VIDEO GAMES AROUND THE WORLD

VIDEO GAMES AROUND THE WORLD

edited by Mark J. P. Wolf

The MIT Press
Cambridge, Massachusetts
London, England

© 2015 Massachusetts Institute of Technology

All rights reserved. No part of this book may be reproduced in any form by any electronic or mechanical means (including photocopying, recording, or information storage and retrieval) without permission in writing from the publisher.

MIT Press books may be purchased at special quantity discounts for business or sales promotional use. For information, please email special_sales@mitpress.mit.edu

This book was set in Gentium Book Basic by Toppan Best-set Premedia Limited, Hong Kong. Printed and bound in the United States of America.

Library of Congress Cataloging-in-Publication Data

Video games around the world / edited by Mark J. P. Wolf ; foreword by Toru Iwatani.
    pages   cm
Includes bibliographical references and index.
ISBN 978-0-262-52716-3 (pbk. : alk. paper)
1. Video games. 2. Video games—Cross-cultural studies. I. Wolf, Mark J. P.
GV1469.3.V565   2015
794.8—dc23
2014034381

10 9 8 7 6 5 4 3 2 1

CONTENTS

Foreword ix
Toru Iwatani

Acknowledgments xv

Introduction 1
Mark J. P. Wolf

Africa 17
Wesley Kirinya

Arab World 29
Radwan Kasmiya

Argentina 35
Graciela Alicia Esnaola Horacek, Alejandro Iparraguirre, Guillermo Averbuj, and María Luján Oulton

Australia 57
Thomas H. Apperley and Daniel Golding

Austria 71
Konstantin Mitgutsch and Herbert Rosenstingl

Brazil 87
Lynn Rosalina Gama Alves

Canada   105
Dominic Arsenault and Louis-Martin Guay

China   119
Anthony Y. H. Fung and Sara Xueting Liao

Colombia   137
Luis Parra and Global Game Designers Guild (GGDG)

Czech Republic   145
Patrik Vacek

Finland   159
Frans Mäyrä

France   175
Alexis Blanchet

Germany   193
Andreas Lange and Michael Liebe

Hong Kong   207
Benjamin Wai-ming Ng

Hungary   219
Tamás Beregi

India   235
Souvik Mukherjee

Indonesia   249
Inaya Rakhmani and Hikmat Darmawan

Iran   271
Ahmad Ahmadi

Ireland 293
Deborah Mellamphy

Italy 305
Enrico Gandolfi

Japan 319
Jennifer deWinter

Mexico 345
Humberto Cervera and Jacinto Quesnel

The Netherlands 359
Christel van Grinsven and Joost Raessens

New Zealand 377
Melanie Swalwell

Peru 393
Arturo Nakasone

Poland 399
P. Konrad Budziszewski

Portugal 425
Nelson Zagalo

Russia 439
Alexander Fedorov

Scandinavia 451
Lars Konzack

Singapore 469
Peichi Chung

South Korea   495
Peichi Chung

Spain   521
Manuel Garin and Víctor Manuel Martínez

Switzerland   535
Matthieu Pellet and David Javet

Thailand   545
Songsri Soranastaporn

Turkey   565
Cetin Tuker, Erdal Yılmaz, and Kursat Cagiltay

United Kingdom   579
Tristan Donovan

United States of America   591
Mark J. P. Wolf

Uruguay   609
Gonzalo Frasca

Venezuela   613
Thomas H. Apperley

Contributors   629

Index   645

FOREWORD

Toru Iwatani

translated by Bryan Hikari Hartzheim

1 On the Global Expansion of Games

Video games today are often played largely for personal use on smartphones, but if we look at the history of gaming, we see that people played games very differently, in arcades—amusement centers that were once a staple of every city. These arcades date back to the coin-operated Edison parlors that first emerged in the United States during the 1890s. In these amusement centers, customers would insert coins into machines for music and visual entertainment.

If we look back at the changes that media such as computers, phones, and video games have undergone, we can see how computers migrated from research centers to the home, where PCs (personal computers) are now just another medium for personal use. As call boxes were once the exclusive domain of companies, telephones became commonplace in the home and are now personal terminals of mobile information. Video games are no different, moving from citywide arcades to homebound console systems so that the current iteration of mobile phone gaming is simply part of the developmental process. The online revolution, in conjunction with the development of gaming on small and portable devices, accelerated the process of popularizing video games, and I believe this is a significant reason for the expansion of video games around the world. In 1972, when the American electronics company Atari introduced *PONG*, which extended inventor Ralph Baer's idea of interactive play on CRT-type televisions (which were, in and of themselves, worthy of admiration), the sensation also created a media revolution.

2 The Role of the Games Industry

Video games differ from other manufactured goods in that they are the rare product to simultaneously be electrical, mechanical, and works of art; they aggregate a range of ideas from the fields of engineering to literature to art to psychology, and they provide society with a necessary cultural tool of "play." Games themselves are formed broadly by the characteristics of play that date back as far as civilization has existed. This human desire to play demonstrates the use of a unique mental instinct, to the point that the Dutch historian Johan Huizinga has claimed of humans, "All are rooted in the primeval soil of play." Therefore, one should not think of games only as fleeting amusements or transient social fads, but as the accumulation of technology and expression, as an emotional medium for players, and as a conscious art that is maturing in its ability to make people respond emotionally. In order to construct the multifaceted support and foundation for new discoveries, which is essential to video games, I strongly feel academia must stake its position very clearly in the years to come as a cultural interpreter for games.

3   To Make Games Is to Understand People's Minds

One can witness people enjoying video games everywhere today. The common feature shared by all these games that become audience hits is the sensitive attention paid by developers to player shifts in mood. Most players appear to be playing games according to a set of rules, but they are actually playing with the developers' *itareritsukuseri*, which in Japanese means "a gracious hospitality that is more fun and kindness than people expect."

Allow me to use my own example of *Pac-Man*, which was released by the Japanese video game studio Namco in 1980, but has spread all over the world and continues to be supported by players young and old, male and female, for more than thirty years. Let me say up front that *Pac-Man* does not excel in any one particular aspect but is constructed to be well balanced in just about every game element.

If we were to enumerate *Pac-Man*'s game elements, they would be:

- Simple game rules and easily understandable goals. Through controlling the protagonist Pac-Man, your goal is to avoid the four ghosts and eat all of the cookies (pac-dots or pellets) inside a maze. When you see the first screen, you should know what to do before you begin playing.
- Simple controls. You can only move Pac-Man in four directions: up, down, left, and right.
- Sense of design and character. The design of Pac-Man is simple, and the design of the colorful ghosts is functional, but Blinky, Pinky, Inky, and Clyde are cute and likable. The designs of both imbue the avatars with character.
- Well-modulated game progress and relief of stress. When Pac-Man eats the power cookies (or power pellets), the hunted becomes the hunter and can chase the ghosts.
- Game design sensitive to the player's shifts in mood. After the ghosts' constant assault leads to a player death, the game's difficulty is lowered upon a restart. Such devices are thoughtfully placed throughout the game.
- Thematic design that is open to all ages and genders. The object of the game is not to shoot down enemies or other such bloodthirsty actions. The main action is to "eat" and becomes a diluted form of attack.

In addition to this list of game elements, little dramatic sketches are interspersed upon the completion of a specific number of stages, relieving the tension and performing the function of relaxing intermissions. Also, the four ghosts are programmed with an algorithm that ensures each of their movements will be different in order to naturally box in Pac-Man and provide thrills.

To sum up, the play in *Pac-Man* constantly changes, appropriately raising or lowering hurdles based on considerations toward changes in the player's psychology. I believe this *itareritsukuseri* game design is why players have continued to play *Pac-Man* for such a long time. If I could describe this spirit in a single phrase, it would be "fun first."

4 Games Research and Vision

Whether from the perspective of the games industry or of the culture at large, video games cannot be content with being forms of play now that they are such an integral part of society. Research and development is urgently necessary for "serious games" such as rehabilitation machines that have the power to develop the interesting aspects of games. The international game research center DiGRA (Digital Games Research Association) was formed in Finland in 2003 for the purpose of conducting academic research on the ever-expanding range of video games. The Japanese digital games association DiGRA Japan was formed in 2006 and is currently conducting vital games research.

In order to discuss the future of video games, we must ask what their purpose is, confront them from the perspective of play design, and examine the sublimation of their status into art. The expression "games are entering us" might sound hazy, but it is a good example of this concept.

5 Conclusion

Through planting my roots in the video game industry, I've been able to make games with the support of many people. I am currently researching and teaching video games based on this experience and my life's belief: "Bring games to everybody."

This belief means, for example, the desire to bring fun games to those who cannot see, or to those who live in remote places where electricity is unstable or unavailable. I know the path to realizing this goal is long, but I want to work together with the people of the world to one day bring games to everybody.

I've arrived at the end, but I would like to express my great appreciation to all the writers from every continent who have contributed to this anthology on the history of video games. And I offer my deepest gratitude to Dr. Mark J. P. Wolf for giving me this forum and indulging my opinionated thoughts. Truly, thank you very much.

Toru Iwatani
Tokyo Polytechnic University
September 2013

Foreword

(*original Japanese*)

1　ゲームがグローバルに拡張した経緯について

今ではスマートフォン(smartphone)でゲームが行われるなど、使用態様は個人ユースとなっているが、ゲームの歴史を振り返ると現在の遊ばれ方とは大きく異なり、街中での遊戯施設であるアーケード(arcades)施設に辿り着く。そのアーケードとは1890年頃に米国(U.S.A)で見られるようになったコインを投入して起動させて映像や音楽を楽しむCoin-operatedのEdison Parlorなどの遊戯施設の事である。

　コンピュータや電話そしてゲームの歴史的メディア態様の変化をみると、研究所や会社にしかなかったコンピュータが家庭用になり、現在ではPC(personal computer)として個人ユースのメディアになっている。電話も同じで、会社や電話ボックス(call-box)などの公共ユースから家庭用となり、現在では個人が持ち歩く携帯情報端末になっている。ゲームについても同様で、街中のアーケード遊戯施設から家庭用のゲーム機で遊ぶように変化し、現在では携帯ゲーム機で遊ぶようになった経緯がみられる。ネット環境の整備と併せて技術的な革新から機器の小型化がゲームの大衆化を促進し、グローバルに展開していった大きな要因と考えられる。米国アタリ社(ATARI)が1972
年に米国アタリ社が発表したビデオゲーム「PONG」は、CRT型テレビを用いたインタラクティブな遊びを発明したラルフ・ベア(Ralph Baer)のアイデアを拡張したものであるが、
それまで鑑賞用であったテレビを遊びの道具に変えるというセンセーショナルなメディア革命を巻き起こした。

2　ゲームの社会に於ける役割

ゲームは電気+機械+著作物が一体となった他の工業製品と違った稀有な製品で、工学や文学や芸術や心理等の様々な学問領域の集合体であり、社会に必要な「遊び」の文化的道具と言える。遊戯としての性格を備えた広義のゲームというものは文明の連綿に等しいほどの歴史があり、人が遊びを欲することはある種の知的本能とでも言うべき行為に他ならず、オランダ(Netherlands)の歴史家ヨハン・ホイジンガ(Johan Huizinga)は「人間は遊ぶ存在である」とまで定義している。それ故にゲームが単なる消費的な娯楽や一過的な社会現象で終わらせてはならず、ゲームは技術と表現の集積であり、プレイヤ(player)に感動を与える存在で、人に感動を与える芸術として成熟を意識し、多角的な支持
と新たな発展を目指す土台を構築するには、学術面に於いてゲームを解釈し、
将来に対して明確に局面を切り拓かなければならないと強く感じている。

3 ゲームを作るという事は人の心を知る事である。

ゲームを楽しんでいる姿は現在世界の至る所で見ることが出きる。そして多くのプレイヤに受け入れられヒットしたゲームに共通して言えることは、それを作った開発者がプレイする人の心の動きを綿密に考え尽くしている点である。一見するとプレイヤはゲームのルールと遊んでいるよう見えるが、開発者である作り手の「至れり尽くせり」のおもてなしの心と遊んでいるのである。

そこで、1980年に日本のナムコ社(NAMCO)から発表されたビデオゲーム「パックマン(Pac-Man)」が何故、グローバルに展開し30数年経っても老若男女に支持され続けているかを、自らであるが一例として示してみる。

「パックマン」は、何か一つの特徴が強烈に秀でているわけでは無く、様々なゲームの要素がバランス良く組みあがって構成されていることが特徴であると最初に宣言しておく。

「パックマン」のゲーム要素を列挙してみると、

・迷路の中の主人公のパックマンを操作して、4匹のゴーストに捕まらないようにクッキーを食べ尽すことがゲーム目的であると、ゲームプレイの前にゲーム画面を見た瞬間に理解できる。(シンプルなゲームルールと目的の分かり易さ)
・パックマンを上下左右の4つの方向にコントロールするだけである。(シンプルな操作性)
・シンプルなデザインのパックマンと適役なのに憎めない可愛らしいゴースト達にキャラクター性がある。(デザイン性とキャラクター性)
・パワークッキーを食べると、逃げる立場から逆転してゴースト達を追いかけることができる。(ゲーム展開のメリハリとストレス解放)
・ゴーストの波状攻撃や、ミスしたあとに難易度を下げて再スタートさせるなどの工夫が随所に施されている。(プレイヤの心の動きを配慮したゲーム設計)
・敵を打ち落とせといった殺伐としたテーマでは無く、「食べる」という行為をメインアクションとして攻撃性を薄めている。(老若男女に訴求する抵抗感の無いテーマ設定)

ここに列挙したゲーム要素以外にも、何面かステージをクリアするとパックマンアニメーション(寸劇)が繰り広げられてプレイ中の緊張感を解し休憩となる。また、4匹のゴーストに別々の動きをさせるアルゴリズムをプログラミングし、4匹のゴーストたちがパックマンを自然に包囲するようにしてスリル感を提供しているなどがある。

総じて言えることはプレイ中に刻々と変化するプレイヤの心理状態を考えて難しさであるハードルの高さを適宜上下させるなど、至れり尽くせりのゲーム設計で構成したからこそ、長くプレイヤに支持され続けていると考えている。一言で言うと「楽しさ第一主義」(Fun First)の精神だと思っている。

4 ゲーム研究と将来像

産業としても文化的側面からも社会的存在意義が大きくなったゲームは、遊びだけにとどまらず「ゲームの持っている興味を持続させるチカラ」を活用したリハビリテーションマシン(rehabilitation machine)などのシリアスゲーム(Serious Game) の研究開発も急務である。このように様々な領域に拡張していくゲームを学術的に研究する必要性から、ゲーム研究の国際学会として2003年にはDiGRA (Digital Games Research Association) がフィンランド(Finland)に設立された。日本(Japan)では2006年に日本デジタルゲーム学会(DiGRA JAPAN)が設立され活発なゲーム研究が行われている。

ゲームの将来像について語るには「ゲームとは何か」を常に自問しながら、遊びをデザインする観点で臨み、アートとして昇華させていくイメージを持っている。煙に巻くような言い方ではあるが、体の中にゲームが入っていくという概念も一例として挙げておく。

5 最後に

私は、ゲーム業界に長く籍を置き、多くの方々に支えられながらゲーム制作を行ってきた。その中で得られたものを活かして、「あらゆる人にゲームを届ける」を人生のテーマにして、現在は大学でゲーム教育と研究を行っている。

このテーマは、例えば目の見えない方々にもゲームが楽しめるように、まだ電気の通っていない様な地域を含めた地球の隅々までゲームが届くようにしたいと言う思いからきている。届けきるには長い道のりが控えているが世界の皆と共に頑張って行きたい。

また最後になりましたが、全ての大陸をカバーした執筆者から寄せられたこの歴史的なビデオゲーム研究随筆集に参画する事が出来た事に大変感謝している。そして私の勝手な話しを載せて頂く機会を与えてくれたDr. Mark J. P. Wolfに深くお礼申しあげたい。本当にありがとうございます。

岩谷　徹

Toru IWATANI
東京工芸大学
Tokyo Polytechnic University

ACKNOWLEDGMENTS

This book was initially inspired by the contributors who wrote essays on countries and regions for my *Encyclopedia of Video Games: The Culture, Technology, and Art of Gaming* (2012). The glimpses they gave of video game history in foreign lands made me want to read more, and I have asked many of them to contribute to this book as well. Many others became involved in the project as it continued and grew, helping to create this mosaic of global video game history. And so I would first like to thank all the contributors, with whom it was a privilege to work: Ahmad Ahmadi, Lynn Rosalina Gama Alves, Thomas H. Apperley, Dominic Arsenault, Guillermo Averbuj, Tamás Beregi, Alexis Blanchet, P. Konrad Budziszewski, Kursat Cagiltay, Humberto Cervera, Peichi Chung, Hikmat Darmawan, Jennifer deWinter, Tristan Donovan, Graciela Alicia Esnaola Horacek, Alexander Fedorov, Gonzalo Frasca, Anthony Y. H. Fung, Enrico Gandolfi, Manuel Garin, the Global Game Designers Guild (GGDG), Daniel Golding, Louis-Martin Guay, Alejandro Iparraguirre, David Javet, Radwan Kasmiya, Wesley Kirinya, Lars Konzack, Andreas Lange, Sara Xueting Liao, Michael Liebe, Victor Manuel Martinez, Frans Mäyrä, Deborah Mellamphy, Konstantin Mitgutsch, Souvik Muhkerjee, Arturo Nakasone, Benjamin Wai-ming Ng, María Luján Oulton, Luis Parra, Matthieu Pellet, Jacinto Quesnel, Joost Raessens, Inaya Rahkmani, Herbert Rosenstingl, Songsri Soranastaporn, Melanie Swalwell, Çetin Tüker, Patrik Vacek, Christel van Grinsven, Erdal Yılmaz, and Nelson Zagalo. I am very thankful that these contributors all speak English and were willing to communicate with me in it (and humbled, too, seeing as I speak *only* English). I would also like to thank Toru Iwatani for his foreword, Bryan Hikari Hartzheim and Doreen Krueger for translation assistance, and others who provided information and contacts, without whom I would not have been able to cover as much of the globe as I have. Thanks go to Francisco Revuelta, Miguel Caballero, Darryn Schneider, Kristin Yates, Ethan Dicks, Audrey Stevens, Stephenie Cahalan, Winston Petty, Federico Peinado, Frenchy Lunning, Koichi Hosoi, Beatriz Helena Rolón Domínguez, Estefania Henao, Kristine Jørgensen, Samantha Hasey, and the press's four anonymous reviewers of the manuscript. Thanks to Doug Sery at the MIT Press for all his editorial assistance and confidence in the project, and ABC-CLIO for reprint permissions for Kasmiya's entries. Thanks also to Judith Feldmann at the MIT Press for her editorial help. Finally, I must thank my wife Diane and sons Michael, Christian, and Francis, who were patient with the time taken to work on this book. And, as always, thanks be to God.

INTRODUCTION

Mark J. P. Wolf

With the enormous growth of the World Wide Web in the last two decades, the rise of mobile platforms and casual games, and an increasing number of game creation programs, the entrance requirements to the global video game industry are lower than ever. Small video game companies are appearing all around the world, each hoping for a hit that will bring it international attention and fame, both of which can grow much faster due to the Internet. With this rise in game production, many more countries have their own video game industries and their own national histories of video games, many of which are only now beginning to be recorded. And yet, thanks to foreign imports and indigenous productions, many national video game histories are already decades old even though the majority of them are not widely known beyond their own borders.

Video game histories typically follow the highlights, innovations, and advances made in the field, and naturally this can only be done using the resources available to researchers. In the United States, the first researchers were journalists or hobbyists looking back on the medium as it grew, such as the multipart "Electronic Games: Space-Age Leisure Activity" by Jerry and Eric Eimbinder, which first appeared in the October 1980 issue of *Popular Electronics*, and books such as George Sullivan's *Screen Play: The Story of Video Games* (1983) or Scott Cohen's *Zap!: The Rise and Fall of Atari* (1984). Interest in video game history picked up during the 1990s and saw the publication of David Sheff's *Game Over: How Nintendo Zapped an American Industry, Captured Your Dollars, and Enslaved Your Children* (1993) and Leonard Herman's *Phoenix: The Fall and Rise of Videogames* (1994). Soon after came Steve Kent's *The First Quarter: A 25-Year History of Video Games* (2000), later renamed *The Ultimate History of Video Games: A History from Pong to Pokemon—The Story Behind the Craze That Touched Our Lives and Changed the World* (2001). As nostalgia grew for video games, mass-market books with color photography appeared, such as Rusel DeMaria and Johnny L. Wilson's *High Score! The Illustrated History of Electronic Games* (2003) and Van Burnham's *Supercade: A Visual History of the Videogame Age, 1971–1984* (2003). In the decade following that, many more books on video game history appeared, including two of my own, *The Video Game Explosion: A History from PONG to PlayStation and Beyond* (2007) and *Before the Crash: Early Video Game History* (2012). Other books have discussed the spread of video games, such as *Gaming Globally: Production, Play, and Place* (2013), edited by Nina B. Huntemann and Ben Aslinger, and an email discussion list, Local Game Histories [LOCALGAMEHIST], has been started by Melanie Swalwell and Jaroslav Svelch as a resource for the global community of game historians.

Most video game history books, however, are produced largely for a North American audience who remembers the games and so the books are thus compiled from the resources available in North America. After the rise of Nintendo following the North American Video Game Industry Crash of 1983, some Japanese history would be included and perhaps some information on Europe. But even in the pre-crash era, video game production and play was occurring around the world as foreign imports and home computers quickly spread. Research in computer science was already underway in multiple countries, with discoveries and developments occurring simultaneously and independently. Games, both amateur and professional, were made and circulated. Hobbyist groups, demoscenes, and communities of enthusiasts formed as early as the late 1970s, and written works including newsletters, magazines, journal articles, reviews, and books appeared, but they typically did not travel beyond their nation's borders. Even into the 1980s and 1990s, national histories of video games would remain largely unknown outside their countries of origin due to language differences and cultural divisions. This continued until the 2000s, when international conferences on video games began to bring together researchers from different countries. The chapters of *Video Games Around the World* attempt to describe the events of various national histories to an English-speaking audience, and many were compiled from primary research when no secondary sources were available. Some of these national histories appear here for the first time in the English language, and some for the first time in any language.

Readers will note that the chapters here vary greatly in terms of length, scope, style, and focus. Although all contributors were given a list of criteria to follow (namely, the history of video games in the country or region, the reception of foreign imports, domestic video game production and exports, indigenous video game culture and how it was influenced by national history, video game company profiles, video game content description, the role of academic video game studies, and the future of video games in the country or region), it was not always possible to cover all these areas. Video game production varies greatly from one country to the next, and where there is less production there is usually even less scholarship on games. While here in the United States one might take for granted that events will be thoroughly documented and that those documents will be safely filed away in an archive, where they will be preserved and made accessible to researchers, this is not the case in many parts of the world (and sometimes not in the United States either, for that matter). Those areas suffering from poverty, political instability, and minimal infrastructure naturally have other priorities, and some cultures do not place the same value on documentation and preservation, or they lack the necessary resources or inclination to do so.

When compiling this collection, there were two approaches that could be taken; I could try to include as much of the world as possible, even though this meant that chapters would vary considerably, or I could include only those chapters written by established academics with all the necessary resources available to write a complete history even if it reduced the global nature of the book's coverage that I originally intended. I have gone with the first option; however, the book described in the second option is still present here, and those who would prefer it can simply set aside those chapters that do not meet their needs. And even the longest chapters here can only touch upon all the various aspects of a national history of video games and suggest an outline of their contours.

As an editor, I have also chosen to allow a variety of voices throughout the book rather than try to homogenize them all into a single style, in order to preserve some of the cultural differences that one inevitably encounters in a project of this scope. Most of the contributors are also natives of the countries they are writing about, so in addition to researching video game history, they have lived through it, experiencing it firsthand, with a deep understanding of the culture in question as well as the prevailing zeitgeist. As noted in the "Contributors" section, many of them are also game designers and founders of game companies, giving them a firsthand perspective on the industries they discuss. Each country or region's unique situations and circumstances affected how their national video game histories developed and the shape that video gaming took; for example, the long history of computer science in the United Kingdom and Scandinavia, the prevalence of mobile games in Africa, Mexico and New Zealand's many arcade games, Brazil's large number of master's and doctoral theses, the Dutch industry's large percentage of serious games, the demoscenes of Eastern Europe, the PC Bangs of South Korea, the warnets of Indonesia, and so forth.

One may also note differences in terminology, which I have, for the most part, included here. For instance, although "video games" is arguably the most common term used to refer to the medium (at least according to Internet search engine results), other terms, such as "computer games" or "digital games," are also used and suggest a different context in which games arose. While people in the United States typically first encountered video games in the arcade or on a home console connected to a television set, leading to the common use of "video" games, in other parts of the world, particularly those where television did not have as strong a presence or had yet to even make any inroads at all, video games were more associated with home computers or even mobile devices, leading to the use of other terminology to describe them (such as "computer games," "digital games," or "electronic games"). I have thus decided not to standardize some of the terminology throughout this book, to allow this cultural variation to manifest itself. As the difference in terminology demonstrates, the level and usage of technology and other resources differs greatly from one place to another, and it is to these that we next turn.

Laying the Groundwork: Infrastructures Necessary for Video Game Industries

A video game industry, even video games themselves, need infrastructures to support them. In the case of video games, we can identify three levels of infrastructure, each built upon the ones below it, which are necessary for a national industry to take root.

At the first level, there are basic needs such as access to electrical power, verbal and visual literacy, and lifestyles that include leisure time for gameplay. Electrical power, needed for the operation of video games, whatever kind they might be, may be distributed by power lines from a central production facility, in the form of batteries, or supplied by solar cells. The first of these three requires a certain level of industrialization, which in turn requires a certain amount of capital and the commercial and governmental means to control and regulate it, while the last two are less centralized but still require an industry to produce them or, at the very least, to import them. In any case, there is an ongoing energy expense for the user beyond the cost of the game and the platform it runs on, and a reliance on a technology that may be considered a luxury. And even electricity is a luxury that many do not have; as recently as 2013, around 1.3 billion people were still without access to it.[1]

Video game play also often requires verbal and visual literacy and some experience with media conventions. Not only do video games have text (at least in manuals and on menu screens, if not embedded within the games themselves), their visuals are often complex and fast-moving and depend on visual conventions borrowed from other media, such as conservation of screen direction and off-screen sound from cinema and television, word balloons from comics (for example, in many Nintendo games), and spatial constructions seen and navigated from a first-person perspective. While most cultures that have video games have already experienced film and television, these media still form the basis for an intuitive understanding of how video games work, and those cultures lacking such experiences are less likely to adopt video games as readily.

Finally, leisure time available for gameplay is necessary for video games, especially those requiring dozens of hours for their completion. Even casual games collectively can take up hours of one's time, hours which may not be available in agrarian cultures or cultures in which the work needed for subsistence takes up much of the day, or among the poor whose long hours in farm fields or factory sweatshops leave little time for relaxation. As mentioned above, there is also the cost of the systems and games themselves, which requires at least a small portion of disposable income. Thus, a certain level of economic development is also at the base of the first level of infrastructure.[2]

Once the first level of infrastructure is satisfied, the next level includes a certain amount of technology, technological know-how, and access to a system of game distribution and marketing. The first of these involves screen technology such as a television set or computer monitor, which the user must own before buying the console or computer system that will be attached to it; or, in the case of arcade games, there must be arcades or other places where the machines can be exhibited and available to the public. In many countries, home computer games are more prevalent than console-based games because of the relatively high cost of consoles (which are more limited in their capabilities) compared to home computers (which are usually cheaper and more multifunctional). Another technology is a system of telephony, either in the form of landlines or mobile phone towers and networks, which are needed for online games and mobile games. These often require Internet service providers (ISPs) as well. And the presence of the Internet does not guarantee the reliability and bandwidth needed to play online games, the requirements of which are a moving target as games continue to evolve.

Naturally, a certain degree of technical knowledge and expertise is necessary for the installation, use, and maintenance of these technologies, which together amount to a technology sector in a country's economy, and this, in turn, requires some degree of training for employees and users alike. Telephony and the Internet almost always involve connections to the rest of the world's communications systems, with all their conventions and protocols, and for many countries this will also include involvement with foreign communications companies.

A system of game distribution and marketing that is accessible to consumers is also necessary for video game culture to emerge. While today the Internet provides a forum for discussions about video games and a venue for their marketing and distribution, in earlier times this was accomplished by events at arcades, home computer clubs, magazines, bulletin board systems (BBSs), advertisements in hobby shops, and gatherings at conventions, demoscenes, warnets, PC Bangs, and so on. In the late 1970s and 1980s, home computer culture made it possible for users to create their own games and trade them, and the rising interest in computer games encouraged the importation of games and console systems, often by companies in related industries. As Melanie Swalwell writes in her chapter about New Zealand, "For some of the companies involved in the manufacture and sale of console (and arcade games), video games were just one of the various electronics design and manufacturing enterprises in which they were involved. Fountain was a well-known electronics company, producing record players, stereo systems, clock radios, and even microwave ovens. David Reid was the name of an electronics store. As far as arcade machines went, Rait made audio amplifiers, whilst Kitronix's products included heat sinks, guitar amplifiers, and printed circuit boards." On the other hand, some companies importing games had little or nothing to do with electronics; as Humberto Cervera and Jacinto Quesnel tell us in their chapter on Mexico, it was a meat-packing company that held the rights to import the Atari VCS 2600 into Mexico.

Once a country has attained the infrastructure of the first two levels just described, a third level of infrastructure becomes possible, which allows for the development of an indigenous game design and production industry. This level includes game designers, developers, programmers, and other professional staff, corporate structures to stabilize and maintain an industry, the necessary investment capital, and a large enough user base to make larger-scale productions financially feasible. Schools and training programs have aided the development of a skilled work force, and since the appearance of the World Wide Web, digital technologies such as game engines and game creation software, as well as the ability to collaborate interactively over long distances, has made it easier for independent producers to work anywhere in the world.

Even when all these things are present, however, the success of the industry will still depend on the successes of individual companies, and those successes will depend on the creation of video games that consumers want to play, if profits are to be made (and they are made not only through game sales, but through membership subscriptions, sales of in-game merchandise, licensing, and advertising). Nothing can guarantee a game's popularity, as history has shown; the balance between conventions and innovations, and familiarity and novelty, along with the right kind of content, is difficult enough for anyone to achieve. The actual creation and marketing of games, as well as their content, is determined by a variety of factors, including the tensions present within the video game industry.

Tensions in National Video Game Industries

Much of video game history (and for that matter, history in general) depicts enduring struggles between opposing forces over time, as well as the effects that produce and are produced by these tensions. The following sections look at some of the tensions present in national video game histories, which vary in importance from one country to the next. Although each of them is described in terms of one thing versus another, they should each be considered as more than just a simple binary opposition; each is a spectrum of possible positions, many of which will be found coexisting alongside each other. As can be seen in the chapters in this book, these forces largely determine the shape of an industry and the games it produces.

Indigenous Production versus Foreign Imports

Every country that has video games has experienced the domination of foreign imports. For all countries except the United States, where they originated, video games arrived as an element of a foreign culture and one that threatened to displace local culture, or at the very least, would need to be assimilated. As Wesley Kirinya writes in his chapter on Africa, "Some of the games that used to exist in the traditional cultures, for example, Mancala and Bao, also faded away as they would be seen as 'backward' by the urban 'elite.'" In places like Africa, which had already experienced colonization by foreign powers, video games could be seen as yet another form of foreign cultural influence. Even in the United States, ever since Nintendo eclipsed Atari following the crash of 1983, foreign home game systems have been dominant, beginning with the importation of the Nintendo Entertainment System (NES) in 1985. Even as of late 2014, only one of the three top-tier video game systems (the Nintendo Wii U, PlayStation 4, and Microsoft Xbox One) is made by an American company. While the crash may have given foreign industries a respite, allowing them to pull ahead, other venues since then such as online gaming and mobile gaming have also reset the stage and given smaller companies the means to reach large international audiences.

Thus, the tension between foreign imports and indigenous production can be found in every country that has video games, and how this tension is managed varies greatly from one to the next. In almost all cases, the entry of foreign imports preceded indigenous video game production, establishing conventions and audience expectations that shaped the country's domestic video game industry and its output. The situation is similar to that faced by the film and television industries of many countries during the twentieth century; the importation of Hollywood cinema and American television programming was cheaper and easier than producing film and television programs domestically, and domestic projects rarely had budgets that could compete with Hollywood. The result was that national cinemas were displaced by Hollywood product to such an extent that governments intervened with quotas (as in postwar Britain), laws keeping foreign income within the country, or laws limiting foreign imports (such as Italy's Andreotti Law), as well as coproductions that were international in scope and were designed for a pan-European audience instead of merely a national one. Similar solutions can be found in national video game industries and are described in the chapters here.

At the same time, however, the distinction between foreign and domestic companies and games is an increasingly problematic one, since both console-producing companies (such as Nintendo and Sony) and major game-producing companies (such as Electronic Arts, Rockstar Games, and Ubisoft) have branches in multiple countries and continents. Each of these branches employs a local labor force, and, as a result, their products may even reflect the national cultural context. Furthermore, branches of foreign companies are often closely linked to local companies.

Just as programmers left Atari and began other companies such as Activision and Imagic, foreign branches of companies can seed national industries by hiring and training local designers who then leave to start their own companies. For example, as Dominic Arsenault and Louis-Martin Guay relate in their chapter on Canada, employees of the Canadian branch of Electronic Arts, known as EA Canada, left to start several of their own third-party studios, greatly enlarging the Canadian video game industry. As we see in Peichi Chung's chapter on Singapore, some governments even encourage foreign studios to enter their countries with the expectation that they will train local talent who will then start their own local companies. However, the opposite can also occur, since successful local companies are sometimes bought by foreign companies—for example, the Australian studios Beam, BlueTongue Entertainment, and Ratbag Studios, which were purchased by Infogrames, THQ, and Midway Games, respectively (see Thomas H. Apperley and Daniel Golding's chapter on Australia). And once the parent companies decide to cut back, some of these national branches are closed, sending professionals back into the job market again.

Sometimes foreign companies outside the video game industry can also play an important role, as in the case of the Timex Corporation in Portugal, which assembled its home computers in that country (see Nelson Zagalo's chapter), or the meat-packing company mentioned above that brought the Atari 2600 into Mexico. Imports don't always make it into countries either; as Melanie Swalwell tells us in her chapter on New Zealand, "Global distribution anomalies appear to explain the fact that some widely available consoles, such as the ColecoVision, Intellivision, and the Vectrex, did not make it to New Zealand." Finally, foreign imports and indigenous production can be combined in such a way as to threaten both industries, in the form of pirated versions of games.

Legitimate Industry versus Piracy

The degree to which piracy is present in a country depends on the efforts made to control it; governmental attitudes regarding patents, copyrights, and their infringements; profit margins; supply and demand; and even cultural attitudes toward copying and ownership. While many countries face software piracy due to the ease of reproducing digital media such as CD-ROMs, DVDs, and other optical discs, some countries also have pirated consoles in their black marketplace as well.

Piracy generally occurs due to the desire for lower-priced merchandise, especially when the population in question already is subject to low wages and economic hardship. Lower-priced pirated merchandise makes it harder for legitimate industry to compete and thus drives away legitimate outlets and companies even as it spreads video game culture among those who would not have been able to afford it otherwise. Pirated goods hurt foreign imports as well and often appear just as fast as the new games do. Consumers, however, can sometimes consider piracy a subversive activity; as Anthony Y. H. Fung and Sara Xueting Liao write in their chapter on China, "While the disregard for copyright is, in general, seen as the piracy of intellectual property, consumers in China (as they argue), see it as resistance against the dominance of the concept of copyright and patents. Local businesses also conceive of it as a counterforce against the giant multinational companies." Governments, then, may use piracy as a way of controlling foreign imports and influence, deliberately limiting them by making it difficult for profits to leave a country. And as Humberto Cervera and Jacinto Quesnel explain in their chapter on Mexico, piracy can lead to brand loyalty that continues into legitimate industry when consumers are later able to afford legitimate products.

Piracy can also help determine the relative balance between the various sectors of the video game industry. For example, online games that require monthly subscriptions and access to a company's servers are less vulnerable to the loss of profits than standalone games that can be copied and sold; thus, in areas with high rates of piracy, companies may find online games more profitable. While statistics do not appear to exist regarding the number of programmers and other technical personnel who get their start in pirated games and later use their experience to make the transition into the legitimate industry, piracy can provide training for local workers who want to get into the video game industry and may make it easier for legitimate companies to find skilled employees. The competition between legitimate and illegitimate producers also mirrors, to some extent, another tension in the industry, between mainstream and independent (or "indie") productions.

Mainstream Industry versus Independent Productions

In the early days of video game production, games were typically created entirely by single individuals, opening production to anyone who was interested, and hobbyists could even begin to form their own companies (for example, in the Czech Republic, as Patrik Vacek tells us in his chapter). As games grew more sophisticated and elaborate, with more complicated hardware, they required larger design teams and game budgets, making it more difficult to enter the industry. By the late 1980s, only larger, mainstream companies were able to program games for state-of-the-art systems. By the end of the 1990s, triple-A game budgets were in the millions of dollars, and by the early twenty-first century, games such as *Grand Theft Auto 4* (2008), *Red Dead Redemption* (2010), and *Star Wars: The Old Republic* (2011) had budgets of USD $100 million or more. Few companies can compete at this level, and even games that are expected to do well due to their ties to existing franchises can eventually fail, such as the MMORPG *Star Wars Galaxies* (2003–2011), which began well but alienated players when changes were made.

Thus, different levels of games and game budgets exist. Since the 2000s and the rise of casual games and mobile gaming, an increasing number of game development tools and software have made it possible for small companies to develop games, and with the World Wide Web, distribution can be done digitally and reach a global audience. In some countries, government subsidies are available to aid the local industry, as Thomas

H. Apperley and Daniel Golding describe in their chapter on Australia, and in others, government policies, programs, and support encourage the local industry growth of small-to-medium-sized companies, as Peichi Chung tells us in her chapter on Singapore. Government involvement can also lead to game censorship laws or to self-regulating organizations supported by the mainstream industry (such as the Entertainment Software Ratings Board [ERSB] in North America and Pan European Game Information [PEGI] system in Europe), or governments may even ignore games and see them as harmless (as in parts of Africa). Such attitudes and initiatives, begun by the mainstream industry, are usually applied to independent productions as well.

While games are usually seen as entertainment by consumers, the recognition of video games' artistic potential can also aid indie productions that are more experimental in nature, calling attention to new directions and possibilities for games. Museum exhibitions also help to highlight innovative games and game technologies, some of which may have commercial potential, and game journalism and academic scholarship can also promote such games. Like the film industry, successful innovations that begin in independent or experimental productions often make their way into mainstream productions, and successful indie companies occasionally grow to join the mainstream as well, blurring the divide between mainstream and indie productions. But gauging the marketplace and attempting to appeal to consumers is no longer something limited to national scope. Especially in smaller countries with limited markets, consideration of the global marketplace has become increasingly important.

National Marketplace versus Global Marketplace

As mentioned earlier, foreign imports often set the tone, conventions, and audience expectations for video games within a national marketplace. Already-established hardware platforms and game engines may place limits or restrictions on game production, marketing, and distribution, and existing game genres may determine how games are perceived and experienced by an audience. Local game development companies are often enlisted to assist with the creation of games for multinational companies and foreign franchises, which have the global marketplace in mind. National audiences, especially in smaller countries, may not be large enough to recoup the cost of more expensive games, leading companies to reduce game content that is considered too nationalistic and thus not as exportable (as discussed in Songsri Soranastaporn's chapter on Thailand).

A wholly national aesthetic, then, may become harder to develop, since games with more nationalistic details may not sell as well on the international market, and the more expensive game development is, the more likely the international market will need to be targeted to return a profit. Like cinema, national languages sometimes give way to more internationally used languages such as English, so as to broaden the potential audience of a game. But whereas a two-hour film can be easily dubbed or subtitled with another language, games requiring dozens of hours of gameplay, with many contingent situations, may have more content to be translated, raising the cost. One of the most extreme examples of this is *Star Wars: The Old Republic* (2011), which reportedly contained more than two hundred thousand lines of recorded dialogue.[3] And yet there are still many games that are strongly rooted in their respective cultures, such as *Adventures of Nyangi*

(2007), *Quraish* (2007), *Capoeira Legends* (2009), *The Spirit of Khon* (2010), and many other games mentioned in the chapters of this book.

Yet while corporations are concerned with selling games as broadly as possible in the global marketplace so as to maximize profits, gamers who are interested in something different may wish to explore games that are rich in localized content, and some games have managed to frame their unique cultural qualities in ways that distinguish them from their competitors, increasing their sales. With international distribution made easier by the Internet, more such opportunities become available as time goes on. Games that rely mainly on graphics as opposed to text, and those that have more recognizable game mechanics, may appeal and be playable regardless of culture and language barriers; and games that find success in their national marketplace may be able to reinvest profits in English-language (or other-language) versions for the global marketplace. Better still, games can be designed from the start to have multiple languages built in to them and their marketing. For example, Afkar Media has an English-language Web page promoting its game *Quraish*, which involves Islamic history (see http://www.quraishgame.com/qe_index.htm), and the game itself features interfaces in both Arabic and English. Even in the area of censorship, which differs from one region to the next, games can be designed with multiple systems in mind. In their chapter on Australia, Thomas H. Apperley and Daniel Golding describe how *Fallout 3* (2008) was altered globally due to requirements for Australian classification. In some cases, governments actively support games that promote national culture, for example, the Indonesian MMORPG *Nusantara Online* (2009). As Inaya Rakhmani and Hikmat Darmawan write in their chapter on Indonesia, "*Nusantara Online* was developed based on sociological, archeological, and historical theories that speak of Indonesia's cultural roots. The characters, social system, architecture, and even the virtual map are consistent with Indonesian cultural history in combination with mythological characters, legends, and folklore." Video games can also be coordinated with other media, and the relationship between them constitutes the last of the tensions to be explored in this section.

Video Games versus Other Media

In the United States, video games became a venue for the adaptation of properties from other media as early as the late 1970s and early 1980s—for example, the Atari 2600 cartridges *Superman* (1979), *Raiders of the Lost Ark* (1982), *E.T.: The Extraterrestrial* (1982), *M*A*S*H* (1983), and *Krull* (1983). Today, video games are another venue for vast, transmedial franchises with works appearing in film, video, books, comics, games, websites, computer software, and other media. Video games have also become a medium of origin for transmedial franchises as well, as so many successful games have demonstrated. Thus, in industrial terms, video games are more often coordinated with other media, rather than simply competing with them, but culturally, video games have displaced other media and cultural activities taking users' time and attention away from them. As a result, tensions can exist between video games and other media, with one media industry driving another or providing content for it as video game industries integrate themselves into the broader media industry landscape.

Since a complete description of video games' place within broader media contexts requires detailed and in-depth discussion of the histories of all the other media in a given national context, this tension can only be touched on within a single chapter. Even so, many of the chapters here show how video games are related to home computer industries, which arose around the same time; film and television industries, from which ratings systems were adapted; and other industries that provided models of regulation and corporate structures for video game companies, addressing some of the same problems such as piracy, censorship, and lack of funding. As with the other tensions discussed, the universal situations faced in every country reveal the forces shaping video game industries, while the specificities of each national context reveal the variety of ways such forces can balance against each other.

A Global Portrait

In the global portrait of the video game industry collectively provided by the chapters of this volume, many overarching trends and connections are present but implied, due to the volume's division by country or region. For example, the availability of off-the-shelf hardware and software tools for game design coupled with the growing market for mobile games and the Internet as a form of distribution and delivery have reduced the overhead necessary for starting a game design company, so much so that they are now possible even in economically depressed parts of the world. Ideas are finding their way into the global marketplace much faster (and spreading their influence faster), while increasing collaboration and border-crossing erodes the traditional boundaries of national industries. Mobile games and MMORPGs can quickly propagate far beyond their national origins and are designed with the global market in mind. Hits such as *Angry Birds* (2009) from Finland and *Pou* (2013) from Lebanon demonstrate that top-selling games can come from companies in smaller national industries. Unlike the older model of an initial national release followed by later releases on other continents, simultaneous worldwide releases offer the lure of greater profits obtained in a shorter amount of time; increasingly, multinational corporations are up to the task. As of this writing, in March 2014, the most recent global blockbuster was *Grand Theft Auto V* (released worldwide on September 17, 2013) by Rockstar Games, a multinational video game developer and publisher with offices in the United States, England, Canada, Scotland, and Japan (and a former studio in Austria). With a reported development and marketing budget of USD $266 million,[4] *GTA5* made $800 million in its first day of sales and reached $1 billion after three days, making it the fastest-selling entertainment product in history.[5] Only with the anticipation of a global audience —and the means for simultaneous worldwide release— are such successes possible.

Collaborations, company ownership, branch office locations, and franchised intellectual property (IP) are crossing national boundaries more than ever before. For example, The LEGO Group, a Danish company, hired Traveller's Tales, a British company, which is now a subsidiary of the US Company Warner Bros. Interactive Entertainment (itself owned by conglomerate Time Warner Inc.), to produce LEGO-themed video games, and to do so, Traveller's Tales outsourced some of the work to the Argentine company Three Melons. The games were then programmed for systems from the U.S. and Japan.[6] Thus, the companies influencing the final form of the LEGO games are located in at least five countries on four continents. This is not unusual; according to a 2008 *Game Developer Research* survey, 86% of game studios used outsourcing for some aspect of game development.[7] So just as many national video game histories are finally being written, the growing shift toward transnational game development is eroding and reconfiguring the very concept of a national industry. What is more, such transnational exchanges also enrich the cultures they impact while at the same time establishing video game conventions on a global scale. The balance between the differentiation needed for novelty versus the homogenization that allows familiarity is something that needs to be examined on a global scale, as games are outpacing the academic study of them. Thus, beyond merely offering a look at neglected areas of video game history, it is hoped that *Video Games Around the World* can provide a foundation for such comparative study, which has only barely begun. Along with the appearance of more national histories, this appears to be the direction that video game history is heading.

A World's Fair of Video Game Scholarship

As the chapters' reference lists in this volume demonstrate, more and more international video game history is being written, but much remains to be translated into English (and other languages as well). While video game history and scholarship has been slow to cross borders, games from around the world are programmed for globally released console systems and home computers, or appear online, allowing these games to be purchased and played virtually anywhere in the world. Many games rely more on images and sound than text, making adaptation easier and quicker, and quite often, video game conventions allow games to be understood and played by anyone, regardless of the language they speak. One such game, *Pac-Man* (1980), is one of the earliest games to find worldwide success and popularity, is one of the most downloaded games of today, and is certainly the most popular arcade game ever made, perhaps even the most popular video game of all time. Thus, the game's creator, Toru Iwatani, is the perfect person to provide a foreword for a volume such as this.

Scholarly work on video games has not spread nearly as fast as the games themselves have, though this is starting to change thanks to publications and conferences, which are more global in scope. The World Wide Web has also sped up dissemination of game communities and game scholarship, and of course, games themselves. Although the exigencies of academic publishing limit this book to grayscale imagery, Internet searches of game titles allow one to easily find many colored images of games and game systems as well as speedruns and walkthrough videos, and most of the companies mentioned in the chapters have an online presence that is easily accessible.

This book attempts to cover the globe as well as can be done within a single volume, and every continent is represented here. North America is covered by chapters on Canada, the United States, and Mexico, while South America is covered by chapters on Colombia, Venezuela, Brazil, Peru, Uruguay, and Argentina. Europe is covered by well over a dozen chapters, including ones on Ireland, the United Kingdom, Scandinavia, Finland, the

Netherlands, France, Germany, Austria, Poland, the Czech Republic, Hungary, Switzerland, Spain, Portugal, Italy, and Russia. Asia is also represented by the chapter on Russia, as well as chapters on China, Hong Kong, Japan, South Korea, Thailand, Singapore, and India. Oceania and the Australian continent are represented by chapters on Indonesia, Australia, and New Zealand. Africa is represented by Wesley Kirinya's chapter, and the Middle Eastern region adjoining it is represented by chapters on Turkey, Iran, and the Middle East. In all, the composite picture of the global video game industry contained in this collection demonstrates a range of innovation and diversification, despite technological convergence and the growing interconnectedness of the industries' players.

These chapters cover the six inhabited continents, but one yet remains, the most sparsely populated one. So finally, for the sake of completeness, we must also ask whether there are video games in Antarctica.

Video Games in Antarctica?

Antarctica, that continent without a country, is a harsh environment populated by only scientific research stations, with no permanent human residents. Still, we would expect its part-time inhabitants to spend most of their time indoors and also to have a certain level of technical proficiency, suggesting that video games might find some popularity there. While what follows here is an anecdotal report limited to stations of the United States, it should be noted that more than two dozen other countries have research stations in Antarctica. Despite their cultural differences, the extreme conditions and isolation experienced by all these stations are no doubt similar enough so that these anecdotes may be representative of more than just the stations described. In any event, I leave a more comprehensive survey of the continent to future researchers. That said, my respondents were from Scott Base, McMurdo Station, and the University of Wisconsin's station at the pole. I have left their emails almost exactly as I received them, only italicizing game titles for greater clarity. All three describe an interest in video games lurking within the scientific culture of the pole, whether openly acknowledged or treated as a guilty pleasure, and one respondent was even a former game designer.

Anthony Hoffman, Winter Engineering Supervisor at the Scott Base Radio Workshop, described (in an email sent May 15, 2011, to the author) some of the limitations faced by residents of the stations:

> Few of the staff here are really into games, though a number of individuals play solitaire on the computer during lunchtime. Would certainly be the most played computer game I've seen.
>
> We work 6 day weeks and some of us are "on call" 24/7, so unfortunately not a great deal of free time.
>
> As we have a limited internet connection, downloading new games and online gaming is out of the question. A few of us do enjoy older games; as an example I played serial linked *Duke Nukem 3D* with the chef the other day. I've seen a few people playing the occasional turn based strategy game such as *Civilization*, *Buzz Aldrin's Race to Space*, etc.
>
> One of the summer staff had a Sony PlayStation which was used a bit in the bar. We have a Nintendo Wii here in the bar with a range of sports games, though it doesn't get a lot of use.

A lot of my own work is based on Amiga computers and I frequently use an emulator on the windows boxes we have here in order to run a range of productivity software plus the occasional game such as *Cannon Fodder*, *Lemmings*, etc, during a bit of time-out.

After hours most of the staff tend to relax by reading, jigsaw puzzles, and watching movies. We have a fairly extensive DVD library on base which gets a lot of use.

Ethan Dicks, who did four winters on the Ice, one at McMurdo Station with Antarctic Support Associates and three at the pole with the University of Wisconsin, echoed some of the same sentiments, writing in an email to the author sent May 16, 2011,

With "summer" at Pole running from, essentially, 1 Nov to 15 Feb (give or take a few days on either end for weather), and the population fluctuation from 150–250 throughout most of the season, the nature of "free time" is very different than the nine-month winter with a population of a few dozen (24–86, depending on which exact year and how much construction/demolition was scheduled—my winters were 75, 65, and 64, IIRC).

During the summer season, with three shifts, the public computer lab is often quite full with a short queue at break times. The in-station personnel (about 50–150, depending on the exact year as the station was under construction) have easy access to computers, but the folks out in "summer camp" (100–150) are pretty much limited to personal laptops if they don't feel like staying in the station late, or making a 15-min walk each way to the computer lab or their work computer).

The winter is a much quieter time—90% of the station is on the same schedule (formally 07:30–17:30, but "grantees" (scientists) work more to the needs of the project and less to the strict tick of the clock), and there are no planes (meaning no cargo, no new people, no departures …) to fill the working hours. There's usually some construction or deconstruction projects going on, the day-to-day data collection by a dozen or more scientists and science technicians, and plenty of evening time for music, arts and crafts, reading, movie watching, and, of course, game playing.

When I was last at Pole (2008), there were a number of PS2 consoles (I had one myself), and a Wii or two, plus an old GameCube and an old Genesis that saw little use. As for PC gaming, there is a US Antarctic Program prohibition of games on the in-station network, so what is common is to set up an isolated network with a few machines and a variety of games loaded on them. One of the favorites in 2008 was one gaming workstation set up with a personally-owned copy of *Mass Effect* that we all took turns at. There were also a few LAN-party games loaded on all machines so we could do CTF and cooperative team-based missions.

One thing that makes modern gaming difficult from the Pole is that the Internet is not a 24/7 resource there. Depending on the exact year and the condition of the satellites in the sky and the ground equipment at Pole, there is anywhere from 4–12 hrs per day (presently about 8, I think) of approx-1Mbps connectivity for all uses—phone calls via VoIP, video conferences, mundane data transfer (science data is via a different, one-way link), then at the bottom of the queue, Internet. It's not viable to play games that must "phone home" for confirmation/registration, and DLC can take days or weeks to pull down. To make matters worse, the equipment there was ordered years earlier (the supply chain is quite long from when anything is ordered to when it shows up on site—up to 18 months), so any games that require fast, fresh, and up-to-date hardware are out of luck. Games that run well on 4–5-year-old hardware are much more likely to run smoothly.

I hope this gives you an overall picture of computer gaming at the Pole. If you have any particular questions, I can try to answer them. I've been a game tester in the past (a "Visioneer" for Activision), have written commercial computer games (back when the Apple II and C-64 were dominant), and I play games to this day, so I'm probably as good a resource as any on the topic.

Finally, McMurdo Station's Inventory Data Specialist, Larry Fabulous, described (in an email to the author sent August 23, 2012) his experiences with video games at the Pole:

> I hope what I write here is helpful for you somehow as it will really just be my experience with video games and not a proper study. ... I have each of the three main video game consoles here as well as a PSPGo, PS Vita, Nintendo DS Lite, Nintendo DSi, and Nintendo 3DS. Well, those and an iPhone and MacBook Pro. I use all for gaming, but the only games I play online are on Facebook (which I don't play regularly at all ... just a social fix once in a great while). Your information is correct about online gaming. The lag is huge and makes most online games unplayable. What I play most on my consoles are RPGs and other single-player experiences. I wish I played more but I find that down here I commit to many more social events than I ever have anywhere else I've ever lived. I still love video games and buy them and read about them and all that but just don't get to them like I used to! I'm embarrassed to say that the only games I've completed while being stationed here are match-three games on my phone and from PSN that I downloaded. Well, also *Dragon Quest IX* but I did have someone to play that with and portable gaming feels so much less cumbersome than console gaming at this point in my life. Or maybe it's just more to get to a game than it used to be when you just hit the POWER button on the NES and the game was on right away. Console gaming feels to me, more so down here than back home, a huge commitment that I rarely feel I have enough time to indulge. I'm sad to say that because there are so many games I want to be playing, lol! I do know there are people here who are NEVER seen in public outside of meals because they game like crazy but I can't bring myself to do that. Isn't that weird? To wrestle with something like that? I am always finding myself in social situations and I fantasize about being left alone so I can just chill on the couch or bed and get lost in a game, haha. Let me know if you have any other questions about video games or anything else. I'll be happy to answer them.
>
> You also asked about film and TV and I'd have to say that yes, far and away, they are more popular than gaming here, at least in the winter. There are ALWAYS events based around watching a series or movies. There are public events, private ones, impromptu ones ... gaming here is kind of like it was in the '80s when it was considered childish and wasteful. So weird.

Video games, then, can be found on every continent and appeal to one of the widest demographics imaginable. The chapters in this book are a testament to their popularity, rapid growth, and great diversity. Yet, with each country or region covered only by a single chapter, this collection demonstrates just how much video game history remains to be written (or, at least, translated into English). There are so many countries that remain underrepresented or even unrepresented here, whose histories remain to be researched and written. As more national histories are compiled and translated, the overarching tale of the global history of video games will emerge, and historians and players may be able to better realize and enjoy the enormous variety of games produced in the world.

Notes

1. According to *World Energy Outlook 2013* (http://www.worldenergyoutlook.org/resources/energydevelopment/ and http://www.iea.org/topics/energypoverty/). Sources vary slightly; for example, the World Bank, quoted in the *Washington Post*, gave 1.2 billion as the figure (see http://www.washingtonpost.com/blogs/wonkblog/wp/2013/05/29/heres-why-1-2-billion-people-still-dont-have-access-to-electricity/).

2. For statistics on leisure time around the world, see http://www.oecd.org/berlin/42675407.pdf, http://www.nationmaster.com/graph/lif_lei_lei_tim_lei_tim_acr_act_oth_lei_act-leisure-time-across-activities-other, and http://www1.vwa.unisg.ch/RePEc/usg/dp2008/DP-14-En.pdf.

3. Apparently, the game currently holds the world record for the "Largest Entertainment Voice Over Project Ever" (see http://en.wikipedia.org/wiki/Star_Wars:_The_Old_Republic).

4. According to Kirsten Acuna, "'Grand Theft Auto V' Cost More to Make Than Nearly Every Hollywood Blockbuster Ever Made," *Business Insider*, September 9, 2013, http://www.businessinsider.com/gta-v-cost-more-than-nearly-every-hollywood-blockbuster-2013-9.

5. See Dave Thier, "'GTA 5' Sells $800 Million in One Day," *Forbes*, September 18, 2013, http://www.forbes.com/sites/davidthier/2013/09/18/gta-5-sells-800-million-in-one-day/; Andrew Goldfarb, "GTA 5 Sales Hit $1 Billion in Three Days," *IGN*, September 20, 2013, http://www.ign.com/articles/2013/09/20/gta-5-sales-hit-1-billion-in-three-days; and Caroline Westbrook, "Grand Theft Auto 5: Game Smashes Records to Become 'Fastest Selling Entertainment Product Ever' after Passing $1bn Mark," *Metro News*, September 21, 2013, http://metro.co.uk/2013/09/21/grand-theft-auto-5-becomes-fastest-selling-entertainment-product-ever-after-passing-1bn-mark-4061933/.

6. In particular, LEGO video games were made for Nintendo systems (Game Boy Advance, Game Boy Color, GameCube, Nintendo 64, Nintendo DS, Nintendo 3DS, Wii, and Wii U), PlayStation systems (1, 2, 3, 4, PSP, and Vita), Microsoft systems (Xbox, Xbox 360, and Xbox One), and the Windows and OS X operating systems.

7. See "Survey: Outsourcing in Game Industry Still on Increase" by the staff of *Game Developer Research*, April 2, 2009, http://www.gamasutra.com/php-bin/news_index.php?story=23008.

AFRICA

Wesley Kirinya

Games in general have been an integral part of African culture, just like any other culture in the world. Most games were created not just for fun, but also as tools to preserve culture and educate the young in the community. They were usually played on special occasions and ceremonies. For example, just before a woman becomes a wife, her father will ask her and her sisters to cover themselves. The groom will then be faced with the "strategic" task of picking out his bride.

There is not much on record regarding games in most African countries for the last half of the nineteenth century and the first half of the twentieth century. Most countries were fighting for independence from their colonial "overlords." Cultures had also been displaced. Education was becoming formal, which meant Africans were learning new tools and cultures. Formal education and culture were founded on the colonizers' cultures; for example, countries colonized by Britain adopted British culture, language, and educational curricula. This was a period of transition that created a deep division in African cultures. Some Africans embraced the colonial culture while others stuck to their traditional culture. This period set the stage for the rest of Africa's future when it came to games and entertainment in general.

Gaming history across Africa is very similar from one country to another after the mid-twentieth century. This is because most African countries had the same colonial overlords, which meant the level of modern development and the issues faced by Africans across the continent were very much the same. In the late twentieth century, Africans living in cities embraced video games. The first video games in Africa were in arcades. Some of the most popular games were PONG (1972), *Space Invaders* (1978), *Asteroids* (1979), and *Pac-Man* (1980). These games were superseded by more graphically appealing games that were still played in game arcades, mainly platform-based street fighting games such as SEGA's *Streets of Rage* (1989). Home-based gaming via consoles started becoming popular in the mid-1990s. Some of the popular consoles were the SEGA Mega Drive, Nintendo Entertainment System (NES), and Super Nintendo Entertainment System (SNES), and later the Sony PlayStation. Handheld games followed closely, with devices such as the Game Boy, Game & Watch, and generic devices that bundled "999 games in 1." The most popular games on handheld consoles were brick games and platform games. The generic devices usually had the famous brick-breaking/block-laying games in different ports. They were essentially the same game but ported in different styles.

The generation that grew up with these consoles is the foundation of today's video game development studios.

Africans living in rural areas did not have access to electricity, let alone video games. Some of the games that used to exist in the traditional cultures, for example, Mancala and Bao, also faded away as they would be seen as "backward" by the urban "elite." Throughout the 1980s and into the twenty-first century, men between twelve and twenty-five years old dominated gaming. These figures are based on my surroundings and other gamers I've talked to around the continent. It's very difficult to get these kinds of statistics because no one has been collecting them.

Influence

Video games were first played by Africans in the 1980s. These were the Africans who lived in urban areas, where there was better infrastructure, such as electricity. These Africans worked in government in their newly independent countries, taking positions previously held by their colonial masters. They had more disposable income, and they embraced modern culture. It is their children who were the first Africans to grow up with video games.

Post-independence, Africans found themselves in a new struggle. They had inherited a new culture from their colonial masters, who were from Western Europe, yet their extended family was living in rural areas with their traditional culture. There was a struggle between the new and the old. National resources were not managed correctly, which led to broken economic promises and civil conflicts. This set the stage for dictatorships, military governments, and false democracies, which in turn led to restrictions regarding electronic media. Video games, however, were seen as harmless. They neither crossed the agenda of the political elite nor had a significant audience. After all, it was just children who played games; adults wanted to be seen as more "serious."

Entertainment—in the form of music, articles in newspapers and books, and plays performed in theaters—however, started becoming a tool to fight historical injustices. West Africa is famous for musicians such as Fela Kuti, and South Africa is famous for the film *Sarafina!* (1992). Comic illustrations, such as the ones drawn by Gado, a famous comics artist from Kenya, became popular. In the late 1990s, entertainment played a role in educating Africans, leading to more democratic processes. Entertainment still continued to be a major tool that was used to reveal to Africans the state of their countries. Comedy on television became popular.

Unfortunately, the state of video games had not changed. There were no African video games largely because there were no workers with the required development skills. The education system was not as advanced as in Western countries at the university level. Well-funded educational programs were mainly professional ones, such as medicine, architecture, and engineering. These were seen as crucial for the development of the continent, and video games had no place in that agenda.

In the early twenty-first century, the children of the 1980s who grew up playing video games set the stage for video game development in the continent. The first efforts can be traced in South Africa, followed by the eastern and western parts of Africa. South Africa had the most modern economy, meaning a higher number of people living in cities. This meant they had more young people who had been influenced by technology at an early age and more advanced higher learning institutions. Therefore, they had more advanced skills in computing.

Today, video game development studios are spread around the continent. Most of the games developed in South Africa are closer, in terms of gameplay and storylines, to Western games, while games developed in eastern and western Africa are more localized to the African context. The game development studios in the northern part of Africa are mainly an extension of Western video game development companies. This mirrors the historic influence on the different parts of the continent. South Africa is one of the African countries that achieved independence quite late, in the mid-1990s. Its colonial country had more time to influence the indigenous African population, and its infrastructure is also much better than other African countries. The East and West African countries share a more common history. They were either colonized by the British or French; for the most part, they gained independence at the same time, between 1955 and 1980, and right after achieving independence, they faced similar economic and political issues such as running a new form of governance that they had inherited from their colonial masters. North Africa has a great Arabic influence. Their close proximity to Europe also allows the region to be more in touch with the Arab and European nations than the sub-Saharan African nations. Because of this, most games developed in the North African region are subject to a greater European and Arabic influence, for example, *F1 Racing Championship* (2000) for the Nintendo 64, developed by Ubisoft Morocco, and *CellFactor: Psychokinetic Wars* (2009), developed by the Egyptian company Timeline Interactive along with Immersion Games of Colombia.

Reception of Foreign Games

Foreign games were well received, mainly because the video game was a medium for children's entertainment that was not seen as interfering with political circles. Foreign games were introduced in the 1980s via arcades. Children would save up coins and play games to their hearts' content. In the 1990s, a few families living in cities would have had game consoles and handheld consoles. During school holidays, children would gather at a friend's home and play video games while the parents were away at work.

In the late 1990s, more graphically appealing games (with three-dimensional graphics) and more powerful game consoles emerged. Games such as *DOOM* (1993) and *Quake* (1996) were quite popular. The games also became more violent in terms of content. However, since parents at this time did not grow up with video games, they were less interested, and so it was still game-on for the teenagers. In the 2000s, the teenagers became adults, some playing fewer games than they used to. They were familiar with games and noticed the aggressive culture in games, and today's parents are more cautious about the games they let their children play.

Foreign games have also influenced how games are played. They drove the transition from arcades to game consoles, to handheld consoles, to home computers, to Yahoo! and Facebook games, and now to mobile games. Foreign games are also setting the standard for games, and the audience expects games developed in Africa to be of a similar quality standard, for example, the *Tomb Raider*, *FIFA*, *Resident Evil*, and *Super Mario* series of games. For African game developers, then, foreign games are both a blessing and a curse. They are a blessing because they are growing the industry in the continent by building a bigger and bigger audience, but at the same time, they are a curse because African development studios lack the resources and skills to develop games to the same standards.

Domestic Game Production and Export

Production of domestic video games began emerging in the early 2000s. One of the first companies to develop games was Clockspeed Mobile from South Africa, founded by Herman Heunis. Clockspeed's attempt to build an SMS-based massively multiplayer mobile game evolved into an instant messaging platform called MXit.

South Africa continued to pioneer video game development when a new studio called Luma Arcade was founded in 2006. Luma Arcade specializes in smartphone games and produces graphics and animation in addition to its video games, which include *Mini #37* (2007), *Marble Blast Mobile* (2008), *Flipt* (2009), *REV* (2009), *The Harvest* (2010), *Racer* (2011), and *Bladeslinger* (2012) (see http://lumaarcade.com).

In 2006, I founded a company in Kenya called Sinc Studios. Sinc Studios developed a 3-D PC game titled *The Adventures of Nyangi* (2007), which sold a few hundred copies. One of the unique aspects of the game was that the game engine development was done locally, and the interest and publicity generated by the game went on to build a good foundation for game development in Africa. The game was featured in leading local and regional television programs and newspapers such as *Africa Journal*, *Art Scene*, *Nation Newspaper* (Kenya), and a Ghanian newspaper. Several blogs also wrote about the game, including The WhiteAfrican blog, which highlights the tech scene in Africa (http://whiteafrican.com/2007/04/27/an-african-3d-adventure-game/).

Around the same time, BlackSoft was cofounded in Ghana by Eyram Tawia, along with Albert Dodoo, Justin Dakorah, and Opuni Asiama. Tawia developed his first video game, *Sword of Sygos*, in 2005, basing it on Greek mythology. It was his great passion for comics that led him to create the story for the game. Due to the small number of video game developers, Eyram Tawia and I decided to form a new video game development company in 2009, Leti Games. "Leti" is a word from the Ewe tribe of Ghana meaning "star." The company focused on the development of multiplayer mobile games, which include *iWarrior* (2009), *Bugz Villa* (2009), *Kijiji* (2009), *Street Soccer Battles* (2011), *Mr. CEO* (2012), *Haki* (2012, in collaboration with Afroes), *Haki 2* (2013, also with Afroes), and *Ananse* (2013) (see http://www.letigames.com). In addition to games, the company also produces comics, instant messaging software, and fun mobile applications.

More game studios have emerged in the late 2000s, including MALIYO Games (founded in 2012) and Kuluya (founded in 2012), both in Nigeria; Planet Rakus (founded in 2011) in Kenya; Afroes (founded in 2009) in South

Figure 1
Games made by Luma Arcade: *Racer* (2011) (top), *The Harvest* (2010) (center), and *Bladeslinger* (2012) (bottom).

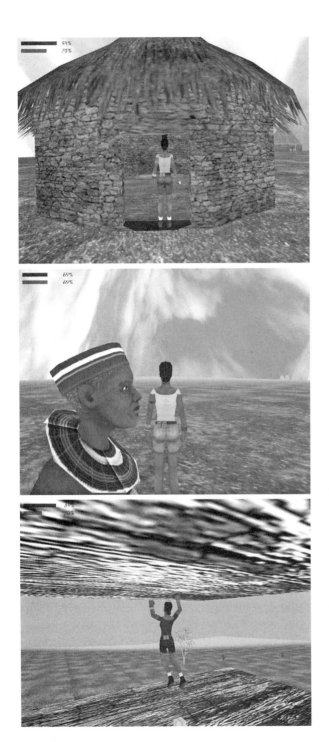

Figure 2
Screenshots from *The Adventures of Nyangi* (2007) by Sinc Studios.

Figure 3
Games made by Leti Games: *iWarrior* (2009) (top) and *Ananse* (2013) (bottom).

Africa, and the French giant Ubisoft's office in Morocco (established in 2005). Indeed, specialization has also started to emerge. For example, Afroes focuses on mobile-based serious games, such as *Champ Chase* (2010), *Moraba* (2012), *Take A Champ* (2012), *Haki* (2012), and *Haki 2* (2013) (see http://www.afroes.com); MALIYO focuses on Flash-based fun and serious games (see http://www.maliyo.com/games); Kuluya on Web browser games (see http://kuluya.com); Planet Rakus on mobile games, such as *M3 Racer* (2012); and Leti Games on purely fun games. Between 2000 and 2012, video game development in the continent grew from one company to almost ten. Each of these companies has played a key role, not just in building the technical skills, but also in building a viable industry. Leti Games leverages its technical strength to build the required technical skills, partnering with Afroes in concept development and production. Other companies in the game space have emerged, such as NexGen, which organizes LAN game parties that video game development companies can take advantage of to popularize video game development among the youth.

The general look and feel of video games developed in the continent depends on the region. South African and North African development studios are most similar to the American and European games in terms of graphics and storylines. This has largely to do with their influence. The game studios in East and West Africa are more localized and are more focused on African-themed content and storylines.

The export of video games occurs via game portals such as Ovi Store, the Google Play store, and others. Planet Rakus's *M3 Racer* (2012) mobile game has had almost one million global downloads. Most games are not developed and shipped as hard copies, since most games are produced for mobile devices and the Web;

Figure 4
Images from the game *Haki* (2012) by Afroes.

therefore, there is no need to produce hard copies. But the revenue from export is quite small, and almost all downloads are free versions of the games. Developers are exploring strategies such as including advertisements in the games in order to boost revenue from foreign players.

Indigenous Game Culture

Video game culture in Africa is mainly centered on game-playing parties with a group of players. Playing a video game is a social and fun event where players enjoy watching other players play. Watching a game being played provides useful insights, and, in a way, the players who are watching are passively playing the game. Game playing sessions are usually held at someone's house or at Internet cafés. The games played are usually competitive games, such as fighting games like those of the *Tekken* and *Mortal Kombat* series. These kinds of games allow players to take turns playing the game, as one player is defeated and another steps in. However, if there are enough game controllers, players will usually play games that take longer to complete, such as action-adventure games.

The culture of playing games in a large social setting extends from other entertainment events in African cultures such as weddings and child-naming ceremonies. Anything fun needs to be shared. The more people are involved, the more interesting things are. The best way to share in the experience is to participate. This is partly why video games have been accepted in Africa.

Video Game Studies in Africa

There is little formal education for video game development in the continent. Most of the formal education institutions are in the southern and northern parts of the continent where video game development had an earlier start. In East and West Africa, video game studios are formed by bringing together teams of talented developers and artists who then work together to understand how their talents can be merged to develop a video game. Usually one member of the team will have an understanding of how everyone's talents can be brought together to develop a game and will take the lead. Most developers have their roots in computer science, while the artists have developed their skills in universities and colleges that train in animation and digital arts. These institutions provide training in digital arts that are geared more toward the advertising industry, but they also touch on game development. One such institution is Shang Tao in Kenya.

There are IGDA chapters in Nigeria and Ghana, which were founded by Benedict Olumhense and Eyram Tawia, respectively. Their goal is to build interest in game development. Eyram Tawia is Ghanian and has had a passion for comics and video games since his childhood. Benedict Olumhense is from Nigeria and is passionate about video game development. He has worked on the Global Game Jam in Lagos and participated in the African Game Conference. His company, CYPHA Entertainment, helps to identify gaming opportunities in Africa.

My Video Game Company Start-Up Experience

At the time of setting up my first video game company (which was also my first company), I only knew of one other video game development company, which was called Luma Arcade and was based in South Africa. I was excited to know this because I saw it as a great opportunity; however, it turned out to be a great challenge, one that taught me great lessons that I treasure and apply today. Below are the challenges I faced when developing my first video game, *The Adventures of Nyangi*.

No Formal Studies

My passion for developing video games had engulfed me to the point that I lived and breathed it every day. I searched for colleges and universities that offered formal game development courses; however, I was disappointed to find none. I therefore decided to postpone my university education and focus on my video game development skills full time.

As time went on, I realized the value I was missing by not being able to learn in a formal institution. Formal studies offer the chance to learn from people's failures and avoid their mistakes; however, I was learning from my own mistakes. The process of developing games was a huge trial-and-error exercise—very tedious and slow, but at the end of the day, it gave me great understanding as to how to develop video games.

Internet Access

Without a source for formal learning, the next best place to gain knowledge was from manuals and articles posted online. The greatest online resource I found was http://www.gamedev.net. It is the largest community of independent video game developers in the world. They are enthusiastic to share their ideas and advice with other independent video game developers.

However, this required Internet access. At that time, the Internet was not as accessible or inexpensive as it is today. I accessed the Internet from public Internet cafés. Each trip involved noting items to research and downloading as much information as I could so that I could review it at home. This was not as convenient as simply launching a browser at any time and getting the right information all from the comfort of a chair.

Skilled Partners

With little video game development studies and a lack of formal studies, there was no community of skilled video game developers with whom I could work. The production of video games not only involves technical work, such as system design and programming, but creative work as well, such as concept writing and artwork. I found myself having to wear both shoes. I learned how to use the programming tools as well as use 2-D and 3-D artwork and animation software. There was also no one to give constructive criticism on how to improve the game. Therefore, I was the concept creator, artist, system designer, programmer, and QA (quality assurance), all in one!

Market Information

Once I completed the game to the point that I thought it was market-ready, I was faced with the next challenge. I had to get the game out there! I recall thinking of all the creative ways to get my game known on zero marketing budget. I used my contacts to build a little buzz, and to my surprise, I got the attention of some of the largest news media companies in the country and the continent. I did TV and newspaper interviews. I thought these would lead to great downloads, but I came to realize it only generated buzz and did little to achieve actual sales and downloads.

I leveraged on this buzz to build contacts who would give me advice about the market. I learned that marketing is very costly and needs careful planning to be effective. I would need to know how many computers in the country could play my game, who my target demographics were, their willingness to buy my game, and many other answers.

Distribution Channels

When I completed my game, my first thought was to use the Internet for distribution. However, after a couple weeks of having the game on the Internet, I realized that this is heavily tied to marketing, so although people were able to get to my game, there were lots of other games available to download online as well.

I contacted a couple of video game portals that had communities of millions of players. However, they declined to host my game on their portals because it was not the kind of game that their visitors played, nor was it up to the quality of games they offered.

I then decided to print a couple of disks and sell them at local stores that sold imported video games. My game was one of the top four best-selling games. It was quite exciting for me! However, I did not have the money to print disks in large quantities. I was also a bit unsure as to how exactly the buyers were responding. Was it because they really liked the game and they recommended it, or was it because they saw it was one of the first video games to be developed in Africa?

No Mentors or Investors

I had spent all my money and energy on my first game. The experience I gained while launching my game taught me that to move forward I needed to find someone to mentor me and invest money into my company. I had made a lot of good contacts, and I even found an investor. However, the contacts did not have time, and the investors did not understand the video game development industry. They knew video games were something growing and interesting, and lots of kids love playing video games, but they depended on me to execute the video game at a standard of international quality.

I managed to add a software engineer and an artist, both of whom were my friends and were inspired by what I had done. However, we realized that the task we had—to develop my game into a multiplayer game with an even richer 3-D experience—was beyond us. Eventually we had to close the company, but everyone walked away with valuable lessons.

Conclusion

My experience from developing my first video game was amazing, risky, and foolish, but at the same time, full of great lessons that no one would have ever given me. The experience has made me a much better entrepreneur and game developer!

The Future of Video Games in Africa

The increase in the number of game development studios in the continent is quite positive. There is a growing number of young people interested in game development in the region. Some of these young people join the game development companies as interns, who later stay on to become a core part of the team. For example, Donald Apoku joined Leti Games as an intern after completing his undergraduate degree in computer science. His thesis was based on creating great ambience in video games. His understanding of the video game development process quickly made him part of the core team, focusing on HTML 5 and Android.

Specialization has also started to emerge in concept development, artwork, technical implementation, and business model development. Afroes in South Africa focuses on concept development for serious games, which are basically games that are aimed at educating the public about social issues.

The number of women playing games appears to be growing. Women entered into the game-playing arena with casual games such as *FarmVille* (2009) on Facebook. They also enjoy casual games on mobile devices. Men are active on the same channels, too. The number of mobile phones in Africa is higher than the number of PCs and consoles combined, thus making them the best platform for game development in the continent.

One of the greatest challenges is electronic microtransactions. Most global gaming business models are built on charging small fees at regular intervals. This is difficult in the continent where most people do not have credit cards and do not put much trust in electronic transactions. But with some of the fastest-growing economies in the world located in Africa, including Ghana, Nigeria, Kenya, and South Africa, the future of game development is bright. More people in the continent are going outside the norm to explore new fields such as video game development. More people also have disposable income. Mobile money services are growing and diversifying, meaning people are able to complete transactions from their mobile phones.

Video games in Africa have a promising future. A great mix of stories, growing income, increased Internet connectivity, and more widespread use of smart phones is creating a fertile ground for the emergence of video game development studios. Five years ago, there were fewer than three video game development studios. By 2013, there were more than ten. Though challenges exist, it's up to the game development community in Africa to seek innovative solutions that will see Africa take center stage in the global video game marketplace.

ARAB WORLD

Radwan Kasmiya

[Editor's note: Originally, in the early planning stages of this book, Radwan Kasmiya, game designer and cofounder of the company Falafel Games (established in 2011 with Vince Ghossoub), had planned to write a chapter on video games in the Arab World; however, due to the political instability of his native Syria and the resulting upheavals that followed, he and his family were forced to flee as refugees, a harsh reminder that the relative peace and tranquility necessary for academic research to occur is still not something enjoyed by all. I had asked Radwan to write this essay since he had written the "Middle East" and "Quraish" entries in my *Encyclopedia of Video Games* (2012), so in order to fill in for the missing essay, ABC-CLIO/Greenwood Press graciously allowed me to reprint those two entries here. The references for both have been combined into a single section at the end.

And so it was until the book was about to go to press; then I finally heard back from Radwan. He had relocated with his family in China and found work in the video game industry there. He was able to do so because Falafel Games has offices in both China and the Middle East. He also reported that recently Falafel Games had become the only developer of online RPGs in the Middle East and that their flagship game, Knights of Glory (2011), had more than a million players (as of this writing in February 2014) and was voted the best browser game in the Middle East for three consecutive years (2011, 2012, 2013). The game's success has inspired MBC, the major TV network of the Middle East, to invest in Falafel Games, and the company is also discussing further growth with other strategic partners in the region.

Although not enough time remained to produce a complete chapter, Radwan did compile some additional, updated information regarding the Middle East, which appears after his encyclopedia entries below.]

Middle East

The fast progress of video game culture in the 1970s and 1980s reached Middle Eastern markets about the same time that it was spreading through households in Japan, the United States, and Europe. Early game consoles were popular in the region, and the senior generation still refers to all kinds of video games as "Atari" games, a result of the dominance of the Atari VCS 2600, socially and commercially. Since the introduction of these game consoles and their succeeding 8-bit home personal computers, such as the Commodore 64 and the ZX Spectrum, the demand for video game consoles with an Arabic-friendly interface grew to a high level, encouraging Al-Alamyyeh, a Kuwaiti company, to start producing an Arabic home computer called the Sakhr (which means "rock" in Arabic) in 1981, which was essentially based on the well-known Japanese MSX. The product was a success and became popular with middle-class families in the Middle East.

During that time, Al-Alamyyeh made an attempt to convert the BASIC computer language into Arabic, and many applications were developed to support domestic users; but in general, the software industry in the region was essentially targeting the corporate applications market. Hardly any of these programs were Arabic video games, and the reason was obvious: nobody would invest in a market flooded with cheap pirated games. Eventually, it was up to the indie developers to start video game production, such as *War73*, developed by Radwan Kasmiya in 1999, and Mohammed Hamza's *Stone Throwers* in 2001. Both were two-dimensional arcade-style games for personal computers and were based on the Arab-Israeli conflicts.

In September 2000, *Under Ash*, a PC first-person shooter (FPS) game, was previewed to the public at a Damascus book fair, and it was considered to be the first commercial video game in the Middle East based on the Palestinian conflict. Created by Radwan Kasmiya and published finally by Dar Al-Fikr in 2002, *Under Ash* targeted Arab gamers, did not contain English subtitles, and was never sold outside the region (limited copies found their way to Europe unofficially). The game received a lot of praise, managing to sell more than 100,000 units within the first six months of release—a considerable quantity, even compared with internationally best-selling titles' sales in Middle Eastern markets. The success of the game naturally led to a sequel, and so Afkar Media was established and became the first independent game studio in the region.

Two years later, *Under Ash II* (also known as *Under Siege*) was released in 2004 but did not hit the retail shelves until late 2005 because the publisher was trying hard to access new markets with little experience in simultaneous releases. *Under Ash II* was received positively because of its realistic storyline, advanced graphics, and artificial intelligence (AI) for the time. The game has been described as a "docugame" because all the levels are based on actual events documented by the United Nations' records from 1978 to 2004. This time around, the game also supported English subtitles.

The publisher decided to split the game into two parts to graze more money from enthusiastic gamers, so they released *Under Siege: Path to Freedom* and *Under Siege: Remnant of Human*. That said, the English version of this game, *Under Siege: Golden Edition*, was, as of late 2010, never released—a strange tactic that was repeated later with *Quraish* (2005), the first Arabic real-time strategy (RTS) game, also developed by Afkar Media. *Quraish* was highly anticipated by Arab gamers because it was based on actual historical (conquest) wars during the early Islamic periods, a highly revered period in the region. The game was finally released commercially in 2008 after spreading its four campaigns on four independent package designs and one "Golden Version" with all the features.

During that time, another Syrian company, Techniat3D, developed *Zoya: A Warrior from Palmyra* (2002), an adventure game inspired by the *Tomb Raider* series. Sadly, the game sold fewer than a hundred copies in the region, and many claimed that the "improper attire" of the female warrior featured on the cover was the reason players avoided the game. A revised version was made by Afkar Media, and the game was re-released as

Victory Castle; however, this version didn't meet with much success either, selling just over a thousand copies. Consequently, Techniat3D was closed down in 2003.

On the edge of the Middle East, Imaginations FZ, LLC, was founded in Dubai (United Arab Emirates) in 2003 and managed to create two games before it shut down in 2005: *Legend of Zord* (2003), featuring stories from the *Arabian Nights*, and *Wadi Basheer* (2005), a poor racing game with a Middle Eastern twist. A group of Lebanese developers came out with their first video game, *Special Force* (2003), which focused on Hezbollah military operations against Israeli occupation forces in Lebanon. The game was quickly adopted by Hezbollah and merged into its propaganda machine. The outcome was a short video game (three levels) with mixed messages of religion and politics. That attitude led to the ban of the game in many Middle Eastern countries; however, this led to greater media attention and developers were rewarded with more resources to come up with a sequel in 2007. The developers established their new studio, Might 3D, and created *Special Force 2: Tale of the Truthful Pledge* (2007), which was based on the 2006 Lebanon War between Hezbollah and Israel. It was presented less than a year after the actual war, and the game was in Arabic with an unofficial patch to add English-language subtitles. Independent critics tend to compare these games with *America's Army* (2002) because both are funded by political parties to promote specific views (*America's Army* is funded by the Pentagon). Apart from the political background of the *Special Force* series, which was controversial but did not boost sales, credit should go to the series developers because they started using the Genesis3D engine for their first game and then managed to develop their own game engine for *Special Force 2*.

In 2003, the game *Jenin: Road of Heroes* was published by Turath (a Jordanian e-book and software publisher) but didn't sell much (4,000–8,000 units), even though it copied the same political concepts regarding the Israel-Palestine conflict. The next game from the same developer was *Wild Races* (2008), a funny animal racing game that featured bareback riding on eight animals at six tracks. It was published by Andalussoft but didn't sell well either, and both developer and publisher stopped venturing any further into hardcore video games. The Jordanian company Quirkat is set to become a main player in the Middle Eastern video game industry and has invested in the localization and refurbishment of well-known international video games. Their first product was *Arabian Lords* (2006), a city-building strategy game from BreakAway Games, published by RED Entertainment Distribution.

Ultimately, it seems that the home video game industry was not generating enough money. Quirkat's next ventures were in new mobile portals and casual games, including *Al-Moosiqar* (2009), a casual online game that lets players play Eastern music on an Oud (an instrument similar to the European lute), and *Tariq's Treasure* (2008), a puzzle/strategy mobile game, published by BreakAway. Quirkat also developed an advergame for HTC Middle East to promote the release of a new handset model. The game, along with the strong HTC brand, attracted thousands of online players, according to its official Web site.

In 2007, the Egyptian studio Al-Khayal was established with support from Egyptian technology funds, and they created *Buha* (2006), a comical adventure game based on a popular movie character (Mohamed Saed); however, that game was the only one the company produced. In 2008, the Syrian company Joy Box was established, which targeted the casual games and mobile games market. Its website has many types of games, ranging from two-dimensional arcade-style games that are basically sold to Al-Majd (a local quiz TV channel) to three-dimensional prototypes of adventure games and racing games.

During the 2009 global financial crunch, Afkar Media had to postpone some of its major game projects, although it did manage to finish *Road to Jerusalem* in late 2009, a comedy/adventure computer game published by Fares al Ghad, still unreleased as of 2014. In other markets, Maktoob, a successful Jordanian company, built its popularity through years of service and managed to get Yahoo! to acquire some of its shares (due to the huge user base). Maktoob has distributed free browser-based games to the Middle East, including *Khan Wars* (2009), neck-and-neck with the company Travian, and made a good profit on it. Alongside localized browser-based games, such as *Damoria* (2009) and *Stardoll* (2009), Maktoob is running a dedicated Web site for casual games and online flash games, with thousands of original Arabic and Middle Eastern titles. Tahadi, based in Dubai, UAE, is a major publisher of online games; it localized and operates famous massively multi-player online role-playing games (MMORPGs) including *Runes of Magic* (2009), *Ragnarok Online* (2002), *Crazy Cart* (2009), and *Heroes of Gaia* (2010). Falafel Games, a new developer with studios in the Middle East and China, is planning to publish a series of local content games starting with *Knights of Glory* (2011), an MMO browser game about Muslim conquests.

What all the titles developed in the Middle East have in common is their release platform; most of them were made for the IBM PC. This is because PCs are a more affordable platform for publishers and also because of console manufacturers' attitudes toward the Middle East video game market. For example, Microsoft does not support Xbox Live services in the region, and although Sony's consoles are popular, many Middle Eastern countries are not even included in their PlayStation Network list of supported countries.

Quraish

Quraish is a real-time strategy game with three-dimensional graphics developed by Afkar Media in 2005, designed by Radwan Kasmiya and published by Dar Al Fikr in 2007. The name comes from "Quraish/Kuraish," a famous Arabian tribe whose descendants include the prophet Muhammad (AD 570–613) and the most famous Muslim leaders, the Omayyad and Abbasid caliphs. The game's story is spread over four historical successive campaigns, covering Middle Eastern history before Islam until early Muslim conquests (AD 600–638), with each campaign consisting of a number of levels. Players must finish each level to be able to play the next one.

The game's first level, "First Encounter," depicts life in Arabia before Islam, the first Arabian tribes' coalition, the defeat of the Persian army at the battle of Dhi Qar in AD 609, and the appearance of Islam. The second level, "Apostasy Wars," is the story of the Wars of Apostasy and battles that reformed the Muslim State. The third level, the "Conquer of Persia campaign," follows the history of the Muslim conquests in southern Iraq and battles with the Sassanid Empire. The fourth level, the "Conquer of Syria campaign," tells the story of Muslim armies sweeping north of Arabia and beating Byzantine armies at the doors of Jerusalem. The narratives of every campaign are performed by non-Muslim characters, allowing players to experience different opinions about Muslim conquest wars. Even though *Quraish* features real historical characters, many top characters were not shown directly due to the sensitivity of showing images of the Prophet Muhammad and some of his major followers in media across the Middle East.

Players can play fictional and customized death matches, choosing from the main four factions of the game: Arabs, Bedouins, Byzantines, and Sassanids. Recommended system requirements for the game include Windows XP or Vista, a CPU speed of at least 1.8 MHz, 64MB of VGA memory, at least 512MB of RAM, and 1.2GB available on the hard drive. Multiple versions of the game have appeared, including four independent Arabic versions, each of which includes a single campaign, and the Golden version, which includes all four campaigns and English and Arabic interfaces. The game is the first Arab/Muslim strategy game and the first to handle this period of history.

Updated Information as of February 2014

From 2010 to 2013, Jordan witnessed a boom in video game industry growth as the government's sudden interest in IT and the software development business generated several serious initiatives. King Abdullah II is said to be a video game fan himself; he even participated in the beta testing of *Operation Arabia* (a first-person shooter on Facebook, see https://www.facebook.com/OperationArabia) created by Wizards Productions (2010–2013). A group of Jordanian game company and service providers including Quirkat, Maysalward, Wizards Productions, TakTek, Crown IT, and Gate2Play created the Jordanian Gaming Task Force to share experience and resources (see http://www.gamingtaskforce.com/home/about-us). Unfortunately, most development studios have closed due to limited funding or unclear strategy, and Mayslward, a mobile games studio, was the only remaining developer in Jordan by 2014 (http://www.maysalward.com).

In Egypt, many independent developers have tried to publish culture-themed games or political-themed games. Among them, Nezal Entertainment is the most successful company, which managed to attain investments of more than a million dollars with *Elmadinah* (2013), an Arabic-style *FarmVille*-like game. *Crowds Vote* (2012) is another distinguished game from Nezal based on the Egyptian revolution of 2012 (see http://nezal.com).

Unearthed: Trail of Ibn Battuta (2013), by SEMAPHORE (established in 2011, see http://www.sema-phore.com), is the first game developed in Saudi Arabia and is an adventure game with settings and locations in the Middle East. It was published on many platforms including PC, PS3, and mobile (see http://www.unearthedgame.com).

One of the most successful games in the Middle East was produced in Lebanon. *Pou* (2013), a Tamagotchi-style pet game by Paul Salameh, is very popular in Europe and the Middle East. The app made the global "Top 5 Paid iOS Apps" list in January 2014, and has experienced between 260,000 and 320,000 free downloads per day on Android. So far, *Pou* has hit the number-one spot in the iPhone kids' games category in ninety countries (see http://www.pou.me). Also in Lebanon, Wixel Studios, a veteran casual games developer, closed down as of February 2014. Its last hit title, *Survival Race* (2013), didn't do well in the marketplace (see http://www.wixelstudios.com).

References

Afkar Media, http://www.afkarmedia.com.

Al-Khayal, http://www.khayalie.com.

Al-Moosiqar, http://fuzztak.com/fuzztak_Public/Fuzztak_public_master_Arabic.aspx?subcatid=69&Lang=1&Page_Id=3095&Menu_ID=13.

Arabian Lords, http://www.arabianlords.com/Public/arabic_main_public_master.aspx.

Damoria, http://damoria.maktoob.com/?aip=topmenu.

Falafel Games, http://www.falafel-games.com.

Joy Box, http://joyboxme.net/main.

Maktoob, http://games.maktoob.com/?utm_source=maktoobhometab&utm_medium=link&utm_campaign=home-testing.

Quraish, http://www.quraishgame.com.

Sisler, Vit. 2006. In videogames you shoot Arabs or aliens—Interview with Radwan Kasmiya, *Umelec/International* 10 (1):77–81, http://www.digitalislam.eu/article.do?articleId=1418.

Special Force 2: Tale of the Truthful Pledge, http://www.specialforce2.org/english/index.htm.

Stardoll, http://stardoll.maktoob.com/ar/?pid=25423.

Tahadi, https://www.tahadi.com.

Under Siege, http://www.underash.net.

Wild Races, http://andalussoft.awardspace.com/WildRaces/WildRaces.html.

ARGENTINA

Graciela Alicia Esnaola Horacek, Alejandro Iparraguirre, Guillermo Averbuj, and María Luján Oulton

In Argentina over the last thirty years, the consumption of technology in general, and video games in particular, has increased due to the new distribution platforms (social networks) and the massive use of mobile devices (such as tablets and smartphones). These "cultural products" (Garcia Canclini 1999) have allowed millions of users to be connected online simultaneously. These global technologies helped digital games become a mass-consumption product.

Argentine research from 2005 to 2011[1] discovered that video games account for 70% of the cultural consumption of youths age ten to eighteen years old. The profile of the typical gamer has changed and is no longer a computer-savvy young boy who spends many hours per day playing video games. Nowadays, video game culture has reached several generations and a broad demographic, and the experience of gaming varies greatly, depending on the type of application, platform, or device chosen. The video game industry developing in Argentina is focusing on this expansion, designing products according to this variety of platforms and user profiles.

Due to the expansion of the global market, activity in Latin America has increased, especially in Argentina, Brazil, and Chile. As a consequence, in 2000, the Asociación de Desarrolladores de Videojuegos of Argentina (Argentine Videogame Developers Association, ADVA) was founded by young Argentine dabblers. This association works actively in one of the most dynamic areas of Latin America: video games. This sector employs young professionals in charge of programming, design, illustration, script, and music.

In the beginning, Argentina's advantage in the development of video games was the low cost of production while keeping its high quality. Nowadays, the Argentine industry has become professionalized, and their services are better valued, achieving creativity and capacity.

Today, video games can be defined as a "multimedia environment of cultural convergence that requires the confluence of disciplines such as cinema, music, video, animation, in an immersive technological system" (Esnaola Horacek 2013). Video games can help connect generational interests among adults and teenagers, professors, and students because everybody shares this ludic environment. Video games are also a venue for transmedial storytelling, mixing fantasy and adventure in worlds where gamers identify constantly with their avatars.

From Gamers to Developers and Game Designers

In Argentina, the video game industry is very new, with rapid development as evidenced by the growth of employment: 156% from 2009 to 2012, with the annual income of the sector increasing 342% over the same period, according to the data of the National System of Cultural Statistics, Argentina.[2] Also, 95% of that amount is for export earnings.[3]

Since 1988, more than a few firms have established offices in Argentina, such as Sabarasa (1996, Javier Otaegui, http://www.sabarasa.com), Evoluxion (1996, Santiago Siri, http://www.evoluxion.com), Cyber-Juegos (1997, Ariel and Enrique Arbister, http://www.cyberjuegos.com), Codenix (1999, Ariel Manzur, Juan Linietsky, Alejandro Iparraguirre, http://www.codenix.com), LatDev (2001, Guillermo Averbuj, http://gamester.com), NGD Studios (2002, Andrés Chilkowski, Nicolás Lamanna and others, http://www.ngdstudios.com), and Moraldo Games (2003, Hernán Moraldo, http://www.games.moraldo.com.ar). Other companies that have established their offices in Argentina include Gameloft (2002), Globant (2003), Three Melons (2005), QB9 (2005), and Metrogames (2008). Some companies work exclusively in the video game industry and are strong exporters: 95% of national production is intended for the international market, and 80% of local companies have customers in the United States.

There are at least sixty-five companies that develop video games in Argentina, in addition to the community of independent developers (organized as the Comunidad INDIE). These companies are producing USD $50 million in sales volume per year and generate more than two thousand jobs for highly skilled professionals. The games created in Argentina are sold to distributors and generate up to USD $500,000 for large publishing houses. As such, Argentine developers are well recognized throughout the world.

Since 2000, the ADVA (http://www.adva.com.ar/) has annually organized an Exhibition of Argentinean Games (EXARGA, http://expoeva.com/). It manages CODEAR (Concurso Desarrolladores de Videojuegos Argentinos), and COREAR (Concurso de composicion de musica para videojuegos), a contest for new developers (CODEAR) and for video-game musicians (COREAR) whose winner receives an award for the best musical composition for video games.

The ADVA is now the leading international reference for any company or individual interested in finding new business opportunities and connecting with the community of game developers in Argentina. According to the ADVA's statistics for early 2013, the video game industry in Argentina has 90% of its games exported to the United States, Europe, and Asia. The ADVA estimates that there are seventy companies (most of which began operating after the year 2000) and 2,000 professionals (including designers, artists, illustrators, programmers, engineers, writers, testers, and musicians), with an average age of twenty-seven, working in different areas within the development process: mobile phones, the Web, advergames, PC games, edutainment, console-based games, and social games. In recent years, developers from other regions of Argentina have begun meeting in small associations linked to the ADVA, such as the Asociacion de Desarrolladores de Videojuegos de Rorario (ADVR).

Also, talented Argentines working at game companies are able to offer creative and innovative ideas to complementary industries such as film, television, and advertising. Now that the video game sector has the support of the government, careers for future game developers can be found in the country.

The Early Years of Video Games in Argentina, 1980–2000

Most arcade games of this period were introduced and licensed by foreign game developers, such as Atari's *PONG* (1972) or Taito's *Space Invaders* (1978). In the early 1980s, Namco's *Pac-Man* (1980) and Nintendo's *Mario Bros.* (1983) were the hottest games and the first ones played by many local players. Their unlicensed merchandise flooded the market. The first domestic production of a video game in Argentina was the popular "card-trick game" *Juego de Truco* (1982), developed by Ariel and Enrique Arbiser, and the 1986 version was updated to have CGA graphics (see http://www.dc.uba.ar/people/materias/pf/truco). This game had a great influence because it was based on a very popular Argentine card game.

Naomi Marcela Nievas, who used the pseudonym "Sharara," designed another early video game. She was one of the first national game developers who developed the game *Scrunff* (1985), programmed in Assembler 8086 on the Z-80 for MSX consoles, and even participated in the creation of a game compatible with the SEGA Genesis console in 1990. Her other works include digital jukeboxes, electronic bingo, and digital slot machines. Nievas was one of the first women to join the video game industry; today, others are working in this sector, but there are only a few women employed at the level of their male counterparts.

Arcade games were popular in the 1980s and 1990s, especially among young males. Some of the most popular games were *Scrunff* (1985–1988), *Guip* (1985–1988), the *Arkanoid*-like game *Killers* (1985–1988), *Asteroids* (1985–1988), *Bichos* (1985–1988), *Gussy* (1990), and *Burbuja John* (1992), a four-color labyrinth game developed for CGA monitors. Other games include *Juego de Escoba 15* (1994), developed by Ariel Arbiser (see http://www.dc.uba.ar/people/materias/pf/truco); *Regnum* (1995) by the Conde Brothers, Paul Zuccarino, and Andrés Chilkowski; Cristian Soulos's fighting game *Nemen vs Llovaca* (1995); *Regnum 2* (1996), which appeared on CD-ROM; and the massively multiplayer online role-playing game (MMORPG) *Regnum Online* (2007), published by NGD Studios (see figure 1). Caimán Company (http://www.caiman.com.ar/) has developed software since 1996, including the *Yo, Matías* series of games (based on the comic strip) and the collection *6 Grandes Juegos Vol. I* (release year unknown), which contains games such as *Mahjong, Memotest, Tetris, El Ordenador, Generala*, and *El Recolector*. A second volume, *6 Great Games Vol. II* (release year unknown), contained games such as darts, submarines, puzzles, Reversi, Domino, *Batalla Naval*, and several educational games.

Among the great local productions, a few titles represent the history of Argentinean video games. One of the most memorable titles, *Fútbol Deluxe 96* (1996) (see figure 1), was designed by Santiago Siri and published by the company Strategy First (http://www.strategyfirst.com/). *Fútbol Deluxe 96* was written in Visual Basic and Flash and was the first Argentine game to be publicized internationally. The game is a tactical football simulator and assumes the role of the trainer, while the player must direct a league team from Argentina. The

Figure 1

Regnum (1995) (top left); *Fútbol Deluxe 96* (1996) (top right); *Malvinas 2032* (1999) (bottom left); and *Argentum Online* (2000) (bottom right).

player can buy and sell players, make all kinds of decisions, and compete in the Copa Libertadores, Conmebol, and Intercontinental for maximum achievement.

Other games of note are *Nuku* (1997), a horizontally scrolling platform game compatible with Game Boy Advance, developed by Juan Linietsky and Pablo Selener (see http://reduz.com.ar); *Bizarreh* (1998), a platform game, and *Edia* (1999), a ship-game for two, both developed by H. Hernán Moraldo; the platform/adventure game *Dapharen's Fear* (1999); *Malvinas 2032* (1999), a military strategy game developed by independent developer Javier Otaegui and rereleased in 2000 by Edusoft (see figure 1); *Cóndor* (2000), a platform game designed by Nicolas Vinacur, Juan Garcia, and Marcelo Rubinstein (http://www.sabarasa.com); and *324—El chofer mercenario* (2001) and *Porko vs Dex* (2002), both designed by Leandro Barbagallo and Diego Trasante. Most of the video games developed in Argentina are registered at http://le_porko.tripod.com/. (See http://members.tripod.com/le_porko/historia/ for examples of games developed in Argentina.)

Argentine history is represented in *Malvinas 2032* (1999), a real-time strategy game in which the player commands Argentine forces and tries to retake the Falkland Islands for Argentina. It was developed by Sabarasa Entertainment. *Malvinas 2032* takes place in 2032, the fiftieth anniversary of the Malvinas War. On that date in the future, there is a deep worldwide oil shortage, and the Argentine government has asked the British government to share oil exploration in the Malvinas. Given the refusal of the British, the Argentine president decides to retake the islands. There the action begins for the player, who will command the troops to destroy the enemy and reclaim the islands. The game has a total of twenty-three stages, which, if met, lead to the ultimate recovery of the Malvinas islands. The first stage is the landing at Puerto Argentino, and as the stages progress, the game's difficulty increases. The game ends when all the land is taken. While the graphics do not achieve excellence, they are still well done, and the game gives a historical account of the Malvinas War of 1982.

One of the most famous local games, *Argentum Online* (2000), a free, online, role-playing game, was developed by Pablo Márquez (alias "Gulfas Morgolock"), at age twenty, aided by Fernando Testa (a fellow university student in the city of La Plata) and Javier Otaegui. This was Argentina's first online game, and the test server used a domestic broadband connection. It was a 2-D medieval RPG drawn with oblique perspective, similar to *Ultima VII: The Black Gate* (1992) (see http://www.argentumonline.com.ar) (see figure 1). Many members of the community of gamers are committed to the project, participating in both the development and in the maintenance of the game. The project was developed with minimal (and personal) resources: computers and spare time. Since 2000, the game has been running continuously on the Internet. For almost three years, the community increased its size a thousand times, and as of 2013, thousands of people enjoy the game. The game takes place in a fantasy world called Argentum, with forests, mountains, rivers, seas, and cities. The fun of the game is that you are not alone; there are hundreds of people who represent characters with whom you interact, communicate, or fight (http://www.argentumonline.com.ar).

The Rise of the Argentinean Video Game Industry

The Argentine Videogame Developers Association (ADVA)

Beyond the first developments, a mailing list named "juegosar" was created in Yahoo! Groups in 2000 for news and discussions of the developing video game industry. Developers who participated in this list included Nicolas Massi, Javier Otaegui, Santiago Siri, H. Hernán Moraldo, Juan Linietsky, Daniel Benmergui, and Nicolas Lamanna. Lamanna also started a Forum named the ADVA, which today can be found at www.duval.vg.

The first administrative committee of the ADVA included Andrés Chilkowski, Javier Otaegui, Hernán Moraldo, Santiago Siri, Nicolás Lamanna, Daniel Benmergui, Guillermo Averbuj, Sebastian Uribe, Juan Linietsky, and Fernando Mato. Also, Santiago Siri designed and Nicolás Lamanna programmed the first interactive website for the ADVA. The second and third versions were created by Santiago Siri, Guillermo Averbuj, and Nicolás Lamanna (for more information, see http://www.youtube.com/watch?v=OttNgR9pklU).

The year 2004 saw the legal foundation of the ADVA (http://www.adva.com.ar), which remains formally registered in its historical articles as an association (see "Historical articles of Association," http://www.adva.com.ar/public/downloads/estatuto.pdf). Another point in this history was the organization of the Video Games Exhibition in Argentina (EVA-ADVA). As of 2013, this event has run for eleven consecutive years, becoming the largest event of the video games industry in Argentina. Since 2003, the ADVA has organized the Argentine Videogames Expo (EVA) in Buenos Aires. Every year, this event is visited by thousands of attendees, who come from all over the world. Like the Game Developers Conference (GDC), the EVA consists of a series of conferences about game development where people gather to learn about the latest advances in programming, game design, art, and business in the industry. The EVA is now a massive success, supported by more than forty sponsors, including Sony, Vostu, Cartoon Network, and Gameloft, among others. Guillermo Averbuj (http://gamester.com.ar) has been the committee chair of the organizing committee, centralizing the production and operative management of the EVA from 2003 to 2007.

The First Historic Steps in the Sector

The history of Argentinean video game companies begins with Sabarasa Entertainment (based in California), which had a studio in Buenos Aires since 1996, with Javier Otaegui as the CEO. Sabarasa produced one of the first computer strategy games in the region, *Malvinas 2032* (1999), and was the first company to be licensed by Nintendo to create games for the Game Boy Advance, Nintendo DS, and the Wii, as well as for the Sony PSP. Their games include *Horizon Riders* (2010), *Save the Turtles* (2010), *ALT-PLAY: Jason Rohrer Anthology* (2010), and others. Inspired by Sabarasa's success, Evoluxion was founded by Santiago Siri in 1996 and became one of the pioneering game development companies of Argentina. Siri directed the production of *Fútbol Deluxe 96*, the first Argentine game to receive a worldwide distribution deal.

Since 1997, the Argentinean Cyberjuegos.com has been a gaming social network for Latin America and Brazil, producing and delivering casual multiplayer games. With Roby Krygel as creative director and CEO, and more than one million registered users, Cyberjuegos has created a strong Spanish-speaking community of social gamers. Another pioneer firm, Codenix (founded in 1999), was a company that participated in the creation of NGD Studios (see below), offering technology and then serving as a consultant to several companies. Currently Codenix is one of the few companies that provide technology and consultancy services. LatDev Games & Tech, founded by Guillermo Averbuj in 2001, has created many websites, games, and software. Today, Averbuj produces events around video games with Gamester and has helped PixOwl develop the first sandbox game designed in Argentina, *The Sandbox* (2012) (see http://www.thesandboxgame.com/team.html).

Other studios established in Argentina since 2000 include Gameloft, which was founded in 1999 in France and came to Argentina in 2002. Gameloft creates games for digital platforms including mobile phones, smartphones, and tablets (including Apple iOS and Android devices), set-top boxes, and connected TVs. Gameloft operates its own established franchises. Another of the historic firms registered is GlobalFun, founded in Sweden in 2000. To support their local markets, GlobalFun also operates in the United States, Argentina, Spain, and Italy. GlobalFun is a publisher and developer of high-quality mobile games, and GlobalFun Argentina has produced games for various franchises including Ben 10, The Powerpuff Girls, Scooby-Doo, The Brak Show, and other games such as *Tank Racer* (2005), *Billy the Kid 2: Hunted* (2006), *Robin Hood* (2007), and *Great Legends: Vikings* (2008). GlobalFun Argentina was the result of Swedish company GlobalFun's purchase of the mobile games division from NGD Studios.

NGD Studios, located in Buenos Aires and founded in 2002, was started by Pablo Zuccarino, Andrés Chilkowski, Eugenio Insausti, Fernando Testa, Matías Pequeño, Nicolás Lamanna, Cesar Guarinoni, and Eduardo Gohyman. They were the creators of the game *Mis Ladrillos* (2002), a game for a construction kit toy similar to LEGO bricks. NGD is best known for *Regnum Online* (2007), the biggest MMORPG made in Latin America, and other games including *Regnum* (1995) and *Regnum Online* (2007), now known as *Champions of Regnum*, and the first Argentinean game made especially for Steam, *Bunch of Heroes* (2011), a cooperative action game (allowing for up to four players at once). *Champions Of Regnum* (http://www.championsofregnum.com/) is a free-to-play 3-D medieval fantasy MMORPG, with the option to pay for premium content. The game is based on battles between realms and brings together a global community of players. The game was published in Germany, Brazil, the United States, and other countries around the world. NGD's other game of note is *Bunch of Heroes* (http://www.bunchofheroes.com/), which is exclusive to the Steam digital distribution platform. Additionally, NGD made a game for Cartoon Network called *Maldark: Conqueror of All Worlds* (2011). NGD is using its extensive experience in building player-versus-player and cooperative gameplay to create intense, engaging, and high-quality games for the PC, consoles, and other platforms.

Globant, founded in 2003, is a Latin American company that creates software products for global audiences. Globant is focused on engineering, art and design, social networks, and game development. Since 2002, Globant has been doing outsourcing for Electronic Arts and has participated in the design of various important games, the most famous of which is *FIFA 12: UEFA EURO 2012* (2012).

Some companies lasted only a few years. Nucleosys Digital Studio, founded in 2003 by Agustin Cordes and Alejandro Graziani, specialized in adventure games. *Scratches* (2006), a graphical adventure game, was well received and later published on Steam, with *Scratches: Director's Cut* (2007) released the following year. On July 15, 2009, it was announced that Nucleosys would be disbanded, and Cordes went on to found Senscape in 2010, with another graphical adventure, *Asylum*, in the works. Moraldo Games (later known as Moraldo Tech), led by Horacio Hernán Moraldo from 2003 to 2008, developed advertising technologies that were used in some of the largest companies worldwide, including mixed and virtual reality technologies, computer vision algorithms, and many single-player and multiplayer games.

Three Melons was founded in 2005, with an office in Buenos Aires and a business development office in Los Angeles, California. They worked on successful *LEGO Indiana Jones* and *LEGO Star Wars* Web-based games, the game *Gardens of Time* (2001) (which won the "Best Social Network Game 2011" Game Developers Choice Online Award), and 3-D MMOGs with top advertisers and media companies, and also developed MelonDaiquiri, a 3-D Flash/Silverlight game development framework. In March 2010, Three Melons was bought by Playdom, which itself was acquired by Disney in August 2010.

Finally, casual game developer QB9 was founded in 2005 and has grown to have a team of more than thirty-five employees designing Web-based games, PC/Mac games, and games for the PlayStation Portable. They have partnered with some of the most successful IP holders, including Comedy Central, VH1, and Shockwave, and have created game design courses at the University of Palermo in Buenos Aires. QB9 is also a member of the ADVA's Education Commission.

New Trends in the Industry

As of 2013, there are several trends regarding specific content and game development within the Argentinean video game industry, including advergaming and social games. Companies such as QB9, MetroGames, and others are developing social games that have been launched on Facebook and Orkut. The massive company Vostu landed in Argentina, betting on the local industry, with six hundred jobs. The first national advergaming event is registered at http://www.advergaming.com.ar. Another area of development is that of edutainment, with educational content used as support for teachers. Events and communities related to educational video games support this trend on the part of the national government, with series of thematic games. This section for players contains a series of events related to the theme, highlighting the International Symposium of Videogames and education in the Model 1-on-1 and the development of international groups of playful pedagogical researchers, such as *Grupo Alfas: Ambientes Lúdicos facilitadores de aprendizaje*, coordinated by Dr. Graciela Esnaola Horacek (the Web page for the conference Video Games in the Model 1-on-1 International Symposium of Edutainment can be found at http://gamester.com.ar) (see figure 2).

Video games have been declared a cultural industry by the National Culture Secretary (http://www.cultura.gob.ar), and nowadays this industry is included in the national events of MICA (Mercado de Industrias Culturales Argentinas/Market of Argentine Cultural Industries) (http://www.mica.gob.ar). MICA (with sectors in art, design, video games, publishing, and music) is the first space in Argentina to bring together different activities of the various cultural industries in order to promote business, exchange information, and show their products to the world. MICA gives producers and artists a chance to meet with leading industries' companies worldwide and open up new business opportunities. These gatherings extend to different zones of the country and promote the meeting of business associations, national institutions, and small- to medium-sized companies.

International game publishers such as Axeso5 (http://www.axeso5.com) and 6 Waves (http://www.6waves.com) have brought games for all platforms to Argentina, and other publishers, including Square Enix, Electronic Arts, and Riot Games, have their games imported into Argentina. This, along with the rise of local development, has led to greater decentralization. The ADVA also promotes the development of groups from other regions of the country outside the Buenos Aires area, including Jujuy, Santa Fe, Mendoza, and Cordoba. The first local group of developers landed in Rosario in the Santa Fe Region. Also, there are websites and initiatives, some of them supported by the ADVA, which run autonomously to provide support or services to the local industry, such as Industria VG (http://www.IndustriaVG.com.ar), Desarrolladores Unidos de Videojuegos de América Latina (DUVAL) (http://www.DUVAL.VG/), VG Map (http://www.VGMap.com.ar), Gamejoint: Connecting Independent Developers (http://www.game-joint.com), Gamester (http://gamester.com.ar), and City Network (http://www.citynetworkhosting.com.ar). News of events and communities within the games sector can also be found at LatinGamers (http://www.latingamers.net), LocalStrike (http://www.localstrike.net), and more.

Music and Video Games

One sector of the video game industry that is still small but growing is that of video game music, which began at the amateur level around 2005. One of the earliest composers is Christian F. Perucchi, who began doing freelance sound and music for games in 2005. Formally trained, Perucchi specializes in music composition and SFX creation for film, video games, movies, television, and other media. He won the Innovar 2008 award for Oniric Games's *Time of War* (2008) and a 2010 ADVA CODEAR award. Juan Cavagnaro, an indie composer who wrote music for Three Melons and currently for Dedalord, scored games including *Running Fred* (2012), which was the best-selling Argentine video game of 2012, and *Boarder X Battle* (2010) for Disney and ESPN through developer Three Melons. He was also the first musician to play live at the Exhibition of Argentina both in 2008 at the San Martín Cultural Center (EVA2008) and at the University of Belgrano the following year (EVA2009). Cavagnaro edited the EPs *Valvular (ECC83)* (2008), *Spontaneo* (2009), and *Steve Hyuga* (2009), which was a tribute to Nintendo's Japanese video game *Captain Tsubasa Vol. II: Super Striker* (1990), the soundtrack of the original game music in *Jazzecity* (2009), *Incestibleach* (2012), *TIC TOC* (2013), and *Songs of Impending Doom* (2012), which featured music from Dedalord's Fred video game saga (which includes *Falling Fred Z* [2011], *Running Fred* [2012], *Super Falling Fred* [2013], and *Fred Skiing* [2013]).

Art and Video Games

The history of art and games goes back to the beginning of mankind, when art and games were part of a magic circle. As time went on, men broke down the circle, but it kept fighting back. Games and art have reunited on different occasions, most recently mediated by technology and under the shape of art games: an art movement that arises from two parallel sides, video games and new media art. In Argentina, we are still taking our first steps. Within the field of video games, the pioneer and main figure is Daniel Benmergui, an independent game developer renowned worldwide for his experimental and art games. His latest game, *Storyteller*, was recognized with the Nuovo Award at the Game Developers Conference in 2012.

The Argentinean indie scene is mostly recognized for developing experimental video games. There are a few developers toying around with the concept of art games, as is evident in such games as *Lumiere and Nycteris* (2010) from Martin Gonzalez; *Rabbit Fable* (2013) from Santiago Javier Franzani; *Panoramical* (2012), a collaborative project from Fernando Ramallo and David Kanaga; or the works of Agustin Perez Fernandez

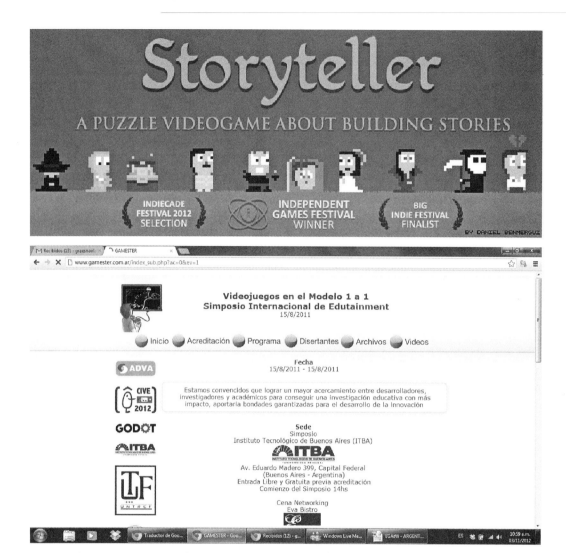

Figure 2
Storyteller (2011) by Daniel Benmergui, a puzzle game about the construction of stories (top), and the webpage for the conference video games in the Model 1-on-1 International Symposium of Edutainment (bottom).

(known as "Tembac"). While there isn't yet a common agreement on the meaning of art games, the term usually rests on the notion of conceptual proposals, meaningful subjects, and disruptive game play mechanics. Perez Fernandez's works are the closest to conceptual art; he usually relies on video games as a tool to convey some sort of statement, whereas Benmergui's works are more related to the history and language of games. As for commercial developments, there have not yet been any games that could be categorized as experimental or artistic.

The potential of video games as art was first explored in Argentina around 2004. The results were usually mods: subversions of existing works or precarious video games designed as sociopolitical criticism. The artists generally saw video games as a new medium, but they lacked real knowledge of the tools; their works weren't as engaging as video games, neither in their game mechanics nor in their aesthetics. However, most recently and since the popularity of tools such as the Kinect camera or the PlayStation Move controllers, there has been a new twist in the works of new media artists. They are creating interactive-ludic installations, new fancy games designed as artistic experiences, and a new wave of works is appearing.

According to the theorist and media artist Monica Jacobo (2012), the first experiment in the field was Ricardo Pons's *Pulqui II* (2004): the reconstruction of the first flight of an airplane prototype commissioned by Peron in the 1950s. Then came *Casa de Juegos* (2004), an online project coordinated by Andrés Oddone and commissioned by the CCEC, and *Batalla del Rio de la Plata* (2005) from Santiago Tavella and Agustín Blanco, an online version of the mythic marine battle. Jacobo herself started producing artworks related to video games in 2004, always creating mods, video installations, and altered photos, and her first work was *4 minutos 44segundos* (2004). Another famous work was *Cartonero* (2006) from Estanislao Florido, an attempt at a video game that conveys a harsh criticism of the economic downfall of 2001. Other examples of new media artists experimenting with video game concepts or technology include the work of Gabriel Rud (a pioneer in machinima), Federico Joselevich Puiggrós, Diego Alberti, Luciano Azzigotti, Joaquin Fargas, Yamil Burguener, Patricio Gonzalez Vivo, and Emiliano Causa and his artistic group Biopus.

Video games have been recently admitted within new media art career programs as another aesthetic possibility, acknowledged as a new tool, ideal for interactive developments. As a result, there is a new wave of young artists' collectives coming up with innovative proposals, such as Proyecto 032, SuperfluoLab, Glitch Studio, and some of the works of CCEBA Media lab. Art games have their own exhibition circuit. In Argentina the first experience was Game on! El Arte en Juego (http://www.gameonxp.com), an exhibition on games, art, and technology, created and run by *Objeto a* (a cultural producer specializing in art and new media). The first edition took place in 2009 uniting indie developers, renowned commercial enterprises, new media artists, digital visual artists, musicians, and a wide range of professionals related to the video game industry. Since its first appearance, Game on! El Arte en Juego has had two more editions and a variety of activities, and has visited different cultural centers and institutions. Different exhibitions around the country have started including video games among their selections. (There have also been some isolated small exhibitions focused only on games, but they didn't prosper.) Some contests such as BridgestoneArte 2013 (http://www.bridgestonearte.com.ar) have included video games as a new category, and in July 2013, Fundación Telefónica launched the first Argentinean digital museum: Espacio Byte (http://www.espaciobyte.org). Among its first six exhibitions, two are covering video games.

But there has been very little theoretical production on the subject of video games and art produced in the country. The most relevant works have been the ones from Baigorri (2004), Diego Levis (2011), and Monica Jacobo (2012). Within an academic context, doctoral candidates are now writing theses that focus on the relationship between art and video games.

Edutainment in Schools

In Argentina, from 2003 onward, the "Digital Agenda for Social Inclusion" (http://www.agendadigital.gob.ar/) has promoted the inclusion of digital media in education, aimed at the general population. In this scenario, mainly in the classroom, the Programa Conectar Igualdad, which is national in scope, contributes to secondary education (http://www.conectarigualdad.gob.ar/) as well as numerous provincial and regional programs. This policy initiative opened up the possibility of including video games in education. Since it is noted that "play" is the main activity of the students when they receive laptop computers, we have discovered two main areas of impact concerning these policy actions: an increase in teachers' interest in including video games in their teaching activities, and the promotion of the national video game design industry. These video games are considered "multimedia resources that promote interactive activities to review and strengthen knowledge which contain various levels of complexity" (http://juegos.educ.ar/) and specifically include language content, mathematics, and education for citizenship.

In relation to this "playful" proposal, we noticed that most of the video games designed for teaching, for the development of skills, or for the acquisition of curricular content ended up losing their "playful" features, resulting in products that children find boring. Edutainment (games with educational aims) is boring for children because the games are often slow and have poor graphics lacking special effects; commercial games, on the other hand, developed with higher budgets and designed as entertainment, are usually more attractive. There is a new point of view in edutainment that considers the educational power of all games, including commercial ones. So, it became necessary for teachers to receive particular training to be prepared for the evaluation and selection of appropriate video games for their students. Edutainment as "learning by playing" is another point of view, which sees video games in school as something to play and enjoy. This is the direction in the latest research in ludic pedagogy (for more information about edutainment and gamification, see the papers at CIVE12 http://www.uv.es/ordvided/ and CIVE 13 http://cive13.blogspot.com.ar/). Another initiative is the Observatory of Videogames (http://portal.educ.ar/debates/videojuegos) from the Argentine government. However, it only offers information and updates on the interactive digital games industry and doesn't generate a teacher training area to include video games in education; this is a weakness in the program.

The latest studies of edutainment argue that digital games can be a form of education that develops complex thinking processes. This is achieved through problem solving and the incorporation of formal and informal education to create interactive environments and develop skills such as the management of technology and digital information. This definition, however, does not consider the educational value implied by the playful proposal, which must be considered "an artistic, narrative pop-up, hypertext" (Horacek Esnaola 2009) in the technological environment. This definition focuses on the fact that in video games' narratives, we must consider the mix of genres and styles as a product of multimedia convergence for the particular structure of immersion and interactive aesthetics. In this context, we wish to stress that games are narratives in the convergence of cultural stories, formats, and devices, typical of the "high-tech pedagogy" (Horacek Esnaola 2009), which allows the management of images and sounds through the digitization of procedures. Video games represent a junction of disciplines including film, music, video, animation, and immersion in virtual environments, thanks to the interactivity that facilitates their technological development (Horacek Esnaola 2009).

These, then, are the two approaches to the inclusion of video games in the classroom. These views explain the increase in the number of researchers and amount of postdoctoral research being done that focuses on the sociocultural narrative of video games and their educational potential.

The Video Game Industry in Argentina Today

The beginning of the twenty-first century has seen the inclusion of more demographic sectors in the Argentine video game market. Until the end of the twentieth century, the typical user profile was a male, ten to fifteen years old. Currently, the audience for video games has expanded to children under the age of seven as well as adult men and women. In addition to their use in Argentina, many games are also exported: 95% of the games made in Argentina are destined for the European market, American market, Japanese market, and the markets in other Latin American countries, according to sources in the industry itself (ADVA Dossier 2004). The devaluation of the peso in 2002, along with the emergence (since 2009) of new venues for gameplay such as social network accounts (such as Facebook) and smartphones explains the increase in local industry and consumption in particular.

Some examples of recent successes of Argentine video games include *Bola* (2004), a football game on Facebook, a *freemium* social game with 100,000 gamers per day; *Más fútbol* (2010), a PC and console game in the FIFA series, which has sold more than 100 million copies in eighteen years and is manufactured in several studios from different countries (in Argentina, the Globant Company develops the menus and tests the final versions of the games before their release); and *Mundo Gaturro* (2010), a virtual world game for children that is played over the Internet. The Digital Media Company (CMD), the Clarín Group with The National Group DRIDKO, and cartoonist Nik with the company's online game development QB9 decided to sign an agreement to launch *Mundo Gaturro*, which found success among Latin American children, with almost three million active users (see also http://www.taringa.net/posts/noticias/12536697/Los-videojuegos-argentinos-conquistan-con-goles-y-heroes.html).

With these products, the Argentine game industry is currently the fourth largest in Latin America—behind Brazil, Mexico, and Chile, according to the ADVA. The video game industry of Argentina already has a workforce of more than two thousand people including designers, artists, illustrators, programmers, engineers, writers, testers, and musicians, who, on average, do not exceed the age of twenty-seven. Currently in Argentina, there are some sixty-five companies linked to video games, the vast majority of which (85%) appeared after 2000. The annual revenue of the sector is around USD $50 million (ADVA Dossier 2004).

Another fact to consider is the emergence of technical careers and other academic initiatives designed to train professionals for the video game industry. This is a sign of local industry growth. Teachers are mostly from developer companies, searching for original talents among their students. However, the participation of academics or consolidated researchers linked to video games is still very low.

The video game industry in Argentina is one of the country's fastest-growing and most dynamic industries. At the Universidad General Sarmiento (UNGS) in Buenos Aires, between September 2010 and February 2011, customized research of companies in the game industry as a whole was done and the following data was obtained:

- The level of formal education of video game industry employees is not only significantly above that of traditional sectors of industry in Argentina, but it also exceeds the average of university graduates who are working in firms engaged in the production of software in general. With the large number of companies that have done training in recent times, this activity identifies employers who value the establishment of connections, while nearly one in every two signatories has been associated with other organizations with technological and/or commercial objectives.
- The difficulties the sector is facing have to do mainly with the availability, rotation, and specific training of human resources. Also, access to financing is difficult for these companies, mainly due to the shortage of resources for acquiring new hardware and licenses for production.

This report is the product of fieldwork carried out at the national level and based on a survey of fifty-five Argentine game companies, covering a wide range of dimensions associated with organizational dynamics and economic and innovative performance. The participation of firms was supported by the ADVA and the CODEVISA (Cluster de Desarrolladores de Videojuegos de Santa Fe). The survey also had the support of the Industry Institute of the National University of General Sarmiento and was funded by the European Union through the EULAKS IdeI-UNGS grant project in 2008. The collected sample of fifty-five cases is very significant because it represents the census of the sector; the original base noted some sixty-five active companies, which formed after the fieldwork expanded in about twenty cases, forming one of the most important sources of information on the topic. The survey aimed to give a realistic view of the sector and for this reason incorporated companies down to the micro-sized ones, which are very important in the dynamics of the studies.

The video game industry in Argentina is mainly made up of small companies. While the average number of employees is thirty-three, half the companies in the sector have eleven employees or fewer. Thus, the companies can be divided into three categories; small-sized, which have between three and eleven employees

(30%); medium-sized, with between eleven and twenty-one employees (40%); and larger (30%), with more than twenty-one employees. The entire sector is growing, and the number of employees has increased in recent years; during 2010, about 1,200 people were employed in the industry—almost 10% of the employment generated by the sector of software and related services in Argentina. In terms of size of the development, the study shows that between 2009 and 2010 there was an average increase of 32% in the employment generated by these companies.

Nearly 85% of the companies emerged after the currency devaluation in late 2002, and there were very few gaming companies prior to 2003 (the oldest company dates back to early 2001). One of the characteristics of this industry is that of exportation: two out of three companies export, while more than half of the companies in this group are strong exporters, selling abroad more than 50% of their products or services. In general, the larger companies in the sample (defined as those with more than twenty-one employees) are more likely to be exporters. In spite of this, it is worth noting that very small firms have also specialized in supplying external markets. The United States is the main destination for exports, with about 65% of the firms that export (to a greater or lesser extent) shipping to that country. Alternative destinations are Mexico (18%) and Spain (18%), which are also important places receiving Argentine exports.

These were the most important conditions that allowed companies to develop games or parts of games (47%) and training for the development of products (51%). These figures express a strong tendency for these companies; in fact, 85% report having made some innovation in products or services between the years 2008 and 2010 (see http://infouniversidades.siu.edu.ar/noticia.php?id=1437). In this sense, the formal level of human resources, training activities, and workplace dynamics from an organizational point of view are essential. Almost 70% of the firms have done some training for their employees between 2008 and 2010. Another area of utmost importance is that of corporate connections, which enhance the organizations' experience and allow the sharing of technological, commercial, and strategic experiences. In the case of the production of video games, companies establish ties with other key firms mainly in three areas: commercial actions (performed by 49% of the companies surveyed), technical assistance (49%), and human resources training (47%).

Among the specific aspects that characterize the companies studied, it is remarkable how the companies need qualified applicants. This generates consequences not only at the level of formal education, but also in higher levels of study. On the one hand, 80% of companies have a high level of formal training. This means that 45% of businesses have at least one full-time employee with university-level education and that 34% have at least one employee with a postgraduate degree. On the other hand, the issues of training for HR and the supply of skilled labor available in the country have been a concern for entrepreneurs in video games. Other concerns are of significant importance: 38% of the surveyed companies mentioned the cost of the licenses and hardware, and 17% mentioned the strengthening of institutional connections (which allow access to promotional tools, public financing of the sector, and relationships among the areas of the sector).

In July 2010, the Center of Studies for the Desarrollo Económico Metropolitano (CEDEM) (see figure 3), which is dependent on the Municipal Government of the City of Buenos Aires, presented the partial results of

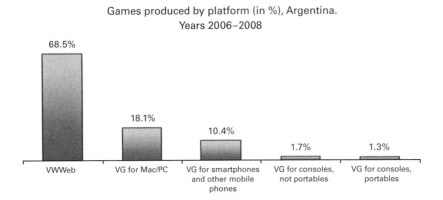

Figure 3
Games produced in Argentina by platform (in %), years 2006–2008 (CEDEM, 2010, http://comex.mdebuenosaires.gov.ar/contenido/objetos/inforvj.pdf).

the survey conducted with the ADVA between June 2009 and February 2010 citing data on games and social networks that were very interesting.[4]

Even though Argentina has seen a great evolution of its gaming events, they are still not very popular. Probably the most important event is EVA (Argentine Videogames Expo), which has been held every year since 2002 and is a great reunion for video game developers in the country. While only 120 people attended the first EVA, it has been growing every year, gathering the most proactive people in video game development and establishing a good place to invite and bring over the most relevant speakers from around the globe, including Ian Bogost, James Portnow, Nick Fortugno, Raph Koster, Chris Taylor, and other business-related personalities from major video game companies such as Sony, Square Enix, and Microsoft. (In 2013, Tim Shaffer was invited to Tecnopolis for an event hosted by the national government). And EVA targets developers more than gamers. Commercial events in Argentina have boosted sales, with special launch events mainly targeting press and gamer communities, with companies such as EA, Square Enix, Riot, and others as hosts. Some of them are very well organized and are becoming more frequent and diverse in their nature and objectives.

Yet, for the general public, the video game scene itself is not enough. Commercial expos are aimed at that audience and are integrated mostly with comic books, manga, animé, and several TV series, although Argentina is far from E3-scale events. Also, the WCG (World Cyber Games) has frequent participation from Argentine players at several techie events. One player, Francisco Sotullo, has won the bronze medal three times in the FIFA competition (2008, 2012, and 2013).

Within the indie scene, Argentina has developed some self-organized events for indie developers like Tembac, who is pushing the indie local scene. These types of events occur mostly in art galleries or cultural clubs and are sometimes integrated with other art contents, such as new media installations or music performances.

One of the main events positioning the video game industry as a producer of culture, focusing on games and their public and presenting video games as art is "Game On! El arte en juego," organized by a local art gallery called "Objeto a." This event showcases games with deep messages and experimental concepts. Several events showcased the games of Daniel Benmergui, the first Argentinean to win the "Nuovo Award" at the IGDA (2012), as well as other local developers.

Recently, the national government has declared the video game a cultural product by adding an area dedicated to it in their series of events called MICA (Culture Industry Market from Argentina). At these events, they gather investors and developers from Argentina (and South America) in business rounds, conferences, and expos. Gamester is one of the local groups that organize and help produce these events.

The Console Market in Argentina

The first game consoles arrived in Argentina in the late 1970s. They were different *PONG*-imitations built by Argentine companies in local factories. Telemacht's version of *PONG* had the largest number of sales. Due to the economic situation in the 1980s in Argentina, there was an extension of the Atari 2600 era but with a local version of the hardware also known as Dynacom, EDU games, and other minor clones. But there is no doubt about which was the leader in the console market in the country over all competitors: the ColecoVision, which had a great TV campaign and was considered a better but much more expensive machine, with plenty of games to play. We have established that for every ten game consoles of that era in Argentina, 7.5 were ColecoVisions.

In the 1990s, some changes in the economy allowed people to finally get the Nintendo Entertainment System (NES), but a version of the Japanese Famicom was already commercially available in the country, better known as "Family Game." It was a big success, exceeding sales of all other consoles before it. There were plenty of versions of "Family Game," outstripping the NES, which had much less success since it was really expensive; the "Family" console was half the NES's price, and so were its cartridges. This led to a strange situation: the people who had the real NES would get an adapter in order to play cheaper "Family" games in their consoles. And it worked!

Without a doubt, the "Family" console began a new video game era in Argentina. Its competitor, the SEGA Master System, was much more expensive, with fewer games, and harder to find. As there were no Mario games, we can easily say that at the time, of the twenty consoles on the market, nineteen were "Family" systems. At the same time that the "Family" console appeared, 16-bit machines reached the stores. Of course, the powerful Super Nintendo Entertainment System (SNES) and the marvelous SEGA Genesis were nothing but expensive in 1992, so it would take a couple years for this generation to reach consumers. Nintendo didn't win the market, and the SEGA Genesis was the leader, although this console never reached the numbers of the "Family" console's success. Indeed, it was still present, dividing the market into a line of cheap and expensive consoles.

During the 1990s, piracy was a key factor in the appeal of the Genesis. This was the first time that Argentina witnessed a generational change: 70% of the people who bought a console got the Genesis, while the rest had the SNES. At the same time, in the 1990s, there were plenty of game consoles even more powerful—such as the 3DO or the Atari Jaguar, or even the SEGA CD—with no success or minimum public awareness due to the astronomical prices and the lack of games available.

The second half of the 1990s was the CD-ROM era. The CD-ROM was everywhere, on both PC and video game machines. In 1995, the most anticipated video game console was the 32-bit SEGA Saturn with a CD-ROM drive. But, as it happened in the rest of the world, the Sony PlayStation quickly surpassed the Saturn. People knew this new console was really expensive, but its innovation level was so outstanding that they didn't care. To the average person, the first PlayStation symbolized the mature version of a video game console. A great version of the most popular soccer game was launched in 1998, and piracy made this 32-bit title available at a much cheaper price than the old 16-bit or even 8-bit games; this was a key to the consolidation of the console in the market. The PS1 was everyone's favorite, even against the more powerful (but ten times more expensive) Nintendo 64, which still had cartridges with a price of about USD $100 each. Because of this, Nintendo had only 20% of the market, the lowest position the company ever had. The SEGA Dreamcast was a middle-aged console, and in Argentina during those times, it was impossible to afford; almost nobody bought it, so it has been seen as a cult console ever since.

In 2000, the PS2 appeared and had a great start, and it became clear that the PS2 would continue the legacy of the PS1 or even surpass it. The PS2 almost eliminated the Nintendo GameCube and the Microsoft Xbox from the market, due to the higher price of those systems and the need to buy original games for them, whereas Sony's system was backward compatible with the PS1 and included new features such as a DVD player. During that period in Argentina (2002–2007), nine out of every ten consoles were PS2s, and the rest were divided between Nintendo's and Microsoft's consoles.

The success of the PS2 was only comparable with that of the "Family" console. In 2008, new consoles were launched: the Xbox 360 was the first, but consumers were concerned about the "three red lights" (a warning sign meaning high alert) issue, mostly because there was no official local technical support. The PS3 was now the new N64, regarding the excessive price of new games. But that wasn't all; the need for an LCD-based TV discouraged buyers, and the older PS2, with the biggest game library available, remained predominant.

Nobody imagined that the creator of Super Mario, which dominated the market in the 1980s, would find its way back inside every house once again, so the Nintendo Wii came as a surprise, captivating the Argentinean market with a great advertising campaign and the possibility of using the Wii in almost every commercial mall and store. The Wii entered into the market fast and furious, with people saying, "Get the Wii! It's better than the PS2." The idea of the gesture-based controller, plus the possibility of getting nontraditionally playful members of the family to play and have fun made the Wii the best option for gamers and new gamers from 2008 to 2012. Around this time, for every ten consoles sold, five were PS2s, three were Wiis, one was a PS3, and one was an Xbox 360. Finally, the PS3 enjoyed a legacy of almost fifteen years of customers' loyalty to the PS1 and PS2, and in the end it turned the machine into a leader within the Argentine console market at the end of 2013.

The Portable Console Market in Argentina

The portable console market was much smaller compared to traditional, TV-based consoles. In the 1980s, Nintendo's Game & Watch series games were very expensive, and only the people who traveled abroad could buy them. In the 1990s, the Nintendo Game Boy found success, but the Nintendo Game Gear did not. In the late 1990s, the Game Boy Color enjoyed a sales bump thanks to the Pokémon game, giving the market an experience that wasn't available on the larger console systems. With Nintendo's monopoly in portable gaming, the fortunes of the Game Boy Advance and Nintendo DS were assured, and in the case of the DS, the possibility of getting free ROMs with a pirate cartridge put the system's sales at high levels. The Sony PlayStation Portable (PSP), the PlayStation Vita, and the Nintendo 3DS are still very expensive in Argentina, and while the 3DS is the clear leader, the portable console market is still behind that of TV-based consoles.

The Most Popular Games in Argentina

People were confused about the difference between arcade games and console games, and this continued until the "Family" console era, when Mario appeared. Some of the most-played games from the first era include *PONG* (1972), *Space Invaders* (1978), *Pac-Man* (1980), *Donkey Kong* (1981), and sports games. These, plus *Pitfall!* (1982) for the Atari 2600, were the only recognizable games for Argentines before "Family" console games, when Super Mario appeared.

In the 16-bit era, Sonic the Hedgehog stole some of Mario's audience, but the most popular genre of the time was fighting games, so *Street Fighter II* (1991) and *Mortal Kombat* (1992) dominated not only the arcades, but also the consoles with versions for the SNES and SEGA Genesis.

In 1994, the first FIFA game became popular, and the "Family" console version couldn't compete with it. This was the start of the great football/soccer video game phenomenon in Argentina. The FIFA series and Winning Eleven (Pro Evolution Soccer) series were the leaders in sales, not only during the first PlayStation generation, but up to the present. There were few blockbuster games, but soon games such as *Call of Duty* (2003), *Grand Theft Auto: San Andreas* (2004), *Guitar Hero* (2005), the LEGO Star Wars games, and a spectrum of great games for the PS2 revived the console and began a more successful era.

The Argentinean market is quite easy to read: the console with the most pirated games is the leader of the generation. As of 2014, the Xbox 360 is at the top. This is the result of the availability of pirated games for the console and picking up the market lost by the PS3.

Publications

Argentina's dedicated magazines and TV shows about video games had moderate success. PC gaming has had dedicated gaming magazines since the 1980s in the Commodore 64 era, while for consoles, the main publication from the 1990s was *Action Games* and was published for almost a decade. Today, there are a couple magazines with relative success such as *Loaded* and *Irrompibles*. On television, an old, cheap cable network known as "Nivel X" had some success worth mentioning.

The Future of Video Games in Argentina

The success of the industry in recent years is due largely to the rise of casual games, which are designed to appeal to a wide demographic, with simple rules and free access. Casual games have expanded the concept of the video game and have brought video gaming to segments of the population that previously had no way to access video games. This new way of conceiving the game as a product of mass consumption was possible thanks to new developments in support platforms and storage capacity, resulting in more complex and faster games, but at the same time, games that are easier to play and which appeal more to the casual player. Such new and increased possibilities for online games have allowed them to grow from small businesses to what may become one of the great industry pillars. Casual games are expected to grow in the global industry in the coming years, along with developments in mobile phones, touch screen interfaces, and games for social networks based on the model of casual players and virtual communities. In the same way, casual games have encouraged new business models, including digital distribution and games that are free to play. In the area of free-to-play games, the key lies in the possibilities surrounding the product (allowing players to pay money to level up their characters or to buy accessories and virtual goods for their characters, and so forth). The large volume of players that these games attract means strong profit potential for the developers.

In the next few decades, it is expected that the Argentine gaming industry will continue to follow a path of development and consolidation, with good prospects for the coming years. Thus, the expansion of global demand for video games, international recognition of the quality and creativity of the Argentine industry, the potential opportunities in mobile gaming and games for social networks, and the flow of foreign direct investment that the industry receives today, all promise an encouraging future for video games in Argentina.

Acknowledgments

Dr. Graciela Esnaola was the essay's main author, with contributions made by Alejandro Iparraguirre and Guillermo Averbuj. The section "Art and Video Games" was written by María Luján Oulton. The authors also wish to thank Emilio Gonzalez Moreira for his contribution to the essay.

Notes

1. Sources consulted: National System of cultural statistics, Argentina: http://sinca.cultura.gov.ar/, Universidad Nacional de General Sarmiento; CEDEM, Center of Studies for the Metropolitan Economic Development (DGEyC—GCBA), and second nationwide survey of developers of Video Games.

2. Ibid.

3. See http://www.eldiario.com.ar/diario/cultura-y-espectaculos/50626-el-pre-mica-centro-en-su-etapa-final.htm.

4. Sources consulted: "Relevamiento a Empresas de Videojuegos de Argentina," Universidad Nacional de General Sarmiento; CEDEM, Centro de Estudios para el Desarrollo Económico Metropolitano (DGEyC—GCBA); and Segunda Encuesta Nacional a Empresas Desarrolladoras de Videojuegos.

References

Baigorri, Laura. 2004. Game as critic as art. 2.0, CD-ROM, *Sextas Jornadas de Artes y Medios Digitales*, Symposium organized by the CCEC (Argentina) and Cordoba University.

Dossier, A. D. V. A. 2004. *Informe Industria Videojuegos en Argentina*. http://www.adva.com.ar/public/downloads/dossierADVA.pdf.

Esnaola Horacek, Graciela. 2009. Video games "Teaching Tech": Teachers of global convergence. The process of docile thought through the macro-ideology cultural and technological convergence. In University of Salamanca, *Revised Theory of Education Volume X*. http://campus.usal.es/~teoriaeducacion/rev_numero_10_01/n10_01_esnaola_horacek.pdf and https://sites.google.com/site/dragracielasnaola/.

Garcia Canclini, Néstor. 1999. *El Consumo Cultural en América Latina*. Bogotá, Colombia.

Historical Articles of Association. N.d. Available at http://www.adva.com.ar/public/downloads/estatuto.pdf.

Industria Argentina de Videojuegos. 2009. http://www.youtube.com/watch?v=0ttNgR9pklU.

Interview Martín Repetto, Secretario de la Asociación Argentina de Desarrolladores de Videojuegos. http://www.youtube.com/watch?v=3jYHY3k0F0o.

Jacobo, Mónica. 2012. *Videojuegos y arte. Primeras Manifestaciones de Game Art en Argentina*, Centro de Estudios en Diseño y Comunicación, Cuaderno 41, June. http://www.scielo.org.ar/scielo.php?pid=S1853-35232012000300006&script=sci_arttext.

Levis, Diego. 2011. *Arte y computadoras. Del pigmento al bit*. Buenos Aires.

Paricio, Guadalupe Gil, and Paulina Seivach. 2010. La Industria de Videojuegos en la Argentina: Segunda Encuesta Nacional a Empresas Desarrolladoras de Videojuegos: Un Diagnóstico en Base a 30 Empresas Encuestadas. August. http://www.observatorioapci.com.ar/archivos/368/a_1330480421.pdf.

AUSTRALIA

Thomas H. Apperley and Daniel Golding

Australia has experienced a relatively long history of local game development, which in recent years has been quite turbulent, but not without promise from emerging models of game development. For many years, Australia also experienced a unique—and arbitrary, possibly draconian—system of digital game classification, which has only recently been changed to measures that are more similar to those found in other Western democracies. It is for these two issues that Australia is primarily known in digital gaming circles. It may therefore appear somewhat paradoxical that a country with high-tech innovative studios—such as *LA Noire* (2011) developers Team Bondi, named after Sydney's internationally famous eponymous beach—also caused a global delay in the release of Bethesda Game Studios' *Fallout 3* (2008) because it was refused classification due to its "realistic" portrayal of drug use (this paradox is noted in Apperley 2008 and Schroeder 2010).

In 1999, the Office of Film and Literature Classification (OFLC) published the lengthy report "Computer Games and Australians Today" (Durkin and Aisbett 1999). Commissioned in 1995, the report responded to the growing significance of video games in Australian life, and at the time it was "one of the largest projects ever conducted into the nature of computer game play" (Durkin and Aisbett 1999, ix). While the focus of the report was on how aggressive content was "experienced and perceived," it also raised other issues, such as "time use, implications for social interaction, gender differences in play and attitudes, and consumers' uses of the classification system for computer games" (Durkin and Aisbett 1999, ix).

More recently, the Interactive Entertainment Association of Australia has commissioned four reports on the Australian video game market. Published in 2005, 2007, 2009, and 2012, these reports are based on large-scale random-sample interviews with Australian households and present a useful macroanalysis of local patterns of video game use and consumption (Brand 2005, 2007, 2012; Brand, Borchard, and Holmes 2009). The most recent report presents the video game as established within Australian culture, with 92% of households containing a device used for playing games (be it a PC, console, handheld console, or mobile device), with most playing for between half an hour and an hour at a time, every other day. The consistent rate of these research reports has also allowed trends to be identified; the latest report notes that between 2005 and 2011, the proportion of women who have identified as video game players has increased from 38% to 47%, which leads the authors to suggest that "female representation equal to males among gamers is imminent" (Brand

2012, 9). Equally, these reports have tracked the average age of video game players in Australia from twenty-four in 2005 to thirty-two in 2012. Furthermore, the Australian Bureau of Statistics released a survey of the Australian video game industry in July 2013 (Australian Bureau of Statistics 2013).

Interest in "the nature of computer game play" also came from Australian scholars. The growing importance of the video game was signaled by the—now expatriate—scholar McKenzie Wark in a 1994 article, "The Video Game as an Emergent Media Form" (Wark 1994). Australian-based scholars working in a number of fields made important contributions to the study of digital games, particularly John Banks's work on the relationship between game developers and players (Banks 2009), Catherine Beavis's research on the connections between gaming and literacies (Beavis, O'Mara, and McNiece 2012), Bernadette Flynn's work on the virtual spaces of games and how gaming reconfigures domestic space (Flynn 2003, 2004), Sal Humphreys's work on community governance in MMOGs (Humphreys 2008), Sue Morris's analysis of the "gaming apparatus" (Morris 2003), Angela Ndalianis's analysis of game aesthetics (Ndalianis 2004, 2012), and Melanie Swalwell's ethnographic work on LAN gaming (Swallwell 2003). Australian-based online journal *M/C—Media & Culture* published two pioneering special issues on video games in 1998 (edited by Paul McCormack) and 2000 (edited by P. David Marshall and Sue Morris). The relevance of game studies was cemented in 2004 with special issues of *Media International Australia* (edited by Chris Chesher and Brigid Costello) and *Scan: Journal of Media Arts Culture* (edited by Patrick Crogan and John Potts) dedicated to video games. In their introduction, Chesher and Costello locate the growth of scholarly interest in video games in relation to a significant growth in academic and scholarly interest in games globally, signaled by the first DiGRA conference, "Level Up," in Utrecht, the Netherlands, in October 2003, and particularly significant for Australian scholarship, the "Digital Arts and Culture" conference in Melbourne in May 2003 (Chesher and Costello 2004, 9).

In this chapter, we first trace the historical development of the digital games industry in Australia and discuss important contemporary developments before moving to the more fraught issue of digital game censorship in Australia. This latter section focuses on how the regulations historically shaped the consumption of games in Australia and outlines the impact of the recent changes to regulations, which are considerably relaxed in comparison to the preexisting framework.

The Digital Games Industry

The story of the digital games industry in Australia begins with the founding of Beam Software/Studios in 1980. Beam was a subsidiary of Melbourne House (Publishers) Ltd., a company founded in the UK in 1978. Originally it was intended that the Melbourne-based company would develop computer books for the UK market, but the core business of Beam soon changed to software development (Donovan 2010, 122). In 1981, the company acquired the license to produce a digital game based on J. R. R. Tolkien's *The Hobbit* (1937) on the proviso that the game was packaged in a joint release with a copy of the book (Australian Centre for the Moving Image n.d.). *The Hobbit* digital game was released in 1982 for the ZX Spectrum, fully exploiting its 48KB capacities (Stuckey and Harsel n.d.). It received the Golden Joystick award for "Best Strategy Game" in 1983 and went on to sell over a million copies (DeMaria and Wilson 2004, 347). Beam released several more games based on Tolkien's world, but they were never able to capitalize on their early successes until the release of *The Way of the Exploding Fist* (1985). This two-player fighting game was designed initially for the Commodore 64, and like *The Hobbit*, it was eventually ported to other commonly available home computers (Knight and Brand 2007; Stuckey and Harsel n.d.). *The Way of the Exploding Fist* won "Best Overall Game" at the Golden Joystick awards in 1985, at the same event Beam was honored with the "Best Software House" award. These early commercial games are still respected today for their innovation and originality, particularly in the areas of graphics, emergence, and marketing.[1]

After Beam's parent company and publisher was purchased by Mastertronic in 1988, the company began to diversify its design operations. They began to design games for the Nintendo Entertainment System, PC, PlayStation, and SEGA Saturn, with considerable success through the 1990s until they were purchased by the French entertainment software company Infogrames in 1999, at a time when that company was undergoing a considerable amount of expansion. It was renamed Infogrames Melbourne House and produced several digital games but was consistently affected by the financial problems of its parent company, which eventually sold them to Australian-owned and Brisbane-based Krome Studios in 2006 (Hinton 2009, 46).

The majority of long-standing Australian game design studios have stories similar to Beam, in which initial success has led to purchase from overseas interests. For example, the Melbourne, Victoria-based, BlueTongue Entertainment was founded in 1995 and met with initial success with the PC Game *AFL Finals Fever* (1996). The company was purchased by THQ in 2004, for whom they designed the critically acclaimed Wii title *de Blob* (2008), but despite the commercial success of the game, THQ closed BlueTongue studios in 2011 as part of a "strategic realignment" (souri 2011). Another example of local success, Ratbag Games, was founded in 1993 in Adelaide, South Australia, and soon developed a reputation for producing high-quality racing games for the PC in the wake of the success of the post-apocalyptic *PowerSlide* (1998). In 2005, the studio was purchased by Midway Studios and was briefly renamed Midway Studios Australia before Midway Studios forcibly closed it later that year. After the closure, many of the experienced staff were rehired by Krome Studios to form the nucleus of its new Adelaide studio (Funky j 2005).

Other early Australian developers have managed to continue operating as independent studios. Sydney-based Strategic Studies Group (SSG) was founded in 1983 and is dedicated to producing strategy games. SSG most recently released *Across the Dnepr: Second Edition* (2010). Canberra-based Micro Forté was founded in 1985 and is noted for producing *Fallout Tactics: Brotherhood of Steel* (2001). Micro Forté was also very active in developing the Australia games industry, taking a lead role in the establishment of professional conferences and training institutes. Micro Forté has very successfully moved into the massive multiplayer online gaming (MMOG) middleware market after working for seven years on the canceled Xbox-only MMOG *Citizen Zero* (Perry 2007).

International companies that have established studios in Australia have enjoyed critical success with their locally developed products. For example, Irrational Games established a second studio in Canberra, Australian

Capital Territory, in 2000. Irrational games was purchased by Take Two in 2005/2006 and the Canberra studio was renamed 2K Australia before merging with 2K Marin in 2010. This studio was primarily responsible for the critical and commercially successful multiplatformer *Bioshock 2* (2010) and the reboot of the *X-COM* series, *The Bureau: XCOM Declassified* (2013).

The labor conditions of the Australian (and indeed global) video game industry were brought into focus in mid-2011 when someone using an anonymous Twitter account began posting allegations regarding the development of *LA Noire* (2011) at Team Bondi in Sydney. Following media interest, eleven anonymous former Team Bondi workers spoke to journalist Andrew McMillen for the Australian edition of IGN.com about their experiences, with one stating, "I never would have thought you could put a sweat shop in the Sydney CBD" (McMillen 2011). The allegations ranged from mismanagement, to employees not being paid for overtime, to verbal abuse, to systematic burn-through of workers. Ultimately, the central criticisms focused on Team Bondi studio head Brendan McNamara and a culture of perpetual "crunch." "Crunch" is an industry term for an intense period of crisis in production, when employees are asked to work long and hard hours (anywhere between 60 and 100 hours per week) in order to reach development milestones. According to the anonymous sources quoted by McMillen, these crunch hours became the norm, with an average week measured at sixty hours, and intense periods jumping to "between 80 and 110 hours per week." Following the publication of these allegations and the release of *LA Noire*, Team Bondi was placed into administration and its assets sold to Australian film company Kennedy Mitchell Miller (KMM), owned by *Mad Max* (1979) director George Miller (Brown 2011). KMM is now one of Australia's only traditional game development studios.

The Australian video game industry has historically received support from both federal and state governments. In the early 2000s, the Victorian and Queensland state governments provided incentives for developers, including development kits, support for overseas travel, and the direct funding of projects and companies. Such support has been haphazard, however, with the schemes criticized for being "piecemeal and disorganized" (McCrea 2013, 205). Equally, the industry has persistently set this support against governmental contributions to Australia's film industry and international schemes like Canada's tax breaks for their video game industry. This conflict came to a head in 2007 when a collection of video game studio heads told the ABC current affairs television show *The 7.30 Report* that "the games industry in Australia could be bigger than the film industry" (Callaghan 2013).

Since 2007, industry bodies such as the Game Developers' Association of Australia (GDAA) and the Interactive Games and Entertainment Association (iGEA), with support from a sector of Australia's enthusiast press, have publicly campaigned for greater government funding and financial incentives for Australia's video game industry. In 2012, this lobbying led to the announcement of an Australian Interactive Games Fund, which comprises AUD $20 million of federal government funding to be allocated over three years from 2013 to 2015. After a brief public consultation process, it appears that the vast majority of allocated funds, as administered by federal government body Screen Australia, will go to existing businesses and established development studios instead of infrastructure or supporting emerging practitioners, a decision that has been criticized as conservative by some (Golding 2013). The federal fund will join three state government funds—the Victorian Digital

Games Fund administered by Film Victoria, the Interactive Media Fund operated by Screen NSW, and Screen Australia's All Media Ignition Program and Digital Sandpit Program—in funding games across Australia.

This governmental support of the video game industry has been framed by the shift toward new modes and structures for video game development both locally and internationally. The most significant development in this respect has been the emergence, and now dominance, of mobile game studios in the Australian landscape, such as Brisbane's Halfbrick, which developed *Fruit Ninja* (2010) and *Jetpack Joyride* (2011), and Melbourne's Firemint, famous for *Flight Control* (2009) and *Real Racing* (2009). These studios drew on the residual talent left isolated by the departure of older studios from Australia and have established themselves as "a new Australian mainstream—one focused on mobile, digital distribution, bite-sized arcade gameplay, and in recent years the possibilities of freemium and in-app purchases" (Callaghan 2012). While mobile development—especially at an independently owned company such as Halfbrick (Firemint was bought by EA in 2011 for an undisclosed sum and was merged with another Melbourne-based EA mobile studio, Iron Monkey, in 2012, to form Firemonkeys)—would have once been labeled "indie" or at least as outside the mainstream for the Australian industry, such companies now form the backbone of Australian game development.

The emergence of a new mainstream has inevitably led to new modes of independent and alternative game production. While new development and distribution technologies have engendered the success of mobile studios such as Halfbrick, they have also opened up unconventional spaces for individuals and small groups in Australia's video game culture. Developers such as Jarrad Woods, who under the pseudonym "Farbs" has developed *ROM Check Fail* (2008) and *Captain Forever* (2009), Alexander Bruce, who developed *Antichamber* (2013), and Harry Lee, who developed *Midas* (2011) and *Stickets* (2013), are all examples of a new kind of independent aesthetic in Australia. Meanwhile, smaller, more specialized studios continue to appear, such as mobile "gamebook" developers Tin Man Games, and boutique independent studio Pachinko Pictures.

Game Regulations

Since the formal introduction of video game classification legislation in 1995, debates over regulation have dominated public discourse surrounding video games in Australia. Central to these debates was the deliberate omission of an adults-only classification level for video games: unlike the classification scale for film (and the video game classification systems in comparable countries such as the United States, the UK, and New Zealand), the highest possible classification for video games under "The Classification (Publications, Films and Computer Games) Act 1995" is MA15+. This meant that from 1995 to 2013, video games with content that the federal Classification Board considered to exceed the guidelines for the MA15+ rating were denied classification in Australia (and by implication, considered illegal to sell or import, and effectively banned). Several video games have been refused classification in Australia on these grounds: most notably, Rockstar Games' *Grand Theft Auto III* (2001) for violence and sex (which was then edited and reclassified [Finn 2003]); The Collective Inc.'s *Marc Ecko's Getting Up: Contents Under Pressure* (2005) for instructing on matters of crime (Apperley 2008), and Z-axis's *BMX XXX* (2002) for standards of decency.

These classification outcomes have meant that regulation has operated as a key discourse surrounding video games in Australia, regularly manifesting itself in mainstream and enthusiast press coverage and culminating in a registered political party with the introduction of an R18+ category as its sole policy platform; a federal government-led public consultation that saw 58,437 submissions, 98% in favor of an R18+ classification; and the largest public petition in the history of Australia, supporting an R18+ classification with 89,210 signatures, exceeding a previous petition to change widely unpopular industrial relations laws.

The initial decision to omit an R18+ classification seems to have been largely related to two concerns: first, that video games could have a higher impact due to their interactive nature; and second, that the absence of R18+ level video games would aid in the protection of children from inappropriate content. "It is one thing to watch a violent video; it is another thing altogether to be involved in the violence," stated member of the Federal House of Representatives Peter McGuarun in the parliamentary discussion of the classification legislation in 1994 (McGauran 1994). Equally, Janice Crosio, a fellow member of the House of Representatives, argued that the classification system "will assist us as parents to make informed choices about the games our children play and ... ensure that our children are not exposed to material deemed unsuitable" (Crosio 1994). These two threads of discussion would continue to reverberate throughout debates over video games in Australia—indeed, the perception that video games are inherently linked to children is a particularly persistent idea, despite evidence to the contrary. A 1996 report by the Australian Bureau of Statistics, for example, found that 50% of Australians playing video games were age eighteen or older (Skinner 1996). In addition, shortly after the introduction of the classification legislation, the Office of Film and Literature Classification (OFLC, the government body that until 2006 housed the Classification Board) commissioned a report on the current research of the effects of video games on "young people" by the University of Western Australia Professor Kevin Durkin. The report, while not discounting the possibility of negative impacts completely, cautioned against alarm: "Contrary to some fears that computers inevitably draw people into solitary activities, one major study has found that social play with computer games is twice as high as social involvement in other media use" (Durkin 1995). This report seemingly did not impact the politics of the time, though it did represent the first of many public indications that such a restrictive classification system may not be effective or necessary. Durkin's report was launched with a press release by Federal Attorney General Michael Lavarch, who took the opportunity to praise the recently enacted classification system, which he claimed had been "well received by parents" (Lavarch 1995).

The greater context of the mid-1990s saw growing public concern over violence in video games, and this may in many ways help to rationalize the low impact of findings like those of Durkin and the ABS. Early in the decade, American video games such as id Software's *DOOM* (1993) and Midway's *Mortal Kombat* (1992) triggered widespread public discussion of video game violence. In 1993, for example, the *Sydney Morning Herald* stated that *Mortal Kombat* was "sick and offensive. Don't buy it. Good luck, though, in getting the kids off your back if you don't" (Camm 1993). Digital Pictures' full-motion video-based horror game *Night Trap* (1993) in particular was so widely criticized for its (now somewhat primitive) scenes of slumber-partying girls being attacked by aliens that it was credited with creating the community and media impetus for the introduction of the

Classification Act 1995. Editorializing on *Night Trap* in 1993, for example, Melbourne's *The Age* noted that the then prime minister, Paul Keating, had moved to seek support for the introduction of a national classification scheme: "All video games should be regulated by the Office of Film and Literature Classification," *The Age* concluded ("Welcome curbs on violent video games" 1993). After the introduction of the classification scheme, the late 1990s cemented the trend of the close regulation of video games. As Terry Flew suggested, the Port Arthur massacre of 1996 meant there was an expectation that the newly elected and socially conservative Liberal Party, under Prime Minister John Howard, would "enact stronger policies toward the censorship and classification of materials deemed to be potentially harmful" (Flew 1998, 89).

Nonetheless, the early 2000s saw the first concerted and public attempt to expand the video game classification system to include an R18+ rating. In 2002, the OFLC conducted a public review of the system and concluded that "there is strong support for an 'R' Classification for computer games both in terms of quantity and quality compared with opposition to it; as such this classification should be introduced with restriction to those aged 18 years or older" (Brand 2002, 32). However, due to Australia's federalist system of politics, any change to the Classification Act 1995 requires unanimous agreement between all state attorneys-general and the federal government. The OFLC's proposal gained in-principle support from all members of the Standing Committee of Attorneys General (SCAG) except for South Australian Attorney General Michael Atkinson and Federal Attorney General Daryl Williams and therefore failed. The annual report of the Standing Committee states, "On balance, not all Ministers were satisfied children would not access games classified as suitable for adults" (Standing Committee of Attorneys General [Censorship], Annual Report 2002–2003).

Bethesda's *Fallout 3* (2008) is a particularly interesting case study of how Australia's classification system functioned without an R18+ level. *Fallout 3* was ruled to be refused classification on July 4, 2008, with the Classification Board report indicating that the game's implementation of drug use was the central concern.[2] In this original version of *Fallout 3*, players could elect to use morphine to mitigate the effects of injuries. Therefore, the Classification Board concluded that the use of a prescribed drug "is related to incentives and rewards ... and is such Refused Classification." As *Fallout 3* was a widely anticipated video game with substantial prerelease advertising and media coverage (in contrast to many refused-classification video games in Australia, which frequently have low public profiles, such as The Farm 51's *Necrovision* [2009] and Gameloft's *Sexy Poker* [2009], both refused classification close to *Fallout 3*), there was substantial public reaction to the classification decision. The Classification Board 2008–2009 annual report indicates that they received 629 complaints about the classification of *Fallout 3* (Classification Board and Classification Review Board 2009, 54). A common complaint in the media focused on the apparent inconsistency of the Classification Board's rulings on prescription drugs: "What are the syringes in *Bioshock* filled with—magic fairy dust?" read one comment on enthusiast Web site Australian Gamer (GamePolitics 2008). Indeed, the Classification Board has classified a number of video games that make similarly explicit reference to and use of morphine both before and after *Fallout 3*, such as Valve's *Half-Life* (1998), classified MA15+ in 1998, and Replay Studios' *Velvet Assassin* (2009), classified MA15+ in 2009.

However, the publishers of *Fallout 3* in Australia submitted an altered version of the game to the Classification Board. This edited edition of *Fallout 3* changed the names of the drugs used in the game to fictional alternatives ("Med-X"), satisfying the board that the game could now be accommodated at an MA15+ level, at which it was classified on August 7, 2008. The alteration and resubmission of refused classification video games is not particularly unusual in Australia—similar processes have been followed for Genuine Games' *50 Cent: Bulletproof* (2005), Valve's *Left 4 Dead 2* (2009), and Double Helix's *Silent Hill: Homecoming* (2008).

What was unusual about the *Fallout 3* edit was that the game's developer, Bethesda, incorporated the changes required for classification in Australia on a global scale. According to Bethesda, the decision was made to alter *Fallout 3* globally to "avoid confusion" ("Censors force Fallout 3 changes" 2008). Thus, we can see a rare example of video game regulation in Australia having a global impact on the creation and reception of a video game—an impact likely not foreseen at the creation of Australia's system of regulation.

The apparent inconsistencies in classification highlighted by *Fallout 3* point to greater practical difficulties in regulating video games in Australia. If the goal of the Classification Act 1995 was to exclude a certain level or type of video game from Australia, its practical effectiveness is worth questioning. Indeed, the very video games that prompted the introduction of the classification scheme have largely been made legally available; *Night Trap* was classified at M for "medium level violence" in 1995, for example. Only a small percentage of video games have been refused classification; in 2009, for example, only four games out of 1,055 were refused classification. It seems unlikely that such a small number of games would be classified R18+ if such a rating were available, since by comparison, seventy-six games were classified MA15+ in 2009. Comparing classified video games across national regulation schemes is also insightful: since 1994, the North American Entertainment Software Rating Board (ESRB) has classified more than 1,700 video games at a "Mature 17+" or "Adults Only" level; since 2003, the Pan European Game Information (PEGI) organization has classified 1,069 videogames as PEGI 18 (though this number includes duplicate certificates for single games across multiple platforms); by contrast, since 1995, only ninety video games have been refused classification (the logical equivalent) in Australia.

Clearly, a significant number of video games has been classified as MA15+ in Australia but identified as being at a mature level elsewhere. Indeed, when the former federal Attorney General Brendan O'Connor declared his public support for an R18+ classification in 2011, he released a comparison of forty-seven video games classified at MA15+ in Australia, only fifteen of which were consistent with international equivalents (O'Connor 2010). Violent and gory video games, such as Epic Games' *Gears of War* (2006), SCE Studios' *God of War* (2005), and *Bulletstorm* (2011) by People Can Fly and Epic Games, are currently classified as MA15+ in Australia, despite being rated Mature 17+ by the ESRB and 18 by PEGI.

In February 2008, six years after the first concerted attempt by the OFLC to implement an R18+ classification, the newly elected Federal Labor government announced that a potential R18+ rating would be discussed at the next SCAG meeting (Moses 2008a). However, as in the 2002 attempt, there was a failure to reach consensus between attorneys general, and the proposal failed. In this instance, the only member of the SCAG opposed to the R18+ was South Australian Attorney-General Michael Atkinson. In his rejection,

Atkinson repeated the two major opposing arguments to an R18+ rating since 1995: that he wished to deny children access to "potentially harmful material" and that "games may pose a far greater problem than other media—particularly films—because their interactive nature could exacerbate their impact" (Moses 2008b).

Nonetheless, an R18+ classification was not dismissed entirely at this point: instead, the attorneys-general agreed to canvass public opinion on the matter and requested a discussion paper be developed. In retrospect, it was this act that reignited public debate on the issue in force. Opposition to the discussion was immediately raised from both ends of the political spectrum, with both a pornography lobby and a Christian group opposing the move.[3] In response, several groups were formed in support of an R18+ classification level, including a political party. This minor political party, Gamers 4 Croydon, was formed to contest South Australian Attorney General Michael Atkinson's parliamentary seat of Croydon in the 2010 state election, taking its name from Valve's *Left 4 Dead 2* (2009), after it was refused classification the previous year. Gamers 4 Croydon received only 3.7% of the votes in Croydon (and 0.8% statewide) (LeMay 2010), but the issue became conspicuous—and heated—enough that Atkinson claimed (perhaps with some election-year hyperbole) that "my family and I are more at risk from gamers than we are from the outlaw motorcycle gangs who also hate me" ("Attorney-General Steps Up Fight with Gamers" 2010). Despite his reelection, Atkinson chose to step down as attorney general only days after the 2010 election.

The public response to the SCAG discussion paper was overwhelmingly in support of an R18+ classification, with 98% of the 58,437 submissions in favor, a figure that Federal Attorney General Brendan O'Connor noted was unusually high (Ramsey 2010). In addition, a number of months after the public consultation process ended, an independently audited petition signed by 89,210 people supporting an R18+ classification was tabled in Federal Parliament. The petition was at the time the largest in Australian history, exceeding a previous petition regarding deeply unpopular Howard-government industrial relations laws. Following these events, by mid-2011, the attorneys general reached an in-principle agreement for the introduction of an R18+ classification, and guidelines for the rating were drafted ("Historic agreement to introduce 18+ videogames" 2011).

In March 2012, legislation to implement an R18+ classification passed Australia's lower house with support from all major parties. Interestingly, the political discourse seems substantially unchanged from 1994. On the basis of the bill's lower-house reading, there remains an impression that video games are a childlike activity (Ewan Jones, MP, claimed that "the best thing about [video games] as a parent is that it keeps them [children] quiet") and that the interactive nature of video games could lead to a higher impact than other media (Golding 2012). Most interestingly, there seemed to be an opinion among parliamentarians that the R18+ classification would allow parents "to provide the necessary protection from material that is likely to harm or disturb [minors]," in the words of Karen Andrews, MP (Golding 2012). This was the same argument used to justify the omission of the R18+ classification eighteen years earlier, suggesting a substantial shift in the rhetoric of Australian video game regulation. Perhaps this is partially related to the perceived effectiveness of classification and censorship in Australia: while it was initially envisaged that disallowing R18+ level video games would make them wholly unavailable, it is now argued that it is more practical to regulate such video games than exclude them entirely. It is worth noting that the perception of danger or potential harm as relating to such R18+ level games has therefore remained largely unchanged.

After passing in all Australian state parliaments, the R18+ classification came into effect on January 1, 2013, with Team Ninja's *Ninja Gaiden 3: Razor's Edge* (2013) becoming the first game rated R18+ under the new system. Furthermore, two new government reports, the National Classification Scheme Review and the Convergence Review, have recommended deeper and more holistic reform across the regulation of all media forms in Australia, such as industry-led regulation for lower classification levels in order to streamline cost and classification workload. However, despite signaling potential for the then anticipated National Cultural Policy to respond to these recommendations, the policy, when released in March 2013, did not incorporate any revision of classification and regulation of media in Australia, indicating that the current system will remain in place for the conceivable future. Tellingly, in June 2013, *Saints Row IV* (2013) became the first video game refused classification under the amended legislation (for "sexualized violence" in the guise of an "alien anal probe" gun) and was closely followed by *State of Decay* (2013) (for the depiction of real-world drugs), indicating that the regular censorship of video games in Australia may well continue into the future.

Conclusion

While over the past decade Australian and Australian-based game studios have been involved in many games that have had considerable commercial and critical success, the traditional studio model of game development has struggled to survive. The smaller teams that have met success by focusing on gaming apps demonstrate considerable promise for the future of game development in Australia. This is encouraging for Australian game developers, as is the renewed interest from the federal government in providing funding to support local game development. The current state regulation of digital games by the government has endured strong criticism and quite vocal and active campaigns against it. While there has been some reform, the regulation of games has not been revised so as to fall under the same rubric as other forms of media content, despite this revision being strongly recommended by government-commissioned reports. The situation as it now stands in mid-2014 suggests that the current system of regulation will remain the status quo. As the popularity of digital games in Australia continues to increase, the government may make more robust formal commitments to supporting local industries and reforming the regulation system. However, as matters currently stand, the future of both these issues is largely in the hands of dedicated developers and consumers who are committed to a wider local production and consumption of digital games.

Notes

1. Australian (and New Zealand) computer games from this period are archived on the *Play It Again* blog http://blogs.flinders.edu.au/play-it-again/. The Australian home development scene for computer games is

discussed in Melanie Swalwell's "Questions about the usefulness of microcomputers in 1980s Australia," *Media International Australia* 143 (2012).

2. Board reports are not usually publicly available, but the *Fallout 3* report was quickly leaked to the media and remains available at http://blogs.theage.com.au/screenplay/Fallout%203.pdf.

3. Both the Eros Foundation—an Australian pornography and "adult industry" lobby group—and the Australian Christian Lobby opposed the introduction of such a classification ("Strange Bedfellows Oppose R18+ Games Debate" 2008).

References

Apperley, Thomas. 2008. Video games in Australia. In *The Video Game Explosion: A History from PONG to PlayStation and Beyond*, ed. Mark J. P. Wolf, 223–228. Westport, CT: Greenwood Press.

Attorney-General steps up fight with gamers. 2010. *ABC*, February 16. http://www.abc.net.au/news/2010-02-16/attorney-general-steps-up-fight-with-gamers/332546.

Australian Bureau of Statistics. 2013. *Film, Television, and Digital Games, Australia, 2011-12*. Canberra, Australia: Australian Bureau of Statistics. http://www.ausstats.abs.gov.au/ausstats/subscriber.nsf/0/E612C796E7A5461FCA257BA50012F64A/$File/86790_2011-12.pdf.

Australian Centre for the Moving Image. n.d. Hits of the 80s: Aussie games that rocked the world: Beam software timeline. http://www.acmi.net.au/.

Banks, John. 2009. Co-creative expertise: Auran Games and Fury—a case study. *Media International Australia* 130:77–89.

Beavis, Catherine, Joanne O'Mara, and Lisa McNiece, eds. 2012. *Digital Games: Literacy into Action*. Adelaide, Australia: Wakefield Press.

Brand, Jeffery E. 2002. A review of the classification guidelines for films and computer games. White Paper, Australia: Office of Film and Literature Classification.

Brand, Jeffrey E. 2005. *GamePlay Australia 2005*. Gold Coast: Bond University.

Brand, Jeffrey E. 2007. *Interactive Australia 2007: Facts about the Australian Computer and Video Game Industry*. Eveleigh, Australia: Interactive Entertainment Association of Australia.

Brand, Jeffrey E. 2012. *Digital Australia 12*. Gold Coast: Bond University.

Brand, Jeffrey E., Jill Borchard, and Kym Holmes. 2009. *Interactive Australia 09*. Gold Coast: Bond University.

Brown, Nathan. 2011. Team Bondi to close. *Edge*, October 5. http://www.edge-online.com/news/team-bondi-close/.

Callaghan, Paul. 2012. The return of the videogame outsiders. *Realtime* 112:25. http://www.realtimearts.net/article/112/10911.

Callaghan, Paul. 2013. Trading dollars for (99) cents. *Metro,* 175:128.

Camm, Mark. 1993. A code for triggering the Gore dimension. *Sydney Morning Herald*, November 15.

"Censors Force Fallout 3 Changes." 2008. *Edge*, September 10. http://www.edge-online.com/news/censors-force-fallout-3-changes.

Chesher, Chris, and Brigid Costello. 2004. Why media scholars should not study computer games. *Media International Australia* 110:5–9.

Classification Board and Classification Review Board. 2009. *Annual Reports 2008–2009.* Sydney, Australia: Commonwealth of Australia, http://www.classification.gov.au/About/AnnualReports/Documents/CBOARDAR%200809.pdf.

Crosio, Janice. 1994. Second reading speech: Classification (Publications, Films and Computer Games) Bill 1994. September 22, House of Representatives. http://parlinfo.aph.gov.au/parlInfo/search/display/display.w3p;query=Id%3A%22chamber%2Fhansardr%2F1994-09-22%2F0020%22.

DeMaria, Rusel, and Johnny L. Wilson. 2004. *High Score! The Illustrated History of Electronic Games.* Berkeley, CA: McGraw-Hill.

Donovan, Tristan. 2010. *Replay: The History of Video Games.* Lewes: Yellow Ant.

Durkin, Kevin. 1995. *Computer Games and their Effects on Young People: A Review.* Sydney, Australia: Office of Film and Literature Classification.

Durkin, Kevin, and Kate Aisbett. 1999. *Computer Games and Australians Today.* Sydney, Australia: Office of Film and Literature Classification.

Finn, Mark. 2003. GTA3 and the politics of interactive aesthetics. *Communications Law Bulletin* 4 (22): 6–11.

Flew, Terry. 1998. From censorship to policy: Rethinking media content regulation and classification. *Media International Australia* 88:89–98.

Flynn, Bernadette. 2003. Geographies of the digital hearth. *Information Communication and Society* 6 (4): 551–576.

Flynn, Bernadette. 2004. Games as inhabited spaces. *Media International Australia* 110:52–61.

Funky, J. 2005. Midway shafts Ratbag Studios … *Five minutes of funk*, December 12. http://blog.funkyj.com/2005_12_01_archive.html

GamePolitics. 2008. Report: Australia's *Fallout 3* ban prompted by in-game drug use. *GamePolitics.com*, July 10. http://www.gamepolitics.com/2008/07/10/report-australia039s-fallout-3-ban-prompted-game-drug-use.

Golding, Daniel. 2012. What our politicians think about videogames. *Crikey*, March 17. http://blogs.crikey.com.au/game-on/2012/03/17/what-our-politicians-think-about-videogames/.

Golding, Daniel. 2013. Videogame makers are essential to a creative Australia. *ABC Arts*, March 15. http://www.abc.net.au/arts/stories/s3716558.htm.

Hinton, Sam. 2009. Gaming nation: The Australian game development industry. In *Gaming Cultures and Place in the Asia-Pacific*, ed. Larissa Hjorth and Dean Chan, 39–57. New York: Routledge.

Historic agreement to introduce 18+ videogames. 2011. *Age*, July 22. http://www.theage.com.au/digital-life/games/blogs/screenplay/historic-agreement-to-introduce-r18-games-20110722-1hs4b.html.

Humphreys, Sal. 2008. Ruling the virtual world: Governance in massively multiplayer online games. *European Journal of Cultural Studies* 11 (2): 149–171.

Knight, Scott J., and Jeffrey E. Brand. 2007. History of game development in Australia. Melbourne, Australia: ACMI. http://www.acmi.net.au/.

Lavarch, Michael. 1995. Computer games may improve family relations: Research. October 4, Press Release: Attorney General, http://parlinfo.aph.gov.au/parlInfo/search/display/display.w3p;query=Id%3A%22media%2Fpressrel%2F5QI20%22.

LeMay, Renai. 2010. Gamers 4 Croydon hails "fantastic" Atkinson resignation. *Delimiter*, March 21. http://delimiter.com.au/2010/03/21/gamers-4-croydon-hails-fantastic-atkinson-resignation.

McCrea, Christian. 2013. Snapshot 4: Australian video games: The collapse and reconstruction of an industry. In *Gaming Globally: Production, Play and Place*, ed. Ben Aslinger and Nina B. Huntemann, 203–207. New York: Palgrave Macmillan.

McGauran, Peter. 1994. Second Reading Speech: Classification (Publications, Films and Computer Games) Bill 1994. September 22, House of Representatives, http://parlinfo.aph.gov.au/parlInfo/search/display/display.w3p;query=Id%3A%22chamber%2Fhansardr%2F1994-09-22%2F0021%22.

McMillen, Andrew. 2011. Why did *L.A. Noire* take seven years to make? *IGN*, June 24. http://au.ign.com/articles/2011/06/24/why-did-la-noire-take-seven-years-to-make?page=1.

Morris, Sue. 2003. First-person shooters—a game apparatus. In *ScreenPlay: Cinema/Videogames/Interfaces*, ed. Geoff King and Tanya Krzywinska, 81–97, London: Wallflower Press.

Moses, Asher. 2008a. R-rated games may be on shelves soon. *Sydney Morning Herald*, February 25.

Moses, Asher. 2008b. R18+ for games? Not a chance. *Sydney Morning Herald*, March 7. http://www.smh.com.au/news/technology/r18-for-games-not-a-chance/2008/03/07/1204780028413.html.

Ndalianis, Angela. 2004. *Neo-Baroque Aesthetics and Contemporary Entertainment*. Cambridge, MA: MIT Press.

Ndalianis, Angela. 2012. *The Horror Sensorium: Media and the Senses*. Jefferson, NC: MacFarland.

O'Connor, Brendan. 2010. International counterparts restrict more video games to adults only. Media release, December 9. http://www.sciencemedia.com.au/downloads/2010-12-9-1.pdf.

Perry, Douglass. 2007. Micro Forte cancels Citizen Zero: Announces secret new spy-themed MMO. IGN.com, March 5. http://au.ign.com/articles/2007/03/01/micro-forte-cancels-citizen-zero.

Ramsey, Randolph. 2010. More than 55,000 R18+ submissions received. *Gamespot Australia*, March 4. http://au.gamespot.com/news/more-than-55000-r18-submissions-received-6252907.

Schroeder, Jens. 2010. *"Killer Games" versus "We Will Fund Violence": The Perception of Digital Games and Mass Media in Germany and Australia*. Frankfurt: Peter Lang Publishing.

Skinner, T. J. 1996. Household use of information technology, Australia, February. Australian Bureau of Statistics.

souri. 2011. THQ Studio Australia and Blue Tongue Entertainment to close this week. *Tsumea*, August 10. http://www.tsumea.com/australasia/australia/news/100811/thq-studio-australia-and-blue-tongue-entertainment-to-close-this.

Strange bedfellows oppose R18+ games debate. 2008. *News.com.au*, March 28. http://www.news.com.au/technology/strange-bedfellows-side-on-games-ratings/story-e6frfro0-1111115915352.

Stuckey, Helen, and Noè Harsel. 2007. Hits of the 80s: Aussie games that rocked the world. *ACMI*, November.

Swalwell, Melanie. 2003. Multi-player computer gaming: Better than playing (PC games) with yourself. *Reconstruction*, 3:4.

Swalwell, Melanie. 2012. Questions about the usefulness of microcomputers in 1980s Australia. *Media International Australia* 143:63–77.

Wark, McKenzie. 1994. The video game as an emergent media form. *Media International Australia* 71:21–30.

Welcome curbs on violent video games. 1993. *Age*, May 16. http://newsstore.smh.com.au/apps/viewDocument.ac?page=1&sy=smh&kw=night+trap&pb=all_ffx&dt=selectRange&dr=entire&so=relevance&sf=text&sf=headline&rc=200&rm=200&sp=nrm&clsPage=1&docID=news930516_0188_2679.

AUSTRIA

Konstantin Mitgutsch and Herbert Rosenstingl

In 2005, at its peak, the Rockstar Games subsidiary Rockstar Vienna was one of the biggest game development studios in Germany and Austria, with over one hundred employees. In the late 1990s, the Austrian game developer Max Design surprised the games market by selling over two million copies of *Anno 1602* (1998), but Rockstar Vienna's establishment heralded a new era of game development in Austria. At this point in time, the future of Austrian video game development seemed bright. Together with other Austrian game development studios such as JoWooD (established in 1989), Pixlers Entertainment (1991), Greentube (1998), Sproing (2001), Xendex (2001), and GamezArena (2002), along with subunits of major publishers including Electronic Arts, Microsoft, Sony Computer Entertainment, Ubisoft, and Nintendo, Austria's location in the heart of Europe appeared promising. A year later, in 2006, Austria rejected a new political initiative for game censorship and regulation with a proactive strategy for handling video games. In 2007, instead of installing game censorship, Austria decided to host a high-profile gaming event in the Vienna Town Hall called Game City. The event brought together representatives of the games industry, nonprofit organizations, academia, and the general public for the discussion of the current state of computer games. The approach was designed to facilitate a rich dialogue about games instead of censorship. At that time in Austrian game history, politicians, game developers, gamers, and the general public were acting in concert; Austria seemed to be a gaming oasis compared to many other countries in Europe (in particular, Germany).

The faith of the Austrian game development movement, however, took an unexpected turnaround when Rockstar Vienna was closed in 2006 and all employees were laid off. As a result, a variety of smaller independent game development studios were established that today create video games, serious games, and social apps. On a political level, Austria stayed unorthodox in its handling of age regulations and censorship. In contrast to other countries, Austrian youth policymakers decided to follow a path of positive assessment for computer and console games rather than rigid legislative regulations as in Germany. In 2003, the Federal Office for the Positive Assessment of Computer and Console Games (Bundesstelle für die Positivprädikatisierung von Computer- und Konsolenspielen, for short: BuPP) was established within the structures of the Austrian Federal Ministry for Youth on a governmental level. While the age ratings of PEGI (Pan European Game Information) and the German USK (Unterhaltungssoftware Selbstkontrolle) appear on game packaging, Austria had, until recently, assessed playability and age-appropriate content through the recommendation of suitable games. The Austrian way of positive assessment on a governmental level is unique and offers an interesting alternative to censorship and youth protection.

To highlight the evolution of video games in Austria, this chapter will outline four important parts of Austrian video game history in greater detail: Austrian game development, the positive assessment of games, research activities in Austria, and Austrian gaming culture. Furthermore, it will examine indigenous cultural initiatives such as e-sports, art and games, and games journalism. Finally, we will discuss the future of video games in Austria.

Austrian Game Development

The roots of Austrian game development led to the first commercial computer games and video games developed by individuals or small independent game studios. For example, based on our research, the first commercial video game developed in Austria was the text adventure game *Der verlassene Planet* (*The Abandoned Planet*) (see fig. 1), created by Hannes Seifert in Austria and released in 1989 by the German publisher Markt & Technik. The first internationally known Austrian game development studio, Max Design, was founded in 1991 and is known today for influencing the creation of the genre "German Business Simulations" with releases such as *Cash* (1991), *1869* (1992), and *Anno 1602* (1998) (see http://en.wikipedia.org/wiki/Max_Design). *Anno 1602* (see fig. 2) sold more than four million copies to date and was the most successful video game ever developed in Austria or Germany in the late 1990s (see http://www.gameswelt.de/red-monkeys/news/red-monkeys-mitbegruender-ueber-vergangenheit-anno-und-was-uns-in-zukunft-erwartet,82897). The *Anno* series was later developed conjointly with Sunflowers Interactive Entertainment Software, which released *Anno 1503* (2003). Although Max Design closed its studio in 2004, Ubisoft has continued the *Anno* series with its latest release, *Anno Online* (2013).

Max Design proved that an Austrian game development studio could significantly impact both the German and international markets, but it also showed the struggle that Austrian developers went through to sustain their business. Looking at the history of the biggest Austrian development studios, one finds there are similarities in their experiences; they started as small enterprises, managed to reach a moderate to high level of success and growth, and then suddenly collapsed for various reasons. The rise and fall of the two biggest Austrian game development studios, Rockstar Vienna and JoWooD, exemplify the struggle to succeed in a global market.

Rockstar Vienna

In 1993, the first Austrian development studio, neo Software, was established by Niki Laber and Hannes Seifert in Hirtenberg. They created their first commercial video games for the Commodore AMIGA with titles such

 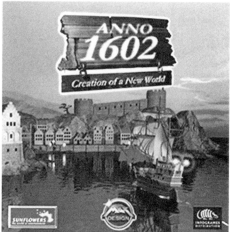
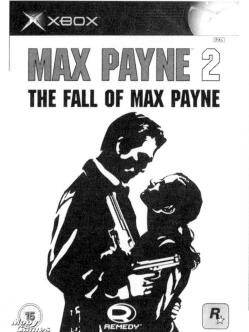 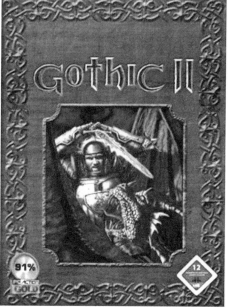

Figure 1

Hannes Seifert's *Der verlassene Planet* (1989) (top, left); Max Design's *Anno 1602* (1998); Rockstar Vienna's *Max Payne 2* (2003) (bottom, left); JoWooD's *Gothic II* (2002) (bottom, right).

as *Whale's Voyage* (neo Software/Flair Software, 1993) and *Der Clou!* (neo Software/Max Design GmbH & Co., 1993), and PC games such as *Die Völker* (neo Software/JoWooD, 1999). In 2001, Take-Two Interactive acquired neo Software, and two years later it joined Rockstar Games under the umbrella Rockstar Vienna (see http://de.wikipedia.org/wiki/Rockstar_Games#Rockstar_Vienna). In this golden era of Austrian video game development, Rockstar Vienna developed the Xbox version of the successful third-person shooter *Max Payne* (2001), followed by Xbox versions of *GTA III* (2003), *GTA: Vice City* (2003), and *Max Payne 2: The Fall of Max Payne* (2003) (see fig. 1). The final project, *Manhunt 2* (2007), was moved to Rockstar London after Take-Two closed Rockstar Vienna overnight in 2006 and fired all its 100 employees without prior notice. In an article on Kotaku.com, Lee Brown (2006) summarized this surprising occurrence.

> The offices Rockstar Vienna—the wispy-mustached lederhose subsidiary of Rockstar Games that was responsible for porting the *Grand Theft Auto* games to the Xbox (and somehow managing to make them look more detailed than the PC versions) as well as porting *Max Payne 2* to the PS2 and Xbox—has been closed.

It was not simply the fact that the biggest Austrian game developer was shut down but the unexpected and vicious way in which the company was closed that shook the Austrian game development scene. While the founders of neo Software announced the formation of a new company—Games That Matter Productions—in 2007, many former employees of Rockstar Vienna left Austria or established their own game studios. Games That Matter Productions was acquired and rebranded as Deep Silver Vienna by Koch Media and closed down three years later. Referencing reports and the online discourse regarding Rockstar Vienna's rise and fall, the studio created an atmosphere of unease and ill will in the Austrian developer scene.

JoWooD Entertainment AG

Two years after the foundation of neo Software in 1995, the publisher and developer JoWooD Entertainment AG was established by Andreas Tobler. In 1997, JoWooD developed and published the successful business simulation *Der Industriegigant,* which sold more than 800,000 copies worldwide. During the sixteen years of JoWooD's existence, the enterprise developed and published more than forty games and sold hundreds of thousands of copies of titles such as *Gothic II* (2002), *Yoga: The first 100% Experience Wii* (2009), *Sam & Max: Season One Wii* (2008), and *Arcania: Gothic IV* (2010). JoWooD also expanded rapidly, acquired different development studios, and was listed on the Austrian stock exchange in 2000. Only two years later, the game development and publishing company narrowly avoided insolvency and struggled to meet the high standards that had been expected of their products and service. Different titles were supposedly shipped incomplete, a legal dispute with an Australian company impacted the production process, and the risk of insolvency remained ever-present. In 2006, after cooperating with the German media enterprise Koch Media GmbH, all development studies of JoWooD were closed, and the company retained only its publishing activities. After further problems with unfinished products, a relaunch in 2007, different company takeovers, and phases of new hope, JoWooD finally filed for bankruptcy in early 2011. JoWooD was acquired by Nordic Games Holding, which uses JoWooD as a publishing label for Nordic Games GmbH.

In 1999, when JoWooD published the successful strategy game *Die Völker*, which was developed by neo Software/Rockstar Vienna, the Austrian game development machine was running smoothly—but within twelve years, both companies had collapsed.

Later Developments

One independent Austrian game development studio that has managed to successfully establish itself on an international level is Sproing Interactive Media GmbH, founded in 2001 by Harald Riegler and Gerhard Seiler. The company has developed more than fifty titles and started with small, online, shoot-'em-up games such as *Moorhuhn* (2004), which were inexpensive to develop but successful in the market (see http://fm4.orf.at/stories/1687659/). Today, Sproing has fifty employees and publishes multiplatform games such as *Scotland Yard* (2008), *Mountain Sports* (2009), *My Fitness Coach Club* (2011), and *Kinect Sports Gems* (2013). One of their recent MMOGs, *Skyrama* (2011), reached more than ten million registered users in 2012, and the company's development approach of addressing different platforms appears promising (see http://www.sproing.com/website/news.html). Compared to Rockstar Vienna and JoWooD, Sproing manages to stay sustainable by focusing on smaller projects, remaining flexible in its development cycle and being realistic with regard to production volume.

In a similar way, the online 2-D and 3-D developer Greentube was able to successfully focus on a niche market of free online games and developed the highly successful sports game *Ski Challenge* (2004) (see http://futurezone.at/b2b/5405-ich-lasse-mir-von-apple-nichts-diktieren.php). Xendex is now a leading developer of mobile games, employing more than forty people with studios in Austria and Poland (see http://www.xendex.com/index.php?page=2&menuid=5).

Besides these middle-size enterprises, the Austrian independent game development scene has been flourishing since 2005. Today, smaller development studios such as Bongfish GmbH (established in 2001), Ovos (2004), Cliffhanger Productions Software GmbH (2005), Cybertime (2005), Team Vienna (2006), fatfoogoo (2006), ClockStone Software (2006), Rarebyte OG (2006), Emoak (2007), Platogo (2008), Mi'pu'mi Games GmbH (2009), Studio Radiolaris (2009), Bobblebrook (2008), Blackish (2008), Socialspiel Interactive Family Entertainment Gmbh (2010), mySugr GmbH (2010), Pro 3 Games (2010), Schein (2010), Broken Rules Studios (2010), Stillalive Studios (2010), ilikescifi (2010), Spielwerk GmBH (2010), All Civilized Planets (2011), Conquering Bytes (2011), White Rabbit Interactive (2011), Game Gestalt (2011), Vienom (2011), V-Play GmbH (2012), and Donkeycat (2012) are reviving the Austrian development scene (see http://www.in-ga.me/branchen-archiv/). In addition, subunits of major publishers including Electronic Arts, Microsoft, Sony Computer Entertainment, Ubisoft, and Nintendo are located in Austria.

The public and political appreciation of video games makes up one further component influencing the video game scene in Austria. When Rockstar Vienna came to life in 2003, another groundbreaking process gathered momentum. That same year, Austria surprised the industry, politicians, and the public with an unorthodox handling of age regulations and censorship.

Video Games and Protection of Minors in Austria

The dramatic rise in popularity of digital games in the 1990s challenged policymakers worldwide. The key issue was, and is, to develop and implement strategies for protecting children from problematic content while supporting the positive potential of video games and digital media in general. Many countries have decided to approach the subject purely from a legislative perspective based on a negative assessment policy; that is, by regulating the distribution of digital games based on rigid rating systems (Mitgutsch and Rosenstingl 2008). However, in a time when anyone can access digital distribution channels outside governmental control, such strategies appear to be ineffective. Furthermore, policymakers are quite often influenced by so-called moral panics. Cohen (1973, 9) defines this phenomenon as a sporadic episode in which something or someone is declared "a threat to societal values and interests." In many cases, novel media formats are victims of these moral panics fueled by mass media. It is therefore not particularly surprising that computer games have triggered a number of moral panic attacks and continue to do so, especially in connection with in-game violence. However, Austria opted for another approach: in 1994, the Department for Youth Policy of the Federal Ministry summoned experts and practitioners to a meeting on the topic of youth and computers. Regarding the subject of digital games—which had become quite popular, especially among young people by this time—there was a common sense about the necessity of measures to be taken by the authorities. The main reason for any call for action was the supposedly harmful content of certain games. The ban of *Wolfenstein 3D* (id Software, 1992) in Germany in January 1994 had provoked a first wave of mass media coverage on violent games. The experts discussed the concept of USK (see http://www.usk.de/en/), a voluntary ratings board of the entertainment software industry in Germany, in comparison with other strategies, and a clear recommendation was made for "white lists" and counseling of parents and pedagogues instead of purely legislative steps and age ratings. The five arguments of the experts and practitioners of this meeting, regarding the different strategies, were as follows (see Rosenstingl and Wagner 2008):

1. *Limitation of access or general ban of violent games.* This is the concept of the German "index of youth-endangering media" (see http://bundespruefstelle.de/bpjm/information-in-english.html). Media products listed in this index are not to be sold to people under the age of eighteen, and any means of advertising, including the display on a publicly accessible shelf in a shop, are prohibited.
    Potential: With no regard to actual harm, all suspicious games are simply to be kept out of the reach of youngsters.
    Shortcomings: It is easy for minors to circumvent this regulation in many different ways, such as downloading hacked versions or gaining access at an older friend's home. Above all, it is to be expected that the

banning of games results in increased attention given to those games, especially among those minors who are already fascinated by violence.
2. *Limitation of access based on an age rating of games, according to a suspected impairment of the development of minors.* This is the concept of the USK in Germany and other rating boards throughout various countries.
Potential: A selective protection that takes age into account.
Shortcomings: Similar to the above solution, circumvention is possible, and an effect of undesirable attention and propaganda might occur. Furthermore, these ratings are easily misunderstood, as they do not take into account any other aspects of adequacy. For example, highly complex economic simulations are rated "6," but no child of this age will be able to play them.
3. *Price increase for violent games through special taxes.*
Potential: Following the costs-by-cause principle, a potential social damage has to be paid by the originator.
Shortcomings: As the experience in other fields (such as cigarettes or alcohol) demonstrates, any moderate increase has only very limited effect and a noticeable, effective increase is hard to argue conclusively.
4. *Recommendation of games according to the results of positive assessment.*
Potential: As a measure of consumer information, this provides substantial support on purchase decisions with possible long-term effects on the market and no undesirable propaganda.
Shortcomings: It is no measure of protection and in this regard only effective if youngsters either accept or follow these recommendations themselves or possess enough competence for framing their gaming behavior accordingly.
5. *Fostering the media competence of youngsters, parents, and pedagogues.*
Potential: This approach impacts all areas of media usage and is independent of any specific media products.
Shortcomings: Critics will always find specific problematic issues that are not being addressed by educational institutions. But, most crucially, it seems impossible to reach all people, especially those with the least competence; therefore, those with the greatest need of support are usually the most "resistant" to accorded offers.

Moreover, the emphasis in media education has long since shifted from "protection" toward media literacy and parental guidance. The aim is to reinforce the ability of children and teenagers to make sensible decisions about the media they use. Of course, this is a learning process, and mistakes will often be made. But young people can learn from their mistakes, and that goes for media use, too. The learning process is even more effective if the right kind of support is available, and nobody is better placed to provide it than well-informed, open-minded parents. Ultimately, there is no solid case for banning games. That would require scientific proof of the harmfulness of computer games, which does not (as yet) exist. No studies of the long-term effects of gaming have produced results that would justify a ban.

The Austrian government subsequently decided to focus on the positive assessment (Argument 4) and fostering strategy, and systematically pursue those further (Argument 5). Instead of passing rigid laws to blacklist specific games, it started to fund initiatives maintaining white lists of recommended games.

Federal Office for the Positive Assessment of Computer and Console Games

Based on the promising feedback for the abovementioned white list, the development and operation of the Federal Office for the Positive Assessment of Computer and Console Games (Bundesstelle für die Positivprädikatisierung von Computer- und Konsolenspielen, for short: BuPP) within the structures of the Austrian Federal Ministry for Youth was authorized in the autumn of 2003. Accordingly, a program was founded in February 2004 with fundamental research on the required catalog of criteria and the necessary considerations to implement these criteria. In parallel, a program for the assessment process and accompanying codes of conduct was defined, tested, and refined. In the summer of 2005, the ministry's department project team had developed a manual for the assessment of games, comprising three sections: a small section with technical aspects and two main sections regarding gaming fun and pedagogical aspects. The latter made pronounced provisions for not only problematic issues, such as violence, but also the positive potential of games. Following the findings of Gebel, Gurt, and Wagner (2004), the manual attempted to map the dimensions of competence stimulation as defined in this study. The process for the evaluation of games was now built around a weekly meeting of the Assessment Commission, in which two assessors who tested the game in question, an external expert, and two representatives of the ministry worked together and discussed each case before finding a majority decision about the awarding or declining of the seal of quality. After several test runs, the BuPP began its official operation in November 2005. Since then, a list of recommended games has been published online at http://bupp.at and is updated almost weekly. BuPP's prime objective is to provide parents, guardians, and others involved in children's upbringing with guidance on the selection of computer games and console games. It does this by giving positive ratings to well-designed games, as well as providing other information and support services. The aim is to help adults to actively engage with their children's computer or console leisure activities.

Since the content of some commercially successful titles is brutal and violent, one of BuPP's core objectives is that of awareness raising and sensitizing parents, teachers, and young people themselves to this problem. Another priority is to exercise a positive influence on the entertainment software market. If a game is awarded a positive rating, this means that it:

1. Delivers superior playability
2. Does not raise any concerns in terms of suitability for young audiences and makes the most of the opportunities for promoting educational outcomes
3. Is technically state-of-the-art

The scheme's credibility with young gamers is underpinned by a strong focus on playability, institutionalized involvement of young people, feedback from media education project work with children, and close interaction with the eSports scene. In conjunction with the activities of BuPP, gaming events such as the annual Game City were initiated to foster a constructive dialog between players, parents, and developers.

Game City Vienna

When a tragic school shooting took place in Emsdetten, Germany, in 2006, political parties in Vienna instigated a new initiative for censorship in Austria. Once more, video games were made the scapegoat for the individual violence of a young man. The authorities of the city of Vienna had to take action and so invited a group of experts to meet. Based on the recommendations of these experts and the exemplary work of the Federal Ministry of Economy, Family and Youth with BuPP, the politicians again chose a strong, proactive way of handling video games: instead of censorship, the city decided to host Game City, a high-profile event bringing together representatives of the games industry, nonprofit organizations, academia, and the general public for the discussion of the current state of computer games in 2007 (see http://www.game-city.at/). Since then, Game City has become Austria's biggest computer and console gaming event, with more than 63,000 visitors in 2012. It is an event for every age group, and admission is free. During Game City, the famous Vienna City Hall, the yard in the heart of the building, and the open air exhibition area around the hall become a gaming area. The gaming room in the main ballroom of the city hall is transformed into a home-style living room and offers exhibitors the chance to present their newest products in this cozy atmosphere. The center yard allows for typical trade fair concepts; the open-air exhibition area has almost no limits concerning the appearance of the exhibitors. In the over-sixteen and over-eighteen areas, the newest action and shooter games can be tested. Game City also provides space for the finals of the Austrian World Cyber Games, and further offers an information zone for children and families, where parents can find professional information about electronic games from experts in the field of pedagogy. The uniqueness of Game City comes from several factors:

1. The authorities of the City of Vienna and of Austria demonstrate commitment by allowing such a gaming event to take place in the city hall.
2. The event brings players, parents, researchers and developers together.
3. The combination of typical trade fair elements—at an extraordinary location—with information for parents, eSports, and an academic conference, is carried out with complete success.

Conferences, Academic Game Research, and Development Schools

One key element of Game City is the Vienna Games Conference—Future and Reality of Gaming (for short: FROG), whose objective is to serve as a public information platform and an international networking event for game researchers of various disciplines. Since 2007, the Vienna Games Conference has attracted hundreds of scholars and visitors to participate in this playful meeting of minds. In 2013, the Vienna Games Conference—now one of the most successful annual games conferences in Europe—was held for the seventh time. The FROG conference brings together international scholars from all over the world and local researchers, graduate students, game developers, teachers, social workers, and media representatives. The results are published in the annual proceedings and in *Eludamos* special editions (Mitgutsch and Rosenstingl 2008; Mitgutsch, Klimmt,

Figure 2
Game City in the Vienna Town Hall (2012).

and Rosenstingl 2010; *Eludamos* 3.1, 6.1; Swertz and Wagner 2010; and Wimmer, Mitgutsch, and Rosenstingl 2012), in archives of recorded videos (see http://www.arimba.com/frog/), and in diverse newspaper articles and news reports. The conference is fully financed by the Federal Ministry of Economy, Family, and Youth and allows for a critical but constructive discourse about the future and reality of gaming.

The annual Vienna Game AI Conference is another international gaming conference held in Austria. The conference is the world's largest event focusing exclusively on artificial intelligence, gameplay, and character animation, and has been hosted in Vienna since 2012 (see http://gameaiconf.com/about/). Another annual conference in Austria that also includes themes in relation to games is PIXEL, the annual conference on computer graphics and animation (see http://pixelvienna.com/). It is organized by the Austrian CG community and features talks and workshops by international computer graphics and animation professionals at different locations throughout Vienna. While other, smaller conferences, workshops, and meetings are taking place on a regular basis, one longstanding and yearlong event that has been attracting local game scholars, developers, and media representatives since 2005 is the SUBOTRON Arcademy, which holds monthly talks about game theory, game design, and development in Vienna (see http://subotron.com/veranstaltungen/arcademy/).

The academic gaming scene in Austria is rich, and most institutions have courses and lectures on video games. Many books and papers on games research have been published in Austria, and all academic areas of games research are covered by a variety of researchers. Besides the Games 4 Resilience Lab at the University of Vienna (see http://kinder-psy.univie.ac.at/home/), the Institute of Design and Assessment of Technology, and the Human-computer Interaction Group at the Technical University Vienna (see https://igw.tuwien.ac.at/designlehren/Site/Welcome.html), different Austrian research institutions focus on video gaming, the impact on players, and game design. In general, there is a strong focus on educational perspectives on games (Pivec, Koubek, and Dondi 2004; Wagner 2008; Mitgutsch and Rosenstingl 2008; Pivec and Pivec 2012; Mitgutsch and Alvarado 2012), game design (Pichlmair 2004; Rusch 2008; Kayali and Purgathofer 2008; Mitgutsch and Weise 2011), cultural and medialized dimensions (Götzenbrucker 2001, 2009; Hipfl and Hug 2006; Wagner 2006; Strebenz 2011; König 2012), game theoretical approaches (Strouhal 1996, 2007), psychological aspects (Stetina et al. 2011; Lehenbauer et al. 2012), arts and gaming (Kayali, Jahrmann, Schuh, and Felderer 2012), and many more.

In terms of the scientific and academic discourse, the Technical University Vienna, University of Vienna, University of Applied Arts Vienna, University of Klagenfurt, University of Salzburg, and Danube University need to be mentioned. On a more practical level, the SAE Institute Wien, FH Hagenberg, FH Salzburg, FH Joanneum Graz, FH Technikum Wien, and the Technical University of Vienna offer students the opportunity to explore and study game development. As outlined in the history of Austria's game development landscape, the graduates from these schools are forced to work in other countries if they are interested in AAA game development. However, for Indie designers, Austria can be an enriching and stimulating place.

Video Game Media

Due to the fact that Austria, with its population of eight million, and Germany, with eighty million people, share a common language, most gaming-related media activities have their home base in Germany. Besides some editorial departments in a few periodicals such as the biweekly *eMedia*, only two gaming media groups with noteworthy impact are situated in Austria: consol.MEDIA publishing Ltd. and MediaXP Ltd.

consol.MEDIA Publishing Ltd.

With *consol.AT*, the Austrian video gaming magazine, and the PC-gaming magazine *Gamers.at*, as well as activities in the online, television, and multimedia markets, consol.MEDIA Publishing Ltd. is the national market leader in media for computer and video games. Its portfolio comprises seven print magazines and eight online sites that provide daily content. The print magazines *consol.AT* and *Gamers.at* each sell for €4.99, with a circulation of almost 30,000 copies. They are published monthly and distributed via standard channels that allow for special cover campaigns at 1,500 points of sale. The consol.MEDIA Network provides websites for all the print magazines and, in addition, products such as GamersMotion and podcasts. The video portal GamersMotion offers news, previews, and reviews as well as entertainment programs, reports, and video clips such as the series "Gaming VS Reality." The podcast portfolio reaches more than 200,000 listeners and provides them with interviews, facts on games, and industry and coverage of events. The consol.Media online network altogether boasts 8,445,000 page impressions, 1,830,000 visits, and 775,000 unique users per month.

MediaXP: Lifestyle and Entertainment Publishing

MediaXP produces, markets, and distributes online and print media as well as innovative Flash-based and corporate publishing products. Latest additions to the portfolio are the German edition of IGN, the world's leading gaming magazine, *Inside Games*, an in-store TV format in cooperation with McDonald's Germany, and the German edition of *Pocket Gamer*. Due to this diversification, a broad target group can be attracted: over 34 million viewers, 1.5 million unique users per month through different websites, and nearly 500,000 readers of the print magazines. The most important print magazine is *GamingXP*. *GamingXP* is distributed free of charge and directly to readers, thus addressing not only established core gamers but also new target groups. *GamingXP* provides a high-quality overview of the latest PC games and video games. The magazine's circulation of 60,000 copies reaches approximately 170,000 readers, making it the farthest-reaching ÖAK-approved Austrian print magazine in terms of gaming, entertainment, and lifestyle. In addition to that, the magazine is available online for all German-speaking readers.

Another print product is *MobileXP*. *MobileXP* features news, previews, and reviews of the latest mobile games and also covers mobile lifestyle, hardware, equipment, and trends. The print magazine is published and distributed in cooperation with Hutchison 3D Austria, and its circulation is 118,000 copies, available free of charge. *OnlineXP* is the first interactive, online game and browser game magazine in the German-speaking market and provides 400,000 readers with previews and reviews of online games as well as reports and news about the gaming community and industry. The magazine is published bimonthly and is available online as an interactive PDF. Further media venues that report on gaming on a regular basis are in print (*The Gap* and *Der Standard*) and online (*Futurezone.at*, *DerStandard.at*, and *FM4.at*).

Gaming Culture, eSports, and the Arts

Austrian gaming culture is rich and supported by different agencies and organizations. The proactive approach to video games is mirrored in publicly funded institutes such as the *spielebox*, which offers a space for children and families to enjoy board games and digital games (see http://www.spielebox.at/). The BuPP and the *spielebox* also established educationally supported workshops and lectures where kids and teenagers can try out recommended computer and console games. Initiatives such as these can be found all over Austria and represent a well-organized informal learning and playing setting for Austrian youth culture.

For a more mature audience, there are workshops, game sessions, and exploration settings hosted by universities and local gaming initiatives such as SUBOTRON, the Austrian Game Jam (see http://www.austriagamejam.org/), the Austrian Chapter of the International Game Developer Association (IGDA) (see http://igda.at/), and events such as Play:Vienna (see http://playvienna.com/). Another example is the so-called Eltern-LAN Workshops (parents-LAN), which are organized by the eSport Association Austria together with the Federal

Ministry of Economy, Family and Youth. These workshops provide an opportunity for parents to gain firsthand experience with games their children like to play. In alternating sessions, the parents can play games such as *Counter-Strike* (1999) or *Track Mania* (2003) to get used to the gaming interface, or hear experts talk about games from a pedagogical perspective.

In 2009, when members of the public once again associated another school shooting with games, a very successful grassroots initiative was born. Using the slogan "Games don't kill—guns do," a few social media activists started to post pictures with statements such as "I have killed monsters in *God of War*. But I have never killed a monster in real life. Not because there are no monsters. There are no guns. Not in my life" (see http://www.flickr.com/photos/36335836@N04/3363595738/in/pool-games-dont-kill/). Within a few days, dozens of similar pictures were posted to Flickr and most of the major media in Austria covered this initiative (see http://derstandard.at/1244460728000/Games-dont-kill-Guns-do).

A further important component of the Austrian Gaming scene is the support of eSports. After years of participating in international eSports tournaments in the Electronic Sports League (ESL) and participation at the World Cyber Games (WCG), the eSport Association Austria (eSport Verband Österreich, for short: esvoe), a superordinate, independent institution for Austrian professional, semiprofessional, and amateur gamers on all platforms (console, PC, mobile, etc.) was established in 2007 (see http://www.esvoe.at/). Esvoe's goal is to reach a future of amateur and professional gamers in Austria and to offer a platform to support electronic sportsmanship and competitions. In 2009, Austria held its first national championship, and the finals of the tournament were exhibited at Vienna Game City with thousands of visitors. Since 2010, esvoe has also offered seminars and workshops on training and development for professional eSport players and established subsections in every Austrian state. As of mid-2014, thirty-five sports clubs have joined esvoe, and eSports in Austria are flourishing.

The Future of Video Games in Austria

The history of video games in Austria, from its unconventional politics and the rise and fall of its two biggest game development studios, to its recent establishment as fruitful ground for conferences, research, and indie development, allows a brief glance into the future. Austria appears to be an ideal environment for games research and the exploration of different game development niches. Although Austria lacks bigger development studios, and governmental support for the gaming industry is minor, the cultural and political discourse about video games is unique. Austria is home to many very interesting smaller developer teams, and the community in Austria manages to attract international attention. However, in the global game developer landscape, Austria is only a small player that might sometimes surprise others. So far, the establishment of AAA studios in Austria seams unrealistic, and history shows that it is flexibility and creativity that allows Austrian game developers to succeed. Finally, Austria can be seen as a model that demonstrates how a proactive approach to video games enriches the public discourse about them.

References

Brown, L. 2006. 100 Rockstar Vienna devs fired. *Kotaku.* http://web.archive.org/web/20120606122352/http://kotaku.com/173328/100-rockstar-vienna-devs-fired?tag=gamingrockstar.

Cohen, S. 1973. *Folk Devils and Moral Panics.* St. Albans: Paladin.

Gebel, C. M. Gurt, and U. Wagner. 2004. *Kompetenzbezogene Computerspielanalyse: In medien + erziehung (merz)* 48 (3): 18–23.

Götzenbrucker, G. 2001. *Soziale Netzwerke und Internet-Spielewelten: Eine empirische Analyse der Transformation virtueller in realweltliche Gemeinschaften, Opladen/Wiesbaden.* Westdeutscher Verlag.

Götzenbrucker, G., and M. Köhl. 2009. Ten years later: Towards the careers of long-term gamers in Austria. *Eludamos: Journal for Computer Game Culture* 3 (2): 309–324.

Hipfl, B., and T. Hug. 2006. Media communities—Concepts, topics, and tendencies in the current discussions. In *Media Communities*, ed. B. Hipfl and T. Hug, 9–32. Münster: Waxmann.

Kayali, F., M. Jahrmann, J. Schuh, and B. Felderer. 2012. Alternate reality games: Persuasion in context. In *Applied Playfulness*, ed. J. Wimmer, K. Mitgutsch, and H. Rosenstingl. Vienna: new academic press.

Kayali, F., and P. Purgathofer. 2008. Two halves of play—Simulation versus abstraction and transformation in sports videogames design. *Eludamos: Journal for Computer Game Culture* 2 (1): 105–127.

König, N. 2012. The play experience: A constructivist anthropology on computer games. PhD diss., University of Vienna.

Lehenbauer, M., O. Kothgassner, A. Felnhofer, and L. Glenk. 2012. Play for change? A psychological view on computer games. In *Applied Playfulness*, ed. K. Mitgutsch, J. Wimmer, D. Rusch, and H. Rosenstingl. Conference proceedings of the FROG Vienna Games Conference 2011. Vienna: Braumüller.

Mitgutsch, K., C. Klimmt, and H. Rosenstingl. 2010. Exploring the edges of gaming. In *Proceedings of the Vienna Games Conference 2008–2009: Future and Reality of Gaming.* Vienna: Braumüller.

Mitgutsch, K., and H. Rosenstigl. 2008. *Faszination Computerspiele.* Vienna: Braumüller Verlag.

Mitgutsch, K., and H. Rosenstingl. 2008. Vom Jugendschutz zur informierten Entscheidung. Altersgemäße Einstufung von Computerspielen. In *Spielen in digitalen Welten*, ed. Winfred Kaminski and Martin Lorber. Munich: Kopaed.

Mitgutsch, K., and N. Alvarado. 2012. Purposeful by design: A serious game design assessment model. In *FDG 2012, FDG '12 Proceedings of the International Conference on the Foundations of Digital Games*, 121–128. New York: ACM Press.

Mitgutsch, K. and Weise, M. 2011. Subversive game design for recursive learning. In *DIGRA 2011 Proceedings: Think Design Play*. http://www.digra.org/wp-content/uploads/digital-library/11310.47305.pdf.

Pichlmair, M. 2004. *Designing for Emotions: Arguments for an Emphasis on Affect in Design*. Vienna: Vienna University of Technology.

Pivec, M., A. Koubek, and C. Dondi, eds. 2004. *Guidelines on Game-Based Learning*. Lengerich: Pabst Verlag.

Pivec, M., and P. Pivec. 2012. *Lernen mit Computerspielen: Ein Handbuch für Pädagoginnen/Pädagogen*. Vienna: Bundesministerium für Wirtschaft, Familie und Jugend.

Rosenstingl, H., and K. Mitgutsch. 2009. *Schauplatz Computerspiele*. Vienna: Lesethek Verlag.

Rosenstingl, H., and M. Wagner. 2008. Towards a positive assessment policy for computer and console games. *Human IT* 9 (3): 1–17.

Rusch, D. C. 2008. Emotional design of computer games and fiction films. In *Games without Frontiers—War without Tears. Computer Games as a Sociocultural Phenomenon*, ed. A. Jahn-Sudmann and R. Stockmann. New York: Palgrave Macmillan.

Stetina, B. U., O. Kothgassner, M. Lehenbauer, and I. Kryspin-Exner. 2011. Beyond the fascination of online-games: Probing addictive behavior and depression in the world of online-gaming. *Computers in Human Behavior* 27 (1): 473–479.

Strebenz, B. 2011. *Genres in Computerspielen—eine Annäherung*. Glückstadt: Vwh Verlag.

Strouhal, E. 1996. *Acht mal acht: Zur Kunst des Schachspiels*. Berlin: Springer.

Strouhal, E. 2007. Comparing games: Chess versus go: Some remarks on the methodology of games research. *Homo Oeconomicus* 24 (1): 95–110.

Swertz, C., and M. G. Wagner. 2010. Game\\Play\\Society. In *Proceedings of the Vienna Games Conference 2010: Future and Reality of Gaming*. Munich: Kopäd Verlag.

Wagner, M. G. 2006. The serious side of eSports. Serious Games Summit, Washington D.C. http://www.gdconf.com/conference/sgs.html.

Wagner, M. G. 2008. Serious games: Spielerische Lernumgebungen und deren Design. In *Online Lernen—Handbuch für das Lernen mit Internet, L*, ed. J. Issing and P. Klimsa. Landsberg: Oldenbourg Wissenschaftsverlag.

Wimmer, J., K. Mitgutsch, and H. Rosenstingl. 2012. Applied playfulness. In *Proceedings of the 5th Vienna Games Conference 2011: Future and Reality of Gaming*, Vienna: Braumüller.

BRAZIL

Lynn Rosalina Gama Alves

In Brazil, the history of video games began in the 1980s when the first video game appeared on store shelves. According to Chiado (2011, 26), the first Atari VCS (1977), "half mounted and half manufactured in Sao Paulo," reached stores in April 1980. Joseph Maghrabi, a 1980s entrepreneur, was instrumental in bringing video games to Brazil and created Channel 3, a pioneering club that manufactured game cartridges. As Maghrabi stated in an interview, "Before creating Channel 3, I founded the Atari Electronics Company. It was for the importing of devices and accessories of the Atari console. We imported the printed circuit boards and joysticks, ordered the supplier to manufacture the plastic box and paperwork (manuals, guarantee, etc.), we mounted the devices and we sold them together with a cartridge to magazines" (Chiado 2011, 26).

The first game produced in Brazil was called *Amazônia*, a text adventure game developed in January 1983 and published entirely by the magazine *Micro Sistemas* in its August 1983 edition (no. 23). As of 2012, *Amazônia* can be found online at the TILT club's website (http://www.clubtilt.net/pags/clubtilt.htm), which gathers tips for game developers. In the game's "Adventures in the Jungle," the player has to solve small problems whose main goal is, according to the game's manual, "to escape the dangers of the jungle after surviving a plane crash and find a paved road. To achieve this goal, the player must construct sentence commands, using a verb (action) plus objects. For example: GET THE BAG." *Amazônia* is considered the classic text adventure game written in Portuguese.

From 1983 to 2010, there were 105 games produced in Brazil. It is difficult to reconstruct the history of game development in Brazil because there is very little written on the subject— mainly only the work of Marcus Chiado and the Gamebrasilis (the Catalogue of Brazilian Electronic Games), both of which were instrumental to the writing of this essay. Researchers and experts in the field of games in Brazil were also consulted, as well as representatives of the two main institutions, the Brazilian Association of Electronic Game Developers (ABRAGAMES) and the Commercial, Industrial, and Cultural Association of Games (ACIGAMES), but unfortunately, neither association has data records about the history of games in Brazil.[1]

Thus, games listed in table 1 were produced from 1983 to 2010.[2] In 2009, there was considerable growth in the production of games in Brazil, mainly due to the action of Tectoy, a Brazilian video game and electronics company, which launched nine games that year (see http://www.tectoy.com.br/). Tectoy was established in

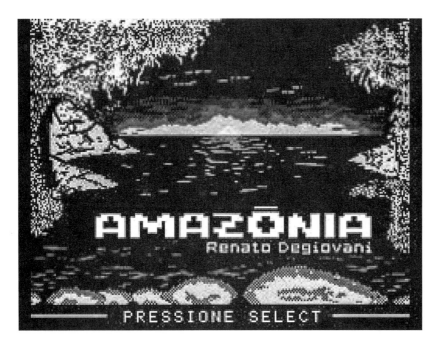

Figure 1
Screenshot from *Amazônia* (1983), the first video game produced in Brazil.

1987 in São Paulo with the purpose of developing and producing high-tech toys and later became the exclusive representative of SEGA in Brazil, making all the consoles the company launched in the West, from the SEGA Master System to the SEGA Dreamcast. In 1998, the company devoted its production to the video game market. In 2005, the Tectoy Móbile division was started, and in 2009 Zeebo Inc., a joint venture between Qualcomm and Tectoy, was created and produced the first Brazilian console, the Zeebo. In May 2011, however, Qualcomm, the leading investor of Zeebo Inc., decided to stop investing in the project, and Zeebo Interactive Studios (ZIS), located in Campinas, was closed.

The year 2009 also saw the release of Donsoft Entertainment's *Capoeira Legends: The Path to Freedom*, which deserves particular mention because it highlights and emphasizes Brazilian culture and history (see the promotional video for the game at http://www.youtube.com/watch?v=WyF8dFMkTlY). The game depicts the lives of black people, white people, and natives in *mocambos* (runaway slave communities) in 1828, and involves *capoeira*, a Brazilian martial art developed by the descendants of African slaves. The game is set in the surroundings of the city of Rio de Janeiro, in 1828. In the game, the black, white, and Indian races that form the ethnicity of the Brazilian people live in communities that were constantly threatened by advocates of slavery. In this context, the capoeira, a Brazilian martial art developed by the descendants of African slaves, emerges as a defense and fighting force.

Table 1

Year	Name of Game	Producer
1983	Amazônia	Micro Sistemas
	Didi na Mina Encantada	Philips
1986	Em Busca dos Tesouros	Micro Sistemas
	O Enigma dos Deuses	Ciberne Software
1990	Zorax	Discovery/Hitek
1991	Mônica no Castelo do Dragão	Tectoy
1992	Ayrton Senna Super Mônaco GP II	Tectoy
1993	Barravento	Discovery/Hitek
1995	Casseta & Planeta em Noite Animal	ATR Multimedia
	Master Multimidia	ATR Multimedia
1996	Desafino	ATR Multimedia
1997	Curupira	Nixtron Interactive
	Guimo	Southlogic Studios
	Planeta Vermelho	Estelar Software
1998	Incidente em Varginha	Perceptum Informática
1999	Hades 2	Espaço Informática
2000	Caxy Gambá Encontra o Monstruário	44 Bico Largo Multimídia
	Gustavinho em O Enigma da Esfinge	44 Bico Largo
	Show do Milhão	SBT Multimídia
	Solaris 104	Apollo Entertainment
2001	Christmas Magic	Espaço Informática
	Hades 2	Espaço Informática
	Micro Scooter Challenge	Perceptum Informática
	No Limite	Continuum Entertainment
	Outlive—A Era da Sobrevivência	Continuum Entertainment
	Putzgrilla	Draft Marketin Esportivo
	Qual é a Música	SBT Multimídia
	Super Mini Racing	Perceptum e Canal Kids

Table 1
(continued)

Year	Name of Game	Producer
	Tainá—Uma Aventura na Amazônia	VAT
	Xuxa e os Duendes 2	Continuum Entertainment
2002	*Aventura na Selva*	Shape CD
	Big Brother	Continuum Entertainment
	Jogo do Banquinho do Raul Gil	Rede Multimídia
	Trophy Hunter 2003	Southlogic Studios
	Vampiromania	Green Land Studios
2003	*Big Brother Brasil 3 D On-line*	Green Land Studios
	Bola de Gude	Icon Games
	Brasfoot	Scoop Software
	Deer Hunter 2004	Southlogic Studios
	Futsim	Jynx Playware
	Impacto Alpha	Oniria
	Matchball Tennis	Espaço Informática
	Sandy & Júnior Ameaça Virtual	Green Land Studios
	Smilinguido em Desafio na Floresta	Continuum Entertainment
	Space Shooter	Oniria
	Trophy Hunter	Southlogic Studios
2004	*Deer Hunter 2005*	Southlogic Studios
	Die Ponyrancher	Preloud
	Dier Pferdebande	Oniria
	Erinia	Ignis Games
	Ryudragon	Decadium Studios Game Developer
2005	*Coca-Cola Super Coach*	Oniria
	Der Pferderennstall	Espaço Informática
	Die Pferdebande—Weiße Stute in Gefahr	Oniria
	Jessy: Ein Zirkuspferd in Not	Bitcrafters Inc.

Table 1
(continued)

Year	Name of Game	Producer
	Outgun	Sorvete Games e Nix & Huntta
	Riding Ground	Preloud
2006	Ayrton Senna Pole Position	Meantime
	Fruzzle	Icon Games
	Fute Bolon Line	Francisco Matelli Matulovic
	Golds of Virtual Boards	Devworks
	Lissy—Und Ihre Freund	Preloud
	Traulian	Canoasoft
	Winguel	Espaço Informática
2007	Bingo de Letras	Oniria
	PillowFight	Oniria
	Torre de Hanoi	Oniria
	Torre Inteligente	Oniria
	Warbots	Délirus
2008	5-3-3 São Paulo Futebol Clube	Tectoy Digital
	8 Segundos	Tectoy Digital
	Lex Venture	Interama Games
	Snail Racers	Icon Games
	Super Vôlei Brasil	Aquiris
	Taikodom	Hoplon Infotainment
2009	Capoeira Legends	Donsoft Entertainment
	Chiaroscuro	Tectoy Digital
	Detetive Carioca	Icon Games
	Dreamer: Musicstar Popstar	Overplay
	Guitar Idol	Interama Games
	Pebolim do Sao Paulo Futebol Clube	Tectoy Digital
	PoChickenPo	Tectoy Digital

Table 1
(continued)

Year	Name of Game	Producer
	Sam Power Footballer	Tectoy Digital
	Super Vôlei Brasil 2	Aquiris
	Zeebo Extreme Baja	Tectoy Digital
	Zeebo Extreme Bóia Cross	Tectoy Digital
	Zeebo Extreme Corrida Área	Tectoy Digital
	Zeebo Extreme Jet Board	Tectoy Digital
	Zeebo Extreme Rolimã	Tectoy Digital
2010	*Invasão ET*	Oniria
	Michael Jackson: The Experience	Ubisoft São Paulo (Nintendo DS e PSP)
	Zeebo F. C. Foot Camp	Zeebo Interactive Studios
2011	*Bubble Up*	Insane Media
	Christmas Jumper	Riachuelo Games
	Combate a Dengue	Icon Games
	Earth Under Siege	Insane Media
	Mahjong Max	Icon Games
	Paperama	Braza Games
2012	*Deuces Wild Casino Poker*	Icon Games
	Monte Bello	Riachuelo Games
	Vaca Maia	Vaca Vitória

These data show that growth in Brazilian game production is still timid, although the international market indicates great potential. In the area of console technology, the twenty-first century saw the birth of the seventh generation beginning in November 2005 with Microsoft's Xbox 360, followed by Sony's PlayStation 3 and Nintendo's Wii in 2006. This generation of consoles is marked by greater storage capacity and a variety of accessories, such as a steering wheel for the Wii and a light gun for the Wii Remote, which accompany these technologies. They also possess greater processing power, graphics, and resolution, and other aspects that favor the creation and development of increasingly photorealistic games, with significant potential for immersion and navigation, increasingly complex narratives, and different degrees of interconnectivity. This whole scenario has favored the emergence of large companies that operate in the development, production,

Figure 2
Screenshots from *Capoeira Legends* (2009), a game about runaway slave communities.

and distribution of technological devices. The Brazilian game industry includes the following players and segments:

- Developers who employ the professionals in charge of programming, art, design, script, and sound effects. In games focused on education, we also find specialists with expertise in this area. These professionals are present mainly in large companies.
- Publishers who are responsible for making possible the sale and distribution of video games produced by small businesses that struggle to reach customers directly, especially with mobile operators who work only with publishers. In Brazil, some of these operators have also been acting as publishers, for example Oi and Claro, which are considered major telecom operators in Brazil, just as Telefonica and Vodafone are in Europe.
- Distributors that are responsible for the dissemination of digital games for consoles and computers. Distribution usually occurs in stores specializing in IT (and/or digital games), big bookstore chains, supermarkets, and on the Web (mainly mobile gaming).

In national and international markets, the industry offers different types of games, from casual games to massively multiplayer online role-playing games (MMORPGs), resulting in an ever-growing typology of games, including those of social networking sites such as Facebook and Orkut, which attract an audience of very different players considered hardcore (experts, veterans, and experienced). Subjects who interact with these video games, especially Facebook's *Farmville* (2009), are typically female, over thirty-five years old, and have no experience as players.[3] But why do these games attract different types of people? Such questions have contributed to the emergence of another player in the field of electronic games: researchers. These researchers have oftentimes been concerned with the processes of interaction between players and games, particularly during the time when a player is immersed in the game world.

Researchers from the humanities and applied social sciences have mainly been working on the different matrices that permeate the relationships established by players and games, which can be seen in the results presented in the CAPES (Federal Agency for Coordination of Improvement in Higher Education) Bank of Theses and Dissertations. CAPES regulates and approves the programs of postgraduate studies in Brazil. After they are defended and approved, abstracts of dissertations and theses are available at the Bank of Theses (http://bancodeteses.capes.gov.br/). (In Brazil, the writing of a master's dissertation precedes the writing of a doctoral thesis, unlike in other countries, where the doctoral work is referred to as a dissertation.) This is discussed in further detail in the following section.[4]

Games Become an Object of Research in Brazil

Investigations into video games began in the mid-1980s, when Atari was at its peak as a major publisher and distributor of games in the world and in Brazil. Elsewhere, early work in this line of investigation was done by Patricia Marks Greenfield (1984), who addressed the development of reasoning in the electronic age, stressing television, computers, and video games. Another important researcher, Sherry Turkle, published *The Second Self: Computers and the Human Spirit* (1984), and later *Life on the Screen: Identity in the Age of the Internet* (1997), pointing out the significant contributions made by games in the area of cognitive development. From this period onward, researchers from Europe and the United States began disseminating research on video games, with issues related to ludology, narrative, learning, and so on. Ludologists such as Aarseth (1997, 2001), Frasca (1999, 2003), and Juul (2005) argued that the study of video games should constitute an autonomous discipline, with ludology distancing itself from discussions that emphasized narrative. From an opposing perspective, Murray (1998), Ryan (2004), and others emphasized issues related to the importance of narrative in games. For narratologists, these types of media have a specific form of storytelling, as do other media including cinema, comic books, and novels. In Brazil, these studies followed the same route as the foreign productions; work was divided between *narratologistas* and *ludologistas*, and more recently, in line with international trends.

Researchers interested in how learning is mediated by games can be divided into two groups. The first considers video games as a space for interaction, particularly early games with two-dimensional environments and simple narratives and gameplay; for example, the work of Antonietti et al. (2002) in Italy. Researchers belonging to the second group examine games that have more complex narratives, with higher levels of interactivity, gameplay, and interconnectivity, and greater graphical realism, all of which encourage greater immersion in the game environment. Also important to this group are games that are part of franchises, such as those of the Harry Potter, Lara Croft, Deus Ex, Counter-Strike, and Warcraft franchises. The work of Gee (2004), Shaffer, Squirre, Halverson, and Gee (2005), and Johnson (2005) in the United States; Mendez, Alonso, and Lacasa (2007) in Spain; Torres, Zagalo, and Branco (2006) in Portugal, and others, deserves special mention for addressing games that do not have explicit educational content.

This work on the international scene has influenced and shaped research in Brazilian universities, allowing the construction of a different look at the cultural phenomenon of video games. In this context, the Periodicals Portal and the CAPES Bank of Theses and Dissertations has significant value for researchers in search of theoretical and methodological support to build the state-of-the-art objects of their research, as well as to disseminate their work. Thus, searching the database of the CAPES Bank of Theses and Dissertations (using the keywords "games," "digital games," "electronic games," and "video games"), it was possible to map the route that Brazilian investigators have followed to make electronic games an object of research, delineating them as a serious and relevant subject that requires theorizing beyond reductionist and Manichean readings.

Although the CAPES Bank of Theses and Dissertations was founded in 1987, Brazilian video game research began in 1994, when it appeared in the Department of Education at the Federal University of São Carlos–São Paulo. In "Game Over: The Child in the World of Video Games," researcher Vivianna Martinez Velasco Carola studied the relationship between children and video games, taking the play as the main activity that humanizes the child as a subject (Carola 1994). Her work constituted an important milestone for research in games and culture.

Ten years later, I defended a PhD thesis in education at the Federal University of Bahia in Salvador that also had "game over" as an expression in its title but with different intent. I used the expression for two reasons.

The first was to paraphrase the authors Corso and Corso (1999), who used it to refer to parents' feelings about the arrival of adolescence for their children—that is, the feeling of powerlessness against the emotions and situations that both generations will experience during this period. The second was because of the relationship of the topic with the dynamic present in video games; with the appearance of "game over," players experience a moment of impotence at the end of a game. In this research, the object of investigation was the violence in the world of games and how players construct meaning for violence. The researcher interviewed the subjects to identify if there was a transposition of violent content in games to real life.

In Brazil, then, the 1990s marked the initiation of research within the universities, with significant growth in the area as revealed by keyword searches (for the terms "games," "digital games," "electronic games," and "video games"). Games became a new object of research, combining *apocalyptic readings* (studies emphasizing only the negative and pessimistic aspects related to games) and *integrated readings* (studies highlighting only the positive aspects of games) into *critical readings* (studies that do not reduce the object to only one point of view but point out positive and negative aspects of games).

From 1994 to 1998, four master's dissertations on video games were defended in education, communication, and computing, and two doctoral theses on video games were defended in the fields of linguistics and sociology. The universities that housed these works are in the southeast (the Federal University of São Carlos [UFSCAR], Catholic University of São Paulo [PUC-SP], Federal University of Minas Gerais [UFMG], University of São Paulo [PUC-SP], University of São Paulo [USP], and the Research Institute of Rio de Janeiro and South Federal University of Santa Catarina [UFSC]).

The beginning of the twenty-first century marked the growth of video game research. From 2000 to 2009, a total of 111 works appeared: 93 master's dissertations and 18 doctoral theses in the areas of education (23 dissertations and 5 theses), communication (15 dissertations and 6 theses), and computing/IT (26 dissertations). Interestingly, there were also papers in the areas of design (10 dissertations and 1 thesis), sociology (1 thesis), electrical engineering (5 dissertations and 3 theses), literature/linguistics (5 dissertations and 2 theses), the arts (7 dissertations), and psychology (2 dissertations). Thus, we can see that education is the field of knowledge that currently has the largest number of papers on the subject (interestingly, only two dissertations from this period, one from 2006 and the other from 2009, actually dealt with the training of teachers). The immersion of students in the world of digital culture, and especially games, may have contributed to the growth of research in this field, and these studies have investigated both Brazilian games as well as foreign imports.

Another important point with regard to the period from 2000 to 2009 is the geographic location where the research involving games was carried out—significant numbers in the south (16 studies) and especially in the southeast (66 studies), with growth in the northeast (25 studies), the midwest (3 studies), and the north (1 study). Such data can be linked to market issues and initial training of the labor force; that is, the southeast region contains a significant percentage of companies and undergraduate (bachelor's and technological courses) in the area of games. The same can be said for the south; in addition to courses, Santa Catarina is now home to the largest "Game Polo" of Brazil—that is, a group of companies and universities that together establish research and business relations. Pernambuco also has a development center, and other cities highlighted in the survey (Salvador, Fortaleza, Teresina, and João Pessoa) revealed timid development, that is, markets that are still expanding. However, the State of Bahia has actively promoted the implementation of a culture of games. These actions, involving the state government, are characterized by three points: The first is the Audiovisual Network created in 2008 by the State Secretariat of Culture, which brings together representatives of all segments of the industry that produce audiovisual products. This network coordinates actions at the level of training, funding, and infrastructure for all segments of audiovisual culture, such as movies, games, and music, among others. The Technology Park (a physical space designed to enhance the relationships between the industry and universities, investing in research and business) opened in 2012 and demonstrates that a demand for professionals and companies that produce interactive content already exists in Salvador, as does the need for qualified professionals to work in game development, simulations, interactive environments, augmented reality, and other fields. This park will be a place to train and absorb this labor. And finally, in September 2009, the Foundation for Research Support of the State of Bahia (FAPESB) created "Game Cluster," which, like the Audiovisual Network, aims to coordinate research and industry in the area of games, a network that brings together research groups from universities and companies working in games. The State of Bahia has allotted a space of 200 square meters for the installation of the Game Cluster in the Technological Park, strengthening the link between companies and universities developing games and games research. Such activity requires training and qualifications, the creation of new companies, and the existence of laboratories with shared technologies that support quality projects. As infrastructure develops, these labs will be used by the institutes of science as well as technology companies.

Parallel to the growth of academic research in the area of games was the emergence of research groups registered in the directory of the National Research Council (CNPQ), which studies video games. The earliest records began in 2002, when two groups were created: in the area of computer science, the Graphic Computer and Digital Entertainment at the Regional University Foundation of Blumenau (FURB) in Santa Catarina, and in the area of communication, Virtual Communities at the University of the State of Bahia (UNEB). Below is a record of the groups found in the directory as of 2010 (Source: Directory of Research Groups registered in the CNPQ).

Research Groups Registered in the CNPQ

Although the list in table 2 includes only the research groups that registered in the CNPQ (using keywords such as "digital games," "electronic games," "games," and "video games"), these data represent a number of doctoral theses and master's dissertations that is much larger, and many of these are not linked to any of the groups listed above. There are theses and dissertations at PUC-SP with an emphasis on semiotics that have games as their object, but which are linked to research groups that are not registered or that are not within the field of game studies.

Table 2

Name of Research Group	Year of Creation	Area	State
Indigent—UFBA	2004	Computer Science	Federal University of Bahia (UFBA)
Visualization, Simulation and Digital Games	2006	Computer Science	Federal University of Mato Grosso do Sul (UFMS)
Games per Computer	2006	Compute Science	Federal University of Piauí (UFPI)
Edutainment	2007	Education	Federal Technology University of Paraná (UTFPR)
IT	2009	Computer Science	Federal Institute of Goiania (IF Goiano)
Research Lab of Graphic Computing and Digital Games	2009	Computer Science	Education and Culture Foundation of Minas Gerais (FUMEC)
Humor, Comics and Games	2009	Communication	Federal University of Paraiba (UFPB)
Augmented Reality Lab, Digital Games and Digital TV	2010	Computer Science	Federal Institute of Bahia (IFBA)

Another important point observed during my research is in regard to alternative places used by researchers to disclose preliminary and end results with the academic community. Beyond journals, researchers exchange knowledge and participate in national events in the area of games, such as the Seminar on Electronic Games, Education, and Communication: Building New Trails, and SB Games (the Brazilian Symposium on Computer Games and Digital Entertainment), which is the largest of its kind in Latin America. Both of these events demonstrate the growth of research in games and make use of the data in the CAPES Bank of Theses and Dissertations.

Theoretical Research Perspectives

Preliminary analysis of research at the levels of master's and doctorate degrees identified in the CAPES Bank of Theses and Dissertations in the areas of communication and education revealed that the authors took as their starting point the concept of play described by Huizinga (2001) and developed by Caillois (1958). According to Huizinga, a game constitutes a universal activity prior to culture itself, since "even in their least restrictive settings, [games] always [presuppose] human society" (2001, 3), and even animals perform recreational activities. Although Caillois (1958) admits the existence of a large number of games in our society, he suggests that they do not have important consequences in the real lives of human beings.

Using these classics to contextualize phenomena, game studies researchers establish dialogue with authors Janet Murray (1998), Espen Aarseth (1997, 2001, 2003), Jesper Juul (2005), and Gonzalo Frasca (1999), revisiting the debate between narratologists (who look at games as narratives) and ludologists (who are more concerned with the playful aspects of games). From 2000 to 2005, this discussion was marked by a reductionist outlook but was later exceeded by the Brazilian researchers siding with Frasca (1999) and Juul (2005), who do not adopt one position or another but understand that both views are complementary and applicable to contemporary games. Also in regard to game studies, we can see the emergence of a state of the art, at least in quantitative terms (since the qualitative analysis of theses and dissertations is still in process), with reference to the CAPES Bank of Theses and Dissertations to contextualize the object of study and indicate the necessity and relevance of the work developed by Brazilian graduate students.

A common feature in discussions about games involves the concept of culture: the convergence culture of Jenkins (2008), with its emphasis on transmedial narrative; the media culture of Santaella (1996); the virtual communities of Rheingold (1997); and games' relationship with cinema, according to Machado (1997a, 2002).

Another interesting point to highlight is that while research in the area of education tends to focus on different learning spaces, studies based in communication have more of a theoretical tendency, outlining bibliographic research but without empirical field investigations. Researchers in the field of education are concerned with issues related to learning mediated by games and have thus established dialogues with Gee (2004), Shaffer, Squirre, Halverson, and Gee (2005), Alves (2005), Moita (2007), Johnson (2003, 2005), Turkle (1989, 1997), Greenfield (1984), and classic authors such as Piaget (1978, 1983, 1990) and Vygotsky (1993, 1994, 2001). The authors Rushkoff (1999), Tapscott (1999), and Prensky (2001) are also often cited in discussions of the characteristics of the digital generation.

Authors Gee (2004) and Shaffer et al. (2005) are often mentioned in discussions about video games and education. For Shaffer et al., "Virtual worlds of games are powerful because they make it possible to develop situated understanding" (2005, 106). Thus, subjects who interact with different media experiences in multiple contexts can understand complex contexts without losing the connection between abstract ideas and the real problems they can solve. In other words, they learn by making sense and deriving meaning through information that emerges from the narrative of games, built in partnership between the game and the player. For Gee (2004), along with Shaffer, meanings in electronic games always have specific situations, with situated meanings, not general meanings.

Taking digital games into a school setting, therefore, does not mean that they are necessarily designed to develop math concepts, language learning, or the teaching of cognitive processes; they may only be designed for entertainment (after all, we cannot overload the kids). This understanding of the technologies, digital media, and their representations is reductionist, contrary to theoretical perspectives that discuss the presence of these elements in different learning environments, mostly at schools. They are also opposed to the classical psychogenetic theories (Piaget, Vygotsky, and Wallon, among others), which have existed for over fifty years and are thoroughly discussed in training courses for teachers. How far do we need to go to understand that play should be present in learning situations? That the school should be a space of pleasure? We need something closer to the semiotic experience of our students.

More work needs to be done regarding the training of teachers in this area, and as we saw earlier, this topic is still one that is underrepresented in the works found in the CAPES Bank of Theses and Dissertations.

Since they deal with children and adolescents, teachers need to immerse themselves in the various semiotic fields that overlay the presence of technologies in contemporary society. Taking digital games to school because they involve our students without any prior exposure to them (and without the construction of meaning), and seeking to fit games into school programs as they are developed, will result in failure and frustration for both teachers and students. In addition, this repeats a path trodden in the 1990s when educational software was introduced into schools in Brazil, including e-books with animation and hypertext, which was soon rejected.

The intention should not be to transform schools into Internet cafés, as they are differentiated learning spaces with different logic. The intention should be to create a space for teachers to identify ethical, political, ideological, and cultural issues within the discourse of games or interactive material that can be explored and discussed with the students, all while understanding the relationships that players—our students—have with these media, while challenging, intervening, and mediating the construction of new meanings for the stories. Through these media, users can learn new ways of seeing and understanding cultural artifacts.

A closer relationship between teachers and developers of games can also help to encourage dialogue between these professional fields, to address differences, and open new perspectives in the production of games for educational use. Finally, understanding video games as cultural phenomena requires a new way of looking at media beyond Manichaean perspectives that associate games with violent behavior, inactivity, addictive behaviors, lack of motivation at school, school failure, and dropout rates. Such uncritical and reductive readings restrict the possibilities of dialogue between teachers, gamers/students, and the universe of games. Thus, the delineation of pathways between digital games and learning requires that once again we hearken to the invitation that Babin and Kouloumdjian (1989) made in the late 1980s when discussing audiovisual culture: facing the new scenario in which games play a part, we must simultaneously immerse as well as distance ourselves to develop critical methods for examining games and taking control of them.

This invitation is open to all those involved in the pleasurable investigative processes that contribute to the consolidation of new knowledge. I conclude this chapter emphasizing the importance of the CAPES Bank of Theses and Dissertations, which enables researchers from different parts of the world, especially Brazil, to build new skills, connect with the research community, and extend further investigations in different fields of knowledge, especially in the case of games. The panoramic view offered by the CAPES Bank of Theses and Dissertations enables researchers to explore new perspectives flagged in the work and go deeper into the studies that have already been initiated.

Finally, there is good news regarding the gaming market in Brazil as of late 2012. According to a survey by US consultancy firm PricewaterhouseCoopers (PwC), in 2011, the game industry in Brazil alone brought in BRL $840 million (USD $328,394,396) and could reach BRL $4 billion (USD $1,564,394,440) by 2016, growing on average 7.1% per year. These data indicate a significant growth in the Brazilian video game industry, compared with industries in countries such as the United States and Mexico.

Foreign companies have also become aware of the Brazilian market's success. Microsoft, for example, opted to nationalize the production of the Xbox console, with production taking place in the Zona Franca de

Manaus since the year 2010, resulting in a 40% reduction in price. Another significant factor is the growth of players: currently one in five Brazilians, or 45.2 million people, are frequent or occasional players, according to the Brazilian Institute of Opinion and Statistics (IBOPE).[5]

I have also highlighted the need for organizations such as ABRAGAMES and ACIGAMES to create communication channels between companies that produce games in Brazil, identifying needs and requirements, and to enhance the process of Brazilian game development as well as the conducting of studies to map the current state of the industry and the subsidizing of new activities and projects. All of these things indicate a very bright future for video games in Brazil.

Notes

More on the history of video games can be found in Lynn Rosalina Gama Alves's book, *Game Over: Video Games and Violence*, New York: Futura, 2005; and the Discovery Channel documentary *The Video Game Era* (2007), which is available on YouTube at http://www.youtube.com/watch?v=P4Iq3ZR5TH8.

1. It is important to emphasize that I had great difficulty accessing data regarding the history of electronic games in the Brazilian industry. Unfortunately, there are no records other than those listed in this essay. ACIGAMES and ABRAGAMES have no data that could support the writing of this chapter.

2. Sources: From 1983 to 2003, Gamebrasilis (the Catalogue of Brazilian Electronic Games) and SENAC (edited by National Service of Commercial Apprenticeship [SENAC], published in September 2003). From 2004 to 2010, game websites and Wikipedia. Information was also provided by Senior Producer André Nogueira and Juliano Barbosa Alves, Oniria's business director.

3. This record is based on empirical observations ratified in Zagalo 2010.

4. This section was built from experience, reading, and research developed by the author, who coordinated a research group that also develops digital games.

5. See http://www.midiamax.com/noticias/820892-mercado+games+cresce+abre+oportunidades+brasil.html and http://www.noticiasbr.com.br/mercado-de-games-cresce-e-se-torna-oportunidade-de-negocios-no-brasil-81347.html.

References

Aarseth, Espen. 1997. *Cibertexto: Perspectivas sobre a Literatura Ergódica*. Lisbon: Pedra de Roseta (2005 translation).

Aarseth, Espen. 2001. Computer games studies: Year one. *Game Studies: The International Journal of Computer Game Research* 1 (1) (July) http://www.gamestudies.org/0101/editorial.html.

Aarseth, Espen. 2003. Jogo da investigação: Abordagens metodológicas à análise de jogos. In *TEIXEIRA, Cultura dos jogos—Revista de comunicação e cultura Caleidoscópio*, ed. B. Luis Felipe, 9–23. Lisboa, Edições universitárias Lusófonas. 2nd Semester2003, No. 4, 9–23.

Alves, Lynn Rosalina Gama. 2005. *Game Over—jogos e violência*. São Paulo: Futura.

Antonietti, Alessandro, Chiara Rasi, and Jean Underwood. 2002. I videogiochi: una palestra per il pensiero strategico? *Rivista: Ricerche di Psicologia Fascicolo* 1:125–144.

Babin, Pierre, and Marie France Kouloumdjian. 1989. *Os novos modos de compreender—a geração do audiovisual e do computador*. São Paulo: Ed. Paulinas.

Bonaviri, Gaetano, Giuseppe Cardinale, Roberta Iervolino, Alice Mulé, and Rosaria Russo. 2006. Aspetti cognitivi e affettivi dell'utilizzo dei videogiochi: Le due facce della medaglia. *Rivista: Psicotech Fascículo* 1:37–48.

Cabreira, Luciana Grandini. 2006. Electronic Games under the perspective of mediators of knowledge—the virtualization of playing from the perspective of teachers of 3rd and 4th grades. MA thesis, State University of Londrina.

Caillois, Roger. 1958. *Os jogos e os homens*. Lisbon: Cotovia (1990 translation).

Carola, Viviana Martinez Velasco. 1994. Game over: The child in the world of video games. MA thesis, Federal University of Sao Carlos.

Chiado, Marcus Vinicius Garrett. 2011. *1983: O Ano dos Videogames no Brasil*. São Paulo: Artes Ofício.

Corso, Mário, and Diana Corso. 1999. Game Over—o adolescente enquanto unheimlich para os pais. In *Adolescência: Entre o passado e o futuro*, ed. Alfredo Jerusalinsky, 81–95. Porto Alegre: Artes Ofício.

de Kerckhove, Derrick. 1997. *A Pele da Cultura*. Lisboa: Relógio D'Água Editores.

Frasca, Gonzalo. 1999. Ludology meets narratology: Similitude and differences between (video)games and narrative. *Parnasso 3*, Helsínquia, 365–371.

Frasca, Gonzalo. 2003. Simulation versus Narrative: Introduction to Ludology. In *The Video Game Theory Reader*, ed. Mark J.P. Wolf and Bernard Perron, 221–235. New York: Routledge.

Gee, James Paul. 2003. *What Video Games Have to Teach Us about Learning and Literacy*. New York: Palgrave Macmillan.

Gee, James Paul. 2004. *Lo que nos enseñan los videojuegos sobre el aprendizaje y el alfabetismo*. Málaga: Ediciones Aljibe.

Greenfield, Patricia Marks. 1984. *Mind and Media: The Effects of Television, Computers, and Video Games*. Cambridge, MA: Harvard University Press.

Huizinga, Johan. 2001. *Homo ludens: O jogo como elemento da cultura*. São Paulo, Brazil: Perspectiva.

Jenkins, Henry. 2008. *Cultura da convergência*. São Paulo: Aleph.

Johnson, Steven. 2003. *Emergência—a dinâmica de rede em formigas, cérebros, cidades e softwares*. Rio de Janeiro: Jorge Zahar Editor.

Johnson, Steven. 2005. *Surpreendente! A televisão e o videogame nos tornam mais inteligentes*. Rio de Janeiro: Campus.

Juul, Jesper. 2005. *Half-Real: Video Games between Real Rules and Fictional Worlds*. Cambridge, MA: MIT Press.

Lemos, André. 2005. *Cultura Das Redes—Ciberensaios para O Seculo XXI*. Salvador: Edufba.

Levis, Diego. 1997. *Los videojuegos, um fenômeno de masas: Que impacto produce sobre la infância y la juventud la industria más próspera del sistema audiovisual*. Barcelona: Paidós.

Machado, Arlindo. 1997a. *Pré-cinema e pós-cinema*. Campinas: Papirus.

Machado, Arlindo. 1997b. Hipermídia: o labirinto como metáfora. In *A arte no século XXI: a humanização das tecnologias*, ed. Diana Domingues, 144–154. São Paulo, Brazil: Fundação Editora da UNESP.

Machado, Arlindo. 2002. Regimes de Imersão e Modos de Agenciamento. In INTERCOM—Sociedade Brasileira de Estudos Interdisciplinares da Comunicação 25th Congresso Brasileiro de Ciências da Comunicação. Salvador/BA, Sept. 1–5.

Mendez, Laura, Mercedes Alonso, and Pilar Lacasa. 2007. Buscando nuevas formas de alfabetización: ocio, educación y videojuegos, comerciales (mimeo).

Moita, Filomena Ma, G. da S. Cordeiro, and F. C. B. Andrade. 2007. *Game On: Jogos eletrônicos na escola e na vida da geração*. São Paulo: Editora Alínea.

Moura, Juliana. 2008. Electronic games and teachers: Mapping pedagogical possibilities. M.Ed. diss., UNEB.

Murray, Janet H. 1998. *Hamlet on the Holodeck: The Future of Narrative in Cyberspace*. Cambridge, MA: MIT Press.

Pecchinenda, Gianfranco. 2003. *Videogiochi e cultura della simulazione—La nascita dell'homo game*. Milan: Editori Laterza.

Piaget, Jean. 1978. *Seis estudos de psicologia*. Rio de Janeiro: Editora Forense Universitária.

Piaget, Jean. 1983. *Psicologia da Inteligência*. Rio de Janeiro: Zahar Editores.

Piaget, Jean. 1990. *Epistemologia genética*. São Paulo: Martins Fontes.

Prensky, Marc. 2001. Digital natives, digital immigrants: A new way to look at ourselves and our kids. *On the Horizon* 9 (5). http://www.marcprensky.com/writing/Prensky%20-%20Digital%20Natives,%20Digital%20Immigrants%20-%20Part1.pdf.

Rheingold, Howard. 1997. *A comunidade virtual*. Lisbon: Gradiva.

Rheingold, Howard. 2003. MUDs e identidades alteradas. In *O chip e o caleidoscópio: Reflexões sobre as novas mídias*, ed. Lúcia Leao, 441–486. São Paulo: Senac.

Rushkoff, Douglas. 1999. *Um jogo chamado futuro—Como a cultura dos garotos pode nos ensinar a sobreviver na era do caos*. Rio de Janeiro: Revan.

Ryan, Marie-Laure. 2004. *Narrative across Media: The Languages of Storytelling*. Lincoln: University of Nebraska Press.

Santaella, Lúcia. 1996. *Cultura das mídias*. São Paulo: Experimento.

Shaffer, D. W., K. D. Squire, R. Halverson, and J. P. Gee. 2005. Video games and the future of learning. *Phi Delta Kappan* 87 (2): 104–111. http://website.education.wisc.edu/kdsquire/tenure-files/23-pdk-VideoGamesAndFutureOfLearning.pdf.

Tapscott, Don. 1999. *Geração Digital—A crescente e irreversível ascensão da Geração Net*. São Paulo: Makron Books do Brasil.

Torres, A., N. Zagalo, and V. Branco. 2006. Videojogos: Uma estratégia psicopedagógica? *Actas, Simpósio Internacional Activação do Desenvolvimento Psicológico*. Aveiro, Portugal, June 12–13.

Turkle, Sherry. 1997. *A vida no ecrã—a identidade na era da Internet*. Lisbon: Relógio D'água.

Turkle, Sherry. 1989. *O segundo EU—os computadores e o espírito humano*. Lisbon: Presença.

Vygotsky, Lev Semynovitch. 1993. *Pensamento e linguagem*. São Paulo: Martins Fontes.

Vygotsky, Lev Semynovitch. 1994. *A formação Social da mente: o desenvolvimento dos processos psicológicos superiores, Org. Michael Cole et al*. São Paulo: Martins Fontes.

Vygotsky, Lev Semynovitch. 2001. *Psicologia pedagógica*. São Paulo: Martins Fontes.

Zagalo, Nelson. 2010. New Models of Interactive Communication in Social Games. Paper presented at the Workshop of Electronic Games, Education, and Communication: Building New Trails. UNEB, Salvador, Bahia, May 6–7.

CANADA

Dominic Arsenault and Louis-Martin Guay

To understand the development and importance of video games in Canada, a few geographical and sociological facts must first be pointed out. Canada occupies the northern part of North America and is surrounded by three oceans and its only neighbor, the United States. While Canada is the second-largest country in the world by total area (9,984,670 square kilometers, which is approximately 5% larger than the United States' 9,526,468 square kilometers), it has a relatively small population of approximately 34 million (Statistics Canada 2012). When compared to the United States' 2010 population of 308 million (US Census Bureau 2010), it is easy to understand why Canada has historically been treated simply as an extension of the US games market.

Asked whether the Canadian games market had any distinctive traits, industry executives typically responded "borders don't matter" and this is a "global business." Canada is widely regarded as a subset of the US market, with differences in sales patterns between the two countries "not of a serious magnitude" (Bertram interview, 2002). The games titles and genres that have dominated the Canadian market over the past five years were roughly the same titles that topped the US and European charts (Dyer-Witheford and Sharman 2005, 194).

Canada's relatively small population, together with the continental proximity, permissive trade agreements, and close cultural exchanges between the two English-speaking countries, have all contributed in minimizing Canada as a distinct territory. The most apparent difference lies in the requirement by law for all products sold in Canada to feature labeling in both English and French. But this difference is not so substantial when considering Canada as a whole because bilingualism, just like the apparent sparseness of the population, is somewhat misleading.

For all its land area, Canada's population is densely situated along the southern border, and even then, in very definite urban centers: 50% of Canadians live in the Quebec City-Windsor Corridor, a narrow 1,200 kilometer-long strip of land along the Saint Lawrence River, and more than one-third of all Canadians live in either Toronto (Ontario), Montreal (Québec), Vancouver (British Columbia), or Calgary (Alberta). This is indicative as well of the population density and uneven economic importance of the ten provinces and three territories that together make up Canada, as these four provinces account for 86% of the total Canadian population. While bilingualism plays an important role in the international perception of Canada, the country is predominantly English-speaking. In eight provinces, francophone minorities represent at most 2.5% (in Ontario) of their population. The provincially bilingual province of New Brunswick has a 30% French-speaking minority, and the provincially unilingual French province of Québec has an 80% francophone majority and 10% Anglophone minority, in stark contrast with the rest of the country.

If there is a sort of leitmotif to be identified regarding the Canadian population, it could be said to be, by location and language, a tendency toward "clustering." *Qui se ressemble s'assemble*; "birds of a feather flock together." This is reflected in Canada's very political structure. As a federation, Canada is headed by a federal government that shares powers and responsibilities with the provinces, which are free to develop unique economic, social, and cultural policies. If we are to understand the gaming culture and industry of the country, we must realize that while some aspects are shared and apply to all cases, every province is unique in other respects, and there are complex interprovincial and provincial-federal relationships at work. Nevertheless, some initiatives have recently appeared in an attempt to bring under one flag all actors across the 5,500 kilometers of Canadian space, such as the Canadian Video Game Awards, established in 2009 to celebrate Canadian-made video games and talent.

An Overview of the Industry

Like almost every other country on the planet, video games entered the Canadian cultural space from the outside before any local industry was set up. The fact that Canada had the United States as its neighbor, though, combined with the absence of linguistic barriers between the two, meant that many home games, consoles, and arcade games were released simultaneously in both countries. Hence, Canada never really had to rely on a private import industry except for Japanese games, which, as in the United States, could appeal to a niche audience. Interestingly, despite the bilingual labeling and instruction manuals, games sold in Canada had their contents only available in English, as the software itself was the same version as for the US market; an import market for French games might have been a possibility in Québec, but it never materialized for two reasons: the games translated for France were produced for the PAL video standard and so would not be compatible with Canadian NTSC televisions; and the games took months or years before being released in Europe and thus would be both outdated and more expensive if they relied on private importation. As this shows, there is good historical reason to consider Canada a subset or auxiliary part of the US market, at least until an indigenous games industry started to emerge.

It is very difficult to obtain valid data on the games industry in Canada, perhaps owing to the relative youth and secretive corporate culture of this sector. A 2009 study prepared for the Entertainment Software Association of Canada (ESAC) best described this reality:

> In November 2007, when ESAC published its first white paper, "Entertainment Software: The Industry in Canada," Canada's entertainment software industry was basically an unknown quantity. There was no aggregate data on the size of the industry in this country, no overall job numbers or economic impact statistics to

provide a snapshot of just how significant the industry was to Canada's economy. The best information that was available either lumped entertainment software in with other forms of content or with other forms of technology. (Hickling Arthurs Low 2009, i)

The same study identified "three primary and seven secondary clusters across eight provinces in Canada. Together, these clusters represent 94% of total employment" (Hickling Arthurs Low 2009, iii). In 2008, the two initial historical hubs around which the Canadian game industry has articulated itself, Vancouver and Montréal, respectively housed 42% and 32% of the total game industry employment. The bulk of the remainder was split among the third primary cluster, the Greater Toronto Area (9%), and the largest secondary clusters, Québec City (4%) and Ottawa-Gatineau (2%). The remaining 11% of total employment was spread across the other secondary clusters, such as London (Ontario), home to Digital Extremes and a handful of small to mid-size game developers, and Edmonton (Alberta), with BioWare and a few surrounding game studios. Very few companies happen to operate outside the clustering approach; Silicon Knights' 120 employees form such an exception, which denies them the advantages of clustering as summarized in the Hickling Arthurs Low (2009) report:

> The reasons why firms cluster are now widely documented: innovation performance among clustering firms can be enhanced as a result of benefits that stem from being in close proximity to market leaders, from being able to access a pool of highly skilled and talented employees, and from the learning and knowledge sharing that comes from being in a community where social interactions can take place inside and outside of office hours. (Hickling Arthurs Low 2009, 7)

The Canadian games industry has seen substantial annual expansions throughout the 1990s and 2000s. Ranked sixth in importance among game-producing nations in 2005 after the United States, Japan, Britain, Germany, and France (Dyer-Witheford and Sharman 2005, 190), it has risen to occupy the third rank (behind the US and Japan) in 2011, with an estimated economic impact of CAD $1.7 billion on the country's economy. With 348 video game companies and 16,000 employees, Canada is the largest nation of game developers on a per-capita basis (SECOR Consulting Inc. 2011, 2).

The strength of the game industry in Canada is attributed to four factors:

1. cheaper cost of living, hence lower average salaries and sometimes operating costs, than its neighbor, the United States
2. the presence and synergistic development of broader high-technology and visual media sectors ("industry clustering")
3. strong government policies and fiscal incentives to favor development of that specific sector
4. the availability of a highly skilled and creative workforce, through long-standing governmental investment in postsecondary education.

The first point, while certainly a valid consideration for multinational studios, hinges on multiple factors that are beyond the control of the games industry itself, such as the exchange rate between the US dollar and Canadian dollar, the presence in Canada of free universal health care and social security measures, labor laws, free trade agreements, and taxation levels and regulations on corporations, among other things. Individual provinces can work out special arrangements or differently implement some of these features, which we cover as a separate point. But first, we will examine the historical clustering and development dynamics of the games sector and larger related industries across the main hubs of game development in Canada.

The Emergence of Entrepreneurs: The Local Players

The first historical artifact of Canadian video game production is the subject of debate. Two games and companies share the claim for the creation of the country's first video game. In 1982, two Vancouver teenagers, Don Mattrick and Jeff Sember, designed a game called *Evolution*, in which the player experienced natural evolution from the first life forms to the appearance of the human race. Published by the Sydney Development Corporation for the personal computer in 1983, the game sold more than 400,000 copies—a striking success when put in proper perspective: it matched Infocom's original *Zork: The Great Underground Empire—Part I* (1980) and exceeds Interplay's 300,000 copies of the smash hit *The Bard's Tale* (1985). Proud of their success, the two creators founded a company in British Columbia called Distinctive Software, with a then unheard of size of sixty employees. All in all, Distinctive Software's development was a relative success story from 1983 to 1991. This came in no small part from their Commodore 64 games. Working closely with Accolade, a well-known publisher at the time, they achieved a good rhythm of production and sales for the rest of the decade thanks to the Commodore 64's high market penetration rate, with sports titles such as *Hardball!* (1985), *Test Drive* (1987), and *Grand Prix Circuit* (1989).

Also in 1983 but 3,500 kilometers away beyond the Canadian Prairies and on the eastern frontier of Ontario in Ottawa, Rick Banks and Paul Butler's Artech Studios created *B.C.'s Quest for Tires* (1983), also published by Sydney, a game in which the caveman player-character must rescue his girlfriend who has been kidnapped by a dinosaur. The next year, the sequel, *Grog's Revenge* (1984), was released. Artech Studios had a diverse production history by making games for practically every platform available in the games industry. This ranged from the early-days multi-platform ports of successful games developed by DSI in Vancouver and Accolade in the United States to a slew of licensed game adaptations such as *Jeopardy* (1998), *Monopoly* (1999), and *Trivial Pursuit Unhinged!* (2004) in the late 1990s and early 2000s. Artech's last original work was *The UnderGarden* (2010), and the studio closed its doors in 2011.

The Beginnings

In 1991, the US developer and publisher Electronic Arts (EA) bought Distinctive Software and renamed the studio EA Canada. The industry grew significantly in the 1990s as employees left the company to start third-party studios such as Radical Entertainment (1991), Relic Entertainment (1997), Barking Dog Studios (1998), and Black Box Games (1998). In the same period of time, other independent game studios led by passionate creators opened their doors. In St. Catharines (Ontario), Denis Dyack founded Silicon Knights in 1992, just after he produced *Cyber Empires* (1991), a DOS-based strategy fighting game. In 1996, the company created the *Legacy of Kain* series, and in 2002 they achieved critical acclaim with *Eternal Darkness: Sanity's Requiem*, one of the most highly acclaimed titles in the survival horror genre.

In addition to Dyack's business, 1992 also saw Québec City entrepreneur Rémi Racine founding Megatoon, which eventually became Artificial Mind and Movement (A2M) and Behavior Interactive. Specializing in the adaptation of popular licenses, the company accumulated numerous sales to become in 2011 the largest independent game studio in Canada. After relocating in Montréal in 2000, they worked on high-profile franchises such as Scooby-Doo Monsters Inc., Ice Age, and Kung Fu Panda. In 2009, they had moderate success with their first original intellectual property, a third-person shooter game called *Wet*.

The year 1993 marked the creation of another Canadian studio, Digital Extremes, by James Schmalz in London (Ontario). Together with US developer Epic Games, they created one of the milestones in the first-person shooter game history, *Unreal* (1998), which received a multitude of awards from the industry and specialized press. The *Unreal* series sold more than fifteen million units worldwide. Following that success, Schmalz and his team have since worked on *Dark Sector* (2008), an original intellectual property (IP), on the PlayStation 3 version of *BioShock* (2008), and on *The Darkness 2* (2012), among others.

In 1995, Alberta made its entrance on the Canadian game development scene through the creation of BioWare. Founded by doctors Ray Muzyka, Greg Zeschuk, and Augustine Yip, and specializing in role-playing games (RPGs) for the PC market, BioWare became one of the most praised companies in the Western world in the early 2000s. At first successful with the Baldur's Gate, Neverwinter Nights, and Knights of the Old Republic franchises, they developed their own intellectual properties with *Jade Empire* (2005), *Mass Effect* (2007), and *Dragon Age* (2009), celebrated RPGs and action-RPGs that have become a reference point for many creators around the world.

The Consolidation of an Industry: The Foreign Giants

In addition to the creation of domestic studios, the 1990s have also seen the establishment of two foreign giants: Electronic Arts Canada and Ubisoft Montréal. From coast to coast, they changed and dictated the business and production of AAA games in the country, and to an extent, in the Western world. With approximately 3,000 employees in 2011, EA Canada became the biggest video game production studio in the world. Their franchises published under the EA Sports and EA Sports Big labels, such as FIFA, NBA, and the Need for Speed series, have sold millions of copies. While EA's Vancouver Studio is the largest center of production in Canada, the firm now spans almost the entire country thanks to the founding and acquisition of satellite studios in Montréal and Edmonton.

In 1997, the Québec government's subsidies convinced the French company Ubisoft to open a sizable studio in the province. After a slow start, Ubisoft Montréal became, fewer than five years later, the second-biggest studio in the country (behind EA Vancouver) and one of the most productive in the world. Thanks to franchises such as Tom Clancy's Splinter Cell and Prince of Persia (particularly the game *Prince of Persia: The Sands of Time* [2003]), the studio acquired both revenue and credibility. Since then, Ubisoft Montréal has grown to become one of the largest development studios in the world, with more than 2,500 employees. In 2007, it launched a new intellectual property called *Assassin's Creed*, designed by the now-renowned creative director Patrice Désilets. In the span of a few weeks, it became the company's biggest seller and one of the well-known games of the 2010s. Followed by *Assassin's Creed II* (2009), *Assassin's Creed: Brotherhood* (2010), and *Assassin's Creed: Revelations* (2011), the franchise sold twenty-nine million units in only four years. From 2005 to 2010, Ubisoft also expanded to open other studios in Québec City and Toronto.

During the first decade of the twenty-first century, more foreign companies entered the Canadian industry, and the already-established firms proceeded with multiple acquisitions to consolidate their presence. The first one to follow was *Grand Theft Auto* creator Rockstar, a subsidiary of Take-Two Interactive that established a Rockstar Canada in Toronto in 1999. The firm then acquired Vancouver studio Barking Dog in 2002, which became Rockstar Vancouver. Some of the important titles that Rockstar Toronto worked on include *Max Payne* (1, 2 and 3), *The Warriors* (2005), *Bully* (2008), and *Grand Theft Auto IV* (2008). Another Vancouver studio, sports game developer Black Box Games, was also acquired in 2002, this time by EA Canada. THQ crossed the border from the US in 2004 with the acquisition of yet another Vancouver developer, Relic Entertainment, whose respectable pedigree of computer games such as *Homeworld* (1999) and *Homeworld 2* (2003), was subsequently improved by *Warhammer 40,000: Dawn of War* (2004) and *Company of Heroes* (2006). Not wanting to be left behind, the US publishing giant Activision grabbed two good independent studios up north as well: Québec City studio Beenox in 2005, a porting house for the PC platform that became a more versatile studio under that new parent company, and Radical Entertainment in 2008 (through a merger with Vivendi Universal). Radical then started work on a new IP, which led to the release of *Prototype* (2009) and *Prototype 2* (2012), the latter being the studio's last game before Activision shut down production there in June 2012.

The influx of new players and increasing concentration of studio ownership also hit Montréal in the 2000s. Electronic Arts settled in with a new studio, EA Montréal, in 2004. The studio worked on some EA franchises such as SSX and NHL, and developed two original IPs, *Boogie* (2007) and *Army of Two* (2008). In 2007, the British publisher Eidos followed suit, with Eidos Montréal developing the next games in the Deus Ex and Thief franchises. THQ then set up a two-team studio in 2010 with a focus on the development of new intellectual properties. It was announced that one of those teams would be headed by Patrice Désilets, who had left Ubisoft Montréal and whose recruitment by THQ led to Ubisoft filing a lawsuit, as it had done against Electronic Arts in 2004, on the grounds that Ubisoft employees had contractual noncompete agreements prohibiting them from working for a competitor for a time following their resignation (Carless 2006). The year 2010 saw yet another studio opening in Montréal, this time by Warner Bros. Games, with a focus on DC Comics properties. The list seems to keep growing, with Square Enix announcing the creation of a Montréal studio in 2012 after acquiring Eidos in 2009.

These developments showcase one thing: how the internal dynamics of the Canadian games industry have also changed over the first decade of the third millennium. At the time of the 2007 Hickling Arthurs Low report, "British Columbia account[ed] for over half of the industry's total employment, followed by Quebec (26%) and Ontario (16%)" (Game Developer Research 2007). But the balance has since shifted, with Québec's video game industry experiencing a 562% growth from 2002 to 2011 (TechnoCompétences 2012), an annual average of 23.4% that resulted in the province being responsible for more than 50% of all video game jobs in Canada (SECOR Consulting Inc. 2011, 8). As the Vancouver cluster in large part evolved out of Electronic Arts, David Godsall's 2011 summary of the situation provides valuable insight.

> EA employed 1,800 people in B.C. at its height in 2007; now only 1,200 remain. ... But at the same time EA was laying people off in Vancouver, it was building an 850-person Montreal studio, with 400 working on mobile games. While EA is investing in the burgeoning non-console market, it is not doing it in Vancouver. (Godsall 2011)

To identify the causes of this eastward shift, we must trace a comparative historical analysis of the clustering dynamics that shaped the Vancouver and Montréal clusters.

Nesting in the Country: Gaming on the Shoulders of Giants

The Canadian game industry's appearance and rise is tied to the presence of various computer graphics, film and television, and visual effects firms. As a high-technology industry that requires specialized workers with a unique combination of technical skills (namely in video game programming, but also in advanced software, hardware and infrastructures) and creativity, the video game industry typically has to emerge out of a milieu that is conducive to this type of mixing through the preexisting presence of related industries. In Toronto, these include ATI Technologies, a firm specialized in graphics processing chips and acquired by AMD in 2006, and Alias Systems Corporation, which was acquired by Silicon Graphics in 1995, merged with Wavefront Technologies, which then released the 3-D modeling and animation software Maya, extensively used by the Hollywood film industry (Townsend 1999). The Vancouver cluster consists of a high number of game development studios created through a process of "firm fission" originating from Distinctive Software Incorporated (DSI). While the area is also home to strong film and TV production traditions (it is referred to as "Hollywood North"), the convergence of these two sectors is only beginning (Barnes and Coe 2011, 270–275): it seems that by having very early successful video game development (through DSI's first game *Evolution*), this sector "evolved" independently from the film and TV industry, spurred by strong internal growth through a sudden and unexpectedly profitable hit title.

In Montreal, the presence of computer-generated imagery (CGI) goes back to the Université de Montréal CGI short films *Tony de Peltrie* (1982, Pierre Lachapelle, Philippe Bergeron, Daniel Langlois et Pierre Robidoux) and *Vol de rêve* (1985, Philippe Bergeron, Nadia Magnenat Thalmann, and Daniel Thalmann), the former leading to the creation of the Softimage firm and its corresponding 3-D modeling and animation software suite that also enjoys widespread usage in the Hollywood film industry. Montréal also saw the creation of two visual effects companies in 1991: Discreet Logic (acquired by Autodesk in 1999, along with Softimage in 2008), and

Hybride Technologies, now a division of Ubisoft since 2008. This latter acquisition shows both the vitality of the games industry and how the technological convergence between film and games has taken on a very concrete economic dimension as well. As journalist Charles Prémont writes, even though the Québec video game industry kicked off in Québec City and through a certain number of small firms such as Strategy First, Megatoon/Artificial Mind and Movement (now Behavior Interactive), and Kutoka Interactive, Ubisoft's establishment in Montréal is what really got the ball rolling:

> According to many key figures of the video game scene, this arrival channeled the expansion and internationalization of the Québec interactive entertainment industry. The creation of a large studio (Ubisoft counted 200 employees in 1997, 700 in 2003, and 1400 in 2005) stimulates the job market, which allows the development of education and training centers. ... Above all, Ubisoft allows Montréal to shine on the international stage and to attract other large game development studios. (Prémont 2009, 16, freely translated)

The Toronto and Montréal clusters illustrate how a video game sector can emerge by tapping qualified personnel from the local talent pool of already-existing film and computer graphics industries. That said, the three clusters differ in many other aspects; 70% of large firms (151+ employees) operate in Québec, while British Columbia is home to around 30% of all Canadian small (6–50) and medium-size (51–150) companies (SECOR Consulting Inc. 2011, 8–9). These numbers reflect the "firm fission" nature of the Vancouver cluster, where a lot of EA employees quit to open their own studios, and the "big business" approach that was favored by Québec's tax incentives and Ubisoft's initial push. In Québec, 90% of industry workers are employed at firms of more than one hundred people, and 72% of the total workforce is concentrated in a handful of studios with more than three hundred employees (SECOR Consulting Inc. 2011). As it turns out, government support may be the one key reason that explains Québec's meteoric rise at the expense of British Columbia and Ontario.

Live or Die by the Grants

The French company Ubisoft may have had many reasons for choosing Montréal as the site of its North American expansion in 1997: the Québec province's unique positioning as a crossroads between North American and European cultures; Montréal's reputation as a hub for film production, visual effects, and computer-generated imagery; and the Québec metropolis's status as the third-largest French-speaking city in the world; but the key reason may have been a lot more down-to-earth and monetary. The Québec provincial government at the time saw in the video game industry a vast potential for the attraction of foreign capital and investments, along with the creation of high-salaried jobs with particular skill sets and expertise, especially for the younger generation. Ubisoft was awarded a record $25,000-per-job subsidy in exchange for a promise to hire 500 people over five years; the headcount has since risen beyond 2,000.

This initial offer was met with a mixture of criticism and disbelief by the already-present multimedia and game software firms, none of whom had received such royal treatment (Tremblay and Rousseau 2005, 308–315). Eventually a policy for the development of the multimedia sector that applied to everyone and covered all the bases necessary for industry growth saw the light of day in under two years, based on multiple incentives that notably included a 37.5% tax credit on employee salaries. The Québec government's involvement provided a key reason for many external firms to set up shop in *La Belle Province*:

> The policy response from the Quebec government, which early-on had identified the growth potential of [the] entertainment software industry, was pivotal to creating a critical mass outside of the Vancouver cluster and gave Canada an international reputation in the industry. The lesson is as much about timing as it is about offering strategic support. Though several other provinces in Canada have since introduced policies targeting the entertainment software industry, they have done so at a different stage of the industry's global development when a far greater number of regions are vying for a position in the global production network of entertainment software. (Hickling Arthurs Low 2009, 25)

These incentives managed to attract foreign investors for years down the road but were routinely complemented by direct subsidies for large-scale projects. In 2010, the Québec government invested $7.5 million for a Warner Bros. operation of 300 employees by 2015, and $3.1 million for a 400-person THQ studio; in 2011, $2 million was given to Square Enix for an Eidos Montréal expansion and the creation of a second Square Enix studio (Brousseau-Pouliot 2011). Square Enix's press release duly acknowledged the importance of these subsidies: "All of this has been made possible by our fantastic partners in Invest Quebec and Montréal International and we look forward to continuing our excellent working relationship with them" (Square Enix 2011). And yet, as important as these government investments are, they can only work if local talent is there as well, something THQ addressed when explaining its 2010 decision:

> Montréal is recognized for a high concentration of media talent and will serve as a hub for the latest location in THQ's expanding studio system. Its selection reflects THQ's strategy to grow development talent in highly skilled locations. (THQ Inc. 2010)

A Challenging Future: Canada's Quest for IPs

As of 2012, future industry growth is expected to rest on the shoulders of the Québec City and Greater Toronto Area clusters. The fate of the Vancouver and Montréal clusters is indeterminate; assuming no changes in government policy and general industry orientations, if the "casual revolution" (Juul 2009) and the rise of mobile and social gaming settles into a long-term market shift, the Montréal cluster's "big business" focus on large firms and AAA traditional console titles might result in an industrial downturn to the profit of other regions where micro- (one to five workers) and small-size developers appear more easily—namely, the Vancouver and Greater Toronto Area clusters. The development of IT and communications as well as the expanding focus on the casual games and nontraditional platforms may lead to a shift away from the clustering dynamic and toward a second type of industrial development identified by Dyer-Witheford and Sharman:

> This geography demonstrates two contradictory spatial dynamics associated with high-tech industry. The supremacy of Vancouver and Montréal shows the importance of "regional innovation milieu," or "clustering" where mutually reinforcing production activity takes shape in one urban locale (Holbrook & Wolfe, 2000). But the success of some small and mid-size developers in the Prairie and Atlantic provinces also demonstrates

"the death of distance" effect (Cairncross, 1997), where the Internet allows businesses to escape established urban centres and to take advantage of lower costs in more remote areas. (Dyer-Witheford and Sharman 2005, 191)

The "death of distance" effect might very well lead to a dissemination of game production in small and micro-enterprises more evenly spaced out, and into a shift away from the clustering dynamics that marked the "big business" AAA game production model. One of the greatest challenges that the Canadian game industry continually faces is the development of original IPs, controlled by Canadian interests. This is a result of the country's massive reliance on foreign investors:

> Most of the video game revenue generated in Canada originates from foreign-owned game publishers like Electronic Arts (U.S.), Ubisoft (France), THQ (U.S.), and Eidos/Square-Enix (Japan). Canada has become an attractive location for foreign publishers, in part, because of generous tax incentives but also because of the clustering of talent and expertise in Vancouver, Toronto, and Montreal. Foreign-owned studios are a significant source of employment for Canadians trained in advanced skills for interactive media, entertainment programming, and new media design but the decisions and controls over content creation and ownership are not Canadian. (Gouglas et. al 2010, 2)

Indeed, there have been relatively few original Canadian intellectual properties. Increasingly, the large firms are investing in this practice in Canada with high-profile game series: Electronic Arts' *Skate* and *Army of Two*, Ubisoft's *Assassin's Creed*, BioWare's *Mass Effect* and *Dragon Age*, and Rockstar Toronto's *Bully* (2006) are testaments to this, but none of them were released before 2006. Historically, original franchises have more often come out of the small and mid-size developers rather than large companies, as the Eternal Darkness, Homeworld, and Unreal franchises illustrate.

The Northern Indies

In this sense, establishing supportive conditions for the successful development and operation of smaller independent studios might be a good way of ensuring a greater appearance of Canadian intellectual properties, given that many good projects have already been developed without established programs to favor them. The new millennium has seen the rise of the independent scene all around the world, and the Canadian creators are no exception. These include Jonathan Mak and his one-man project *Everyday Shooter* (2007), which won the Design Innovation award at that year's Independent Games Festival (IGF). At the same event, the Seumas McNally Grand Prize was awarded to Alec Holowka from Winnipeg (Manitoba) and Derek Yu for their game *Aquaria* (2007). Later on, another independent game designer from Montréal, Phil Fish, won the same award at the 2012 IGF for his game *Fez*, released in 2012 after five years of development chronicled in the Canadian documentary film *Indie Game: The Movie* (2012).. In fact, the Grand Prize was named thus to honor Seumas McNally, the programmer and president of Toronto-based independent studio Longbow Digital Arts, who won that prize in 2000 with *Tread Marks* (2000) before dying shortly after. It may be that Vancouver's entrepreneurship and Québec's "big studio" approaches can be contrasted with a "cultural" concern in

Toronto. The high numbers of micro and small independent studios that have elected to make their home in this cluster are also complemented by related initiatives, such as Syd Bolton's 2005 opening of the Personal Computer Museum in Branford (Ontario). This venue is meant to preserve the history and present state of the video game industry in the world, with a special emphasis on Canadian content. As of late 2010, more than 11,000 software pieces are exhibited in the museum, a number that will only increase as the industry continues to prosper in Canada.

As this brief and incomplete tour shows, independent game developers are active in Canada. It would make sense for the various levels of government to diversify and invest more in this type of production for an important reason: keeping the revenue stream domestic, thus giving the province and country its due share of revenues through the taxation of profits. As foreign investors typically set up development shops to work on the games but still retain control of publishing in their home country or in other places abroad, publishing revenues escape the local governments. Because the advances in digital distribution lower the barrier of entry for smaller developers, independent game designers typically eschew the traditional industry's revenue-splitting mechanics, which often leave most of the profits to the publisher. Those clear advantages, though, are offset by very practical problems. One of Canada's weaknesses lies in the difficulty for companies to access venture capital (Canadian Chamber of Commerce 2012). In these situations, it is typically local governments that are called upon to invest first so as to spur a new type of economic activity, but the highly risky nature of computer game financing makes such targeted investments difficult to implement properly. One key type of partner in this undertaking may be universities and colleges that train students seeking a career in the game industry.

The Research Pipeline

In 2010, a group of researchers led by Sean Gouglas produced a report for the Social Sciences and Humanities Research Council of Canada on the role of universities in promoting innovation in the computer game domain. In the introduction, they set the tone for what could be the biggest challenge for the future:

> The key to Canada's digital economy, therefore, is to sustain economic growth while encouraging cultural innovation in the video game industry. The diversity of video game and video game technology producers is remarkable, ranging from studios that produce multi-million dollar titles to individual companies that produce distinct game assets to sole proprietorships that create simple (yet often addictive) casual games. (Gouglas et al. 2010, 2)

The development of a stronger economy will come through collaboration between companies, the government, and the higher education sector. Some bridging efforts have already been made in recent history, with universities and game developers partnering for the development of knowledge, training, and research. Brock University in St. Catharines (Ontario) had an informal partnership with the now-defunct Silicon Knights, notably through the biannual Interacting with Immersive Worlds conference series that began in 2007. The

University of Alberta's GAMES Group similarly uses BioWare's game engines to develop path-finding algorithms and scripting languages, and invites industry contacts to provide feedback in game-related courses. In Québec, Ubisoft launched a Ubisoft Campus project from 2005 to 2010, bringing together different universities, colleges, and video game development programs in complementary disciplines in a single location. In this particular case, the government's financial participation in the project was decried both by the opposition political parties and Ubisoft's competitors, who felt it was not the government's role, through taxpayers' money, to invest in the training of employees for a specific company.

According to Gouglas et al.'s report, one of the biggest problems in university-industry collaboration is the conflict in terms of the needs and wants of both groups. University instructors teach abstracts and fundamentals, reasoning that the tools may change and that attaining comprehension of the underlying principles will ensure a long-term capacity of adaptation; in contrast, the game industry needs employees with very definite, narrow, but in-depth skills in certain tools (particularly in large firms where worker specialization is more valued), who are able to efficiently carry out plans by respecting the tight deadlines and immediate needs on which the industry is built. As the report said, "Many of the people we interviewed suggested that formal programs at colleges and universities would not, and in fact, could not, prepare most for a career in the gaming industry" (Gouglas et al. 2010, 39). The feeling may even appear to be mutual: there is a growing number of universities that offers courses, programs, and research projects revolving around video game studies or experimental development, but these initiatives do not necessarily mesh together with the industry. For instance, the Canadian Game Studies Association has had very few speakers or attendees from the industry at its yearly conferences since it started operating in 2007. While the four provinces that lead the games industry have more than a dozen universities offering some measure of game-related education, these do not necessarily offer practical, hands-on game development programs, and many of them are not partnering with studios to the extent that internships or conferences by game professionals have become the norm. While it may appear to be something of a Canadian cliché, academia and industry may be said to act as two solitary entities. The forging of a deeper relationship between the two might spur a second wave of game entrepreneurship but also a larger cultural take on games that would complement the industry's large-scale industrial production and provide a more diverse future for video games in Canada.

References

Barnes, Trevor, and Neil M. Coe. 2011. Vancouver as media cluster: The cases of video games and film/TV. In *Media Clusters Across the Globe: Developing, Expanding, and Reinvigorating Content Capabilities*, ed. C. Karlsson and R. Picard, 251–277. Cheltenham: Edward Elgar.

Brousseau-Pouliot, Vincent. 2011. Jeu vidéo: 350 nouveaux emplois chez Eidos. *La Presse Affaires.* http://affaires.lapresse.ca/economie/technologie/201109/01/01-4430609-jeu-video-350-nouveaux-emplois-chez-eidos.php.

Cairncross, Frances. 1997. *The Death of Distance: How the Communications Revolution Will Change Our Lives.* Boston: Harvard Business School Press.

Canadian Chamber of Commerce. 2012. Tackling the top-10 barriers to Canadian competitiveness. http://wpmedia.fullcomment.nationalpost.com/2012/02/top10barriers.pdf.

Carless, Simon. 2006. Electronic Arts, Ubisoft clash on Montreal hiring. *Gamasutra.com*, http://www.gamasutra.com/php-bin/news_index.php?story=7985.

Dyer-Witheford, Nick, and Zena Sharman. 2005. The political economy of Canada's video and computer game industry. *Canadian Journal of Communication* 30:187–210.

Entertainment Software Association of Canada. 2011. 2011 essential facts about the Canadian computer and video game industry. http://www.theesa.ca/wp-content/uploads/2011/10/Essential-Facts-2011.pdf.

Game Developer Research. 2007. *2007 Game Developer Census.* http://gamedeveloperresearch.com/game-developer-census-2007.htm.

Godsall, David. 2011. Vancouver's ailing video game industry. *BCBusiness.* http://www.bcbusinessonline.ca/profiles-and-spotlights/industries/media-arts-and-entertainment/vancouvers-ailing-video-game-industry.

Gouglas, Sean, Jason Della Rocca, Jennifer Jenson, Kevin Kee, Geoffrey Rockwell, Jonathan Schaeffer, Bart Simon, and Ron Wakkary. 2010. *Computer Games and Canada's Digital Economy: The Role of Universities in Promoting Innovation.* Report to the Social Sciences & Humanities Research Council Knowledge Synthesis Grants on Canada's Digital Economy. http://circa.ualberta.ca/wp-content/uploads/2010/03/ComputerGamesAndCanadasDigitalEconomy.pdf.

Hickling Arthurs Low. 2007. The entertainment software industry and its impact on Canada. http://www.theesa.ca/documents/ESAC_whitepaper2007.pdf.

Hickling Arthurs Low. 2009. Canada's entertainment software industry: The opportunities and challenges of a growing industry. http://www.theesa.ca/documents/ResearchReport09.pdf.

Holbrook, J. Adams, and David Wolfe, eds. 2000. *Knowledge, Clusters and Regional Innovation: Economic Development in Canada.* Montréal and Kingston: McGill-Queen's University Press.

Juul, Jesper. 2009. *A Casual Revolution: Reinventing Video Games and Their Players.* Cambridge, MA: MIT Press.

Prémont, Charles. 2009. *Guide de l'Industrie Jeux Vidéo, Première Édition.* Montréal, Québec: Le Lien Multimédia.

SECOR Consulting Inc. 2011. *Canada's Entertainment Software Industry in 2011.* Report Prepared for the Entertainment Software Association of Canada. http://www.theesa.ca/wp-content/uploads/2011/08/SECOR_ESAC_report_eng_2011.pdf.

Square Enix. 2011. Square Enix to expand further in Montréal. http://www.square-enix.com/eng/news/2011/html/f59a62fddbf4cff95811a41d5cde6676.html.

Statistics Canada. 2012. 2011 Census: Population and dwelling counts. http://www.statcan.gc.ca/daily-quotidien/120208/dq120208a-eng.htm.

TechnoCompétences. 2003. Développement de la main-d'oeuvre des entreprises québécoises de production de jeux électroniques. http://www.technocompetences.qc.ca/sites/technocompetences.qc.ca/files/uploads/industrie/etudes-et-rapports/2003/Developpement%20m-o%20jeux_0.pdf.

TechnoCompétences. 2012. L'emploi dans l'industrie du jeu électronique au Québec en 2011: Un portrait sommaire de la situation. http://www.technocompetences.qc.ca/sites/technocompetences.qc.ca/files/uploads/industrie/etudes-et-rapports/2011/Rapport2011_Jeu_VFR.pdf.

THQ Inc. 2010. THQ opens Montreal development studio; announces industry heavyweight Patrice Désilets to join THQ studio system. http://investor.thq.com/phoenix.zhtml?c=96376&p=irol-newsArticle_pf&ID=1484058.

Townsend, Emru. 1999. Along the banks of the St. Lawrence... *Animation World Magazine* 3 (12), March. http://www.awn.com/mag/issue3.12/3.12pages/townsendcanada.php3.

Tremblay, Diane-Gabrielle, and Serge Rousseau. 2005. The Montreal multimedia sector: A cluster, a new mode of governance, or a simple co-location? *Canadian Journal of Regional Science* 28 (2): 299–328. http://cjrs-rcsr.org/archives/28-2/7-Tremblay-Rousseau.pdf.

US Census Bureau. 2010. 2010 census data. http://www.census.gov/2010census/data/.

CHINA

Anthony Y. H. Fung and Sara Xueting Liao

Video games have become an increasingly popular activity in everyday life, especially with the strong growth of online games, which are now the cause of a worldwide mania. In 2010, the revenue of the global game market was USD $52 billion, and it is expected to increase to USD $70 billion in 2017 (DFC Intelligence 2012). A research report from Gartner also foresees that the global gaming industry will even exceed USD $74 billion in 2011 and possibly reach USD $112 billion by 2015 (McCall and van der Meulen 2011).

In China, as in other Asian-Pacific areas such as Japan and Korea, video games drive a huge flow of capital and bring striking profits, gradually forming a pillar industry in the economy. The 2011 game report by the CGPC (China Game Publication Committee) and IDC (International Data Corporation) demonstrated that the Chinese game market generated USD $6.98 billion (RMB 44.61 billion) in revenue that year, increasing 34% from the revenue in the previous year and amounting to one-tenth of the global revenue in the video game industry. China's market share in the global game industry has become more prominent as well.

The Emergence of Youth Culture in China and the Rise of the Chinese Game Market

The increasing appeal of the video game market is attributable to a growing young culture in China. Youths tend to consolidate their identity and reinforce their social roles by appropriating cultural practices that are widely shared among their peers (Hebdige 1979), and it is not difficult to imagine that gaming has become a popular phenomenon among youths in China. Illustrating the idea of youth engagement in political and social movements, Clark (2012) highlights how new media technology and the Internet have inevitably become pivotal vehicles and platforms for public participation and even intervention. With more video games going online, they will readily become more embedded in the social lives of Chinese youths. When playing video games online, Chinese youths are likely to build up a "virtual me" online and will more readily identify themselves with other youths in virtual worlds (Cao and Downing 2008). In his study of Chinese youths, Clark (2012) also points out that players isolate themselves while gaming or "playing" in front of their computers, but ironically, at the same time they are joining in with the global trend when they participate in video game play.

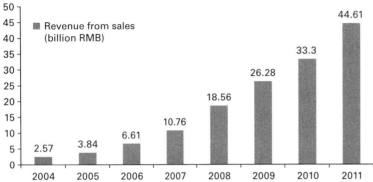

Figure 1

Chinese video game market revenue (2004–2011). Source: CGPC and IDC. "2011年中国游戏产业调查报告(摘要版)" (2011 Chinese Game industry Report [Abstract]), http://www.cgigc.com.cn/201201/120087820889.html.

Moving Online and Becoming Mobile

In China, relatively isolated video game activity is complemented by online games in which players are connected to other players. There are multiple reasons, including improving Internet connections and infrastructure, domestic industrial protection, policy and regulations, and issues of piracy. Among those it seems that the availability of Internet access has fundamentally changed youths' lifestyle in play and leisure. According to the China Internet Networks Information Center, despite other possible factors affecting the accuracy of counting the number of Internet users, China has the world's largest online population, with an increasing number of online users. That number reached 513 million as of the end of 2011 (Kan 2012). Even though game companies such as Blizzard Entertainment encountered difficulty in establishing business in China due to the country's stringent criteria for outside investors, foreign investment is increasing annually and many video game projects have been designed or even published in China. Nowadays, online games in China constitute the biggest part of production and revenue for game companies (Niko Partners 2011). Recently, video games on mobile devices have become increasingly popular due to the social networking function of games. Premature as it is, the gaming market on mobile devices and social network services provides opportunities for the changing video game business in terms of its operation and profit model.

Constraints Due to State Ideology

On the one hand, noticing the potential influence of video games, the Chinese authorities are playing devil's advocate, restricting, deterring, and slowing down the dissemination of Western games and the development of the Western video game industry in China. As a result, from production and introduction to circulation and consumption, all sectors of the video game industry are highly controlled. For example, Blizzard Entertainment attempted to publish *World of Warcraft* (2004) in China, but citing the game's violence as the reason for rejection, the state strictly denied its approval. Blizzard then had to rely on a local distributor to localize its product, and to be censored (the local collaborator is either a state-related enterprise or a company that confirms to the state's censorship system, and thus the collaborator also takes up the role as censor). This becomes the indirect way that the locals are "made" to get involved in the game market. On the other hand, the state also acknowledges the economic potential of the game industries and thus allows domestic capital to become involved in video game production in different forms. Of course, under authoritarian ideology, local developers cannot avoid joining in with the ideology of the central government and creating societal harmony. However, the local production of video games also produces side products: pirated video games, which to a certain extent cause tension among game companies and the state, among domestic game industries, between countries exporting their cultural products, and between users and the state. While the disregard for copyright is, in general, seen as the piracy of intellectual property, consumers in China (as they argue), see it as resistance against the dominance of the concept of copyright and patents. Local businesses also conceive of it as a counterforce against giant multinational companies.

A Brief History of China's Video Game Market

The history of video games in China began in the 1990s. Yet, it is not much less than the history of video game industry development in most Western countries, which started in the early 1970s. Its swift, worldwide spread is tied to the invention of personal computing, which had a breakthrough in speed and storage in the 1970s—the same golden age of the video game industry, which saw the large-scale introduction of console games, arcade games, and PC games. Later, the availability of the Internet also boosted the online game business and the mobile game industry (Lowood 2006). The ability to accommodate multiple players (allowing a bigger market) and the development of gameplay in PC and online games also explained why they slowly superseded early console and arcade games as the most popular kinds of games.

Xiaobawang (meaning "big brother") was the first game console released in China (in the late 1980s).[1] It is a relatively simple electronic device with slots into which video game cartridges are inserted; two game controllers are then connected to the keyboard for players. When the console is connected to a television with cables, users can play games read from the cartridges. Game cartridges published for Xiaobawang contain pirated games of foreign imports.

Exorbitant growth of console games in China occurred from 1988 to 1991. During that time, consoles were imported from Japan and more than a dozen local imitations rushed into the market, creating a highly competitive environment. Though big international companies and their systems (such as the Nintendo Entertainment System [NES], the NEC PC-Engine, the Super Famicom [the Super Nintendo Entertainment System] and the like) eagerly knocked on the door of the Chinese domestic market, they could not afford to pay the tremendously high tariff. These companies were required to pay a 35% import tax for specific "preferable" items and 130% for regular importation. When it was deemed unrealistic to try to sell these expensive imported consoles to the relatively low-income Chinese users, pirated consoles were made locally in China and met the market need. Amid the keen competition, the Xiaobawang outsold other competitors due to its low cost and relatively high quality. It soon occupied almost the entire Chinese console game market. To play games, users only needed a pirated console and a pirated cartridge containing exactly the same content, say, of an NES cartridge. Later in the market, the Tianjing-based company Xinxing promoted a series of games called *Street Fighter*, a pirated version of the worldwide game *Street Fighter* series from Capcom, and it soon become a market hit. As *Street Fighter* could be played on SEGA's MD platform (the 16-bit Mega Drive), a locally cloned console, the SEGA-based 5th Generation, was manufactured. This new, higher standard (16-bit) console was soon prevalent across the entire market and created a new wave of competition among local console manufacturers. Outraged by the public theft of its copyrights, SEGA took legal action, but in vain. Counterfeit video games by Xiaobawang, Xinxing, and other pirating companies were still the local winners.

At the height of the video game console era, piracy was not confined to China. In various developing countries, including Russia, South Africa, Poland, Brazil, and Taiwan, game production companies all "mimicked" the production technologies of those flagship game publishers in developed countries in order to fulfill the desire of local markets where consumers probably could not afford the high price of a legitimate console system. In Brazil, for example, a pirated game console based on the SEGA Mega Drive called "Phantom System" was released in 1991 by a company called Gradiente (see http://en.wikipedia.org/wiki/Nintendo_Entertainment_System_hardware_clone#Brazil). In Taiwan, Micros Genius (its Chinese name is Xiao Tiancai, literally meaning "small genius"), was a brand name used by the TXC Corporation for making, selling, and spreading clones of Famicom consoles in Southeast Asia. A modified pirated console named "Multicarts" was also developed in Taiwan and could accommodate many different games in one cartridge (see http://en.wikipedia.org/wiki/Micro_Genius).

Taiwan later played a pivotal role in China's video game development. In the 1980s, Taiwan was a major Asian hub for PC game development, with companies that operated in the game business and R&D, publication, promotion, and circulation. In 1983, Taiwan's Soft-World International Corporation was established. It became the first company to distribute foreign video games in authorized Chinese versions. In 1988, the first professional Chinese video game development company, Softstar Entertainment Incorporation, was set up in Taipei. During this period, Taiwan was the only legal exporter of games to China. In 1995, Softstar's *The Legend of Sword and Fairy* (仙剑奇侠) for the Windows 95 operating system was published, and it was very successful in both Taiwan and mainland China.

The mid-1990s could be called an incipient period for China in terms of publishing video games. The earliest Chinese video game, *Condor Rush* (神鹰突击, 1994) was released by Jinpan Electronic Corporation. In 1995, giant game companies including Qiandao Software (前导软件, which went bankrupt in 1998), Western Hills Residence Studio under Kingsoft Company, and Object Software Limited were founded. Of the earliest games created and distributed, *Jianxiaqingyuan* (剑侠情缘, 1997) by Western Hills Residence Studio could be considered the greatest success, and it established a reputation for the company. Subsequently, the studio promoted a series of *Jianxia* games from 1997 to 2009.

The year 1997 was a milestone for the development history of the Chinese video game industry; many companies were established in this wave of development, and they published series of famous games that drew public attention. After that, there was a period of fast development in which Suntendy Interactive Multimedia Co., Ltd. (新天地互动多媒体技术有限公司, which went bankrupt in 2005) played a big role. Suntendy pioneered an innovative business model: it became a local agency for the importing and localizing of foreign games and managed the operation of a proxy server under the authorization of the host game company. More prominently, it was the legal agency of an adventure game, *Tomb Raider*, issued by Eidos Interactive. In 1998, Suntendy distributed the Spanish RTS (real-time strategy) video game *Commandos: Behind Enemy Lines* (1998), while producing a meticulously Chinese-subtitled version that was released in 1999. Getting beyond the obsolete image of Chinese game companies as dens of piracy, such localized cooperation with foreign companies was definitely a big step forward for the Chinese game industry. Soon afterward, many companies such as Thor (世纪雷神) and Popwan (天人互动) emerged and acted as distributors for imported as well as domestically developed video games, opening up a new epoch for video games in China. Well-developed games included *Heroes in Marshes* (水浒英雄传, 1997) (see http://www.china.com.cn/info/games/2011-03/22/content_22190653_2.htm), the *Xuan-Yuan Sword* series (轩辕剑系列, 1998–2002), *The Three Kingdoms: Fate of the Dragon* (傲世三国, 2000) (see http://www.china.com.cn/info/games/2011-03/22/content_22190653_2.htm), *Heroes in Wulin* (武林群侠传, 2001), *Sakura Wars* (樱花大战, 2001), and *Millionaire* (大富翁, 1998–2002) (see http://article.ali213.net/html/1880_2.html). Translated foreign games such as *StarCraft* (1998), *Counter-Strike* (1999), and *Civilization III* (2001) were also very popular in China.

Chinese video game development soon encountered severe crises, however. The huge profits made on domestic video games brought on another wave of piracy, this time pirating famous local content. While new companies soon found a lack of corporate experience and resources, others failed to provide quality products, which resulted in the withdrawal of investors even from legitimate companies such as Qiandao, which went bankrupt in 1998. Besides piracy, the rise of the online game industry also challenged the sustainable development of traditional video game companies. Game development was shrinking, with less capital input and investment. Investors were afraid of the rampant growth of the pirated game market and the perceived likelihood of the gradual shift from PC games to the more attractive and profitable online games.

At the same time, the world of online games has flourished globally. With the largest population in the world, China has become a mass market for foreign investors and developers. With less than a decade's development, according to iResearch, in 2010, the annual turnover of the Chinese online game industry reached USD $5.14 billion (RMB32.74 billion), increasing 21% from the previous year (see http://www.chinadaily.com.cn/bizchina/2011-02/14/content_12007639.htm). China is now regarded as one of the three biggest markets (along with the United States and South Korea) in the global online game market, accounting for 27% of the pie. In retrospect, the development of Chinese online gaming could be dated back to the success of MUDs

(Multi-User Dungeons) and *Ourworld* (1997) during the Asian financial crisis. (*Ourworld* is an online virtual world that contains a range of online causal games and activities, allowing for socialization and interaction among players.) *Swordmen's Legend* (1997) represented the best of text-based MUDs, and *Ourworld* released its online game platform called *Ourgame Online Game World* in June 1998, providing five online chess and board games: *Weichi*, *Xiangqi*, *Chinese Checkers*, *Tuolaji*, and *Gongzhu*.

China's online game market has a unique characteristic: domestic developers lagged behind the market, so the first generation of online game players comprised, in fact, fans of *Ultima Online* (1997), which became the forerunner of global MMORPGs and which came to China in 1998. Though operating on private servers and run by anonymous operators, *Ultima Online* quickly crystallized a huge generation of Chinese online players in the late 1990s. It was only afterward, in July 2000, that the first genuine graphical MUD, *King of Kings*, the predecessor of Chinese MMORPGs, was released. Table 1 summarizes the key events in the development of the Chinese online game industry.

The Emerging Model of Game Business in China

In China, the early development of video games, like other major industries, was largely constrained by the technology available. Expensive hardware and consumers' low income rendered insufficient demand in the video game market. Even those who were the first to acquire computers might also have been unaware of their potential for entertainment, instead focusing on the computer's value in the areas of education and urbanization. The first generation of video game consumers included those whose families could afford a console system or whose classmates or friends' houses had a computer that could be used to play games.

As China's economy was rising, however, computer and video game innovations diffused from top to bottom. The state supported the industry and regarded it as a policy and economy issue, enhancing the flourishing of game companies. Digital technology grew quickly under the exceptional investment made by the government. The education system, controlled by the central state, also formed another channel to promote computer techniques as essential and necessary skills for the new generation. As former leader Deng Xiaoping said in 1989, "The spread of computer literacy should start with children" (see http://it.sohu.com/20090916/n266774054.shtml), an idea that was incorporated into primary school textbooks and that triggered rushes to the newly introduced media as symbols of modernization.

Under the pro-computer ecology, video games, as a kind of content for the computer, also became popular, especially among the youth population. Unlike the early days of arcades, more and more families could afford personal computers at home, which made it possible for video games to enter people's everyday lives and homes. Outside the household, the mushrooming boom of game centers and cybercafés all over the nation also nourished and amplified the fad of computer entertainment.

Except for computer hardware, however, China had no favorable conditions for game companies to grow. In Chinese society, games are regarded as perverting children and adolescents' mental and physical health

Table 1

Important events in the Chinese online game industry, 2000–2011.

2000.07	The first genuine Chinese graphical MUD, *King of Kings*, is released. The developer starts the first Chinese company, Larger, which applies computer graphics technology to online entertainment.
2001.01	Waei (currently Wayi) operates *Stone Age*, which included avatars in a cute and cartoon-like style, differing from the bloody and violent nature of previous games. This was the first game to be seriously affected by Waigua (a third-party assistant program). Waei also employs a new subscription method for game users called WGS (Web Gold System). Users need to exchange virtual currency with real money in order to purchase, upgrade, and use the service provided by the server.
2001.11	Shanda operates *Mir II*, a Korean online game. It becomes the most popular MMORPG in China in 2002 and 2003, with over 250,000 simultaneous users being reported. Meanwhile, *Mir II* has more private servers than any other online game in China. In 2003, Wemade and Actoz (*Mir II*'s developer and mother company) cease their contract with Shanda unilaterally, bringing the conflict between game developers and operators into public sight.
2002.06	The Lanjisu net bar (or cybercafé) fire in Beijing causes twenty-one deaths. The government starts to regulate these so-called Black Net Bars (unlicensed bars).
2003.05	TQ Digital Entertainment is one of the exhibitors at the 2003 Electronic Entertainment Expo, with its product *Monster & Me*. It is the first time a Chinese developer appears at such a high-level international game exhibition.
2003.09	Online games are brought into the "National 863 Project," and the government allocates RMB5,000,000 in support of the development of original, native, online games.
2003.12	The "first case of virtual property" (involving the game *Red Moon* [2000]) is closed with the plaintiff winning, bringing the problem of the protection of virtual property into the public agenda.
2004.08	The company Game-Orange announces that *Great Trader* (2004) will be free to play, and it becomes the first Chinese game to be supported solely through the sale of virtual items.
2004	Chinese online game companies are listed on NASDAQ, including Shanda and The9 Limited.
2007.07	The "Preventing Addiction System of On-line Games for Youth" initiative is implemented, formally limiting minors' daily game time.
2008	Web-based games become a new form of online game, and because these games are free, some players remain online all day long. These games grow even more popular in 2009.
2009.06	The Ministry of Culture and Ministry of Commerce together issue "Notice Governing Transactions Involving Virtual Currency in On-line Games," attempting to keep the virtual property market in order.
2009.06	The proxy dispute regarding *World of Warcraft* leaves The9 Limited with a 90% loss of profit, indirectly bringing to light the conflict between General Administration Press and Publication of PRC (GAPP) and the Ministry of Culture on the legitimate authority regarding the governance of online games.
2010.06	Five Chinese online games, *Wanmei*, *Tianlongbabu*, *Menghuanxiyou*, *Zhangtu*, and *The World of Legend*, appear on the *Forbes* list of great successes in 2009.
2010.10	A couple who used an illegal programming Waigua in an online game to "level up" their avatars is sentenced to prison. It is the first court case in China to condemn players who destroy equality and balance in an online game platform.
2011.01	The China On-line Games Copyright Protection Alliance (COGCPA) is established, cofounded by sixteen online game companies, with support from the government. It shows that online games, as a form of entertainment, are officially protected by the state.

and are therefore often rejected in formal settings. As for the government's policy, video games as a medium are always subjected to government control, which might eventually reduce them to propaganda. Thus, early online games were only imported from Japan and Korea, the game design of which appeared more relevant to the Chinese audience due to geographical and cultural proximity. Early online start-ups in China actually turned to those nations to learn how to operate, manage, and develop games. Chinese companies soon established profits through martial arts (or wuxia) games, a genre that is unique to China's culture and history. Prominent companies such as Softstar, Domo Studio, Qiandao, and Western Hills Residence Studio released successful and influential titles including *Xuan-yuan Sword* (1990), *The Legend of Sword and Fairy* (1995), *Jianxiaqingyuan* (1997), and so on. With such games, these companies drew upon indigenous martial arts and, with the addition of enthralling storylines, were able to cultivate the tastes of game consumers of that generation.

Shanda Interactive Entertainment Limited introduced *The Legend of Mir II* (2001) to the Chinese market from Korea by creating a proxy model—that is, it licensed a foreign game, operated it domestically, and then shared its profits with the producer. This model proved successful, and other companies soon followed suit. Even nowadays, the proxy model is still an important revenue source for some companies, but as one can imagine, such foreign collaboration has inevitably led to increasing disputes between the foreign game developers and local agents. To overcome these conflicts, China needs a new model. In November 2005, Chinese online game companies, led by Shanda, introduced the come-stay-play revenue business model, which allows users to play the basic functions of locally developed online games for free and to purchase in-game value-added services, including in-game items and premium features that enhance the game experience. Shanda began to profit after two years of this experiment. In 2007, the highest quarterly number of active paying accounts (APA) reached 3.47 million for Shanda's MMORPGs and 2.38 million for their casual games (see the *Shanda Interactive Entertainment Limited 2007 Annual Report*, available at http://www.snda.com/Upload/20091/43e31431d3b345d4ae5cca75f954a9f9.pdf). Combined with in-game advertising revenue, this business model helped Shanda dominate the Chinese game market for several years. In 2007, the number of formally operated online games in mainland China was 203, among which 154 adopted the model, making it the mainstream model of the Chinese online game industry (Popsoft Editorial Office 2008).

Nowadays, the free-to-play model is still dominant in the game market, including both PC and online games. However, several changes have appeared recently. The popularization of Chinese social networks such as Renren and Weibo (a Twitter-like service), and the widespread use of portable and mobile devices for mobile games, particularly those in social network applications, are gradually eroding the PC game market. Popular online activities distract consumers, who are spending more and more on online games and in-game items. Game companies are attempting to become an entertainment platform, not only to produce games but also to provide places for small-to-medium-sized businesses to operate their games and to create and develop a longitudinal industrial chain of different digital entertainment offerings around a core of video games (actually, the core can be in any medium so long as there is substantial content). These can then be expanded to cover other product lines, for example, game fictions, game expos, films, TV series, cartoons, peripheral products, and so forth.

The Politics of Control

Aside from problems in the market, video games nowadays are also regulated by the state and bureaucracy. The rise of game arcades, consoles, and offline computer games in the 1990s engendered public controversy about the potential impact of video games. Academic studies have documented debates surrounding both negative and positive effects of gameplay (Lee and Peng 2006). Simply put, scholars argue that games can arouse users' aggression with their violent and sexual elements, and these effects may lead to various health problems and game dependence, upset the socialization process, and cause gender stereotyping, among other things (Provenzo 1991; Griffiths 1999; Brenick et al. 2007; Jenkins and Cassell 2008). Due to the negative news coverage and academic coverage of game playing, including child truancy due to games, game addiction, and family conflicts due to gameplay, the government took strict measures to control video games as a response to public outcry.

Though no concrete measures were laid down to regulate the emerging industry, back in the late 1980s and early 1990s, a video game had to go through rigorous censorship before it could be distributed in the market. The logic behind the government's stance was that it protected adolescents from antisocial, violent, obscene, superstitious, and pseudoscientific media content, and imposed a social moral standard, if not ideological control. In the year 2000, the most stringent control over video games, the Ministry of Culture's "Special Administration Proposal for Video Games' Operational Places," was released by the General Office of the State Council (see http://www.chinaculture.org/gb/en_artists/2003-09/24/content_42919.htm). In article 6 of the announcement (see [2000] 国务院办公厅转发文化部等部门关于开展电子游戏经营场所专治理意见的通知 at http://www.people.com.cn/item/flfgk/gwyfg/2000/11201000059N.html), the state commanded that all companies and individuals must stop producing, distributing, and circulating video game devices domestically. This was the infamous ban of console games and arcade games in China, and as of early 2013, the selling of consoles in the national market is still illegal. The entire industry is affected by this policy, which has forced all the production, distribution, and circulation of video games to move online. In 2000, the total sales of online games was around RMB38 million (USD $4,591,312), accounting for 9.8% of the wholesale industry. In the year following the announcement, online game revenues sharply increased to RMB500 million (USD $60,412,009), accounting for half of the game industry's income. In other words, the government was the "invisible hand" that disbanded the console game and offline PC game market in favor of the more profitable and promising online games. Figure 2 illustrates the decline of the offline PC game market.

In line with the ban of video games, the Chinese authorities also released and implemented the following laws and regulations to control them or other related media:

- (2004) *Advice on Strengthening and Developing the Moral Construction of Adolescents from the State Council of CPC Central Committee* (中共中央国务院关于进一步加强和改进未成年人思想道德建设的若干意见). The announcement suggested that new video games should be strictly censored for the good of adolescents' development.

Figure 2

The offline PC game market in china, 2001–2010. Source: CGPA and IDC. See "2005 年中国游戏产业报告" (2005 Chinese Game Industry Report, http://www.cgigc.com.cn/201009/79650588472.html), and "2010 年中国游戏产业报告" (2010 Chinese Game Industry Report, http://www.cgigc.com.cn/201109/109468513257.html).

- (2004) *Notice Forbidding the Broadcasting of PC/Online Games Programs* (广电总局通知:禁止播出电脑网络游戏类节目). Issued by the General Bureau of Radio, Film, and Television, the announcement prohibited traditional media programs that were video game-themed.
- (2007) *Regulation on Digital Publications* (电子出版物管理规定).
- (2008) *Regulation on the Publishing of Digital Publications* (电子出版物出版管理规定).
- (2009) *The Administration of Software Production* (软件产品管理办法). These three rules pertained to the censorship of video games and were enforced in the domestic market. At the same time, the announcements emphasized that copyright laws protected those digital publications.

The final analysis shows that China's reaction to video games is out of an ideological concern. Such new technology is highly related to the permeation of Western capitalist ideology into Chinese society. It is not just a political concern. Chinese society is always vigilant of excessive consumption, or in this case, the addictive nature of games and the money spent on them. Second, China is still quite defensive when considering foreign competitors and investors in their opening market. They realize that the infant domestic market would soon be eroded and conquered by large, competitive, international companies mainly coming from the Western world, pushing local companies to the margin of the market. Eventually, this means economic colonization. Thus, we can see such regulation of video games as not just control over social negative effects, but a matter of economic protectionism, which is always a political agenda in the background.

The Politics of Piracy

The state's policy and regulation not only affected the survival of the video game, but also became a function of the growth of piracy in China. The relationship between piracy and the ban of video games is very similar to the chicken-and-egg relationship; it is hard to say whether the strict control of the market leads to uncontrolled piracy as a consequence of unsatisfied consumer desire, or whether the widespread piracy of video games causes societal anxiety and concern, which results in the state taking charge of the industry. In any case, the central policy hinders the development of copyright law, resulting in lack of protection for the video game product against piracy.

It is difficult to study the early pirated market of the 1990s. Early console companies such as Xiaobawang and Xinxing simply left no formal public record of finance and audit. Comments online include rough estimates of the profits based on calculations of the sales of the Chinese cloned consoles and cartridges: assuming 25 million consoles might have been owned by domestic players (see http://club.pchome.net/thread_1_15_383208___171357.html) and supposing that every console costs RMB 100, one cartridge costs RMB 10, and two cartridges will be purchased along with one console, we then can estimate a total turnover of around RMB 7.5 billion (roughly USD $1 billion). As early as 1996, pirated video games had already spread across China. During that time, every legitimate copyrighted video game cost more than RMB 100, a price that most adolescent players could not afford. From the point of view of the users, a pirated CD-ROM with the same content that cost RMB 5–20 was more affordable. Besides, in the late 1990s and early 2000s, only a few families owned personal computers with hardware that could run those games. Cybercafés and video game centers then became places where video games were played, but it was also these places in which piracy spread. Owners only needed to buy one copy of a game and then offer it to dozens of players. This sharply reduced the profit of legal PC game developers.

During the golden age of Chinese PC games, from 1997 to 2002, legal video games did exist in China on a very small scale. According to Federal Software, the market scale of legal video games in 1998 reached around RMB80 million (about USD $12 million), and most of these games were made by companies affiliated with foreign game enterprises. Among them, about 75% of the video games distributed had no more than 5,000 copies in circulation; however, the breakeven point estimated was around 10,000 copies. Apparently, most of the legal operations did not make a profit at all. This also explains why few Chinese companies would be willing to legally release original video games. Since 2005, ChinaLabs has documented the annual rate of video game piracy in China. A ratio of the number of illegal copies to the total number of installed legal copies is displayed in figure 3, which shows that the piracy rate has, in fact, diminished every year, from 68% in 2005 all the way down to 26% in 2010.

Yet the figure should be interpreted with caution. On the surface, we can see the decline of pirated video games in China; however, the study might not completely account for the total number of pirated games in the market, and it is possible that quite a large number of pirated games went undetected and reached the users directly by clandestine channels. Many personnel in the field even said that the number of pirated video

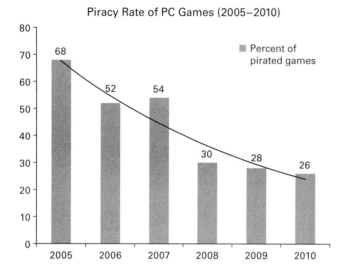

Figure 3

Piracy rate of PC Games, 2005–2010. Source: ChinaLabs, "2010年度中国软件盗版率调查报告" (2010 Chinese Piracy Rate Report).

game copies far exceeded the number of legal copies. Besides, as more people owned PCs and had Internet access in the late 1990s, there would be no way to trace the number of illegal video games downloaded from the Internet.

The severity of piracy basically deterred further investment, game development, and operation of games or proxy servers in which games are stored. Companies had to take necessary measures to curb piracy and increase the confidence of investors and developers. Two strategies were developed and practiced: low pricing strategies and encryption techniques. In 1998, Qiandao released a game called *Heroes of the Marshes: Tales of Upholding Justice* (水浒传:聚义篇), which was sold for only RMB48 (USD $5.80). Later in the same year, a Beijing-based company, Soft-World International Corp., distributed a game named *Fengyun* at RMB19 (USD $2.30). These actions threw the market into disarray, since the prices of legal games were getting close to the prices of counterfeit ones. Game companies originally could not rely much on the income from game sales, and now profits shrank to a level of unsustainability. In 1997, Kingsoft Company's *Jianxiaqingyuan* was sold at RMB128 (USD $15.47), but by the year 2000, the price of *Jianxiaqingyuan 2* (2000) was further reduced to RMB38 (USD $4.59). This pricing war, named after Kingsoft's 1999 campaign, was known as the "Low-Priced Legal Software Campaign" (*Hongsezhengbanfengbao*, 红色正版风暴). It sparked long and popular discussions over the strategy and its implications for the Chinese software market. In the end, however, the outcome of the pricing war was disappointing. The cutthroat strategy forced game companies to operate at a large scale, while only earning meager revenues. It also gave consumers a false expectation that copyrighted games should be sold at prices comparable to illegal ones. The popularity of low-priced games also reshuffled the market. Many small-to-middle-sized companies were ousted because of the huge operation cost, while many that survived turned to publishing foreign games rather than investing in their own local game development.

Another common strategy to tackle piracy was adding encryption to games, including both console and PC games. The situation in PC games seems more serious, as consumers bought a single copyrighted copy of a game on a floppy disc or CD-ROM and copied it onto other discs and CD-ROMs, sharing them with friends. As the potential profit in piracy became evident, large-scale pirated production appeared. Manufacturers even compiled multiple games onto a single CD and sold it at RMB 5, a price just slightly higher than a raw CD. After the year 2000, this form of piracy was replaced by free illegal downloads online. For the users, piracy was now more convenient and less expensive than purchasing CD-ROMs. In response, game companies started to search for stronger encryption techniques for software copy protection. Famous encrypting brands such as StarForce, SecuROM, and DNA were once mainstream and top-ranking copy protection mechanisms that were adopted by many game companies. However, despite stronger encryption, in the end, no single video game could avoid eventual decryption and online sharing. Besides, the complicated and redundant procedures could discourage players who had to decrypt their licensed copies first before playing, further driving more people to turn to easily accessed pirated games (see http://jandan.net/2011/12/06/why-people-pirate-games.html).

The authorities never seem to have a high-profile position in the fight against video game piracy as they do in the fight against the piracy of movies and computer software. Ostensibly, the state responded only to the public call for morality and social health, checking on the growth of video games at the expense of the industry's profit. This is quite an exception when compared to the other so-called creative industries in China. For example, the state came down heavily on industries related to the piracy of movies because of external pressure. The ambivalent attitude specifically toward video game piracy has kept video games from consideration as a legitimate entertainment and cultural product, just like movies, TV dramas, animation, and music, despite the fact that the youth at this time have been captivated by this new and exciting product from the West. Thus, the state's laissez-faire policy over the market actually indirectly led to the formation of a pirated video game market. Such practices are not very different from the emerging practices of the Chinese authorities toward new Western cultural forms or platforms. If the state insisted on the Western conception of a piracy-free market, pirated games would not be so easily found by Chinese search engines or at websites such as Baidu or Tudou.

Rather than tackling piracy, the state suppressed consumption. On June 16, 2000, the party published an article about the potential harm of games on *People's Net*, an official website that represents the ideology of the party pertaining to the curtailing of gaming (see http://www.people.com.cn/GB/paper53/807/105357.html). The article demonstrated how certain students were addicted to playing arcade games, console games, and PC games in Internet cafés and video game centers, arguing that video games were equivalent to "electronic heroin" and spiritual pollutants for adolescents, taking a position that resolutely opposed this "societal scourge." The ultimate effect was a general social panic about games in society. Under great pressure from the party, both central and local authorities soon took concrete measures regarding video game restriction and game center censorship. Several provinces and cities issued rules to prohibit establishment of video game centers (see http://news.sohu.com/91/07/news146340791.shtml and http://news.xinhuanet.com/society/2009-07/29/content_11793240.htm). Using the fire tragedy at the Lanjisu net café in 2002 as justification (see http://news.xinhuanet.com/misc/2002-06/17/content_444681.htm), the state formally imposed age requirements for people in Internet cafés and limited the circulation of video games (Jenkins 2002). Finally, the state simply decided to sacrifice the entire offline video game industry and supported the development of online games by launching the "National 863 Project" in 2003 and investing RMB 5 million (USD $604,120) in online games. The logic is clear; exports of online games that are made in China and that carry traditional cultural Chinese content enhance China's international soft power.

The different treatment given to online games demonstrates that Chinese authorities' ban of offline video games was not entirely based on social concern. As domestic companies are never strong on video game development, apart from cloning and importing foreign video games, the industry ban was largely implemented to halve the proliferation of foreign elements in China. In short, it was done for the purpose of economic and national protectionism.

Gaming Culture and the Audience: Hegemony and Control

The last intriguing question related to China's video game industry is in regard to the role of gamers. It is often said in China that the demise of the video game industry and the piracy issue are essentially due to a unique player culture and context in China. Players in China belong to the Net generation, which grew up with home computers, games, but most importantly, the Internet. Born in the very unique historical period in which China is politically restrictive but economically free, this generation is keen to use digital media and skillful in navigating information, technologies, and platforms. While the formal mass media is tightly controlled, the Net generation relies heavily on the Internet for information and news—and even feels empowered to challenge the ideological control of the state in online forums, something that the state-owned media channels would not allow. Everyday online activity now is simply a ritual for this generation. Thus, it is quite natural that they would make friends online, shop online, and seek entertainment online. Console games or video games that serve as standalone activities might gradually lose their appeal. Besides, as this generation is free to use the Internet for various cultural, social, economic, and political activities, they are not accustomed to paying for video games, but more willing to switch to the free-to-play online games. That explains why online games have replaced offline video games as the major gaming activities of this generation in China.

The trend seems irreversible. The number of "Chinese netizens" is growing (Hauben 2010), and China has already become the world's largest Internet market. According to the Chinese Internet Network Information Center (Kan 2012), in late December 2011, among the huge Internet population, the generation between 10 and 29 years of age reached 55% (26.7% are aged 10 to 19, and 29.8% are aged 20 to 29). It is then quite reasonable to predict that online games will remain the main trend.

In addition, the Net generation also subscribes to the participatory culture of the Internet. This participatory culture in the digital era, as argued by Jenkins (2006), is shaped at the intersection of three trends: (1) new tools and technologies that enable consumers to archive, annotate, appropriate, and recirculate media content; (2) a range of subcultures that promote Do-It-Yourself (DIY) media production, and a discourse that shapes how consumers have deployed those technologies; and (3) economic trends favoring horizontally integrated media conglomerates that encourage the flow of images, ideas, and narratives across multiple media channels and demand more active modes of spectatorship (pages 135–136). Such culture requires them to reuse and reappropriate the content and resources made available by new information technology—a new digital logic that, in effect, challenges the traditional notion of intellectual property rights. The rapid development and spread of these new technologies is often associated with users' violation of copyright; even companies, institutions, and conglomerates increasingly encourage consumers to trespass on the boundary of copyright protection through the promotion of participatory culture and interactive consumerism (Jenkins 2006). For the Net generation, should they adhere to the logic of intellectual properties, thereby refraining from pirating games, they will have to sacrifice the pleasure they derive from video games and other games that they have been playing. They will also feel obstructed and kept from participating in the culture. In the new cultural, economic, social, artistic, and technological contexts, originality and creativity appear to be opposed to protection of copyright, which are regarded as just the monetary interest of multinational companies in trade (Mertha 2007). Thus, fewer players will have the moral concepts of buying and playing non-pirated video games. As stand-alone video games that must be purchased or licensed fail to meet the expectations of the Chinese audience, this business model is doomed to fail in China.

Acknowledgments

This work was fully supported by a grant from the Research Grant Council of Hong Kong Special Administrative Region (Project no. 4001-SPPR-09).

Note

1. For images of the Xiaobawang console system, see https://www.google.com.hk/search?hl=zh-CN&safe=strict&q=%E5%B0%8F%E9%9C%B8%E7%8E%8B%E6%B8%B8%E6%88%8F%E6%9C%BA&bav=on.2,or.r_gc.r

_pw.r_cp.,cf.osb&biw=1280&bih=649&um=1&ie=UTF-8&tbm=isch&source=og&sa=N&tab=wi&ei=aR5cT_bWL-6aiQeeqNSXDQ.

References

Brenick, A., A. Henning, M. Killen, A. O'Connor, and M. Collins. 2007. Social evaluations of stereotypic images in video games: Unfair, legitimate, or "just entertainment"? *Youth & Society* 38 (4):395–419.

Cao, Y., and J. D. H. Downing. 2008. The realities of virtual play: Video games and their industry in China. *Media Culture & Society* 30 (4):515–529.

Clark, Paul. 2012. *Youth Culture in China: From Red Guards to Netizens*. New York: Cambridge University Press.

D. F. C. Intelligence. 2012. Report predicts pc-based game revenues to lead industry, driven by growth in digital distribution among core gamers. June 6. http://www.dfcint.com/wp/?p=338.

Griffiths, M. D. 1999. Violent video games and aggression: A review of the literature. *Aggression and Violent Behavior* 4:203–212.

Hauben, Ronda. 2010. China in the era of the Netizen. *blogs.taz.de*, February 14, http://blogs.taz.de/netizenblog/2010/02/14/china_in_the_era_of_the_netizen/.

Hebdige, Dick. 1979. *Subculture: The Meaning of Style*. London: Routledge.

Jenkins, Henry. 2002. The Chinese Columbine: How one tragedy ignited the Chinese government's simmering fears of youth culture and the Internet. *MIT Technology Review*, August 2. http://www.technologyreview.com/web/12913/.

Jenkins, Henry. 2006. *Convergence Culture: Where Old and New Media Collide*. New York: New York University Press.

Jenkins, Henry, and Justine Cassell. 2008. From quake grrls to desperate housewives: A decade of gender and computer games. In *Beyond Barbie and Mortal Kombat: New Perspectives on Gender and Gaming*, ed. Yasmin Kafai, Carrie Heeter, Jill Denner, and Jennifer Y. Sun, 20–35. Cambridge, MA: MIT Press.

Kan, Michael. 2012. China's Internet users cross 500 million. *PCWorld*, January 15. http://www.pcworld.com/article/248229/chinas_internet_users_cross_500_million.html.

Lee, Kwan Min, and Wei Peng. 2006. What do we know about social and psychological effects of computer games? A comprehensive review of the current literature. In *Playing Video Games: Motives, Responses, and Consequences*, ed. Peter Vorderer and Jennings Bryant, 327–345. Mahwah, NJ: Erlbaum.

Lowood, Henry. 2006. Game studies now, history of science then. *Games and Culture* 1 (1):78–82.

McCall, Thomas, and Rob van der Meulen. 2011. Gartner says spending on gaming to exceed $74 billion in 2011. *Garnter Newsroom*, July 5. http://www.gartner.com/it/page.jsp?id=1737414.

Mertha, Andrew C. 2007. *The Politics of Piracy: Intellectual Property in Contemporary China.* Ithaca, NY: Cornell University Press.

Partners, Niko. 2011. 2010 Chinese online games market annual review and five-year forecast. http://www.nikopartners.com/reports/Niko%20Chinese%20Online%20Games%20Report%20TOC%202010.pdf.

Popsoft Editorial Office. 2008. *The 7th China Computer Game Industry Report.* Beijing, China.

Provenzo, E. F. 1991. *Video Kids: Making Sense of Nintendo.* Cambridge, MA: Harvard University Press.

COLOMBIA

Luis Parra and Global Game Designers Guild (GGDG)

The history of video games in Colombia begins in the 1980s and 1990s, when a group of enthusiastic young people, motivated by classics like the Mario Bros. franchise, started programming short game experiences, recreating games like Tejo, a traditional sport in Colombia. Later on, they would write games using Symbian and Java for the mobile game market. Colombians have had access to the latest games and consoles almost as they appeared in the US market, giving the young enthusiasts the right motivation and role models to push their efforts. These young developers later became the pioneers of an entertainment industry that, although still considered to be in its early stages, is continually learning, growing, and becoming established, and is on its way to achieving international acknowledgment.

From Exploration to the Birth of a New Industry

Two kinds of developers may be identified at the beginning of this story: those who worked at enterprises and developed projects for foreign game studios, responding to their demands, and those who, with great efforts, were able to carry out their own projects independently, making games that in most cases were highly creative but went almost unnoticed. During this first stage, production was naive and disintegrated, without a productive corpus and a representative name.

Some of the Relevant Studios in Colombia

As of 2014, Colombia had roughly fifty registered video game studios; below are short descriptions of some of the most relevant ones.

Immersion Games

The beginnings of the video game industry in Colombia are linked to the curiosity of college students and their research groups. Such is the case of Eivar Arlex Rojas, founder of Efecto Studios, who along with Jesús Cardona created a research group in 2000 at the Universidad Autónoma de Cali, which focused on the study of issues related to virtual reality, augmented reality, and stereoscopy. Their student group developed 3-D applications, specifically, the design of a flight simulator for the Colombian Air Force, a project which, given its magnitude, became the origin of a company focused on the development of virtual interactive services. Immersion Software and Graphics was formed initially by Rojas, Ernesto Galvez, Miguel Posada, and Julian Castillo. The company would later shorten its name to Immersion Games.

Because of the graphical quality of its first projects, Immersion developed a promotional demo that caught the eye of the famous US company Artificial Studios, developer of the "Reality Engine." Thus was born the first phase of the game *CellFactor: Combat Training* (2006), which used Ageia PhysX physics acceleration cards, and thanks to them, Immersion managed to make an excellent technical demonstration that allowed the company to participate in the Game Developers Conference (GDC). Between 2005 and 2006, Immersion developed a small set of six levels called *CellFactor: Revolution* (2007), which can be considered the first representative of the emerging video game industry in Colombia. Once internationally recognized, Immersion had the opportunity to develop *Monster Madness: Battle for Suburbia*, a game for the Xbox 360 and PS3 that, despite the problems encountered because of their relative inexperience, was completed and released in 2007.

In 2008, Immersion helped to develop *CellFactor: Psychokinetic Wars* (2009) along with the Egyptian company Timeline Interactive and with the support of Ubisoft as a publisher. In 2009 and 2010, Immersion worked on the project *Lucha Libre AAA: Heroes del Ring* (2010) for PS3 and Xbox 360. Finally, due to financial problems, Immersion was liquidated. However, with the participation of Eivar Arlex Rojas, Julian Castillo, and Miguel Mateo Rojas Posada, a new game development company, Efecto Studios (http://www.efectostudios.com), was formed, which currently develops games for PC, PS3, Xbox 360, and iOS. The studio's latest creation is *Grabbity*, which was released on August 6, 2012, and published by Mexican media company Televisa.

NDiTeravision

Teravision Games was founded by Enrique Fuentes in 2005 in Venezuela, where it quickly became one of the main developers for entertainment companies such as Disney and Nickelodeon in Latin America, later creating products for video game companies including Atari and Namco, as well.

In 2010, Teravision Games moved its operations to Bogotá, Colombia, where its development center is currently located, and in 2012 a joint venture with the Canadian company NDi Media took place. Currently both companies operate under the name NDi Teravision (http://www.nditeravision.com). The company currently focuses on creating mobile and Web games, with an emphasis on the development of character-based intellectual properties, as it is also an interactive partner for entertainment companies such as Breakthrough, PBS, and Nickelodeon.

C2 Game Studio

C2 Game Studio (http://www.c2gamestudio.com) began operations in 2008, providing simulation services, entertainment, advergames, serious games, and motion-capture. From 2008 to 2011, the studio performed work for some of the largest companies in Medellin, Colombia, including EPM, Nutresa, ISA, El Condor, and SENA—always with the illusion of making games, but waiting for the right time to bet on this industry.

In 2011, a strategic shift took place, as the company decided to produce and market its own games. The first commercial game from C2 Studio was *Cowboy Guns*. It was rated the best iPad game in its launch week and won the national design award "lápiz de acero." In 2012, the game *Nitro Chimp* was released, which received good reviews and was a finalist for several awards including App Circus and "lápiz de acero." Both games were published by Chillingo.

Currently, C2 Studios is developing a new game based on the intellectual property of *Nitro Chimp*. This is the first self-published game and was designed from the start for the free-to-play business model.

Flamin' Lab

On August 1, 2010, Flamin' Lab came to life with just two members, a programmer and a designer, Hermann Vallejo and Felipe Rodriguez (http://www.flaminlab.co). After creating the company Web page and its first Unity engine game, about a character that had to catch a small flame that appeared randomly in a space, Flamin' Lab was officially launched on October 3, 2010.

In 2011, the studio developed six different interactive experiences for one of the most important museums in Colombia, "El Museo del Agua" of the EPM company in Medellin, followed by the development of five other experiences for three more museums. In order to get involved in the iOS market, in November of 2011 Flamin' Lab launched *MagTrap*, its first video game for the iPad, with most of the downloads being made from China. Since then, Flamin' Lab's main goal has been to create video games and mobile apps, and so in March of 2012 it launched *Roach n Roll*, its second game for iOS, which is available for iPad and iPhone.

Finally, in July of 2013, Flamin' Lab became part of the sixth round of the "YetiZen Accelerator Program" in San Francisco, motivating them to move the operation to the United States and continue with its objective of internationalization, expansion into other markets, and growth in the video game industry.

Press Start Studios

Another case study is that of the company Press Start Studios (http://pressstart.co), created in 2011 by Luis Ernesto Parra, Ivonne Tovar, and René Serrato, and initially focused on casual games for Facebook, publishing four titles and reaching over two hundred thousand players in more than fifty countries. Press Start Studios' first published game was *Doña Gloria: The Game* (2011), and despite being based on a local event (see http://www.youtube.com/watch?v=c4XlFdlLddI), the game set a record of reaching one hundred and sixty thousand players in two weeks, all without any marketing budget. In 2012, the company changed its focus to the mobile gaming market with a project called *Drive Me Bananas*, and with this change, Press Start got its first

Figure 1
The logos for Press Start Studios, Flamin' Lab, and the Global Game Designers Guild.

angel investment. In 2013, Press Start opened offices in San Francisco, California, and was accepted into the prestigious video game accelerator YetiZen, which meant an excellent opportunity for recognition in the US market and growth for the Colombian gaming industry.

TN3STUDIO

TN3STUDIO (http://tn3studio.com/en/) is an interactive studio founded by Alvaro Felipe Bacca in 2010. It was not until 2011, however, that new members entered the company and began its consolidation as a developer of digital content in the city of Cali, Colombia. In October 2012, the studio became one of the companies winning funding the CREA DIGITAL grant, organized by the Ministry of Culture and ITC (Information Technologies and Communications). The result was the interactive story "Andrés y la Ballena," which is part of the reading plan of the Ministry of Culture and was published in 2014 for tablets. The studio's beginnings were marked by the development of serious games, educational games for private platforms, interactive digital content, and games with Kinect technology.

Currently, TN3STUDIO works with production companies and public projects guided by the mission statement: "We are based on local awareness and the reality of the country, decid[ing] to make content with added value, focused on fun and creativity, rather than resorting to violence, the easiest argument and one of the most stigmatized for Colombians" (see http://tn3studio.com/en/). This is one of the main reasons why their portfolio highlights interactive stories, adventure games, puzzle games, and casual games in general. Some of the projects released in 2014 were: *Andrés y la Ballena* (for Colombian public libraries, in February 2013, and for tablets, in April 2014), *Kick it: Road to Brazil* (a casual game for the iPhone released in March 2014), and *Groove the Worm* (a PC game released in April 2014).

Boron Studios

Boron Studios (http://boronstudios.com), located in Bucaramanga, Colombia, in the Andes Mountains, was founded in 2012 by Sebastian Castilla, Carlos Arenas, and Oscar Salazar. The studio is currently working on *Aluna*, its first game and first IP, designed for mobile platforms. Recent work by the studio includes work-for-hire for the development of the application *Arkis*, aimed at children between two and four years old.

Kanazú

Kanazú S.A.S. (http://kanazu.org) goes back to Julio Enrique Aguilera Fandiño as coordinator of technology and instructional material at Maloka, an entertainment-learning center in Bogotá, where he directed the production of toys, video games, and interactive learning materials. In 2008, Julio Enrique won the "Lápiz de Acero" award for toy development at Norma and represented Colombia in the first "Bienal Iberoamericana de Diseño." In 2011, Kanazú was nominated for the "Lápiz de Acero" award in the digital category for the video game *Mission Bicentennial* (available online at http://malokapro.org/juegos/fac/). In 2012, thanks to the program of Based Technology Entrepreneurship (EBT) from Colciencias (a government agency that oversees higher education and academic and scientific research), Kanazú S.A.S. was established and that same year won the CREA DIGITAL grant from the ITC Ministry to develop the video game *Edutainment Croanak* (available at http://www.croanak.com). This coproduction features a frog that fights to save the environment.

Kanazú specializes in creating and sharing stories that materialize through video games and toys for education and entertainment. During 2014 and 2015, the studio will launch a line of interactive toys with augmented reality components. It is also developing mobile versions of video games made previously.

A Little Bit of Hardware

Searching for the origins of this industry, we should pause the story in order to mention the case of Carlos Anzola, an engineer who, in the 1980s, began programming small video games on the Commodore 64 but made his real contribution in the area of hardware, with his creation of the HiE-D in 2007. His vision was of a system ahead of its time; a gesture-based interface that recognized body movements without the need for a physical, handheld controller. This new system was named HiE-D (Human Interface Electronic Device), which can be considered a predecessor of the future Project Natal or Microsoft Kinect (see http://www.youtube.com/watch?v=FqNPruJhei0). Indeed, some suggest that Microsoft, which requested a prototype of the HiE-D in 2007, may even be infringing on Anzola's patents with the Kinect (see Benchoff 2011). Since then, Anzola has been improving his prototype in order to obtain the necessary funding to fulfill his dream and put the HiE-D into mass production.

Various Sectors with the Same Goal

Government support for the development of the game industry has been viewed positively by members of the gaming community. However, it is important to note that it is an accompaniment to various initiatives in many areas, specifically public relations and creating events, contests, and other activities. Thus, "This kind of support is invaluable to the growth of the game industry in Colombia," according to Jairo Nieto of Brainz Games, the developer of successful games such as *Vampire Season* (2012).

The year 2006 saw Loop, the first festival of animation and video games, organized by Oscar Andrade, who is a self-described game programmer by hobby and filmmaker by profession. This popular festival was one of the first entertainment media events that allowed the sharing of experiences through the showcasing of products and developments, and it helped to showcase the various players in the emerging leisure industry in 2009. With extensive experience in the audiovisual industry, Andrade highlighted his vision of designers, developers, and the work developed in the country, and it has been an excellent venue for the dissemination of industry experience.

The Colombian game industry received a boost in 2009 thanks to the roundtable discussions of the Telecommunications Research Center (Cintel), a center of technological development in information technology (ICT), which freely exchanged information and whose main objective was supporting the development of the ICT industry through consensus formulation of recommendations on technology, regulations, and business.

In one of these roundtable discussions, an analysis of the digital content industry called "Business Opportunities for Content Generation on Different Platforms" was presented, shaped by different actors in the industry and academia such as Cintel, the Ministry of Technology Information and Communications, the National Television Commission, the Ministry of Culture, Jorge Tadeo Lozano University, and Microsoft Colombia (Cintel 2011). The members of the roundtable discussed the potential for development of the video game and animation industry sectors in Colombia. Ernesto Galvez, the founder of Immersion, was chosen as a guest speaker at the first Forum of Digital Content for this reason. This first forum opened the door to several small studios, allowing them to see the positive growth of gaming and to consider the creation of a particular industry in this sector. From 2011 onward, the industry grew into various areas such as education, government, and industry.

From the perspective of Jairo Nieto of Brainz Games, however, the video game industry really began to evolve when companies such as Apple and Google opened their mobile platforms to independent developers, giving them the opportunity to work without a publisher, allowing Colombian developers to display their software directly to the world.

Associations in Colombia

As for the presence of the associations of developers in our country, it is important to highlight three guilds that have facilitated the strengthening and dissemination of the Colombian industry nationally and internationally: SOMOS, the Game Developers Association in Colombia (IGDA), and the Global Game Designers Guild (GGDG).

SOMOS

SOMOS is a Creative Industries and Digital Content Guild founded in 2009 to establish connections with the government, education, and business sectors. It was founded by leaders of companies including Jaguar Taller Digital, Creative Connection, Caterpillar, ZIO Studios, and Naska Digital. Since its inception, SOMOS has promoted important initiatives, such as the first CONPES of Creative Industries, Digital Content Policy, the creation of the Short Films and Feature Films category of the FDC, and the CREA fund. SOMOS is also in the process of strengthening the industry and has participated in the creation of the Coalition of Creative Industries and Content and other activities that promote the development of the industry. As of October 2013, SOMOS had thirty affiliated companies, six chapters, and was continuing to strengthen its regional presence along with the industry.

IGDA Colombia

The Game Developers Association in Colombia is a well-known nonprofit association established in 2010 through the joint efforts of several companies, among which stand out Brainz (formerly part of ZIO Studios), XOR, Colombia Games, Immersion Games (now Efecto Studios, http://www.efectostudios.com), and NDiTeravision. IGDA Colombia brings together developers, academia, and government offices around the topic of video games and their applications.

The GGDG

The Global Game Designers Guild (GGDG) was created in 2012 by entrepreneurs Luis Ernesto Parra, Ivonne Marcela Tovar, and Rene Serrato from the vibrant company Press Start Studios. The goal of the guild is to discuss, learn, connect, and report on issues related to game design and the game industry in general. These

Colombian entrepreneurs decided to start several projects to create community in Latin America, including newspapers, newsletters, the first award for women in games in Colombia in 2013, brought four startups to San Francisco to a one month educational Bootcamp with YetiZen and another five startups took it online, where part of the leading team of the first LATAM gathering during GDC and podcasts with international guests. They also used social networks to broadcast specialized information about video games, specifically about the design process. This is coupled with an interest in promoting business opportunities for current and future gaming studios and start-ups. GGDG also has various chapters in different regions of Colombia including Bogotá, Medellin, Cali, Bucaramanga, and Barranquilla, and international chapters in Mexico, Australia, and France.

Government Support

Since 2009, these measures of internationalization of the Colombian video game industry have been supported by Proexport Colombia, a government entity, and in January 2012, the Ministry of Information Technologies and Communications of Colombia included the sector in the Vive Digital Plan (Proexport 2013), which seeks to improve infrastructure. Finally, it can be said that the industry's progress has made it possible to create different dynamics of interaction between video game development companies and various government agencies, resulting in sustained growth.

Acknowledgments

The authors would like to thank Eivar Arlex Rojas Castro of Efecto Studios, Mateo Rojas Borrero of Efecto Studios, Óscar Andrade of Jaguar Taller Digital, and Jairo Nieto of Brainz Games and Enrique Fuentes, The Product Guy cofounder/partner at NDiTeravision. We would also like to thank Álvaro Felipe Bacca, Luis Correa, and Julio Enrique Aguilera Fandiño.

References

Benchoff, Brian. 2011. Did Microsoft Steal the Kinect? *Hack A Day*, July 14. http://hackaday.com/2011/07/14/did-microsoft-steal-the-kinect/.

Cintel. 2010. Mesa sectorial "Generación de contenidos." March. http://cintel.org.co/.

IGDA. 2011. *Estudios de desarrollo de videojuegos en Colombia*, en Directorio de empresas. http://igdacolombia.co/desarrolladores/.

Proexport. 2013. *Servicios de Proexport para el Sector Videojuegos*, July. http://www.proexport.com.co.

CZECH REPUBLIC

Patrik Vacek

Though relatively small in geographic terms and population, the Czech Republic has, since the 1990s, played an increasingly important role within the video game industry of the territory known as Central and Eastern Europe. Czech game designers have not only produced commercially successful mainstream titles and received considerable critical acclaim, but they have also been instrumental in creative efforts ranging from mobile phone games to underground or independent titles to modifications of well-known game products.

A history of video games in the Czech Republic is not complete without a closer look at the socio-historic and cultural conditions of the country and wider region of Central and Eastern Europe. Unlike the situation in the United States, Japan, or Western Europe, the Czech Republic (which was an integral part of what was formerly known as Czechoslovakia from 1918 to 1992, a federation of the Czech and Slovak Republics) had been politically separated from the development of the Western world, which had profound effects on the development of popular culture as a whole.

Things we usually tend to consider as taken for granted (such as freedom of press and gathering or traveling abroad) were out of reach for most citizens of Czechoslovakia and—most importantly—the usual consumer goods and things of everyday use and enjoyment such as TV sets, VCRs, cars (other than those of the Eastern bloc), soft drinks, chocolate, clothes, and computers, were distributed in very limited amounts through Tuzex (a contraction of two words, "tuzemský" [domestic] and "export"), a special sort of upmarket chain of retail outlets owned and controlled by the state with its own artificial and relatively hard to obtain currency that served to drain hard currency out of circulation. Ubiquitous grayness and visual monotony (though not so severe as in North Korea or the Soviet Union) did not captivate consumers' senses and served as vehicles of state propaganda and its persuasive communication strategy, full of mediocre visual clichés and ideological conformity.

In everyday life, consumers were never tempted by the visually striking merchandising of renowned department stores, the originality of boutiques (as private retail activity was almost completely suppressed), or other visual splendors of public space. Television commercials and music videos were almost nonexistent or very naïve both in formal structure and the way their subject matter was portrayed. The few areas in which painters, graphic artists, or illustrators could excel were limited to film posters, book illustrations, and typography, and even the most common items, such as food and beverages in supermarkets, automotive spare parts, and shoes, were presented in packages that were far less attractive and less technologically developed than the same stock in the West. While it may sound rather curious, during the mid-1980s, the supply chain of the communist economy was in such an unsatisfactory state that even the most trivial appearances of consumerism such as empty Sprite/Fanta/7-Up cans or single-use colored shopping bags from Western high-street retail could become collectors' items among Czech and Slovak teenagers, simply because of their visual attractiveness and otherness.

This and many other factors contributed greatly to the situation that can be best described as "a popular culture gap"—a situation indicating the state when most of the population experienced very limited exposure to such Western cultural fundamentals as comics and video games. In such a situation, the first appearance of video games was considered a sort of revelation. The real problem was the sparse availability of Western technologies, and as Czechoslovak foreign exchange reserves were precious, companies were encouraged to draw technological solutions from other Central and Eastern European countries. The hardware designs once successful in the Soviet Union, Poland, or Bulgaria frequently served as a base for further development or modification within the rest of the territory.

It is also important to differentiate "video games" from "computer games." While in the United States and Japan the video game is traditionally played on a single-use device such as a gaming console attached to a TV set, Czechoslovakia and the Czech Republic have traditionally preferred the European term "computer game," as the vast majority of gamers have played their games on personal computers (both PCs and older 8-bit and 16-bit devices).

The sole exception to this terminological rule can be seen in an early phase of gaming at the beginning of and during the mid-1980s when trucks were crisscrossing Czechoslovakia packed with older coin-operated machines that attracted great numbers of players from all age ranges, though predominantly teenagers and their fathers. This very first mass manifestation of video games was considered a sort of curiosity and was not encouraged by state officials who had shown little or no respect or enthusiasm for the medium, which was generally thought of as a mere protrusion of Western capitalist culture. The caravans with arcade games traveled mostly along with traditional circuses or were a part of permanent fairs. The cost for a single game of *PONG* (1972) or a classic 2-D platform game was relatively expensive—about CZK 2—while a scoop of ice cream was priced at about CZK 0.70 and an afternoon cinema screening for children at CZK 2. Although the social and cultural status of this activity was very low, it enjoyed an immensely intensive following among young people.

A slightly more advanced option for those who wished to play was some form of hobbyists' or enthusiasts' club formally covered by its technological, science, or programming focus. These extracurricular activities were often supervised by a local socialist organization called The House of Children and Youth (in Czech, Dům dětí a mládeže, or DDM) and presided over by an older student chief. Besides serious tasks such as amateur programming, the members could also use the computers for limited gameplay. The devices used in such groups included mostly Czech or adapted hardware sets such as IQ151 (which served as a de facto standard in the Czech schools) or PMD 85 (which represented the same in Slovakia). Among other game-related computers produced and distributed in Czechoslovakia were the Consul 2717, manufactured by Arsenal Brno, the PP-01 (of which the latest generation was the PP-06, a PC XT clone with two 360KB floppy disk drives), and the Didaktik Gama and its alterations (formerly an improved version of the PMD 85 and later the first machine of a whole series of Sinclair clones made by the Slovak company Didaktik S), which became highly successful and sought after by Czech gamers. At the very pinnacle of the Czech and Slovak gaming experience during the 1980s was the possession of a home computer. Those lucky enough to have bought imported 8-bit computers at Tuzex (such as the Atari 800XL, which became relatively available in the mid-1980s) or one of the ZX Spectrum clones, enjoyed a high social status among their schoolmates or colleagues.

This changed rapidly with the fall of communism after 1989, when Czechoslovakia (and, after 1993, the Czech Republic as a sovereign state) transformed itself into a relatively successful free market economy with the influx of current IT. Still, personal computers, which were increasingly purchased for businesses, schools, and government bodies, were relatively expensive items and so were games. Not only was it necessary to build up the basic hardware infrastructure that would keep up with global development and reduce technological lag, it was also necessary to implement a legal model of software distribution.

As the 1980s in Czechoslovakia represented a period where there were no game producers, distributors, or retailers, piracy was the only option and an everyday occurrence. Those who had legally imported or smuggled in games, mostly stored on cartridges or tapes, provided them quite liberally to players they befriended who gathered at various informal clubs. There even existed special parties (called "copy parties") of gamers who copied the games extensively for this purpose, using a vast array of relatively incompatible tape recorders and self-made utilities.

The Reception and Distribution of Video Games

Video games received a great deal of respectability due to the Czech game magazines that emerged in the early 1990s. Their editors in chief and contributing editors steadily pointed out that the original games—however expensive they may seemed to both parents and children—had clear advantages over black market copies, such as the standard guarantee period, original packaging and manual, and most of all, the continuous user support from the publisher. Another frequent feature in such articles was an opinion proclaiming that by purchasing original games, players could significantly contribute to the inception and further advancement of domestic game designers and their creative efforts. This careful persuasion could be found in editors in chief's forewords, essays, and columns, and was influenced by the interests of newly established game distributors who advertised in the newly incepted game magazines. It is certain that most games were—at least at the beginning of legal distribution in the Czech Republic—far beyond the financial capabilities of a majority of the population. The brand new big-budget and highly anticipated games were sold in different price bands, ranging from CZK 999 to 1,999, which in 1993 represented about a fifth to a third of an average monthly income before taxes (see http://www.czso.cz/cz/cr_1989_ts/0503.pdf, data provided by Czech Statistical Office).

A few months later, the prices of less successful titles started to fall, and after a year or two, the retail prices of the same games were only a half to a quarter of what they were originally, while very successful games maintained their high prices for a prolonged period of time. Such a situation did not favor the legal distribution model, and it improved significantly after the Czech population had satisfied their growing (and previously denied) consumer appetites and salaries started to increase.

During the first half of the 1990s, the ownership of at least one personal computer within a family (most often a secondhand PC XT or relatively expensive AT 286/386) was still more a privilege than a norm, which was why starving gamers tended to use every opportunity to play. As widespread use of the Internet was still a few years in the future, local networks in schools and homes flourished. The number of secondary schools equipped with sufficient game infrastructure rose swiftly, and while most schools used their personal computers and newly founded LAN networks almost exclusively for education in recently established courses of computer and information literacy, there were also after-hours sessions that attracted many passionate players. It is also important to note that the level of school hardware equipment varied widely from big cities such as Prague or Brno to smaller settlements in the countryside. Some schools even permitted students to play in information literacy classes; however, they only encouraged playing those games that were believed to develop the students' sensorimotor and logic skills and knowledge base, such as turn-based strategy games, flight simulators, or adventure games. Contrarily, playing *Wolfenstein 3D* (1992), *DOOM* (1993), and other three-dimensional first-person shooter (FPS) games often provoked more than a teacher's raised eyebrow (because of possible keyboard and mouse damage) while not being directly banned or significantly limited.

Another instrumental place of mass game admiration in the early 1990s was INVEX (the International Fair of Information and Communication Technologies),[1] which was held in Brno, the second largest city in the Czech Republic and a traditional venue for international fairs in Czechoslovakia with a significant international outreach. During the 1990s, this event served as a showcase of the world's latest advances in information technology, and for a few years it was the third largest in the world just after the German CeBit and American Comdex. Although the event was far more concerned with general attendance and business professionals than gamers, a minor part of the show served as an annual catalyst for the video game industry in the Czech Republic. It gradually became a place where the gaming public meets game journalists, designers, and local distribution representatives, while at the same time it also incorporated the top tier of the Czech pro-gaming circuit. Some of the gaming public's favorite events included the throwing of highly attractive new games into the frenzied crowd, massive LAN competitions, and other promotional activities. As time went by and exceedingly rare technological novelties (such as the first Pentium processors) were becoming common and widely available through online retailers, INVEX's importance lessened. Nevertheless, an image of adolescents paired with distributors in suits, both occupying cutting-edge laptops and workstations and loudly enjoying themselves while playing the latest three-dimensional first-person shooter, was a thing of peculiar beauty.

By the end of the 1990s, the structure of game distribution in the Czech Republic was firmly established and a few importers dominated the market, supplying all standard sale channels. With the advent of a new millennium and increasingly widespread use of the Internet, the online retail segment has gained a considerable portion of the game market in the Czech Republic.

The beginnings of systematic reception of game culture in the Czech Republic can be attributed to the early 1990s, when dedicated game journals came into existence. The first significant game periodical, *Excalibur*, started the wave that—at its height—included no fewer than five monthly journals (*Score, Level, GameStar, Official PlayStation Magazine*, and *Doupě* [in English, *Hideout*]) severely competing with each other.[2] Printed journals have become and still are the most preeminent and publicly visible advocate of gaming in the country, while their circulation has been decreasing sharply in favor of a growing number of dedicated websites often backed up by big game distributors or nationwide newspapers.

As media, video games experienced very limited critical exposure before the 1990s, when they were treated mostly as a bizarre technological novelty and something very insignificant. During the 1980s, hobbyist titles such as *Elektronika* (*Electronics*) or *Mikrobáze* (*Microbase*) had noted video games' existence but continued to treat them in a very cautious manner as can be seen in a skeptical editorial in *Elektronika* magazine (no. 4, 1986, translation by the author):

> It is certain that having a Sinclair on your table can lead to a late night until dawn, when one is a scuba diver and chasing pearls or getting eaten by a sea creature, skiing down slopes, or taking the helm of an F1 racecar. Such a program can even be amusing, but it is necessary to realize it originated in completely different conditions. It would pale in a while.

Such unfavorable comments were in accordance with state doctrine, and it could develop to the opposite extreme, lovely examples of which were present in the writing style of video game journals of the 1990s.

The first video game journals were made by players and for players. The editors in chief and contributing editors were recruited from the very core of the first generation of gamers, from the pool of the most experienced and technologically advanced players of the time, who suddenly became influential opinion makers representing two distinctive authors' styles. While the first group favored being highly sophisticated, using the British journal *Edge* as a model, the latter pioneered the aggressive local variant of gonzo-journalism, which—quite frequently—became an extreme of an author's self-exhibition. On the pages of *Score*, a magazine begun in 1993 and the leading Czech game journal of the 1990s, the roles were divided evenly between the editor in chief (Jan Eisler) and Andrej Anastasov (the deputy editor in chief and the most productive author of the magazine). There, the latter quickly gained the indisputable status of a gamer's god (or an Übermensch, at least, whose game proficiency highly outranked that of mere mortals) much adored by readers, while the editor in chief was almost equally recognized for his sophistication and a more family-oriented approach. Behind the leading duo, a group of less eccentric contributors followed, most of them specializing in a certain game genre and continuously cementing their reviewer's expertise in that field.

Such an extravagant authorial style (emphasizing the heavy use of jargon, first-person address, impish smiles, omnipresent digressions, redundant appraisals, nonlinear narrative, and personal impressions of actually reviewed games) was tolerated by publishers since it was perceived as one of the indicators of a newly established democracy, and quite accordingly, it pleased most teen readers while intimidating or even discouraging the remainder—both parents and other journalists.

Potential fears, critical remarks, or accusations of amorality were skillfully articulated on the pages of the "Letters to the Editor" section of every magazine, where they invoked three basic levels of interaction among the readers and "their" magazine. The largest group of writers demonstrated unconditional admiration for contributing editors with an ambition to achieve the same level of expertise. The second category of readers was composed of adults or parents who were expressing their concerns or even horror after having read an article from their son's "innocent game magazine." That group—slowly growing in its importance to the publisher—had more often than not provoked a sort of prankish response both from the editor and other readers who were avowedly showing their reluctance to follow any guidelines, and even more so, the guidelines set by parents. The last group, preferred by the magazine's staff, championed a thoughtful and critical approach to video games, stressing the importance of games while not omitting prospective threats. Such readers were usually rewarded (through the "Letter of the Month" subsection) with warm editorial appraisal and/or a free annual subscription to the magazine or one of the newest games. Quite comically, and to the great astonishment of many elders, *Score* magazine had reappropriated the verb "to booze" (incorporating a noun "boozer" for a passionate player; the Czech verb is "pařit" while the noun is written "pařan"), which served for decades as a colloquial term for a casual drinker in the domestic context.

It is necessary to emphasize that—at least for a few years after 1989—the Czech game scene was considered to be a predominantly male domain (mirroring all gender and status stereotypes) and a sporadic woman's interest, or advanced proficiency in video games was perceived as an absolutely unique and much-applauded affair. This was manifested not only on the pages of game journals, where both editorial staff and readers zestfully welcomed every female gamer but also at other places of social interaction (such as school classrooms and game conventions).

Today, the majority of the Czech population is, however, still somewhat cautious about video games, while newspapers and weeklies tend to be on the protective side, manifesting their attitude through general news coverage (an old myth about copycat crime, the health risks of excessive gaming, etc.), to more or less knowledgeable investigative reports, to some educated contributions from academia or game industry representatives.[3]

A striking example of a biased contribution to video games' agenda and the most discussed journalistic article to emerge in the Czech context was Radovan Holub's polemic published in *Reflex* in 1999, one of the leading Czech weeklies covering social and political affairs. When reading "The Culture of Cripples" ("Kultura kriplů"), one cannot doubt the author's prejudice and incompetence.[4] Holub writes,

> It is most interesting to observe the reviews. In a review it is essential to understand and assess the work's meaning. The review should be attached to an aesthetic ideal. Where we can, however, find an aesthetic ideal in a flood of Elexis Sinclair's [sic] swollen lips, beefy guys with skulls at waistlines and wry mutants? I answer simply: nowhere. And so, the reviews become just an enumeration of technical gadgetry. ... Reviewers address

an indistinguishable, half-imbecile, half-infantile crowd in spite of the fact they gibbet everything moronic, juvenile and embarrassing.

Understandably, the article provoked an enormous response, but we have to add here that there also exist—quite naturally—far less heated features aimed at games and their prospective benefits.[5]

Game Industry Indicators

In accordance with the trend in which game industry or national policy bodies poll their customers on the subject, the Czech Republic has also initiated a representative study of gamers. This was conducted by the Association of Game Industry of the Czech and Slovak Republic (which serves as an umbrella for both importers and distributors of video games in the Czech Republic, with Electronic Arts, Microsoft, and Sony being members) and published under the title *Game Industry in 2011: Basic Information On The Videogame Market In The Czech Republic*.[6] As such, it can be considered an up-to-date, professionally executed, and thus relevant data sheet on the games and gamers within the country.

The report, which is based on a survey of 1,002 respondents, represents the Czech population fifteen years of age and older.[7] The principal data acquired from the research say that 27.8% of the Czech population play video games on personal computers, consoles, or mobile phones, while 63.1% are males and 36.9% females, with age cohorts divided as follows: 15–17 years (9%), 18–29 years (36.2%), 30–44 years (31.2%), 45–59 years (13.3%), 60 years and older (10.3%). Rankings based on families' equipment (the question was: "Does your family own any device which allows the playing of computer games?") have provided these results: 80% of interviewed families are in possession of a high-performance PC, 17% of families have a game console, 19% own a smartphone, 6% use a handheld device for playing games, and the remaining 17% do not own any gaming machine. When asked about the number of games the families have in their possession for all game devices, a slight discrepancy can be observed in the results being gathered because the majority of families (58%) claim they do not own any games (25% of the respondents own between one and five games, 12% have from six to ten games, and just 8% revealed they had purchased between twenty to fifty computer games). If we take a relatively high rate of software piracy into account, this conclusion is something to consider.

The data presented above show that the Czech game population is almost as well-developed as the rest of the gaming world, with a significant dominance of personal computers instead of game consoles. Also, the impression of a relatively high piracy rate is confirmed by further results, which reveal that 24.7% of gamers do not purchase video games (another 14.7% of players do download them from the Internet). When asked about their game habits, Czech gamers' behavior is revealed to be dominated—quite understandably—by online modes of gaming (51%), while offline single-gamer experiences are still quite common (41%).

The frequently discussed issue of games' possible negative impact on children was also reflected in the survey, revealing that a full third (34%) of families allow their children to play fewer than five hours per week. One-fifth of children younger than fifteen years are allowed to play between five and twenty hours per week, and only one in every ten children is allowed to play as much and as often as he or she wants. A mere 2% of children are banned from playing games, while a third (32%) are not interested in gaming.[8] In addition, only 14% of parents are aware of actual PEGI (Pan European Game Information) classifications and use them when purchasing computer games. A vast majority of parents (78%) are not familiar with this rating system.

The report also presents up-to-date information on the state of the game industry in the Czech Republic. From a global point of view, the numbers can seem very low thanks to a limited market size; however, it is important that they continue to grow. While the Czech game market showed an annual turnover of CZK 1.623 billion (about USD $81 million at 2013 exchange rates) in 2008, it reached CZK 1.871 billion (USD $93.5 million) one year later and concluded with CZK 2.189 billion (USD $109.5 million) in 2010.[9] Nevertheless, the sales of game software made up only a third of the total sum in each respective year.

Video Games within the Academic World

Along with the general public and gamers' interest in computer games, digital game culture has also attracted the attention of the Czech scientific community—be it within a basic research realm at universities or in more applied fields on various levels. While the latter area's representatives belong mainly to the fields of psychology, psychiatry, or educational practice, which tend to approach the topic with a certain measure of watchfulness and more often than not point out possible threats and notorious risk factors (gaming being considered a part of gambling, excessive gaming, and poor educational outcomes), an emerging generation of researchers aged between their mid-twenties and late thirties and based at leading Czech public universities (Masaryk University in Brno, Charles University in Prague, and Czech Technical University in Prague) has embraced the opportunity to investigate computer games with enthusiasm.

As is the case with all new fields of inquiry, the role and academic prestige of game studies within Czech academia has yet to be codified. The limitations are, as one would expect, similar to those of the Czech game developers: the small local market and language, a lack of tradition in the discipline, and decades of interrupted or severely limited development in the social sciences and humanities. And while game developers have almost seamlessly joined the world game scene and routinely publish their creative efforts in the English language, Czech game studies have only established themselves and still do not have even a single monograph written in English.[10] It is a pity, then, that Czech game scholars have also experienced quite a limited attendance on the world's game studies conference circuit, a problem originating mainly from a lack of substantial funding provided mostly by the government grant agencies.[11]

After all, despite being unsystematic and sparse, certain successes have been achieved in Czech game studies. There exist authors who produced studies ranging from cultural and sociological overviews of game culture (including Jaroslav Švelch, see http://differentgaming.blogspot.cz/), to intercultural and religious understanding (for example, Vít Šisler—see http://uisk.jinonice.cuni.cz/sisler/), to computer games serving as a medium of expression.[12] Besides academic publication outcomes, there are also a substantial number of works related to knowledgeable authors from other academic disciplines. Their positively biased efforts have often concentrated on promoting computer games as a part of one's development or useful and attractively packed resources suitable for informal education.

A few years ago there also appeared the first courses solely, or almost exclusively, covering the topic. At best, they were incorporated as elective courses within the existing university curriculum, and they adjoin the more traditional programs of film and television studies, visual studies, social history or theory, and the history of interactive media. These courses had, quite naturally, attracted large followings of students, which accelerated the development of the field through their own critical outcome realized primarily through bachelor's and master's theses.[13]

Czech Game Development

From big studios with global aspirations to small creative efforts with local familiarity, the game market in the Czech Republic has it all. Nevertheless, the beginnings were quite unassuming, and Czech game developers had to start from scratch. During the 1980s, creating video games was an occasional activity, a product of pure enthusiasm that lacked both professional skills and commercial background. Among the Czech video games of the era are those that try to imitate or spoof classic titles and —most of all— text adventures that came to life through programming activities in informal clubs or in homes.[14] Some of the best-known titles included text adventures by pioneers of the Czech game culture, including František Fuka (Fuxoft),[15] who, along with Miroslav Fídler (Cybexlab) and Tomáš Rylk (T. R. C.), made up the Czech game programming group Golden Triangle, which created the ZX Spectrum game *Belegost* (1989) based on the novels of J. R. R. Tolkien (see http://www.worldofspectrum.org/infoseekid.cgi?id=0006003&loadpics=3).

The first professional Czech video game was *Tajemství oslího ostrova* (*The Mystery of Donkey Island*), a graphical point-and-click adventure developed by Pterodon Software and distributed nationwide for personal computers by Vochozka Trading in 1994. As the title suggests, it was intended as a parody of the classic *Monkey Island* series, and as such, it attracted favorable if somewhat overrated critical and gamer responses, due primarily to its being the very first Czech professional video game. The former goal of the authors (Jarek Kolář and Petr Vlček) was only to prove their programming skills and to show that Czech game designers were able to create a video game comparable to those being imported. At the same time, there started to emerge other minor creative efforts, but most of them vanished without leaving any significant impact on the domestic video game market.

It was Jarek Kolář who, in cooperation with Michal Janáček, established the firm Pterodon—a game studio that developed titles such as *7 dní a 7 nocí* (*7 Days and 7 Nights*, also known as *7?*) (1994), the second point-and-click adventure that was published in the same year as the first one, and was later released as freeware.[16] The game was produced within six months by two principal programmers and a number of external collaborators.

While not so dissimilar to *The Mystery Of Donkey Island,* the slightly frivolous and ambiguous title (frequently compared to the *Leisure Suit Larry* series) succeeded to generate a sequel called *6 ženichů a 1 navíc* (*6 Grooms and 1 More*) (2003). In 1995, there followed a photo adventure (using scanned photographs) *Ramonovo kouzlo* (*Ramon's Spell*), a slightly educational game with a dungeon-like interface set in the small Czech town of Nové město nad Metují.

In 1998, after a few minor titles such as *Colony Wars* (1996), *Acte Europa* (1997), and *Hesperian Wars* (1998), to which Pterodon game developers had only partially contributed, the firm began working with Petr Vochozka's Illusion Softworks, which resulted in *Flying Heroes* (1998), an action FPS that would be published worldwide by Take-Two Interactive and sell one hundred thousand units. This was, however, just a preparation phase for the studio's greatest success, which came with *Vietcong* (2003) and *Vietcong 2* (2005), solid tactical FPS games with a strong response from gamers (the first game sold more than one million units). After producing *Vietcong 2*, most of Pterodon's staff went over to what has gradually become the most successful and visible game studio in the Czech Republic, now known as 2K Czech.

2K Czech (previously known as Illusion Softworks, now a subsidiary of Take-Two Interactive) is a company based in Brno and Prague that was set up in 1997 and has contributed enormously to the good reputation of the Czech video game industry. After the *Vietcong* series' success, 2K Czech's greatest critical and popular success came with the innovative tactical first/third-person shooter *Hidden & Dangerous* (1999) and *Mafia: The City of Lost Heaven* (2002), the latter selling more than two million units, while the *Mafia* franchise as a whole is said to have shipped about five million units, making it the best-selling video game to have ever emerged from the Czech Republic (see http://www.gamasutra.com/view/news/37228/Grand_Theft_Auto_IV_Passes_22M _Shipped_Franchise_Above_114M.php#.UJAQJmd_3wU). The third major game developer of Czech origin is Bohemia Interactive, a member of IDEA Games and an independent computer and console game developer based in Prague. The firm, founded in 1999, gave life to the highly praised tactical shooter and battlefield simulator *Operation Flashpoint: Cold War Crisis* (2001), which was even adapted as a special combat training simulator called *VBS1* (Virtual Battlespace Systems 1) (2002), of which the United States Marine Corps was the first military customer.

An image of the Czech game industry would not be accurate without mention of the creative efforts that lie beneath both mainstream media and the game community's focus. There exist studios and projects that are either successfully established or about to release their first titles. Also, almost completely outside of business practice stand those who try to use video games for other than commercial purposes (for instance, education, self-amusement, or as a vehicle for delivering some sort of ideology).

It is safe to say that mobile telephone game design in the Czech Republic is flourishing. Qualified sources confirm that there exist between ten and fifteen Czech mobile phone game developer teams that cater to the needs of the global market while successfully maintaining their own existence.[17] A fine example of an independent Czech game developer is Amanita and its *Machinárium* (2009), a game that has sold more than one million units as of 2013, the majority of which were digital downloads.[18] As for its second title, the visually enchanting *Botanicula* (2012), the creators first let players set their own price that they would like to pay, and after two weeks it was fixed by the studio at CZK 200 (USD $10) per download. Within the realm of social networks, a Prague-based company of twenty people called Geewa has earned more than two million daily active Facebook players with its *Pool Live Tour* (2010) (see http://www.geewa.com/). Other growing studios developing video games for mobile devices or networking communities include CraneBalls from Ostrava and Arsenal Brno's Madfinger Games, while all mobile phone game developers rely on distribution channels including the Apple store, Android Market, and Facebook.

In the Czech Republic, video games have existed since political and economic conditions became stable and favorable. The country's game output cannot be as overwhelming or at least as plentiful as it is in some big countries with widespread influence and a significant tradition of industry. However, when talented and creative people received access to the technology and the world market, they tried to offer their best. Some of them succeeded, indeed brilliantly. From a global point of view, twenty years of world game development are equivalent to a century's worth of development in more traditional and less-frenzied fields. Nonetheless, it is good to know that Czech games and gamers are back on track now and enjoying good health.[19]

Notes

1. The convention started in 1990 and peaked both in general audience and exhibitor numbers in the mid-1990s. The last exhibition was held in 2008. A direct successor and a much smaller event has been held annually since 2009 (http://www.bvv.cz/en/invex-forum/).

2. To be precise, the very first Czech video game journal that was founded, even before *Excalibur* was born, was *ZX Magazín*, published in 1988. On October 15, 1989, there appeared a single release of a magazine named simply *Počítačové hry: Informace pro uživatele mikropočítačů* (*Computer Games: Information For Microcomputer Users*). Its circulation reached 72,000, and the issue had thirty-six pages. As such, it was a striking example of amateurism that manifested itself in a letter from the editor that proclaimed the magazine staff had no exact idea about what topics the readers might be interested in. True to this attitude, the "journal" does not include any reviews and the volume is composed of essays, commentaries, and various hardware news and programming hints. The magazine is now available online at http://www.oldgames.sk/mag/pocitacove-hry-1/.

3. In spite of these usual difficulties, video games have enjoyed a dedicated television presence since 1994. First they appeared through the British show *Bad Influence* with limited success because TV NOVA (the first commercial TV station in the Czech Republic) broadcast a year-old series dubbed into the Czech language. The focus of the show was aimed predominantly at SEGA and Nintendo's consoles, neither of which was a gamer's standard in the Czech Republic. It was quickly replaced by the Slovakian show *Level majstrov* (*Level of Masters*) (http://www.youtube.com/watch?v=2Jj8lIaXaPQ), and the development concluded in *Game Page*, a weekly TV magazine broadcast by nationwide and public Czech Television (http://www.ceskatelevize.cz/porady/1095870977-game-page/). The show ranks among the most successful programs, and as of 2013, its twenty-six-minute run is divided into standard features such as news, previews, reviews, and competitions.

Game Page also reflects the traditional distribution model of game platforms in the Czech Republic, while three times a month it covers games for personal computers, and the remaining episode focuses on game console releases.

4. See *Reflex* 33 (1999). The Czech article in its entirety as well as subsequent debates can be found at http://doupe.zive.cz/ohlednuti-za-kulturou-kriplu/a-20046.

5. One of the best efforts to date is a brief anthology covering various popular culture topics within high school classes, with one of the chapters focusing on the use of strategic games in history—a classic example and a good model to follow. See Milena Bartlová, *Pop History—O historické věrohodnosti románů, filmů, komiksů a počítačových her*, Praha: Lidové noviny, 2003, 165 pages.

6. Available for download in the Czech language at http://www.herniasociace.cz/hlavni-stranka/servis-pro-media/herni-trh-v-cr/.

7. Conducted by Factum Invenio. Data was collected in April 2011 through CATI (Computer Assisted Telephonic Interview).

8. The wording of the question, aimed at parents bringing up children fifteen years of age and younger, was: "What is your and your children's attitude towards playing computer games?" And even if most parents supervise their children's game habits and a third of parents proclaim that their children are not interested in gaming, this opinion is not, however, shared by the parents who themselves play computer games. Only 20% of those parents who are also active gamers believe that their children are not interested in computer games.

9. The total numbers include the turnovers realized through sale of computer games for personal computers, the game consoles PlayStation 2, PlayStation 3, Wii, and Xbox 360, and the DSi and PSP consoles. Also, the sales of game hardware platforms are included, with the exception of personal computers.

10. See http://gamestudies.cz/. It is symptomatic that the Web page considered to be a vehicle for promoting game research activities is still lacking an English version. This contrasts with other "non-English-speaking" countries such as Denmark, Sweden, and Finland. However, it is important to note that those countries have a well-developed system of foreign language education and were—contrary to the case of the Czech Republic—spared the Russian language monopoly, whose instruction was mandatory in former Czechoslovakia for forty years of communist rule.

11. It is necessary to say that some researchers succeeded in getting better funding for their game research projects. For example, the author of this essay received funding of about US $2,500, but 90% of the sum was assigned for purchasing literature, while the use of the money to acquire games was not allowed. But researchers had to first have solid argumentation pretending to focus predominantly on the negative aspects of computer games or at least incorporate computer games into existing media research. Second, they had to have good luck with government evaluators. Thus, getting basic game research properly funded is still more the exception than the rule in the Czech Republic.

12. See http://www.iluminace.cz/index.php/cz/article?id=95. As far as I am aware, this is the first volume of any Czech peer-reviewed journal aimed solely at the topic of computer games.

13. The aggregate website www.theses.cz can serve (besides its function as an anti-plagiarism tool for Czech higher education) as a relevant insight into students' game research interests and fascinations. Its bilingual interface provides—through the translated titles and short abstracts only—a brief look at the computer game topics that Czech bachelor's and master's students are investigating in their theses. The access policy differs in accordance with every single university's policy, so whereas some of the Czech universities choose to make their students' final works accessible to everyone, others choose not to do so.

14. For a good overview of Czech text adventure games of the 1980s, see Jaroslav Švelch's study at http://web.mit.edu/comm-forum/mit7/papers/Svelch_MiT7.pdf.

15. See http://www.fuxoft.cz/english.htm. Besides his programming activities, Fuka also authored three short books (*Počítačové hry I a II* [*Computer Games I and II*], Zenitcentrum, 1987 and 1988; and *Hry pro PC* [*Games for PC*], Cybex, 1993) dedicated to video games. The first two can be downloaded at http://www.fuxoft.cz/tmp/dl/Pocitacove_hry_1.pdf and http://www.fuxoft.cz/tmp/dl/Pocitacove_hry_2.pdf.

16. See http://www.pterodon.com/. I mention it here only because of historical accuracy; the site seems not to have been updated since at least 2009 (as is apparent in its news section).

17. According to Martin Bach (vice president of the Game Industry Association of the Czech and Slovak Republic), Czech game developers are going through a sort of renaissance. The proof of this is the fact that domestic developers are producing games for big commercial markets. For instance, the Prague-based Rake in Grass studio pushed its *Archibald's Adventure* (2008) for the iPhone.

18. Even three years after its release, *Machinárium* still sells very well according to the authors, and this title alone would bring in enough income to keep the studio alive—all this without usual forms of advertising, while the authors are said to keep from 80% to 90% of revenue. See http://amanita-design.net/.

19. As in every region, even the Czech Republic has its share of projects that can be considered (from a mainstream perspective) something special, unusual, bizarre, or even weird. Under such somewhat fuzzy labels, we could include some educational games or games aimed at specific needs of a certain community, religion, interest group, or video games promoting various ideologies. As these games are worthy of special research attention and some of them have already been investigated in various students' essays, I mention here only a handful of titles belonging to the group:

   1. *Europe 2045* (2008), according to its website, is "a multi-player, on-line strategy game, designed to be supporting educational material for social science courses, attempting to familiarize players with political, economic and social issues in a united Europe and the present-day world" (while for others it can represent a further example of the European Union's burst into society), is a nice example of a "serious" game developed by experts who also authored a scholarly paper on the subject (http://uisk.jinonice.cuni.cz/sisler/publications/ACM_MindTrek_Europe_2045.pdf).

2. While the Czech Republic is one of the most atheistic countries in the world, the only religious video game was created in Slovakia, the Czech Republic's eastern and more spiritually anchored neighbor. The game, named *David: In The Name of Crowds* (2010), is based on the Biblical account of the battle between David and Goliath and its development was supported and promoted by the conference of Slovakian bishops. A sample of the adventure game, aimed at children aged eight to twelve, can be found at http://www.david-hra.sk/video.

3. The last but not least example of a game developed with a specific purpose is *GhettOut!* (2008) (http://www.ghettout.cz/en/), a title aimed at illustrating the social exclusion of a minority.

FINLAND

Frans Mäyrä

Finland, a sparsely populated Nordic country with 5.4 million inhabitants, has a fast-growing and disproportionately large video game culture and industry. Combining high technology with the spirit of experimentation, Finnish game design has managed to join other Finnish success stories in high technology, such as the Linux operating system and Nokia mobile phones. On the other hand, the public perception of gaming in Finland has often been conflicted. The studies conducted have nevertheless revealed Finns of all ages to be rather active game players.

The earliest game playing in Finland relates to the rich tradition of folk games. Ranging from verbal puzzles to physical outdoor games, this folk tradition also attracted the attention of scholars as early as the eighteenth century. The research carried out in the field of folkloristics also forms the early basis of Finnish game studies (Sotamaa 2009). The rich mythology of oral poetry, transmitted from generation to generation and finally compiled and published as the Finnish national epic *Kalevala* (1835/1849) should also be mentioned, as it has had a long-standing influence on the Finnish cultural imagination. After the Second World War, more Finnish card games and board games started to appear. The most famous of these is *Afrikan Tähti* (*The Star of Africa*), a board game designed by Kari Mannerla and published in 1951. However, there is no direct continuity from the folk game traditions or traditional board game publishing to Finnish electronic games and video games.

The first Finnish electronic game, *Nim*, was designed and constructed by Hans Andersin, one of the team of engineers responsible for implementing the first Finnish mainframe computer project, ESKO, in the 1950s. Inspired by a newspaper article about an earlier American electronic version of the same game, Andersin described the game rules in mathematical logic and constructed the *Nim* game system using relays and diodes in 1955 (Paju 2003).

The earliest popular gaming machines in Finland were pinball machines installed in bars, cafés, and amusement parks. However, the culture of gaming arcades never properly established itself in Finland, possibly partly due to the relatively short history of urbanization and public amusements in Finland. The early video arcade games were installed in bars alongside pinball games and *pajatso* slot machines (Saarikoski 2004, 217). The dominant cultural ethos, rooted in a long religious tradition, was also culturally antithetical toward games and joyful leisure of any kind; the northern Protestant ethic was traditionally very work-oriented and perceived gaming and gambling of any kind as "sinful" (Alho 1981). In 1976, a law was passed that granted the gambling monopoly RAY the exclusive right to control access to the "leisure automatons," which included arcade video games. The aim was to put public gaming under tighter regulation, and in discussions, the rationale for this was based on the supposed addictive nature of gaming and the possibly immoral character of video games. As gaming arcades waned, the door was opened to home computers and home console video games to dominate the Finnish gaming scene.

The cultural tension between work and leisure, or entertainment and utilitarian purposes, surrounded the early experiences with computers and video games in Finland. Jaakko Suominen, a historian of technology, has described how computers and electronics moved from the domains of administration and science to become parts of the increasingly electronic landscape of modern Finnish homes (Suominen 2003). *PONG* (1972) and other early video games in the 1970s were presented as tools for activating passive television audiences; advertisements portrayed games as engaging interactive media that would bring entire families together in the living room. The role of video games and computers proved to become particularly central to the culture of Finnish boys and young men; girls participated in video gaming from early on, but the male dominance in computer clubs and other venues where gaming enthusiasts gathered was often noted (Suominen 1999). Games scholar and developer Sonja Kangas has noted how the gender roles clearly mark technology in Finland as belonging to the male domain, so that most girls learn to shy away from displaying their interest and competencies in games and in information technology in general. However, Kangas also notes that particularly from the mid-1990s onward there have been significant cultural changes in Finnish computer and gaming culture that have begun to allow more flexibility and room for constructing identity as a female gamer, male gamer, or computer enthusiast (Kangas 2002).

Petri Saarikoski, another Finnish cultural historian, has identified three key periods in the formation of Finnish computer culture (Saarikoski 2004, 411):

1. The Microprocessor Revolution (1973–1981)
2. The Home Microcomputer Boom (1982–1990)
3. The Recession Years (1991–1994)

The period of the Microprocessor Revolution was dominated by (male) electronic hobbyists, radio amateurs, and business users who were the early adopters of computers in Finland. Some users utilized assembly kits to build their own computers such as the Sol-20 or Telmac 1800, which were introduced by Telercas, a Finnish company. Games were often perceived as programming exercises during this time. It was during the Home Microcomputer Boom when computers entered Finnish homes and began their process of cultural domestication. Video games were a key element in the popularization of home computers, and young boys in particular emerged as the new "virtuoso" group who developed skills for mastering their full potential. The years of economic recession in the early 1990s in Finland were also the years when personal computers and the Internet became elements in the everyday lives of wider demographics. Dedicated video game consoles started to become popular in Finland only in the mid-1990s when the Sony PlayStation became available. Surveys carried out in the early 1990s reported that only 4 to 7% of Finnish gamers owned a video game console system (Saarikoski 2004, 289). Among the early home video game consoles, at least the Nintendo Entertainment System (NES) gained some popularity in Finland, but as Finnish video game collectors point out, there were great numbers of different, individual video gaming devices that had reached Finland in one way or another in smaller numbers. These included such early systems as the Magnavox Odyssey (1972), Atari Home PONG (1975) and Atari 2600 (1977), as well as later ones including the Mattel Intellivision (1979) and the GCE/Milton Bradley Vectrex (1982) (see, e.g., the collections documented at http://pelikonepeijoonit.net/collection/consoles.php?sort=4). There are no systematic records of which arcade video games have been popular in Finland, but there are mentions of at least games produced by Atari (including *Fire Truck* [1978], *BattleZone* [1980], and *Centipede* [1981]), Namco (including *Pac-Man* [1980], *Ms. Pac-Man* [1982], and *Phozon* [1983]), and Nintendo (including *Mario Bros.* [1983] and *Super Mario Bros.* [1985]) being played in the Finnish gaming arcades during the 1980s and 1990s.

The Commodore 64 Years

The single most important turning point in the history of Finnish game culture was the introduction of home computers, particularly the Commodore VIC-20 (1980), the Commodore 64 (1982), and the Commodore Amiga (1985) in the Finnish markets. The early computer hobbyist and hardware hacker cultures started evolving at that point into a programming and gaming-oriented computer subculture. Finland at this point already had a century-long history in highly ranked education (as later witnessed by the top positions in the international PISA studies), and excellence in such fields as engineering, design, and the fine arts. Computer games and video games appeared as an area where all these strong traditions could be united.

The Finnish home computer culture also supported the evolution of specialized publications such as the wide-ranging, leisure-oriented IT magazine *MikroBitti* (1984) and a dedicated games magazine *C-lehti* (1987), which was later rebranded as *Pelit* (1992). Both publications soon built up the largest circulations of their kind in the Nordic region. The illegal copying, sharing, and cracking of copy protection of video games was an important element in the creation of gamer and computer enthusiast communities; the crack intros (computer animated title screens) designed by the computer software cracker teams soon evolved into demos, or noninteractive real-time presentations of computer art. The gatherings of this demoscene grew into important breeding grounds for the Finnish video game developers. The most important event of this kind, Assembly, has been organized annually since 1992 and still brings together thousands of coders and gamers with their computers, requiring the use of the largest available sports arenas (see http://www.assembly.org).

The first Finnish video games were programmed in the early 1980s to the popular Commodore home computers. Publishing printed programming code in magazines such as *MikroBitti* was an early form of game publishing. The pioneering commercial developers were single, multitalented individuals such as Stavros

Fasoulas and Jukka Tapanimäki, who both achieved international distribution of their Commodore 64 games in the late 1980s. From early on, there was a broad range of genres that Finnish developers were exploiting, including text adventure, action adventure, shooting games, and platform games. Copying or liberal borrowing from successful foreign games (such as *Tetris* [1984], *Elite* [1984], or *Boulder Dash* [1984]) were in ample evidence, but so also was original creativity.

The games of Jukka Tapanimäki provide good examples of early Finnish video game design and programming. Tapanimäki, originally an aspiring designer and a literature student at the University of Tampere, published his first commercial game, *Octapolis* (1987), through a small company called English Software. Prior to that, Tapanimäki had published games as printed programming code in *C-lehti* and *MikroBitti*, where he also wrote articles about game programming.[1]

Octapolis is a genre hybrid that involves both shooting-game and platform-game modes. The backstory displays familiarity with the conventions of science fiction and revolves around the challenge of a sole pilot (representing Galactic Imperium), who alone attacks the eight cities of planet Octapolis. The player must first navigate his or her way into the cities by flying a small spaceship in a split-screen mode, destroying obstacles while simultaneously moving the ship forward and backward, up and down, and left and right (by holding the fire button down while moving the joystick). The platform game mode, which occurs inside the cities, is focused on evasion strategies, since the space pilot is not able to harm any enemies except the "evil eyes," which grant bonus points when shot. The game was positively reviewed at the time (see, e.g., *Zzap!64* 33, p. 28 [Christmas Special, 1987]). The main point of comparison was Stavros Fasoulas's *Sanxion* (1986), a futuristic side-scrolling space shooter published a year earlier. Both shooter and platform game elements were nothing new, but as a demonstration of skills in art and programming by a single young developer, a game like *Octapolis* was a forerunner and set an example for the next generation of Finnish game makers.

Commercial Success: PC and Online Games

The combination of programming and audio-visual design excellence was established as the trademark of the best Finnish video games. Marketing, however, was not the forte of Finnish developers, and commercial success was relatively rare for early Finnish video games. The stereotypical view of Finns being introverted and highly technically skilled yet lacking in social skills is partly a myth, yet it has also been established as a part of the public perception of Finland abroad (Moilanen and Rainisto 2008). Individual game developers working mostly alone during the 1980s rarely had access to major marketing efforts. The first Finnish commercial game development companies, Terramarque and Bloodhouse, emerged from the demoscene and were both established in 1993. The companies developed several games for the Commodore Amiga home computer but without major commercial success. The two companies merged together to form a new firm, Housemarque, in 1995 and focused their joint energies on the PC market that was emerging as an important game development platform at the time. That investment paid off, as Housemarque was able to release

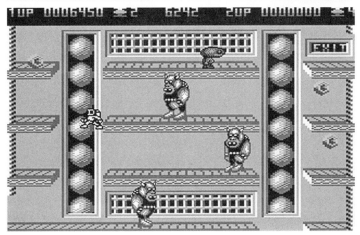

Figure 1

Screenshots from *Octapolis* (1987). (Source: Mobygames.com.)

an internationally best-selling video game, *Supreme Snowboarding,* at the end of the 1990s (Saarikoski and Suominen 2009).

Supreme Snowboarding (Infogrames, 1999; retitled *Boarder Zone* in the US market) was among the first snowboarding games for PC to take full advantage of the 3-D graphics cards that were quickly becoming the norm among PC gamers of the late 1990s. Housemarque designers also utilized their Finnish "snow-how" and demoscene background to design high-speed tracks with visually impressive, dynamic lighting effects. Reviewers praised the game for its realistic snowy environments and special effects, including multiple shadows, reflective surfaces, and weather effects.[2] The game was also easy to control and enjoyable to play, yet repeated criticism was that the game had too limited a number of tracks and was therefore lacking in content when played for longer periods of time. The game music and brand-name snowboards were all designed to tap into the fashionable snowboarding subculture. As a technologically advanced game, *Supreme Snowboarding* was also adopted by hardware manufacturers to demonstrate the potential of the latest PC processors and 3-D graphics cards at various industry events from 1998 to 1999.

The great Finnish IT success story of the late 1990s and early 2000s was Nokia, which established itself as the global leader in mobile phones during that period. The public perception of Finnish game development has also largely focused on games for mobile handsets. There were several mobile gaming companies founded

Figure 2
Screenshot from *Supreme Snowboarding* (1999). (Source: Housemarque.)

during the late 1990s and early 2000s, but historically, sales and international visibility have actually been greatest in the areas of PC and console games. Housemarque, for example, has continued to develop games for multiple platforms, including *Transworld Snowboarding* (2002) for the Microsoft Xbox, *Floboarding* (2003) for the Nokia N-Gage mobile platform, and *Super Stardust HD* (2007) for the Sony PlayStation 3. The close relations to Nokia proved to be a mixed blessing for several game companies, since commissions of games and technology demos from Nokia provided welcome revenue but at the same time, tied their resources sometimes to nonoptimal development environments, including the failed N-Gage.

The most successful release of the early 2000s in Finnish games, or game-related services, was *Habbo Hotel* (2000) by Sulake. *Habbo* is a teen-oriented virtual world focused on chatting and playful interaction in virtual rooms such as restaurants, dance clubs, or in customizable "guest rooms." There are eleven online communities and users logging in from over 150 countries, with the reported average number of unique visitors exceeding ten million per month (Sulake 2012). User-created content is the driving force behind *Habbo*'s success, as young users are active in organizing events and competitions with each other or in inviting each other over to take part in self-designed games or quests (Ylisiurua and Durant 2009). While *Habbo* is free to access, the virtual furniture (or "furni") that is necessary to set up and personalize one's room is purchased using different means such as credit cards, pre-paid cards, or SMS payments. The virtual environment has also been used for promotional activities including visits by musicians such as Gorillaz and Avril Lavigne. In addition, some nonprofit youth organizations have set up their own operations inside *Habbo* (such as "Elämä on parasta huumetta," the Finnish antidrug association). The actual video games implemented inside *Habbo* are typically social and casual, such as *SnowStorm* (2006), a virtual snowball fight. The average age of *Habbo* users ranges from thirteen in Belgium to sixteen in Peru; it appears that in the northern countries, young people prefer to move on to other, more "adult" games and services at an earlier age than in southern and developing countries (KZero 2011). The visual style of *Habbo* and its avatars is based on the 8-bit era of 1980s video games and appears today as an ageless "retro" or nostalgic style, looking back to the origins of digital culture and virtual worlds.

The role of national cultural background and mythology has not been particularly strong in those Finnish video games that have gained international acclaim. It appears more appropriate to characterize Finland as a country whose developers have been skillful in adopting diverse cultural influences and sometimes successfully creating new, technically advanced interpretations of international themes and content.

Remedy Entertainment's third-person shooter game *Max Payne* (2001) is a good example of this. The game is most famous for its slow-motion "bullet time" gameplay mode, where the perception of time is slowed down so that it is possible to avoid bullets and gain advantage over enemies. The visual influences of bullet time can be tracked through the science fiction film *The Matrix* (1999) to the Hong Kong martial arts cinema. The adaptation of slow-moving "battle ballet" to a shooter game was, however, an original innovation by Finnish developers. Even while there was an earlier game with the bullet-time effect (Cyclone Studios's *Requiem: Avenging Angel* [1999]), the adaptation of slow-moving "battle ballet" to a shooter game was carried out in such a stylistically distinctive manner that bullet time in games is today usually associated with *Max Payne*.

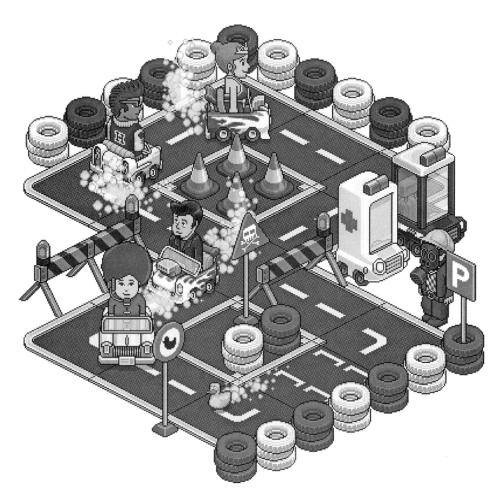

Figure 3
Car racing in *Habbo Hotel* (2000). (Source: Sulake.)

Figure 4
Screenshot from *Max Payne* (2001). (Source: Remedy Entertainment and Rockstar Games.)

At the level of milieu and themes, *Max Payne* is similarly indebted to several international influences. The plotline focuses on Max, a policeman whose wife and newborn daughter were brutally murdered and who is carrying out a personal vendetta, trying both to gain revenge and solve the mystery behind the crimes. There are obvious visual and narrative references to the storylines of hard-boiled detective novels as well as to *film noir* style of crime dramas. The video game also makes use of graphic novel conventions, replacing narrative cut-scenes with comic book-style panels. In addition to the multiple popular cultural references, there are certain mythical undertones in the game; most importantly, the Norse myth of Ragnarök, the end of gods, is used to infuse the game world with an apocalyptic sense of doom.

Max Payne was a major success and was used as the foundation for a franchise including games for multiple platforms as well as a Hollywood film. The publisher, Take-Two Interactive, reported that the game series had sold over 7.5 million copies by 2011.[3] The sequel, *Max Payne 2: The Fall of Max Payne* (2003), was developed by Remedy Entertainment, but the next game in the series, *Max Payne 3* (2012), was developed by Rockstar Vancouver while Remedy focused on developing new intellectual property around *Alan Wake* (2010). *Max Payne* was rewarded with multiple "game of the year" awards and has been discussed by game scholars as a notable example of a video game, which although it perhaps fails to reach the status of "great art," is nevertheless an important artistic achievement (Smuts 2005; Løvlie 2005).

The Rise of Mobile Games and Digital Distribution

Even while Finnish game development's greatest success stories were usually related to PC games, mobile games have remained an important, though often difficult, area to work in. One of the main challenges for mobile game developers in the early 2000s was the fragmentation of the handset market. With variable hardware specifications such as screen sizes and processors, and many different mobile phone operating systems, Finnish mobile game developers had to develop, test, and release each game in literally hundreds of different versions. Access to consumers was channeled through network operators, who decided which games were given the best positions in the operator decks, typically favoring familiar brand names. Consequently, the complex and risk-averse mobile game ecosystem was highly problematic for any real innovation to emerge (Costikyan 2004). There were, nevertheless, some interesting and innovative releases such as RedLynx's *Pathway to Glory* (2004), a turn-based tactical strategy game that allowed real-time mobile multiplayer gaming over the network.

Pressured by their fragmented development and publishing environment and associated risky revenue models, Finnish mobile game developers were eager for alternatives, and Apple was the first to provide a unified environment for developing and distributing mobile games with its iPhone and iOS App Store, the latter released in 2008. Another way of minimizing operational risks for small game studios was through mergers, and four leading Finnish mobile game developers, RedLynx, Mr. Goodliving, Universomo, and Sumea, were all acquired by larger foreign companies between 2004 and 2011. But there are risks involved in being owned by foreign capital, as Universomo and Mr. Goodliving were both closed down by their parent companies sometime after their acquisition. Industry analysts have pointed out the lack of venture capital in the Finnish game industry (Lukkari 2011), but, on the other hand, Tekes, the Finnish Funding Agency for Technology and Innovation, has been active in supporting promising game developers. New digital distribution models constituted the most radical change for video game publishing and sources of revenue in Finland during the latter part of the first decade in the twenty-first century. The iOS App Store, Google Play, Valve's Steam, and the online delivery services run by major console manufacturers have all opened up new possibilities for Finnish games to bypass geographical marginalization and reach worldwide audiences (Sotamaa and Karppi 2010).

No other example captures the full potential of this new distribution model and business ecosystem better than *Angry Birds* (2009), the game series and franchise created by Rovio Entertainment. The strategic puzzle game is focused on the physics of flying trajectories, as the player is challenged to sling different birds toward fortifications of pigs in game levels featuring increasingly fantastic designs. *Angry Birds* was among the first video games to effectively make use of the possibilities opened up by the touch screen interface. The casual enjoyment provided by various physics puzzles is coupled with cartoon-style design and a theme that makes the game and its characters into a distinctive and identifiable brand.

In March 2012, it was reported that different versions of the game had been downloaded more than 700 million times, and that more than 30 million people played *Angry Birds* each day (Barnett 2012). Rovio has expanded its franchise aggressively, and there have been releases of plush toys, a board game, T-shirts, and

Figure 5
Screenshot from *Angry Birds* (2009). (Source: Rovio.)

even dedicated retail stores for various *Angry Birds* branded products. In April 2012, the first *Angry Birds*-themed activity park opened in Tampere, Finland. *Angry Birds* has been called the most successful mobile application in the world, but Rovio has been careful to point out that it created fifty-two mobile games before having a huge hit with *Angry Birds* (Holthe Eriksen and Abdymomunov 2011). Rovio's game was based on the development and marketing expertise groundwork laid by more than a decade of mobile video game business carried out in Finland—for example, the classic *Snake* game for Nokia mobile phones programmed by Taneli Armanto in 1997 (Saarikoski and Suominen 2009, 30–31).

The rise of Rovio and *Angry Birds* is the most well-known success story in the new generation of Finnish game development. At the same time, it also demonstrates the renewed interest in mobile game development. In 2011, it was estimated that almost 40% of all Finnish game developers were designing games for mobile devices. At that point, there were around seventy game development companies in Finland, with an estimated annual turnover of €165 million (Neogames 2011).

Researching Finnish Gaming Culture

Academic video game research in Finland has been active and multidisciplinary and has included theoretical, analytical, and descriptive approaches to games and player experiences, as well as to the social and cultural aspects of video games. Numerous studies have focused on game design, the game industry, and gaming technology issues. Finnish game researchers have also been active in establishing conferences, publications, and international initiatives such as DiGRA, the Digital Games Research Association. Multidisciplinary teams of game researchers have grown in some universities, including the universities of Tampere, Turku (especially the Pori unit), Jyväskylä, Oulu, and Aalto University. In addition to the anthropological, folkloristical, and cultural-historical research mentioned above, work done in educational research and psychology on children's play and games has formed its own tradition in Finland (Piironen 2004).

The overarching view of Finnish gaming culture to date is provided by the Player Barometer series of survey studies (Karvinen and Mäyrä 2009; Kuronen and Koskimaa 2011; Karvinen and Mäyrä 2011; Mäyrä and Ermi 2014). It is based on an earlier series of qualitative interview studies of Finnish game players (Kallio, Kaipainen, and Mäyrä 2007; Kallio, Mäyrä, and Kaipainen 2011) and is aimed at capturing the game-playing interests and behaviors of Finns between ten and seventy-five years of age. Including card games, board games, sports games, and gambling games, in addition to computer games and video games, the survey data reveals active interest in game playing among Finns. In the representative sample of more than 3,300 respondents, some form of game playing was reported by almost everyone (about 98%), and active game players who played a game at least once a month formed a clear majority of 89% of respondents. Digital game playing was most popular among the younger (under forty years of age) generation, but overall more than half of Finns appeared to play some form of video game at least once per month. In addition, during the three-year period, there was a statistically significant increase of active digital game players (from 51% in 2009 to 56% in 2011).[4] Also, the oldest surveyed age group, those between seventy and seventy-five years of age, had significantly increased their game playing, so gaming appears to be growing more popular among older generations as well as the young (Karvinen and Mäyrä 2011).

There is no single video game genre that dominates Finnish video game culture. The most popular game that is mentioned most often as a recently played game in the Player Barometer surveys is solitaire; apparently its ready availability (due to its being installed along with the Windows operating system) explains much of its popularity. Also, puzzle and classic games such as *Mahjong*, *Tetris* and Sudoku are high on the list of games that are played often. Sports games and action games are popular particularly among boys and men, and in 2011 their favorites included the *NHL* series, the *Call of Duty* series, and the *Grand Theft Auto* series. The popular Facebook game *FarmVille* was also among 2011's most-played games in Finland, but the one game that clearly peaked in the 2011 popularity listing was *Angry Birds* (Karvinen and Mäyrä 2011).

The public reaction to video games in Finland has been somewhat divided. On the one hand, the older discourse of worry and negative media effects has proven to be resilient even while society as a whole has increasingly adopted game playing. The other discourse popular in the media has focused on games and the game industry as a potential major national success story. There were two highly publicized school shootings in Finland in the early 2000s—in Jokela in 2007 and in Kauhajoki in 2008—and the media reported that the killers had played first-person shooter (FPS) video games. However, no major media outlet has exclusively blamed video games for the shootings, suggesting that negative perception no longer dominates the public discourse regarding video games. It is already expected that every Finnish young person will have played some form of video game, since the culture and technology of video games have become "domesticated" (Haddon 2003).

As a whole, Finnish gaming culture can be seen as both rooted in national and Nordic cultural history and also immersed in international influences. Enthusiasm for exploring the possibilities for new technologies is a pervasive thread in this development, as well as the increasingly permissive attitude toward new media and popular culture. Tensions surround the rise of Finnish video game culture, but video game development, game playing, and game studies have nevertheless managed to achieve high visibility in Finland.

Notes

1. For more about Tapanimäki's career, see *Wikipedia*, available at https://en.wikipedia.org/wiki/Jukka_Tapanim%C3%A4ki.

2. For reviews, see *GameSpot*, available at http://www.gamespot.com/boarder-zone/reviews/boarder-zone-review-2538262/, and *IGN*, available at http://pc.ign.com/articles/162/162128p1.html. The PC hardware manufacturer relations are mentioned on *Wikipedia*, available at https://en.wikipedia.org/wiki/Housemarque.

3. As reported by *Gamasutra* news, available at http://www.gamasutra.com/view/news/37228/Grand_Theft_Auto_IV_Passes_22M_Shipped_Franchise_Above_114M.php. See also *Max Payne* on *Wikipedia*, available at https://en.wikipedia.org/wiki/Max_Payne.

4. In statistical significance testing, this change produces $p=0.025$ (p values relate to the probability of observed results arising by chance, and values below 0.05 are commonly regarded as statistically significant).

References

Alho, Olli. 1981. Uhkapelit. In *Pelit ja leikit*, ed. Pekka Laaksonen, 104–116. Kalevalaseuran vuosikirja 61. Helsinki: Suomalaisen kirjallisuuden seura.

Barnett, Emma. 2012. Angry Birds "Space" app tops 10 million downloads. *Telegraph*, March 27. http://www.telegraph.co.uk/technology/news/9169117/Angry-Birds-Space-app-tops-10-million-downloads.html.

Costikyan, Greg. 2004. Soapbox: Has mobile game innovation ended already? *Gamasutra*, November 16. http://www.gamasutra.com/view/feature/2163/soapbox_has_mobile_game_.php.

Ermi, Laura, and Frans Mäyrä. 2014. *Pelaajabarometri 2013: Mobiilipelaamisen nousu*. TRIM Research Reports 11. Tampere: Tampereen yliopisto. http://urn.fi/URN:ISBN:978-951-44-9425-3.

Haddon, Leslie. 2003. Domestication and mobile telephony. In *Machine That Becomes Us: The Social Context of Personal Communication Technology*, ed. James E. Katz, 43–55. New Brunswick, NJ: Transaction Publishers.

Holthe Eriksen, Eirik, and Azamat Abdymomunov. 2011. Angry Birds will be bigger than Mickey Mouse and Mario: Is there a success formula for apps? *MIT Entrepreneurship Review*, February 18.

Honkela, Timo, ed. 1999. *Pelit, tietokone ja ihminen = Games, Computers and People*, Suomen tekoälyseuran julkaisuja 15. Helsinki: Suomen tekoälyseura.

Kallio, Kirsi Pauliina, Frans Mäyrä, and Kirsikka Kaipainen. 2011. At least nine ways to play: Approaching gamer mentalities. *Games and Culture* 6 (4): 327–353.

Kallio, Kirsi Pauliina, Kirsikka Kaipainen, and Frans Mäyrä. 2007. *Gaming Nation? Piloting the International Study of Games Cultures in Finland*. Hypermedia Laboratory Net Series 14. Tampere: Tampereen yliopisto, available at hypermedialaboratorio.

Kangas, Sonja. 2002. Tytöt bittien ja pikselien maailmassa. In *Tulkintoja tytöistä*, ed. Sanna Aaltonen and Päivi Honkatukia. Helsinki: Suomalaisen kirjallisuuden seura. http://souplala.net/show/girlIT.pdf.

Karvinen, Juho, and Frans Mäyrä. 2009. *Pelaajabarometri 2009—Pelaaminen Suomessa*, Interaktiivisen median tutkimuksia 3. Tampere: Informaatiotutkimuksen ja interaktiivisen median laitos, Tampereen yliopisto.

Karvinen, Juho, and Frans Mäyrä. 2011. *Pelaajabarometri 2011: Pelaamisen Muutos*. TRIM Research Reports 6. Tampere: University of Tampere.

Kuronen, Eero, and Raine Koskimaa. 2011. *Pelaajabarometri 2010*. Jyväskylä: Jyväskylän yliopisto, Agora Center. https://www.jyu.fi/erillis/agoracenter/julkaisut/verkkosivut/pelaajabarometri2010.pdf.

KZero. 2011. Avg user ages by country for Habbo. http://www.kzero.co.uk/blog/avg-user-ages-by-country-for-habbo/.

Løvlie, Anders Sundnes. 2005. End of story? Quest, narrative, and enactment in computer games. In *Proceedings of DiGRA 2005*. Vancouver, Canada: University of Vancouver. http://www.digra.org/dl/db/06276.38324.pdf.

Lukkari, Ulla, ed. 2011. *Luova raha: Näkökulmia luovien alojen rahoitukseen*. Hermia Oy:n julkaisuja. Tampere: Hermia. http://www.tem.fi/files/29724/Luova_Raha_Nakokulmia_Julkaisu_2011.pdf.

Moilanen, Teemu, and Seppo Rainisto. 2008. Suomen maabrändin rakentaminen. Helsinki: Finland Promotion Board. http://www.imagian.com/kuvat/Raportti_SMKT_2008.pdf.

Neogames. 2011. *The Finnish Game Industry 2010–2011*. Tampere: Neogames.

Paju, Petri. 2003. Huvia hyödyn avuksi jo 1950-luvulla: Nim-pelin rakentaminen ja käyttö Suomessa. *Wider Screen* 2–3. http://www.widerscreen.fi/2003-2-3/huvia-hyodyn-avuksi-jo-1950-luvulla/.

Piironen, Liisa, ed. 2004. *Leikin Pikkujättiläinen*. Helsinki: WSOY.

Saarikoski, Petri. 2004. *Koneen lumo: Mikrotietokoneharrastus Suomessa 1970-luvulta 1990-luvun puoliväliin*. Jyväskylä: Jyväskylän yliopisto.

Saarikoski, Petri, and Jaakko Suominen. 2009. Computer hobbyists and the gaming industry in Finland. *IEEE Annals of the History of Computing* 31 (3):20–33.

Smuts, Aaron. 2005. Are video games art? *Contemporary Aesthetics*, 3, http://www.contempaesthetics.org/newvolume/pages/article.php?articleID=299.

Sotamaa, Olli. 2009. Suomalaisen pelitutkimuksen monet alut. In *Pelitutkimuksen vuosikirja 2009*, ed. Jaakko Suominen, Raine Koskimaa, Frans Mäyrä, and Olli Sotamaa, 100–105. http://www.pelitutkimus.fi/wp-content/uploads/2009/08/ptvk2009-09.pdf.

Sotamaa, Olli, and Tero Karppi. 2010. *Games as Services—Final Report*. TRIM Research Reports. Tampere: University of Tampere.

Sulake. 2012. Habbo Hotel—Check in to check it out! http://www.sulake.com/habbo/?navi=2.

Suominen, Jaakko. 1999. Elektronisen pelaamisen historiaa lajityyppien kautta tarkasteltuna. In *Pelit, tietokone ja ihminen = Games, Computers and People*, ed. Timo Honkela, 70–86. Suomen tekoälyseuran julkaisuja, Symposiosarja 15. Helsinki: Suomen tekoälyseura. http://www.tuug.fi/~jaakko/tutkimus/jaakko_pelit99.html.

Suominen, Jaakko. 2003. *Koneen kokemus: Tietoteknistyvä kulttuuri modernisoituvassa Suomessa 1920-luvulta 1970-luvulle*. Tampere: Vastapaino.

Ylisiurua, Marjoriikka, and Peter Durant. 2009. Virtual tourism of Habbo UK users. MIT6 Media in Transition Conference, Boston, Massachusetts, April 24–26. http://web.mit.edu/comm-forum/mit6/papers/Ylisiurua__Durant.pdf.

FRANCE

Alexis Blanchet

Consideration of the history of video games in France prompts the historian to wonder about the French history of video games and the history of French video games. These two histories cross paths but proceed from different dynamics: on the one hand, around the start of the 1970s, the arrival in France of a new form of entertainment and the development of a globalized industrial, economic, and cultural sector; on the other hand, the emergence of national structures (production studios, editors, a specialized press, and cultural institutions such as museums and libraries) that—dedicating themselves to the creation of video games and to the formation and promotion of the sector—participate in the excitement of this industry and provide a structure for the practices that it generates. What I propose to do in the pages that follow is to weave the two histories together in order to understand how it took only a few years for the video game sector to become a fixture in the French communications and cultural landscape, developing along the way an industrial, communications, and institutional environment on a national scale.

Before the Importation of Video Arcade Games into France: The Awareness of an Absence

The latest research into the video game sector in France during the 1960s indicates that it was virtually nonexistent. According to a 1968 article in the newspaper *Le Monde,* the number of computers in service during 1967 numbered no more than 2,850—and that constituted a 43% increase from 1966. The use of computer technology in governmental and business administrations certainly expanded over the course of the postwar years—its function being to streamline the bureaucratic tasks in which both public and private enterprises were involved—but this does not involve any diversion or appropriation of hardware or software for the purpose of entertainment. There are multiple explanations: the number of computers is considerably smaller in France than in North America, and access to them is strictly limited to a handful of qualified engineers.

The situation was quite different in the United States, where the college-campus culture permitted students to have access to computers costing up to several million dollars. This free access to computers, which made possible the appearance in the US of the earliest sorts of video games, had no parallel in France, no doubt on account of a lack of confidence on the part of the powers that be, in French youth. The war cry of the student uprising of May '68 was all about asserting that young people have a place of their own at the heart of French society.

There were a few rare cases of game programming, but the aim was always to show off the computer's capacities. Paul Braffort, a French mathematician working for EURATOM (The European Atomic Energy Community) in the early 1960s, wrote a *Go-Bang* (or *Go-Maku*) simulation in 1960 on an IBM 1620 and 7090. This is the very first game application seen in France dealing with artificial intelligence research. In a French television documentary (produced by the ORTF, short for Office de Radiodiffusion-Télévision Française) from January 1963, the reporter Roger Louis plays a game of *Marienbad* (a version of *Nim*) against Antinéa, a computer at the Centre National d'Etudes des Télécommunications (CENT). After Louis makes his first move, the computer does likewise and then announces, "I've already won, but you can keep on playing." Likewise, a section of the television magazine *Eurêka* devoted to science news shows a technician from the French electronics company Thomson-Brandt playing *Morpion* (Tic-Tac-Toe) with the help of an optical pencil on a cathode ray tube monitor. When its adversary wins, the machine rewards him by displaying a few sonnets on its screen.

The testimony offered by French programmers who worked during the 1960s and 1970s confirms this impression that computer games were next to nonexistent in France. The arrival, during the latter decade, of the programming language BASIC permitted French programmers to record lists of code and to indulge in games of ballistics simulation, resource management, and arithmetical calculation; one of the programs is called *Le Compte est bon* (*It Adds Up*) and mimics the very popular television game show *Des Chiffres et des Lettres* (*Numbers and Letters*).

L'Automatique de Divertissement: The Emergence of the Video Game Arcade in France (1973–1993)

The French equivalent of the American arcade industry is called *l'automatique de divertissement.* It is an umbrella term for fairground-stall games, pinball machines, and other electromechanical games. These entertainment machines are found chiefly in cafés and at traveling fairs and charity bazaars. The first mention of a video game in the specialized *automatique de divertissement* press dates back to the summer of 1973; in various professional magazines, one finds advertisements for *TV Ping-Pong,* an American clone of Atari's *PONG* (1972) by Amutronics. Offered by various distributors such as Salmon S.A. and Bussoz S.A., it was presented as a "sensational American novelty."

After only a few months of ad campaigns for *PONG* and its clones, arcade video games disappeared from these professional magazines. It seemed to be the release of Atari's *Breakout* (1976) that revived and entrenched a lasting enthusiasm for arcade video games in France. The automated game sector, which up until that point marketed the jukebox, pinball machines, table football, and automated billiards, had a new product: *le vidéo*.

At first, arcade video games were imported from the United States, where an integrated circuit cost the equivalent of five to six thousand francs, and then from Southeast Asia, where copies of games were marketed at the equivalent of eight hundred to a thousand francs. The importation of these counterfeit versions accounts in large part for the rapid increase in popularity of arcade video games in France. At the end of the 1980s, American makers even sent an FBI agent, Robert C. Fay, to France in a futile attempt to dismantle the French network of counterfeit game importers.

Thus at the beginning of the 1980s, the French market was dominated by American imports. In December 1983, the number of arcade games was estimated at 100,000—the same as the number of pinball machines. These games generated a billion francs, or 20% of the total for all automated entertainment platforms: pinball machines, automated billiards, foosball, and jukeboxes. On the European field, the French market for automated games held third place, behind those of West Germany and Great Britain, even though in 1981 there were more arcade video games in France than in Germany.

The French automated entertainment market is represented by authoritative bodies—such as the Confédération Française de l'Automatique (CFA), created in 1982—publishes various professional reviews, and holds annual conventions: for instance, the Forainexpo/Amusexpo was held at the business center of Le Bourget, near Paris, from 1971 to 1999, and was capable of accommodating 10,000 visitors. In 1981, the network was faced with a new tax, totaling between 1,000 and 1,500 francs per machine, imposed on games in all entertainment venues such as game halls, bars, and fairgrounds. Between 30 and 40% of machines—somewhere between 50,000 and 100,000—had to be withdrawn from use. Because the cost of warehousing them was prohibitively high, in the case of some operators, this meant not just warehousing the machines but destroying them (in the presence of a bailiff) in order to earn exemption from the tax.

France remained, nonetheless, one of the most dynamic markets in Europe. The makers of American video games (Atari, Bally, Midway, etc.) and Japanese ones (SEGA, Taito, Namco, etc.) negotiated exclusive deals with various distributors. Beginning in 1975, Atari invested in a factory in the Franche-Comté region, near the Swiss border, in order to shorten the distribution channels and circumvent the time-consuming process of importing its products from the United States. The French makers Jeutel and Stambouli had their own factories where digital circuits were assembled, but it was not until 1982 that the first French attempt at programming an arcade video game was made by the company Valadon Automation. Inspired by the first platform games (such as Nintendo's *Donkey Kong* [1981]), *Le Bagnard* (*Bagman*) was a one- or two-player game that proved to be quite popular in Japan as well as in France. Innovative for its time, *Le Bagnard* produced a series of digitized French exclamations, such as "Aïe aïe aïe!" and "Et hop là," when the player succeeded at certain actions.

In 1992, Alain Prost, the French Formula One champion, joined forces with SEGA, the publicist Philippe Gimond, and three other partners to create the company La Tête dans les Nuages (Head in the Clouds), a network of video game halls based on the concept of Japanese "game centers." (One of the halls, on the Boulevard des Italiens, in the heart of Paris, remained in operation as of summer 2014.) In March 1977, the company went public, presenting to investors a dynamic economic development: 160 halls, income amounting to a

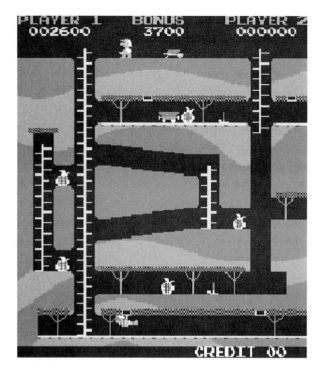

Figure 1

Le Bagnard (Valadon Automation, 1982) was the first (and rare) case of an arcade video game developed in France.

billion francs, and profits of one hundred million francs, in exchange for 35% of the capital. Two years later, the stock had lost 90% of its value, dropping from 170 to fifteen francs. By 1980, the company had lost eighty million francs and had closed five of its twenty-seven halls. In 1999, the company declared bankruptcy. It was one of the rare French attempts to establish a nationwide network of video game halls.

Reduced to bars and cafés, the domain of French automated games was hit hard by the gradual diminution in the number of such establishments in France: from 120,000 in 1980 to 30,000 as of summer 2013. In 1990, 3,500 distributors were prospering; now, in 2013, fewer than 700 remain in operation.

The Home Video Game Market in France, 1976–1984: Dominated by Atari

The first traces of the sales of the first home video game console, the Magnavox Odyssey, date back to 1974, when it seems that several units were imported into France. In a January 1978 report published by the Institut national de l'audiovisuel (INA), it is estimated that 16,000 units—all various models of Atari's Home PONG (1975)—were imported during the year 1976 (Cregut et al. 1983, 29–32). In 1982, the sales volume of the consoles came to 250,000 units; the next year, the total increased by 10%, to 275,000 (Cregut et al. 1983, 33). The American materials designed for NTSC (National Television System Committee) televisions, at the time the dominant analog television system in North America (but also used in some South American and Asian countries) had to be adapted to conform with the French color-encoding system SECAM (short for *Séquentiel couleur à mémoire*), which is different from PAL (Phase Alternating Line), used elsewhere in Europe. This requirement that American consoles be adapted to conform to French norms helps to explain why their importation into France was so belated.

One of the first French video game and game console manufacturers was the Société Occitane d'Electronique. Founded by Jean-Luc Grand-Clément in 1976, this company had a turnover of eight million francs in 1977 and of twenty million in 1978, thanks to a range of products from the Home PONG type such as Occitel, to a microcomputer, Model X1. A Saudi consortium, Tag International, took control of the company by purchasing 51% of its shares; a French bank, Société Générale, purchased 13%, and the founder kept the remaining 36% for himself.

Toward the end of the 1970s, the Dutch consortium Philips took over Magnavox and decided to commercialize its Odyssey^2 console throughout Europe under the brand name Videopac, using various manufacturers as of 1980—the year before the arrival of Atari's French-market consoles. The Odyssey's games were not imported but produced by a Philips subsidiary, La Radiotechnique, in one Paris suburb (Suresnes), and the cartridges in another (Les Châtelets). In 1982, Mattel began to market its Intellivision in France, followed by Coleco's Colecovision in 1983. In the summer of 1982, Atari consoles accounted for 65% of the market, and Philips's Videopac the remaining 35%. The next year, Atari and Mattel dominated the market.

Beginning in 1983, the French home console market, like the American one, experienced a crisis. The arrival on the French market of the first microcomputers —of American, but also European (including French) makes— was seen as the end of the home video game era. The marketing of these machines in France was patterned after the marketing in North America, vaunting their potential as a teaching tool, a medium of artistic creation, and a personal-management system. In 1982, 70,000 microcomputers were on the market; the following year, the number more than doubled to 180,000. In 1983, Thomson's French computers, the TO7 and MO5, were equipped with an optical pencil, offering users a more intuitive relationship with the screen than the usual keyboard.

The French government itself was seduced by microcomputing's potential in regard to education and in 1985 set in motion an ambitious project, "Informatique pour tous" ("Computers for schools"). A budget of two billion francs permitted the purchase of 120,000 Thomson computers to furnish elementary school introduction-to-computer workshops, particularly in BASIC and LOGO programming. The range of Thomson computers (TO7, TO7-70, MO5, and TO8) was also capable of running many video games programmed exclusively for the French market. For example, the video game company Infogrames produced games inspired by the then current success of a certain sort of American sci-fi movie, *Troff* (1983). *Troff* was a futuristic motorcycle race game based on Steven Lisberger's *Tron* (1982), and *I.L. L'intrus* (1984) was a harrowing space adventure game strongly inspired by Ridley Scott's *Alien* (1979).

Computer Games: The First French Software

With the emergence of microcomputing, French video game creation at long last found support for developing its own software. Structures for the development and release of new titles in French for the French and francophone (Belgian and Swiss) markets were created in response to a problem for many French families: the fact that Anglophone software dominated the market. The year 1983 seemed to be when not only the French, but also the francophone video game industry took off: Bruno Bonnell and Christophe Sapet created Infogrames; Laurant Weill and Marc Bayle established Loriciel(s); while Bertrand Brocard and Gilles Bertin joined forces to create Cobrasoft; and Emmanuel Viau and Philippe Ulrich did likewise with their Ere Informatique. In 1984, Elliot Grassiano founded Microïds, a software publisher and development studio. France Image Logiciel (FIL), a software and video game publisher and distributor created in 1985, stood out from the crowd in that it produced translations of foreign titles and handled their distribution in Italy as well as in France. In 1986, in the distribution and game development sector, the five Guillemot brothers, originally from Brittany, created Ubisoft Entertainment. Last but not least, Jean-Luc and Béatrice Langlois and Bruno Gourier found Lankhor, noteworthy for its two bestsellers, *Le manoir de Mortevielle* (*Mortville Manor*) (1987) and *Vroom* (1991).

These French companies enjoyed success with their game adaptations of French and francophone films and comic strips: *Les passagers du vent* (*The Passengers of the Wind*) (Infogrames, 1986), adapted from a François Bourgeon comic strip; *Les Ripoux* (*The Crooked Cops*) (Cobrasoft, 1987), adapted from a popular police comedy directed by Claude Zidi; Infogrames' *Bob Morane* games, adapted from the novels of the Belgian author Henri Vernes; and *Nord et Sud* (*North and South*) (Infogrames, 1989), adapted from the Franco-Belgian comic strip *Les Tuniques bleues* (*The Blue Tunics*) of the writer-artist duo Cauvin and Lambil.

Many of the first French programmers were adolescents and young adults learning BASIC on a microcomputer. At that time, Sylvain Karpf, programmer of *I.L. L'intrus* (1984), was in high school, where he was a member of the computer club. It took him only three weeks to program his game, modeling it on American text adventure games such as *Zork* (1979). After an attempt at local, small-scale distribution in the northern city of Lille, the game was picked up by Bruno Bonnell, who wanted to build a catalog for his future video game company, Infogrames. He reworked the packaging and took charge of the nationwide distribution of the title. Thus, the publishing process for a French video game was modeled on that of book publishing, where the authors met with the publisher and presented a finished product; all that remained to be done was reach an agreement about its distribution.

The "French Touch" (1987–1995)

Toward the end of the 1980s, with the establishment of a French-market oligopoly composed of one British and three American microcomputer makers (Amstrad, Apple, Atari, and AMIGA, respectively), a generation of game designers emerged in France and made a name for itself on the international scene. The emergence

Figure 2
Another World (Delphine Software, 1991), Éric Chahi's most important contribution to the "French touch," deals with sci-fi and fantasy worlds.

of these French talents owed a great deal to Paul de Senneville, a music producer and investor in the video game sector beginning in the late 1980s. Independent creators such as Paul Cuisset, Éric Chahi, and Frédéric Raynal collaborated with de Senneville until the mid-1990s on, for example, *Les Voyageurs du temps* (*The Time Travelers*, 1989), one of the first French games in the point-and-click adventure genre; *Another World* (*Out of This World*, 1991); and *Little Big Adventure* (1992).

The first fruits of the "French touch" are probably found in the games developed by ERE Informatique under the guidance of Philippe Ulrich, who was trained as a musician. His interest in computer science dates from the appearance of the first microcomputers in France. Beginning in 1984, his studio produced its first games for the Sinclair ZX81 and then the ZX Spectrum and the Oric Atmos. Equipped with a team of creative talents including Rémi Herbulot, Didier Bouchon, and Michel Rho, ERE Informatique enjoyed success with, for example, *Macadam Bumper* (1985) and *L'Arche du Captain Blood* (*Captain Blood's Ark*) (1988). The latter won awards around the world and sold very well on the American market. While very strong on the creativity front, ERE Informatique did not pay sufficient attention to its French and international distribution networks, an element vital to its economic viability, and in 1988, the company was bought out by Infogrames. Ulrich moved on to Virgin Games USA, where he and Herbulot collaborated on an adaptation of *Dune* (1992), a combination adventure-strategy video game.

Another unusual video game pioneer was Muriel Tramis, originally from Martinique, one of the very few women involved in video game production. Her trademark, as it were, was adventure game tie-ins to Creole culture. *Méwilo* (1987), the first adventure game that she programmed, in collaboration with the writer Patrick Chamoiseau and the graphic designer Philippe Truca, was set in the Martinican town of Saint-Pierre on the eve of the famous 1902 eruption of Mount Pelée, which obliterated the town. The player incarnates a renowned parapsychologist who has come to investigate a case involving zombies in the residence of settlers knows as *békés*. Also coauthored with Patrick Chamoiseau, her second game, *Freedom: les Guerriers de l'ombre* (*Freedom: The Shadow Warriors*) (1988), was a tactical role-playing game that addressed the issue of slavery in the context of an eighteenth-century plantation. Tramis continued her work on the development front at the studios Coktel Vision and Tomahawk; she explored various game genres, from erotica (*Emmanuelle* in 1989, *Geisha* in 1990) to puzzle adventures, such as *Gobliiins* (1991) with Pierre Gilhodes, before specializing in educational games later in the same decade.

In 1988, Paul de Senneville founded Delphine Software International (DSI), Delphine being the name of his first daughter; his codirector, as well as lead designer, was Paul Cuisset. The company set up shop in Paris near the Champs Élysées. DSI, which belonged to the group Delphine, was a publisher primarily of software for microcomputers. During the previous two decades, de Senneville had made a career for himself as a composer and producer, collaborating with Olivier Toussaint on songs for French pop singers such as Michel Polnareff, Dalida, and Mireille Matthieu. Toward the end of the 1970s, he shifted to instrumental music, composing for the pianist Richard Clayderman. These compositions sold well: one hundred million copies worldwide.

The studio's first game, *Bio-challenge*, created in 1989 by Paul Cuisset for Atari ST, was singled out for its technical excellence, its polished soundtrack, and its sci-fi atmosphere. Delphine Software achieved fame later the same year with the point-and-click *Les Voyageurs du temps* (*The Time Travelers*), the product of a collaboration between Paul Cuisset and the very young Eric Chahi, who already had some experience in video game development and had worked for the account of Loriciel and Chip, two French producers. The two men had a taste for erudite and sophisticated science fiction, such as Frank Herbert's *Dune* saga and Stanley Kubrick's *2001: A Space Odyssey* (1968). The project was guided by their desire to produce a very cinematic game. The emphasis was on *scènes cinématiques*, or cut-scenes—Cuisset had just established a department at Delphine devoted to them—and on game backgrounds designed by Chahi. The critical reception was enthusiastic, and the studio built on it, little by little.

In 1991, the studio achieved international renown with Erich Chahi's *Another World* (*Out of This World*). The studio offered Chahi two years' time to create an adventure game that featured more technical innovations than ever, for the purpose of conveying a melancholy, refined aesthetic. He chose to use vector graphics to maximize the fluidity of the animation, and at the same time created an instantly recognizable visual style. *Another World*, with its physicist hero and its parallel universe, was just the sort of fanciful fiction favored by the studio. An international bestseller with 400,000 copies sold, the game was offered in formats adapted to platforms other than computers, including Nintendo and SEGA's 16-bit game consoles. The vast distribution of the game contributed to Chahi's international renown; many esteemed Japanese game designers, including Hideo Kojima (*Metal Gear Solid*, 1998), Fumito Ueda (*ICO*, 2001), and Goichi Suda/Suda 51 (*Killer 7*, 2000), cite *Another World* among their games of reference and Eric Chahi's oeuvre as a major influence on theirs.

Delphine Software attempted to sustain this international success with *Flashback* (1992), heir to *Another World*. Produced by Cuisset, this new game incorporated technical options that were very innovative for the time, such as 3-D cut-scenes and the use of digitized voices in the game's footage. To improve the animation's realism, the Delphine team applied the technique of rotoscoping—previously used by Jordan Mechner on *Prince of Persia* (1989)—for the animation of the characters. With 1.25 million copies sold, *Flashback* was the biggest French video game hit of the early 1990s.

In 1988, Frédérick Raynal made a name for himself with his graphics for *Pop-Corn,* a *Breakout*-style game, distributed for free. He met Infogrames' Bruno Bonnell, who hired him to produce the horror-adventure game *Alone in the Dark* (1992), inspired by H. P. Lovecraft's stories. Raynal circumvented the technical limitations of the period's microcomputers by choosing to animate elements worked out in 3-D against independently prepared 2-D backgrounds. This technique required that the camera angle be fixed differently for each background, a constraint that, paradoxically, intensified the visual drama. With *Alone in the Dark*, Raynal invented a trick that inspired the Japanese survival horror genre of the second half of the decade. Following a disagreement with Bruno Bonnell, Raynal left Infogrames in 1993 to found (with the support of Paul de Senneville) Adeline Software International, where he was joined by a number of other Infogrames veterans: Yaël Barroz, Didier Chanfray, Serge Plagnol, and Laurent Salmeron. In 1994, Adeline Software came out with a game of its own, *Little Big Adventure,* which sold 350,000 copies.

After the success of *Another World,* Eric Chahi, along with a dozen other former Delphine software employees, created Amazing Studio in 1992. His new project, titled *Heart of Darkness,* was six years in the making, and, therefore, when at long last it appeared on the video game production scene in the late 1990s, it was an anachronism: a classic 2-D platformer in an industry that three years prior had begun the transition to 3-D graphics. Chahi, exhausted by the production of the game and the management of a major team, decided to absent himself for a few years from the world of video game creation. The days of small-scale developments in computer technology had come and gone.

Michel Ancel was the last important representative of this generation of software creators; he reached his acme in 1995 with the release of the first *Rayman*. His creative talents noticed in the late 1980s by the Guillemot brothers, Ancel started programming games for Ubisoft. In the early 1990s, he became interested in the computer graphics technique of ray tracing, and his research into character animation culminated in the creation of *Rayman*, the name an allusion to the aforementioned technique. The game, with a budget of 15 million francs, was developed under the supervision of Serge Hascouët in Montpellier for the Atari Jaguar; it was released on the Sony PlayStation in 1995, at the time of the latter's North American and European launch. The critical reception in both Japan and the United States was enthusiastic, and by the end of 1995, 400,000 copies had been sold in Europe. By the end of its career, the total number was in the millions, and its bouncy little eponymous hero was renowned worldwide.

The expression "French touch" was used by the Anglophone press, first in English reviews and then elsewhere, to describe a quality particular to video games that emerged from French studios. One may wonder whether its application to French video games was not due to its use somewhat earlier, at the start of the

1990s, in reference to French design and electronic music. The term refers to characteristics that are at one and the same time aesthetic, technical, and thematic. The French video games of the 1990s offered sophisticated visual worlds, a firm cinematic anchorage, and a technical mastery in matters of programming and technological innovation. Its universes of predilection were those of science fiction and fantasy, guaranteeing them worldwide distribution. The specificities of French national culture were pretty much avoided in the "French touch" productions of this period. Paradoxically, this international appreciation of the "French touch" of French video games arrived in tandem with a progressive internalization of the French studios.

France and the Globalization of Video Games: Infogrames and Ubisoft

In 1992, Philippe Ulrich, along with Jean-Martial Lefranc and Rémi Herbulot, founded the company Cryo Interactive Entertainment and developed hit games such as *Megarace* (1993), *Lost Eden* (1995), *The Raven Project* (1995), and *Versailles 1685: Complot à la Cour du Roi Soleil* (*Versailles 1685*, also known as *Versailles: A Game of Intrigue*) (1996), a historical adventure video game the aim of which was to make cultural history appetizing by adding the sweetener of gameplay. The game was produced in collaboration with the Réunion des musées nationaux (RMN), a publicly funded organization in charge of all the national museums, including the chateau of Versailles. Cryo's series of historically themed point-and-click games—*Egypte* (1997–2004) and *Chine* (1998)—provided yet another opportunity to collaborate with the RMN.

These historical games, with their aim of whetting players' appetites for their national heritage, constituted one of the rare attempts to introduce French cultural content, including museum collections, into video games. In 1999, after a successful initial public offering on the French stock market—in a few months the price increased twentyfold—the company fell victim to the bursting of the Internet bubble in 2002 and was obliged to slash its workforce by 80%. In 2007, after a buyout by JoWooD, the last remnants of Cryo's holdings were liquidated. As for Coktel Vision, in 1992 it became a subsidiary of Sierra On-Line, which marketed its titles in the United States. In 1999, after a buyout of Sierra by Havas, Coktel Vision was incorporated into Vivendi Universal Games. In 2005, Mindscape bought the company from Vivendi, at which point the studio's activity as a game developer ended.

In the mid-1980s, Infogrames had undertaken an ambitious acquisition policy, buying up two British companies (Ocean Software and Gremlin) and five American ones (Accolade, GT Interactive, Paradigm, Hasbro Interactive, and Shiny Entertainment). These buyouts constituted a heavy financial burden for the group; in 2002, with a debt of 580 million euros, it saw its stock's price—at 5,106 euros in June of 2000—plunge to 80 cents. And with the purchase of Hasbro Interactive on May 7, 2003, Infogrames bought what remained of Atari.

In 1997, Adeline Software International was picked up by SEGA and rechristened No Cliché. Between 1999 and 2001, under the direction of Frédérick Raynal, the studio developed three games for the Dreamcast, most notably *Toy Commander* (1999), marketed in Europe as a launch title. When, early in 2001, SEGA decided to cease production of Dreamcast, No Cliché was obliged to close shop, thus abandoning an ambitious survival horror project, *Agartha*. The game aimed to deal with moral questions, the player having to choose multiple ways between good and evil to complete the adventure.

Also in 1997, French game designer David Cage (his real name is David De Gruttola) founded the Paris studio Quantic Dream. In the course of its fifteen years of existence, the studio worked on only four projects with four different producers but managed to earn an international reputation on account of its mastery of the process known as motion-capture and the thought that the game designers put into the emotional aspect of the games. Late in 1999, Cage released his first project, *The Nomad Soul* (*Omikron: The Nomad Soul*) published by Eidos. Adapted from a novel that he wrote, it is set in an open, futuristic world. The game is noteworthy in that David Bowie lent not only his voice to the soundtrack but also his features to two of the characters. Six hundred thousand copies of the game were sold, prompting plans for a sequel, but it was never completed.

Cage's next project, the adventure game *Fahrenheit* (*Indigo Prophecy* in the United States), was released by Atari in 2005. Here, Cage indulged his ideas regarding interactive fiction: the video game as a medium for adults, and one allied more closely than ever with movies. The soundtrack was entrusted to Angelo Badalamenti, a film composer most notable for working with David Lynch; the voices used in the French-language version are the very ones used for the dubbing of the Hollywood stars Keanu Reeves, Angelina Jolie, and Will Smith. *Indigo Prophecy* is a paranormal thriller set in New York City. The game is a cinematic interactive action game that engages the player in an investigation of his own identity.

In the first decade of the new millennium, Quantic Dream begins to work exclusively for Sony and its PlayStation 3, first on the game *Heavy Rain* (2007), then on *Beyond: Two Souls* (2013). To render facial expressions and body movements, the studio made use of full performance-capture and motion-capture. In 2013, after several resounding pronouncements regarding not only the deplorable state of contemporary video game creativity but also his own conventionality and immaturity, Cage defended the ideal of restoring the art of acting to the heart of video game creation in order to provide new forms for emotional expression. Ellen Page and Willem Dafoe were among the actors cast in *Beyond: Two Souls*, the studio's latest game.

A Case of International Development: Ubisoft

Following the international success of *Rayman*, Ubisoft began undertaking ambitious new projects, all of them in conjunction with various pop culture media: movies, television, and comic strips. Ubisoft started with French cinema, from which it drew various popular characters and universes: *Les Visiteurs* (Jean-Marie Poiré, 1992) and the *Taxi* series, produced by Luc Besson, became the basis of video games made for consoles and PCs. The *Taxi* series had a triple advantage: it met Hollywood's requirements for an entertainment movie; its car chase theme transferred smoothly to the video game medium (and more specifically, the racing game genre); and it had in place a European distribution network, which offered games access to territories beyond the French market.

From 2000 on, the acquisition of movie franchises became not only more deliberate, but also reached far beyond France's borders—more specifically, to the output of Hollywood's studios: animated cartoons for family consumption (such as *Gold and Glory: The Road to Eldorado* [2000], and Disney's *Dinosaur* [2000] and *Tarzan Freeride* [2001]), foreign movies that were American box office successes (*Crouching Tiger, Hidden Dragon* [2003],) action films and other blockbusters produced by American studios (*Charlie's Angels* [2003], Tim Burton's *Planet of the Apes* [2001], and *The Sum of All Fears* [2003, tied in with Tom Clancy's franchise]), and runaway production on a vast scale (such as Oliver Stone's *Alexander* [2004]). The French producer multiplied his offensives in the realm of popular movies. Alongside these movie franchises, he drew content from television series with ever-growing audiences: from *Sabrina, the Teenage Witch* (2002), and *V.I.P.* (2000–2002) with Pamela Anderson, this acquisition policy widened its scope to include *CSI: Crime Scene Investigation* (2003) and its two spinoffs, *CSI: Miami* (2004) and *CSI: New York* (2008). In parallel, Ubisoft became involved in the production of video games based on popular European comic strips, such as *Largo Winch* (2002) and *XIII* (2003).

At the start of the new millennium, a milestone was reached when Ubisoft acquired adaptation rights for Peter Jackson's film *King Kong* (2005). Up until that point, the French producer had been unable to compete with his American rivals, such as EA and Activision, when it came to rights to the biggest box office movies of the start of the decade; but the reputation of Ubisoft's teams of developers, managed by Michel Ancel, seems to have persuaded Jackson himself to entrust the adaptation of his film to Ubisoft. The game *King Kong* will be the first film adaptation born to a new generation of game consoles, thanks to its having been developed for Microsoft's Xbox 360. In order to intensify the sensation of immersion in the fiction, the development teams decided to eliminate from the screen any and all graphic interface providing information as to the player's state of health or ammunition supply. Furthermore, the video game formula used for this adaptation, a mixture of two genres, first-person shooter (FPS) and 3-D action, invites the player to incarnate, in alternation, Jack and Kong. It is the aim of the adapters to render every facet of the film, especially its ability to keep the audience's emotional identification flip-flopping between man and beast.

King Kong and Ubisoft's international expansion placed the company in an excellent position for winning adaptation rights for other hit movies, which in turn guaranteed that it would reach its ambitious sales goals. The company is particularly fond of 3-D animation, making close adaptations of many Sony Pictures Animation titles: not only *Open Season* (2006), *Surf's Up* (2007), and *Cloudy with a Chance of Meatballs* (2009), but also *Beowulf* (2007) and *Arthur and the Revenge of Maltazard* (2009). In an increasingly competitive environment, and especially one in which video game producers are offering more and more titles, the adaptation is a value haven, guaranteeing as it does a certain number of international sales.

In 2009, Ubisoft acquired adaptation rights for the most ambitious movie project of the decade, *Avatar* (2009), directed by James Cameron. The movie was considered a major event on account of both Cameron's reputation and its technological sophistication; within only a few weeks its worldwide box office sales put it ahead of all the competition, including Cameron's own *Titanic* (1997). Several models of game elements, such as vehicles and sets created by Ubisoft for the game, have been incorporated into the film. The concept of a fictional universe that owes its existence equally to movies and to video games seems to find a new and solid method of collaboration, whereby graphic and visual elements circulate between the movie and the video game. The video game's sales, however, were disappointing, and so Ubisoft decided to shift its center of balance to the marketing of its own franchises (including *Prince of Persia*, *Far Cry*, and *Assassin's Creed*) in a variety of formats, such as television, graphic novels, and comic strips (both print and digital).

Since the start of the decade and in parallel with this aggressive policy regarding movie industry intellectual property (IP), Ubisoft has undertaken to develop its own franchises. In recovering ownership of *Prince of Persia*, in 2003, the company relaunched a cycle conceived in its Montreal studios. The company also contracted with Tom Clancy for five video game franchises: *Rainbow Six* (1998–2013), *Ghost Recon* (2001–2012), *Splinter Cell* (2002–2013), *EndWar* (2008), and *Hawx* (2009–2010). In addition, Ubisoft's Montreal branch handled two franchises important to the studio, assuming responsibility for the *Far Cry* series in 2005, and in 2006 creating the *Assassin's Creed* franchise under the aegis of Patrice Désilets.

In 2003, Michel Ancel released the game *Beyond Good & Evil*, a big critical success but only a moderately commercial one. He also participated in the creation of one of the launch titles produced for the Nintendo Wii, and *Rayman*'s spin-off, *Les Lapins Crétins* (*Raving Rabbids* [2006], which became an independent series and a casual gaming favorite. Ubisoft's interest in the Wii since 2006 led to a partnership of mutual trust: every Paris visit on the part of Nintendo designer Shigeru Miyamoto, the official purpose of which is to promote Nintendo IP, is also an opportunity to meet with Ubisoft's directors at their Montreuil headquarters. Ubisoft's current policy seems to be focused entirely on the transmedial marketing of its own IP, converted into print and digital comics, novelizations, animated movies, and television programs.

Henceforth, Ubisoft figures among the top third-party video game producers worldwide. By the end of the decade, the company had twenty-six studios in nineteen countries, from France to Canada by way of Morocco, Romania, and China, which represented 7,450 employees, 6,250 of them in production. Since 2007, the company's turnover has been, on average, around a billion euros (Ubisoft Entertainment 2011).

The Diminution of French Video Game Creativity, 2000–2010

By the end of the 1990s, the French video game sector, consisting of 1,200 businesses employing 25,000 people, was the fifth-largest video game industry in the world. In May 2002, tech stock prices plunged worldwide, and the bursting of the Internet bubble took with it, in a matter of just a few months, three of the most prestigious French studios: Kalisto Entertainment (*Dark Earth* [1997], *Nightmare Creatures* [1997]), No Cliché (*Toy Commander* [1999], *Toy Racer* [2000]), and Lankhor (*Vroom* [1991]).

Between 2003 and 2013, the French video game industry shrank by 80%; with only 240 companies and 4,800 employees left, it dropped to seventh place worldwide in terms of overall size and strength. According to calculations made by the Syndicat National du Jeu Video (SNJV), French studios employ on average twenty people. The sector is obliged to maintain this post-bubble modus operandi, one of extreme austerity, because the majority of the studios develop material for virtual platforms: only 7% are involved in the fabrication of video game hardware.

In the totally digital mobile device and social games market, France is the world's third-ranking producer, with modest budgets and modest turnovers, but greater profitability than that of the video game console market. Bulkypix (*Hysteria Project* [2009], *My Brute* [2009], *Saving Private Sheep* [2010]), Pastagames (*Maestro* [2009], *Pix'n Love Rush* [2010], *Rayman Jungle Run* [2012]), and The Game Atelier (*Flying Hamster* [2010]) are among the decade's standout studios. On the other hand, along with their numbers and their size, the viability of French studios has diminished; according to the SNJV, after five years of activity, the number of closures is 30% higher in France than in England and Germany. There are complaints about how the heavy tax burden puts a brake on structural development, limits game developers' compensation, and drives them out of the country. And, in fact, many students trained in France, as well as young professionals, do emigrate, primarily to francophone Canada—more specifically, to Montreal.

Despite these economic constraints, however, by the start of the next decade, French studios signed more and more contracts with international producers: Capcom signed DontNod's AAA-game project *Remember Me* (2013) (introducing a futuristic neo-Paris), SEGA produced the Paris studio Arkedo's game *Hell Yeah! Wrath of the Dead Rabbit* (2012), and Nintendo granted Ubisoft privileged-partner status (*ZombiU* [2012]). Likewise, American producers bet on French projects: Bethesda picked the Lyon studio Arkane and its FPS *Dishonored* (2012), while THQ/Disney and Microsoft went with the Bordeaux studio Asobo (*Ratatouille* [2007], *Toy Story 3* [2010], and *Kinect Rush: A Disney Pixar Adventure* [2012]). Moreover, if the "French touch" label is no longer trusted as a guarantee of creativity *à la française*, many French studios—such as Eugen Systems, with its real-time games of strategy (*Act of War* [2005] and *R.U.S.E* [2010]); Cyanide, with its cycling-management videos (*Le Tour de France* [2011] and *Pro Cycling Manager* [2005–2013]); and Ankama, with its MMO (massively multiplayer online) games (*Dofus* [2004] and *Wakfu* [2012])—remain the internationally respected standard in the realm of simulation.

The Specialized Media: The Press, TV, and the World Wide Web (1983–2012)

The first French magazine devoted to video games was *TILT*, published by Edition Mondiale; the first issue hit the *kiosques* in September 1982. *TILT* covered microcomputer games and game console news. In 1984, the editorial staff organized Tilts d'Or, the first French award ceremony to honor video game creators. From 1991 to 1997, Jean-Michel Blottière, *TILT*'s editor in chief, served as both producer and host of the television station France 3's *Micro's Kids* (1991–1997), one of the first regular shows about video games on French television.

The success of video game consoles on the French market in the early 1990s prompted the growth of specialized magazines, including *Player One*, *Joypad*, and *Console +*, written and edited for the most part by young men —some still in high school— who signed their articles with pseudonyms drawn from pop culture sources (video games, movies, novels, and rock music). The spread of these magazines is accompanied by a strengthening of press relations between game designers working for French video game maker subsidiaries and editors alike. While their primary focus is video games, these magazines also pay attention to events in the movie world and to the latest comic strip releases, and for certain readers, they are an import vector when it comes to the popularization of manga (distinctively Japanese comic strips) and animé (a Japanese graphic genre featuring hand-drawn as well as computer-based animation).

During the same period, the video game press, crowded with French publishers—such as Média Système Édition (*Player One, Nintendo Player, PC Player*), Hachette (*Joystick, Joypad*), and Pressimage (*Génération 4, Banzzaï*)—was the target of numerous takeovers by international consortia. This was especially true of Future France (renamed Yellow Media in 2009), which maintained a virtual monopoly on the video game press with titles such as *Jeux vidéo magazine,* created in 2000, and various official magazines tied in with individual game makers (Nintendo, Playstation, and Microsoft) born or reborn (that is, bought out) since 2000. However, the press suffered a crisis caused by the rising power of specialized online sites. At the end of 2012, Yellow Media, having meanwhile become Mer7, declared bankruptcy and sold all of its titles. The independent magazine *Canard PC*, founded in 2003 by former *Joystick* staff members, was one of the rare ones to enjoy an increase in sales during this decade, thanks to an increase in the number of its subscribers.

French Web sites specializing in video game information, having grown in strength toward the end of the 1990s, delivered a fatal blow to the print press. Over the course of the next decade, these sites, partly the creation of video game enthusiasts, were swallowed by giant media conglomerates. Thus, Jeuxvideo.com is the most frequented francophone video game site; according to the audience measurement specialist Mediametrie, in November 2011, it ranked fourth among such sites in terms of number of visits. Created by Gameloft, a company based in Aurillac specializing in Web design, the site was bought out in 2006 by Hi-Media, a specialist in the placement of online advertising and micropayment systems. In 2007, Gamekult.com, founded in 2000, was bought out by the American conglomerate CNET and the following year was incorporated into CBS Interactive. As for Lyon-based Jeuvideo.fr, it became the property of the French television station M6. And last of all, also in 2007, a group formerly employed in the specialized paper press sector founded the pure player (that is, exclusively online) site Gameblog.fr.

Institutionalization of Video Games (Exhibitions, Conservation, Legitimization)

Interest in video games on the part of public cultural institutions went through several phases. In 1992, the Bibliothèque National de France was the first to assemble a collection of digital documents, including video games, open to accredited researchers in the library research division. The most important French association concerned with preservation of the national video game heritage was MO5.com. Founded in 1996 by collectors who wished to pool their funds, it was one of the foremost associations in France to promote video game culture, organizing traveling exhibitions of hardware and software and holding conferences throughout the country.

Figure 3

Michel Ancel, Frédérick Raynal, Minister Renaud Donnedieu de Vabres, and Shigeru Miyamoto, Chevaliers dans l'Ordre des Arts et des Lettres. (From Open Access material by Ministère de la Culture [http://www.culture.gouv.fr/culture/actualites/images/videotous.jpg], courtesy of Farida Bréchemier/MCC.)

Cultural institutions began to acknowledge that the video game medium was one worthy of their attention by the end of the 1990s. The Minister of Culture accords the rank of Chevalier in the Ordre des Arts et des Lettres to several game designers, not all of them French: Philippe Ulrich in 1999; Shigeru Miyamoto, Michel Ancel, and Frédérick Raynal in 2006; Peter Molyneux, Eric Viennot (*In Memoriam* [2003]), and Antoine Villette (*Cold Fear* [2005]) in 2007; David Cage in 2008; and Paul Cuisset and Anthony Roux (Ankama) in 2009. Moreover, in 2005, La Poste (the French postal service) printed three million stamps featuring images of numerous video game characters, including Pac-Man, Mario, Lara Croft, Donkey Kong, the Prince of Persia, and the Sims.

In 2004, *Game On,* a touring show created by London's Barbican Centre that recounted the history of video games, came to the Tri Postal, La Poste's sorting center in the northern city of Lille. In 2010, the Conservatoire National des Arts et Métiers played host to *Museo Games,* designed in collaboration with MO5.com and installed by the plastics designer Pierre Giner. It was a different sort of exhibition in that visitors were permitted to play the games on display. The same year, the Paris suburb of Evry's Agora theater created Arcade, a selection of video games endowed with a stripped-down, vector graphics aesthetic (as found in *Rez* [2002], *Geometry Wars: Retro Evolved* [2005], and *OSMOS* [2009]). The exhibition traveled throughout Europe and the Middle East. For three months in 2012, a wing of Paris's Grand Palais, usually a venue for grand art exhibitions, hosted *Game Story*; allowing such a popular medium into one of the temples of highbrow culture served as a form of legitimization. Since 2010, the Gaité Lyrique, a converted nineteenth-century theater now devoted to computer art, has provided video games with a permanent display space. In 2012 it hosted the exhibition *Joue le jeu*, which was organized around the theme of independent video game design, with media artists and video game designers Lynn Hughes, Heather Kelley, and Cindy Poremba serving as cocurators.

In 2003, the French prime minister, Jean-Pierre Raffarin, announced the creation of the first national video game school, the Ecole Nationale du Jeu et des Médias Interactifs Numériques (Enjmin), its home Angoulême, a city in the southwest that since 2008 has hosted the Cité nationale de la bande-dessinée et de l'image (the National Comic-Strip and Image Center). The Enjmin is the outcome of a partnership, dating from 1999, between the CNAM and the universities of Poitiers and La Rochelle. Since 2005, it has offered (to a student body steadily increasing in size) a two-year course of study in any one of six subjects, leading to a master's degree. It was on December 21, 2006, that an Enjmin student was accorded the first master's degree in computer games and interactive media.

The status of video games, and the perception of video games by the public and by public institutions, seems to have come a long way since the 1990s. The aging of the video game player population has been accompanied by a change in the nature of the medium and hence a modification in its reputation, since it can no longer be considered a form of entertainment exclusively for youngsters. It was not until the beginning of this century that French universities began to open their doors to video games by underwriting doctoral research on the subject in the fields of sociology, education, the information and communication sciences, and education, as well as film and media studies. As for the artistic aspect of video games, by now it has been asseverated by French creators, the specialized press, and numerous cultural institutions. In the 1920s, cinema was deemed the seventh art and defined the between-the-wars era. Little by little, the video game is becoming the tenth art and doing the same for ours.

References

Audureau, William. 2012. La France, ce petit pays du jeu video. October 25. http://www.gamekult.com/actu/la-france-ce-petit-pays-du-jeu-video-A105352.html.

Blanchet, Alexis. 2010. *Des pixels à Hollywood: Cinéma et jeu vidéo, une histoire économique et culturelle*. Châtillon: Editions Pix'n'Love.

Centre National de la Cinématographie et de l'image animée. 2010. *Les entreprises françaises du secteur du jeu vidéo*. Paris: CNC.

Cregut, Françoise, Hélène Monnet, Jean François Delas, et al. 1983. Jeux vidéo, le degré zéro de l'informatisation. Culture Communication CESTA, No. 6, December.

Donovan, Tristan. 2010. *Replay: The History of Video Games*. Lewes: Yellow Ant Media.

Fichez, Elisabeth, and Jacques Noyer, eds. 2001. *Construction sociale de l'univers des jeux vidéo*. Université Charles de Gaulle-Lille 3, coll. Collection UL3: travaux et recherches, Villeneuve-d'Ascq.

Gaume, Nicolas. 2006. *Citizen Game*. Paris: Editions Anne Carrière.

Gorges, Florent, and Isao Yamazaki. 2009. *L'Histoire de Nintendo, 1983–2003: Famicom/Nintendo Entertainment System*. Châtillon: Editions Pix'n'Love.

Ichbiah, Daniel. 2011. *La saga des jeux vidéo*. Triel-sur-Seine: Editions Pix'n'Love.

Ichbiah, Daniel, and Sébastien Mirc. 2011. *Michel Ancel: Biographie d'un créateur de jeux vidéo français*. Triel-sur-Seine: Editions Pix'n'Love.

Le Diberder, Alain. 2002. La création de jeu vidéo en France en 2001. *Développement Culturel*. Ministère de la Culture, Paris: Juillet.

Maneuvrier, Arno. 2009. *Du tilt à l'extra ball: Vingt ans d'automatique en France*. France: Devoldaere.

Ubisoft Entertainment. 2011. *Rapport de gestion Ubisoft*. Rennes: Ubisoft.

GERMANY

Andreas Lange and Michael Liebe

In describing the computer gaming culture in Germany, one must deal factually with two different countries at least until 1990, when the GDR and FRG were united.[1] This is not an uninteresting chapter in Germany's history; while computer games and video games were unavailable to most people in East Germany, in West Germany, as in all industrialized countries, games were quickly developing into a popular medium. This makes it possible to compare the development of computer games in two entirely different systems, which differ both socially and politically, yet have a common history and culture. This will be taken into account in this chapter, and the computer game culture in East Germany will have its own dedicated section. Since video game culture in West Germany continued uninterrupted through reunification, all other representations of video game culture from 1990 in West Germany will be of the reunified Germany.

One of the common roots of contemporary video game culture in both Germanys is the long tradition of strategic games. In the late eighteenth and early nineteenth centuries, the "war game" (*Kriegsspiel*) was first developed by Johann Chr. L. Hellwig and later refined by Georg Leopold von Reisswitz and his son Georg Heinrich von Reisswitz. It was the first complex simulation of strategic battles—a sort of extended chess, in which every unit type had individual features. These features were defined by numeric values and covered more than mere rules of movement. However, they also fixed values such as firepower and strength as well as variables such as health, ammunition, and so on. It was the first tabletop simulation war game, although the "players" were military generals who did not actually use it for entertainment but for the education of officers as well as strategic planning of real-life battles. It set the foundation for game mechanics that are now defining for any modern computer-based strategy game. A "computer" at the time was a human calculator who literally computed the results of actions undertaken by the "players," who then could estimate the outcomes of different strategies in battle. The playfield consisted of standardized square blocks, which represented different elements of landscapes and which could be arranged according to real or virtual scenarios.

The development of computer technology began before the 1949 postwar separation of East and West Germany. In 1941, Konrad Zuse was the first to build a fully automatic apparatus that ran by programs that were not fixed but which could actually be designed by coders: the Z3. It used binary numbers and is considered to be Turing-complete; that is, to be computationally universal and capable of simulating every real-world design for a computing device. Certainly, the development of the computer suffered setbacks due to what Germany lost in World War II. However, Zuse was already designing a chess program between 1942 and 1945 in his own higher programming language "Plankalkül," the implementation of which happened only decades later within the framework of a research project for computer history.

Games Germans Like to Play

Games have a long tradition in Germany. Already in 1795, Friedrich Schiller formulated that "the human being only plays where he is in the complete definition of the word 'human being'. And he is only completely human where he plays" (Schiller [1795] 1879, 15). Even if Schiller meant the aesthetic play of the theater, "trivial" games also enjoy great popularity.

Regarding digital games, typically German products—in the sense that the genre is either more popular in Germany than in other countries or that the games are produced in Germany—are strategy games such as the *Anno* series (also known as *A.D.* series, since 1998) by Sunflowers Interactive Entertainment Software, or the *Settlers* series (which started with *Serf City: Life Is Feudal* [1993] by Bluebyte Studios, not to be confused with another very typical German game, *The Settlers of Catan* [1995], a board game by Klaus Teuber). These games all follow a similar principle: players have to collect and manage resources, build up an army or similar forces, and defend, explore, and conquer new territories. Thanks to a string of research possibilities, trade options, and other strategic variables, the range of configuration parameters can be quite extensive. These game mechanics are also predominant in the many browser games made in Germany, such as Gameforge's *OGame* (2002), Travian Games' *Travian* (2006), and Bigpoint's *Dark Orbit* (2006) and *Farmerama* (2009).

As soccer is the most popular sport in Germany, a lot of video games played in Germany focus on it. Yet on the production side, German soccer games feature a mechanic similar to that of strategy games. Management simulations are particularly popular; titles such as KRON Simulation Software's *Bundesliga Manager* series (begun in 1991) or Ascaron's *Anstoss* series (from 1991 to 2006) are good examples. In these games, the range of possible actions even goes as deep as individual player training, buying and selling on a transfer market, and changing tactics or strategies during simulated matches.

Another successful genre in Germany is that of simulation games, ranging from train, subway, and farm simulators, to dredge and road sweeper simulators. The German publisher Astragon is rather active in this genre, and the *Farming Simulator* series (since 2008) is its biggest success with more than one million copies sold.

Also popular are fantasy role-playing games (RPGs) such as Piranha Bytes' *Gothic* series (from 2001 to 2010), Attic's *Das Scharze Auge* series (from 1990 to 1997), and Radon Labs' *Drakensang* series (2008, continuing as Bigpoint's *Drakensang Online* since 2011). *Das Schwarze Auge* and *Drakensang* are pen-and-paper games originally designed by Ulrich Kiesow in 1984. As of 2013, the most recent game in the *Drakensang* franchise was published by Bigpoint Berlin and adapts the complex role-playing mechanisms to browser game technology and a free-to-play business model. Yet it no longer holds the rights to the original brand, *Das Schwarze Auge*. International brands such as Bethesda Softworks' *The Elder Scrolls* series (since 1994) and the Polish titles CD Projekt RED's *The Witcher* (2007) and Reality Pump's *Two Worlds* (2007) also sell successfully in Germany.

Presumably the best-known German action games are Rainbow Arts' *Turrican* (1989) and *The Great Giana Sisters* (1997), both programmed by Manfred Trenz. They both are classical side-scrolling platform and jump-'n'-run games. Notably, both games gained international acclaim through their soundtracks. Thanks to these hits, the musician Chris Hülsbeck was invited to compose music for new releases of *R-Type* and *Star Wars* games. The first-person shooters *Far Cry* (2004) and the *Crysis* series (since 2007) by Crytek, and *Spec Ops: The Line* developed in 2012 by Yager Development, all became international hits.

Introduction to the German Video Game Industry

Although Germany has ranked among the world's great markets as a location for developers, it has remained a *developing country* for a long time. In a different way from German board games, which are well-known internationally for their special strategy-oriented multiplayer designs and sold under the label "German Games," digital games from Germany have only recently received international recognition. One of the reasons for this is the fact that no relevant internationally operating publisher existed in Germany. The large foreign publishers found it difficult to establish German productions and market them internationally. In technical circles, the reigning opinion was that German studios were not able to make easily marketable games and instead concentrated too much on many unnecessary details and techniques.

Nevertheless, Germany was, and is, well-regarded for its engineering qualities. Crytek, with its CryENGINE (available since 2004), provided the technically most advanced game engine of the industry, a tool that today is widely used for pre-visualization in architecture and movies. Bigpoint stretched the capabilities of standard Web browsers to their limits by pioneering massively multiplayer games played in 3-D without needing any additional client software installed, a feature that in the long run was also the basis of success for other German browser game companies such as Gameforge, Innogames, and Travian. Consequently, the latest edition of the game development tool, CryENGINE 3, includes real-time rendering in browsers.

The industry sector of game services also has some typical German characteristics. Because of the current structure of the German full-price market, the localization industry is rather large. On the one hand, almost every video game published in Germany is translated into German and, in most cases, has to be adapted to the system for the protection of minors or other cultural needs. On the other hand, browser games roll out internationally from the first day of their release, making localization for external markets an integral part of development. Also, thanks to the success of online gaming and online payment systems, community management and digital distribution play a major role in the industry. Gamegenetics, for example, takes already-successful browser games and connects them with popular Web platforms and fosters their market reach to a large degree. By doing so, they established a service tailored for the needs in an ecosystem mainly based on self-publishing distribution models. Sponsorpay offers micropayment solutions, and HitFox Game Ventures has focused on investing in young, online, and mobile publishers.

History of the German Game Industry

With the beginning of the home video and computer game industry in the 1970s, the German market was mainly built on international imports, especially from the United States. Also, the first Atari VCS 2600 consoles were initially imported piece by piece by the Hamburg-based company WEA Musik GmbH. In 1980, the company flew to the United States and bought the original packages, which were to be sold in US stores. The package included the VCS console and the game *Space Invaders*, as well as two joysticks. They took out the joysticks and the game and sold these separately to local German vendors. Following the instant success of these imports, Atari Germany was founded on March 6, 1981, by the Hamburg-based entrepreneur, Klaus Ollmann. With the introduction of the Atari VCS, the third-party publisher, Activision, also expanded to Germany and established a local branch in 1985 (founded by Winrich Derlien).

All internationally successful game consoles were also released in Germany. The Famicom was introduced as the Nintendo Entertainment System (NES) in 1986, followed by the Super Nintendo Entertainment System (SNES) in 1992, the Nintendo 64 in 1997, the Nintendo GameCube in 2002, and the Nintendo Wii in 2006. The SEGA Mega Drive was introduced in 1990 and for a brief time was more successful than the Nintendo consoles. The Sony PlayStation was introduced in 1995, the PlayStation 2 in 2000, and the PlayStation 3 in 2007. The Sony consoles soon became the most successful consoles in Germany. The Microsoft Xbox was released in Germany in 2002 and the Xbox 360 in 2005. Before 2005, most consoles were released approximately one year later in Germany than in Japan or the US. This changed with the simultaneous rollout of the Xbox 360. In 2012, the console game market made 819 million euros (including hardware). Apart from the hardware developments in the GDR described below, only one console system was developed in West Germany. It was called the VC4000 and was produced by the Cologne-based company Interton from 1978 until 1982. During this time, around forty games were published for the console, which was built around the Signetics 2650A CPU. But because the quality of the games was lower than that of its competitors, the VC4000 never became more than a niche product, mainly benefiting from its cheap price. Traditionally, the PC has been the most popular gaming platform in Germany. Software sales alone made 464 million euro in revenue for PC gaming. One of the first internationally relevant developer companies from Germany was Kingsoft, founded in 1983 in Aachen by Fritz Schäfer. The company, which was acquired by Electronic Arts in 1995, built its success on the Commodore computers C16, C116, and Plus/4. A gaming boom in Germany started when the Commodore 64 was introduced in Germany in 1983. It was the first popular home computer in Germany and sold more than three million units. Its successor, the Amiga 500, sold more than one million units from 1987 until 1993 (Commodore Germany 1994). Also very popular for gaming were Sinclair computers, the ZX81 (introduced to Germany in 1981), and the Spectrum. Within the first two years, the ZX81 sold more than two million units.

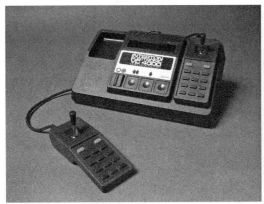 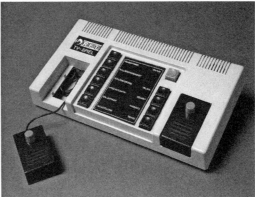

Figure 1
Early German console systems: the Interton VC4000 (left) and the RFT TV Spiel (right).

IBM-compatible PC systems (especially using the Intel 80286 processor of 1984) were used for gaming to a large degree as well.

Computer game development for the international market started quite late in Germany. One of the milestones in that respect was *Urban Assault* (1998), which was published by Microsoft and developed by Terratools, founded in 1993 in Potsdam near Berlin. According to Terratools, *Urban Assault* was the first game from Germany to be published in thirty countries simultaneously. Also, in regard to gameplay, *Urban Assault* was a milestone since it was the first game to combine real-time strategy with a 3-D first-person perspective.

Computer Games in East Germany

Due to the limited spread of computers and game consoles in the GDR, there was no official video game industry and relatively few players. In spite of that, computer games enjoyed great popularity just as in the Western world. Owing to the necessity of having to solder and program themselves, computer gamers in the GDR gained greater technical capabilities and computer knowledge than the players in West Germany.

In order to comprehend the situation, it is important to understand the difficulties of the states of the Warsaw Pact regarding the production and supply of IT. Because of the CoCom (Coordinating Committee for Multilateral Export Controls) embargos, Western companies were not allowed to export computer technology to the states participating in the Warsaw Pact. Therefore the GDR, in contrast to West Germany, could not simply import all the necessary computer technology it needed, so people were forced to develop their own hardware and software to a great extent. Even if the development of the GDR lagged behind that of the West by about five years due to supply problems, there were developments and products worth mentioning, and those relevant to gaming are described here. Although they were not simply an obscure area of focus—rather, quite the opposite, as one of the primary communist planning measures—these achievements were still born of the shortage of importable technology.

For this purpose, so-called computer stations were installed in schools and youth centers where interested children learned to program games and other applications. Also, the *Gesellschaft für Sport und Technik* (GST) responsible for *Wehrsport* was expanded into the area of computers. The concept "computer sport," therefore, developed in the environment of the GST. Under this label, many championships and competitions took place in the 1980s. For example, the participants in the 1987 Programming Olympics held in Berlin had the task of organizing the main memory of a database, developing the algorithm to access it, and implementing them in BASIC.

Even in the realm of scientific conferences such as "Computer Use in Non-educational Activities," held on October 20, 1988, the possibilities of the application of computer games were discussed in detail. Dr. Gerd Hutterer came to the conclusion that "computer games possess the objective tendencies which acquire the ideas and values of socialism" (Hutterer 1988, 7–18).

Against this background, it is plausible that the first video game was produced in the GDR about 1980. It concerned the BSS 01 (Bildschirmspielgerät 01), a screen game device that contained a *PONG*-like home game, which was built around the General Instruments AY-3-8500 chip, like many comparable Western devices, and which, by this time, would have used the remaining stock. These chips could be acquired reasonably and did not fall under the CoCom Embargo. Although the *PONG*-like console was actually built in a foundry in Frankfurt on the Oder for private use, like so many other products, it rarely made it into stores. Its price of 500 marks was almost commensurate with the average monthly wage. The device was sold preferably to educational institutions and youth centers and was displayed openly for young visitors. While in West Germany the medium spread as a subculture that was not publicly promoted, computer games in East Germany were

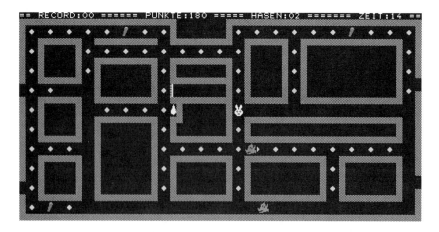

Figure 2
Screenshot of *Hase und Wolf* (*Hare and Wolf*, 1986), one of the eight games on the *Poly Play* arcade machines.

an integral part of the official cultural and educational policy from the beginning. Additionally, they quietly provided proof that the GDR was on the same scientific and technical world-class level as the West, another reason why people at the time were proud of computer and video games.

In any case, the BSS01 was not to be long-lived. The system's name hints at the possibility of a second series, and according to the information from the former project leader, the BSS02 was almost identical in build, technically improved, and probably in color before it was canceled due to the production of clock radios. The game tradition was not continued until 1986 with the production of *Poly Play* arcade machines. They were produced up until the reunification at the VEB Polytechnik and offered a selection of up to eight games. The selection consisted mainly of reproductions of Western games such as *Pac-Man* (1980). In the communist *Poly Play* version, Pac-Man and the ghosts were replaced by the Russian fairy tale characters of the Rabbit and the Wolf. Besides the game consoles, home computer development was primarily used for games. For many East German computer fans, the Z1013, which was available beginning in 1984 as a construction kit, was their first personal computer, and as such was available for normal citizens at a price of about 1,000 marks. To purchase it, one had to "simply" submit a written order, wait about a year, and personally pick it up at a specialty store. With its 16mb main memory, it had little capacity, but it could be expanded with its TV and cassette interface device and with the help of soldering irons and some creativity. However, the KC Series of the VEB Microelektronic Mühlhausen was more interesting. This computer, produced as of 1987, was based on the Z80-Nachbau-Chip U880D. The computer's concept from the beginning was a modular system capable of upgrading, and it commanded an acceptable graphic and sound output. Even though there were only 50,000 copies produced, the KC computer was similar to the Commodore 64 used in the GDR.

Not only was hardware in short supply; it was almost worse in the area of software. On the one hand, there was a focus on the replication of Western programs. This is how the "inferior to MS-DOS" operating system, the CP/M, experienced a renaissance in the early 1980s, albeit under a difference name, SCP. Even MS-DOS was eventually copied and hastily renamed DCP. On the other hand, computer users in the GDR had little choice but to write their own programs, which, of course, was also true for many game programs.

Distribution was normally done hand to hand, whereby one program was gladly exchanged for another. Another possibility of program distribution, which was almost visionary by today's standards, existed in broadcasting by radio. On October 16, 1986, a computer program was transmitted for the first time over the ether within the framework of the science magazine of Radio DDR 2. The sounds, which were undetectable by the human ear, had to be recorded on a cassette and then loaded onto the computer. Because of the overwhelming success (about 50,000 listeners) the broadcast, "BASIC 1x1 des Programmierens," including software transmission, was broadcast beginning in 1987 on a regular basis. Two years later, *REM das Computermagazin* emerged from this concept. The half-hour moderated program enjoyed great popularity in the computer scene until the reunification *Wende*. Games were a regular, integral part of these mass downloads.

Although it was forbidden, according to the CoCom embargos, to export even small computers into the states abiding by the Warsaw Pact, an increasing number of Western computers surfaced in the GDR. As a result of the embargos, this involved mostly privately smuggled products (the expression was "Grandma from the West"). Since the GDR also had ongoing interest in importing Western computer technology, customs regulations were accordingly conciliatory concerning the influx of Western computers. Each person could receive one computer from the West duty-free. One only had to smuggle it past customs on the Western side. Each subsequent computer would cost about 6,000 marks in customs duty, which correlated to the market price of a computer in the GDR.

Eventually, the private computers were said to be sold in special East German "Intershops," even though they were insufficiently available. It is estimated that about 200,000 home computers produced in the West found their way into the inner East German border. Inside the GDR, a Western computer scene developed where the most frequently used computers were the Sinclair and the Commodore 64. Interestingly, the two communities, those with the Western computers and those with the Eastern, rarely mingled. While the followers of Eastern computers built and programmed by themselves to a great extent, the scene around the Western computers developed more into one of games and copying. With all the open-mindedness of the official GDR regarding computer games, one must not forget that this was entirely purpose-driven, for in East Germany, they could not miss an opportunity within the framework of propaganda to denounce capitalistic computer games. In November 1989, an article entitled "Bombing, shooting, slaughtering ..." appeared in the regular rubric of Imperialism in the magazine *Funkamateur* (*Amateur Radio Operator*), which was significantly involved in the spread of computers in the GDR. In the article, the capitalistic profit-seeking of unscrupulous gaming companies such as Sublogic was castigated. With violent games such as *Jet* (1985) or *Falcon* (1987), tens of millions would be earned at the expense of the development of the youth ("Bomben, ballern, metzeln ..." 1989).

After unification, it was easy for the East Germans, who were used to East German computers, to continue with their hobbies and/or professions. Because of the shortage of IT, the East German government decided, in accordance with the Russians, to copy Western standards such as IBM and DOS. Therefore, the technical knowledge of how to program even Western computers was already present; for example, the abovementioned game company Terratools (founded in 1993) was mainly based on staff with an East German background.

The German Game Industry Today

As of summer 2013, Germany had the biggest market for digital games in Europe. In 2012 its estimated value was around 1.85 billion euros. In respect to Triple-A console and PC games, import numbers are traditionally higher than export or domestic production. Yet in respect to casual games (mainly browser-based strategy games), the industry has successfully managed to find and exploit a niche market. According to the latest market research data, sales of virtual items alone reached close to 226 million euros in 2011, while revenue from subscription and premium accounts reached 124 million euros. Around fourteen million Germans play online and browser-based games (Gesellschaft für Konsumforschung [GFK] 2013).

Every international publisher—including Nintendo, Sony, Electronic Arts [EA], Take-Two Interactive Ubisoft, and Vivendi—has an office in Germany. CDV (established 1989, insolvent 2010) and Frogster Interactive Pictures (established 2005, now Gameforge) are game publishers with a German origin, but neither of the two focuses on Triple-A console titles; in 2008 Frogster had shifted its portfolio completely to online games (including an adaptation of Runewalker Entertainment's *Runes of Magic* [2009] and Bluehole Studio's *The Exiled Realm of Arborea* [2011]), establishing close ties to South Korea. Publishers in Germany are represented by the industry association, BIU (Bundesverband für Interaktive Unterhaltungssoftware).

Koch Media, founded in 1994 by Franz Koch and Klemens Kundratitz and based in Planegg, near Munich, is an exception to this trend. Today it produces and markets games under the Deep Silver label mainly in Europe and North America. Deep Silver has published more than 200 games since 2003, including its own franchises such as *Dead Island*, *Sacred*, *Secret Files*, and the *Risen* role-playing games.

Established in 1999, Crytek, the maker of the original *Far Cry* (2004) as well as the *Crysis* series, is currently the biggest Triple-A developer in Germany. Lesser known internationally is Blue Byte (established 1988), although it has a longer tradition. Its most successful games include *Battle Isle* (1991), *The Settlers* (1993), and *Anno 1404* (2009). Yager (established 1999) received a contract to develop *SPEC OPS: The Line* (2012) for the international publisher Take-Two Interactive, and hence has become one of the big studios. Surfing the Facebook trend, Wooga became one of the leading social media game developers of the world, with *Monster World* (2010) and *Diamond Dash* (2012) being two of their most successful titles. In the mobile games sector, numerous new players have entered the market, with Chimera Entertainment (established 2006), Flaregames (established 2011), and Infernum Productions (established 2011) among the most successful. In 2010, Bigpoint, Europe's biggest browser games developer, started a massive mobile strategy further pushing the market shift toward smartphones. Developers in Germany are represented by the G.A.M.E. Bundesverband der Computerspielindustrie (Association of Game Developers).

Germany is host to the largest European video game fair, Gamescom, in Cologne, which had more than 370 exhibitors and 275,000 visitors in 2012. (Gamescom ran under the name "Games Convention" in Leipzig from 2002 until 2008.) During the fair, the biggest European game developers' conference, GDC Europe, takes place. Since 2007, the DGT, "Deutsche Gamestage" (German for "game days"), has taken place in Berlin (since 2014 it continues as International Games Week Berlin). It encompasses the developer's conference Quo Vadis, the A MAZE. Indie Connect, as well as the recruiting event Making Games Talents, the Gamefest at the Computerspielemuseum, and several other matchmaking and convergence events. Other big events in Germany were the Browser Games Forum in Frankfurt/Main and Casual Connect in Hamburg.

A Leader in Browser Games

A German success story was enabled through the shift from retail to digital distribution in the video game market. In the late 1990s and early 2000s, games such as Phenomedia's *Moorhuhn* (1999), Alexander Rösner's

OGame (2002), and Denys Bogatz's *space pioneers* (2004) were the first signals of a wave of browser-based games that would spread worldwide.

Browser games do not need client software to be installed, and they run entirely through standard Web browsers such as Firefox or Internet Explorer. Not only the data but also most of the processing power is provided by the game servers, hence technical requirements for the players are minimal, making them attractive to a whole new audience of less tech-savvy or less economically strong computer users worldwide. For most browser games, gameplay is simple and intuitive, allowing them to be played during work or at short breaks during the day. *Moorhuhn*, for example, was a simple point-and-shoot game where the player has to hit as many grouses as possible in a certain period of time. Yet, most other games defining the genre were strategy games that were slow in pace, allowed players to check in briefly every couple hours, and did not need constant engagement. Most of the players' input is based on giving long-term orders; in *space pioneers*, players had to build and manage an empire of planets, spaceships, and resources. Gameplay was rather long-term, yet casual. For example, to build a spaceship on one of the planets, players set the parameters and then had to wait for several hours until the construction was completed—making it ideal to just check in every now and then during a long workday.

Furthermore, guild-building and teamwork are of great importance in these games. The early strategy-based browser games were supported by an active open-source community and initially had a noncommercial approach. Still today, registration in most cases is free, but by asking for micropayments, a new basic game mechanic was added: pay for pace. Construction or other tasks can now be sped up or weapons and tools can be made efficient through investments in in-game shops, making gameplay faster and easier for those players willing to buy the power-ups.

Companies such as Bigpoint (established in 2003 as m.wire, renamed Bigpoint in 2007), Gameforge (established 2003), Wooga (established 2009), and several more are new but potent players on the international browser game and social media game market. In early 2012, Wooga overtook EA in the Facebook games market and followed directly after Zynga in the world rankings. Bigpoint had more than 220 million registered users in 2011, making it the fastest-growing game company in Germany. Like Wooga, they do not address core gamers but attract a rather casual gaming community. More German companies are successful in the online casual game market: GameDuell (established 2003) is Europe's leading casual competition platform, and Intenium is successful with http://www.deutschland-spielt.de (established 2003). Innogames runs *TribalWars* (2003), Funatics runs *Cultures Online* (2011), and Travian Games runs the internationally successful village management game *Travian* (2004*)*.

Video Games as Cultural Artifacts and Their Perception in Society

Since the rise of computer games in the 1970s, public skepticism has grown and the medium as a whole was widely rejected as dangerous for the health and education of children and adolescents. In 1984, the "law for protection of minors" was updated, and arcade games were banned from public spaces open to minors, putting them into the same realm as "one-armed bandits" and other gambling machines, which are intended for adults only. Also, the "Bundesprüfstelle für jugendgefährdende Schriften" (Federal Department for Media Harmful to Minors) started policing games more and more. The first game banned for minors was Activision's Atari VCS 2600 game *River Raid* (1982) because of its military content.

The increasing sales numbers of video games in Germany led to the foundation of a new body taking care of the ratings. In 1994, the USK (Unterhaltungssoftware Selbstkontrolle/Entertainment Software Self-Rating) rating system was initiated. It was implemented by the Foerderverein fuer Jugend und Sozialarbeit e.V. (FJS) in cooperation with the first German industry association VUD, which was dismantled in 2004. The principles of the USK basically follow the process established for movie ratings, but until 2003, on a voluntary basis. Since 2003, the ratings for computer games have to be done on a legal basis due to changes in the law for the protection of minors. Since the beginning, the rating process has been funded by the publishers, while the games were rated by independent experts. Since 2003, the system is officially certified and embedded in the German legal system and the different age ratings, even the ones below an age-eighteen rating, have to be controlled and respected by every sale of a game. Games that are not rated by the USK are not allowed to be advertised, yet can be sold. Since 2007, the USK has been owned by the two industry associations, G.A.M.E. and BIU, and in spite of it, every age rating has to be accepted by the German "Bundesländer," which is responsible for the protection of minors in federal Germany.

The strongest legal action against games, or any other media, is a complete ban, which makes the possession of such works illegal for everyone. One main criterion for banning content in Germany is gameplay that glorifies violence extensively; for example, through reward systems tightly connected to actions that are violent just for the sake of violence; Rockstar Games' *Manhunt* (2003) was a game that fell into this category. Another "don't" in German media is racist ideology and discrimination against minorities. It is also forbidden to depict the swastika or any other obvious Nazi symbols. Id Software's *Wolfenstein 3D* (1992), for example, is forbidden for both reasons.

As a reaction to these restrictions, many international publishers have decided to exchange the human enemies in their international version for androids or zombies in the German version. To do so, they simply manipulate the overarching narrative and turn the color of the blood to green, stating in the game manual that the game characters are aliens or robots; Virgin Interactive's *Command & Conquer* (1995) is one of the more prominent examples of this.

The contents of video games have led to numerous discussions on a political level in Germany. Publicly, the most visible were the debates following incidents in schools (for example the shooting in Erfurt in 2002 and in Winnenden in 2009), after it was discovered that the perpetrators had played *Counter-Strike* (1999) and other first-person shooters. Whereas first-person shooters in Germany are traditionally called "Ego-Shooters," the new name "Killerspiele" (literally, "killing games") was coined in the context of the violence debate around computer games. Studies in criminology were well-funded and supported by the public opinion that games were dangerous for youth. One of the opponents of game critics, Prof. Dr. Christian Pfeiffer, published several studies confirming the thesis that an extensive consumption of violent computer games has a negative influence on minors. In recent years, the debate about the negative influence of violence in computer games has taken over the discussion about game addiction, especially with the great success of *World of Warcraft* (2004) and reports of players who have lost control of their gaming habits. Today, institutions such as the ambulance for gaming addiction ("Spielsucht Ambulanz," founded in 2008) at the University of Mainz provide support for players with problematic gaming behavior.

Another milestone in these debates was the declaration of video games as cultural and potentially artistic artifacts in 2007 by the "Deutscher Kulturrat" (German Culture Board), which is the top organization for all artistic professionals. This declaration was very helpful in establishing a more differentiated public opinion of the medium, slowly minimizing the general suspicion shown toward computer games. As a consequence, the developer association G.A.M.E. became a member of the German Culture Board.

Following this declaration, several federal governments have institutionalized game content funding within their cultural support programs. In 2007, Berlin was the first to do so, and North-Rhine Westphalia the last to follow in 2012. Like the age-rating system, the model copies the system established for the film industry (although with a far smaller budget). These programs mainly contain funding for prototypes or smaller productions, generally ranging from 10,000 to 100,000 euros, which can be applied for at the regional media funding bodies in Baden-Württemberg, Bavaria, Hamburg, and North-Rhine Westphalia. Berlin and Brandenburg as well as the three federal states of Saxony, Saxony-Anhalt, and Thuringia all have joint industry support programs for their respective regions.

Another program supporting cultural or high-quality game content made in Germany is the "Deutscher Computerspielpreis" (German video games award) established in 2009 in cooperation with the "Bundesbeauftragter für Kultur und Medien" (BKM, Federal governor for cultural affairs and media) and the industry associations BIU and G.A.M.E. as well as the IT-industry association Bitkom (although the latter ended its engagement with the award in 2010). Games winning this award receive up to 75,000 euros in prize money, allowing smaller developers to start new projects. The prize is awarded in several categories: Best German Game, Best Kids' Game, Best Youth Game, Best Mobile Game, Best Serious Game, Best Browser Game, and Best Newcomer Concept. The award is based on a statement of the German Parliament from November 14, 2007:

> Computer Games, including other interactive entertainment media (video consoles, online, and cellphone games) have continuously gained cultural impact in the last years. They became an important economic, technological, cultural and social factor in Germany. ... Computer games transport social images and contain their own cultural topics. Due to that, they have become an important part of the cultural landscape of our country and are formative for our society. (Deutsches Parlament 2007)

An Educational Landscape

Video game research and education in Germany has diversified into various branches, with empirical studies and quantitative research on one hand, and fundamental, philosophical studies and qualitative research on the other. Cultural research first started as journalism, and the work of Florian Rötzer, Konrad Lischka, and Mathias Mertens had a noteworthy influence on video game studies in the early 2000s. Game studies in particular had its origin in the "Arbeitsgemeinschaft Games" (SIG Game), a chapter of the "Gesellschaft für Medienwissenschaft" (Society for Media Studies), founded in 2000 by Britta Neitzel and Rolf Nohr. Today, the Digital Games Research Center (DIGAREC) of the University of Potsdam, and the Game Lab Cologne (GLC) are major centers in the field. Private schools such as the Games Academy (Berlin, Frankfurt/Main), the Mediadesign Hochschule (Berlin, Dusseldorf, Munich), and QANTM/SAE Institute (Berlin, Munich) have established design-oriented education programs. Several universities are also offering bachelor's and master's studies, which are approaching games in different ways.

Game Culture and Art

In the time of floppy disk cracking, Germany grew a vibrant demoscene. Demos developed from the short trailers (called cracker intros), which hackers added as a sort of splash-screen before a game starts, showing the credits of the cracker of the copy-protection. As storage capacity was very limited (high capacity disks such as the 3.4" double-density floppy disks had around 1.44 MB), the animations had to be code-only (no graphics or sound files and everything generated through algorithms). These coded animations became so popular that events and competitions formed around these technical skills in which not only the aesthetics of the animation but also the elegancy of the code were rewarded. Its biggest yearly gathering, Evoke, has taken place every year in Cologne since 1997. Prominent organizations in this field of game art and culture are the A MAZE. festival (established 2008), the "Videospielkultur e. V." (Association for video game culture) in Munich (established 2006), the Transmediale (since 1997), as well as the Club Transmediale (since 1999) in Berlin, the "Zentrum für Kunst und Medien" in Karlsruhe, and the Next Level Conference in Cologne (since 2010).

The Computerspielemuseum (Computer Game Museum) in Berlin is a unique German institution. Founded in 1997, it is the world's first publicly accredited museum dedicated exclusively to video games. With the support of the Lotto Stiftung, the European Regional Development Fund, and Medienboard Berlin-Brandenburg, it opened a new permanent exhibition in 2011, which displays video game history and culture with more than 300 exhibits and 80,000 visitors every year. It also hosts regular events celebrating video game culture, while its collection contains more than 22,000 games and software programs, over 2,300 hardware devices and art installations, as well as around 10,000 magazines. The museum is also a co-initiator of the European Federation of Game Archives, Museums, and Preservation Projects (EFGAMP), which aims to advance the conditions of preserving and accessing our common digital heritage, which is surely becoming a challenge, both technically and legally.

In all respects, from the industry to public institutions, and artistically and culturally, computer games in Germany have slowly but steadily established themselves as mainstream media.

Acknowledgments

The authors would like to thank Dr. Doreen Krueger of Concordia University Wisconsin for translation assistance with certain sections of this chapter.

Note

1. If not explicitly expressed, the term "computer games" is used generally for digital games, without distinguishing between different platforms.

References

Bomben, ballern, metzeln ... 1989. *Funkamateur* (November): 528–529.

Commodore Germany. 1994. Press release, Frankfurt am Main.

Deutsches Parlament. 2007. Antrag: "Wertvolle Computerspiele fördern, Medienkompetenz stärken." *Bundesdrucksache* 16(116):2007.

Gesellschaft für Konsumforschung (GFK). 2013. Market Research Data, Berlin: Bundesverband Interaktive Unterhaltungssoftware (BIU).

Hutterer, Gerd. 1988. Computernutzung in der außerunterrichtlichen Tätigkeit. In *Proceedings of the Conference "Erkenntnisse und Standpunkte zur Nutzung des Computers als Arbeitsmittel in der außerunterrichtlichen Tätigkeit."* Pädagogische Hochschule "N.K. Krupskaja" (Halle/Köthen).

Schiller, Friedrich. [1795] 1879. *Briefe über die ästhetische Erziehung des Menschen.* Stuttgart: G. Cotta'sche Buchhandlung.

HONG KONG

Benjamin Wai-ming Ng

Hong Kong, a free port city with a population of more than 7.1 million, is one of the major game consumption centers in Asia. More than being a mainstream young male culture, gaming has become a citywide entertainment, actively consumed by people of different ages, genders, and socioeconomic backgrounds. Children and adults are as ardent as teenagers in gaming, and the number of female players has increased tremendously. There is no obvious difference between white-collar and blue-collar patterns of game consumption; wherever you go, you will always find someone playing games. At home, people play console games, online games, and computer games; outside the home, they play arcade games, handheld games, mobile games, and tablet computer games. Game consoles are considered a household and personal necessity, and gaming is both personal entertainment and a social activity. Gaming has had a very strong impact on the daily life of Hong Kong people. Game jargon and terminology have been incorporated into spoken Cantonese. Gaming has enriched Hong Kong comics, movies, pop songs, and dramas by providing stories, characters, and ideas. Hong Kong game players consume games in their own ways, creating new rules, skills, and jargon.

This chapter is a historical study of gaming in Hong Kong, from its beginnings in the 1970s to the present, outlining the history of popular game consoles and software, tracing the transformation in game consumption from Japanese games to non-Japanese games and from handheld game consoles and family game consoles to online and mobile games. The chapter identifies important trends and characteristics of gaming in Hong Kong and discusses the impact of gaming on the region's culture and society.

The First Wave of Gaming: Arcade Video Games, Late 1970s to Late 1990s

The first generation of game players in Hong Kong were arcade game players. In the late 1970s and early 1980s, small-size game centers began to appear in the urban areas of Hong Kong Island and the Kowloon Peninsula. These centers usually had only between twenty and forty coin-operated machines and were dark, dirty, and smoky. However, they provided a new form of entertainment at affordable prices (from 50 cents to one Hong Kong dollar per game). During this time, very few families could afford an Atari VCS 2600 or other home console system. Arcade games were thus overwhelmingly well received.

Most arcade games of this period were introduced and licensed by US game developers such as Atari and Midway but often designed by Japanese companies such as Namco and Taito. The first arcade game introduced to Hong Kong was Atari's *PONG* (Atari, 1972), but the arcade games that created a citywide craze were Taito's *Space Invaders* (1978), Namco's *Pac-Man* (1980), and Nintendo's *Mario Bros.* (1983). In the early 1980s, *Pac-Man* and *Mario Bros.* were the hottest games and the first ones played by many local players. Their unlicensed merchandise flooded the market. The Hong Kong animated film *Old Master Q, San-T* (1983) even featured the protagonist playing *Pac-Man* at a game center, and the live-action Hong Kong film *Future Cops* (1993) borrowed characters from *Mario Bros.*

Arcade games were popular in the 1980s and 1990s, marked by the rise of game centers and game players. Playing arcade games became a popular form of entertainment among young males. Hundreds of game centers were open from morning until midnight and were most commonly frequented by players between the ages of sixteen and thirty. In principle, students with school uniforms were not allowed to enter, and smoking was prohibited, but these rules were never forcefully implemented. The image of game centers during this period was somewhat negative, often associated with gang fighting, drug trafficking, and gambling in the public imagination and the media. The government issued *The Amusement Game Centres Ordinance* (1993) to regulate this business, but the condition remained largely unchanged.

The popularity of arcade games in Hong Kong during this period owed much to the overwhelming success of two Japanese combat games, both of which started game franchises: Capcom's *Street Fighter* (1987) and SNK's *The King of Fighters* (1994). Launched in 1992, *Street Fighter II* created an unprecedented commotion in the history of arcade video gaming in Hong Kong. The queue for this game was always long, and some game centers had nothing but *Street Fighter II*. In the early 1990s, it was not uncommon to find stationery shops or grocery stores putting one or two handmade wooden arcade consoles outside or inside their shops where students could play *Street Fighter II*. The game's merchandise, both licensed and pirated, sold like hotcakes. The game was repeatedly featured on the front page of local game magazines and was adapted into eighteen comics and two movies in Hong Kong. For instance, *Future Cops*, an unlicensed film adaptation of *Street Fighter II*, borrowed the characters, storyline, and fighting tactics from the game. *Jietou bawang* (*Street Fighter*, one hundred issues, 1991), a popular Hong Kong comic by Xu Jingchen, was an unlicensed comic adaptation of *Street Fighter* and added elements of Chinese martial arts and Hong Kong comics to *Street Fighter*. Showing a high level of localization in game consumption, Hong Kong players created their own jargon and rules for the *Street Fighter* series. For instance, during three-round games such as *Street Fighter*, the first-round winner would lose the second round on purpose to his opponent so that both players could play the final round and the weaker player would not feel offended (Ng 2009). *The King of Fighters*, introduced to Hong Kong in 1994, replaced *Street Fighter* as the most popular combat game at game centers. Its popularity reached its peak in the late 1990s, having a strong impact on Hong Kong comics, youth fashion, and movies. At least thirty local *kung fu* comics based on this game were published.

Arcade gaming has been declining since the turn of the millennium for the following two major reasons: First, most people own home consoles and handheld consoles, and thus they do not have to go to game centers to play games. Many popular arcade games have been adapted for home consoles or handheld consoles. Second, most game centers have been forced to shut down due to the surging rental market in the last decade. Street-level shop rents in Hong Kong's Causeway Bay rank second highest in the world after those of Fifth Avenue in New York (Li 2011). In 2002, there were 413 game centers in Hong Kong; by 2010, only 310 remained, and by 2012, there were only 291 (as of January 31, 2012).[1] Despite the downturn, arcade gaming will not be sidelined. Arcade game players are usually hardcore gamers who look for the level of excitement that other game platforms cannot provide. For this reason, large-size simulation games (such as SEGA Rosso's *Initial D Arcade Stage 4* [2008]) and combat games (such as Capcom's *Gundam vs. Gundam Next* [2009]) are particularly popular. Game centers also have social functions, serving as hangout places for friends and couples. Hong Kong has many hardcore arcade gamers, and most are males in their twenties, thirties, or forties (Xu 2010). They are very active at game centers and on the Internet. The largest arcade gamers' organization is Fight Club. Founded in 2001, it has 30,000 members, runs a very popular discussion forum on the Web, and organizes game gatherings and competitions three or four times per month on average, at which about 200 to 300 members participate (*Next Magazine* 2004). It seems that although the number of arcade game players is shrinking, the remaining players are very ardent.

The Second Wave of Gaming: Home Consoles and Handheld Consoles, Mid-1980s to Late 2000s

Home consoles and handheld consoles have been the major forms of gaming in Hong Kong since the mid-1980s. Nintendo and Sony have been the two key players in the market, and gaming in Hong Kong would not have been the same without them. Nintendo jump-started the age of the home game consoles, and its games created crazes one after another. Sony joined the competition in the late 1990s, and at times outperformed Nintendo.

The introduction of Nintendo systems marked the beginning of the age of home consoles. The Atari VCS 2600 failed to create a video game revolution in Hong Kong in the late 1970s. Although it had popular games such as *Pac-Man* and *Space Invaders*, it was not attractive to Hong Kong people due to high prices. Nintendo popularized home consoles by introducing the Nintendo Entertainment System (NES, commonly referred to as "Famicom" in Asia or the "red and white machine" in Hong Kong) in 1985. The home console version of popular arcade games such as Nintendo's *Mario Bros.* and *Donkey Kong* (1981), Konami's *Contra* (1987), home console games such as Nintendo's *Super Mario Bros.* (1985) and *The Legend of Zelda* (1986), Square's *Final Fantasy* (1987), and Hudson Soft's *Bomberman* (1987) contributed considerably to the overwhelming success of the NES. Ironically, hardware and software piracy also boosted the popularity of the NES in Hong Kong. Most players added a disk drive to their NES so that they could play pirated games.[2]

The age of 16-bit home consoles began in 1990 when SEGA launched the SEGA Genesis (also called Mega Drive) in Hong Kong. At its peak, the SEGA Genesis and the NES each controlled half of the home game console market. The hottest game for the Genesis was *Sonic the Hedgehog* (SEGA, 1991), and other popular titles were *Crack Down* (arcade game 1989; Genesis version, 1990) and *ATP Tour Championship Tennis* (SEGA, 1994). In 1991, Nintendo regained its leading position by introducing the long-awaited 16-bit console, the Super Nintendo Entertainment System (SNES, commonly referred to as "Super Famicom" in Hong Kong and Asia). The SNES had many excellent games including Nintendo's best-selling *Super Mario World* (1991), and the SNES versions of *Donkey Kong, Final Fantasy, Dragon Ball,* and *Street Fighter II* were all well received. The SNES was the dominant home game console for the next five years, and many middle-class families owned one. For those who could not afford to own one, game shops offered SNES rental services. It was the golden era of home video game consoles in Hong Kong.

With the introduction of the Sony PlayStation (PS) and the SEGA Saturn in 1996, Hong Kong entered the age of 32-bit game consoles. In the same year, Nintendo launched its 64-bit console, the Nintendo 64 (N64), in Japan, and it was shipped shortly after its release in Japan through parallel imports. It achieved limited success, as most local players had not yet digested the 32-bit technology, and N64 had few pirated games due to its cartridge-based technology. The Sony PlayStation was the winner, replacing the SNES as the most popular game console in Hong Kong. In addition to having better graphics and sound effects as well as an overwhelming number of PS games (such as Namco's *Tekken 2* [1996] and *Time Crisis* [PS version, 1997], Capcom's *Biohazard 2* [1998] and Square's *Final Fantasy VIII* [1999]), piracy played an important role in the popularization of the Sony PlayStation. When a customer bought the PS console, the shopkeeper often inserted a chip into it so that it could read pirated CD-ROM software. Sometimes the shopkeeper also gave several pirated CD-ROMs to the customer for free.

Sony continued its dominance of Hong Kong home game consoles by introducing the 128-bit PlayStation 2 (PS2) in December 2001. The PS2 was perhaps the most successful and long-lasting game console in the history of Hong Kong gaming, outperforming Microsoft's Xbox (launched in Hong Kong in November 2001), Nintendo's GameCube (launched in Hong Kong in 2002), and SEGA's Dreamcast (launched in Hong Kong in November 1998) by a large margin. Ironically, the great success of the PS2 could be partially explained by the availability of pirated software; hence, its sales figures are not available. Its backward compatibility with PS software, quality customer service, and functional DVD player were additional bonuses for players.

Hong Kong entered a new era of next-generation home consoles in the mid-2000s, following the release of Microsoft's Xbox 360 in March 2006, Sony's PlayStation 3 (PS3) in November 2006, and Nintendo's Wii in December 2008. At the beginning, the Wii led the market over Sony PS3 and Microsoft's Xbox 360. People appreciated its user-friendliness, innovative ideas (in particular the remote controller), Nintendo's high-quality games (such as *Mario Kart Wii* [2008], *Wii Sports Resort* [2009], and *Super Mario Galaxy 2* [2010]) and highly competitive prices (approximately HKD $1800, which is about one-third the price of the PS3 in 2008; the Wii sold for HKD $1500, and the PS3 for HKD $2000 in 2012). The PS3 and the Xbox 360 have been catching up since then, providing more games and functions (in particular, much stronger online connectivity) than the Wii. As of spring 2012, the PS3 is leading the market, as Sony cut its prices and released high-quality games (such as Namco Bandai Games' *Saint Seiya Senki* [2011] and *Gundam Extreme VS* [2011]) exclusively for PS3.

Handheld games have been as influential as home consoles in Hong Kong. Showing a similar pattern of development, these two game platforms have been reinforcing each other, and many handheld games have been adapted from games for home consoles. Japanese game makers dominated the handheld game market for decades until the rise of mobile games and tablet computer games in the late 2010s.

Nintendo launched its first handheld game devices, the Game & Watch series, in Hong Kong in 1980 and nearly monopolized the market for a decade. Each of these non-cartridge devices could only play one game and display time. The Game and Watch series included top NES titles from franchises such as *Mario Bros.*, *Donkey Kong*, and the *Zelda* series. Originally designed to allow game players to play their favorite games outdoors, the series successfully brought many females and children into the world of gaming.

In 1989, Nintendo introduced its first handheld game console, the Game Boy, in Hong Kong, marking the beginning of the golden decade of handheld game consoles. The Game Boy became the most popular and best-selling game console in the history of gaming in Hong Kong. Local players usually brought Taiwan-made pirated cartridges to play Nintendo games, and each cartridge carried thirty games. Nintendo continued its dominance in the market by launching upgraded versions of Game Boy including Game Boy Color in 1998 and Game Boy Advance in 2003.

In 2005, Hong Kong entered a new era in handheld game consoles, as Nintendo and Sony released their next generation consoles, namely the Nintendo DS (NDS) and Sony's PlayStation Portable (PSP). The consoles have been competing shoulder-to-shoulder since then. Local players see the NDS as a handheld game console but regard the PSP as a multimedia device. Thanks to brilliant NDS games such as Nintendo's *Nintendogs* (2005) and *Mario Kart DS* (2005), as well as backward compatibility with Game Boy Advance cartridges, the NDS outsold the PSP in the local market by a small margin in the beginning. However, in the last few years, the PSP has overtaken the NDS because it has more games than the NDS. With the rise of new forms of portable gaming in recent years, however—in particular mobile games and tablet computer games—traditional handheld game consoles are struggling to thrive.

The Third Wave of Gaming: Online Games and Mobile Games, Late 2000s to the Present

Online games and mobile games have been the fastest-growing areas in Hong Kong gaming over the last decade. Japan is no longer the name of the game in these areas. South Korea, Taiwan, the United States, Europe, Japan, and Hong Kong are all sharing the market.

Online gaming began in the early 2000s and has been growing steadily over the years. In 2003, Hong Kong had about 200,000 online gamers (*Hong Kong Economic Times* 2003), a number that increased fourfold within a decade.[3] People play online games mainly at home and in Internet cafés. At home, people play online games with their PCs, using a CD-ROM (purchased from convenience stores or game shops), or programs downloaded from official websites. They also buy bonus cards, tools, and game currencies to play the games. At home, players typically prefer role-playing games (such as Square's *Final Fantasy Online* [2002] and Blizzard Entertainment's *World of Warcraft* [2004]), which are more compatible with PCs.

Many students and youth prefer playing online games in Internet cafés. The rise of these cafés in Hong Kong has sped up the decline of game centers. The year 2001 marked the beginning of the online game boom in Hong Kong, and that year, about 100 Internet cafés were founded in Hong Kong (*Next Magazine* 2004). At its peak in the mid-2000s, about 300 Internet cafés were operating in Hong Kong.[4] People found them more flexible (open twenty-four hours per day with no age restrictions) and more affordable than game centers.[5] Some saw them as places to kill time or even spend a night.

Compared to arcade gamers, patrons of Internet cafés are younger and many are students. Game addiction and gambling have caused concern among teachers, parents, and social workers.[6] War games such as Nexon Corporation's *Counter-Strike Online* (2008) are very popular in Internet cafés. In recent years, owing to high rents and minimum wages, the number of Internet cafés has been dropping, and many have moved from the ground floor to the second or third floor to cut costs. There were 180 Internet cafés in April 2011, and the number has been dropping (*Shingpao* 2011). About half (54%) are owned by two chains, namely i-One (thirty-seven Internet cafés, founded in 1999) and Msystem (twenty-eight Internet cafés, founded in 2002).

Foreign online games have been dominant in Hong Kong. South Korea, the leader in the market, has provided a large number of games, many of which have large followings and have jump-started the online game culture in Hong Kong. NCSOFT's *Lineage* (1998), released in Hong Kong in February 2000, was the first online game for many local players. In 2002, GRAVITY's *Ragnarok Online* (2002) created another commotion. South Korea has continued to release hot titles one after another, including T3 Entertainment's *Audition Dance Battle Online* (2004, released in Hong Kong in 2006), Rhaon's *Tales Runner* (2005, released in Hong Kong in 2006), and Softmax's *SD Gundam Online* (2009).

Taiwan is another major player in online gaming, having released many games based on Chinese martial arts novels or comics. Popular titles include Chinese Gamer's *Jinyong Online* (2007) and *Chinese Hero Online* (2009). The Taiwanese online game company Gamania set up a very large branch in Hong Kong in 2000 and is the agent for many popular online games in Hong Kong including *Counter-Strike Online* (2008), Eyedentity Games' *Dragon Nest* (2010), and Square Enix's *Fantasy Earth Zero* (2006, released in Hong Kong in 2008). In 2010, Gamania purchased Firedog, an award-wining game developer in Hong Kong. Firedog Computer Entertainment was founded in 1999 by the famous *dojinshi* (self-publishing) group Firedog. Its debut game, *Cupid Bistro* (2001), was named the most popular PC game in the Asian Game Show and received the bronze prize at the Hong Kong Digital Entertainment Excellence Awards in 2002 and 2003 respectively. In June 2002, *Cupid Bistro* was adapted into a Japanese Xbox game and became the first Hong Kong game distributed in Japan. Its Japanese PS2 version, *Cupid Bistro 2*, was launched in August 2003. The company was renamed Firedog Studio when it became a subsidiary of Gamania in August 2010. Gamania and Firedog Studio worked together to develop a 3-D online game called *Tiara Concerto* (2011).

The United States and Japan, two gaming giants, maintain a respectable presence in online gaming. The United States has not released many online games in Hong Kong, but several titles are very popular and long lasting, including Blizzard Entertainment's *World of Warcraft* (2004, released in Hong Kong in November 2005), *Diablo II* (2000), and *StarCraft II* (2010, released in Hong Kong in 2011). Japanese game companies are not particularly strong in the area of online games, but they are catching up quickly. The first Japanese online game that made an impact in Hong Kong was Square Enix's *Cross Gate* (2001). It was released in Hong Kong two months after its formal launch in Japan. Many Japanese online games are adapted from other game platforms including Capcom's *Monster Hunter Online* (2007, released in Hong Kong in 2010) and *Fantasy Earth Zero*.

Mobile gaming has been around for more than a decade, but it did not become a social phenomenon until the popularization of the smartphone in the early 2010s. People usually download game apps to their iPhone or Android phone for free and play them while commuting. Many mobile games can also be played on tablet computers, and people can play Facebook games on their mobile phones. Rovio Mobile's *Angry Birds* (2009) has been the most popular mobile game, and Hong Kong people have a strong demand for its merchandise. Maxims, a Hong Kong cake shop chain, has even acquired the license to make *Angry Birds* cakes and mooncakes. Other hot mobile games include Gameloft's *GT Racing Motor Academy* (2009), HalfBrick's *Fruit Ninja* (2010), City Games LLC's *Shoot Bubble* (2011), XMG Studio's *Cows vs. Aliens* (2011), and Best, Cool & Fun Games' *Bunny Shooter* (2011). It is interesting to note that mobile game developers are from different parts of the world; for instance, Rovio Mobile is Finnish, Gameloft is French, HalfBrick is Australian, City Games LLC is American, XMG Studio is Canadian, and Best, Cool & Fun Games is Brazilian. Gameloft has an office in Hong Kong, its first office in Asia. Thus, Japan and the United States have no advantage in this area. Mobile gaming is not designed for hardcore gamers, who often find mobile games too simple. However, for the same reason, many young females and children are fond of playing mobile games (see Grundberg and Hansegard 2014).

Characteristics of Gaming in Hong Kong

Hong Kong has a relatively long history of gaming. From the late 1970s to the present, playing video games has been a mainstream fixture in youth culture. It has gone through three stages of development—from arcade games, to home consoles and handheld games, and finally to online games and mobile games. The characteristics of gaming in Hong Kong can be summarized in five main points.

First, game piracy is the key factor in popularization. Regardless of its legal and moral problems, game piracy has existed since day one in the history of gaming in Hong Kong and has been prevalent in the areas of home consoles, handheld games, computer games, and mobile games. Game piracy has various forms including software piracy, hardware piracy, and digital piracy. Local game players buy pirated software made in China, Taiwan, Hong Kong, or Malaysia and download games for free from websites in China. Many have never purchased any licensed game software. Pirated home consoles (mostly made in Taiwan) are shipped to Hong Kong, and many Hong Kong game shops and players reprogram them by inserting an unlicensed chip so that the machine can play pirated games.⁷ Hong Kong businessmen have produced some unauthorized game accessories and merchandise such as the motion controller for PS and PS2 shooting games, the loudspeaker for the Game Boy, and *Angry Birds* T-shirts and lanterns. This condition, however, has gradually improved in recent years; the market has become more mature and the Hong Kong government has been more determined in cracking down on game piracy. In addition, game piracy is not a serious problem in next-generation home consoles. Players have to reprogram their consoles to play pirated software, but reprogrammed consoles cannot be connected to official servers to play games online.

Second, Japanese game consoles and software have been dominant in the Hong Kong market, leading their competitors by a large margin in all areas except online games and mobile games. Japan has no competitor in arcade games, providing the machines and the games for game centers in Hong Kong. Namco has been running game centers in Hong Kong since 1977, and now its Namco Wonder Park (six game centers) is one of the largest arcade chains and runs game-related events frequently. In home consoles and handheld consoles, Japan has almost monopolized the market for three decades, with Nintendo, Sony, and SEGA comprising the big three in the market. (Started by the American David Rosen in Tokyo in 1954, SEGA [short for "Service Games", and formerly Rosen Enterprises] has been localized in terms of management and ownership over the decades. In 2004, it was renamed SEGA SAMMY when SAMMY, a Japanese pachinko company, became its largest shareholder.) The only non-Japanese contender is Microsoft, which is lagging behind Sony in the market of next-generation home consoles. Although not the leader, the Japanese maintain a significant presence in online games and mobile games. Many of these games are adapted from such existing popular game series as KOEI's *Sankokushi* and Square Enix's *Final Fantasy*. In mobile games, Japanese companies like to develop card game tie-ins to generate revenue, such as those for SEGA's *Virtua Fighter Cool Champ* (2012) and Bandai's *One Piece ARcarddass* (2012).

Third, Hong Kong is a mature consumption market for gaming. While the Chinese and Koreans are fond of online games and the Japanese prefer home game consoles and computer games, Hong Kong players are more comprehensive and balanced in game consumption.⁸ Different game platforms, including arcade games, home console games, handheld games, computer games, online games, and mobile games, have large followings, and many players play games on multiple platforms. At home, they play handheld games, home console games, computer games, and online games, whereas outside the home they play arcade games, handheld games, and mobile games. The maturity of the market can also be seen in the presence of foreign game developers in Hong Kong and the near-simultaneous introduction of new game consoles and software. Major game developers including Sony, Nintendo, SEGA, Microsoft, Namco, Konami, and Bandai have offices in Hong Kong to promote their game products. Due to geographical, commercial, and legal reasons, foreign game developers often release their game products in Hong Kong soon after their launch in their home countries. Before the formal release, Hong Kong game shops and electrical appliance shops often import the latest game products through parallel importation, making near-simultaneous consumption possible.

Fourth, gaming has had a strong impact on Hong Kong culture and society. It is an integral part of youth consumption culture. Game-related jargon has enriched Cantonese and is used commonly among young people and Internet users. Even Donald Tsang, the chief executive of Hong Kong, used game-related jargon in public to show his affiliation with grassroots society. Game-related jargon also appeared in a public examination organized by the Education Bureau of the Hong Kong government.

Playing video games is a popular form of entertainment. The majority of Hong Kong people are game players, and they play wherever they go. In 2003, Arthur Li, the chief of the Education Bureau, was found playing *Bejeweled* (2001) during the Legislative Council. Hong Kong popular culture has been inspired by gaming, and elements of gaming have been incorporated into Hong Kong movies, songs, comics, TV dramas, children's programming, magazines, and newspapers.[9] There are about one hundred Hong Kong comics adapted from Japanese video games, and two Japanese combat games, *Street Fighter* and *The King of Fighters*, have been made into at least sixty-five Hong Kong comics. Video games are used at school for teaching and learning, in hospitals for therapy, and at community centers as entertainment for the elderly. Universities offer courses on video games, and Microsoft is working with Hong Kong Polytechnic University and the University of Hong Kong to train students to develop games for the Xbox 360. On a darker note, however, some social workers, teachers, sociologists, educational psychologists, and psychiatrists have blamed addiction, gambling, and reclusive behavior on the negative influences of gaming.[10]

Fifth, the Hong Kong game industry is still underdeveloped. Due to the lack of capital, technology, and talent, Hong Kong games are often considered inferior in quality. In the past, Hong Kong game companies developed some computer games, home console games, online games, and mobile games. With a few exceptions (such as Firedog Studio's *Cupid Bistro* [2000], Fore One's *Derby's Tycoon Online* [2001], and Celestial Digital Entertainment's *Stargate Online* [2004]), most were not very well received.[11] In recent years, many game companies in Hong Kong seem to be more interested in serving as the agencies for foreign online games than developing their own games, although some do develop online games and mobile games on the side. Gameone, the largest local game company, is one such example. Gameone has launched a number of very hot foreign online games including *SD Gundam Online* (Softmax, South Korea), *Demi-Gods and Semi-Devils Online* (Sohu, China), *Audition Dance Battle Online* (T3 Entertainment, South Korea), and *Realm of Magic Online* (Net Dragon, China), but it has also developed its own online games such as *Young and Dangerous Online* (2002) and *Dream Gulong Online* (2009). Although the Hong Kong government has identified the game industry as a core area in the creative industry to promote, neither long-term strategic plans nor tangible logistics support has yet been made available.[12] The social perception of gaming remains somewhat negative. It seems that the Hong Kong game industry is unlikely to make any major breakthroughs in the near future, and Hong Kong will remain a consumption center for foreign games.

Notes

1. For the 2002 figure, see *Mingpao Daily News* (Hong Kong), July 11, 2002, page A3. The 2010 and 2012 figures were acquired from the Television and Entertainment Licensing Authority of the Hong Kong government through an e-mail enquiry on February 21, 2012.

2. Regarding the NES revolution in Hong Kong, see Yat-fai Tsang's *A Study of the Game Console Market in Hong Kong* (1991).

3. According to Anthony Fung's estimate, there were about 800,000 online game players in Hong Kong in 2010. That number has been on the rise (see Fung 2010).

4. There were 290 Internet cafés in 2002. See *Mingpao Daily News* (Hong Kong), July 11, 2002, page A3.

5. From August 2003 onward, in principle, people under sixteen years of age were not allowed to patronize Internet cafés after midnight. See *Apply Daily* (Hong Kong), July 15, 2003, page A13. This law has, however, never been forcefully implemented.

6. Regarding the social perception and impact of online gaming in Hong Kong, see Chew 2006.

7. Starting April 24, 2008, following the amendment of the Copyright Ordinance 2007, the modification of game consoles to play pirated games became a criminal offense in Hong Kong.

8. For a comparative study of gaming in Asia, see Ng 2007, 211–222.

9. For examples of Cantonese pop songs inspired by video games, see Yaowei and Jieshi 2005, 53.

10. There are several reports and studies on the negative impact of gaming on young people in Hong Kong, including Leung Nga-man 2010, and Hanhui and Jingyun 2006. For a relatively positive analysis of the online community in Hong Kong, see Fung 2006, and Cheung 2009.

11. The award-winning *Cupid Bistro* and *Stargate Online* were released in Japan in 2002 and 2004 respectively.

12. As a gesture of support, Donald Tsang, the chief executive, visited Gameone in 2009. In 2011, the Hong Kong government spent 4 million Hong Kong dollars to sponsor the Asia Online Game Awards. CEPA (Mainland and Hong Kong Closer Economic Partnership Arrangement) has also provided a framework for Hong Kong game companies to set up its business in China.

References

Apply Daily (Hong Kong). 2003. July 15, A13.

Cheung, Meily Mei-fung. 2009. The role of video games in the cultivation of literacy: A medium perspective. PhD diss., Hong Kong Baptist University.

Chew, Matthew. 2006. Policy implications of massively multiplayer online games for Hong Kong. Occasional Paper No. 176, Hong Kong Institute of Asia-Pacific Studies, Chinese University of Hong Kong.

Fung, Anthony. 2006. Bridging cyberlife and real Life: A study of online communities in Hong Kong. In *Critical Cyberculture Reader*, ed. David Silver, Adrienne Massanari, and Steven Jones, 129–139. New York: New York University Press.

Fung, Anthony. 2010. *2010 nian Xianggang ji Aomen wangluo youxi chanye diaocha shichang baogao (Report on Online Game Industry in Hong Kong and Macau in 2010).* School of Journalism and Communication, Chinese University of Hong Kong.

Grundberg, Sven, and Jens Hansegard. 2014. Gaming no longer a man's world. *Wall Street Journal*, August 19.

Hanhui, Mo, and Zhang Jingyun. 2006. *Participating in Online Gaming: Young People's Views on Avoiding Committing Offences and Protecting Personal Cyber Property.* Hong Kong: Hong Kong Federation of Youth Groups.

Hong Kong Economic Times (Hong Kong). 2003. June 23, A11.

Leung, Nga-man. 2010. Online game playing and early adolescents' online friendship and cyber-victimization. Ph.D. diss., Chinese University of Hong Kong.

Li, Sandy. 2011. Shop rents hit all-time high. *South China Morning Post*, July 29.

Mingpao Daily News (Hong Kong). 2002. July 11, A3.

Mo, Hanhui, and Zhang, Jingyun. 2006. *Participating in Online Gaming: Young People's Views on Avoiding Committing Offences and Protecting Personal Cyber Property.* Hong Kong: Hong Kong Federation of Youth Groups.

Next Magazine (Hong Kong). 2004. December 1.

Ng, Benjamin Wai-ming. 2007. Video games in Asia. In *The Video Game Explosion: A History from PONG to PlayStation and Beyond*, ed. Mark J. P. Wolf, 211–222. Westport, CT: Greenwood Press.

Ng, Benjamin Wai-ming. 2009. Consuming and Localizing Japanese Combat Games in Hong Kong. In *Gaming Cultures and Place in Asia-Pacific*, ed. Larissa Hjorth and Dean Chan, 83–101. New York: Routledge.

Shingpao (Hong Kong). 2011. April 27, A8.

Tsang, Yat-fai. 1991. A study of the game console market in Hong Kong. MBA thesis, Chinese University of Hong Kong.

Xu, Mingfang. 2010. *Dianwan wuxianwan* (Infinity of Gaming). Documentary TV program, Radio Television Hong Kong.

Zhu, Yaowei, and Chen Jieshi. 2005. *Xuni houleyuan: Toushi diannao youxi wenhua (Virtual Paradise in the Backyard: Seeing Through the Culture of Computer Games).* Hong Kong: Enrich Publishing.

HUNGARY

Tamás Beregi

When the average foreigner is asked about Hungary, he or she can usually mention only a few names from the twentieth century, such Béla Bartók, the composer and pianist, Ferenc Puskás, the legendary football player, and Ernő Rubik, inventor of the famous Rubik's Cube, which led to a worldwide fever in the early 1980s. The cube, just like the Russian game *Tetris* (1984), became the symbol of Eastern European creativity, which flourished even behind the Iron Curtain, a symbol of a game that is free in its abstraction and that follows strict mathematical laws, yet still is immensely variable.

The operation of the cube is based on the laws of mathematics. This was a discipline in which Hungary gained a worldwide reputation in the second half of the twentieth century. It is enough to mention the names of two Hungarian immigrants, who made their career in the United States: John von Neumann, a pioneer in modern quantum mechanics, functional analysis, and one of the fathers of the digital computer, and John Kemény, who was one of the inventors of the BASIC programming language. In some of the most elite Hungarian educational institutes, for example, the famous Fazekas High School, the Technical University of Budapest, and the Eötvös Lóránd University (ELTE), mathematical education was always very strong. This didn't change with the spreading of computers, but the lack of technology, as it so often happens, inspired creativity. Thus, in the 1980s, a new industry was born in Hungary in the shadow of the Rubik's Cube. The golden age of Hungarian video game development can be characterized by the strange mixture of technical brilliancy, artistic creativity, and charming amateurism. This is even stranger if we take into account the fact that Hungary became one of the leading video game exporters in Eastern Europe during the mid-1980s (Humphrey 1984, 25).

From the Tail of the Cat to the Fin of the Dolphin

Looking at the ancestors of Hungarian computer games, we must talk about the cybernetic games of Mihály Kovács, the Piarist priest and first Hungarian teacher of computing in high schools. Kovács's games were the hardware themselves; his "thinking machines" were soldered from components obtained from old radios, telephone centers, and technical accessories obtained from shops. For models and inspiration, he used Western technical catalogs and the works of Claude Shannon and others. His most famous game is probably the "agrarian-simulation" game *The Farmer, the Wolf, the Goat, the Cabbage* (1967), which could be played on the cybernetic building kit called the Mikromat. The Mikromat appeared in shops in 1967 and could be assembled at home (Képes 2009, 112–116). The first Hungarian *PONG* (1972) clones, such as the Videoton Sportron 101 console of 1977 and the horrible green-colored plastic Electronic TV Game console of the TV manufacturing company Videoton, appeared relatively early, at the end of the 1970s and beginning of the 1980s. However, the real technical and economical breakthrough happened between 1982 and 1983 as a result of the activities of the company Novotrade, which was founded by an extremely innovative man, Gábor Rényi. Rényi's company was one of the first Hungarian pseudocapitalist ventures, being a joint-stock company that could carry on intensive exportation-importation activity, completely unusual in the communist economy (or "Comecon") countries. Due to relatively liberal and Western-friendly Hungarian economic politics in the 1980s, and with the backing of several banks, Novotrade had high export-import activity with computers, software, medical instruments, and even agricultural machines.

The dawn of the Hungarian video game "industry" can be marked with a competition that was announced through the media in 1983. Anybody could enter a synopsis or a concept for a computer game, and the best ones would be programmed by the Novotrade team or by the applicant on a computer provided by the company for the period of the work. Many synopses were sent by enthusiastic students who didn't have a personal computer and saw the competition as a good opportunity to learn programming. Novotrade was overwhelmed by game ideas, many of which were realized shortly thereafter. Among the earliest Hungarian computer games were *Caesar the Cat* (1983), in which players had to catch mice with a huge cartoonesque cat in a room full of furniture; *Dancing Monster* (1983), in which players had to shoot the body parts of a dancing elephant-like grotesque monster; *Buffalo Roundup* (1983), in which players had to round up buffalo into the fold with a nude cowboy, riding a horse; and the *Chinese Juggler* (1984), in which players had to juggle plates as a Chinese acrobat.

Some of these games were presented at the show held by the Hungarian Trade Commission in London, November 16–18, 1983. This was the first exhibition of software by any Eastern Bloc country, so it received great media attention. The magazine *Popular Computer Weekly*, for example, featured the exhibition on its cover, emphasizing two games, *Caesar the Cat* and *Dancing Monster*. The article praised the movement of the monster and generally the originality and freshness of Hungarian games. It also noted that Commodore 64 and VIC-20 machines were especially popular in Hungary and that Novotrade was the country's most prolific game creator. Another company, Andromeda, based in London and led by Robert Stein, a Hungarian Jewish dissident, made a contract with Novotrade to distribute twelve games every two months. According to the article, the programs are first written on the Commodore 64 and then are rewritten for the Spectrum and BBC machines. It also mentioned that "all of the programs are written by freelance programmers, employed by Novotrade, working from the ideas developed jointly by Novotrade and Andromeda" (Kelly 1983, 5).

In the mid-1980s, various game developing studios were already working with Novotrade. Though the company's headquarters were in the city center, programmers usually worked in rented flats on computers provided by the software house. These young freelancers' real dream salaries were often paid in currency (a privilege usually enjoyed only by diplomats), not to mention the fact that they could learn professional programming. Novotrade and Andromeda had distribution contracts with companies including Mirrorsoft (a division of *Daily Mirror*), Commodore, Quicksilva, Ocean, Mastertronic, Virgin, Activision, and Sierra On-Line. In Novotrade's heyday, as many as fifty projects were in development. "The programmers were, without an exception, geniuses in their twenties. Their average IQ was, in my opinion, between 130 and 140, including the porter," said Tamás Révbíró, who was the artistic director of the company for years (Beregi 2010, 303).

Leading one of the most important developing divisions of Novotrade was Donát Kiss, who was one of the developers of *Caesar the Cat*. "The movement of the cat seems to be trivial," explained Kiss in an interview, "but if I ask you in which order it moves its legs, I am not sure you will be able to answer. András Császár, the other developer of the game was lying on the ground beside parking cars to examine the movement of the cats, and making notes. This seems to be a minor problem, but you cannot imagine how silly it would look if we wouldn't do this quite realistically" (Beregi 2010, 288). Kiss also remembers that because of an unsolvable problem (the tail of the cat started to flash), they had to start all the programming over from the beginning after two-and-a-half months of development.

Kiss's division developed the very popular time-traveling text adventure game, *Eureka* (1984), which was ordered by the British company Domark. In this game, players had to travel through five ages (a prehistoric age, ancient Rome, King Arthur's Britain, Germany during the Second World War, and the modern Caribbean) searching for the missing parts of a talisman that could save the world from an upcoming catastrophe. *Eureka* was born of a commercial idea: Domark offered £25,000 to the first person to complete the game, which involved collecting secret codes (hidden in riddles) and assembling a telephone number to be dialed.

"Robert Stein calls me one day," Kiss tells the story, "and says there is a huge project. There are two guys from the city, with a lot of money, but they don't know anything about the computers. However, they want an extremely difficult game with a very huge award at the end, and with a very big marketing campaign. One guy was called Dominik, the other one Mark; that is why the company was called Domark." According to Kiss, in the spring of 1984, Domark contacted various British software houses with a September publication in mind, but was refused everywhere because of the close deadline. Then one agent suggested Andromeda, where Robert Stein told them that Hungarians usually undertake such crazy tasks. Hearing the deadline, András Császár suggested a really huge salary, thinking he would be refused. "After a few days, Robert Stein called us, saying, 'Listen guys, get visas and travel to England at once, our offer was accepted'" (Beregi 2010, 289).

Eureka's story was written by Ian Livingstone, famous for his *Fighting Fantasy* role-playing gamebook series. One of the first stages for the programmers of Kiss's studio was the development of an interpreter program, a virtual machine similar to the famous "Z-machine" of the American text adventure game company Infocom. With the help of this interpreter, Kiss and Császár didn't have to waste time programming the game on both the Commodore 64 and the ZX Spectrum; it was enough to write the program only once and then, with a press of a button, the codes for the two machines were ready immediately. During the nonstop work, the Commodore and Spectrum teams were connected by telephone and often visited each other's studios. "Imagine two guys sitting in the back of the taxi at 3 a.m., and arguing loudly about the question, whether the strength of the slap of the Vestal Virgin shouldn't be turned off at such-and-such part of the game … Poor taxi driver, what he could think? He saw that we were not drunk, there were no drugs that time in Hungary, but we lived our lives at that time with such problems," laughs Kiss (Beregi 2010, 289–290).

The penalty for each day of lateness was £660, and a paper with the frightening amount on it hung on the wall of the studio. In the last weeks of development, Kiss and Császár were literally living in the office. "Our wives brought the lunch in 3-day turns, and sometimes they even took our children in to let us to see how they were developing," explained Kiss. The development process was especially risky because Ian Livingstone worked on the story in parallel with them. For example, in the final "James Bond" scene, he wanted some "extras" that were impossible because of the deadline. The game was finished at the last second. "The master was sent on a tape. We were in such a hurry, that the man responsible for delivering the tape to the airport had to come to our studio at the block of flats," laughs Kiss while his retro-style moustache jumps up and down. "I didn't even have the time to take down the tape in the stairwell, as he would miss the plane. So I threw the tape out the window of the building. He caught it, jumped in the taxi, and just reached the plane. We knew that the next day 10,000 copies of the game would be produced. If there was one bug that would make the game impossible to finish, we would be dead men, and would be sued" (Beregi 2010, 290).

Fortunately, the game was alright, but almost a year and a half passed after the game's publication until somebody finally completed it. The game was solved first by a teenager who was so shy that he called Domark's secret telephone number three times before he could speak and give the solution.

Thanks to the game's great publicity, by the end of 1984, vast numbers of magazines in England had written about Andromeda and Novotrade. *The Times*, for example, said that Western shops had cleared their shelves to make room for Hungarian games: the journalist compared the success of these games to the success of the Rubik's Cube and mentioned Gábor Rényi's optimistic speculation that Hungarian computer games would soon rival American ones (Humphrey 1984, 25). *The Guardian* had a lengthy article about Robert Stein's career, comparing him to Jack Tramiel, Commodore's founder, who was also an immigrant Jew. The article gave an account of Andromeda's new office in Budapest and in the US. "We could establish a completely new industry in Hungary," Stein said at the end of the article, "turning intellectual property into dollars" (Kelly 1986, 13). Journalists all praised the originality of Hungarian games. This originality can be explained not only by the talent of the programmers, but also by the fact that these programmers were not really familiar with Western video games at that time. The first-class animation of these early games, on the other hand, can be connected to the high quality of Hungarian cartoons and animated films of the 1970s and 1980s. The Pannonia Film Studio (like the Russian Soyuzmultfilm), for example, ranked at that time among the five major cartoon studios along with Walt Disney, Toei, and Hanna-Barbera (Lendvai 1998).

It is impossible to give an account of all the successful games developed by Andromeda and Novotrade: most of them are characterized by new and interesting ideas and their technical realization.[1] For example,

Ariolasoft's *Scarabaeus* (*Invaders of the Lost Tomb* in the US) (1985) was an action-adventure game inspired by sci-fi films and the Indiana Jones movies. In this game, the player's cosmonaut and his pet explored the mazes of an ancient pyramid while solving puzzles and capturing various demons. Technically the game was ahead of its time: players moved continually in a first-person view, and the labyrinth was depicted with raster graphics instead of vector graphics, which was an extremely high technical achievement, and gave a very strong feeling of claustrophobia. This beautiful "pre-*DOOM*" game was awarded an overall score of 96% in issue 8 of *Zzap!64* magazine, and was later even ported to the Nintendo Entertainment System. Quicksilva's *Traffic* (1984) was a traffic-simulation game, where you could control London traffic by manipulating traffic lights to avoid traffic jams. The game, written by three university math students, had five traffic maps, ragtime music, and even synthetic sounds. This was the one and only Eastern European (and probably European) game that was licensed at that time in Japan by Sony and was converted to the MSX standard. In another early, simple, yet brilliant game, Commodore's *Arctic Shipwreck* (1983), players had to balance a three-dimensional ice floe with the movements of a mammoth, keeping the tiny wreck survivors from falling into the icy water. Epyx's *Alternative World Games* (1986) was a humorous Olympics-style program: the various sports ranged from pillow fighting on a gondola in Venice to logrolling in Scandinavia to sumo fighting. Novotrade produced Epyx's *Impossible Mission 2* (1988), the sequel to the famous Commodore 64 game *Impossible Mission* (1984). Just like the first game, it was situated in the high-tech fortress of the evil professor Atombender where the hero, capable of amazing somersaults, had to avoid robots and assemble security combinations. But in the sequel, platforms now had a solid three-dimensional feel, and to reach the professor's secret control room, the player had to tie together musical sequences with the help of a recorder. The game, programmed by József Szentesi, István Cseri (who was also one of the developers of *Scarabaeus*), and Zoltan Kanizsai, was almost as brilliant as the original game and received some rave reviews. *Zzap!64* magazine, for example, gave the game a 96% rating and the prestigious Golden Medal Award, writing that it was "more super a sequel than anyone could have hoped for" ("Test: Impossible Mission II" 1988, 21).

It is largely unknown to the world that probably the most famous and cultic Commodore 64 game, System 3's *The Last Ninja* (1987), was partly developed by Hungarian programmers, a team called SoftView. The game centers on the adventures of Armakuni, the last ninja, who wants to get back the stolen sacred scroll of his sect from the evil shogun. Armakuni has to fight his way with various ninja weapons through six huge worlds (a wilderness, palace, prison, and so on) but also has to solve various spatially based puzzles, which are now commonly found in action-adventure games (how to open a closed gate, how to climb a wall, etc.). The heart of the game was the amazing, detailed, isometric graphics—the work of Bob Stevenson, Hugh Riley, John Twiddy—and also the beautiful musical score, composed by Ben Daglish and Anthony Lees. *The Last Ninja*, with its gardens full of blooming cherry trees, fountains, waterfalls, beautifully animated animals, and with its dynamic yet haunting musical score, was a real audiovisual feast.

According to an interview with Zoltán Ádám, one of the Hungarian programmers of the game *The Last Ninja* was developed in strong collaboration between the Hungarian and English teams (Szerdahelyi 2010, 196–207). System 3 sent designer Mark Cale's main concepts to the teams along with the English artists' graphical work,

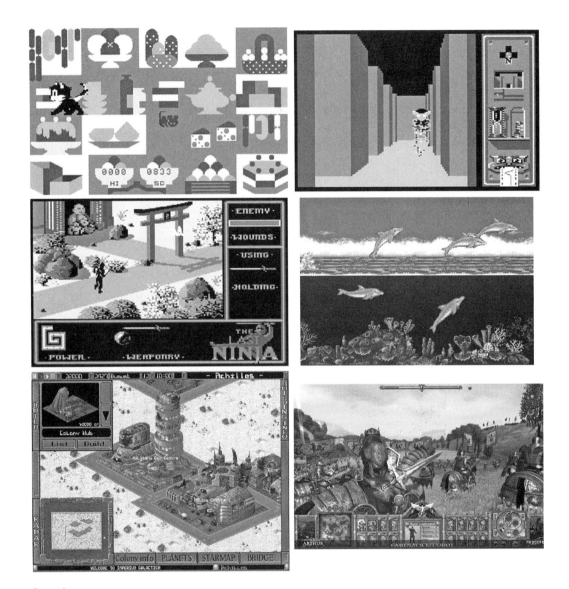

Figure 1

Caesar the Cat (1983), the first internationally successful Hungarian computer game (top, left). The pseudo-3-D graphics of the marvelous action-adventure game *Scarabaeus* (1985) were groundbreaking for the game's time (top, right). *The Last Ninja* (System 3, 1987), one of the most successful Commodore 64 games ever, was partly designed and programmed in Hungary (center, left). The SEGA Mega Drive hit *Ecco the Dolphin* (SEGA, 1992) clearly demonstrates the cartoonesque and peaceful nature of most of the Hungarian video games from this bygone period (center, right). *Imperium Galactica* (Digital Reality, 1997) helped to popularize the space real-time-strategy genre (bottom, left). *King Arthur: The Roleplaying Wargame* (2010) was an interesting hybrid of turn-based and real-time strategy, as well as role-playing games. Its gorgeous visuals and music were praised by critics (bottom, right).

and then came the Hungarian ideas for the world details, graphics, puzzles, the small yet charming episodes, and SoftView's technical virtuosity. SoftView's leader was Ferenc Ruzsa, who was very much into Eastern culture and philosophy; the main programmer was József Szentesi, and István Bodnár, Dániel Erdély, and Zoltán Ádám were among the crew. *Retro Gamer* magazine wrote in detail about the history of the game's development, and thus we can get a glimpse of the almost superhuman fight to squeeze this marvel into the memory of the Commodore 64 ("The Legend of The Last Ninja" 2005). This was possible only with the use of the graphics control program Integrator, which was developed by Ruzsa's team and made possible the use of isometric graphics. Integrator stored the graphics in a huge bitmap image, various sections of which were put onto the screen at any given time. But the beautiful graphics had drawbacks as well: the pictures were drawn before the player's eyes, and the hero could move only on a certain path on the screen.

"Integrator was born when there was a need for a graphics program that could provide a diverse set of images, put them easily on the screen, condense them, allow them to be edited by an artist, and allow them to be painted superbly," explains Zoltán Ádám. "It was like you were seeing a painting, without those repetitive, built-up, stereotyped backgrounds that were characteristic at that time in Commodore games. In my opinion, Integrator is one of the precursors of the modern graphics engines. It was developed to be easily adapted to other computers (for example the Spectrum and Atari) as well" (Beregi 2010, 200).

Programming *The Last Ninja* for a relatively close deadline was an almost superhuman task for an extremely talented yet relatively inexperienced team, and the continual delaying led to tension between System 3 and SoftView.[2] Finally, the program was recalled by Cale (according to Ádám, it was 85% finished at that time), and the program was finished by System 3. The Hungarian team scattered soon afterward. The game was published in 1987 and became one of the greatest hits in the history of the Commodore 64.[3] Sadly, however, the names of the Hungarian programmers are not even listed in the game, and they have yet to receive recognition for their work. Some of these talented young people still had nice careers: the artist, Dániel Erdély, became famous for his geometric form that he called a "spidron."[4] As of late 2012, Ferenc Ruzsa and Isván Bodnár were both at the University of ELTE, the Institute of Philosophy, Department of Ancient and Medieval Philosophy; Ferenc Ruzsa is associate professor, specializing in Indian philosophy, while István Bodnár, specializing in ancient philosophy, is chairman of the department.

The story of the development of another Hungarian megahit game, *Ecco the Dolphin* (1992), is fortunately not so sad. The game was developed by the company Appaloosa, one of the successors of Novotrade. The game's concept was conceived by Ettore Annunziata, and the game was programmed originally for the SEGA Mega Drive by László Szenttornyai (a game development veteran, with conversions of MicroProse simulation games from 1987), József Molnár, and Zsolt Balogh (the graphic artist nicknamed "Talent"). Later conversions included the work of László Mérő (the programmer of *Buffalo Roundup*, and currently a famous psychologist) and József Szentesi (one of the programmers of *Impossible Mission 2* and *The Last Ninja*). This beautiful "dolphin simulation" action-adventure game is a typical example of the pacifist nature of Hungarian games. Ecco is the descendant of Caesar the Cat and of all the animal protagonists that so often appear in Hungarian animated films. In the game, the player has to save his or her dolphin family, which has been sucked into a giant whirlpool. Besides swimming fast, Ecco could eat fish, get to the surface to breathe, use his sonar, and perform amazing jumps. The game, like Sierra's *EcoQuest: The Search for Cetus* (1991), had an ecological message; however, it also incorporated sci-fi elements (time travel) into its story. It became probably the most successful video game ever developed almost entirely by a Hungarian team, and had many sequels and conversions (for example, for Game Gear, Windows, Wii, Xbox Live Arcade, the iPhone, and the iPad).

We must also mention some interesting indie games from the late 1980s and early 1990s. István Rátkai's games were aimed at players who were interested in graphic text adventures but didn't know the English language well. These games, like *Időrégész* (*Time Archeologist*) (1987) or *A bosszú* (*The Revenge*) (1988), were the work of one man and didn't have very beautiful graphics or a parser as good as those found in similar games made by Magnetic Scrolls or Level9, but they were still charming and very enjoyable, with interesting stories and a lot of humor. "I was 17 or 18 when I went to the Ministry of Education," recalls Rátkai about his game *Time Archeologist*, "and said that I want to publish a game privately. They just couldn't do anything with it. They saw that it would be on a tape, but was not music. It had pictures and text, but everything was on a tape. They felt that this project belonged to them, and they were very helpful, but they just couldn't put me anywhere. I still remember the Kafkaesque scene when we wandered the lifeless corridors from one department to the other. Finally, I got the permission for private publication. I drew a cover, hurried to the press on my bicycle, and ordered 300 copies. I bought 5-minute tapes, copied the games to them, folded up the covers, and started to sell the games" (Beregi 2010, 298).

Although Rátkai even suspended his studies at the university for three years, he couldn't live for long away from programming. When I asked him if he tried to sell his games abroad, he replied that there were English versions of both *Time Archeologist* and *The Revenge*, but both were financial failures. "I was twenty-five when one of my acquaintances said that if I had immigrated to America and done the same there, I wouldn't have had to work the rest of my life," he moped. "If I had published *Time Archeologist* in the U.S., it would have sold 300,000 copies, not 300. But the Hungarian market was just too small" (Beregi 2010, 298).

Abandoned Places (1992) is an important Hungarian RPG from the late 1980s to early 1990s. The game was developed by a small team of young programmers, István Fábián, Ferenc Staengler, György Dragon, and Miklós Tihor, who formed an indie company called ArtGame. The game was influenced by famous first-person dungeon crawlers such as *Dungeon Master* (1987) and *The Eye of the Beholder* (1991) but was, in some aspects, more innovative than many of the similar RPGs from the period. For example, players could not only wander in the dungeons but in huge open areas as well, and there were many nonlinear quests. In 1990, with a demo version in hand, ArtGame started to look for a Western European publisher. "We contacted Electronic Arts as well," recalls György Dragon, "and they were interested. But when we sent the demo we got the answer: Sorry, they received a very similar game from another company just one week ago, and they are already negotiating with them. It is true that our game is excellent, but they just don't want to publish two similar games. If we had sent the game one week earlier, we would now be as famous as Raven Software" (Beregi 2010, 300).

An interesting game from this period is the strategy game called *S.U.B.* (1994), which was one of the last programs published by Thalamus Europe. However, due to bad marketing, it didn't become well known. In this game, players had to build an empire of submarines, mining the seabed, researching new technologies, and fighting rival companies. Gábor Forrai, one of the developers of the game, still works in the game industry and is one of the founders of the company Catmoon Productions.

The traditional 8-bit and 16-bit game genres were popular in Hungary even in the mid-1990s, and a good example is the indie adventure-RPG *Newcomer* (1994), which was finished for the Commodore 64, an almost-dead platform already at that time. After fifteen years of development, the new version of this unique nonlinear game (now available in English as well) was published digitally in 2010. Since that time, tens of thousands of downloads have been registered.

An Exclusive Club, but Not Only for the Members

Unlike in Western Europe, in communist countries home computers were extremely expensive for the public, so machines were bought usually in Germany and Austria, and they were sometimes even smuggled in the trunk of the famous Comecon cars, such as the Zhiguli, Skoda, Dacia, Wartburg, or Trabant. Those who could afford them could buy these computers for currency in the diplomatic shops of their home countries, and there was another back door: the famous German and Austrian VAT (value-added tax) called *Mehrwertsteuer*, which could be claimed back later by the consumer. Vienna and its famous shopping street, the Mariahilfer Strasse, were literally swept over by the flood of Hungarian customers at the end of the 1980s. These people were ravenous for modern technical goods, including color TVs, Hi-Fi players, and home computers. The more members of your family traveled together, the more goods you could bring in without paying duty, so many people weren't even ashamed to take their grandma and great-grandpa along on the shopping trip. It is said that some of the biggest Austrian shopping centers at that time even had depositories for elderly people, and there are urban legends of families sitting up their dead grandmothers in the back of the car at the border for the purpose of reducing the duty.

While the computers sold in shops in Hungary were too expensive for most people, the newest software and games were impossible to obtain legally. Thus, these games got into Hungary in the form of pirated copies from Danish, Dutch, and Norwegian hacker groups, through countries such as Germany, Austria, the Czech Republic, Poland, and Yugoslavia. Eastern Europe, and especially Hungary at that time, was the Mecca of the pirate trade. The pirated copies were distributed to various computer game clubs, which became the most interesting and important centers of the Hungarian computer gaming culture.

While in the Western part of the world, hobbyist computer clubs were scenes of innocent technical tinkering and colloquy, the East European clubs were real pirate dens, where social life was sparkling around the sizzling floppy drives and tape recorders. There were no software police at that time in Hungary, so copying games was not a crime at all; it was literally the only way to obtain the programs. The most important computer game clubs were, strangely, not in the city center, but on the outskirts of Budapest in various community centers where the new "supply" arrived on Fridays. These clubs became the most important points for meeting and exchanging information. In socialist countries, it was much more difficult to obtain the newest information about games (such as previews, reviews, walkthroughs, and cheat codes); thus the newest issue of a computer magazine, brought in from the West, was extremely valuable; gamers sometimes literally came to blows with each other to have one or to get a glimpse of a screenshot of an upcoming Amiga marvel. Also, if a game arrived with bad tracks on the floppy or was demagnetized by the train or the underground on the way home, the player had to wait sometimes for weeks to get a good version.

Today, in this time of information hypertrophy and the Internet, it is difficult to imagine the hunger of computer gamers back then and the power that a new game or a new magazine could provide for its owner. According to Tamás Révbíró, the art director of Novotrade,

> I had hundreds of Western computer games, developer software, and user software for the Commodore 64; it was my duty to know and to manage as much as I could. ... Once, a big black car appeared in front of the studio, and one of the leaders of Novotrade told me to sit in it and to take my software bag, stuffed with games. We didn't stop till Balatonaliga, where the car, after a strict inspection, took a turn towards the holiday home of the communist party. Here we were welcomed by someone well known from the TV news: Yegor Ligachev, the second most important man in the Soviet Union [after Gorbachev]. He was on holiday at Lake Balaton with his sixteen-year old grandson, Andrei. In Moscow, at that time, there was probably only one Commodore 64, and he, Andrei was the owner of it. But a computer without programs is just a lifeless ornament in the room. If there is no partner to exchange with—and in Moscow there wasn't—then there are no programs. Fortunately, Hungary was a Great Power of software, and Andrei's grandpa could arrange that I was transported there, with my bag full of games. I let him copy 2–3 dozen floppies (I took the empty floppies as well), and meanwhile we were talking to each other. The young man was very well-informed, and inquiring, and spoke English exquisitely, so I spent very enjoyable hours with him in the holiday home. Meanwhile, my colleague, who finished his university in Moscow, started to argue vividly with comrade Ligachev. He tried to persuade him, the orthodox communist, the main adversary of Gorbachev, about the benefits of the market economy. I doubt he was successful. (Beregi 2010, 304–305)

The center of the Hungarian computer gaming culture was a legendary community house called Csokonai. As of late 2012, the director of the community house was still Lajos Toth, who started the club in the mid-1980s after obtaining a ZX Spectrum. "I knew that computers and computer games would spread with difficulty in the outskirts of the town," he remembers. "This was a poor region, so I knew there will be certainly great need for computers here. ... Then we found out. ... that we will teach children on that miserable computer. We started with an educational program called *Turtle*. Then we told the parents: we must ask them for money to buy another computer. Then came the idea to organize a club for the owners of computers. But this was not so easy under socialism. So we decided to organize an English-type of a club. ... If it is difficult to get in, everybody will want to become a member" (Beregi 2010, 270).

Although on Saturdays the club was opened to the general public, the private club operated on Friday afternoons. For the privilege of becoming a member, you had to have the recommendation of two club members. "Imagine, there were people who waited an entire year to become a member!" laughs Tóth. "And I even got a phone call once from the Ministry of Foreign Affairs saying that there was a diplomat from the

German Democratic Republic who wanted to come here with his son, and was asking my help to get in" (Beregi 2010, 271).

Today, in the time of online communities that are organized spontaneously and Web 3.0, it is very difficult to convey the feeling of what it was like to be a member of this club. Everybody found his second home here, from the elementary school boy, to the university student, to the father in his fifties—from a taxi driver to an engineer. This community was forged by the passion for a hypnotic world, which appeared through games, demoscene programs, and electronic circuits. Spectrum and Commodore users were sitting under the neon lights, around the worn tables, but in the background often had bloody feuds—for example, the walls of public toilets were often vandalized with graffiti abusing the rival computer. There were other computers popular in Eastern Europe, such as the Enterprise, the Commodore Plus4, and the Hungarian machine Primo. The console culture was very limited in Hungary, probably because of the relative difficulty of copying cartridges, so to own a Nintendo or SEGA machine was a real privilege. The most important role of the club was that it created a vibrant community for the gamers and the hobbyists. Moreover, with the help of a program and a mailing database, people with similar interests could find each other throughout the whole country. Not only were friendships born through correspondence, but those living in the countryside could also get games.

When I asked Lajos Tóth if he had any problems because of the copying of the software in the club, he replied, "Never. Although somebody called me once from Novotrade, saying that they will sue me, since one of their programs appeared in our club. I said to him: 'Okay, sue me. Think it over: what can I do if he doesn't protect his programs? I am the head of a public institution where thousands of young people come, so it is natural that the program will appear here sooner of later. I can guarantee one thing: if in this building, somebody buys or sells a program for money, I will forbid him or her to enter the club. If children can get the program, this is their fault, not mine.' They were fuming, but didn't call me anymore" (Beregi 2010, 272).

The most important merit of the Csokonai Club was probably the organizing of the annual show "Computer Christmas." This show, also known as the *BIT-LET* or *Mikrovilág* (*Micro-World*) Christmas (the names of the two most important Hungarian computing magazines at that time, who were the sponsors of the show), was the biggest annual meeting of computer game enthusiasts and hobbyists for more than a decade. The program originated in the club, but around 1985–1986, it moved to the great hall of the Hungarian Technological University. It was a completely crazy, chaotic, yet harmonious event—the real symbol of Hungarian computer gaming culture in the 1980s. In the basement, thousands of people swarmed around the computers, copying pirated games, while on the first floor professional companies sold their products and held presentations. This show excited the genius math student with huge plastic-framed eyeglasses just as much as the family man, with old-fashioned, worn-out nylon bag in his hand, seeing an AMIGA computer for the first time in his life.

"In 1987, after the Budapest International Fair, it was already the biggest computer show in Hungary," Tóth explains, whose thick hair even stands up as though blown by the wind of nostalgia. "It was just unbelievable. Five to ten thousand people were waiting in the queue. We even had to close the gates once or twice because there was no room for more people. Once a man was sent from a cultural institute in Moscow to see what is going on here. In 1990–1992, people from the countryside were already coming in buses, but we had visitors from Novi Sad with computers in their backpacks. It is amazing what kind of a life was there" (Beregi 2010, 272).

A unique event of the 1986 *BIT-LET* Christmas Show occurred when the technicians of the Hungarian Post transmitted computer programs from the top of the building, which could be received by computers throughout the whole town. Meanwhile, Hungarian Radio also transmitted programs, coded as audio sounds. It was actually a simple technique: personal computers stored programs on tapes in audio format. Thus it was possible to record a program that was transmitted on a certain frequency. As we learned from *BIT-LET* magazine, these programs were sometimes even recorded with a microphone near the TV's speaker (Angyalosi 1987). Although it was far from being a perfect technology (for example, Spectrum owners could record programs more easily than the C64 owners), specialists were already speculating as to the future of the technology that was called RADIOSOFT and its future use in education, advertising, and communication. Games, stored in similar audio formats, were so popular at that time in Hungary that one of the most famous pop bands, KFT, included a ZX Spectrum game as a final track on one of its records.

Besides Csokonai there were, of course, other clubs in Budapest and the rest of the country. Many of them were organized with a strong educational purpose by the John von Neumann Computer Society. There was a club for the Spectrum owners in an old factory called "Szellőzőművek" (Ventilation Works) as well, where a dozen computers were connected to a central tape player. When the long-awaited games arrived, the operator of the tape player pressed the button and blew a whistle so others could start to record the programs at the same time.

In the golden age of Hungarian video gaming, not only was club life very special, but the computer game press was as well. While in the Western world the first magazines specializing in home computer games (such as *CRASH* and *Zzap!64*) appeared in the mid-1980s, in Hungary there were a few years of delay. The first gaming magazines, *Spectrum Világ* (*Spectrum World*) and *Commodore Világ* (*Commodore World*), were the brainchild of Lajos Rutz, and appeared very late, only in 1987 and 1988 respectively. On a technical level, these magazines were far below their Western counterparts and were laughably amateurish. The black-and-white screenshots were barely recognizable, and the covers were fabricated at home manually. However, the amateurish appearance, as it often happens, was compensated for by very good content, due to a cult figure known as CoVboy, whose real name was László Kiss. Together with Lajos Rutz and László Székely, Kiss is the father of Hungarian computer game journalism and became famous for his entertaining style, graphomania, sarcastic humor without any respect for authority, and his capacity to consume any quantity of alcoholic liquids. He is a real gonzo type of journalist, as he was called in a 2006 article in *Tomorrow* magazine (Stöckert 2006). Kiss recalls the founding of *Spectrum World* in the following way:

> We didn't care about the minor problems, that at that time you couldn't start a private magazine: to publish a periodical you needed some kind of a social organization behind it. So we decided to go to this and that ministry for the permission. I arrived at the meeting from a punk concert in a bomber jacket and heavy boots; furthermore, I had a hangover, and also drank some alcohol on top of it. In the Ministry of the Education, in

an upholstered room full of palms, some keeper-of-a-seal was sitting, a lady called Pamela. Her main concern was that there was not a social organization behind us. However, I had put the Constitution of the People's Republic of Hungary in my bomber jacket the previous evening, and taking it out, I asked Pamela: please, show me where it is written down that I, László Kiss, cannot publish a magazine. She said there is still a need for a social organization. I said: Okay, the organization is called *Spectrum World*. But then she just kept on asking, who is behind this organization? I replied: here sits Mr. Székely, he is the Spectrum; and here I am sitting, I am the World. And I added, dear Pamela, I will sit here every morning if you don't give us that f***ing permission. So, in the end, she said that the magazine could be published monthly, as a book in 12 parts. (Beregi 2010, 283–284)

Another anecdote recalls the making of the cover of the magazine's first issue:

We didn't have the slightest idea how the press works, we only knew how to hand down a copy. For example, the color breakdown looked like the following: Lajos Rucz appeared with five liters of wine, and then we cut out the parts of the cover from colored cartoons completely drunk, and then assembled them together with glue. The pen writing "Spectrum" was made by Lajos ... but we were drunk and we didn't have a ruler, so you see that the lines are not straight. So this was the first issue, it wasn't a real samizdat, but wasn't a real magazine as well. The later issues were made with the use of an excellent Commodore 64 editing program, called GEOS. (Beregi 2010, 284)

If we compare these magazines to their Western counterparts, besides the obvious technical differences, we immediately recognize some differences in content and in the style of the writing as well. There are fewer reviews, but they are longer and much more personal in style. The reviewer addresses the reader continually, and the articles sometimes include the humorous comments of the editor as well. While in Western computer game journals the correspondence column (letters to the editor) usually has a marginal status, here it was the heart of the whole magazine. The correspondence column was illustrated with naïve, caricaturized drawings by Mihály Müller, writing as Getto Kis, and editor CoVboy communicated in a fatherly yet very sarcastic voice with his readers, who, instead of taking offense at his style, enjoyed his parodistic verbal abuse and indoctrination. The correspondence column editor thus became a real guru, the teller of the "truth" in the digital world. His cult status in Hungary can probably be explained by the previously mentioned fact: obtaining information about games was really difficult here, so the owner of the information became a real master. The magazine tried to solve the problem of the lack of information by publishing the mailing addresses of some of the most important computer game companies in Europe. Moreover, the headquarters of Rucz and CoVboy became a gathering place for friends, where anyone could always find the coolest new programs and the coolest bottles of beer.

In the wake of the Commodore 64 at the beginning of the 1990s, new computer game magazines appeared, including *Computer Mania*, *576 Kilobytes*, and *Guru* (originally in floppy disk format), which were aimed at Western technical quality. But this was the beginning of commercialization, the dawn of the golden age of club life. It was more risky to copy games, and from the second half of the decade—almost fifteen years late—the newest game software on CD-ROMs could finally be bought officially in shops. There were still half-legal ways to get around the software policy; for example, you could borrow the game from the shop for a few days, then bring it back (of course, only after copying it at home), but these maneuvers were soon banned as well.

The End of an Era, or the Beginning of a New One?

From the mid-1990s, gaming culture in Hungary went through significant changes. The closing up of the Western world brought professionalism on one hand, but on the other hand, the special charm and unique subculture that was characteristic of the 1980s completely disappeared. Thus, the difference between the 8-bit and 16-bit era and the new PC/CD-ROM era was probably much stronger in Hungary than in the Western world. An original idea, enthusiasm, and technical virtuosity in game development were no longer enough. The Hungarian gaming industry just couldn't keep up with foreign developments of millions of dollars and so almost completely lost its status in Europe. But there are of course some important games from this period as well. For example, the company Digital Reality, led by Gábor Fehér (who as of late 2012 was the director of Disney Online, Budapest) became well known for its real-time space strategy games, the *Imperium Galactica* series, the first part of which appeared in 1997 (its successor, *Imperium Galactica II: Alliances* [1999], was even more successful internationally and won the 2000 BAFTA Interactive Entertainment Award in the "Music" category). These were the forerunners of other Hungarian real-time strategy games, such as Philos Laboratories' pseudo-Aztec-themed *Theocracy* (2000) or StormRegion's parodistic *S.W.I.N.E.* (2001), about a bloody conflict between pigs and rabbits. Invictus Games' *1NSANE* (2001), an off-road, multiplayer, car racing game, became an international hit, gaining many awards, and became popular even in South Korea. There was an amazing, epic steampunk tactical-adventure game, Philos Laboratories' *Escape from Alcatraz*, in development; however, the company went bankrupt in 2004, and only a rip-off of the game was published. But Neocoregames' *King Arthur: The Roleplaying Wargame* (2009) gained a nice reputation for its Hungarian programmers, as did Digital Reality's beautiful shooter *Sine Mora* (2012).

Although the time when Hungarian games flooded the shelves of the Western shops seems to be long gone, some new companies emerged from the ashes of the titans. These companies usually develop games for mobile platforms (e.g., Catmoon Productions, Artex Studios, and Two Fish) and work with graphics developments (such as Other Side Outscoring) for various companies, proving that Hungarian animation is still world class. The most famous of these animation companies is probably Andrew G. Vajna's Digic Pictures 3D, founded in 2001, which produced videos not only for games including *Warhammer: Mark of Chaos* (2006), *Assassin's Creed II* (2010), and *Mass Effect 3* (2012), but also worked on some special effects on the movie *Terminator 3: Rise of the Machines* (2003). Their game trailer for *Assassin's Creed III* (2012) also won the "Best Trailer" award of G4TV in 2012 (see http://www.g4tv.com/thefeed/blog/post/724761/x-plays-best-of-e3-2012-winners-announced/).

The nostalgia factor, however, is extremely strong in Hungary. For many years there has been an annual retro-gaming exhibition in the Csokonai community center, organized partly by Gábor Szakács (Sakman), who is one of the biggest retro game collectors in Hungary (his collection, more than 1,000 pieces of old hardware and software, includes even some rare handheld games). One of the most interesting museum exhibitions of the year 2010 was the computer game history exhibition *Űrhódító* by the Hungarian Museum of Science, Technology and Transport. This exhibition, organized by Gábor Képes, Head of the Department of Technological Museology, and a curator of the computer science collection, saluted the fortieth anniversary of the game *Space Invaders* (1978) and the history of its Hungarian variants. In the same year, the *PixelHeroes* party was held in Budapest. This party, organized on the occasion of the publication of *PixelHeroes: The First Fifty Years of Computer Games*, became one of the biggest retro-computer game and chiptune parties in Europe, with exhibitions, a roundtable talk, and concerts with the participation of one of the most iconic Commodore 64 music composers, Jeroen Tel. The concerts were organized by *Budapest Micro*, which is a chiptune community concentrating on the use of the legendary Commodore 64 SID chip, and the Nintendo musical chip in modern electronic music. The retro-themed articles of Gabor Stöckert on the blog IDDQ are also extremely popular (see http://iddqd.blog.hu/), and it seems that the culture of video games is starting to become part of media studies at universities as well.

The time is ripe for all of this, as the golden age of video games is a real goldmine not only for nostalgic people, but for popular media researchers as well. The unique Hungarian computer games of the period, just like the cube of Ernő Rubik, are the embodiment of one of the most ancient and sacred desires of *Homo ludens*, and to top it all off, they speak a language that can be understood without interpreters or dictionaries.

Notes

1. An interesting chapter of the history of Andromeda concerns the discovery of the Russian game *Tetris* (1984); this story could be a topic of an entire essay. The game was discovered accidentally by Robert Stein during a visit to the Hungarian Computer Technology Coordination Institute (SZKI). Stein immediately realized the potential of the game; however, the obtaining of licensing of the program became something almost similar to a spy movie. Finally, Spectrum Holobyte, Atari, and Nintendo were fighting for the licensing of the game. See the documentary film *Tetris: From Russia with Love* (2004) by Magnus Temple, described at http://www.imdb.com/title/tt0409371/.

2. Hugh Riley says in an interview with Kai Spitzley that he "spent 3 weekends in Budapest (still behind the Iron Curtain) where it became apparent that their programmers were having difficulties. They were programming in Forth (used for washing machines) so back in London, John Twiddy was brought in to rewrite the code and we worked together in Mark Cale's (Mr. System 3) front room and various other places." From http://lastninja.lemon64.com/old/index.html.

3. Mark Cale says in an interview that approximately four million copies were sold for the Commodore 64. From "The Legend of the Last Ninja," 57.

4. Originally, the spidron was modeled in 1979 by Erdély as homework for Ernő Rubik's design class at the Hungarian University of Arts and Design.

References

Angyalosi, László. 1987. Sugárzunk? *BIT-LET* 40:17.

Beregi, Tamas, ed. 2010. *PixelHeroes: The First Fifty Years of Computer Games*. Budapest: Vince Kiadó.

Humphrey, Peter. 1984. Hungary leaps into high technology. *The Times*, November 27, 25.

Kelly, David. 1983 Hungarian dancing monster. *Popular Computing Weekly*, December 1–7, 5. http://www.zxsoftware.co.uk/8bitgamer/?p=25.

Kelly, David. 1986. Hungary silence. *Guardian*, August 28, 13.

Képes, Gábor. 2009. A way to the modern informatics: Mihály Kovács and his disciples. *HiperGalaktika* 3:112–116.

Lendvai, Erzsi. 1998. Animated cartoons in Hungary. *Filmkultúra*. http://www.filmkultura.hu/regi/articles/essays/anim.hu.html (English translation available at http://web.archive.org/web/20080208021418/http://www.filmkultura.iif.hu:8080/articles/essays/anim.en.html.)

Stöckert, Gábor. 2006. "I was a new voice": Interview with László "CoVboy" Kiss. *Tomorrow Magazine*, August. http://cov.szpeti.hu/oldal.php?oldal=extra/covboy_tomorrow.

Szerdahelyi, Márk. 2010. Interview with Zoltán Ádám, one of the Hungarian developers of *The Last Ninja*. In *PixelHeroes: The First Fifty Years of Computer Games*, ed. Tamás Beregi, 196–207. Budapest: Vince Kiadó.

Test: Impossible Mission II. 1988. *Zzap!64* 38 (June): 21.

The Legend of the Last Ninja. 2005. *Retro Gamer* 18: 54–58.

INDIA

Souvik Mukherjee

India is the sleeping giant of the video game world. Recent developments in the industry and the entry of new gaming consoles, however, mark a significant shift in the culture and reception of video games. As game designer Ernest Adams comments, "India has the talent, the resources, and the attitudes required to become a major player in this industry. All [they are] lacking is experience, and that will come with training and time" (Adams 2009). Adams's optimism is echoed by Thomas Friedman in *The World Is Flat* with the warning, "So today India is ahead, but it has to work very hard if it wants to keep this position. It has to never stop inventing and reinventing itself" (Friedman 2007, 28). Friedman's comment follows from his conversation with the CEO of India's pioneering game development company, Rajesh Rao of Dhruva Interactive. Rao himself catches the spirit of India's ambition to be a global player in gaming through a ludic metaphor: "I think today [the] rule is about efficiency, it's about collaboration and it is about competitiveness and it is about being a player. It is about staying sharp and being in the game" (Friedman 2007, 1).

Almost in answer to Rao's hopes, the Indian IT and BPO monitoring body NASSCOM (National Association of Software Services and Companies) stated in its 2011 report that "Casual Gaming in India is at a nascent stage but is expected to grow significantly at a CAGR (Compound Annual Growth Rate) of 32% over [the] 2010–15 period" (National Association of Software and Services Companies 2011, 4). To put this into context, casual gaming revenue is expected to *quadruple* in the next four years. Gaming, already a popular activity, is growing hugely as a cultural activity in India, and it is interesting to examine the various facets of gaming culture in the country.

A Ludic History

Play has always formed a central part of the sociocultural experience of Indians. *Lila*, or "divine play," is a key concept in Hindu philosophy and religious practices. As Richard Schechner comments, "The *maya-lila* notion of playing describes volatile, creative-destructive activities that are transformative, less bounded, less tame, and less tightly framed in time and space than Western play" (Schechner 1993, 36). The god Krishna is associated with divine *lila,* which is also the way in which the universe is managed, in a state of play, as it were. Ludicity informs the very core of Hindu cultural practices; indeed, theatrical performances and dance are connected to the *Ras lila,* the ritual dance celebrating the love of Krishna and his consort Radha. In other religions and social communities, too, play has formed a significant part of Indian culture. The Mughal rulers of India were patrons of sports such as wrestling and pigeon flying; games such as chess and *Pachisi* (called *Parcheesi* in the United States) were extremely popular in court, and often royalty played the latter using color-costumed members of their harems as pieces on large outdoor boards such as the one preserved in Fatehpur Sikri. The history of both the games and their roots in ancient India is too well known to merit a separate mention. Chess (then called *Chaturanga* in its earliest form) and *Pachisi* both date back to circa AD 600. Western games and conceptions of play were introduced in colonial India by the European settlers, and as such, cricket, football, and hockey, alongside indigenous games such as *kabaddi* and *danda gilli* (also referred to as "the poor man's cricket"), add to the rich tradition of play. Indoor games have also formed an interesting mix, including *Pachisi,* which has made a comeback in Indian homes, *Ludo,* and games such as *Monopoly* getting their own Indian variants. Video games form part of this very rich ludic heritage.

Representations of India in Video Games

Indian places and characters have often appeared in video games. For example, Dhalsim, the yogic fighter in *Street Fighter II* (1991), is Indian. Some of the action in *No One Lives Forever 2* (2002) takes place in the bylanes of Calcutta, and *Hitman 2: Silent Assassin* (2002) has India as its setting. In *Fallout 3* (2008) there are some Indian name references; the "Brahmin" cattle is one of the more controversial ones (Microsoft did not release the game in India because of "cultural sensitivity" issues). More recently, *Call of Duty: Modern Warfare 3* (2011), has a mission set in Himachal Pradesh, India, where American soldiers battle Russians—surprisingly, without any intervention by the Indian army! A similarly implausible representation of India is in *Age of Empires III: The Asian Dynasties* (2007), which has Brahmin healers riding elephants and an infantry comprised of Rajputs, Gurkhas, and Sepoys. For those not familiar with Indian culture and history, this can be misleading: the Sepoy, unlike the Rajput and the Gurkha, is not an ethnic community but the standard name for a soldier in the East India Company's time. The word itself comes from *Sipahi* or *Sipah,* which was a generic term for infantry soldiers in the Mughal and Ottoman armies. Finally, elephants were traditionally used by the warrior class or the *Kshatriyas*; Brahmins, or the priestly class, would seldom be seen near them.

Two very prominent borrowings from Indian (more specifically Hindu, Jain, and Buddhist) philosophy are the concepts of the *avatar* and *karma.* The former, loosely translated in video game terms, denotes the virtual presence of the player in the game whether as her own "self" or in character. In Hindu philosophy, however, *avatar* is the incarnation of a god such as the ten incarnations of Vishnu in which He descends (*avatar* originates from the Sanskrit roots *ava* and *tri* meaning "descent") on earth to rid the world of all evils.

The term was imported into video game parlance in 1986 by Chip Morningstar, designer of the online world *Habitat* (1986), who tied the Sanskrit term "avatar" to "[one's] real-time presence in an online world" (Israel 2011). *Karma*, as it is used in video games, is also very different from the original usage in Indic philosophy. Instead of the vast chain of deeds and their complex consequences that form complex networks through many rebirths, the video game *karma* (interpreted as the consequences of in-game action) is instantaneous and simple. Although India is not entirely absent from the discourse within video games and games' stories, the above examples reveal a very Western impression of India. Mediations of Indian culture in gaming are still few and far between, but to get any idea of what they are, one needs to delve into the history of video gaming in the country.

From *Parcheesi* to *Huebrix*: The History of Video Games in India

The history of video games in India is difficult to document. The early days are largely unrecorded and not many people remember earlier games such as the game *Bhagat Singh* (2002), a *Wolfenstein*-like 3-D first-person shooter (FPS) that received rather poor reviews on gaming sites. This was an attempt at a video game adaptation of the then-popular films on the Indian freedom fighter, Bhagat Singh. The first and only attempt at a postcolonial video game, albeit unwitting, *Bhagat Singh* is an adaptation that is technically wanting in many ways. The other game that vies for the oldest game slot is *Yoddha: The Warrior* (1999), developed by Indiagames following the wave of patriotism that followed the Kargil War with Pakistan. According to the gaming magazine *Skoar!*, the game was more popular for its music video than anything else (Sharma 2009).

With the coming-of-age of mobile technology in 2003 and the advent of consoles such as the Xbox 360 in 2006, there was a major turn toward sophisticated and world-standard game design. Indiagames itself has grown hugely in recent years and has been involved in developing games such as *Spiderman* (2003) for mobile phones, *BioShock* (2008), *Garfield* (2004) and *Ra-One Genesis* (2011) (based on the video game-themed blockbuster Bollywood film). Another pioneering entrant in the Indian video game industry, Dhruva Interactive, launched its own game engine based on Direct 3D and aimed at the mass market. In 1998, it collaborated with Infogrames (now Atari) to build the *Mission Impossible* (1998) game for PCs. Dhruva now has a slew of major titles under its belt—collaborative projects such as *TOCA Race Driver 3* (2006) and *Operation Flashpoint: Red River* (2001) (with Codemasters, UK), and independently developed games such as *Maria Sharapova Tennis* (2005). Other major players have emerged in different parts of the country, such as Lakshya Digital in Gurgaon (near Delhi), which started operations in 2004. With one of India's gaming pioneers, Manvendra Shukul, as its cofounder, Lakhsya is now a major player, having had the experience of working on more than seventy-five titles and having offices in Delhi, Pune, and San Diego.

In 2009, FX Labs, a Hyderabad-based company, developed *Ghajini*, the game version of the eponymous Bollywood box office hit from 2008, itself loosely based on the 2000 Oscar-nominated Hollywood thriller *Memento* directed by Christopher Nolan. The developers have claimed this as "India's first 3-D game" (Reuters 2009), but given the unmapped beginnings of game development in the country, this claim is debatable. The game received mixed reviews; as a reviewer on Techtree.com describes it,

> *Ghajini* ... is a worthy effort from FX Labs, the developers of the game. Now let me clarify a few things before there is a major flame war. As a gamer I think the game needs a lot of fine-tuning and technically the game needs at least a year of refinement, but I also feel that FX Labs in on the right track. (Prakash 2009)

Indigenous games, drawing on Indian ethos, first arrived on the PS2 platform with *Hanuman: Boy Warrior* (2009), developed by Aurona Technologies and launched by Sony Computer Entertainment Europe. Another PS2 game with an Indian connection is *Singstar Bollywood* (2007), which is available in Hindi, and there are plans to include more Indian languages. Other PS2 and PSP games with a local flavor are Trine's *Street Cricket* (2010), and Immersive Games' *Chandragupta: Warrior Prince* (2011), which is based on the historical battles of the Indian emperor Chandragupta Maurya. The website GamingExpress.com gives an indication of the growth of this aspect of the industry in India.

> Adding some sales figures, *Hanuman Boy Warrior* has shipped more than 90,000 PS2 copies in India. *Desi Adda*, on the other hand, sold about 80,000 copies on the PS2 and 40,000 on the PSP whereas the Quiz games sold about 10,000 copies till now. Atin expects 'Street Cricket' to surpass all these figures and set a benchmark for upcoming IPs in the future. (Shirke 2010)

The article quotes Atindriya Bose, country manager for PlayStation Sony Computer Entertainment, stating,

> The market is maturing very fast in India and the product quality has improved significantly. All our developers till now, Immersive, Gameshashtra, and Trine, have done considerable development on the PlayStation Network and have done substantial outsourcing work as well. They are utilizing their skills sets optimally and it is only a matter of time for our developers to start developing for the PS3 as well. (Shirke 2010)

In July 2012, Bose announced that Sony would be manufacturing PS3 games in India and that the company was also in talks with some third-party publishers (Desai 2011). Besides Sony, other companies such as EA, Capcom, Warner Bros., SEGA, and Namco Bandai currently manufacture PS2 games in India. The development of the console market will be interesting to watch in the coming years.

While development continues for other platforms, from 2010 onward there has been a surge in the development of games for mobile platforms such as Android and iOS all over the country. According to NASSCOM's study on casual gaming in India, the total mobile gaming revenue in 2011 was USD $84 million and is expected to be USD $113 million by the end of 2012 (National Association of Software and Services Companies 2011, 13). This is far higher than the other platforms such as PCs and consoles (grossing USD $13.82 million and USD $4.74 million respectively in 2011) and is expected to have a higher growth rate. Some popular titles in mobile games are *Super Badminton 2010* (2010), *Dhoom 2* (2007), and *Touch Squash* (2009). The Indian daily newspaper, *The Economic Times*, reports on an iOS success by an Indian developer:

> In June this year [2012], a mobile application developed by an Indian gaming company was the toast of the mobile world as it emerged as the topmost application across the US and UK, notching up 10 million downloads by users of Apple devices. Created by Mumbai-based Games2Win, the app *Parking Frenzy* is a mobile

game where the player's objective is to park his car in challenging spots within a city, thus mirroring the parking challenges individuals face in their day to day lives. (Syal 2012)

The developers declared that they were making $3,000 per day from advertising alone on the game. Other companies such as Rolocule and Sourcebits have also seen big successes. Rolocule is another case in point for the recent Indian gaming success stories. Rohit Gupta returned to India from New York after completing his MS in computer graphics at Columbia University and resigning from Electronic Arts. There he teamed up with Anuj Tandon (who left his job at Infosys) to found Rolocule. Rolocule, with its *Flick Tennis* (2011), won the "People's Choice Award" at the eighth IMGA awards ceremony held in the Mobile World Congress 2012 in Barcelona; it was the only Indian game developer to do so.

Development on the indie front has also seen some sophisticated game design in recent years. Eminent indie game designer, Shailesh Prabhu, comments,

> There are few individuals as well as few teams in India who are "Indie." We run a small We run a small group on Facebook called Local Indie Game Devs, also there is another group called Indie Game Development India Community. They are small but very dynamic groups with very interesting developers from all aspects of game development. However, there are very few completed Indie games out in the market from India. (pers. comm., 2011)

Prabhu is the CEO of Yellow Monkey Studios, which has designed innovative game concepts such as *Finger Footie* (2010), a top-view flick-based football game; *It's Just a Thought* (2011), a more philosophical game about a symbolic journey of a thought inside the player's mind; and *Huebrix* (2012), an unconventional puzzle game. The indie surge is relatively recent: Yellow Monkey Studios was founded around 2008; Kinshuk Sunil, founder of Hashtash, set up the Indie Game Development Community India in 2011; and MPowered, another indie studio, launched its game *Carmelia* in the same year. As Prabhu says it, indie developers in India aim at doing "what we feel has potential, and is fun for people to play" (pers. comm., 2011).

Compared to the obscure and one-off game productions of the early days, the present-day scenario has become one that is much noisier and busier. Gaming websites such as Zapak.com and iBibo.com are becoming popular for online gaming. Interestingly, both of these contain special sections called "Games for Girls" and "Games for Boys." Popular titles are flash-based adaptations of card games, chess, and badminton. Although not yet as popular as it is in the leading gaming nations, direct-to-home (DTH) solutions are predicted to rise in 2015, and Telecom companies such as Airtel are using their relationships with casual gaming companies to offer DTH services.

Besides the earlier collaborations between Bollywood and the video game industry, video games have recently started making headway into popular culture. Bollywood superstar Shah Rukh Khan starred in *Ra-One* (2012), one of India's most expensive films (whose production cost approximately USD $8.3 million). *Ra-One* is the story of a game designer who manages to design a game where the antagonist is more powerful than the protagonist. The game boss, Ra-One (a pun on the demon king of the Indian epic, *Ramayana,* called Ravan) comes to life and causes havoc in the real world. It is up to the player, playing as G-One (which is a pun on the word "Jiwan," which means "life" in many Indian languages) to defeat Ra-One and restore order to the world. Although the film received mixed reviews, the introduction of gaming concepts into India's most popular entertainment institution, Bollywood, certainly reflects the growing acknowledgment of gaming as a cultural element.

The Players

A significant percentage of gaming in India is based on imports, and international titles are extremely popular. *Counter-Strike* (1999) is perhaps the most played game and is supported by the LAN gaming culture in cybercafés, universities, and gaming events. According to a 2009 study by the Internet and Mobile Association of India (IAMAI), although the penetration of massively multiplayer online games (MMOGs) is not high in comparison to other countries, the high profile launches of the Microsoft Xbox 360, Kinect, and the Sony PlayStation 3 in 2009 have added a new dimension to gaming (IAMAI 2009). The study also states that the gamers prefer single-player games and multiplayer single-session games. A small-scale pilot survey conducted for this chapter on the IndianVideogamer.com forum still shows similar trends. More people preferred shooters and adventure games, with racing, sports, and real-time strategy games following suit. The survey also revealed a preference for consoles and handheld devices, as well as the fact that all the respondents were male. This supports the IAMAI's conclusion that the typical Indian player is a college-going male from an affluent family in one of the eight major metropolitan cities. The study also shows a prevalence of consoles as preferred gaming devices; whether this is an indicative statistic remains to be seen. An ASSOCHAM (Associated Chamber of Commerce and Industry in India) survey conducted in 2011 among 2,000 teenagers and 1,000 parents in major Indian cities showed a significant growth in gaming habits, which it identified as cause for concern (*The Indian Express* 2011). More than 82% of the teenagers surveyed said that the average time they spend playing video games is around fourteen to sixteen hours per week; parents and researchers expressed concern at what they felt was time spent away from schoolwork and the possibility of interacting with adults unmonitored over the Internet. The ASSOCHAM statistics claim that children aged between twelve and eighteen years have the following gaming preferences:

- 82% play on consoles.
- 71% play games on a desktop or a laptop computer.
- 61% use portable gaming devices such as a Sony PlayStation Portable or a Nintendo Game Boy.
- 58% use a cell phone or handheld organizer to play games. (ASSOCHAM 2013).

The five most popular games were *Guitar Hero* (2005), *Halo 3* (2007), *Dance Dance Revolution* (1998), *Madden NFL*, and *solitaire*. For most teens, gaming was seen as a social activity and a major component of their overall social experience. Obviously, any assumption that teenagers form the majority of gamers in India is problematic, and such surveys should extend to other age groups. More than 54% of India's population is under twenty-five years of age, and, therefore, the country presents a large consumer base for video games. However, the penetration of video games in the market is only around 3%, as the newspaper *Daily News and Analysis* (DNA) reports (Dhomse 2011). The player distribution is not clearly recorded, although according to the *Economic Times*, most of the players are based in the metropolitan areas and have an average gaming experience of eighteen months.

Online gaming is small in comparison to countries such as the United States, Korea, Japan, and China. According to the website CGTantra.com, much of the small online gaming revenue is generated through cybercafés (Mistry 2013). Of the 140,000 cyber cafés in India, 97% fall into the "unorganized" category, often having outdated systems, poor bandwidth, and pirated software. Competitive gaming is also in a nascent stage, but the E-sports Council of India, new websites, and major events such as the BYOC (Bring Your Own Computer), show signs of promise. As Vishal Dhupar, managing director of NVidia Southeast Asia, states, "Gaming has begun to be seen as a sport in its own right—an 'e-sport' if you will" (Bali and Rossi 2011).

The Industry

Reports on the Indian video game industry are relatively difficult to locate, but the statistics published by global research organizations show much promise. A report on the market for computer games and video games states:

> The Indian games industry grew to Rs 15 billion ($277m) in 2012, recording a growth of 16% over the previous year, the Indian Media and Entertainment Industry Report 2013 has stated. The report also estimates that the games industry in India will grow at a 22% CAGR to reach Rs 42 billion ($776m) by 2017. (Desai 2013)

The promise made in reports such as this one is increasingly being made good by the Indian gaming industry. According to NASSCOM's game studios directory, more than 106 studios are listed as building games across India; some of these are in relatively remote towns. At this point, NASSCOM's unique and extremely vital role in the future of video games in India needs to be discussed. The NASSCOM Gaming Forum was set up in 2006 to coordinate the needs of and skill sets within the video game industry. Today, it has chapters in seven different cities with quarterly meetings, and it also hosts an annual Game Developers' Conference (NGDC). The NASSCOM Facebook group has 2,100 members, and that membership is growing quickly. Besides facilitating interaction among the designers and the player community, the group regularly publishes white papers on the industry and regularly interfaces with the government to seek support for the industry. It also coordinates nationwide and global game jams (events for a diverse set of developers to come together and prototype games).

Despite all the promise and the efforts of NASSCOM and its members, the industry still faces many obstacles. A spokesperson for NASSCOM's outreach program articulated some of these fears to the newspaper the *Times of India*:

> One of the key challenges being faced by Indian gaming developers is the stiff competition from international content which often tends to have better quality. However, given the success of localized games we have seen

in the past, the Indian game developers appear to be improving content that is on par with international standards. The lack of casual games based on Indian environment, society, lifestyle, and culture also inhibit the adoption in Tier-2 and Tier-3 cities. The majority of the games available today is based on international themes and story-lines and invokes limited interest from Indian consumers. Also the absence of viable alternative [payment] mechanisms for online and mobile ... games has hindered innovation [for] game developers. Infrastructure related issues like high cost of consoles and even game titles, inadequate bandwidth, low awareness of gaming as it is a relatively ... newer form of entertainment and not integrated into the Indian culture and the prevalence of piracy are some of the major challenges faced by the Indian game developers and industry. All the major metropolitan cities have their own hubs for pirated software. According to the newspaper, *Times of India*, "New Delhi's famous Nehru Place market has been placed among the top 30 notorious IT markets of the world that deal in goods and services infringing on intellectual property rights." (Athavale 2012)

The non-Indian cultural element, the better quality of the international competition, and the ever-problematic infrastructural issues are the big hurdles that need to be overcome. Game developers, therefore, seem to want to shift toward mobile gaming so as to avoid many of the above problems, although payment mechanisms still remain unaddressed. A NASSCOM study on casual gaming sees an increase in the percentage of mobile gamers from 45.6% to 61.4% from 2010 to 2015. The industry is, therefore, more focused on games for the mobile platforms such as the Android and the iOS. The statistics for games published from 2011 to 2012 show that fifteen titles were released for the iOS, eight for the iOS and Android, and two exclusively for Android; this can be contrasted with seven titles for the PC and five for consoles (PS3, PS2, and PSP combined).

Recent years have seen many new development studios appear in all parts of India. The major concentration of companies is around Mumbai (thirty-four), Bangalore (seventeen), Hyderabad (fifteen), and Delhi and the National Capital Region (seven). Many of these companies consist of relatively small teams of between eleven and twenty-five members, but some of these such as Rolocule (mentioned earlier), The Awesome Game Studio (TAGS), and Yellow Monkey (also mentioned earlier) are already quite well known in the industry. Yellow Monkey is an indie studio and won the Best Original Game award at HOPLAY 2011. The key players are, however, companies such as Dhruva Interactive, Indiagames, Trine, and Lakshya, among Indian companies, and those such as Zynga, Ubisoft India, Electronic Arts, and Sony, among international companies.

Dhruva and Indiagames are the two oldest game developers in the country. Established in Bangalore in 1997 by Rajesh Rao (see above), Dhruva now has between 201 and 300 employees and provides a full range of content creation, game art, and game development on iOS, Android, Flash, and HTML5. Dhruva has worked with illustrious names such as Microsoft, Capcom, Electronic Arts, Ubisoft, Disney, THQ, and Codemasters, among others. Besides winning numerous awards, Dhruva is also noted for being the first studio to get a million-dollar outsourcing contract on a single game title. Set up in 1999 by the other pioneer of Indian game development, Vishal Gondal (see above), Indiagames has a base of between 301 and 500 employees and builds games for mobile devices, social media, and the PC. Gondal's team has also won numerous national and international awards and is credited with building the first Indian FPS. Lakshya Digital, arguably the main player in the country's capital region, also features among the oldest and the largest in India's game development history. According to NASSCOM, Lakshya has an employee base in the range of 301–500 employees and has worked with major game developers such as Ubisoft, Rockstar, and EA. With offices and development centers in Delhi, San Diego, Tokyo, Pune, and Singapore, the company is "leading the game development movement in India" (National Association of Software and Services Companies 2012, 37). Lakshya has also recently formed a joint venture with the US-based console companies Zeebo and Educomp "to sell devices with educational content in India" ("Educomp inks JV with Zeebo, Lakshya" 2011). Zeebo is a low-cost console aimed at the BRIC (Brazil, Russia, India, and China) countries.

The other leading segment of the industry is composed of international companies that have opened up studios in India. Ubisoft Pune is a case in point: the company, globally famous for titles such as the Assassin's Creed, Prince of Persia, and Splinter Cell series makes games for consoles and PCs as well as mobile devices. Electronic Arts, based in Hyderabad, is the other international game industry giant in the country. Identifying the leaders within a growing industry is easy; however, there are hundreds of companies and studios that are starting up in central and southern India. It is toward these that one needs to look to anticipate the development of the gaming industry.

In a short survey conducted for the purposes of this essay, members of the NASSCOM gaming forum on Facebook (both game developers and players) were interviewed. Of the twenty-two respondents, almost everyone (barring one who did not wish to disclose his/her gender) was male. The majority (61%) was in the age group of twenty-five to thirty-five, and 73% said they develop games for Android; but the developers for the PC and the iOS were also a sizeable percentage (68%). When asked how the industry is faring, using a Likert scale ranging from "very poorly" to "very well," 22% said "very poorly," 39% said "poorly," 35% said "neutral," and 4% reported "well." While 52% thought that the industry would take at least five to ten years to reach international standards, 35% were more optimistic and thought that three to five years would suffice.

In the absence of recent and targeted statistics, the survey was deemed the best way to capture the popular opinion about the industry. The demographics clearly indicate the predominance of male developers, and it is evident that designing for mobile platforms is very much the norm in India. There is also a general realization that the industry needs to do a lot better than it is at present but that it is poised to achieve such a goal. The developer community in India is generally considered a "boys' club"; NASSCOM's panel on women in gaming was a new attraction at its 2012 conference. A report by MCV India summarized the panel's opinions:

> The wide-ranging discussion mostly revolved around the challenges faced in educating women about the fact that gaming is a viable career option and a good alternative to pursuing a career in the IT industry. … [A developer] spoke about pressure from family members and the need to educate them about the gaming industry and its long-term stability. With the growth of casual and social gaming, all panelists agreed that having women in the team would help in the development process, especially for games that are also targeted at women. (Bali 2012)

While the skewed gender ratio in the industry is a big area in need of change, other industry trends are apparent. One respondent in the above survey said, "Most significant positive is growth—mushrooming number of studios, indie devs, talent. Biggest negative—most games out of India still are very poor in terms of polish and finish. We lack in terms of execution—which IMO is more important than 'ideas'" (Mukherjee 2012). Another person stated that "everybody wants to be rich thinking Angry Bird [*sic*] can be developed in two months. All are living with [a] Me Too approach. Guys like … are glorifying outsourcing services which can never put you on the global map. Go and ask someone in USA if they know about THAT company. Investors are corrupt because people used first mover advantage. Positive is that there are at least 2–3 companies which want to do good stuff which has never been the case before" (Mukherjee 2013). Other responses highlight the high quality of art in Indian gaming and the quick pace of growth in the industry; however, they are critical of factors such as the development costs, lack of governmental support, immature gaming audience, piracy, and lack of game-related education. The other topics have been touched on earlier; at this point, it might be useful to look at education and gaming.

Game Design Education and Games in Education

Game design is now a popular topic in computing and multimedia courses all over the world, and Indian academic institutions are slowly heading toward hosting design programs. The most prominent name in game design education is that of DSK Supinfocom in Pune. The institute is a global collaboration between the DSK Group, India, and the Chamber of Commerce, Grand Hainault, France. The students of this institute have won many awards within the country and abroad. Another name, from the Art and Design angle, would be that of Srishti School of Art, Design and Technology in Bangalore. Established names such as BITS Mesra among others are also beginning to set up courses related to video game animation and design.

Scholarship regarding game design is still rather scant in India, and there is even less research regarding gaming cultures or the game studies angle. For example, Pramod K. Nayar's book on New Media, *Digital Cool: Life in the Age of New Media* (2012), contains a chapter on game studies. Nayar addresses some of the older debates and issues in game studies and offers an explanation as to why game worlds are "cool":

> Gameworlds combine aesthetics (the interface, but also the graphics, the virtual environment), history, geography (Gamespaces but also the world they represent), social and cultural codes (race, gender, nation), economics (profits, advertising and promotional culture), psychology, pedagogy (as learning environments) and media studies. A medium and practice that spans so many aspects of the social and cultural domains is surely worth examining. *Games are the newest, and "coolest" form of social and cultural practice.* (Nayar 2012, 62, italics in original)

Nayar's comment is a start, and video game scholarship is still at an introductory stage in India. However, the first research in game studies was started as early as 2001, when the field was relatively new all over the world. Souvik Mukherjee, arguably one of the first to complete an MPhil (from Calcutta) and then a PhD (from

Nottingham, UK) in the area, has pursued game studies research from the early days of the field. However, his work on the Indian aspects of video games, especially that on the concepts of *karma* and the *avatar* in gaming vis-à-vis Indian philosophical discourses, is more recent. More students from the leading Indian universities are pursuing PhDs in video game cultures in India and abroad. International game design and theory experts such as Ernest Adams and Trip Hawkins have conducted workshops in the country.

Although the use of games in schools and higher education institutions is virtually unknown (except for recent cases such as my own use of *Assassin's Creed 2* [2009] to teach the Renaissance to undergraduates at Presidency University, Calcutta), interest in games as a teaching medium is on the rise. E-learning companies, such as Tata Interactive Systems, have their own game-based learning unit, and large game development studios, such as Lakshya Digital, also produce games for education.

Game-Related Publications

Besides actual game titles and game design schools, the media has an important role to play in shaping video game culture in the country. *Skoar!* is the only gaming magazine published in India. Started in 2003 as a gaming supplement of Jasubhai Media's *Digit* magazine, *Skoar!* has seen many ups and down in its publication. Earlier a full-fledged magazine, it is now a bimonthly supplement that is brought out by *Digit*, and part of the company 9.9 Mediaworx. *Chip* magazine also brings out a bimonthly video gaming supplement called *Insider: Your Guide to Gaming.*

Other than the print and online versions of *Skoar!*, gamers in India can access online journals such as Indianvideogamer.com and gameguru.in, both of which run features on current games, reviews, and interviews. MCV (Market for Computer and Videogames) India is another website that provides news on the gaming industry. Both Indianvideogamer.com and MCV India are edited by Sameer Desai, a prominent figure in Indian video game reporting.

Concluding Comments

It is hard to offer a conclusion, or if we are to think ludically, *conclusions* to such a study. Indian video gaming, in the early years of the last decade, was a largely uncharted area. Today, with a far more organized approach, zones of dialogue, and conferences announcing the "rise of the indies," video gaming shows more promise both as a cultural phenomenon and as a career option. Nevertheless, this is, arguably, only the incubation period, and there is scope for much more development. The industry needs the support of both the government and the public, but it also needs to pay greater attention to the needs of the Indian playing community and to incorporate Indian contexts in its designs. Whether India can be called a sleeping giant is perhaps a moot question, but its potential as a creative zone, a market, and a playing community is huge and undeniably significant.

References

Adams, Ernest. 2009. The promise of India. *Designer's Notebook*, November 7, http://www.designersnotebook.com/Lectures/India/india.htm.

ASSOCHAM. 2013. 85% metros kids play with technology gadgets: ASSOCHAM Survey. ASSOCHAM India. September 13. http://www.assocham.org/prels/shownews-archive.php?id=4174.

Athavale, Dileep. 2012. Indian computer game developers appear to be improving content. *Times of India*, November 2, http://articles.timesofindia.indiatimes.com/2012-11-02/software-services/34877317_1_game-developer-conference-game-developers-captive-centres.

Bali, Avinash. 2012. NGDC 2012: Women in gaming. *MCV India*, November 5. http://www.mcvindia.com/news/read/ngdc-2012-women-in-gaming/0105797.

Bali, Avinash, and Fernandes Rossi. 2011. An Interview with NVIDIA. *Tech 2*, November 12. http://m.tech2.com/features/gaming/an-interviewnvidia/256902.

Desai, S. 2011. Round 6: Interview with Atindriya Bose, Playstation India. *IndianVideoGames*, July 2. http://www.indianvideogamer.com/features/round-6-interview-with-atindriya-bose-playstation-india/8626.

Desai, S. 2013. Indian video games industry grew 16% in 2012. *MCV India*, April 17. http://www.mcvuk.com/news/read/indian-video-games-industry-grew-16-in-2012/0114217.

Dhomse, Himanshu. 2011. Videogame penetration is a mere 3%. *DNA,* May 15. http://www.dnaindia.com/india/report_video-game-penetration-in-india-is-a-mere-3pct_1543478.

Educomp Inks JV with Zeebo, Lakshya. 2011. *Indian Express*, January 11. http://www.indianexpress.com/news/educomp-inks-jv-with-zeebo-lakshya/736103.

Friedman, Thomas L. 2007. *The World Is Flat [A Brief History of the Twenty-first Century].* Further updated and expanded, release 3.0, New York: Audio Renaissance: Distributed by Holtzbrinck Publishers.

IAMAI. 2009. Report on gaming in India. Internet and Mobile Association of India, December 29.

Israel, S. 2011. How gamers made us more social. *Global Neighbourhoods*, March 23. http://globalneighbourhoods.net/2011/03/sm-pioneers-farmer-morningstar.html.

Mistry, Daneesh. 2013. Gaming gets serious—gamers get professional. http://www.cgtantra.com/index.php?option=com_content&task=view&id=285&Itemid=33.

Mukherjee, Souvik. 2012. Videogames in India: A research survey. *Ludus ex Machina*, March 7. http://readinggamesandplayingbooks.blogspot.in/2013/03/videogames-in-india-research-survey.html.

National Association of Software and Services Companies. 2011. *Casual Gaming in India*. New Delhi: NASSCOM.

National Association of Software and Services Companies. 2012. *Indian Game Studio Directory*. New Delhi: NASSCOM.

National Centre for the Performing Arts (India). 2006. *The Art of Play: Board and Card Games in India*. Mumbai: Marg Publications (on behalf of the National Centre for the Performing Arts).

Nayar, Pramod K. 2012. *Digital Cool: Life in the Age of New Media*. New Delhi: Orient Blackswan.

Prakash, Navneet. 2009. Ghajini The Game (PC review). *Techtree.com*, February 7.

Reuters. 2009. "Ghajini" inspires India's first 3D videogame. Reuters India. January 22. http://in.reuters.com/article/2009/01/22/idINIndia-37594820090122.

Schechner, Richard. 1993. *The Future of Ritual: Writings on Culture and Performance*. London: Routledge.

Sharma, Sriram. 2009. Indiagames to release 2611 based FPS game mission. *Skoar!* November 25. http://skoar.thinkdigit.com/Gaming/Indiagames-to-release-2611-based-FPS-game-Mission_3739.html.

Shirke, Rohit. 2010. Sony Computer Entertainment and Trine launch "Street Cricket" for PS2 and PSP. *GamingExpress.com*, November 9. http://www.gamingxpress.com/newsitem/items/sony-computer-entertainment-trine-launch-street-cricket-for-ps2-psp.html.

Syal, Sudhir. 2012. Mobile apps developed by Indian ventures notching up millions of downloads on global stores. *Economic Times*, July 27. http://articles.economictimes.indiatimes.com/2012-07-27/news/32889364_1_mobile-application-mobile-devices-mobile-game.

Too much gaming makes kids aggressive. 2011. *Indian Express*, July 18. http://www.indianexpress.com/news/too-much-gaming-makes-kids-aggressive-violent-survey/819030/.

INDONESIA

Inaya Rakhmani and Hikmat Darmawan

This chapter provides a general overview of the video game industry and practices in Indonesia. The term "video game" is often used to mean arcade games, console games, computer games, and online games that have become popular over the past twenty years.[1] However, we have found through interviews and literature reviews that console and online games have had the most significant economic and cultural implications in Indonesia, a country of 240 million citizens, 130 million of whom live on the island of Java, where physical and information infrastructure are significantly more developed compared to the rest of the country.[2]

With 45 million Internet users, it is estimated that there are seven to ten million game users in Indonesia and that between the years 2010 and 2012, this number increased by 33% each year (Ika 2011a; Firman 2010). If the trend continues, by 2015 Indonesian video gamers will amount to approximately 30 million. In a country with a monthly GDP per capita of 2.5 million Indonesian rupiahs (approximately USD $278, as of 2012) (Wibowo and Pratomo 2012), the average gamer spends approximately 100,000 to 1 million Indonesian rupiahs per day (approximately USD $11 to $111) during the holiday break (the average four-week school break between mid-June and mid-July) (Ito 2009), with an average of two to ten hours per week throughout the year (Kusumadewi 2009). Among urban Indonesian youth, console games and online games have replaced both more traditional, physical games and television viewing (Barliana 2004, 4). A recent study on video game playing habits revealed that 95% of its respondents enjoy playing, spending an average of eight hours per day doing so (Heryadi 2008, 411).

Video games have not only gradually become a promising industry, but have had complex social and cultural impacts on their users. We discuss these implications throughout this chapter and have tried to describe how video games provide a space for critical expression and the shaping of new communities. Despite Indonesian scholars' growing interest in the study of video game culture, researchers have failed to address the limitations of each other's findings, due to general disorganization in the Indonesian academy. This study is among the first attempts to fill the void by providing a general overview and description of video game development in Indonesia. By borrowing from political economy and cultural studies approaches, we attempt to provide a descriptive portrayal of Indonesia's video game industry and practice.

In this chapter, we provide a history of video games in Indonesia and map out the video game development that is generally divided into three main waves: console games, personal computer games, and online games. Second, we reveal how "gray market culture" (de Kervenoael and Aykac 2008) is the basis of the video game industry and practice in Indonesia. This includes the distribution of console games, the industry's financial "losses," and their implication for gamers. Third, we also describe the government's stance and protectionist policy-making regarding the handling of piracy in the software industry. After failures in law enforcement to support the regulation, the government later expanded its regulatory approaches to include investing in creative economies in which they have included local game developers. Fourth, we describe video game communities based on the largest number of members and briefly explain the interaction within each community to provide baseline data on the structure of Indonesian online gaming communities. Finally, we set forth a more localized context of video game practice in Indonesia through an ethnographically inspired case study on the sociocultural implications of online gambling. Considering the roles of the industry stakeholders, the government, and gaming communities, we elaborate on the current state and issues related to the field and provide suggestions as to how scholarship can facilitate the rapid growth of Indonesia's video game industry.

The Development of Video Games in Indonesia

Asep Kurniawan, editor in chief of *Game Station*, one of the leading Indonesian game magazines with a monthly circulation of 140,000, claims that since nothing has been written about video game history in Indonesia, the exact date of when video games arrived in the country depends on the collective memory of its users. (Since statistical data on gaming in Indonesia is difficult to obtain, if not nonexistent, communities and industry workers alike rely on testimonies of gamers and trend precedents.) Remy Fabian, editor in chief of *Video Games Indonesia*, on the other hand, estimates that Indonesia's video game industry and market dates back to the 1970s with the arrival of arcades and consoles. This development is followed by personal computer (PC) games, handheld games, and mobile games (Remy Fabian, interview, December 23, 2011). Based on both interviews, the development of video games in Indonesia can be traced as far back as the mid-1980s. Because only the upper-middle class could afford to buy video games, the development of video game culture in Indonesia has culturally been separated into two categories. The first is the isolated video game practices of those owning video game consoles, and the second is the "unregistered" or "gray market culture" video game rentals that made video games accessible to lower-middle class gamers. In this chapter, we distinguish between levels of gaming practice by describing how they have significantly different implications because of their class-culture specificity.

Kurniawan divides the most popular games in Indonesia into three general types. The first is console games, in which he includes everything from arcade games to personal console machines such as the Super Nintendo Entertainment System (SNES) (early 1990s), Sony PlayStation, and Microsoft Xbox (late 1990s). The second type is computer video games that later developed into online games, most notably *StarCraft* (1998),

Counter-Strike (1999), and *Ragnarok Online* (2002). Although some computer video games may be available on both consoles and home computers, players predominantly play them on home computers. The third type of video game emerged during the 2000s, marked by the booming of smartphones in Indonesia. Along with the rise of smartphone use, game applications (which are usually compatible with tablet computers) also rose in popularity.

Console Games

In the early 1980s, the Atari VCS 2600 became the first console system available in Indonesia, followed not long after by Nintendo's Game & Watch series. Gamers who have their own consoles practice similar habits to gamers in other countries. Home console users can afford their own consoles, meaning they typically come from the upper-middle class, and they have a more isolated gaming practice among players—playing games not as part of an interaction with other gamers, but individually (Remy Fabian, interview, December 23, 2011). According to Kurniawan, in the early 1980s, the Game & Watch series became a common pastime among students ages seven to fifteen in large cities such as Jakarta, Bandung, Surabaya, and Yogyakarta. In response to both its rising popularity and exclusivity, street vendors began providing rental services for *Game & Watch* games in the late 1980s. These vendors were illegally stationed in front of schools in large cities, offering students affordable rates and a chance to participate in the rising video game culture. Parallel to these *Game & Watch* vendors and mobile rentals, Atari and Nintendo home console system rentals emerged in arcades (sometimes alongside pinball machines), and both were commonly present in Dingdong centers. These places were often mentioned among gamers as "Dingdong" places, most commonly found in shopping centers in large cities.[3] Dingdong places received a negative reputation because of their association with juvenile delinquency such as bullying and school dropouts.

Not all arcade centers were seen negatively by the public. When the biggest arcade company in Indonesia, Timezone (PT Matahari Graha Fantasi), opened its first arcade center in Bali in January 1995, arcade games became more integrated into mainstream culture, compared to the subculture status characteristic of Dingdong places. Timezone quickly became the owner of the largest network of arcades in the country with more than 104 centers all across Indonesia, with 1,000 permanent employees and more than 9,000 arcade machines (see http://www.timezone.co.id/about/profile/). The second biggest arcade company, Amazone, was established in August 2001 and owns sixty-two arcade centers located in large cities all across Indonesia. In the 1990s, shopping malls contributed to the popularity of arcade games through the cooperation between Timezone and the country's cinema monopoly, Studio 21 (Asep Kurniawan, interview, February 15, 2012).

Around 1995, Studio 21 cinemas across Indonesia included a Timezone arcade section where moviegoers could kill time before their films began. Because of its extended arcade, Studio 21 cinemas became a hub in which upper-middle class moviegoers congregated with lower-middle class arcade gamers. Because arcade games were the earliest form of video games in Indonesia, and since they do not compete technologically with home console games and computer games, they have a stable and increasing market (Heriyanto 2011a). According to Akihiko Arai, an arcade game designer from Namco Bandai, Indonesia houses the largest number of arcade machines in the Asia Pacific region. Of the 1,500 arcade machines distributed by Namco Bandai in the Asia Pacific, 1,000 are currently operating in Indonesia. The legal status of arcades made way for them to be integrated into mainstream urban culture, namely through their use in the cinemas. These legitimized arcades and helped sustain their gaming practices and consumption despite the development of more technologically advanced games in Indonesia.

Personal Computer Games

Computer games, on the other hand, marked the second wave of video game development in Indonesia. Although single-player games did not arrive in Indonesia until the 1990s, by the early 2000s, PC games were already growing rapidly in Indonesia. The lag in the popularity of PC games was due to the high cost of personal computers, and this was addressed by the emergence of Internet cafés (*warung Internet* or *warnet*) in 1996, first on the island of Java (Lim 2005). By 2002, there were already 2,500 warnets all over Indonesia (Widodo 2002) and by 2007, this number doubled to 5,000 (Association of Indonesian Internet Cafés [AWARI] 2007). The huge number of warnets led to fierce competition, resulting in market segmentation. According to the head of the Indonesian Warnet Association (*Asosiasi Warnet Indonesia*/Awari), there are three main markets: the first is made up of business center *warnets* that specialize in office-based operations such as printing and photocopying. The second is made up of general warnets in which customers usually browse, check e-mail, and visit social networking sites. The third is made up of online game warnets, where most customers are gamers. The latter type of warnet is comparable to South Korea's PC Bang in that it facilitates physical interaction between gamers, particularly with computer games that are designed to be played collectively, such as massively multiplayer online games (MMOGs).[4]

Fabian and Asep mentioned that the popularity of computer games rose with free-to-play (FTP) games around the early 2000s, such as *Counter-Strike* (1999), *Ragnarok Online* (2002), and *Ayo Dance* (2008) (an Indonesian version of South Korea's *Club Audition* [2007]). The main reason for their popularity, according to Fabian, is the fact that they are free and made even more accessible with the multiplication of online game warnets in large cities. In this sense, the multiplayer nature of their content coincided well with the preexisting collective culture of urban youths that was not facilitated by physical public spaces (Pal 2010).

The majority of these games are developed in South Korea and distributed by local publishers. One of the largest local publishers in Indonesia, Lyto (PT Lyto Datarindo Fortuna), was established in 2003 by Andi Surya. This company, which serves Indonesia, Malaysia, and Singapore, introduced the prepaid card called Game-On, designed to make transactions easier for online games (see http://www.lyto.net/about/). Because these games are distributed by local publishers and are aimed at a local market, they localized the content of imported computer games. An example of this localization, according to Fabian, is that one or two levels of the computer game *Special Force* (2004) are set in Indonesia's capital city of Jakarta. In these levels, the city's landmark, *Monas* (*Monumen Nasional*/National Monument) is visible. The practice of localizing the content of video games in Indonesia is a strategy often taken by non-Indonesian game developers, not only for console games but also mobile games such as *Harvest Moon* (1996) and *Angry Birds* (2009).

Online Games

The third wave of video game development in Indonesia arose with the increase of smartphone use in Indonesia. The Nielsen Southeast Asia Digital Consumer Report stated that in 2011, 48% of Indonesian Internet users accessed the Internet from their smartphones (Ika 2011b). This increase in Internet access from smartphones is related to two factors. The first is the increasing popularity of social network sites such as Friendster and Facebook. In 2010, the number of Indonesian Facebook users reached 69 million (Firman 2014). The second is related to the rise of local game developers. According to Fabian, Indonesian game developers are more focused on mobile online games because they require less capital to finance their development. While console game and PC game development requires at least 100,000,000 Indonesian rupiahs (USD $11,111), mobile online games require barely 10,000,000 Indonesian rupiahs (USD $1,111). These games are furthermore integrated into Facebook, Ovi Store, iTunes, and the Android Market Place. Two of the largest local game developers are Altermyth Studio, established in 2003, and Agate Studio, established in 2009. Agate Studio was set up by sixteen students of the Bandung Institute of Technology in their spare time. In only two years, they produced more than one hundred online games, 20% of which are specifically designed for smartphones, and their workforce rose in number from sixteen to sixty people (Purwanti and Wahyudi 2011).

One week after it was launched, more than one million users had downloaded Agate Studio's online puzzle game, *Earl Grey and This Rupert Guy* (2010) (Herdiyan and Sufyan 2011). Indonesia's mobile phone industry also responded to the growing online game market. One of Indonesia's largest mobile providers, XL, claimed that social media access makes up 60% of their income in value-added service. In response to this, XL embraced local game developers, including Agate Studio, and encouraged them to develop online games for their game portal KotaGames (Purwanti and Hidayat 2012). In 2012, XL expanded the strategy to include online games specifically designed for mobile technology. According to XL's technology director, Dian Siswarini, XL sees a great business opportunity in online games. There are at least 40 million Facebook users in Indonesia and 20 million Facebook gamers, not to mention mobile gamers and online gamers, and this growing market has expanded in parallel with an increase in local game developers (Galih and Ngazis 2012).

A similar strategy has been adopted by Nokia Indonesia. The company is trying to expand its business by cooperating with national entrepreneurs by helping them develop local game applications to be integrated into Nokia mobile phones ("Nokia Siapkan Game 'Smash Mania' [Nokia is preparing 'Smash Mania' game]" 2011). Nokia is currently cooperating with several universities, including the Bandung Institute of Technology, Bina Nusantara, and Universitas Indonesia, by providing scholarships to students who have the skills to develop applications. Global game developers have also recognized Indonesia's market potential. Rovio, the developer of *Angry Birds*, chose Jakarta as the city to officially launch its Facebook application because Jakarta is, according to Rovio, "the global capital of Facebook" (Purwanti and Hidayat 2012). With its expansion into Indonesia, Rovio expects to develop *Angry Birds* on Facebook, including in it "Indonesian attributes" such as batik, the Garuda bird, and the Komodo dragon. At the same time, Rovio is also cooperating with Nokia to organize an idea competition for mobile game content development in Indonesia.

The mobile industry's approach toward identifying a mobile-based gaming market shows that the sector has increasing market value. Content development by local game developers who are currently being supported by the mobile industry also shows that gaming practices in Indonesia are expanding in response to social media usage. The mobile-based online game *Farmville* (2009), for instance, is popular among forty-year-old urban housewives (Zachra 2011). This means that the third wave of video games in Indonesia, unlike the first two waves, is extending beyond its initial base of young, urban users.

Video Games, the Gray Market, and Its Implications

According to Fabian, it is difficult to estimate the number of games imported into Indonesia because dependable data regarding the number of publishers, networks, and official stores does not exist (Remy Fabian, interview, December 23, 2011). Fabian can only estimate that around 65% of the games produced in the US and Japan do enter Indonesia. These games reach Indonesia a mere two days after the games are launched in their countries of origin, as piracy-based gaming businesses in Mangga Dua and Kelapa Gading, Jakarta (the core distribution spots), approach suppliers to preorder.[5] In fact, according to Fabian, the stores selling pirated games are able to provide the newly launched games even before the legitimate stores do. Therefore, although Indonesian console games have a large market, piracy has made it undesirable for game developers. Fabian claims that such video game piracy was not always the norm.

In the early years of the PlayStation 3, 80% of the games played were distributed by authorized dealers, which Fabian related to the fact that pirated video games cannot be played online. However, once piracy was on the increase, authorized dealers withdrew from distributing in Indonesia. Because of the scarcity of these games, their prices rose from 200,000 Indonesian rupiahs (USD $22) to 500,000 Indonesian rupiahs (USD $55 dollars) in 2011 (Remy Fabian, interview, December 23, 2011). Now these games can be either downloaded illegally online or bought at stores that sell pirated video games for a mere 10,000 Indonesian rupiahs (USD $11) per game; both facts push game developers even further from opening authorized dealerships in the country.

The growing phenomenon of gray markets—"in which a firm's products are sold or resold through unauthorized dealers" (Antia, Bergen, and Dutta 2011[4])—in the information technology sector is not exclusive to Indonesia. In 2004, countries such as India, China, and South Korea suffered an estimated loss of USD $59.5 million, USD $510 million, and USD $349 million respectively due to copyright piracy in the entertainment

Table 1

Estimated trade losses due to piracy in Indonesia and levels of piracy, 1998–2002.

INDUSTRY	2002		2001		2000		1999		1998	
	LOSS	LEVEL	LOSS	LEVEL	LOSS	LEVEL	LOSS	LEVEL	LOSS	LVEL
Motion Pictures	28.0	0.9	27.5	0.9	25.0	0.9	25.0	0.9	25.0	0.9
Records and Music	92.3	0.89	67.9	0.87	21.6	0.56	3.0	0.2	3.0	0.12
Business Software Applications	102.9	0.99	63.1	0.88	55.7	0.89	33.2	0.85	47.3	0.92
Entertainment Software	N/A	N/A	N/A	N/A	N/A	0.99	80.4	0.92	81.7	0.95
Books	30.0	N/A	30.0	N/A	32.0	N/A	32.0	N/A	30.0	N/A
TOTAL	253.2		188.5		134.3		173.6		187.0	

software industry, under which video games are categorized (International Intellectual Property Alliance 2009).

Pirated video games have also affected gaming practices in Indonesia. First, Fabian argues that cheap pirated video games have made Indonesian gamers take game developers for granted. With an array of cheap games available, players are less committed to finishing games and throw them away when they no longer play them. This habit has also, according to Fabian, made them unsupportive of local game developers because they have generally been indulged by an abundance of choice due to cheap products. Kurniawan confirmed this assumption by saying that even console systems, such as the Xbox, are popular not because of their features or technology, but because most of their content can be illegally downloaded through the Internet (Asep Kurniawan, interview, February 15, 2012).

Second, video games that are popular in Indonesia, according to Kurniawan, are defined more by what kinds of games are available for piracy (both distributed through gray market stores and/or downloaded online). This is also related to the fact that gamers can request new pirated video games at illegal stores, and thus this limited demand leads to specific games being widely distributed through unauthorized channels. Gamers are known to browse popular games through websites and customer reviews then order games through stores. The rise in popularity of *Counter-Strike* and *Crossfire* (1999) in Indonesia is widely accepted among gamers to have emerged in this way (Richard Samuel Manuhurapon, interview, April 1, 2012).

Third, although Indonesia is home to a large gaming market, its gray market nature and its instability have discouraged foreign investors. Microsoft and Nintendo refuse to open authorized branches in Indonesia because of the high level of piracy (Remy Fabian, interview, December 23, 2011). Instead, foreign companies choose Singapore for better protection and regulation. Indonesia is currently the country with the eleventh-highest amount of software piracy, according to International Data Corporations (IDC) in 2010 (Setiawan and Ngazis 2011). The same research reveals that Indonesia's software piracy constitutes 87% of the market, with an increase of 1% in 2009. Business Software Alliance (BSA) Indonesia estimates a loss of 1.3 billion Indonesian rupiahs (USD $144,444).

Indonesia's gray market gaming culture is a paradox. On one side, distribution through unauthorized channels has pushed foreign investors away from tapping into the country's thriving market. But on the other, it is these illegal channels that have helped the cultural adoption of game content. Users consciously choose to play games that they can obtain for free, making games that are available for piracy the most popularly adopted games. This reality includes the unsustainability of Indonesia's gaming sector as an industry and the sustainability of Indonesia's gaming culture as facilitated through gray market methods. Interestingly, the government has attempted, and failed, to regulate the practices of gray market distribution, which we explain further in the next section.

The Government and the Video Game Industry

Various policy-making efforts by the government in relation to video games have focused on trying to curb piracy with some attention to cultural content. In 2004, the Indonesian government attempted to protect intellectual property rights, particularly copyrights.[6] The law obliged a production code for optical discs, aimed at curbing piracy, specifically with the increase of optical disc production facilities that are alleged to be illegal and assumed to be the basis of the production and distribution of pirated optical discs in Indonesia. The same law was operationalized to include local law enforcement to raid *warnets* and survey their software.[7] The fine for warnets operating with pirated software can reach 500 million Indonesian rupiahs (USD $55,555) or the confiscation of PCs. However, AWARI questioned the effectiveness of these raids; in practice, the warnets pay the field officers between 15 million and 50 million Indonesian rupiahs (USD $1666 to $5555) to avoid confiscation or fining (Gandhi 2005). The method is often mentioned as *jalan damai*, or the "peaceful way," in which field officers help warnets avoid administrative and monetary sanctions in exchange for bribes.

The field realities and ineffectiveness of the regulation resulted in a continuing increase in piracy. This pushed the government to issue another law under the Ministry of Trade stating that all imported production machinery, raw materials, and finished goods in relation to optical discs must be filtered by the government (that is, they are held at customs and checked by an Indonesian surveyor), which added to the unit cost of the end value of the optical discs.[8] The implementation of this law does indeed, in design, hinder piracy in attaining production facilities of optical discs, but the effectiveness of the law depends greatly on field officers—which is why the number of piracy cases continues to increase (Setiawan and Ngazis 2011).

In 2008, the government developed regulations that pertained to online games.[9] The Electronic Transaction Act was established, which aimed to regulate online transactions and to monitor the downloading of illegal games. Under this law, individuals who deliberately and without permission transfer any pirated product through the Internet can be jailed for a maximum of nine years or fined a maximum of 3 billion

Indonesian rupiahs or USD $333,333. The law also allows the government to close down websites that provide services that violate copyrights, such as shortcuts and cheats for gamers, which is even specified in verses 3, 32, 33, and 35 of the law. However, gamer forums criticize the law by stating that hackers will know how to work around it (Yohanlie 2009). The blocking of file-sharing websites, especially those for which the regulation is not intended, in fact motivates hackers to find more advanced methods of entering blocked websites through anonymous proxy, hiding IP addresses, and accessing foreign websites with the assistance of free online translation services.

In 2012, the Ministry of Tourism and Creative Economies established a program to develop creative economies based on cultural arts and media, design, and ICT (Information and Communication Technology) with a budget allocation of 900 billion Indonesian rupiahs (USD $100,000,000). In this program, interactive games are among the fourteen subsectors to which funding is allocated. The program proves that the Indonesian government sees video games as a promising sector, and instead of regulating the spread of video games through illegal channels, the government embraces the industry through investments. Recent data estimates that creative games and animation contribute to 7% of the country's GDP (Kurniawan and Agus 2011). The program also responds to findings that reveal that 50% of Indonesia's video game market still consumes foreign products.

Aside from protectionist regulations and funding allocation to stimulate the video game industry, the government has also noted the production of interactive games that attempt to promote national culture. A cooperative effort between state enterprise PT Telekomunikasi Indonesia (Telkom) and game developer Sangkuriang Studio and PT Nusantara Wahana Komunika, the MMORPG *Nusantara Online* (2009) was launched to target 20% of Telkom's Internet users (Yono 2011). Telkom also distributed free installer CDs to warnet members of AWARI through Telkom's Internet users and the public. The theme of the game is Indonesia's national history, interwoven with the culture of past kingdoms including Majapahit, Pajajaran, and Sriwijaya.[10] *Nusantara Online* was developed based on sociological, archeological, and historical theories that speak of Indonesia's cultural roots. The characters, social system, architecture, and even the virtual map are consistent with Indonesian cultural history in combination with mythological characters, legends, and folklore (see fig. 1).

After three years of game development, Indonesian President Susilo Bambang Yudhoyono launched the game during the 2009 Indonesian Creative Products Week (*Pekan Produk Kreatif Indonesia*/PPKI). PPKI 2009 is an event designed to stimulate the development of Indonesia's creative industry by supporting local entrepreneurs in the culture and technology sector. In his speech, the president expressed his desire to increase creative economies centered on the nation's culture. The event also mentioned *Nusantara Online* as the product of Indonesia's sons and daughters (*anak bangsa*), an obvious symbol of the state's idea of national cultural product. Through officially endorsing *Nusantara Online*, the Indonesian government's approach toward the gaming industry can be compared to that of China's.

According to Zhong (cited in Cao and Downing 2008, 525), the Chinese state strives both to educate young people to resist increasing foreign content in cultural industries and to cultivate a generation that identifies with the state's ideology." Based on PPKI 2009's objective to develop creative economies and the president's

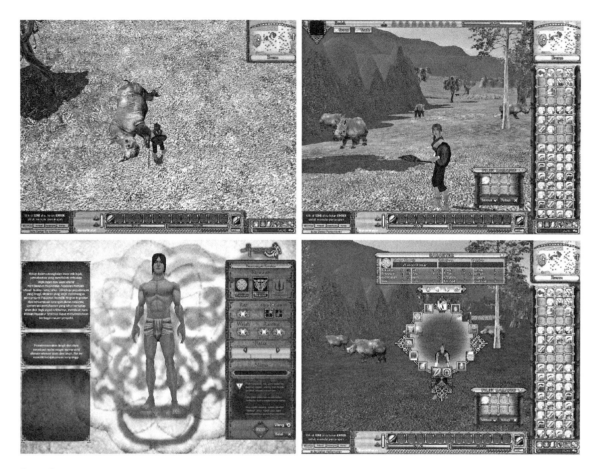

Figure 1
Screenshots from *Nusantara Online* (2009).

speech, which called for authentic Indonesian culture in technology-based cultural products to differentiate the nation from others, however, we argue that *Nusantara Online* is more about commodifying national culture to increase market value than it is about state ideology, as argued by Zhong. Ministers that were present at the PPKI 2009 were of the Ministry of Industry, Ministry of Trade, Ministry of Culture and Tourism, and the State Ministry of Cooperatives and Small and Medium Enterprises. Unlike Zhong's argument that games are used to cultivate a generation that identifies with the Chinese state ideology, the fact that the Indonesian government did not have the Minister of National Education present suggests that the government's interest in the gaming industry works under capitalistic ideology.

This raises the question as to why the political and cultural image of the nation as a "family" (Bulkin 1984), in which local game developers are the nation's sons and daughters and the country is positioned as their father figure, is encouraged by the current government in the context of Indonesia's video game industry. On one hand, Indonesia's current transitional democratic state ideology differs greatly from that of China's authoritative state. On the other hand, Indonesia's authoritarian past, particularly within the Suharto regime (1968–1998), shaped a corrupt system and false capitalism, which employed state ideology as means to rationalize the said corrupt system.

Although there is some attention given by the government toward the local gaming industry, however, it has not been received with much enthusiasm by Indonesian gamers. In 2011, *Nusantara Online* had 17,530 forum members and a total of 26,000 players (Iqbal 2011), there are only 320 threads and 2,442 posts in its forums, as compared to *Ragnarok Online*'s Indonesian forum, which houses 807 threads and 46,444 posts.[11] VGI's thread on local game developers, for instance, has only 2,000 views in comparison to PlayStation's 70,000 views and 8,000 replies. This discrepancy is caused by gamers who are much more interested in US, Japanese, and South Korean games, particularly in their visual quality and taste (Remy Fabian, interview, December 23, 2011). Indonesia's gaming industry, based on the number of players subscribing to MMORPGs, is dominated by Japanese, South Korean, and US game developers that have shaped the visual taste and gaming logics of Indonesian gamers. Although the Indonesian government shows some capitalistically motivated interest in developing the local gaming industry, their strategy will not succeed if there is no effort to recognize that the market is currently dominated by foreign taste and that appealing to the Indonesian gaming market through national culture is not an effective strategy considering the current market taste.

We argue that Indonesia's national rhetoric is moving in a different direction than the consciousness structure of the urban youth, which includes gamers. The lack of structural support for the construction of national ideology provided the opportunity for Indonesian gamers to identify with global taste instead. The development of video games in Indonesia since the first wave has been through imports. The adoption of global gaming culture among Indonesian urban youth is unparalleled by a growing local industry shaped by Indonesian gamers whose tastes, as consumers, are in turn shaped by their consumption of imported games. Among these gamers, online gaming communities have formed, each of which have unique ways of interacting and information-sharing, and these are explained in the next section.

Video Game Communities

The definition and development of gaming communities have been studied as a consequence of the importance of communication in online console games, online games, and virtual worlds in which interaction with other players is vital (Cao et al. 2008). According to Cao and colleagues (2008), there are two different forms of communication in gaming communities; the first occurs in the communities in which players discuss new game features, argue about problems they've experienced while playing, and exchange advice about strategies, while the second occurs during gameplay within games that provide the possibility to communicate while playing. This interaction transcends the online world; the gaming community becomes the most important reason for the player to stay in the game (Cao et al. 2008; Friedl 2003).

Table 2

A list of the largest Indonesian video game communities.

No.	Name of Gaming Community	Year Founded	Members	Number of Threads	Number of Posts
1	Indogamers	2003 Web: 2006	235,079	484,382	6,348,563
2	Liga Game	2007	187,528	64,072	4,387,699
3	Video Games Indonesia	1998 (mailing list), 2002 (forum)	49,303	64,749	2,872,897
4	Kotak Game	2008	32,578	N/A	N/A
5	Egamesbox	2001	17,657	5311	1,037,518
6	Indowebster	N/A	N/A	7,650	997,464
7	Komunitas Gamers Indonesia	N/A	569	767	2062

The three largest Indonesian gaming communities are respectively Indogamers, Liga Game, and Video Games Indonesia (see table 2). The largest Indonesian gaming community, Indogamers, was founded in 2003 and was first established to provide a space for its members to socialize and exchange tips and gaming strategies in online games. The space for warnet members to list warnet prices on the front page of the Indogamers' website is an example of such socialization. As it developed into a Web-based community, threads expanded to include music, sports, film, photography, and other hobbies.

The second largest gaming community in Indonesia is Liga Game (Game League). Liga Game is a professional body that organizes gaming competitions. The objective of this forum is to motivate gamers to become professional players and to shape a positive public perception of gaming as a professional skill. As part of this objective, Liga Game attempts to organize professional competitions that help develop the gaming skills of its members. In order to moderate its members, Liga Game applies the "karma" system, in which the prosocial behaviors of its members (such as helping fellow members) are rewarded with Good Reward Points (GRP) and antisocial behavior (such as deception) is punished with Bad Reward Points (BRP). These points are accumulated and translated into the credibility of each user. Karma points can be transferred to fellow users by those who have posted more than 1,500 times. In the past four years, Liga Game has organized Liga Game Awards that assess good moderators, game administrators, and threads. The winners of the Liga Game Awards are chosen based on voting and are awarded with GRPs. Members of Liga Game usually have offline contact during gaming competitions, which are often sponsored by the industry. They also meet through cash-on-delivery purchases, as the forum includes threads in which members buy and sell firsthand and secondhand computers and gaming equipment.

The third largest gaming community, also the longest-running gaming community in Indonesia, is Video Games Indonesia (VGI). Established in 2002, the community developed in 1998 from a mailing list that centered on the exchange of advice on video games. The mailing list evolved into a forum and finally a Web site and currently has 45,000 members, of which anywhere from 2,000 to 3,000 post actively. While other Indonesian gaming communities such as Kotak Game, Streetgame, Gameqq, and Liga Game concentrate on online games for PCs and are available in warnets dominated by South Korean games, VGI focuses on console games from international publishers such as Sony and Microsoft. Although VGI hosts a large gaming community, the interaction between VGI's members is comparable to Kaskus, one of Indonesia's largest forums, because VGI also includes threads on food, film, cars, news, and other topics.[12] The members' offline interaction is also organized by individuals within the forum, not by VGI as a community.

Differing from the interaction found on Indogamers and VGI, the interaction of Liga Game's members is shaped by the common interest of its members in sharing information about professional gaming and does not allow the sharing of cheats. Around 90% of Indonesian professional FPS gamers, such as *Counter-Strike* players, are members of Liga Game (Richard Samuel Manuhurapon, interview, April 1, 2012). The professionalism and sportsmanship shared among Liga Game members has shaped a community that is trusting and fair. This is demonstrated by the members' preference for purchasing gaming equipment through the community, despite the fact that the price may be cheaper through other forums. This is because Liga Game members are known to be individuals who are serious and reputable in the gaming community; thus, unlike in other forums where online frauds are common, transactions via Liga Game are verified through the culture of the community as well as its GRP system. The unique character of Liga Game that encourages a professional gaming community, as opposed to the communities of Indogamers and VGI that depend on their members' commitment to share information (particularly cheats), suggests that Liga Game is resisting the more dominant gray market culture of Indonesian gamers, although further research is needed to confirm this assumption.

A Closer Look at Gaming and Gambling

Although in this essay we have focused on the larger structure of the gaming industry by describing the waves of video game development, communities, its gray market culture, and the role of the government, it is important to recognize that parallel to the "formal" gaming industry exists unique gaming practices that are left out of literature reviews and statistical data. To strengthen our argument, we take a closer look into how gaming culture among youths has affected their economic and social status in their personal lives and communities. First we acknowledge the growing trend of gambling in relation to video games. In South Jakarta, for instance, there is a growing gambling culture among PlayStation players made possible through gray market PlayStation rentals that proliferate in the area. Gambling culture developed from console games is comparable to sports gambling in which gamers seek the thrill of the competition between players.

The increase of online gambling, which occurs in several large cities in Indonesia, is an interesting trend. Incidents may seem isolated from each other, but news coverage has reported three police raids in three different cities in Indonesia within a four-month period—Jakarta in September 2011 (Pitakasari 2011), Makassar in December 2011 (Pius 2011), and Medan in March 2012 (Portibi DNP 2011). Online gambling in Indonesia is organized through warnets and uses simple online games; examples include free online penguin games and online poker (Farhan [pseudonym], interview, February 28, 2012). It is estimated that warnets have generated hundreds of thousands of US dollars daily (Portibi DNP 2011).

One of the authors of this chapter, Hikmat Darmawan, conducted ethnographically inspired research in Mampang, South Jakarta, to understand the implications of online gambling and how video games are central to it. He studied a community typical of Jakarta in which traditional values are sustained in contrast to the city's urban setting. Most of the community's inhabitants are "native" Betawi by ethnicity, practicing Muslims, and lower-middle class.[13] The residents of the community usually live closely to each other in small houses accessible through narrow alleyways that are separate from the practices of Jakarta's modern society.

In Mampang, several youths are heavy gamers and spend almost every night at the local warnet, which also organizes online gambling. An interview with one of the gamers revealed that through online poker, particularly Facebook's poker, a gamer can make around USD $2,000 per evening, with all exchanges done electronically (Farah [pseudonym], interview, February 28, 2012). What is interesting about the typical online gambling in Mampang is that the area's youths, despite their strict Islamic upbringing, which considers gambling unlawful, readily admit that the money generated from online gambling has made routine office work very undesirable since they can generate in one night of online gambling what they would generate in a month of working at a conventional job.

The practice of online gambling has also made parents more permissive toward their children, allowing them to play games in warnets for hours because they are able to generate income doing so. Their economic independence, made possible through online gambling, has also given them the autonomy to challenge the existing social hierarchy in the community. In one instance, the local Muslim preacher gave a sermon on how the youths of the area have become Internet addicts and online gamblers. One of the gamers took offense and walked out of the mosque, an occurrence that would not have occurred without the existence of a parallel subculture in an otherwise devout community.

The gaming behavior of lower-middle class gamers has been blamed for the abandonment of religious ideals, challenges to parental authority, and even challenges to the state's authority through illegal practices. Although further research needs to be conducted to determine whether or not this is a general trend, the fact that online poker's virtual money has real monetary value, due to organized gambling, shows that economic pressure has transformed gaming behavior into economic practices.

Conclusion

This essay has included an overview of the development of video games in Indonesia in terms of their adoption, their gray market characteristics, the role of government and video game communities, and has taken a close look at gaming's implications in a community in South Jakarta. Our descriptive analysis shows how

Indonesia's video game industry is heavily influenced by its gray market characteristics. We have also described how the government ineffectively attempts to curb unauthorized distribution through regulation and enforcement. The reason for this ineffectiveness, we argue, is that the spread of video games as a cultural practice in Indonesia, particularly in larger cities, has been divided into two classes: the upper-middle class with isolated gaming practices, who became gamers through their role in the international market, and the lower-middle class, who localized gaming practices through unauthorized yet authentic channels. The latter resulted in various consequences that include but are not limited to unauthorized game-rental places, the increasing demand and supply for piracy, and the increasing popularity of online gambling. The Indonesian government has recognized, however, that protectionist regulation does not address the issue. Instead, the government, inspired by South Korea's robust creative economies in which video games are central, has strategized to stimulate the development of Indonesia's video game industry through investment in local game developers.

We have noted the growing trend of online games in Indonesia, confirmed by our recent observations. The mobile phone industry's interest in local game developers, alongside the government's strategy of developing local creative economies, will stimulate the production of local gaming content, which is also made possible by the increase of smartphone use among Indonesians in large cities. We also noted the Liga Game community's attempts to develop gaming as a professional skill through organizing competitions, which will respond to the mobile phone providers' market interest. Thus, future research should take a closer look at how gaming communities respond to the mobile provider's recent interest in the gaming market and how the government, particularly the Ministry of Tourism and Creative Economies, will attempt to profit from the interaction.

Another potential research area is how video games, both popular console and online games, are incorporated into the local context. Indonesia is a developing country with two hundred ethnic groups and six religions acknowledged by the state. The case we described in one area, Mampang, South Jakarta, is exemplary of many other cases across the country where video games have influenced the local social hierarchy. We also emphasize the importance of decentering the focus from larger cities to rural areas. In writing this chapter, the first of its kind, we have inevitably become focused on the developments in large cities as available literature and data are also focused on the urban market. Little literature is available that analyzes trends in the video game industry, let alone literature that deals with more localized context on video game practices in Indonesia. Therefore, this chapter is descriptive in nature and provides a stepping-stone for future research that investigates the complexity and impact of the rising popularity of video games in Indonesia based on empirical findings.

Acknowledgments

The authors of this chapter would like to give special thanks for the assistance of Marsya Anggia throughout the process of data collection.

Notes

1. Arcade games are usually referred to by scholars as coin-operated machines, which allow short action games (Rockwell 2002). Console video games are personally owned machines that allow display on television screens (Wolf 2007). The term "computer game" refers to any game played on a personal computer but may also involve multiple players interacting (Tausend 2006).

2. In this chapter, we recognize that when we describe the development of video games in Indonesia, we inevitably speak of the urban Indonesians who predominantly live in Java—although this is nowhere near representing the trends in other cultural contexts throughout the country (such as urban/rural, regional/local, etc.).

3. We presume that the term "Dingdong" was phonetically inspired (as onomatopoeia). Another term that became popular because of its sound is "kretek," which refers to the Indonesian clove-spiced cigarettes (Reyes 2006).

4. For more on PC Bangs, see Choudrie and Lee 2004.

5. Mangga Dua and Kelapa Gading are the core distribution spots of pirated CDs and DVDs, which include film, music, and games, in Indonesia's capital city of Jakarta (Fahmi 2010; Pertiwi and Asdhiana 2012).

6. See Government Regulation No. 29, Year 2004, regarding High Technology Production Facilities for Optical Discs, http://rulebook-jica.ekon.go.id/english/4384_PP_29_2004_e.html.

7. See Copyright Law Article 72, Verse 3, http://www.iipa.com/rbc/2014/2014SPEC301INDONESIA.PDF.

8. See Minister of Trade Regulation No. 05, Year 2005, on the Importation of Machinery, Raw Materials, and Optical Discs, http://rulebook-jica.ekon.go.id/english/213.MDAG%2005-0405.eng.qc.html.

9. See Information and Electronic Transaction Law No. 11, Year 2008.

10. Majapahit, Pajajaran, and Sriwijaya are arguably the three largest dynasties in pre-independence Indonesia. For further reading, see Miksic 2002 and Ekadjati 2005.

11. *Ragnarok Online* is the most popular MMORPG in Indonesia (Taufiqurrakhman 2010).

12. For more on Kaskus, see Abdillah 2011.

13. Betawi is perceived to be the native ethnicity of Jakarta (people of *Batavia,* Jakarta's colonial name), but their presence has been somewhat diluted by major immigration from Java (Cribb 2009, 14).

References

Abdillah, Muhammad S. 2011. E-Commerce: Forum Jual Beli (FJB) Kaskus [E-Commerce: Kaskus' Trading Forum]. http://chalief.blogspot.com/2011/01/e-commerce-forum-jual-beli-fjb-kaskus.html#axzz1UaF0kIGh.

Anggia, Marsya. 2011. Interview with Remy Fabian. Recorded December 23.

Anggia, Marsya. 2012. Interview with Richard Samuel Manuhurapon. Recorded April 1.

Antia, Kersi D., M. Bergen, and S. Dutta. 2004. Competing with gray markets. *MIT Sloan Management Review* 46:63–70.

Barliana, Syaom. 2004. Pengaruh Siaran Televisi dan Video/Computer Game terhadap Pendidikan Anak: Implikasi bagi Pengembangan Teknologi dan Strategi Pembelajaran. Paper for Persidangan Antar Bangsa UPI-UPSI: Pendidikan dalam Dunia Pesat Berubah: Menilai Semula Proses Pengajaran dan Pembelajaran, Kuala Lumpur, Malaysia.

Bulkin, Farchan. 1984. Negara, Masyarakat dan Ekonomi [State, society, and economy]. *Prisma* 8, ed. Ruslan Burhani (January 19), *Kemenparekraf Target Kembangkan Ekonomi Kreatif Berbasis Seni* [Kemenparekraf is aiming to develop art-based economics]. http://www.antaranews.com/berita/1326991252/kemenparekraf-target-kembangkan-ekonomi-kreatif-berbasis-seni.

Cao, Yong, and John D. H. Downing. 2008. The realities of virtual play: Video games and their industry in China. *Media Culture Society* 30 (4): 515–529.

Cao, Y., A. Glukhova, R. Klamma, D. Renzel, and M. Spaniol. 2008. Measuring community satisfaction across gaming communities. In *International Workshop on Interactive Digital Entertainment Technologies (IDET 2008), Proceedings of the First IEEE International Conference on Ubi-Media Computing and Workshops*, 414–419. Lanzhou, China, July 15–16.

Castillo, Nieves P. 2010. Perceived quality of massive multiplayer online games in mobile environments. Master's thesis, Universidad Politecnica de Madrid; Vienna, Austria: Vienna University of Technology.

Choudrie, J., and H. J. Lee. 2004. Broadband development in South Korea: Institutional and cultural factors. *European Journal of Information Systems* 13 (2):103–114.

Cribb, Robert. 2009. *Gangsters and Revolutionaries: The Jakarta People's Militia and the Indonesian Revolution 1945-1949*, 13. Singapore: Equinox.

D., Y-aldo Bakas Samudra. 2011. Cara Membuka Situs yang di block [How to open the blocked websites]. *Samudra's Blog Community*, November 20. http://anak-game.blogspot.com/2011/11/cara-membuka-situs-yang-di-block.html.

Darmawan, Hikmat. 2012. Interview with Asep Kurniawan. Recorded February 15.

Day, Irwin. 2007. Angka-angka di Bisnis Warnet [Numbers in the Warnet business]. *Iwin Notes-Warnet dan IT* (blog), February 12. http://irwinday.wordpress.com/2007/02/12/angka-angka-di-bisnis-warnet/.

de Kervenoael, R., and D. S. O. Aykac. 2008. Grey market e-shopping and trust building practices in China. In *Trust and New Technologies: Marketing and Management on the Internet and Mobile Media*, ed. T. Kautonen and H. Karjaluoto. Cheltenham: Edward Elgar.

Dimas. 2009. Di Kota Kecil, Warnet Tumbuh Pesat (In small towns, Warnets are growing rapidly]. *Tempo.co*, December 14. http://www.tempo.co/read/news/2009/12/14/072213761/Di-Kota-Kecil-Warnet-Tumbuh-Pesat.

Ekadjati, Edi S. 2005. *Kebudayaan Sunda: Zaman Pajajaran*, Jilid 1. Garut: PT. Pustaka Jaya.

Eyles, Mark, and Roger Eglin. 2007. Ambient Role Playing Games: Towards a Grammar of Endlessness. Paper for the Women in Games Conference, Newport, South Wales.

Fahmi, Wahyudin. 2010. Polsek Kelapa Gading Sita Ribuan Keping Cakram Video Bajakan [Kelapa Gading police officer confiscates thousands of pirated video discs]. *Tempo.co*, July 26. http://www.tempo.co/read/news/2010/07/26/064266268/Polsek-Kelapa-Gading-Sita-Ribuan-Keping-Cakram-Video-Bajakan.

Firman, Muhammad. 2010. Pemain Game Indonesia Naik 33% per Tahun [Indonesian gamers are rising by 33% every year]. *VIVAnews*, July 5. http://teknologi.vivanews.com/news/read/162371-pemain-game-indonesia-naik-33--per-tahun.

Firman, Muhammad. 2014. Pengguna Facebook di Indonesia Naik 6 Persen [Indonesian Facebook users increase by 6%]. http://www.tempo.co/read/news/2014/06/29/072588907/Pengguna-Facebook-di-Indonesia-Naik-6-Persen.

Friedl, M. 2003. *Online Game Interactivity*. Newton Centre, MA: Charles River Media.

Galih, Bayu, and Amal N. Ngazis. 2012. Kembangkan Games, XL Gaet Developer Lokal [Developing Games, XL Attracts Local Developer]. *VIVAnews*, March 13. http://teknologi.vivanews.com/news/read/295911-kembangkan-games--xl-gaet-developer-lokal.

Gandhi, Grace S. 2005. Asosiasi Warnet Pertanyakan Razia Pembajakan Perangkat Lunak [Warnet Association questions software piracy sweeping]. *Tempo.co*, June 8. http://www.tempo.co/read/news/2005/06/08/05662175/Asosiasi-Warnet-Pertanyakan-Razia-Pembajakan-Perangkat-Lunak.

Guttenbrunner, M., C. Becker, and A. Rauber. 2010. Keeping the game alive: Evaluating strategies for the preservation of console video games. *International Journal of Digital Curation* 1 (5): 64–88.

Herdiyan, and Muhammad Sufyan. 2011. Agate Studio, Pencipta Game Football Saga yang Mendunia [Agate Studio, Football Saga game creator going global]. *Bisnis Indonesia*, October 4. http://bisnis-jabar.com/index.php/berita/agate-studio-pencipta-game-football-saga-yang-mendunia.

Heriyanto, Trisno. 2011a. Jumlah Gamer Arcade Indonesia Terbesar di Asia [Indonesian arcade gamers are the biggest in Asia]. *Detikinet*, September 29. http://inet.detik.com/read/2011/09/29/150328/1733367/654/jumlah-gamer-arcade-indonesia-terbesar-di-asia.

Heriyanto, Trisno. 2011b. Jumlah Pengguna Facebook di Indonesia Lampaui Inggris [The number of Facebook users in Indonesia exceeds the UK]. *International Business Times*, January 11.

Heryadi, Yaya. 2008. A case study on habit, motivation, and perception of urban high school students toward computer games. Paper for the Fourth International Conference Information and Communication Technology and System, Solo, Indonesia.

Ika, Aprillia. 2011a. Game Online Asing Masuki Pasar Indonesia [Foreign online game enter Indonesian market]. *Indonesia Finance Today*, March 24. http://www.indonesiafinancetoday.com/read/5155/Game-Online-Asing-Masuki-Pasar-Indonesia.

Ika, Aprillia. 2011b. Nielsen: Jumlah Pengguna Internet dari Handset Meningkat Signifikan [Nielsen: Number of handset Internet users increases significantly]. *Indonesia Finance Today*, July 7. http://www.indonesiafinancetoday.com/read/11008/Nielsen-Jumlah-Pengguna-Internet-dari-Handset-Meningkat-Signifikan.

Iqbal, Muhammad. 2011. Nusantara Online Targetkan 52 Ribu Member Tahun Depan [Nusantara Online is targeting 52 thousand members for next Year]. *Indonesia Finance Today*, November 24. http://www.indonesiafinancetoday.com/read/18460/Nusantara-Online-Targetkan-52-Ribu-Member-Tahun-Depan.

International Intellectual Property Alliance (IIPA) 2009. Indonesia. 301 Report on Copyright Protection and Enforcement. http://www.iipa.com/rbc/2009/2009SPEC301INDONESIA.pdf.

Ito, Budi. 2009. Game Online Makin Melejit [Online games are skyrocketing]. *inilah.com*, July 2. http://www.inilah.com/read/detail/122753/game-online-makin-melejit.

Kurniawan, Aloysius B., and Mulyadi Agus. 2011. Bisnis "Game" dan Animasi Yogyakarta Mulai Bergairah [Yogyakarta's game and animation business start to pump]. *Kompas.com*, September 7. http://regional.kompas.com/read/2011/09/07/2040209/Bisnis.Game.dan.Animasi.Yogyakarta.MMula.Bergairah.

Kusumadewi, Theodora N. 2009. Hubungan antara Kecanduan Internet Game Online dan Keterampilan Sosial pada Remaja [The relationship between online gaming and social skills among adolescents]. Bachelor's thesis, Fakultas Psikologi Univeritas Indonesia.

Law of The Republic of Indonesia Number 19 Year 2002 Regarding Copyright. 2002. United Nations Educational, Scientific and Cultural Organization. http://portal.unesco.org/culture/en/files/30382/11424187703id_copyright_2002_en.pdf/id_copyright_2002_en.pdf.

Lee, H., B. O'Keefe, and K. Yun. 2003. The growth of broadband and electronic commerce in South Korea: Contributing factors. *Information Society* 19 (1):81–93.

Lim, M. 2005. *@rchipelago Online: The Internet and Political Activism in Indonesia*. Bandung, Indonesia: TSD-University of Twente.

Miksic, John. 2002. *Indonesian Heritage: Sejarah Awal*. Jakarta: Grolier International Inc. and Buku Antar Bangsa.

Nokia Siapkan Game "Smash Mania" [Nokia is preparing "Smash Mania" game]. 2011. MediaIndonesia.com, October 23. http://www.mediaindonesia.com/mediagadget/index.php/read/2011/10/23/2463/3/Nokia-Siapkan-Game-Smash-Mania.

Pal, Sarmistha. 2010. Norms, culture, and local infrastructure: Evidence from a decentralised economy. *Discussion paper series/Forschungsinstitut zur Zukunft der Arbeit*, 5281.

Peraturan Menteri Perdagangan RI No. 05/M-DAG/PER/4/2005 Tahun 2005 Tentang Ketentuan Impor Mesin, Peralatan Mesin, Bahan Baku dan Cakram Optik. 2005. Gaikindo. http://gaikindo.or.id/download/industry-policies/b-kebijakan-dep-perdagangan/06-PM_05_MDag.pdf.

Peraturan Pemerintah Republik Indonesia Nomor 29 Tahun 2004 Tentang Sarana Produksi Berteknologi Tinggi Untuk Cakram Optik (Optical Disc). 2004. BPKP, http://www.bpkp.go.id/uu/filedownload/4/61/955.bpkp.

Pertiwi, Ni L. M., and I. M. Asdhiana. 2011. Hotel Harris Jeli Melihat Peluang [Harris Hotel sees opportunity sharply]. *Kompas.com*, June 24. http://travel.kompas.com/read/2011/06/24/12161584/Hotel.Harris.Jeli.Melihat.Peluang.

Pertiwi, Ni L. M., and I. M. Asdhiana. 2012. Panduan Wisata Belanja Mangga Dua [Mangga Dua's shopping guide]. *Kompas.com*, February 16. http://travel.kompas.com/read/2012/02/16/21021536/Panduan.Wisata.Belanja.di.Mangga.Dua.

Pitakasari, Ajeng R. 2011. Satu Lagi, Perjudian Berkedok Warnet di Jakut Kena Gerebek. *Republika.co.id*, September 10. http://www.republika.co.id/berita/nasional/hukum/11/09/10/lralpi-satu-lagi-perjudian-berkedok-warnet-di-jakut-kena-gerebek.

Pius, Romualdus, ed. 2011. Penyedia Jasa Judi Online Ditangkap Polisi. *Serambi Indonesia*, December 6. http://www.tribunnews.com/regional/2011/12/06/penyedia-jasa-judi-online-ditangkap-polisi.

Portibi DNP. 2011. 3 Warnet Judi Online Beromset Miliaran Rupiah Digrebek. Portibi DNP, September 26. http://portibionline.com/berita-2340-3-warnet-judi-online-beromset-miliaran-rupiah-digrebek.html. Retrieved March 29, 2012 (website no longer available).

Purwanti, Tenni, and Reza Wahyudi. 2011. Agate Studio, Pengembang Game Dengan Modal Awal "Passion" [Agate Studio, game developer with "passion" as initial capital]. *Kompas.com*, December 13. http://tekno.kompas.com/read/2011/12/13/09563339/Agate.Studio.Pengembang.Game.Dengan.Modal.Awal.Passion.

Purwanti, Tenni, and Wicaksono S. Hidayat. 2012. Angry Birds untuk Facebook Akan Diluncurkan di Jakarta [Angry Birds for Facebook will be launched in Jakarta]. *Kompas.com*, January 26. http://bola.kompas.com/read/2012/01/26/15383495/Angry.Birds.untuk.Facebook.Akan.Diluncurkan.di.Jakarta.

Purwanto, Didik, and Wicaksono S. Hidayat. 2012. Tawarkan Game Rp 500-an, XL Gandeng Pengembang Lokal [Offering Game Rp 500-ish, XL collaborated with local developer]. *Kompas.com*, March 12. http://tekno.kompas.com/read/2012/03/12/1627548/Tawarkan.Game.Rp.500-an.XL.Gandeng.Pengembang.Lokal.

Reyes, Danilo Fransisco M. 2006. Selling symbols, keeping one's face: Iconography, identity politics, and globalization in Indonesia and Thailand. In *Negotiating Globalization in Asia*, ed. N. Fictions. Quezon, Philippines: Ateneo Center for Asian Studies (ACAS).

Rockwell, Geoffrey. 2002. Gore galore: Literary theory and computer games. *Comparative Computers and the Humanities* 36:345–358.

Setiawan, Aries, and Amal N. Ngazis. 2011. Kerugian Akibat Pembajakan Software Rp 1,3 M [Software piracy losses due to Rp. 1.3 M]. *VIVAnews*, August 24. http://teknologi.vivanews.com/news/read/242490-pembajakan-software-telan-kerugian-rp1-3-m.

Stewart, J. F., C. Mallery, and J. Choi. 2010. A multilevel analysis of distance learning achievement: Are college students with disabilities making the grade? *Journal of Rehabilitation* 76 (2):27–39.

Suryadhi, Ardhi. 2010. Microsoft Bangun "Markas" di Sarang Software Bajakan [Microsoft is building a "headquarters" in the nest of piracy software]. *Detikinet*, March 2. http://inet.detik.com/read/2010/03/02/144802/1309472/398/microsoft-bangun-markas-di-sarang-software-bajakan.

Syam. 2009. Warga Semakin Tidak Nyaman di Kota Besar [Citizens in big cities are getting more uncomfortable]. *Pos Kota*, December 17.

Taufiqurrakhman, Ahmad. 2010. Ragnarok World Championship 2010 Digelar di Jakarta [Ragnarok world championship 2010 held in Jakarta]. *Okezone.com*, October 2. http://techno.okezone.com/read/2010/10/02/326/378391/ragnarok-world-championship-2010-digelar-di-jakarta.

Tausend, Ulrich. 2006. Casual games and gender. Paper presented for Advanced Seminar: Economic Sociology, Ludwig Maximilian, University of Munich, Germany.

Undang-Undang Republik Indonesia Nomor 11 Tahun 2008 Tentang Informasi dan Transaksi Elektronik. 2008. BATAN-Badan Tenaga Nuklir Nasional. www.batan.go.id/prod_hukum/extern/uu-ite-11-2008.pdf.

Wibowo, Arinto T., and Harwanto B. Pratomo. 2012. Pendapatan Per Kapita Naik 13,8%, Kabar Baik? [Per capita income increases by 13.8%, is it good news?]. *VIVAnews*, February 6. http://fokus.vivanews.com/news/read/286054-kelas-menengah-dan-pendapatan-per-kapita.

Widodo, S. 2002. Pengguna Internet Indonesia Mencapai 4,2 Juta User [Indonesia Internet users reached 4.2 million]. *Detikinet*, January 3. http://www.detikinet.com/berita/2002/01/03/20020103-174809.shtml.

Wolf, Mark J. P., ed. 2007. *The Video Game Explosion: A History from PONG to PlayStation and Beyond*. Westport, CT: Greenwood Press.

Yohanlie. 2009. Saran Kepada LYTO Demi Pengembangan ROHAN [Suggestion to LYTO for the development of Rohan], message 35. May 7, message posted to http://www.kotakgame.com/forums/showthread.php?t=3757&page=4.

Yono, Yossie. 2011. Nusantara Online: Game Online Bertema Budaya Nusantara [Nusantara Online: Archipelago culture-themed online games]. *Chip Online ID*, May 10. http://chip.co.id/news/read/2011/05/10/657424/Nusantara.Online.Game.Online.Bertema.Budaya.Nusantara.

Zachra, Ellyzar. 2011. Gara-gara iPhone, Nintendo Hilang Pamor. *SWA*, June 20. http://swa.co.id/updates/gara-gara-iphone-nintendo-hilang-pamor.

IRAN

Ahmad Ahmadi

A roaring flood of video games from the Western world reached Iran when its first generation arrived during the Iran-Iraq War in the early 1980s. The Atari VCS 2600 was among the first generation of game consoles that officially entered Iran, and *Pac-Man*, *Pitfall*, and *Missile Command* were some of the best-selling games at that time. When the war's status was "white" and the ominous "red" alarm was off, groups of children found opportunities to sit in front of the TV and shook their new black-and-gray joysticks up and down, taking turns and sometimes even fighting over them.[1] Now, after three decades of fast technological development and an increase in public access to the tools needed to make them, especially in the last decade, video games have attracted thousands of Iranian users. Since the early 1990s, the market for video games has experienced a boom in Iran, and this led to the production and introduction of the first generation of Iranian video games in a serious way.

When the first game from the *Prince of Persia* series was released in 1989, Iranian gamers hurried to play a game that carried their country's name. But there was always a question in their minds: why was there no sign of Iranian language, architecture, characters, or culture in a game named *Prince of Persia*? All the content elements, for example characters' clothing, appeared in Arabic style. Another question was: is it possible to produce a game that contains original Persian culture?

Although the video game industry in Iran emerged in 2010, the main process of its formation was in the middle of the last decade of the twentieth century. During this time, the Iranian video game industry experienced three generations of game developers. When we talk about three generations, we do not mean three generations regarding age or time; what we mean here involves the evolution of game development experience and knowledge. Some of these companies had years of experience in IT but did not have any experience in the game industry or game production.

In late 1995, along with the beginning of movements in game development, Kanoon, a semi-governmental cultural and educational organization, began its research activities in games and participated in making a few simple educational computer games. Kanoon's activities in game development will be dealt with separately.

The game *Tank Hunter* (1996) was the first 2-D game by Hanifa developed using a simple programming language. The first generation of developers didn't have enough information about game design, level design, engines, and game development software. Without access to game design resources, they developed their first games by examining the foreign games played at that time. *Cobra Operation* (1998), *Rostam* (2000), and *Hell's Gate* (2001) were other titles that were developed the same way. So the first experiences of game development for the first passionate generation of game developers were formed. *End of Innocence* (2006), *Driving in Tehran* (2007), and *The Warrior* (2007) were outstanding 3-D titles of that generation, which, despite being primitive, were accepted by Iranian gamers and fueled the motivation for the continued development of more sophisticated games in Iran. Most first-generation developers were from student groups that produced their games as university projects; and while most of these developers are not working now, their initial efforts opened the way for game development in Iran. Sami Rayan Pardaz, the developer of *Driving in Tehran*, is the only first-generation company that still remains active.

With the establishment of the Iran Computer & Video Games Foundation (http://www.irvgame.com) in 2007, an increase in game development knowledge, and a better understanding of game structure standards, a new spirit entered the body of the industry. The release of games such as *Lotfali Khan Zand* (2008) and *Nader's Sword* (2009), based on the life stories of two great kings and founders of powerful Iranian dynasties of the seventeenth century, showed that game production in Iran was entering a new phase, and the start of the second generation of game development. A significant increase in the quality of these games (compared to those of the first generation), accompanied by the new movement of game studios (such as Fanafzar Sharif, Espris Pouya Nama, Ras Games, and RoyaGaran, all supported by the foundation), created a situation that led to production of games that achieved success in the domestic market as well as the international one.

Garshasp (2010), by Fanafzar Sharif, was the first title with high enough quality to make its way to the international market, and it increased Iranian gamers' trust in local productions. *Garshasp* is based on ancient Persian mythology, bringing to life the epic battles of the mythological hero Garshasp against evil monsters. The gamer takes the role of Garshasp, the strongest man in Persian mythology, on a journey to revenge during which he must reclaim the legendary mace of his ancestors, which has been stolen by the Deevs, and along the way, more significant mysteries are revealed. The main mechanics of the game revolve around melee combat, platforming, and exploration. By July 2013, more than 250,000 copies of the game had been sold.

Age of Pahlevanas I (2009) and *Mithra's Planet* (2010) by Ras Games, *Mir-Mahna* (2010) by Espris Pouya Nama, *OutLand* (2010) by Fotros, *Gando* (2011) by Pasargad Game Studio, *The Mountaineer* (2010) by Niv Studio, *Hidden Treasure* (2010) and *Black Gold* (2011) by Royagaran, *Speed in the City* (2011) by Arvin Tech, and more than forty other games are the outcome of the second generation. At the same time, production of the first group of Iranian Web browser-MMO games began, and games such as *Asmandez I* (2009) by Simulator Developer Co. and *Speed Up* (2010) by Noavaran Persian World, were developed and released. *Asmandez* is an MMO science fiction city-building game, in which players save humankind from attacking robots. The game's backstory is that in the year 2808, a great war begins between mankind and robots on all the planets of the solar system. The earth is then in danger of total destruction by the machines and robots, so the earth's inhabitants must immigrate to other systems. The spaceship reaches the gates of a new system, and hereafter the players are responsible for accommodating this ship.

The success achieved by *Garshasp*, *Mithra's Planet*, *Nader's Sword*, and *Black Gold* in the international marketplace was good news for the Iranian game industry and proved that global success was possible. The success of these second-generation games, along with significant improvements in the global video game industry (as demonstrated by the emergence of new platforms and consoles), encouraged more people to enter the field of game development. Although first generation games didn't have considerable sales, second generation games attracted many Iranian gamers. *Garshasp* set the record, selling 250,000 copies. After *Garshasp*, *Speed in the City*, with 150,000 copies, and *Mir-Mahna*, with 120,000 copies, were the next bestsellers.

The combination of this new group of developers, with a greater number of industry pioneers, was the starting point for the third generation of development in 2012. Game studios such as BaziResane, PardisGame, Raspina, and Amitis are the leading companies of this generation. A variety of teams and companies of different sizes have created a dynamic atmosphere encouraging the emergence of new titles with unique ideas. Since the third generation has started working, more than 150 different games have been finished or are in development.

Most notable about the third generation of games is their variety in genres and versions for different platforms. In the past, most games were PC-based, and typically action genres (shooting games and role-playing games). This was due to these genres' numerous fans within the local marketplace. But developers now take market segmentation into account, with the production of casual and online games, various games for different platforms, tablets, and smartphones, and also games for specific customer segments such as women and children.

The success of the third generation's productions within the global marketplace is a testament to their improved quality and a better understanding of market demands on one hand, and accounts for the recognition and approval of Iranian game studios by international publishers and distributors on the other. Successful titles such as *Shaban* (2012) by Peta Games, *Murder in Tehran's Alleys 1933* (2012), and *Love Path* (2013) by RSK Entertainment, *Legends of Persia (Siavosh)* (2013) by Sourena Game Studio, *The Dark Phantom: Dawn of the Darkness* (2013) by Pardis Game Studio, *Hate The Sin, Love The Sinner* (2013) by New Folder, *Awakening* (2013) by Bazi Resaneh Studio, *Asmandez II* (2013) by Simulator Developer Co, *E.T. Armies* (2013) by Raspina, and dozens of other games released in Apple's App Store or in the Android Market, indicate the progress of Iranian games in the global market.

Domestic Video Game Production

The year 2008 was the real starting point of the Iranian video game industry. With the relative success of some of the second-generation producers, a sort of self-confidence was born in Iranian developers, and it seemed that the acceptance of Iranian games by foreign publishers and gamers was not out of reach. The fear that contracts with international publishers were impossible due to the difficulties imposed by sanctions and other political limitations was finally put to rest. "Produce the game well, and be sure that the

Figure 1

Asmandez II (2013) (top, left), *E.T. Armies* (2013) (top, right), *Garshasp* (2010) (bottom, left), and *Mir-Mahna* (2010) (bottom, right).

distributer shows you the contract"; this was what one of the famous international publishers told me at GamesCom2011.

Although the release of *Garshasp* through Steam's website did not result in a dazzling increase in sales for the game's producer, the fact that it was bought by gamers from more than thirty countries and was sold at the most reliable digital sales website was counted as a great success for the young Iranian game industry. So despite all the problems in the production process, the sanctions' effects, and the impossibility of any contact or formal contracts with other companies as a result of the governments' political issues, Iranian entrepreneurs and small companies could now look for a ray of hope to be assured they were walking the right path; the success of *Garshasp* was this ray.

Among second-generation games, *Nader's Sword*, *Mithra's Planet*, and *Garshasp* all entered the international market. The worldwide release of Royagaran's *Black Gold* (2011), a strategy management game about the oil industry, introduced a totally new subject and revealed the different view of Iranian producers regarding game structure.

As mentioned earlier, successful experiences from 2008 to 2011 made the third generation of game producers more confident about their ability to produce popular games in the international market. In addition, some companies took casual and online game production more seriously. After the success of *Shaban* (2012) at the Bigfish website and its distribution by other casual game publishers, casual game productions in Iran flourished, and dozens of games are now available in Apple's App Store and in Google's Android Market. Before that, the aim of most Iranian game producers was to create a hard-core game, and they defined success as the completion of such a game. Causal games such as *We Need a Hero* (2013), *Hate the Sin, Love the Sinner* (2013), *Naughty Sheep* (2013), and other titles show the success in understanding the changing trend and focus on casual games.

Contracts for the distribution of games, such as *Murder in Tehran's Alleys 1933* (2013) and *Love Path* (2013), were other successful experiences that introduced Iranian adventure games into the global marketplace. This genre is a good one for Iranian game production studios because of its ability to convey stories and its use of characters and environmental design, all of which can be taken from Iranian cultural history. The first game of Bazi Resaneh's *Awakening* series, *Awakening: Burning Ashes* (2013), about the Tunisian revolution of 2010–2011, is an example of an Iranian adventure game of the highest quality.

Although the production of online games requires specific production knowledge of this genre and its revenue models, more than ten different online games, most of them simple, have been produced and released inside the country. *Asmandez II* (2013), however, was so successful that it gained contracts for distribution in Southeast and Middle Asia, Russia, East Europe, and continental America and was warmly welcomed in the competitive environment of online games.

Note that the global success of Iranian games was more than just in international sales; *Hate the Sin, Love the Sinner* received the "Best Project Award" of 2013 at Game-Connection Asia in Shanghai, while *Asmandez II* was selected as the best indie game in 2013 and received the "MMO of the Year" award. At Indies Crash E3 2013, two Iranian games, *E.T. Armies* and *The Dark Phantom: Dawn of the Darkness* (2013), ranked fourth and seventh respectively among ten selected indie games by users' votes. *Parvaneh: Legacy of the Light's Guardians*, of Beardedbird studio, was also nominated for "Best Console & PC Hardcore game" at the Game Connection Europe 2014. These achievements, in addition to gaining respect in the global video game society, point to the abilities of Iran's young studios.

Game production in Iran may still be in its infancy, but it's important to consider what we have now is the result of an industry that started its official activity in 2008, only five years ago as of this writing. All of these products have been produced within the limitations of economic sanctions on Iran and with limited interaction in the international and global marketplace. Iranian producers have tried to participate and gain knowledge and experience in international exhibitions many times, but their attempts often failed due to a lack of travel visas from certain embassies. These circumstances and the different limitations the Iranian gaming industry faces make these successes all the more valuable and important.

Kanoon Parvaresh Kodakan

The Institute for the Intellectual Development of Children and Young Adults (known as "Kanoon") was the first investor and pioneer organization in the development of games in Iran. From the time of its founding in 1961 through production of artistic works, the organizing of festivals, discussion panels, competitions, and various conferences, as well as a variety of practical measures and consulting services, the IIDCYA has answered an extensive part of the cultural and developmental needs of the children and young adults of Iran. All this has been carried out in two main directions: activities *for children*, addressing children and young adults directly, and activities *about children*, with the purpose of promoting adults' knowledge about the peculiarities of the world of children. With high average annual statistics in the area of the production of books, tapes, and toys, and a considerable investment in the fields of recreational and literary activities, these colossal tasks of culture-making and the consolidation of the cultural infrastructure are being pursued. Today, 519 libraries and cultural centers with more than 2,000 tutors throughout the country help to realize these goals.

In late 1996, Kanoon decided to develop games for the indirect education of children. This was the first movement for development of "serious games" in Iran. In 1997, the first game by Kanoon, *Moamaye Jorchin*, was developed by DOS and released on floppy. Then between 2002 and 2003, six other simple games of this kind were developed. Games such as *BaraneSetare* (2002), *Khaneye Doust* (2003), *Shahre Asal* (2003), and *Majaraye Saat* (2003) relate to the years in which they appeared, and were under the management of Ms. Roshanak Noori. In 2004, with game development companies starting to appear, Kanoon began to outsource some of its projects. In 2005, with the cooperation of Kanoon, Royagarn, and C++Software, the first Iranian game development engine, Nutriga, was created, and the game *Shatranj* was developed using it. This organization has established an SBU (strategic business unit) called "Kanoon Game," which follows game development projects. Up to 2013, this institution has developed about twenty-two games with the cooperation of other companies.

The Iran Computer and Video Games Foundation (IRCG)

In early 2007, a group of Iranian game developers submitted a plan to the Ministry of Culture and Islamic Guidance to formalize game development activities and establish video game industry basics in Iran. The plan also involved getting intellectual and financial support from the government and other organizations, and proposed the establishing of a nongovernmental foundation or organization with the aim of supporting the video game industry in Iran. This proposal resulted in the establishment of the Iran Computer and Video Games Foundation (IRCG) in late 2007. This foundation is an Iranian nongovernmental, nonprofit, cultural artistic institution. It is an independent legal entity and is located in Tehran, the capital city of Iran. The main objectives of the IRCG are:

- Supporting game development in Iran.
- Enhancing computer and video game knowledge in Iran.
- Establishing the first Games Education Institute and designing scientific and academic framework to educate a new generation of game designers, developers and artists in Iran.
- Holding festivals and exhibitions to increase the enthusiasm for gaming and forming communities with active members.
- Attending regional and international exhibitions and presenting the achievements of Iranian video games.

The foundation has played an effective role in the stability and development of the video game industry in Iran, and almost all high-quality video games made after 2008 have been with the foundation's intellectual and financial support. Financial support comes about when the design and production plan of a game is approved by the foundation's Production Council, which covers the main costs of the project. This process not only reduces production risk for game studios but also encourages more game enthusiasts to enter the industry. Of course, not all design plans are accepted, as the financial resources of the foundation are limited. After the production council's confirmation, financial support is dedicated to the project, which amounts to between 10% and 80% of the total development cost. According to the contract, the game development company is committed to repay the money to the foundation from profits made from the game's sales, allowing the foundation to be able to continuously sponsor future projects. The government and organizations such as universities, research centers of ministries, and large private companies provide the foundation's budget. Based on budget, the IRCG annually supports about twenty to thirty games. In recent years, some private companies have invested in game development to enter the industry.

The foundation's programs are not limited to supporting the production process. The establishment of the Iranian Game Development Institute in 2012 and the Iranian Game Development College in 2013 were other important actions taken by the foundation. The shortage of a professional workforce in this industry is one of the biggest challenges faced by game development. The establishment of the first game development institute and college in Iran, which is the biggest in the Middle East, was an important step toward training and educating video game specialists. During the two semesters of 2012, about 300 students registered at the institute.

To introduce Iranian games into the global marketplace, the foundation has taken part in some reputable international game exhibitions and has set up booths at Gamescom Germany (every year from 2009 to 2014), Game Connection France (from 2009 to 2014), Casual Connect Singapore and Seattle (2012) and Shanghai (2013), Tokyo Game Show (2012), G-Star Korea (2011 and 2012), and Dubai Game Expo (from 2009 to 2011). These events are good opportunities for presenting the potential of the Iranian game industry as a hub in the Middle East and exchanging experience and knowledge with other countries.

Over the years, Iranian game studios have faced many challenges that most foreign companies have never encountered. When the need to consider impacts resulting from political sanctions that threaten the future of your company always overshadows your business and performance, such unwanted problems impose grave limitations on your work. Many of these problems arise from political issues between Iran and other countries that result in sanctions, limited access to information flows, and less free trade with the outside world. These issues lead to problems in accessing information and the legal use of game engines, software, and game development kits by Iranian game developers, and prevent easy release and direct sales in the global marketplace.

Iran, the second most populous country in the Middle East, has the biggest video game market in the area; in 2012, there were more than 20 million gamers in the country. Such a large market has been an important stimulus for the industry's increasing growth. Developing games in different genres for different platforms in recent years has been a significant boost to the industry in Iran.

Tehran Game Expo & Festival

To increase the understanding of the importance of the game industry in today's entertainment world and to contribute to its growth in Iran, the IRCG has held the Tehran Game Expo & Festival annually since 2011. At the first expo, forty companies took part in the exhibition and 150,000 people visited it. The second and third expos, in 2012 and 2013, experienced considerable growth both in quality and quantity of attendees. In 2013, nine universities and more than 120 companies participated in the expo and more than 450,000 people visited during its five days. Along with the expo, each year game development competitions and educational workshops are held that attract many students, game enthusiasts, and professionals. In addition, each year's festival of video games includes the judging of games submitted by participants. During the third festival, more than 100 games were judged in two main categories of casual and hard-core in different fields such as game scenario, music, character design, and cinematic (cut-scenes). The jury announced the winners of each area and selected the Game of the Year. Winners received an official statuette and cash prizes. The fourth Tehran Game Expo was held in 2014. *Parvaneh* received the prize for the best game of 2014 in Iran.

Considering the number of participating companies and visitors, the event is the biggest video game expo in the Middle East. Although the aim of executives is to hold it as an international expo, only a small number of invited foreign companies have participated in it. Part of this is due to the political situation in the country, which overshadows international relations. But there is hope that with some political changes, the Tehran Game Expo & Festival can solidify its position in the region and also appear on foreign companies' calendars.

Game Distribution and Ratings

The lack of sufficient knowledge about Iran's market size and capacity is to some extent due to political issues and sanctions; they do not allow the global video game community to have a clear image of this large, totally dynamic, and intelligent market. The problem with copyright law and its confluence with the special political situation of Iran in the world have put limitations on foreign videogame producers and publishers inside the country. This, in turn, has caused the Iranian video game market to be somewhat outside of the worldwide competitive corporate environment.

The lack of acknowledgment of copyright law in Iran does not mean there are no intellectual property laws in Iran. If an Iranian company has the official and legal license to distribute a game from the company that owns it, the legal authorities support the company against the illegal distribution and selling of that game by unlicensed companies. Because of the lack of legal limitations and pursuit of intellectual property rights, as in some other countries, distribution companies in Iran download the games from the Internet or import them from abroad, unlock them, and then copy and sell them at the lowest price.

These situations make it possible for Iranian gamers to buy the world's newest games at the same time that they are released worldwide. Although the existence of these unlocked versions sometimes causes problems for players' systems and consoles, this kind of buying is inevitable as long as there is demand for a game. Of course, this does not mean that all the games are easily distributed and without any limitations in Iran's market.

According to research done by the Iran Computer and Video Games Foundation, Iranians tend to play games developed by American and European companies. From 1980 to 1995, when the first generations of consoles and games entered the country, games were played without it being important where they were made; gamers were interested in testing all of them. Over time and an increase in the variety of games and different platforms, they found their favorites in Western genres. Game and level design, characters, setting, and story are important factors that Iranian gamers found interesting in Western-made games; it can be said that Iranian gamers' interests are close to those of Western gamers.

Interaction and communication in gamers' communities are the main factors in the increase in sales in the market. There are more than 2,000 active websites, forums, and blogs that review, comment on, and share information about games. These virtual communities not only provide the most up-to-date news about the latest and future games, but they also review their technical features and provide a challenging space for the free exchange of ideas; of these, http://news.dbazi.com and http://pardisgame.net/ are among the best-known and most popular websites.

In addition to these digital spaces, about ten newspapers, magazines, and periodicals exclusively cover video game industry news and information. I remember at Gamescom 2011, when I showed one of these magazines (*DonyayeBazi*) to the marketing team of EA, they were interested in various content and were astonished at the print quality of their latest game images and reviews.

One of the IRCG's programs was to organize and rate the content of the games for different age groups, so the Entertainment Software Rating Association (ESRA) was formed in 2008. The requirements to distribute and sell a game in Iran include getting a distribution license and an ESRA rating. The ESRA rating system is the only rating system among Muslim countries.

To determine a game's rating, the ESRA considers four characteristics:

1. Physical (motional)
2. Intellectual (mental)
3. Emotional
4. Social

The age groups are:

- The first rating (Kids) 3+ years old
- The second rating (Childhood) 7+ years old
- The third rating (the beginning of adolescence) 12+ years old
- The fourth rating (the end of adolescence) 15+ years old
- The fifth rating (adult, single) 18+ years old

Seven kinds of content are considered when ratings are assigned:

Violence
This refers to the display of violence, as when a behavior is shown to harm someone or something, including the destruction of property.

Tobacco and Drug Use
The virtual use of drugs and tobacco in games can erode the player's internal social taboo against using them in real life.

Sexual Stimuli
Games involving sexual perversity, sexuality out of socially acceptable norms, and so forth can lead to social and physical harm related to the sexual needs of players and their social situations.

Fear

Fear is an internal feeling based on insecurity and a lack of trust. This feeling, at different ages, can lead to chronic stress, stuttering in children, pessimism, cardiac and respiratory diseases, and fearful and timid behaviors in social situations.

The Violation of Religious Values

This refers to the violation of religious values in accord with Islamic principles. Two important elements of it are as follows:

1. The breach of the basic principles of religious beliefs (how heaven or hell is being depicted)
2. Sacrilege in holy places (sacrilege in a mosque or church)

The Violation of Social Norms

The use of vulgar words and behaviors that lead to the breaking of social norms are among the social harms to which children and adolescents should not be exposed.

Hopelessness

This refers to the content in games in which players feel that they must do something that makes them feel sinful.

The ESRA's vision is to establish a place where it can help define the best age ratings for games by helping the growth and development of players, and through the avoidance of the potential mental and physical harm that games may cause or promote. The ESRA, in the National Association of Computer Games, will continue to support the study and research of video games using new perspectives such as communication science, educational science, and social psychology. Designing the criteria for the identification and classification of harmful material was among the association's important research activities, as well as developing the literacy regarding it, which was designed and programmed in five research phases by the ESRA. The extraction of gameplay examples and their analysis in order to assign age ratings is another part of the process. The training of the workforce of human experts for these tasks is difficult, expensive, and requires the ESRA's full capacities and resources.

While there will always be debates regarding what kind of content is appropriate at what age level, another kind of content that is encouraged in all games is that which references the country's long history.

The Influence of National History on Video Games

When one considers countries that are known for their ancient civilizations, there is no doubt that the spiritual beliefs of a nation help to form its long national history, which in turn contains the themes for a treasure trove of stories and biographies that make up a culture. As narrative and an internal message provide the heart and spirit of a video game, Iran's ancient civilization and long national history have provided the largest and most enduring sources for the creation of interesting video game storylines. For example, the story of Sourena's action RPG *Legends of Persia (Siavosh)* (2013) is derived from *Shahnameh*, *Bondahesh*, and *Avesta* (ancient Persian books), and begins when Siavosh, a Persian prince, is killed by Afrasiab, the king of Tooran. Keykhosro, the son of Siavosh, starts to reign after he is transferred to Persia by Giv. After years of searching, he encounters the powerful Afrasiab and defeats him in a severe war, avenging his father's blood.

Iran, commonly known as Persia, is among the few countries that do not have an Independence Day; but during its long history, the country has been invaded by many foreign nations. These events have been effective in creating national historical stories of myths, heroes, and dynasties, as well as tales of individuals' gallantry and bravery in defending their homeland. Iran's history is also divided into two eras: pre-Islam and post-Islam. Pre-Islam (4000 BC to AD 1000) refers to the glorious and great empire of ancient Iranians. Powerful ruling dynasties conquering new territories, interaction with other nations, music, art, monuments, and Persian architecture have provided a rich source for creation and development of games for Iranian and even non-Iranian developers; as in, for example, the *Roman Empire* series.

Post-Islam refers to the beginning of a new era in Iranian literature, cultural, and religious values, national Iranian identity, and creation of masterpieces, which open a window for considering the importance and role of Iranian identity in the formation of the Islamic world's history and its value commonality with other Muslim countries. By keeping these things in mind, developers can create games that will be accepted by gamers in other Muslim countries. Iranian game developers have produced games based on the spirit of patriotism to present their national history. Shooting games such as *Mir-Mahna* (2010) and *Cry of Freedom* (2013) depict uprisings in southern Iran against Portuguese forces and the East India Company invasions of regions and islands south of Iran in the nineteenth century. *Mir-Mahna* is based on Mir-Mahna, an amir who fought against Dutch colonial forces in Iran during the 1740s. The game was based on Iranian literature and features a soundtrack of traditional Iranian music.

Lotf Ali Khan (2008) and *Nader's Sword* (2009) surround the life events of two kings who founded the dynasties Zand and Afsharieh in Iran in the nineteenth century. *Torkamanchay* (2012) depicts the wars between Iran and the Russian empire. *Hidden Treasure* (2010), a strategic management game, involves Iranian national cultural history and its artifacts in famous museums around the world and introduces them as symbols of glorious Iranian history to gamers. Players must find ancient Iranian artifacts in museums around the world and bring them back to Iranian museums. Other games such as *Dowrane Eftekhar* (2013), *The Last Bullet* (2013), *Shekaste Hasr* (2011), the shooting game *Demolition Operation* (2009), and dozens of other war games depict the period of Iranian national history during the Iran-Iraq War.

Dowrane Eftekhar (*Age of Honor*, 2013) is a first-person shooter game that depicts important events in the Iran–Iraq War. The first episode takes the player to two important operations in southwest Iran to defend his homeland with other Iranian soldiers against the Iraqi military. The game's events were designed based on veterans' memories and photographs of regions engulfed in war, and the game was warmly welcomed by gamers and publishers.

Black Gold (2011) is about the history of the oil industry and tells gamers about how the oil industry became nationalized in Iran. Iranian game studios take pride in their national history, and the storylines of their games, whether from ancient history and myths or contemporary events, have been the core of many sophisticated games in different genres.

Video Game Culture

Culture and video games can be considered in two ways: the first is the culture among Iranian gamers and developers, and the second deals with how Iranian culture affects game development. The culture among Iranian gamers is based on a culture of consumption and fun. Video gaming is one of the most important entertainments among Iranians. For most of them, playing games that give them maximum excitement is a priority. That is why action games, RPGs, and race games have the most fans in the country. Even at game websites of online communities, these genres get more comments from users. Game nets, video game websites, and online forums are important sources for game enthusiasts to get the latest news from the world of video gaming. However, culture among game developers is different somehow; they try to bring a new taste to gaming by modeling successful games after Iranian stories, with cultural and historical characters.

Game development in Iran, then, has been affected by national culture. As with other media, video games represent society and daily life. However, games are somehow different because our relationship with games is twofold. Computer games act as an "environment," which the gamer occupies vicariously, and as a result, we can speak of an "investment in the game" (Malaby 2006). Formation of a virtual self in computer games results in a new interaction between the person and himself or herself. On the other hand, the effects, outcomes, and messages of the game (moral, emotional, ideological, and cultural) are important. While playing, the gamer is reading and interpreting, and the addressee can choose from three positions (accept, oppose, or negotiate) in this confrontation. To investigate the reciprocal relationship between Iranian culture and the game industry in Iran, the relationship between space, narrative, and identity should be considered using Paul Ricoeur's concept of "narrative identity" (Ricoeur 1991). According to Ricoeur, identities are not fixed or preexisting things but are made within a narrative's flow. Human identity is made in society and formed with the help of media and narratives. The narratives intervene between the man and the world (preferentiality), between people themselves (communality), and between the person and himself (self-understanding). Games are a new form of mediation to aid in the formation of identity, and several studies have considered the formation of new identities based on narratives (Schott and Horrell 2000), which have the ability to change sexual, racial, and historical identities.

The main question here, however, is, "What is the effect of Iranian identity on the Iranian game industry and vice versa?" One can understand Iranian identity as a phenomenon affected and formed by thousands of years of history and culture. Moreover, ancient Iranian artworks, some of which are preserved in famous museums around the world, show the importance of Iranian culture and civilization as a piece of the puzzle of human civilization.

Iranian culture plays an influential role in four dimensions in video games. The first dimension is the land (Iran) and contains three elements: cultural heritage, the position of the monarchy and mythology, and the cultural norms of humans. Iran's thousands of years of civilization and artists' artworks are applied in designing environments, spaces, and objects, and this makes players aware of Iranian art and cultural heritage. Also, this dimension narrates stories of myths like those of Kave Ahangar, Rostam, Sohrab, Esfandiyar, Siavosh, and so forth, which are the themes of many video games. Those who are involved in gaming have heard of *Prince of Persia*, but those who are interested in history undoubtedly have heard of the philanthropy and the liberalism of Cyrus, the Persian king. Since the beginning of the game industry in Iran, many games have been designed based on ancient Iranian myths including those of *Garshasp, Legends of Persia (Siavosh)*, and others. This dimension of Iranian culture emphasizes concepts such as land, Iran's ancient civilization, cultural heritage, valiancy, bravery, honor, and the fight against evil. The key point in the story of these myths is the difference between evil's nature, its definition, and its position in Iranian myths versus Western myths. In Iranian stories, there are no creatures such as vampires or zombies, but there are demons—supernatural characters that are not human beings but rather devil-like spirits.

The second dimension is Iranian identity and faith. Ideological structures in Iran, whether based on Zarathustrian thought, on theosophical individualism, indulgence, lenience, and tolerance, or on the Islamic principles of justice, anti-oppression, and philanthropy, have elements of unity, equality, friendship, and moderation. This dimension has great potential for the narration of stories with the focus on ideological and religious culture in video games, especially Islamic culture, which is dominant in other nations in the Middle East. Games incorporating the Islamic lifestyle are popular among Muslims in the Middle East. Considering Islamic priorities in designing game levels, characters, setting, story, and narration can add to a game's attraction for the markets of Muslim countries. Iranian developers understand these priorities and not only observe them but also try to earn a competitive advantage for their games in the markets of Muslim countries (for example, *Dropfun* is working on games involving the basic concepts of Islamic philosophy and gamification).

The third dimension is that of traditions, customs, and rituals. This dimension contains concepts such as architecture, art, music, painting, sculpture, traditions, customs, and rituals, and plays a role in the artistic design of environmental architecture and objects, space, characters, and music within games. Iran has always been home to diverse ethnicities and cultures living in peace and a mutual sense of respect. The respect of other cultures and religions is present in Iranian games and is basically part of their conceptual beauty. Poems and stories, decorative arts and crafts, and music have a very special place in Iranian culture. For example, the large rock relief gravure of the women playing harp in Taq-e Bostan belonging to the Sassanid era and religious poems in parts of Avesta indicate the importance of music in ancient Iran before the Arabs' attack. Although Iranian music has weathered many problems during its history, it still has its grace and identity and has impacted the music of other nations. The multiplicity of instruments and twelve basic modes (seven primary modes [dastgahs] and five secondary modes [avazes]) have given Iranian music a diversity that cannot be found in Western music. Sculpture in Takht-e Jamshid or Takht-e Rostam is evidence of these arts among the ancient Iranians. Other arts (such as carpet weaving), crafts, and Islamic architecture also have an international reputation.

To have a better understanding of the depth and characteristics of Iranian developers' ability in design, the Persian CG Community website (http://www.cgart.ir) is helpful. As representatives of young Iranian designers, the managers of this site have participated in SIGGRAPH Asia (2011 and 2012), and their artworks have been entered in international competitions, gaining exposure to professionals all over the world. On the site's "Wall of Inspiration" (http://www.cgart.ir/inspiration), you can find the comments of famous artists such as Scott Robertson, John Howe, Don Hohn, Paul Hellard, Denis Zilber, and others. A quote from Raphael Lacoste even reads: "Iran was a great inspiration when we developed *Prince of Persia*."

Carpets, handcrafts, symbols, statues, and artwork from ancient Iranian empires are widely known around the world. When designing an environment, a development team uses these symbols along with literary or historic stories. This way, Iranian games take on their own identity, like East Asian games that have their own special environment and characterization. As the number of Iranian games increases in the future, there will be a new content style dedicated to Iranian games among other countries' productions.

The fourth dimension is modernity. Modernity in Iran started with the continuous comparison between Iran and Europe, which was the reason Iran began to be seen as backward in many aspects and resulted in a move toward Western European values, including Western aesthetics and norms. The most important elements of this dimension are a person's free will, the rule of law, and freedom of speech. These characteristics, present in modernity, help to create a bridge between ancient and modern culture in the world of Iranian game producers and are a factor that introduces creative ideas into the Iranian video game industry and marketplace.

In Iranian culture, much energy is spent on the production of joy, creating art, and increasing the level of excitement in life; Iranians are not afraid of taking risks and wanting a higher level of excitement in their daily activities. In general, these behaviors and characteristics of Iranian culture can be seen in the works of the third generation of Iranian game producers. As one author put it, "The future of the videogames is also the future of the storytelling" (Shapiro 2013); it could be argued that storytelling is not known in other nations the way it is in Iran. The richness of Persian literature supports this claim. In the stories of HezaroYekShab, storytelling is a technique used to deal with bad-tempered people. Iranian stories and myths have been translated into other languages and have many fans around the world. The presentation of common concepts in a new way in video games can be interesting to gamers from other countries. For example, in America, translations of the great Iranian poet, Molavi Rumi, have been popular. This shows how the conceptual differences of Iranian culture can be interesting for audiences in other countries.

This believability, along with delicate educational and ethical points, clever exaggerations, and comedic fantasies, make these stories right for the production of love-mythical games. Other types of Iranian stories are those narrated by animals, and their purpose is to express social and ethical values in the form of animals' behaviors. The game *Hate the Sin, Love the Sinner* is a new experience of the "love" concept from the viewpoint of an Iranian game development team about the common subject of love. In games such as *Donkey and Crow* (2002), *The Fox Who Followed the Voice* (2011), and *The Mouse and the Snake* (2012), with allegorical expression of the characteristics of each animal (which are popular in other nations, too), concepts of life and social relations are presented.

Iranians like poetry, and they use and have used it to transfer their messages with symphonic language. The reputation of Persian language in the world is not accidental. Molana, Khayam, Hafez, Saadi, and Ferdosi are among the most famous poets in the world. Although the Persian language is one of the bases of Iranian identity, it has influenced territory beyond the Iranian plateau. For instance, before India was colonized by England, in the era of the Gurkani government (the successor to the Teymuriyan emperor in India), Persian was the formal language of that vast empire. Apart from the fact that Persian was the primary language of Iran even among different races, it had a great influence on the neighboring countries.

This issue is important for the Iranian Game Industry from two perspectives. The first is that Persian language and culture is completely recognized in the Middle East, Indian subcontinent, and central Asia, and has even impacted the culture in some of these countries. For example, with the coming of spring, Iranian "Nowruz" (the Persian New Year) is celebrated on March 21 by 300 million people in more than thirteen countries. The second is that Iranian culture, recognized and rooted in other countries, provides a situation in which Iranian games will be familiar and acceptable for their people.

Epic events during the war with Iraq are the other elements that have formed themes for video games; several games were produced or are under production based on events of the Iran-Iraq War. According to Boellstorff (2006), "Game culture" is one of three possible relationships between cultures and games, along with "Cultures of gaming," and "the Gaming of cultures." Game culture means to look at a game as a world of meaning, or a cultural world that is created in relation with the real world. This world, like the real world, is constructed through the perceptions and aspirations of communities. From this perspective, Iranian literature describes a utopia, and cultural productions, especially games, are not created in a vacuum. Any producer inevitably uses signs and cultural codes to produce his games, most of which are discovered in the cultural treasures of the nation. Inspired by the "world of art" concept, it can be said that all game production elements use this cultural treasure in their production in one way or another, in different aspects of a game, including character, geography, space, subject, storyline, plot, and so forth; myths, heroes, and cultural archetypes are among the most usual subjects of this sort. This use is not only unrelated to the history but is also a common feature of popular culture.

Video Game Company Profiles

Besides regulatory and support organizations such as the Iran Computer and Video Games Foundation and Kanoon Game, there are more than 100 companies and institutions active in different sectors of the video game industry, including those of production, publication, and distribution. Also, in recent years, universities from all over the country have begun to pay special attention to the entertainment industry, particularly game development. Sharif University, Tehran University, Amirkabir University, Iran University of Science & Technology, and Kashan University, for example, have started specialized groups to conduct different projects involved with game production and the organizing of game development contests.

According to statistics from the Iran Computer and Video Games Foundation, in 2013 there were ninety game development companies and fifteen game publishing companies in the country. There were also more than forty-five online and print publications that exclusively covered video game news. Some of the more prominent game development companies are described in the rest of this section.

Fanafzar Sharif (http://fanafzar.ir)was founded in 2000 by a group of Sharif University graduates. The company offers services in various fields of software and hardware design. In 2008, the company started activities in game development. Their first experience in the new field was producing the game *Garshasp*, which became one of the best and most popular hard-core games in Iran and was one of the first Iranian games to make its way into foreign markets. The company continued the project with an expanded version, *Garshasp 1.5: Temple of the Dragon* (2012), and is presently working on *Garshasp 2*.

Royagaran Narmafzar (http://www.royagaran.ir) was established in 2000 as a software company specializing in developing software systems. Graphics programming and computer game programming are among the main activities of the company. After some successful coproduction titles with Kanoon, the company developed the game *Black Gold*, the first construction and management simulation game in Iran, which attracted simulation game fans around the world.

Ras Games (http://www.rasgames.com) was founded in 2000 by a group of Sharif University graduates. Since 2003, the team has focused on game development and has produced some successful titles, such as *Mithra's Planet*, which was selected by *EDGE* journal as one of the ten best indie games of 2010. At present, the company is working on *Thrateon*, the next episode of the *Age of Pahlevans* series. Ras Games' mission is to provide high-quality entertaining titles to both the Middle East market and the international market.

Simulator Developer Co. (http://simulator.ir) was established in 2001 and started with educational simulators, and computer games. After years of experience, the company decided to develop the first Iranian online game. In 2011, the company introduced *Asemandez* to online gamers inside and outside the country. In 2013, *Asemandez* became a huge success and was selected as one of the best MMO games, receiving Game Genetics' "MMO of The Year 2013" award. It also received the audience prize for the best indie MMO game.

Raspina (http://raspina.co) has been working in graphics and game development since 2009. The studio is now focused on production of computer games and animation. *E.T. Armies*, the company's latest game, has had very good feedback both inside and outside the country. *E.T. Armies* (Extraterrestrial Armies) is a first-person shooter game in a unique science fiction world, designed to show Persian historical themes and rich environments. The game was among ten indie games that were selected to join E3 2013. The game was also selected as the best upcoming indie game of the year in indie DB 2013 by players' choice.[2]

Pardis Game Studio (http://www.pardisgamestudio.com) is an indie game development studio that was founded in 2011. *The Dark Phantom*, a third-person shooter, was the studio's first project and was among ten indie games selected to join E3 2013. The game has attracted many people both inside and outside the country. It was the second indie game selected as best upcoming indie games of 2013 by players choice.[3]

Dropfun (http://www.dropfun.com) was founded in 2012 and is the first knowledge-based company specializing in the conception, development, and marketing of video games in Iran. Business consulting, gamification, serious games, and game publishing are the areas the company provides consulting services for both domestically and internationally. With the increasing importance of serious games and gamification in all aspects of life, the company is doing projects on health, education, industry, news games, life, commerce, marketing, and work in the country.

RSK Entertainment (http://www.rsk-ir.com; ResanaShokoohKavir game studio) was established in 2007. During a short period, RSK offered popular and successful adventure games including *Diamond of Maranjab* (2007), *Ghajari's Cat* (2011), *Murder in Tehran's Alleys 1933* (2013), *Forgotten Sound* (2011), and *The Fly of Dowran* (2013). The game *Murder in Tehran's Alleys 1933*, whose focus was an old detective story, made its way to other countries' markets, such as the UK. At present, the company is working on a sequel to *Murder in Tehran's Alleys 1933*.

The Industry Overall

The Iranian video game industry is arguably the largest in the Middle East. The number of game productions, companies, and game studios, Iranian interest in technology, the size of the domestic market, and the great amount of financial exchange in this market all support this claim.

To maintain a clear image of the industry, the IRCG has done studies that are updated annually. These studies include topics such as the consumption market, production market, useful and harmful effects of games, consideration of Iran's special cultural, religious, and social characteristics, and trends of the industry in the Middle East and other countries. According to the IRCG, as of 2014 there are more than 120 active companies, studios, and gaming companies in Iran. Of course, we should consider that the size of these companies and their human forces are completely different from those of large companies in other countries. Active gaming companies in Iran are usually funded by entrepreneurs or strategic business units (SBUs). Also, according to the IRCG, the volume of production has had a 100% increase as of July 2013, compared to figures from 2012. Presently, on average, seventy games are produced every year. About ten companies are active in online game production, and more than half of the rest produce hard-core and casual games. Among these active companies, all the members of the companies Fanoos and HamrahanFajr, and the main members of the NewFolder developer team are women.

Iran's game market is the largest in the Middle East. According to IRCG, it was made up of twenty million gamers in 2012 within the age range of five to sixty-five years old. The IRCG reported that in 2012, 57% of

Table 1

Amount of playing time by gamers, 2012 (Source: the IRCG).

Time	Playing Percentage
Every day	36.6%
2–3 days per week	26.9%
4–5 days per week	11.4%
A few days per month	11.4%
Irregularly	13.7%

players were between five and twelve years old; 21.5% were thirteen to fifteen years old; 16.8% were sixteen to twenty-nine years old; and 4.1% were thirty years old and above. It is also interesting to note that 42% of gamers are women. Based on IRCG data, gamers spend forty million man-hours per day playing games.

Action games and shooting games are the favorite genres of Iranian gamers. Of course, the *FIFA* series and *PES* series are at the top of the list, far ahead of their competitors; in fact, *FIFA 2014* (2013) and *PES 2014* (2013) sold more than one million units each in their first two weeks of release. After two months, each game had sold four million units on all platforms combined. PC games have the most players, and the consoles are in second place, but online games have the fastest growing percentage of gamers.

As mentioned earlier, a lack of up-to-date production knowledge and a lack of game design professionals are the biggest problems faced by the Iranian game industry. Also, an absence of investment companies to fund projects has caused serious problems for the producers. With the establishment of the first Game Development Institute in Iran in 2012 and the first gaming university in 2013, the IRCG is trying to train experts to compensate for this lack. The number of girls and boys registered in these two educational centers was so high that all the educational sessions were at capacity.

The Future of Video Games in Iran

Iranians welcome new ideas in the entertainment world and are interested in the latest technologies. The emergence and rapid growth of communication and entertainment technologies, the Internet, social media, and video games have created a new definition of leisure activities among Iranian users as well as those around the world. With the introduction of new, expensive platforms and best-selling games to the market, the demand has continued to increase, even during the economic recession and despite price increases.

The number of Iranian Internet users has grown from thirty-seven million in 2011 to forty-five million in 2012, and mobile phone penetration in Iran rose 110% in the first half of 2013. There was a 120% growth in mobile smartphone sales in June 2013 compared to 2012 (according to an interview with the director of Tehran Mobile Mall), and a 40% increase in video game consoles imported in 2012 (according to an Iran Customs

Table 2

Internet access in Iran, 2012 (Source: The National Internet Development Center [MTMA]).

Type of Access	Members	Users
Dial-up	3,467,380	6,934,760
WiMAX	810,830	2,027,075
ADSL	3,274,600	8,186,500
Mobile	22,629,809	22,629,809
Fiber Optic (ADSL)	2,656,000	6,106,000
Total	32,838,619	45,884,144

Report), all of which indicate an increase in venues and demand for video games. The increase in demand has convinced more companies to enter the video game industry, resulting in a more dynamic industry in the country. With rapid growth of video games in the technology and entertainment world, universities in Iran have started to plan game development courses. Three universities are planning to organize MS courses in game development engineering in 2015.

Verbal communication and text messages are important in Iranian society, and this has motivated VAS (value-added service) companies to cooperate with game development teams to launch SMS (short message service) game systems. Applying the principles of games to contexts other than that of entertainment, the concept of gamification has expanded the use of video games in the world. Recently, the production of games in different fields such as health, education, the military, urban design, and the automobile and oil industry has indicated the entering of a new phase for the video game industry in Iran. With the increase in game design knowledge and experience, game studios proficient in different genres, and the reliance on rich historical, literary, artistic, and cultural resources, we will see a kind of conceptual evolution and new combinations of content and technique in Iranian games in the near future, creating a special place and a new market for the Iranian game industry.

Notes

1. The statuses "white" and "red" were common during the Iran–Iraq war. White was used when the danger of air raid had passed. Red was when air raid danger was approaching and the siren was sounded.

2. http://www.indiedb.com/events/2013-indie-of-the-year-awards/features/best-upcoming-indie-game-of-2013-players-choice.

3. Indie DB.

References

Becker, Howard. 1982. *Art Worlds*. Berkeley, CA: University of California Press.

Boellstorff, Tom. 2006. A ludicrous discipline? Ethnography and game studies. *Games and Culture* 1 (1): 29–35.

Malaby, Thomas. 2006. Parlaying value capital in and beyond virtual worlds. *Games and Culture* 1 (2): 141–162.

Ricoeur, Paul. 1991. Narrative identity. *Philosophy Today* 35 (1): 73.

Schott, Gareth R., and Kirsty R. Horrell. 2000. Girl gamers and their relationship with the gaming culture. *Convergence* 6:36.

Shapiro, Jordan. 2013. The future of video games is also the future of storytelling. *Forbes*, June 17. http://www.forbes.com/sites/jordanshapiro/2013/06/17/the-future-of-video-games-is-also-the-future-of-storytelling/.

IRELAND

Deborah Mellamphy

Although video game play has long been an important part of popular culture in Ireland, video games were never treated seriously by the Irish government or Irish society in the past, which is demonstrated by the enormous lack of information on the early history of video games in the country; the Irish government and media never recognized the potential until recently when global video game companies began to develop operations in Ireland. It is surprising that Atari established a manufacturing base in rural Tipperary in 1978, employing just over 200 people in their plant, manufacturing Atari arcade cabinets, which were then shipped to various other European and worldwide locations for arcade games including *Asteroids* (1979), *Missile Command* (1980), and *Centipede* (1981). Considering Atari's worldwide success in the 1970s, this demonstrates Ireland's long relationship with the global gaming industry (see http://www.gamedevelopers.ie/features/viewfeature.php?article=45). The plant was taken over by Namco and Warner in 1984 and was eventually closed in 1998 after changing hands several times.[1] In addition, Norwegian company Funcom established Funcom Dublin Ltd. in 1994, initially employing twenty people. Jørgen Tharaldsen, Funcom product director, remembers that

> we decided to set up an office in Dublin because of several factors, one of the most important being the art resources in the city. We saw the need to have a department within Funcom which could specialise in console games. While looking for interesting places we naturally explored every detail from recruitment to rent, wages, taxes, and more. We chose Ireland in the end ... there were mostly advantages, as we saw it, compared to many other places. (Barter 2004)

The company developed *Speed Freaks* (1999) (released as *Speed Punks* in the US in 2000) when they were briefly located in the Sandyford Industrial Estate in Dublin from 1994 to 2001 (for more information and a concise history of Funcom Dublin, see Barter 2004). The office closed in 2001 after Funcom Norway decided to focus on the development of *Anarchy Online* (2001), a massively multiplayer online game. This chapter will discuss the history of video game development and play within Ireland, focusing on the findings of Aphra Kerr's 2009 survey of the industry in Ireland, as well as examining some of the most significant indigenous Irish game developers and foreign companies that have established operations in the country. I will also comment on the use of Irish characters and mythology in video games released globally.

October 11, 2011, remains a central date in the development of the gaming industry in Ireland, as *Taoiseach* (the Irish Prime Minister) Enda Kenny announced plans to introduce new packages, including tax incentives and project grants, to attract more game developers to Ireland. This followed a lengthy campaign by Paschal Donohoe of *Fine Gael* (one of Ireland's main political parties), Deputy for Dublin Central, after he emerged as a staunch supporter of the development of the games industry in Ireland. Donohoe, who is unofficially nicknamed Minister for Games, developed a policy paper titled "Helping Ireland Win in the next wave of Digital Entertainment: Digital Gaming in Ireland Fine Gael Policy Recommendations" (see http://paschaldonohoe.ie/?p=3984), which was approved by Enda Kenny and fellow TD (*Teachta Dála*, a member of *Dáil Éireann*, the Irish government) Richard Bruton, Ireland's Minister for Jobs, Enterprise and Innovation, and was incorporated into the political party's general election manifesto. In the paper, Donohoe specifically notes the worldwide financial success of *Call of Duty: Modern Warfare 2* (2009) and *Grand Theft Auto IV* (2008). He also cites the ongoing success of *World of Warcraft* (2004) and discusses the various opportunities for Ireland to develop and promote jobs, demonstrating the increasing interest in the sector and its future importance to the Irish economy as it attempts to overcome its current period of recession.

So great is the growth potential in the sector that Richard Bruton asked Forfás (the national policy advisory board in the Republic of Ireland that advises the Irish government on investment and enterprise in the areas of trade, technology, and science) to produce a report on the projected future of the games industry in the country in order to best support and optimize potential areas, skills, and organizations. The resulting report, "The Games Sector in Ireland: An Action Plan for Growth," was published in October 2011, and, rather than being a policy document, it is a "road map that lays out suggestions for the future" (Barter 2011) as it "talks about developing an international cluster, building international visibility, accelerating creative content development, and delivering next generation broadband" (Barter 2011), illustrating the current government's recognition of the financial and cultural value of the industry. In the report, Donohoe calls for more tax incentives for start-up companies, but this is unlikely due to the current climate and so far has not been recognized. It was also on the above date that David Sweeney, CEO of the Interactive Games Association of Ireland (IGAI) (or Games Ireland),[2] proclaimed that

> the video game industry is the most dynamic player in today's entertainment industry. Its unique mixture of creativity, technology, interactivity and fun gives it a head start on the competition. ... Ireland's rich reservoir of talent, technique and tenacity mean that it is now perfectly placed to play a leading role in Europe's video game industry, [sic] The Forfás report is a clear signpost for the kind of Games Incentive Package that Ireland will develop and offer to the fast-changing industry to become the best place to develop games by 2016. (Weber 2011)

It is clear that Ireland's international reputation as a significant gaming hub is developing with Enterprise Ireland[3] supporting and developing the growth of the industry in the country and with the current government aiming to increase game jobs from the current number of approximately 1,500 jobs to 2,500 by 2014 and to 5,000 by 2016. With a population of approximately 4.5 million, Ireland contains a wealth of IT knowledge. Although still considered at the periphery of the global video game industry in relation to larger countries including the United States and the UK, and despite the current global recession, the Irish IT sector has maintained its viability in the post-Celtic Tiger period[4] and is quickly establishing itself as a Digital Hub.[5] In 2011, the IGAI emerged at the same time as the new Irish government following the Irish general election in February 2011 and was established to represent Irish-based companies and to lobby for the industry in terms of government investment and development. The IGAI operates via a membership scheme and is open to companies and bodies associated with the games industry in Ireland. The company's website also importantly advertises jobs in the sector, and the company organizes industry events throughout Ireland. One such event was GIG (Games Ireland Gathering) 2012, which took place in February in Dublin and was attended by Activision Blizzard, DemonWare, PopCap Games, Big Fish Games, and Electronic Arts, and included master classes, workshops, and demos to boost video game development and community in Ireland. Before I discuss the sector in detail within the contemporary context, I will first refer to the findings of Aphra Kerr's 2009 survey, which provide a concise overview of video games in Ireland today.

"The Games Industry in Ireland 2009"

In "The Games Industry in Ireland 2009," Aphra Kerr (National University of Ireland Maynooth) and Anthony Cawley (University of Limerick) used Survey Monkey to conduct a comprehensive survey of twenty-one game companies in Ireland, both indigenous and foreign. The report begins by referring to Kerr's 2002 report, "Loading ... Please Wait: Ireland and the Global Games Industry," the first study of the video game industry in Ireland, which was based on fifteen interviews in which Kerr found that there were just over 300 people employed in the game industry, most of whom were employed in localization.[6] The 2009 study demonstrates the significant growth of the industry in Ireland as it found that there were 1,277 full-time permanent employees, 170 contractors, and twenty-two freelancers, highlighting a growth of more than 400% in seven years. It found that localization had decreased by half and that customer support had experienced the greatest increase. Significantly, it found that foreign companies employed a greater number of people than Irish companies, and just over 60% reported that their core function was game development, while 30% reported that it was game publishing. The most popular platform was PC/Mac (almost 60%), followed closely by mobile/iPhone. In employment demographics, women made up 13% of the total number employed and tended to be employed in either online customer support roles or in areas including quality assurance, administration, management, and localization. Kerr and Cawley's study also found that the greatest percentage of employees (43%) were aged between twenty-six and thirty-five and that most people employed in the sector in Ireland were German, with the number of Irish employees the second highest nationality. Thirteen of the companies were located in the greater Dublin area with nine in Dublin city center, five in Munster, and three in Ulster. Sixty-six percent located their headquarters in Ireland, and almost half had offices outside Ireland in Europe, the United States, Canada, and Asia (except Japan). The report also reveals the companies' motivation for locating in Ireland:

While availability of skilled labour was the most significant reason for almost half of respondents, this was closely followed by an ability to attract talent, even if it wasn't available locally, i.e., access to Europe and an English speaking workforce. In addition, four companies cited grants and financial incentives, and one identified links to universities. (Kerr 2009)

The findings show that the industry is growing and indicate the need to continue monitoring its growth. They also indicated the industry's reliance on international networking, as 86% of the companies surveyed were selling into the European market, with almost 70% each to the Irish and the US markets. As reported in Kerr and Cawley's findings, it is the sectors of middleware services and technology that have particularly grown and continue to grow exponentially in Ireland. Ireland's importance in the global gaming world is twofold: first, Ireland has its own indigenous middlewear companies including Havok, Demonware, and Jolt, to name but a few, and second, Ireland has seen the influx of foreign investment in recent years, as well as the opening of Bigfish, Zynga, and Popcap offices.

Indigenous Irish Companies: Havok, Demonware, and Jolt

Havok

Founded in 1998 by Hugh Reynolds and Steven Collins, Havok began as a Trinity College Dublin start-up company in the computer science department and now has offices in Dublin, Copenhagen, Montreal, and Munich. Despite the downturn in the Irish economy, Havok expanded operations to Tokyo and San Francisco in 2011, with the largest research and development team in Ireland. The company provides cross-platform support and technology for the Xbox 360, PlayStation 3, Windows, PlayStation Vita, Wii, Wii U, Nintendo 3DS, Android, iOS, Mac, and Linux. It helped in the creation of characters and environments for about 350 games including *Assassin's Creed* (2007), *Uncharted 2: Among Thieves* (2009), *Fallout 3* (2008), *Left 4 Dead 2* (2009), and *LA Noire* (2011). In 2007, the company was sold to Intel for a reported €76 million and is now one of the leading middleware providers for the global video games industry. Havok provides several tools and technologies:

- Havok Physics: Launched in 2000, it is software that allows for real-time 3-D collisions, contributing to the creation of a more realistic virtual game environment, making the physics involved in gameplay more realistic and a bigger part of the gameplay.
- Havok Animation: This software enhances the quality of character animation, making the gameplay experience more interactive and more aesthetically realistic.
- Havok Behavior: This software creates more realistic character performances, making gameplay more cinematic.
- Havok Cloth: This software, which has been used in *DJ Hero 2* (2010), simulates the movement of flexible materials including textiles and hair.
- Havok Destruction: This software has been used in *Battlefield: Bad Company 2* (2010) in the design of the destruction of structures and more realistic action sequences.

- Havok AI: This software creates more plausible character movement and behavior; for example, non-player characters that navigate realistically around obstructions.
- Havok Script: This software contributes to the easy use of the programming language Lua in console games.
- Havok Vision Engine: In August 2011, Havok acquired Trinigy, a leading 3-D game engine provider. Trinigy's Vision Game Engine was rebranded and retitled Havok Vision Engine, which is a multi-platform technology that enables large and complex environments to run at smooth frame rates.

The technology developed by Havok for games has also been used in films including *Charlie and the Chocolate Factory* (2005), *Harry Potter and the Order of the Phoenix* (2007), *Quantum of Solace* (2008), *Watchmen* (2009), and *Clash of the Titans* (2010). Havok has seen the need to recruit from outside Ireland due to the lack of suitably qualified Irish candidates and has worked with Engineers Ireland to promote the subjects of math and science at the high school level. Reynolds and Collins left the company in 2007 to set up New Game Technologies and Kore Virtual Machine. The company has continued to develop and has won several awards; each year between 2008 and 2011, Havok Physics was the winner of *Game Developer*'s Front Line Award for Best Middleware, and the company was also awarded an Emmy by the Academy of Television, Arts & Sciences in 2008 for its physics engines and its realistic simulations.

DemonWare and Jolt

DemonWare is a middleware company that specializes in online support and software for gaming, working with game development studios and supporting more than ninety games, including the *Call of Duty* and *Guitar Hero* series. The company has offices in Dublin and Vancouver and will soon open offices in Shanghai. Founded in 2003 by Dylan Collins and Sean Blanchfield, the company was purchased by Activision Blizzard in 2007 and is now a wholly owned subsidiary. Mike Griffith, CEO of Activision, said that

> the acquisition of DemonWare will enable us to eliminate many of the challenges associated with on-line multiplayer game development, reducing development time and risk, and allowing us to deliver consistent, high-quality on-line gaming experiences. In addition to increasing our talent pool of highly skilled engineers, DemonWare's suite of technologies combined with Activision's own library of tools and technologies will enable us to easily share on-line development capabilities on multiple platforms across our development studios. (Fahey 2007)

The main products developed by DemonWare are:

- DemonWare State Engine, which uses a C++ programming framework to code multiplayer games.
- Matchmaking+, which provides services such as matchmaking and user profile management and statistics for multiplayer games including the *Call of Duty* series.
- DemonWare's customers include Activision, Ubisoft, Atari, SEGA, and THQ. In addition to DemonWare, Dylan Collins also established Jolt Online Gaming in Dublin in 1999, demonstrating that indigenous companies are not limited to console gaming. The Jolt Online Gaming Network hosts and publishes free-to-play

browser-based games including *Utopia* (2008), *Legends of Zork* (2009), and *NationStates 2* (2008) and was acquired by OMAC Industries in Dublin in 2008.

Foreign Companies: Activision Blizzard, Zynga, Big Fish, and PopCap

Activision Blizzard

A number of international companies have developed operations in Ireland in recent years, including Activision Blizzard in Dublin and Cork. The Cork branch, which was established in 2007, employed approximately 700 people.[7] Originally, this office was only to employ 100 people, but the number quickly grew. The operations in both Cork and Dublin are European customer support centers.

Zynga

Founded in 2007 in the US, Zynga, the world's largest social game developer, established its largest European office and multilingual operations center in Dublin in 2010. The office comprises customer support, community management, IT, human resources, business intelligence, and CS content management in languages including French, Italian, German, Spanish, Portuguese, Turkish, Indonesian, and Korean. Speaking at the office's launch, Zynga's chief operating officer, Marcus Segal, said that Ireland's Industrial Development Authority (IDA) helped to develop the deal. "They found us in San Francisco and they reached out early and often. But that's only part of the story—there are lots of countries that have IDA functions, but nobody did as great a job as the IDA did on really partnering [with us] and showing us the way" (Millar 2011). Barry O'Leary, CEO of IDA Ireland, also stated that "Zynga is a very welcome addition to the digital media industry cluster in Ireland with many of the world's leading 'Born on the Internet' companies having established significant operations here. With an ever increasing number of such companies locating in Ireland, we are well established as the Internet capital of Europe" (Kennedy 2011).

Big Fish Games

Big Fish Games, one of the leading casual games developers, has regional offices in Vancouver, Luxembourg, and Cork, Ireland. In early 2012 it expanded its cloud gaming research and development initiative in its Cork branch. Jeremy Lewis, president and CEO of Big Fish Games, Inc., explained that, "Big Fish Games chose Cork because of the skilled and multilingual workforce we require for the development and growth of our worldwide business" (Richmond Recruitment 2009). Barry O' Leary, CEO of IDA Ireland, also said that "Ireland's competencies in areas such as cloud computing and software engineering, together with its vibrant digital media portfolio, make it a perfect fit with innovative companies in this space" ("Big Fish Games Expanding Cloud Game Research in Cork, to Create 30 Jobs" 2012).

PopCap Games

A subsidiary of Electronic Arts, PopCap Games, which is based in Seattle, opened its international headquarters in Dublin. The office works in product localization, mobile games development, marketing, sales, and business development. Significantly, it has partnered with Trinity College, Dublin, Carlow Institute of Technology, Dublin City University, and Dublin Institute of Technology to provide students with six-month work placements. Most of these students are then employed, in full-time or part-time roles, when they finish their university courses. From early 2011 to early 2012, the number of employees in the Dublin office almost doubled, growing from forty-five to eighty people. The 2012 GAME British Academy Video Games Awards (BAFTA) named PopCap Games' *Peggle HD* (2011) the best game in the Mobile and Handheld Category, a game that was developed at the company's Dublin studio. The company was acquired by Electronic Arts in 2011 and has a worldwide staff of more than 500 in Seattle, San Francisco, Dublin, Seoul, Shanghai, and Tokyo.

Irish Culture in Video Games

There is a very healthy video game culture in Ireland that relies on foreign exports, with the *Call of Duty* series being the top-selling game series between 2009 and 2010[8] Despite the interest and investment in the area in the past number of years, Irish culture has influenced elements of certain games, yet it remains difficult to find distinctly "Irish" games or games that are set in Ireland. Only one Gaelic sports game, *Gaelic Games Football* (2005), has been released, and it was only released in Ireland on the PlayStation 2. Barry O' Neill discussed the domestic market saying that "there are only so many games you can make about Celtic mythology or the GAA" (Barter 2011). Although few video games are set in Ireland and Irish characters are rarely featured, Irish and Celtic mythology does influence aspects and characters in a range of multi-platform games. For example, *Final Fantasy XII* (2006) features a character named *Cúchulainn*, the Celtic hero, and several games in the series contain a weapon called the *Gáe Bulg*, *Cúchulainn*'s legendary spear. *Cúchulainn* is one of the foremost heroes in Irish mythology and features in the stories of the Ulster Cycle (one of the branches of Irish mythology that is set in what is now Eastern Ulster). The figure of *Cúchulainn* is now associated with both sides of the conflict in Northern Ireland, as he has been adopted by Irish Nationalists and Ulster Unionists. He is considered the most significant Irish hero and has come to symbolize the Irish nation and culture. By contrast, Ulster Unionists consider him a symbol of Ulster unionism as he successfully defended Ulster from enemies. *Selkie* are monsters in the *Dungeons & Dragons* games who have the ability to transform into humans, as in Celtic mythology.

Other games that feature Celtic mythology include *Tir Na Nog* (1984) and its prequel *Dun Darach* (1985), both released on the Spectrum and the Amstrad. *Tir Na Nog*, which literally translates as "the land of youth," aka *Arabesque* (1984), is also based on the Irish mythology surrounding *Cúchulainn* and his battles in ancient Ireland. The game is populated by the *Sidhe*, a race of people who resemble monkeys. *Cúchulainn* wanders the land of the youth searching for the four pieces of the Seal of Calum to reunite them in order to defeat the Great

Enemy. The booklet boasts that the game is a "computer movie," as the camera angle can be altered between fore, middle, and background, and the game features parallax scrolling, creating a sense of real movement—revolutionary aspects of the game for the time. In the background, clouds roll, birds fly, and smoke billows from distant volcanoes. The game was released by Gargoyle Games in the UK on the ZX Spectrum 48K and was created by Gargoyle's head designer Greg Follis and programmer Royston Carter. The game was influenced by the pair's mutual love of science fiction and fantasy and was particularly influenced by J. R. R. Tolkien's writings. "I also used to play *Dungeons and Dragons* and *Tunnels and Trolls*. And real-life mythology of course and—all right, I confess!—I still read *Imagine* magazine."[9] The pair settled on the influence of Celtic mythology when they realized that it was not only Irish but spanned across European mythology. They also saw the television series *Robin of Sherwood* (titled *Robin Hood* in the US) (1984–1986), which was heavily influenced by Irish mythology and imagery and featured music by Irish traditional band Clannad. According to Greg Follis, "It was lovely. It had that super soundtrack by Clannad and it was a clincher for a Celtic game" (Bourne 1985). *Tir Na Nog*'s prequel, *Dun Darach* (1985), which translates as "Old Fort," was also heavily influenced by Celtic mythology and is set in a Celtic metropolis. Skar, a sorceress, has imprisoned *Loeg*, *Cúchulainn*'s friend. Chris Bourne even calls the game and its prequel "two of the best games ever seen on the Spectrum" (Bourne 1985).

Numerous other contemporary games feature Irish characters or characters with Irish heritage, though most are portrayed stereotypically as violent criminals. Although he only makes a very brief appearance, *Fallout 3* (2008) features Colin Moriarty, a character with a strong Irish accent who cusses profusely. A villain, the character is a local bar owner, drug dealer, and crime lord in Megaton City. Likewise, Sean Devlin in *The Saboteur* (2009) embodies the Irish stereotype as he brawls, drinks, smokes, swears, is short-tempered, and is a womanizer. *Far Cry 2* (2008) features Frank Bilders, who was, before the game begins, a terrorist in Belfast, involved in the IRA. He was imprisoned in The Maze, a notorious prison in Northern Ireland, where he killed four other inmates and was released from prison after being shot in both knees. He then turned informant and disappeared in Africa. The character wears a green tracksuit, winks a lot, and is pale-skinned even though he now lives in Africa, representing stereotypical Irish physical traits. In *Grand Theft Auto IV* (2008), Patrick (Packie) McReary, the employer of the game's main character Nico, is Irish-American. Other characters that are less obviously Irish include Nina Williams in the *Tekken* series (1994–2012) (although there is nothing Irish about her; the game's writers claim she is Irish, but she has an American accent and will be played in the upcoming movie adaptation by South African model and singer Candie Hillebrand), and Jack "Tiny" O'Hara in the *Commandos* series (1998–2006).

The most positive representation of an Irish character in a video game is featured in *Clive Barker's Undying* (2001), a first-person action/adventure game, in which the main character, Patrick Galloway, is Irish and is portrayed as a very good-natured hero. The game takes place in Ireland and involves the Covenant family, who are English. The environment includes navigation through a lavish Irish manor house, an outdoor area filled with Rune-carved standing stones/Ogham stones, a cottage, and a mausoleum. During WWI, Galloway survived what is referred to as a "mystical incident," during which he attained the Gel'zibar stone, a green gem that hangs around his neck and is imbued with magical powers. Patrick uses weapons throughout the game, but he can also cast spells and summon audible moments from the past, overhearing past conversations and arguments. He can also summon the power of the Gel'zibar stone and can "scrye" or see visions of the past. He even goes back in time at one stage to retrieve the Scythe of the Celt in order to execute siblings of Jeremiah, the main villain of the Covenant family. All other Irish characters in the game are minor, as they are servants in the manor and play no great part.

Tomb Raider: Chronicles (2000) is another game set in Ireland, although only for one level. The game also portrays the country as a mystical environment, as an Irish Catholic priest, Father Patrick Dunstan, a very stereotypical religious representation of an Irish character, describes apparitions that have been seen on Black Isle, the location of the level. The level takes place mostly outdoors and features small villages and cottages with phantoms and ghosts, again playing on Ireland's Celtic mythology.

Conclusion

As is clear, the video game sector is one that will grow in economic and cultural significance in the future on a global scale, and specifically in Ireland, as the Irish government now understands its importance to the country's financial recovery. The government now grasps the opportunities that are available to become more progressive and competitive in the digital era. One way that this must be done is through greater advertisement of existing funds that are available for both Irish and foreign game companies. Another sector that needs considerable attention in Ireland is academic research. Although there is a range of programming courses, there is a lack of video game theory and history courses at the university level.

Notes

1. "After being established by Atari/Warner Communications from 1978 to 1984, the Tipperary factory was run as a joint venture between Namco and Warner from 1985 to 1990. Ownership reverted back to Warner from 1990 to 1995 until Chicago based Midway Games purchased the plant in 1995. The plant was purchased the following year by Namco Europe and would close after 20 years in 1998" (McCormick 2008).

2. "Established on an ad-hoc basis in 2010 and incorporated in 2011, Games Ireland (the Interactive Games Association of Ireland) represents companies in Ireland involved in the creation, development, publishing and distribution of interactive games. It is an advocacy group which seeks to drive sustainable growth in the industry at a crucial time for this country. Members include international stakeholders such as Activision Blizzard, Big Fish Games and Popcap Games to local pioneers Havok and Demonware" (*Games Ireland* 2013).

3. Enterprise Ireland is a government body set up to aid the growth and development of Irish business ventures (see http://www.enterprise-ireland.com/en/About-Us/).

4. The Celtic-Tiger era is regarded as the period between 1995 and 2007, approximately, when the economy of the Republic of Ireland experienced rapid economic growth. Since 2008, as a result of the global recession, the country has experienced a decrease in economic growth.

5. In fact, Aphra Kerr stated that the Celtic Tiger years bred growth in the industry.

6. "Game localization or game globalization refers to the preparation of video games for other locales. This adaptation to the standards of other countries [or one's own country] covers far more than simply translation of language. There are different areas such as linguistic, cultural, hardware and software, legal differences, graphics identity and music" (Davis 2011).

7. It was reported in February 2012 that 200 jobs were to go at the Cork operation, from a workforce of 900. The redundancies were blamed on global restructuring.

8. *Call of Duty: Modern Warfare 2* was the best-selling game in Ireland in 2009, and *Call of Duty: Black Ops* the best-selling game of 2010.

9. See Bourne 1985. Follis created the 14-bit walking man, a fourteen-part animation of a man walking across the screen, a revolution in the 1980s, on which the rest of the game was built.

References

Barter, Pavel. 2004. Fun anyone? *gamedevelopers.ie*, December 16. http://www.gamedevelopers.ie/features/view.php?article=25.

Barter, Pavel. 2011. The Irish games industry levels up. *gamedevelopers.ie*, December 1. http://www.gamedevelopers.ie/features/viewfeature.php?article=5648.

Big Fish Games expanding cloud game research in Cork, to create 30 jobs. 2012. *Silicon Republic*, February 9. http://www.siliconrepublic.com/careers/item/25725-big-fish-games-expanding-cl.

Bourne, Chris. 1985. The Gargoyle Speaks. *Sinclair User*, November. http://www.nvg.ntnu.no/sinclair/industry/publishers/gargoyle_su1185.htm.

Davis, Alan. 2011. PS&L Ireland test. http://www.localisation.ie/resources/courses/summerschools/2011/AlanDavis_Test.pdf.

Fahey, Rob. 2007. Activision confirms demonware acquisition. *Games Industry International*, March 6. http://www.gamesindustry.biz/articles/activision-confirms-demonware-acquisition.

Games Ireland. 2013. "About" page. http://gamesireland.ie/wordpress/gamesireland/.

Kennedy, John. 2011. "Facebook effect" and IDA ingenuity inspired Zynga expansion in Dublin. *Silicon Republic*, October 6. http://www.siliconrepublic.com/reports/item/22147-behind-zyngas-expansion-in/.

Kerr, Aphra. 2009. The games industry in Ireland 2009. *gamedevelopers.ie*, October 29. http://www.gamedevelopers.ie/features/view.php?article=47.

McCormick, Jamie. 2008. Atari and Ireland. *gamedevelopers.ie*, November 13. http://www.gamedevelopers.ie/features/viewfeature.php?article=45.

Millar, Angela. 2011. Social games company Zynga launches in Dublin. *BBC News Technology*, June 11. http://www.bbc.com/news/technology-13739404.

Recruitment, Richmond. 2009. About Big Fish Games Cork. *Idaireland.ie*. http://www.idaireland.com/news-media/press-releases/big-fish-games-expected-t/.

Weber, Rachel. 2011. Ireland woos industry with new games incentive packages. *Games Industry International*, October 11. http://www.gamesindustry.biz/articles/2011-10-11-ireland-woos-industry-with-incentives.

ITALY

Enrico Gandolfi

It's quite hard to try to get a coherent overview of the reality of video games in Italy. On one hand, we are referring to a well-established market with numbers similar to the most important European countries. On the other hand, we observe a borderline presence of national producers and a lack of institutional attention to the medium.

In the history of digital entertainment, in the past as well as the present, Italy usually appears in terms of characters and settings: from Super Mario to Ezio Auditore passing through the mafia's topoi, the stereotypes and the artistic patrimony of this country were exploited with great success by the most famous producers. Following a chronological perspective, it's also easy to note some excellent productive cases with Italian influences: Andrea Pessino and Massimo Guarini are well-known names, but they are affirmed and legitimated only within foreign contexts. In the end, Italian creators and products remain marginal in the productive landscape of digital games. The reasons are several and due to both social and economic factors. Moreover, the cultural issue about the status of digital games is central and recurrent according to the debates regarding their effects and the related artistic legitimacy.

According to the Associazione Editori Sviluppatori Videogiochi Italiani (AESVI) (2012b), in 2011, the Italian video game market (excluding online games and mobile games) had a total income of 993 million euros with an annual decrease of 7.1%, representing the fourth largest market in Europe (though the distance from the third, France, is ample). The console hardware sector brought in 393.6 million euros (home console systems represent 67.2% of the market) while software was worth 599.5 million euros (with 94.6% of it for home console systems). The small size of the PC sector is partially due to the rising importance of digital delivery in the country as well as worldwide, not reported in AESVI analysis. The penetration of home console systems in 2011 reached 45.5% of Italian families, whereas personal computers reached 58.8% of them; such data confirms the good diffusion of the former.

The NewZoo report (2012) tells us that in 2012 there were 18,600,000 active players in the country (about 30% of the population), with overall 1.8 billion euros spent and an average of 3.9 game platforms for each player. The growth in terms of payers (19%), players (16%), and money paid (6%) is higher than the average European growth (respectively, 17%, 8%, and 3%), and Italians seem to be more used to playing than their European and American colleagues: 23% of them usually play on their computers and personal (mobile), entertainment (TV), and floating (portable devices) screens. In this chapter, conscious of the fact that perfect completeness is an ideal rather than a fixed goal, I will try to summarize the significant traits of these scenarios, attempt to point out some guidelines, and also consider the future of video games in Italy.

History

The history of video games in Italy follows the main tendencies of the history of video games in Europe, but with some peculiar traits to report. The first element to consider is that Italy produced one of the earliest consoles, the Ping-O-Tronic, by Zanussi (a furniture company still active in several businesses) in 1974, and was licensed to run the official version of *PONG* (1972). The company followed it with a new version in 1977, the Play-O-Tronic. The overall diffusion of the latter was appreciable if we consider the time and the context; the Play-O-Tronic sold almost 22,000 units in a few months, even with a German version published by Quelle. The foreign influences of arcade culture were nationally diffused by names such as Sidam (which held the license for *Space Invaders* [1978], renamed *Invasion* in Italy) and Bertolino (Atari's distributor). The former was the protagonist of a legal controversy involving the latter and Atari, ending with a sentence significant for Italian copyright law: Sidam was judged guilty of cloning famous Atari games such as *Asteroids* (1979) without permission, and, indirectly, video games officially started to be perceived as intellectual property. Other important producers were Midcoin and Zaccaria, active in flipper and coin-op sectors and also in exporting their creations to foreign markets.

As in the rest of Europe, the North American video game industry crashes of 1977 and 1983–1984 affected the Italian national scene, which was less mature and more focused on home computers. After releasing the world's first desktop computer, the Programma 101, in 1965, the Italian company Olivetti promoted and developed Olivetti computer culture during the 1960s, 1970s, and 1980s, with personal computers like the Olivetti M20 and Olivetti M24, which, after 1983, competed with the strong diffusion of the Commodore 64. These machines, and to a lesser extent the Commodore Amiga, were used as game machines and played a strong role as learning tools, whereas Atari platforms were less successful due to their high costs and insufficient promotion.

In the 1980s, magazines such as *Video Giochi* (which covered Atari VCS, Intellivision, ColecoVision, and coin-op games), *MCmicrocomputer* (covering informatics), *The Games Machine* (covering home computers), *K* (covering all), and later on, *ZZAP!* (covering Commodore 64, AMIGA, etc.)—all Italian versions of the more famous English journals— exercised a strong influence in creating a solid consciousness and a communitarian feeling around the medium. They were the best, and often the only, way to access information for garage programmers. They also represented a native point of view, due to the fact that their editorial teams became independent in content creation. Important names in this field, such as the game designers and programmers

Bonaventura di Bello and Francesco Carlà, and the researcher and journalist Matteo Bittanti, supported and grew up in those contexts, and many hobbyist programmers started following their advice.

Furthermore, Nintendo and SEGA consoles reached a good diffusion despite Mattel's mediocre management of Nintendo in Italy, and because of Giochipreziosi's good management of SEGA in Italy, which together resulted in more equality between the two brands. However, it was Sony's first console, the PlayStation, that was the real turning point. The PlayStation became synonymous with video games in only a few years, aided by the gigantic pirate market that exploded during that period and is still operating. Thus, young people could obtain more titles than ever, becoming part of what journalists have called the "PlayStation generation." Sony's supremacy is still evident today since the PlayStation brand has the highest sales in the country; according to Sony Italia in 2009, the overall number of Sony machines sold reached the goal of 10,000,000. Sony's lead is on par with European trends, though it continues to fight with Nintendo and Microsoft, both of which are now stronger than ever.

National Producers, Foreign Affiliations, and Institutional Actors

Even if during the "Commodore era," from 1983 to the early 1990s, the Italian "demoscene" was active, certain game designers such as Enrico Colombini, Carlo Landolfo, and Bonaventura di Bello were prolific (with the concomitant rise of the "compilation" model of distribution)[1] and influential, and the first Italian software house to achieve substantial success was Simulmondo. Established in 1987 and closed in 1999, Simulmondo has produced and published almost 150 titles for multiple platforms, from the Commodore 64 to PC and Game Boy (for example, *F1 Manager* [1989] and *Time Runners* [1993]). The games developed by Francesco Carlà and his team were basically sports simulations and comics adaptations of Bonelli characters (for example *Tex* [1993] and *Dylan Dog: Attraverso lo specchio* [1992]). Another significant name was the Dardari Bros., who focused on the sports genre (with games such as *Italy '90 Soccer* [1988] and *Over The Net* [1990]).

Moreover, the diffusion of the Commodore Amiga during the first half of the 1990s inspired teams that formed around the Amiga's machine language, who then went on to export their knowledge to other platforms. The beat-'em-up game genre was particularly popular; for example, Light Shock Software released the high level *Fightin' Spirit* (1996) and *Pray for Death* (1996), and NA.P.S. Team was responsible for *Shadow Fighter* (1994) and *Gekido* (1999) for the first PlayStation. Another genre popular in that period was that of adventure games, including Prograph's *Tony Tough and the Night of Roasted Moths* (1997) and Trecision's *The Watchmaker* (2001) for PC, which demonstrated quality superior to the national average, even if they were often produced by short-lived software houses.

One of the most creative studios of the period was Milestone S. r. L., which was established in 1996 by Antonio Farina, originally under the name Graffiti. Over the years, Milestone was able to ally with Virgin, EA Sports, and Capcom, and was incorporated in 2002 by Leader Group, the main publishing actor in Italy. With almost 100 members and a triple-A production focused on racing simulation games for all the relevant platforms (*Screamer* [1995], and the *Superbike* and *MotoGP* series), it remains the most important software house in Italy. Moreover, it has been a training ground for programmers and creative talents now employed at different companies, such as Massimo Guarini (who collaborated on several projects at Ubisoft Studios Milan) and Giovanni Bazzoni (now CEO at Digital Tales).

Other important but smaller independent software houses also publish their products abroad; for example, Spinvector, a versatile company that produces mobile games and augmented reality, had important partnerships with major corporations such as Intel and Microsoft; Artematica Entertainment, which specializes in advergames, casual games, and licensed games such as *Diabolik: The Original Sin* (2009) and *Pinocchio the Game* (2008); TiconBlu, responsible for several products under the tag "serious entertainment" (*GuidaTu* [2007], *The System* [2013]); Raylight, with relevant connections to Nintendo; and Digital Tales, a dynamic company with an impressive and growing portfolio of products.

As evident from their profiles—and the first AESVI report (2012a)[2] about Italian developers—beyond the core goal of creating entertainment, these companies operate in several other areas as well, including edutainment, advergames, technological support, and communication; thus, they configure themselves as service providers, trying to exploit the potential of convergent trends of the industry, from social networks to gamification (and not strictly in creative terms). In other words, the core business is fundamentally a hybrid, and often business to business.

The same report stated that more than a third (37%) of the seventy-two Italian companies interviewed active in game development employ only one or two people. More than a quarter (27%) show annual revenues under 1,000 euros, 12% made between 1,000 and 10,000 euros, 27% from 10,000 to 100,000 euros, 27% from 100,000 to 1 million euros, and only 7% crossed the six-zero sum (AESVI, 2012a). The entire sector employs almost one thousand people, even if their efforts are often rhapsodic: according to the AESVI report, the majority of companies (62%) were established in the last three years, and 90% are mostly self-financed; only 21% have some contact with publishers, and 16% have a holding company. Thus, even if the failures are recurrent and the names redundant, we are observing a fluid and dynamic scenario, where start-ups and little creative crews try to reach success in Italy and abroad (59% of developers interviewed were thirty-five years old or younger).

Even if the digital divide in Italy and the inadequate penetration of high-speed Internet access remain serious problems to solve, the expansion of the online market and huge mobile market are strong drivers of a proliferation of developers. According to NewZoo data (2012), Italians in 2012 spent USD $300 million on games on social networks and casual game websites, USD $190 million on mobile games, and USD $200 million on MMOGs. Unsurprisingly, the mobile market is central for 37% of Italian companies, due to the lower costs of production and the ease of diffusion (including the reintroduction of older games), followed by the PC market (24%) and console market (15%). However, several young software houses succeeded in emerging from this *mare magnum*, even if restrained: *ANNA* (2012) by Dreampainters is an acclaimed PC adventure that also found an international publisher, Kalypso. Santa Ragione (*MirrorMoon* [2013]), Studio Evil (*Syder Arcade* [2012]), Darkwave Games (*Master of Alchemy* [2012]), HeartBit Interactive (*Doom and Destiny* [2011]), and

Urustar (*Zwan* [2013]) have obtained remarkable results, often showing significant artistic skills. *MirrorMoon*, for example, was the 2012 Official Selection of the Indiecade, an important international festival of independent games, and *Doom and Destiny* received positive reviews from foreign critics (3.5/5 on hardcoredroid.com and 3.9/5 on mashthosebuttons.com).

Furthermore, from a national perspective, certain individuals have become more famous than entire groups. For example, Paolo Pedercini of Molleindustria is a significant contributor in the worldwide serious games scene; Fabio Viola (2011), with his company DigitalFun, is one of the lead people associated with the phenomenon of gamification; Stefano Gualeni is a qualified game designer who is also active in academia; Matteo Bonvicino is an important author of machinima and the figure beyond BNV Entertainment, a dynamic creative force. Matteo Stanzani is a transmedia and game expert who has worked with Disney, Ferrero, and Ferrari; and Leonard Mechiari and Mattia Traverso are part of the team behind the controversial *RIOT* (seeking funding on Indiegogo). Some of these names have to go abroad in order to find the right recognition, as well as be defined as the "exiled talents" in the AAA perspective: people such as the ex-Blizzard Andrea Pessino, co-founder of American Ready at Dawn Studios, Riccardo Zacconi, CEO at King Digital Entertainment plc, and Massimo Guarini from Ubisoft and Grasshopper Manufacture, the second now back in Italy and directing Ovosonico, a promising software house that already has a strong relationship with Sony concerning *Murasaki Baby* (2014) for PSP Vita.

The overall situation tells us that Italian producers are late in gaining respect in other countries, even though there is a relative acceleration in recent times. According to producers and associations, this condition is due to the lack of private investments and the absence of state support. The proposed law n.5093 (presented April 1, 2012, in the Italian Parliament), which provides tax credit and tax shelter for this sector, is a positive signal even if it does not become a law. We can find a partial solution in the supporting measures concerning new technology-oriented start-ups, even if sometimes their nature is regional and their focus is not on this specific sector.

Another problem is the inadequacy of formative and educational institutions concerning the development of productive skills. Even though there are several masters and private schools that try to offer such training, Italian universities, with few exceptions (such as the University and the Politecnico of Milan), have a poor presence on this front. Unsurprisingly, many Italian developers are self-made programmers, often arriving from technical institutes (67% of them didn't attend a university). Moreover, the fragmented productive landscape has a poor connection with the academic world. However, Italy is one of the countries where the Global Game Jam is held annually, usually due to efforts of individuals and associations such as E-Ludo.

International majors in Italy are increasing their operations with affiliated companies in order to oversee their logistics and distribution (though there are consolidated Italian publishing companies such as 505 Games and Digital Bros, or the now-defunct CTO S.p.A.). Electronic Arts was the first third-party publisher to establish a branch in Italy in 1996, followed by Ubisoft in 1998, which immediately began to recruit Italian programmers for its Game Boy and DS dedicated production: now the studio Ubisoft Studios Milan, the other big team in Italy with Milestone, works on Wii and Kinect technologies (some of their products are *MotionSports*

[2010], in collaboration with Barcelona studios, and *We Dare* [2011]). Other important presences are Activision-Blizzard Italia, Koch Media Italia, and Disney Interactive, and of course the three main hardware producers, Nintendo, Sony, and Microsoft.

In recent years, associations were established to represent all these actors and possibly to improve this condition and industry network. The first is AESVI, established in 2001 and a member of the ISFE (Interactive Software Federation of Europe). It represents both international and Italian companies almost in their totality, as the main force in institutional and cultural relations and monitoring efforts from detailed reports to annual events. These events include the Dragone d'Oro (Dragon of Gold), a sort of Italian Oscar of video games, Games Week for the public, and the practitioner-oriented Italian Game Developers Summit and Game Forum. Particularly interesting is its program called AESVI4Developers, designed to connect and bolster Italian creators of games through several initiatives. The second and more marginal program is the Institution of Italian Producers within Confindustria (the national industry aggregation),[3] established in 2009, which organizes the annual IVDC (Italian Videogame Developers Conference) with AIOMI.

One specular event is SvilupParty, which is organized every year by the veteran practitioner Ivan Venturi with the support of the Cineteca of Bologna and the software houses Studio Evil and TiconBlu. SvilupParty is a meeting for experts that is gaining importance with every conference. Finally, on the Internet, the portal Gameprog.it is a place where Italian developers can meet and share information.

The Role of National History and Ludic Habits in Digital Game Consumption

Settings from Italian history are popular with global audiences and are exploited by many foreign products (from *Assassins' Creed* [2007] to *Dante's Inferno* [2010] and, of course, *Rome: Total War* [2004]), whereas Italian video games make poor use of them, if we exclude specific educational programs. The only two examples specifically referencing our national past, and both supervised by Raoul Carbone, were *Il rosso e il nero—The Italian Civil War* (2004), about the civil conflict between partisans and fascists during the second World War, and *Gioventù Ribelle* (2011), which concerns the battle for national independence in the nineteenth century. The awful quality of the second, aggravated by the institutional sponsorship and the media attention it received, represented a sort of boomerang for the entire national industry, as well as a negative international echo.

Another product that refers to local culture is 10th Art Studio's *Shadows on the Vatican* (2012), a four-episode adventure (as of late summer 2013, only the first two were published), which puts players within Vatican intrigues, while also depicting Roman culture. But it is hard to find further connections or relations between recent national history and the Italian game industry. The governmental indifference (aggravated by the current political instability) and the low profile of the national producers have meant a sort of lack of communication between the two sides.

If we analyze local products that are able to cross national boundaries, the Italian connotation is still present: the simulation games developed by Simulmondo and later by Milestone are exemplary in employing the typical Italian love for cars, racing, and sport. Furthermore, and referring to recent pop culture, several licenses exploited and diffused by Italian developers now, as in the past, are of famous Italian comics, novels, settings, and games, such as *Diabolik: The Original Sin* (2009), *Sine Requie* (2003), *Bang!* (2011), and *Nicolas Eymerich: Inquisitor* (2013). Forge-Reply's *Lone Wolf* (an old series of fantasy book games), is an interesting case of a foreign IP acquired by an indigenous publisher, and is a brand still popular in the country. For the rest, production follows international and generalist topoi (in order to reach a wider audience), if we take as significant examples the recent manga culture-based *Dengen Chronicles* (2013) by Mangatar, nerd-culture-based *Doom and Destiny* (2012) by HeartBit Interactive, or the older fantasy-action RPG *ETROM: The Astral Essence* (2006) by PM Studios. For some games, their educational vocation means the presence of values linked to such things as the environment (such as PM Studios' *Eco Warriors* [2008]) and legality (such as TiconBlu's *The System* [2013]).

If we analyze the overall consumption and sales rankings, observing the weekly and monthly charts compiled by AESVI with the support of GfK Retail and Technologies (which only refer to boxed software, and not in quantitative terms; see http://www.AESVI.it), the importance of national culture is evident through the popularity of football. Every year, current and past versions of *Pro Evolution Soccer* (*PES*) and *FIFA* are at the top of the charts, both in general and single platform classifications (for PCs, another enduring presence is *Football Manager*). In addition, products partially linked with this sport, such as the *Inazuma Eleven* series, are successful. We can read this picture as a radicalization of the European trends through the national culture.[4]

The rest of the charts are dominated by what we can define as "institutional turning points" acclaimed by critics, the press (specialized as well as generalist), hard-core communities (such as those surrounding *The Last of Us* [2013] and franchises including *Call of Duty*, *Diablo*, *Assassin's Creed*, and *Uncharted*), and transgenerational brands such as *Pokémon*, *Mario*, and *The Sims*. Also, family games and casual games are quite widespread; for example, *Just Dance* (2011) has experienced increasing popularity, which has grown quite vast. These preferences explain why in 2011, 18% of games sold were classified PEGI 18, while 53% were PEGI 7 or less (AESVI 2012b). Furthermore, on the latest PC sales charts there is a significant presence of MMORPGs: in 2012, *Guild Wars 2* (2012) and the evergreen *World of Warcraft* (2004) achieved important results: for example, the former was first in August, fourth in September, and fifth in October, while the prepaid card of the latter, respectively second, sixth and fourth (with *World of Warcraft: Mists of Pandaria* [2012] at the top of the chart in September).

According to software charts (used as indirect lenses due to the absence of detailed data about hardware and software sales), the leading console remains the PlayStation 3. Sony versions of games are always in first place. For example, on the charts for December 2012 we find:

1. *FIFA 13*—PS3
2. *Just Dance 4*—Wii
3. *Call of Duty Black Ops II*—PS3
4. *Assassin's Creed III D1 Version*—PS3
5. *FIFA 13*—Xbox 360

In March 2013 we find:

1. Tomb Raider—PS3
2. God of War: Ascension—PS3
3. Naruto Shippuden Ultimate Ninja Storm 3 D1—PS3
4. Tomb Raider—Xbox 360
5. Gears of War: Judgment—Xbox 360

And in June 2013 we find:

1. The Last of Us—PS3
2. Animal Crossing: New Leaf—3DS
3. Donkey Kong Country Returns 3D—3DS
4. FIFA 13—PS3
5. FIFA 13—Xbox 360

(These rankings also confirm what is written above regarding software trends.) This supremacy probably is also due to the heredity of the revolutionary impact of the first PlayStation. In addition, the Nintendo DS brand seems to be popular compared to the weaker but still solid presence of the PSP (according to CfK, there were 243,700 3DS units sold in the first forty-four weeks after the launch, which was a European record). It is difficult to analyze Italian consumption, however, because these data are not released in detail, and often we have to ponder estimates rather than real numbers. Even if we cannot be totally sure about the accuracy of AESVI and CfK reports, and even though their value is strictly indicative of the market trends (also because they are concerned only with boxed software), the solidity and the uniqueness of these institutions (more in-depth CfK surveys are sold for a fee) legitimate them and their data as primary free sources of information concerning the Italian market.

Turning to the practice itself, the Italian attitude toward sports is also evident in so-called e-sports. Established in 1998 and still active as of 2013, the avant-garde company NGI is an Internet provider totally focused on online gaming and being "pro-gamers." Nowadays, e-sport culture is being improved and expanded by e-sport associations such as FNIV (Federazione Italiana Videogiocatori), Personal Gamer, and ELS Italia, by an increase in the number of tournaments (from the Personal Gamer Italian championship to the itinerant show Videogames Party), and by the diffusion of games such as *League of Legends* (2009), the *Call of Duty* series, and the *Battlefield* series. Such trends culminated in the CONI (Comitato Olimpico Nazionale Italiano), the most important sports association in Italy, officially recognizing digital games as a sport in 2013. Thus, the most famous Italian athlete, Simone "Akira" Trimarchi, who specializes in real-time strategy (RTS) games, was the winner of several titles at the World Cyber Games (whose 2006 edition was held in Monza, near Milan), and is now a famous columnist and opinion leader.

Even though the video game industry has a solid foundation, the Italian pirate market is considerable and has a cultural dimension. Italy is one of the nations that the American government has put on its so-called "watch list" (USTR 2013) due to the absence of concrete regulations regarding copyright. According to the

Business Software Alliance (BSA) (2012), 48% of the software used by Italians on their PCs is illegal, much higher than the European average of 33%. We are referring to almost US $2 billion in value terms for 2011. Of course, this situation also concerns digital entertainment, and in reference to this, AESVI has recently launched All4Games, a project aimed at public awareness videos in which Italian developers explain their jobs and their passion for what they do, communicating how the black market damages their efforts.

Another point worth mentioning is the feeling of being a second-class market, which Italian players have strongly felt, especially in recent years. Along with the average delay in publishing foreign games, another reason for this feeling is the lack of an Italian translation of famous RPGs such as *Final Fantasy VII* (1997) and *Planescape: Torment* (1999). While this might seem only a secondary problem, in Italy there is a solid tradition of dubbing, with the participation of famous professionals including Luca Ward (the Italian voice of several actors, from Russell Crowe to Antonio Banderas) in blockbusters such as the *Splinter Cell* series. However, a positive consequence of this situation was the reaction of the gaming community, which began producing fan-made translations. These translations were often developed in teams (by famous translation groups such as ITP and IAGTG) and then shared; sometimes the results were excellent and better than the professional ones. In a very different manner, Hive Division represents another important example of Italian generative fandom. In 2009, this team of *Metal Gear Solid* aficionados produced *Metal Gear Solid: Philanthropy*, a live-action movie that received positive reviews from players and critics due to its high quality; also, the original creator of the series, Hideo Kojima, appreciated it.[5]

The Cultural Debate about Video Games, from Artistic Status and Entertainment to Media Pillories

The representation of video games in Italy is quite controversial and influenced by opposing approaches. On one hand, efforts to legitimate video games as an art form are diffuse and sometimes concrete. On the other hand, attacks by the media are frequent, and prejudice among old generations is still strong, as is the connection between video games and electronic gambling games (for example, slot machines).

Among those working to legitimatize video games as an art form are Luca Traini and Debora Ferrari, art critics and experts who have organized several events and museum exhibitions about the relationship between art and games, with important supporters such as Ubisoft and Nintendo. Their activity has taken them to the prestigious fifty-fourth edition of the Venice Biennale of 2011, with the Neoludica: Game Art Gallery project (Ferrari and Traini 2012). Their synergic partner has been the Sicilian company E-Ludo (composed of E-Ludo Lab and E-Ludo Interactive) of Salvatore Mica, established in 2009 with the managing of the first Italian version of the IGDA Global Game Jam (Chapter Catania, now in its fourth edition) and strongly dynamic from a cultural, formative, and productive (supporting the regional developers) point of view.

A recent addition to the overall game culture is VIGAMUS: The Videogame Museum of Rome, the largest museum of video games in Italy, which opened at the end of 2012. It is the fulcrum of several activities, from PR events to conferences, and the main activity of AIOMI (Associazione Italiana Opere Multimediali

Interattive), the first Italian association for the legitimation of the medium, an association founded by the journalist Marco Accordi Rickards in 2008, which has since merged with the museum. Other similar centers are the Archivio Videoludico of the Cineteca of Bologna (established in 2009 and more archivist and academic in nature), and La Mecca del Videogioco (a small institution established the same year as VIGAMUS).

Beyond these efforts aimed to affirm the "high culture" profile of video games, the growing relevance of the medium is already appreciable in important events such as Lucca Comics and Games (whose related space was dramatically expanded in recent years), Cartoons on the Bay, and the previously mentioned Games Week. From a generalist perspective, the Italian edition of *Wired* magazine and journalists and intellectuals such as Jaime D'Alessandro, Ivan Fulco, Luca Maragno, and Matteo Bittanti have played and continue to play an important role in the proliferation and domestication of the medium. The fashion of "geekiness" (or nerdiness in the national context, as connected with gaming practices; the term "geek" in Italy means "cool" and is now a sort of popular tag) is growing among young people, as is the comprehension by older Italians that to new generations, the playing of video games is something natural. Furthermore, broadcast (with television shows such as *Gamerland*) and editorial (ISBN edizioni books) attention to the medium has grown in recent years, with products directed at gamers as well as mainstream audiences. Concerning the past, the retrogaming landscape is still alive due to the efforts of several associations and portals such as Gamescolletion.it (which also supports events and exhibits), Oldgamesitalia.it (covering a wide scope, from the rescue of old games to the legitimation of the medium), and *Tilt!* (which specializes in pinball as well and has an archive recognized by UNESCO).

As mentioned earlier, attacks on digital entertainment are still frequent and are often launched by generalist media, politicians, and parental associations such as MOIGE (Movimento Italiano Genitori). These attacks are often awkward attempts to find scapegoats rather than legitimate efforts toward change. Certain cases are particularly illustrative. For example, *Rule of Rose* (2006), a Japanese survival horror game, was described by the weekly magazine *Panorama* in an exaggerated manner that emphasized its "perverse" connotations. The condemnation of the game was so glaring that it brought on a strong reaction from players, even if they practically ignored the game itself. The second case was a TV special by the TG1 (the newscast of RAI 1, the first channel of the Italian broadcasting public service [RAI]), in which digital games were correlated with the July 22, 2011, tragedy in Oslo, implicated as one of the things that pushed Anders Behring Breivik to kill. The "Movimento contro la disinformazione sui videogiochi" (in English, "movement against the disinformation about video games") meant a quick, skilled, and collaborative response from the gaming community developed on the Internet but also through conferences and meetings. Some deemed it a "call to arms" and a "rediscovered consciousness," and we can partially take for granted these tags. It is an important example of a constructive approach, avoiding every militant and extreme action, often not productive. This aggregate opinion was organized and directed by the crew of multiplayer.it, the leading informational website about digital games, which also produced publishing guides and related novels.[6] But we should remember that another way video games gain legitimacy is through academia and the rise of video game studies—a field that is still in the making in Italy.

Italian Game Studies

When discussing Italian game studies, it is easier to talk about profiles of single individuals rather than refer to organized structures. Furthermore, only at the end of the last century did scholars from various fields begin to show interest in this topic. From a disciplinary point of view, semiotics remains the most developed field in Italy. From Massimo Maietti's influential book *Semiotica dei videogiochi* (2003) to the efforts of scholars such as Ruggiero Eugeni, Patrick John Coppock (who is also interested in philosophy), and researchers such as Dario Compagno, Gabriele Ferri, and Agata Meneghelli (2011), the contributions to game studies were, and remain, significant. And we must mention the work of young academics from related disciplines such as Riccardo Fassone, Mauro Salvador, Paolo Ruffino, Marco Benoît Carbone, Valentina Rao, Alessandro Canossa, Roberto Semprebene, Ivan Mosca, Stefano De Paoli, and Roberto Dillon. Furthermore, education studies has produced some relevant works, such as Rosy Nardone's *I nuovi scenari educativi dei videogiochi* (2007) and Damiano Fellini's *Videogame Education: Studie percorsi di formazione* (2012). The University of Udine, with Cristiano Poian, Giovanni Caruso, and the Postcinema group, is active in research, especially concerning the relationship between cinema and digital games. Moreover, a strong network of researchers involved with the medium was built up at the Politecnico of Milan, at the University of Modena and Reggio Emilia, and at the University of Bologna in terms of communication, humanism, design, and semiotics. A result of all these presences was the 2011 establishment of *G|A|M|E*, which is the first academic journal about game studies in Italy, even though it follows a multidisciplinary and international vocation.

Moving to profiles of individuals, the main name is the abovementioned Matteo Bittanti, called the "philosopher." During his activity, beginning in the late 1980s, at first in journals and magazines and later in an academic context, he has made important contributions to a wide range of research fields, from critical studies to semiotics to media studies (see Bittanti 2002, 2008). Together with Gianni Canova, a famous exponent of Italian film studies, Bittanti is the curator of Ludologica, the preeminent book series about video games in Italy. Furthermore, he has dedicated meaningful efforts to the artistic status of the medium, along with media art expert Domenico Quaranta, also within the Neoludica project. Other people directly interested in this field and in ludology are the communication expert Peppino Ortoleva (2012), the sociologist Gianfranco Pecchinenda (2003), the economist Paola Scorrano (2008), and the methodologist-designer Luca Giuliano, a hybrid figure between theory (1997) and practice with his theatrical tabletop role-playing game *On Stage!* (1995). In addition, the recent work of Luca De Santis (2013), focused on gender and LGBT themes in video games, is worth mentioning for its distinctiveness. In recent years, the situation has gotten better, although the absence of continuative structures that could improve and consolidate such knowledge has forced many scholars (including some of those mentioned above) to work totally or partially abroad.

For practitioners and insiders, another important contribution is that of Emiliano Sciarra (2010), the creator of the board game *BANG!* (2002), which had a ludologic framework with certain innovative elements. Concerning digital games, authors have made significant attempts to communicate, label, and analyze the peculiar traits of the medium: Francesco Carlà (1996) and Ciro Ascione (1999), from a more popular and mainstream position, and the abovementioned Fabio Viola, Francesco Alinovi (2011), and Roberto Genovesi (2006), from a more theoretical perspective.

The Future of the Italian Game Industry

Studying the Italian videoludic landscape results in conflicted feelings. From one perspective, institutional structures are weakened by the lack of private investments and the indifference of the state, and the road that the national industry must still travel seems long and hard. From the other perspective, the increasing efforts of associations such as AESVI, in coordination with others (for example, E-Ludo Lab and *Wired* magazine), and the strong market business and the creative Italian talents, even those of small teams, are encouraging signs. The potential of the Internet, the mobile market, and the new politics for independent developers help their production and freedom even though it remains restricted. Furthermore, in recent years, public opinion regarding video games has passed from a common prejudice to a sort of naïve curiosity, which is also driven by the evident economic success of the medium and its diffused consumption. People begin to see digital games as a cultural medium rather than simply a new technology with strange effects on younger generations (Wolf and Perron 2003). In addition, the cultural background of core gamers is present and ready to be activated and stimulated. The rise of cultural initiatives is synergic and propaedeutic, referring to a new awareness that must be developed with an integrated attitude. Only when institutions start to give the proper attention to the medium will it be possible to have an organic discourse about video games as well as academic framing that avoids every enthusiastic or catastrophist claim about them. In other words, we need a mature consciousness about the topic and knowledgeable gatekeepers facilitating the related cultural and social debates. At the national level, we need a "level up" concerning the cultural discussion regarding video games: they are neither the devil nor God. Simply, they are a pilaster of the current popular culture, and besides that, a fundamental business to exploit. A pragmatic lens is useful here, or maybe more than that, a perspective based on the simple question: "Are digital games art?" Now we are observing only potentialities and single, excellent exceptions. Our hope is to see something more in the future, a cohesive "Made in Italy" brand of video games.

Notes

1. These are thematic compilations of games (usually adventure games) that were periodically published with magazines on newsstands and sold at a low price (for example, *Next Strategy* [1985] and *Collana Viking* [1987]). An important aspect of this phenomenon was the "interactive comics" collections (*L'uomo ragno* [1993], *Diabolik* [1993]) released by Simulmondo, with high-frequency, chapter-based publications. This model of business found great success in the 1990s, with tens of thousands of units sold, as reported by Francesco Carlà, the founder of Simulmondo.

2. The data remain representative even if they are not strictly current.

3. Also, AESVI recently entered Confidustria in the Confindustria Cultura Italia group.

4. From discussions on forums and specialized websites, there appears to be a hard-core group that has an aversion to *FIFA* games and other sports games. According to the Cultural Studies approach (Hall 1997), this "hate" may be interpreted as a reaction to the mainstream popularity of sports simulations in order to defend a sort of niche gaming identity that characterizes and distinguishes regular (and often long-time) gamers from other gamers. We are talking about a fluctuating attitude indeed; unsurprisingly, something similar is recently happening with the *Call of Duty* series and its hegemony in the charts.

5. As reported on www.mgs-philanthropy.net/main/?p=1241&lang=en. Some members of Hive Division had a meeting with Kojima in 2013.

6. Other important websites are, for example, Spaziogames.it, Everyeye.it, Nextgame.it, Indievault.it (focused on independent products), Italiatopgames.it (a sort of national gamerankings.com), it.IGN.com (the Italian version of IGN.com), gamesearch.it, and eurogamer.it (the Italian version of eurogamer.com). The "paper press" is less relevant than in the past, with a few exceptions such as the historical *The Games Machine* and the multiplatform *Game Republic*.

References

AESVI. 2012a. 1°Censimento dei game developer italiani. http://www.aesvi.it/cms/download.php?attach_pk=1411&dir_pk=902&cms_pk=1929.

AESVI. 2012b. Rapporto annuale sullo stato dell'industria videoludica in Italia 2011. http://danielelepido.blog.ilsole24ore.com/i-bastioni-di-orione/files/aesvi_rapporto-2011.pdf.

Alinovi, F. 2011. *Game start! Strumenti per comprendere i videogiochi.* Milan: Springer Verlag Italia.

Ascione, C. 1999. *Videogames: Elogio del tempo sprecato.* Rome: Minimum Fax.

Bittanti, Matteo, ed. 2002. *Per una cultura dei videogames: Teorie e prassi del videogiocare.* Milan: Unicopli.

Bittanti, Matteo, ed. 2008. *Intermedialità: Videogiochi, cinema, televisione, fumetti.* Milan: Unicopli.

BSA (Business Software Alliance). 2012. Shadow market of pirated software grows to $63 billion. http://globalstudy.bsa.org/2011/.

Carlà, F. 1996. *Space Invaders: La vera storia dei videogames.* Rome: Castelvecchi.

Compagno, D., and P. J. Coppock, eds. 2009. Computer games between text and practice. In *E/C: Rivista dell'Associazione Italiana di Studi Semiotici* 5. http://www.ec-aiss.it/monografici/5_computer_games.php.

De Santis, L. 2013. *Videogaymes: Omosessualità nei videogiochi tra rappresentazione e simulazione (1975–2009)*. Milan: Unicopli.

Fellini, D. 2012. *Videogame education: Studi e percorsi di formazione*. Milan: Unicopli.

Ferrari, D., and L. Traini, eds. 2012. *Neoludica Art and Videogames 2011–1966*. Milan: Skira.

Genovesi, R. 2006. *L'ABC dei videogiochi*. Rome: Dino Audino Editore.

Giuliano, L. 1997. *I padroni della menzogna: Il gioco delle identità e dei mondi virtuali*. Rome: Meltemi.

Hall, S., ed. 1997. *Representation: Cultural Representation and Signifying Practices*. London: Sage.

Herz, J. C. 1997. *Joystick Nation: How Videogames Ate Our Quarters, Won Our Hearts, and Rewired Our Minds*. Boston, MA: Little, Brown.

Maietti, M. 2003. *Semiotica dei videgiochi*. Milan: Unicopli.

Meneghelli, A. 2011. *Il risveglio dei sensi: Verso un'esperienza di gioco corporeo*. Milan: Unicopli.

Nardone, R. 2007. *I nuovi scenari educativi dei videogiochi*. Bergamo: Edizioni Junior.

New Zoo. 2012. *2012 Country Summary Report: Italy.* http://www.newzoo.com/wp-content/uploads/Italy_summary_deck_new1.pdf.

Ortoleva, P. 2012. *Dal sesso al gioco*. Turin: Espress.

Pecchinenda, G. 2003. *Videogiochi e cultura della simulazione, la nascita dell'homo game*. Rome-Bari: Editori Laterza.

Sciarra, E. 2010. *L'arte del gioco*. Milan: Ugo Mursia Editore.

Scorrano, P. 2008. *Competitività, collaborazione e valore nelle network industries: Un'analisi nel settore dei videogame*. Bari: Cacucci.

USTR. 2013. *2013 Special 301 Report*. Office of the United States Trade Representative. http://www.ustr.gov/about-us/press-office/reports-and-publications/2013/2013-special-301-report.

Viola, F. 2011. *Gamification: I videogiochi nella vita quotidiana*. Pisa: Arduino Viola.

Wolf, Mark J. P., and Bernard Perron, eds. 2003. *The Video Game Theory Reader*. New York: Routledge.

JAPAN

Jennifer deWinter

Japan is a densely populated island country with a population of 127.52 million as of 2012 ("Statistical Handbook of Japan" 2013) coupled with an active US military population of 39,222 (Department of Defense 2011). Since the late 1970s, Japan has had a strong role in the international computer game market, developing both successful hardware and software brands. This success is of particular note when compared to the rather weak-performing IT industries in Japan (Casper and Storz 2012; Azuma et al. 2009). The Japanese game industry's success may be attributed to the rather strong influences of related entertainment industries, such as manga and animé (Kohler 2004; Allison 2006; deWinter 2009), as well as a rather long history of gambling in the form of pachinko and *Hanafuda*. Japan was a dominant force in the computer game and related industries throughout the 1980s and 1990s, leading to the catchy phrase "gross national cool" (McGray 2002) to describe their entertainment exports; however, the 2000s saw a rise in anxiety regarding the "collapse" of the Japanese game industry as strong game competitors began emerging in the United States and Europe while a strong entertainment competitor began rising in South Korea. Nevertheless, export data and critical reviews suggest that the Japanese computer game market continues to be a strong force in both Japanese and global markets.

Challenges to Studying Computer Games in Japan

Japan is the third most powerful computer game market after the US and Europe as a collective whole, based on sales figures (as can be seen in the 2012 aggregate data presented in VG Chartz, as well as in the annual reports of Nintendo, Sony, Namco, and Capcom). However, the Japanese game industry does not work in a national vacuum. The Japan External Trade Organization (JETRO) 2007 report, which traces Japan's international connections through import/export sales and company buyouts and mergers, has collected data that illustrate the international nature of the Japanese computer game industry. Japanese companies often have overseas offices (such as Nintendo in the United States), making it more difficult to demarcate the national boundaries of ownership and historical exigency. Further, Japanese hardware and consoles have had a strong presence in the gaming industry from very early on, affecting the distribution of international games through hardware specifications and distribution agreements ("Japanese Video Game Industry" 2007). Indeed, Nintendo got its start in the video game industry in 1974 by becoming the distributor of the Magnavox Odyssey in Japan.

In addition to the international dialectic of the Japanese gaming industry, the well-documented fact that this industry is intimately connected to other entertainment media often makes it difficult to talk about games in isolation. For example, the dialectical histories of animé, manga, toys, and computer games often mean that creations in one medium are used to sell creations in other media. In its simplest form, this means that properties are being adapted from one medium to another. In other cases, they are used in a type of serial narrative that ends in one medium and picks up in another (deWinter 2004). In addition to this media-vampirism, there is the propagation of gamic memes in other media. For example, in the animé *Excel Saga* (Koshi 1999) and *Magical Shopping Arcade Abenobashi* (Akahori 2002), computer game interfaces become the metaphors for expressing emotions and relationships, such as the four-part option sequence (in response to a girl asking you whether you will go to the school fair, you can answer: (A) Ignore her; (B) Tell her yes, with another girl; (C) Tell her yes; (D) Tell her yes and ask her if she would like to come with you) commonly found in dating sim games. As a result, there will be a certain amount of bleed throughout this chapter in discussing Japan in both a national and global context as well as Japanese computer games as one part of a complex media landscape.

Who Plays Computer Games?

According to an Enterbrain, Inc. report (reported in GMOCloud 2012), 56% of console gamers are male and 44% are female. Of those numbers, 36% are younger than eighteen years old, 28% are between eighteen and thirty-five years old, and 36% are older than thirty-six (GMOCloud 2012). Further refining these numbers shows that there is a trend of women gamers over fifty years old playing console games, thanks in large part to DS games such as *Brain Training* (2005) (*Brain Age* in the US) and *Nintendogs* (2005) (Leyton 2006; Nouchi et al. 2012). These numbers vary slightly for Japanese social gamers. According to 2011 financial report data from Gree, DeNA, and Mixi, 54% of social gamers are men and the other 46% are women. The demographic demarcations provided indicate that 19% are nineteen years old and younger, 41% are between the ages of twenty and twenty-nine, and the remaining 41% are thirty years old and above. The trend, according to DeNA's Mobage (2011), is for the demographic to shift toward older players; in 2009, for example, 26% of players were over thirty versus 41% in 2011 (Toto 2011).

The over-fifty gamer set will become more important to the Japanese gaming industry as the population continues to age. Japan's National Institute of Population and Social Security Research (2012) anticipates a steady increase of the above-sixty-five population to climb from more than 30 million in 2012 to a peak in 2042 with 38.78 million, only to stabilize at 34.64 million by 2060. In other words, by 2060, approximately 40% of

Japan's population will be above sixty-five years old. This shifting demographic data is affecting the breadth of the industry, from console gamers, social gamers, and arcade gamers. Kyoko Matsuda, a spokesperson for SEGA SAMMY HOLDINGS, explained that they are already courting older clients at their arcades with senior days and low-tech frequent player cards (Lah 2012).

Japan's Early Game Industry: SEGA, Namco, Taito, and Nintendo

The Japanese computer game market emerged in the 1970s while the country enjoyed the discretionary wealth and vitality that accompanied the post-war reconstruction of Japan. However, the Japanese government did express doubt that game centers, and later computer games, were a good cultural investment in a society that worked 6.5 days per week. David Rosen, a founding member of SEGA (SErvice GAmes), argued successfully with the Ministry of International Trade and Industry (MITI) that coin-op arcade games would provide an emotional release, and in 1957, was granted a license to import coin-op games (Pettus 2012, 4). According to Pettus (2012), Rosen was right: "The Japanese took to the coin-op arcades in droves—even more so than the Americans—and both games imported from America and produced locally raked in profits for all of the industry players" (4–5). Evidence appears to prove this point. For example, the Japanese company Nakamura Manufacturing Company (later shortened to Namco) first started importing *PONG* (1972) games after Masaya Nakamura bought the rights for Atari Japan from Nolan Bushnell in 1973. Nakamura Manufacturing was already in the business of amusement rides and amusement machines, such as pachinko, so his distribution channels and contacts were compatible for the new computer game medium. Simultaneously, Taito began to import *PONG* clones for distribution into the Japanese market. Increasing demand for computer games, coupled with the connection to pachinko parlors and gaming halls, ensured that the early history of computer games was intertwined with the yakuza, the Japanese mafia. Indeed, because the yakuza manufactured and marketed many of these early games, both Taito and Nakamura Manufacturing faced market pressure that required them to respond. In the case of Nakamura Manufacturing, Masaya Nakamura decided to break his contractual agreement with Atari—an agreement that ensured all manufacturing rights belonged to Atari alone with Nakamura Manufacturing acting as the Japanese distributor—and manufacture and distribute his own cabinets of *Burokku-kuzushi* (the Japanese name for Atari's *Breakout* [1976]). Nakamura later lost a lawsuit brought against him by Atari.

Meanwhile, Taito was developing games with microprocessors rather than integrated circuits, which game designer Tomohiro Nishikado used to create his wildly successful game *Space Invaders* (1978). According to Kohler, this game introduced a sense of narrative rising tension with a manga-like art style. Further, the success of this game in Japan caused a 100-yen coin shortage, forcing the government to quadruple production of the coin. Positively, store owners reported that this game paid for itself within a month (Kohler 2004, 19). During this time, Toru Iwatani was addressing a gender disparity that he saw in the game centers (Japanese arcades) and started designing games for Namco that targeted female players. This led to his 1980 global sensation, *Pakku-Man* (*Pac-Man*). In addition to Taito and Namco's commercial successes, SEGA saw national and international success with *Frogger* (1981).[1]

While Taito and Sega concentrated on developing games for the Japanese market and Namco relied on importing and distributing Atari games, Nintendo saw an opportunity to simultaneously develop hardware and software that was portable and cheap to manufacture. This history of Nintendo, like its competitors, already had close ties to older entertainment and gambling media. Nintendo, which literally means "place where luck is in the hands of heaven," started as Yamauchi Nintendo Playing Cards, the Kyoto-based *Hanafuda* playing card manufacturer. While their work in playing cards—both Japanese *Hanafuda* and Western-style decks called *toranpu* (for trump)—was lucrative for the company because of yakuza-based high-stakes gambling, Hiroshi Yamauchi saw an opportunity to expand into computer games by becoming the distributor of the Magnavox Odyssey in Japan in 1974. Later, in response to the commercial success of *Space Invaders*, Gunpei Yokoi of Nintendo created the Game & Watch series, the first entry of which combined a digital clock with the computer game *Ball*, in 1980. The success of the Game & Watch series led to 30 million units sold in the series' first eleven years (Donovan 2010, 155). While computer technology allowed for the commercial success of games, the importance of the game designer's role was about to emerge, and the dual emphasis on story and technology would spur computer games to become a more immersive medium. No designer's name is more important in this early era than that of Shigeru Miyamoto. Miyamoto was hired originally as an artist who had a knack for hardware and visual design. He was given a chance to design a game when Minoru Arakawa, the head of the American branch of Nintendo, asked for a game for their newly developed *Radarscope* (1980) cabinets in an attempt to reuse the hardware of that failed game. Miyamoto created *Donkey Kong*, arguably the first game that began its design process around a story (Kohler 2004, 39), launching Nintendo into the corporate company of Taito, SEGA, and Namco.

In the early 1980s, Yamauchi was poised to replicate Atari's success in the home console market. He ordered the development of the Nintendo Famicom (short for "family computer"), asking designer Masayuki Uemura to "create a console that was not only a year ahead of the competition in technology but also a third of the price of the Epoch Cassette Vision" (Donovan 2010, 158). The Nintendo Famicom was released in July of 1983 and by the end of the year had sold more than a million units. Due to the pressures of their success, Yamauchi created a new licensing agreement for other game publishers—an agreement that still dominates the computer game market to this day. In essence, publishers had to pay cash up front to create a game and then share a cut of the profits with Nintendo. Likewise, Nintendo could veto the release of any game, ensuring quality and content control. As a result of this, sexual games that were popular on PC platforms (such as *Night Life* [1982] and *Lolita Syndrome* [1983]) were banned from the Nintendo market, ensuring their commitment to the "family computer" of entertainment. While the reason for their success is multifaceted, Izushi and Aoyama attribute Nintendo's ability to displace earlier competing products from Tommy and Bandai (both Japanese companies) to "price competitiveness, its ability to deliver original mega-hit software such as *Super Mario Brothers* (1985), and its initial alliances with best-selling arcade video games" (1847)—in other words, a dual focus on hardware and software, with an eye toward an eager consumer market. Thus began the rise of the home console market in Japan.

Home Consoles and Handheld Devices: A Nation That Plays Games

The sales figures for home consoles in Japan are staggering in relation to worldwide sales figures. Japan constantly rates in the top three countries tracked, along with the United States and the UK, indicating that both their productive power and consumption of this medium has tremendous influence both in local and global markets. While there were dedicated computer game systems and early consoles before the release of Nintendo's Famicom (first, Nintendo acting as a distributor for the Magnavox Odyssey, then Nintendo's Color TV Game 6, 1977; Color TV Game 15, 1978; Color TV Racing 112, 1978; Color TV Game Block Breaker, 1979; Computer TV Game, 1980; and Bandai Super Vision 8000, 1979), it was not until the 1980s and the beginning of the 8-bit era that Japan's console market had significant market penetration with the releases of Nintendo and SEGA's consoles. Nintendo, SEGA, and Sony (since the 1994 introduction of the Sony PlayStation), are the key players in what is commonly referred to as Japan's console wars or system wars (Finn 2002; Herman 2007). Indeed, this time in hardware history has permeated Japan's cultural narratives; *Aoi Sekai no Chuushin de* (2007), a fantasy manga about the war between the Segua Kingdom and the Ninterudo Empire in the land of Consume, was adapted to anime in 2012.

Nintendo

With the release of the Famicom in 1983, Nintendo was strongly positioned in the impending console wars (see table 1, best-selling Nintendo console sales as of March 31, 2013). Throughout the early years, they instituted exclusivity clauses into their agreements with third-party developers, effectively choking the competitive market early (Ryan 2011, 110). Nevertheless, SEGA's release of the Sonic franchise destabilized Nintendo's dominance in the computer game industry. In response to this new market threat, Masayuki Uemura of Nintendo decided that the Super Famicom (or Super NES) would be a 16-bit game system that would emphasize sound and graphics, a combination that would allow for what Kohler calls the cinematics of Japanese games. Accompanying this release was *Super Mario World* (1991), produced by Miyamoto, and the sheer size of the game (seventy-two levels) coupled with the secrets sprinkled throughout, ensured game immersion based on time-on-task along with a satisfying reward system. The following year, Nintendo released *Super Mario Kart* (1992), a simple racing game based on the characters from the Mario franchise. This game went on to be a resounding success, so much so that in 2009 *The Guinness Book of World Records* named it the most influential computer game (an assessment based on initial reactions and longitudinal impact) (Ivan 2009). Important to note here is the positive impact that the console wars between Nintendo and SEGA had on Nintendo: previous to SEGA's entry into the home console market, Nintendo was essentially synonymous with video games,

Table 1

Best-selling Nintendo console sales as of March 31, 2013. (Source: Nintendo Co., Inc., Historical Data.)

Home Consoles		Units Sold
1983	Famicom/Nintendo Entertainment System (NES)	61,910,000*
1990	Super Famicom/Super Nintendo Entertainment System (SNES)	49,100,000*
1995	Virtual Boy (Table Top Console)	770,000**
1996	Nintendo 64	32,930,000*
2001	Nintendo GameCube	21,740,000*
2006	Wii	99,840,000*
2012	Wii U	3,450,000*
Handheld Consoles		Units Sold
1989	Game Boy	118,690,000*
2001	Game Boy Advance	81,510,000*
2004	Nintendo DS (including Lite, DSi, and DSi XL)	153,870,000*
2011	Nintendo 3DS (including 3DS XL)	31,090,000*

* *Source:* Nintendo Co., Inc., Consolidated Transition by Region.
** *Source:* "Nintendo Company Statistics," 2012.

controlling approximately 85% of the market share. However, the two-party competition SEGA introduced helped retailers feel more in control, making them less antagonistic toward Nintendo. Further, the competition helped Nintendo sales, and Nintendo responded to this by allowing its vendors to make SEGA games, which effectively meant that SEGA would have problems providing exclusive content for its system.

Simultaneously with these console developments, Nintendo continued to produce portable gaming devices. Building from early trials of Game & Watch, Nintendo released the Game Boy in 1989 for the suggested retail price of ¥12,500 (approximately US$90), and its two signature launch titles, *Tetris* and *Super Mario Land*, are still the best-selling games in Japan for this platform. Handheld game devices do well in Japan, and a number of suppositions have been posited to make sense of this. For one, Japan is a tiny place, about the geographical size of California with about half the US population in residence. The majority of that population is concentrated in the major urban areas such as Tokyo and Osaka, where apartments are small and people rely on public transportation. These small handheld devices take up little room in living spaces, and they are easy to play on commuter trains and the subway systems (Nakamura 2012; Y-N 2006). Finally, the sociological study of Japanese families finds that many families are child-centered, leading to the "child-king" and buying that child whatever he wants. While this is beginning to change, this was strongly the case in the late 1980s and early 1990s (Atou 2001).

Game Boy's early market success had to do with Gunpei Yokoi's design philosophy of *Kareta Gijutsu no Shuhei Shikou*, or "lateral thinking of withered technology" (Donovan 2010, 205). Yokoi used existing technologies, such as monochromatic LCD screens, to keep the device light and also to help conserve battery life. This also helped to keep the price down. Further, Yokoi applied his years of designing for the Game & Watch series and successfully made the argument that the Game Boy should be a cartridge-based system, ensuring longer monetizing possibilities. Thus, even though there were color-based competitors on the market, the Game Boy entered the market and sold 118.5 million units over its lifetime, second only to the Nintendo DS. This same year, the *Japan Economic Journal* named Nintendo the best company in Japan, and according to David Sheff (2011), "By 1991 Nintendo had supplanted Toyota as Japan's most successful company, based on the indices of growth potential, profitability, penetration of foreign and domestic markets, and stock performance. Nintendo made more for its shareholders and paid higher dividends between 1988 and 1992 than almost any other company traded on the Tokyo Stock Exchange" (5).

Handheld devices continued to be the means of playing casual games and simple puzzle games for the most part. But this changed with the introduction of Satoshi Tajiri's 1996 game *Poketto Monsutā Aka Midori* (*Pocket Monsters: Red and Green*), or *Pokémon*. Inspired by his childhood playing with beetles, *Pokémon* allowed players to collect and battle small monsters, and each battle allowed the monsters to level up. Ultimately a nostalgia game in Japan, *Pokémon* was resoundingly successful in its home market, revitalizing the dying Game Boy; further, when released into the international market two years later (with the support of complementary media), *Pokémon* went on to become, according to Joseph Tobin (2004), "the most successful computer game of all time" (7).

Nintendo already had experience managing the Mario brand, licensing the character into toys, lunchboxes, cartoons, a movie, and so forth. *Pokémon* appears to be Nintendo's cross-media franchise *par excellence*. Buckingham and Sefton-Green (2004, 19) posit that *Pokémon* was planned as a cross-media enterprise from a very early stage. One look at the release dates alone instantiates their claim: In 1996, based on the computer game for the Game Boy, Media Factory publishes the Trading Card Game—a game that will be remediated into the computer game *Pokémon Trading Card Game* for the Game Boy Color in 1998 and includes a limited edition *Pokémon* card. The manga *Pocket Monsters Special* begins in 1997, providing character development and narrative story arcs that were unavailable in the game space. Indeed, Satoshi Tajiri, *Pokémon*'s game designer, is on record saying, "This is the comic that most resembles the world I was trying to convey" (see "Pokémon Adventures" 2002).

Also in 1997, the *Pokémon* animé began airing on TV Tokyo and was then licensed for international distribution in 1998. The first of sixteen movies (as of June 2013) showed in Japanese theaters in 1998. The small, electric Pikachu appeared on myriad products, from the mundane lunchbox to the Pokémon Jets of All Nippon Airways. The wild success and almost immediate diversification of this brand led the three copyright holders of the franchise—Nintendo Co., Ltd., Creatures Inc., and GAME FREAK Inc.—to create the Pokémon

Center Company in 1998, now called The Pokémon Company, as of 2000. The role of this Nintendo subsidiary company is brand and community management. Tsunekazu Ishihara, the company's president, explained the purpose thus: "A more accurate description of what it means to produce Pokémon is that we play a leading role in making Pokémon products as appealing as possible and that we make every effort to achieve the best possible business results when those products come out" (Ishihara 2013). These business results are impressive: each game in the series has sold more than six million copies. Thus, according to Benzinga (2013), "During the late '90s, when Nintendo's console [the Nintendo 64] was failing and its handheld was thriving, Hiroshi Yamauchi (the company's president) reportedly told the Japanese press that the only game Nintendo needed was Pokémon."

The combination of handheld consoles and *Pokémon* products provided enough sales and revenue to enable Nintendo to winter the storm of poor Nintendo 64 and GameCube sales—systems that lost to the emergence of Sony during the console wars—providing enough market share to produce and release the Wii. Originally named the Nintendo Revolution, Nintendo announced in April of 2006 that the system would be renamed the Nintendo Wii. "While the code-name 'Revolution' expressed our direction, Wii represents the answer. Wii will break down that wall that separates game players from everybody else. Wii will put people more in touch with their games … and each other" (quoted in "E3 06: Revolution Renamed Wii" 2006). With its release in November 2006, the Wii went on to become the top-selling next generation console, beating both the Xbox 360 and PS3. (Interestingly, the PS2 is still selling well and remains the top-selling console, as of June 2013.) There are a number of reasons cited for the success of the Wii: it appeals commercially to the casual gamer with its innovative nunchuk controller, it provides families with a family console, it has innovative controls, it successfully captures the female demographic with casual games (and more importantly, *Wii Fit* [2008]), the Wii is accessible to persons with disabilities and elderly players, and finally, the Wii doesn't break that often (especially compared to the PS3's "yellow light of death" or the Xbox 360's "red ring of death").

Regardless of the Wii's success, Nintendo didn't like being relegated to the casual gamer. In the interview series "Iwata Asks," archived on Nintendo's website, Nintendo's global president, Satoru Iwata, interviews key creators in the company, and in the Nintendo E3 2011 interview, Miyamoto and Iwata reflect on the success and limitations of the Wii and the possibilities presented by the Wii U. Iwata states, "Shortly after the Wii console was released, people in the gaming media and game enthusiasts started recognizing the Wii as a casual machine aimed toward families, and placed game consoles by Microsoft and Sony in a very similar light with each other, saying these are machines aimed towards those who passionately play games." To this, Miyamoto agrees, noting that the only reason that the Wii is not considered a core gamer system is that it wasn't designed around the specifications of HD television. The Wii U would overcome this visual and psychological limitation while still appealing to its wider demographic through backward compatibility and future development projects. Yet, against these hopes, sales figures have not been robust, with only 390,000 units selling during its 2012 Christmas-season debut. In the interview, Iwata is quoted as saying, "We feel deeply responsible for not having tried hard enough to have consumers understand the product." The consumers, it appears, think of the Wii U as a Wii with an additional controller, something that Nintendo has had difficulty dispelling. However, Keza McDonald (2013) from IGN is more forgiving of the Wii U launch, calling its moderate success respectable in the face of lackluster marketing. Regardless of the moderate success of the Wii U, it is safe to conclude that Nintendo has been a heavyweight in the industry ever since it entered the console wars in the 1980s, defining new technologies, games, and industrial practices.

SEGA

In 1983, at the same time that Nintendo released the Famicom, SEGA released the SG-1000, which was followed two years later by the SEGA Master System. SEGA already had a long history as a successful arcade game provider, so this move into the home entertainment market seemed the next logical step. However, sales were lackluster at best. In his book *Service Games: The Rise and Fall of Sega*, Sam Pettus (2012) probably best summarizes the reasons for this. Not only did Nintendo have name-brand recognition, it also coerced software developers into exclusivity agreements: "When you own 90% of the world's largest videogame market, then you don't have to play fair. That only leaves 10% for your competition, which in theory means they never should be able to catch up with you no matter what they put out" (26). Add to this SEGA's initial decision to distribute through Tonka, a toy company with no experience in video games, and the choked entry of SEGA into the console wars makes sense (see table 2, SEGA consoles with units sold 2012).

Table 2
SEGA consoles with units sold as of 2012.

Home Consoles		Units Sold
1983	SG-1000	unknown
1985	SEGA Master System (SMS)	10,000,000*
1988	SEGA Mega Drive (US: SEGA Genesis)	28,540,000**
1994	SEGA Saturn	8,820,000**
1998	SEGA Dreamcast	8,200,000**
Handheld Consoles		Units Sold
1990	SEGA Game Gear	10,620,000**
1994	SEGA Mega Jet	unknown
1995	SEGA Nomad	1,000,000***

* *Source*: Forster 2011, 80.
** *Source*: "Game Platforms," VGChartz 2012.
*** *Source*: SEGAtastic 2009
(Note: The SEGA CD is not a console but rather an add-on, and it is not included in the list above.)

The SEGA Mega Drive (or the SEGA Genesis in the US) enabled SEGA to produce market competition while helping to drive the quickly iterated development of new system technologies (Akagi 1992). Nintendo was still developing 8-bit games, and while they were in the development process of a 16-bit system, the Super Famicom would not be released until 1990. SEGA already had the SEGA 16 System running in its arcade games based on the 16-bit Motorola MC68000 CPU, and CEO Hayao Namayama decided it was time to translate that technology into the home market (Pettus 2012, 33). The SEGA Mega Drive layered the 16-bit technology onto the already-developed SEGA Master System, allowing SEGA to develop what was essentially a PC engine in the disposable medium of entertainment hardware. What should have been a smooth launch met with competition from an unexpected source: NEC entered the arena and released the PC-Engine: Core Grafx 8-bit system (renamed the NEC Turbografx 16 in the US) in 1987. It looked and sounded better than Nintendo's Famicom, but that was its only marketing strength. SEGA rallied and sent a rebranded SEGA Genesis into the US, where they found a market hungry for the product. However, SEGA didn't have great games to play, especially when compared to Nintendo games. SEGA's licensing agreements were highly unfavorable to third-party developers until Trip Hawkins of EA asked some of his engineers to reverse-engineer the Genesis. They then approached SEGA and negotiated a deal that allowed EA to make as many games as it wanted for the system, in such a way that both parties profited (a milder success than the one that Hawkins was initially negotiating for). Hawkins explained in an interview later that his developers were resistant to designing for the Genesis; they wanted to design for the more high-powered home PC. However, as Hawkins puts it, "If the customer buys a Genesis, we want to give him the best we can for the machine he bought and not resent the consumer for not buying a $1,000 computer" (Ramsay 2012, 9).

Now that SEGA had inked partnerships with game developers, it needed to have a mascot as recognizable as Mario. Up to this time, SEGA had been using Alex Kidd, but Nakayama wanted a more strategic approach to the system's mascot. According to Pettus, "His instructions were quite specific. The new mascot would have to be as easily recognizable as Mario, yet as unlike him as possible. The new mascot would have to be a rather unorthodox character, and the game developed for him would have to reflect this" (51–52). Yuji Naka, Naoto Ooshima, and Hirokazu Yasuhara, the Sonic triumvirate, took to this challenge, creating a series of characters before settling on a hedgehog (rather than a rabbit or armadillo). The design choice was to limit actions to a one-button jump and to incorporate elements of pinball (Ko et al. 2012). Sonic made his first appearance at the 1990 Tokyo Toy show and then as character placement on the J-Pop band *Dreams Come True*'s tour bus during their 1990 national tour. The Sonic game was released in 1991, and US players especially preferred the game with this cocky, fast hedgehog over the newly released *Super Mario World*. Then, with the dual packaging of Sonic with the Genesis, SEGA's market share showed they could provide real competition and reduced the projected release sales of the Super Famicom/SNES, leading Sheff to conclude that SEGA was like a "sobering cold shower" (Sheff 1994, n.p.). This shower, according to Sheff, led Yamauchi of Nintendo to start forming secret alliances with complementary industries for projects in the pipeline. Thus, even though SEGA had Sonic, and Sonic was a global sensation, Nakamura decided to scrap the Genesis in favor of the Saturn, a gamble that ended poorly for SEGA.

The 1994 SEGA Saturn came out the same year as Sony's PlayStation. Now SEGA had two major competitors to face. Further, because the PlayStation had a 1994 release date, Nakayama pushed up the Saturn's release date four months against protests and a complete lack of finished games. The PlayStation had the advantage, too, because Sony employed C for programming, whereas developers had to learn both the hardware and programming languages of the Saturn. SEGA redesigned the console, but the result was a higher off-shelf price for the consumer: ¥44,800 for the Saturn versus ¥39,800 for the PlayStation in Japan, and $399 for the Saturn versus $299 for the PlayStation in the United States. While the launch of the Saturn was accompanied by the acclaimed *Virtua Fighter,* the console had no other games to boast of, providing Sony ample time to win over a market of critics and consumers. The financial failures of SEGA led then-CEO Nakayama to seek a merger with Bandai, a Japanese toy and entertainment company; however, the management of Bandai resisted the merger, and in May of 1997, the merger was declared a failure (Pettus 2012, 240). The following year, the SEGA Dreamcast launched, and while the system had innovative hardware and a large number of important games designed specifically for it, the Dreamcast could not compete with the emergent PlayStation 2 with its backward compatibility and integrated DVD player. Couple that with the rise of piracy and bad marketing, and SEGA had to mark an exit from the home console market in the next millennium. SEGA was perceived as so weak during this time that rumors circulated about Nintendo acquiring it, which Nintendo denounced in a December 12, 2000, corporate news release, translated roughly as "We 100% deny the acquisition of SEGA."[2]

This is not to say that SEGA does not remain an important player in computer games. In 1994, Sammy, a pachinko company, acquired SEGA to become SEGA SAMMY HOLDINGS. The profits from pachinko and pachislot helped to fund some of SEGA's ventures. In 1995, SEGA paired with ALTUS Co., Ltd. to release Print Club, colloquially known as purikura (for purinto kurabu), a wildly successful arcade-like photo booth that printed sheets of small stickers. Following this, in 1996, SEGA opened Tokyo Joypolis (a brand that started in Yokohama), providing a large-scale venue for a SEGA-themed amusement park with rides and arcades. SEGA SAMMY now defines their group as "highly complementary due to the structure of being a group of comprehensive entertainment companies" (SEGA SAMMY HOLDINGS 2013). These complementary companies continue to include pachinko, arcades, amusement centers, and now "consumer business," which includes software development for consoles, PCs, and mobile phones with its toy and video business. While the earnings reports from 2008 and 2009 show a marked loss of revenue, SEGA SAMMY's Fiscal 2012 Business Results Report indicates that the company is recovering from the downturn and is still a ¥58.3-billion operating company (based on operating income).

Sony

The history of Sony's entry into the game market is tightly knit with Nintendo in its earliest instantiation. Ken Kutaragi, a Sony engineer, convinced Sony to collaborate with Nintendo to develop CD-ROM technology for home game systems, a partnership that was announced at the 1991 Consumer Electronics Show in Chicago.

However, Donovan argues that Nintendo became paranoid that Sony might be trying to muscle its way into the game market (and it was already struggling against the Sonic competition), so it canceled the collaboration (Donovan 2010, 265). This, apparently, angered Nario Ooga, who then created Sony Computer Entertainment and put Kutaragi in charge. The team created a CD-ROM system with a 3-D graphics card coupled with a controller that molded to a player's hand rather than being a flat board. At the beginning, there was some trepidation about the system: 3-D graphics had not been explored fully and developers thought that the new system, because of the cost of the technology, would be too pricy on the consumer market.

And they were right to be concerned. In the US, Trip Hawkins of EA released the 3DO system with Panasonic; however, the launch price was $699.95 (compared to the PlayStation's $399 launch price), which severely limited its adoption by the consumer market. Further, the launch of SEGA's *Virtua Fighter* (1993) showed developers what could be done with three-dimensional graphics, inspiring British-based Core Design to create *Tomb Raider* (1996) for the PlayStation system (and PC platforms). Then, as luck would have it, 1997 saw the release of *Gran Turismo* (1997) and *Final Fantasy VII* (1997), two of the top-selling games for the console. *Gran Turismo*, a racing game, took five years to develop and sold a total of 10,850,000 units with 2,550,000 of those units sold in the Japanese market (Gran Turismo Series Software Title List 2013). This launched the franchise and the follow-up game *Gran Turismo 2* (1999), went on to be the second-best-selling game for the PlayStation, with 9,370,000 total units sold. The publisher Square (now Square Enix) broke away from Nintendo after publishing its previous six titles and created a 3-D CD-ROM-based game for the PlayStation. According to Tetsuya Nomura, director and coauthor of the game, "There were already several 3D games in America that were accepted by fans. My fear had been that the Final Fantasy franchise might be left behind if it didn't catch up to that trend" ("Afterthoughts" 2005). The game's release met with immediate critical and commercial success,

Table 3

Sony PlayStation consoles with units sold.

Home Consoles		Units Sold
1994	PlayStation	104,240,000
2000	PlayStation 2 (PS2)	157,680,000
2006	PlayStation 3 (PS3)	77,450,000
2013	PlayStation 4 (PS4)	No Data
Handheld Consoles		Units Sold
2004	PlayStation Portable	79,100,000
2011	PlayStation Vita	5,150,000

Source: "Game Platforms," VGChartz 2012.

garnering multiple awards and nominations and selling 9,720,000 units worldwide, with 2,470,000 of those in Japan (or 25.4% of total sales) (Final Fantasy VII 2013).

The PlayStation and the games developed for it adhered to Sony's expanded vision of gamer demographics, which positively affected their console market penetration (see table 3, Sony PlayStation consoles with units sold). Nintendo had been associated with children and family gaming, and SEGA followed on that trajectory for the most part. Sony, on the other hand, directly targeted both children and adults from the beginning of its launch ad campaigns, featuring children, young professionals, older management, and retired people. The slogans represented this as well: "All the games are collected here," "Let's go to one million units," "It has become great," and "For good children and good adults."[3] From the moment of the launch, Sony made good on the promise to diversify its gamer base, attracting child players with *Barbie: Race and Ride* (1999) and *Arc the Lad* (1995), and attracting adult players with gritty titles such as *BioHazard* (US, *Resident Evil*) (1996) and *Silent Hill* (1999).

Sony was well poised to enter the next generation of consoles with the PlayStation 2, competing directly with the Dreamcast. While SEGA rushed the Dreamcast, Sony was able to use technological developments in other company business units to integrate emerging DVD technology into the PS2. In addition to this, Sony developed the Emotion Engine, a CPU that integrated eight discrete units: a CPU core, two vector processing units, a graphics interface, a 10-channel DMA unit, a memory controller, an imaging processing unit, and an input-output interface—all of which enabled backward compatibility. These moves read the market well, for as reported in a 2001 *Economist* article, "when Sony's PS2 was first released in Japan in March 2000, its initial sales relied significantly on the console's ability to play digital video discs (DVDs) as well as games" ("No Laughing Matter" 2001). The PS2 DVD player was, effectively, the cheapest DVD player on the market, marking a transition in console technologies and expectations that game systems become the center of home entertainment. While the DVD player provided the initial gimmick needed to defeat the Dreamcast and overshadow the release of the Nintendo GameCube, the continued success of the PS2 remains remarkable. In a 2005 press release from Sony, the company celebrated its fast rise to 100 million units sold, stating that "PlayStation 2 is strongly supported by a wide range of users as well as from third party developers and publishers, and by maintaining backward compatibility of software titles for its predecessor system for the first time in the history of computer entertainment, PlayStation 2 has become the standard home entertainment system with over 14,000 PlayStation and PlayStation 2 titles playable, of which 6,200 titles are for PlayStation 2 alone" ("PlayStation 2 Breaks Record" 2005, 1). The PS2 was still active on the market until 2012, when Sony stopped manufacturing it, well into the life cycle of the PS3. The demand for and value of the console strongly continues in Japan. As *Kotaku*'s Brian Ashcraft (2013) reports, "Market value site Kakaku.com tracks the PS2 as spiking from an average price of ¥20,000 ($213) late last year. For most of this year, the PS2 has been priced on average at around ¥30,000, which is $320. (Keep in mind, the manufacturer's suggested retail price is still ¥16,000!)."

Compared to the success of the PS2, sales of the PS3 were disappointing to Sony. The launch was plagued with delays, and more importantly, the US had developed a strong competitor in the Microsoft Xbox. Thus, the competition provided by the Wii and Xbox 360 led to anemic sales; Sony consistently lagged behind its two major competitors. Part of the problem had to do with the small library of launch titles, and the other oft-cited problem had to do with developers; they simply found the architecture too difficult to program for. CEO Kazuo Hirai explained in an interview that this was intentional—he wanted a console that would have longevity for developers and anticipate many of the emerging technologies (similar to the argument Ruggill and McAllister [2011] make about games as anachronistic). Hirai explained, "We don't provide the 'easy to program for' console that [*developers*] want, because 'easy to program for' means that anybody will be able to take advantage of pretty much what the hardware can do, so then the question is what do you do for the rest of the nine-and-a-half years?" (quoted in Purchese 2009).

Sony worked to develop the PlayStation Network (PSN), an online content provider and multiplayer network. This network is accessible via the PS3 and the two handheld consoles PSP and PS Vita. Here again, Sony has struggled; in 2011, the information on its network was hacked, and the account information of 77 million users was compromised (Caplin 2011). As a result of this, Sony had to shut down the network for an "indefinite" amount of time and didn't disclose the problems that it faced for seven days. And while a survey conducted by GameSpot shows that the majority of PlayStation gamers will remain loyal to the company, they did report a 9% increase in the sales of Xbox 360s at the time. The promise of the PS4, on the other hand, may help Sony claim the strong place that they once held in the early 2000s. GameSpot reports that already 600,000 people have signed up for the "First to Know" list (Makuch 2013). Further, as Gaudiosi (2013) writes in *Forbes*, "When it comes to PS4, things are looking very good for Sony's PS4 launch. Twenty-nine percent of all social media chatter around the PS4 were gamers announcing their intent to buy the new console."

Game Centers, Game Parlors, and the Performance of Gaming

Japan has approximately 4,650 game centers, or video arcades, as they are known in the United States. This number is 3.64 per 100,000 of the population. According to a *Statistics Japan* entry on "Video Arcades" (2010), the largest concentration of these is found in Okinawa—three times the national average—followed by Tokyo. The only statistical correlation that suggests itself in these numbers is that there are a greater number of dedicated game centers where US populations tend to be denser. The total number of game parlors, those that support gambling games such as pachislot and pachinko, is 8,625 as of 2007, or 6.77 per 100,000 of the population, and pachinko parlors continue to have a close relationship with the yakuza (Son 1998; Johnston 2007), almost double that of game centers. Add to these numbers the observation that Ashcraft makes in *Arcade Mania!* (2008) that game cabinets are crammed into any corner that a child might see, and the landscape of Japan seems to be painted with pixels.

Japan has a long history of arcades, weathering the arcade crash of the early 1980s well and going on to develop innovative and novel arcade interfaces. One reason for this historical success may have to do with the currency of games; early Japanese arcade games standardized the 100-yen coin (the approximate equivalent to the US dollar), which, Brian Crecente argues, allowed the arcade market to thrive in Japan while the quarter-based arcade market struggled and eventually crashed in the United States. Additionally, game centers were and continue to be colocated with train and subway stations, thereby providing a few minutes of distraction for patrons who are waiting.

Early in the life of Japanese game centers, they were seen as dangerous places. As Kusahara (2006) reports, high school boys would often skip school and hang out in these centers. When younger children arrived to play, they were often bullied, which motivated many PTAs (parent–teacher associations) to ask mothers to patrol the game centers (Kusahara 2006, 169). Thus, the introduction of the home console was a welcome relief. Game centers responded by developing more innovative and technologically advanced games—games that often had specialized interfaces that emphasize the performativity of gameplay as part of the spectacle. Crane games were hugely popular, dispensing the cute Kawaii characters so popular in Japan. Then, in the 1990s, arcade game makers started to develop games in response to the success of *Beatmania* (1997), Konami's DJ game in which the controller was a simulated turntable. Based on the success of this game, Konami created *Dance Dance Revolution* (1998), a full-movement dancing game in which players had to step on a controller pad. The game was successful in both its home market and abroad. Indeed, according to Wong (2000), "Konami's line of music-based video games—primarily *DDR* and *Beatmania*—has sent the company's net income for the fiscal year ending March 31 soaring nearly 260% to 18.3 billion yen, or $173.6 billion, up from 5.1 billion yen, or $48.4 billion." Japanese people, from young children to after-work salary men, could be seen on the streets of Tokyo, dancing on the dance pads of the arcade machines placed in front of shops and game centers. These music/dance-based games are not the only ones to offer novel interfaces; the infamous game *Boong-Ga Boong-Ga* (2001) (a Japan-only release by a Japanese-trained Korean developer) simulates Kancho, an adolescent game wherein people put their hands together and point up with their two forefingers to poke someone in the backside and simulate an enema.

In addition to these performance games, the 1990s saw the introduction of purikura. Purikua machines take pictures of people in a booth and then allow people to theme the borders, draw pictures, and add aftereffects to the image. Then, for the small cost of ¥300, a sheet of small stickers is printed on the spot. These photography machines are not new, yet purikura machines could be found all over Japan, from game centers to the local convenience store. According to Okabe et al. (2006), the reason for the success of these machines probably had to do with ease of use and their sociability. Furthermore, they allowed people to capture memories of themselves with friends in a location-based entertainment form (4).

Game centers did not weather the release of the Wii as well as they weathered the arcade crash of the 1980s. In a *New York Times* article, Gibbs (2008) notes that the release of the Wii, with wide-screen TVs, enabled game center experiences to be replicated at home. This pushed game designers to innovate the game center approach. As Jun Higashi, president of Namco, explained at the arcade expo in Tokyo, "We need to innovate, especially in the realm of games where people move their bodies. ... We also need to develop games that can't be played at home" (quoted in Gibbs 2008). While Higashi sees in the future in movement games (an area of interest to Japanese scholars as well—see, for example, the Ritsumeikan Center for Game Studies at http://

www.rcgs.jp), Japan has been developing novel approaches to its IC card games. These games are a hybrid between arcade games and trading card games. Based on technology that was patented in the mid-1990s (see US Patent 5,498,860), IC cards can store information about characters, special abilities, unlocked levels or courses, and the like. On average, they cost ¥300–500 and are limited to one function; thus, it is not uncommon to see arcade players arrive with a binder of carefully organized IC cards for play. Not only do these cards store particular information that a player can access at a later date, they also provide artwork to fans of the franchise, which entices many players (Kierkegaard 2006; Stone 2012). IC card-based arcade games have increased the level of player-developed content, enabling players to design clothes and looks, and in some games, those designs go to social media where they are voted on and the winner's designs are produced for fashion shows or cosplays.

Social Gaming in Japan: DeNA, Gree, and Mixi

According to the Yano Research Institute (and reproduced in GMO Cloud 2012), Japanese social games are experiencing rapid growth, from ¥45 million in 2008 to ¥1,171 million in 2011 (118). When Enterbrain announced that the game market grew for the first time in five years, the numbers provided led Serkan Toto, a leading consultant in Japanese mobile games, to conclude that "the sizes of the video and social gaming markets in Japan are now very similar" (Toto 2013). Further, Toto summarizes the Nomura Research Institute's market predictions, which anticipate that Japan's social gaming market will likely be worth ¥393.5 billion by 2016, almost double current market value (Toto 2012). This may have to do with the expanding demographic of social gamers in Japan: 67.9% of the population in Japan has played social/mobile games (CA MOBILE, quoted in GMO Cloud 2012, 75). Of that, GREE, DeNa, and Mixi reported on their websites (2011) that the demographics are strongest between twenty- and twenty-nine-year-olds (between 36% and 53.8%) and second strongest among ten- to nineteen-year-olds (between 9.4% and 31%), and also that the demographics tended to favor male players (47.9% to 60%) over females (40% to 53.3%). The four most common genres of games are pet breeding, city building, farming, and restaurant management (GMO Cloud 2012, 104). Furthermore, Japan-specific genres do as well as mobile and social games, such as manga-based games, social J-RPGs, dating simulations, and card-collecting games.

Particular to Japanese social and mobile games is the controversial game mechanic "*Kompu Gacha*" or "Complete Gacha." The game mechanic is simple enough: a player pays for a randomized attempt to get a virtual reward. To this, the game adds a collecting element, and according to Brown (2012), oftentimes the best, most rare reward is locked until all other collectable awards are unlocked. This mechanic was big money for mobile and social games and was able to sidestep the very strict Japanese gambling laws. However, a series of 2011 and 2012 news articles highlighted the social and monetary ills of this mechanism. According to Hori (2012) in his *Diamond IT & Business* article, he interviewed housewives who would spend ¥100,000 per month or ¥1.5 million per year as part of a fully developed gambling addiction. This press coverage compelled many of the major mobile and social game developers—GREE, Mobage, DeNA, Mixi, CyberAgent, Dwango, and NHN Japan—to enter into a self-regulating agreement on May 9, 2012. Following this, Peterson (2012) reported in *Games Industry International* that Gree and DeNA's stock plummeted 20%. However, these companies showed impressive foresight, for on May 18, 2012, the Consumer Affairs Agency of Japan released a press release that declared kompu gacha illegal under the "Act against Unjustifiable Premiums and Misleading Representations" and the "Law for Preventing Unjustifiable Extra or Unexpected Benefit and Misleading Representation." At the end of 2012, in response to the kompu gacha laws, Japanese social game developers banded together to form the Japan Social Game Association (JASGA). The group describes its mission thus: "Enhance the environmental improvement for users by conducting written examinations for platform operation corporations, patrolling social games, and dealing with information provided by customers." While kompu gacha disappeared as a result of this, gacha as a mechanic is still around; however, the mobile game providers must disclose the probability of winning and the industry is putting caps on how much people can spend on content, especially for younger players. These are attempts to ensure that social gaming will be able to develop robustly and with little government oversight, a development model that is credited for the earlier computer game success of Japanese console games.

Censorship and Rating Systems

Japanese computer games may seem to be outside of the more stringent and formal control that is common in other countries; however, this is not the case. From early in Japanese computer gaming history, Japan has imposed a series of laws and ordinances on Japanese games and game providers. The earliest example of this is from 1985, when Japan revised the Entertainment Establishments Control Law to include game centers. According to The Ministry of Foreign Affairs of Japan (1999) in their report on the "Situation of Japanese Women," "In Japan, in the early part of the 1980s, new types of adult entertainment trade emerged one after the other and this had an enormous amount of negative influence on the good morals of society and sound development of the juvenile." The Entertainment Establishments Control Law, then, is intended to control prostitution and the sex trade, and this is the law unto which the game centers must adhere. Therefore, according to Ono (2013), it is difficult for Japanese arcades to survive legally.

These laws, naturally, do not extend to the home market, and this might be one of the reasons that the home console market flourished in the 1980s and 1990s. Satoshi Tajiri, the creator of *Pokémon*, may have too idealistic an outlook regarding the reason for this. In his interview with *Time* magazine (1999), Tajiri states, "In Japan, violence in games is pretty much self-regulated. In the 1980s, there was a game called *Bullfighter* where the matador stabbed the bull and red blood squirted out. The day after it was released, they changed the blood to green." Yet as Kelley (2010) rightly notes, "By the mid-1990s, the rapid expansion of the domestic market for video games in Japan was accompanied by a growing speculation about the potential negative effectives that this relatively new and booming media might be having on the population of (primarily) youthful users" (146). In 2002, the Computer Entertainment Rating Organization (CERO) was formed as a branch of the Computer Entertainment Suppliers Association (CESA). CERO rates for age appropriateness: A for all ages, B for twelve-year-olds and above, C for fifteen-year-olds and above, D for seventeen-year-olds and above, and Z for eighteen-year-olds and above (CERO 2013). They rate by evaluating sex-related expressions, violence, antisocial acts, and language and ideological expressions (CERO 2013). The evaluative process is very much culturally driven; for example, Ashcraft (2011) of the *Japan Times* notes that *Shin Megami Tensei: Persona 4* (2008) received a B rating in Japan (twelve-year-olds can play it) yet an "M" ("mature," for ages 17 and up) rating in the United States.

While CERO attends to the majority of released games, the Ethics Organization of Computer Software (EOCS) in Japan oversees PC games, and, more specifically, they regulate the adult game industry. In Japan, dating sim games have a long history, from the popular high school dating game *Graduation* (1992) to the controversial *RapeLay* (2006), a game that has effected the greatest change of EOCS policies. *RapeLay*, a Japan-only release, simulates a man who has been rudely rebuffed by a woman. This man can follow the woman onto a train and rape her and her two daughters. Rape fantasies in media are not uncommon in Japan, something that Cather (2012) notes in her book *The Art of Censorship in Postwar Japan*, and while Cather is interested in manga censorship, the landscape of sexual entertainment is similar across media. So it was that *RapeLay* did not garner any attention in Japan until it was picked up by a third-party distributor and released in the UK and subsequently made accessible in Europe and the United States. EOCS received pressure from Equality Now, a women's rights group, to ban rape games, and while it appeared that it would in 2009 (see Caoili 2009), the group has since retracted its intentions for further action against games of this genre.

"Our Game Industry Is Finished": The Present and Future State of Computer Games in Japan

In his infamous outburst at the 2009 Tokyo Game Show, Keiji Inafune declared that the Japanese game industry was finished (Kohler 2012)—not that it was in decline, but that it was done. While Inafune's words of doom were extreme, they articulated a growing anxiety among Japanese game companies and developers who see the sun setting on their global dominance. Neither the competition introduced into distribution channels (such as the Xbox and Steam), nor the way social games have undermined conventional games (Boxer 2012), can fully account for recent sales figures. In 2011, Japanese hardware sales were up 2.4%; software sales, on the other hand, were down 13.7% (Rose 2012). That year also saw a number of game journalists commenting on the lackluster presentations at the Tokyo Game Show, with dwindling numbers of booths and little-to-no new content (Ryckert 2011; Gilbert 2011). However, while the number of booths appears to be shrinking, the number of attendees broke records in 2011 and 2012, which indicates that the game industry is not finished; it's only adjusting.

This is something that Japanese game developers are beginning to comment on. For example, Yosuke Hiyashi, head of Team Ninja of the *Ninja Gaiden* franchise, explains, "There was a time when Japanese games were number one, but that's changed. We need to figure out and focus on the core audience and that would get into all kinds of games. We need to be able to deliver a game that they'll definitely enjoy" (quoted in Robinson 2012). Robinson (2012) argues strongly in his *Eurogamer* article that while Japanese games may not be as globally prominent as they once were, to accuse them of lacking creativity (a charge being leveled against Japanese developers) is a mistake. He points to *Demon's Souls* (2009) and quotes Gavin Moore as saying, "You think that could have been made in the west, but no one would have the balls to make that in the west. No one would make it that difficult—it's only that Japanese mentality, that old-fashioned sense of clearing the game. And everyone in the west thought that game was awesome" (Robinson 2012). *Demon's Souls* went on to sell 1.7 million units, 51% of those in the United States ("Demon's Souls" 2013). Indeed, the sales numbers from the first quarter of 2013 indicate that Japanese game developers and publishers are still performing well in more traditional computer game markets; seven of the top ten are Japanese games that went on sale between November 2012 and March 2013 (D'Angelo 2013). In addition to console games, social and mobile games continue to expand into new markets, and on the development side, the Japanese gaming community is fostering new talent through the Tokyo Game Show Sense of Wonder Night for indie game developers. Regardless, the future of Japanese computer games will be a dual future of national gaming identities and policies alongside global partners and markets.

Notes

1. Like many of these companies, it's difficult to speak of SEGA as an exclusively Japanese company. It was started by ex-patriot US citizens and was eventually bought by Gulf & Western in 1970. However, while SEGA was traded on the New York Stock Exchange, the company, maintained in Japan with its own name and logo, was eventually acquired by Sammy to become the SEGA SAMMY HOLDINGS (Pettus 2012).

2. My translation. The original Japanese reads: "弊社がセガ社を買収することは100%ないことを表明します."

3. My translation. The original Japanese is "全てのゲームは、ここに集まる" (subeteno geemu wa, koko ni atsumaru), "いくぜ、100万台" (Ikuze, Hyakkuman dai), "すごいことになってきた" (sugoi koto ni natte kita), and "よい子とよいおとなの" (yoi ko to yoi otona no). The original ad campaigns have been collected on the No Problem LCC site, maintained by Koshimo, one of the advertisers hired for these campaigns. See http://www.noproblem.co.jp/works/koshimo/ps.html.

References

Afterthoughts: Final Fantasy VII: Interview with Yoshinori Kitase and Tetsuya Nomura from Electronic Gaming Monthly. 2005. *Electronic Gaming Monthly*, issue 196, October. http://www.ff7citadel.com/press/int_egm.shtml.

Akagi, Teppei. 1992. *SEGA vs. Nintendō: Maruchimedia Wōzu no Yukue.* Tokyo, Japan: Nihon Nōritsu Kyōkai Manejimento Sentā.

Akahori, S. (writer), and H. Yamaga (director). 2002. *Magical Shopping Arcade Abenobashi/Abenobashi mahout shoutengai* (Television series). Tokyo, Japan: Gainax and Madhouse.

Allison, A. 2006. *Millennial Monsters: Japanese Toys and the Global Imagination.* Berkeley, CA: University of California Press.

Ashcraft, B., with J. Snow. 2008. *Arcade Mania! The Turbo-charged World of Japan's Game Centers.* Tokyo: Kodansha International.

Ashcraft, Brian. 2011. Video games now have the same U.S. protection as books and films. *Japan Times*, July 6. http://www.japantimes.co.jp/life/2011/07/06/digital/video-games-now-have-the-same-u-s-protection-as-books-and-films/#.UbKlZ4L3hJs.

Ashcraft, Brian. 2013. Believe it or not, the PlayStation 2 is getting *more* expensive in Japan. *Kotaku*, February 19. http://kotaku.com/5985252/believe-it-or-not-the-playstation-2-is-getting-more-expensive-in-japan.

Atou, M. 2001. Very low fertility in Japan and value change hypothesis. *Review of Population and Social Policy* 10:1–21.

Azuma, Tetsuya, Marton Sziraczki, Nobutaka Takeo, and Satoshi Yamada. 2009. The video games cluster in Japan. http://www.isc.hbs.edu/resources/courses/moc-course-at-harvard/Documents/pdf/student-projects/Japan_Video_Games_2009.pdf.

Benzinga, Louis. 2013. Can Pokémon X and Y save the Nintendo 3DS? *NASDAQ*, January 10. http://www.nasdaq.com/article/can-Pokemon-x-and-y-save-the-nintendo-3ds-cm206308.

Boxer, Steve. 2012. Is Japan's development scene doomed? CVG visits the Japan Foundation in London to find out. *Computers and Videogames*, March 2. http://www.computerandvideogames.com/337455/features/is-japans-development-scene-doomed/.

Brown, Nathan. 2012. Complete gacha a "disaster of historic proportions." *Edge*, May 10. http://www.edge-online.com/news/complete-gacha-disaster-historic-proportions/.

Buckingham, D., and J. Sefton-Green. 2004. Structure, agency, and pedagogy in children's media culture. In *Pikachu's Global Adventure: The Rise and Fall of Pokémon*, ed. Joseph Tobin, 12–33. Durham, NC: Duke University Press.

Caoili, Eric. 2009. Japanese organization to ban sale of rape games. *Gamasutra*, May 28. http://gamasutra.com/php-bin/news_index.php?story=23803.

Caplin, Nick. 2011. PlayStation Network and Qriocity outage FAQ. *PlayStation.Blog*, April 28. http://blog.eu.playstation.com/2011/04/28/playstation-network-and-qriocity-outage-faq/.

Casper, S., and C. Storz. 2012. Diversity in the emergence of new industries: Public sector engagement and networking in the emergence of the Japanese, Korean, and US game industry. Paper presented at the SASE Annual Conference 2012, MIT, Cambridge, MA. http://citation.allacademic.com/meta/p568395_index.html.

Cather, K. 2012. *The Art of Censorship in Postwar Japan*. Honolulu, HI: University of Hawai'i Press.

Crecente, Brian. 2011. How one coin saved arcades in Japan and another killed them in the U.S. *Kotaku*, February 22. http://kotaku.com/5767303/how-one-coin-saved-arcades-in-japan-and-killed-them-in-the-us.

CERO. 2013. What is the "Age-appropriateness Rating System"? Computer Entertainment Rating Organization. http://www.cero.gr.jp/e/rating.html#05.

Consumer Affairs Agency, Government of Japan. 2012. 「カード合わせ」に関する 景品表示法(景品規制)上の考え方の公表 及び 景品表示法の運用基準の改正に関するパブリックコメントについて. http://www.caa.go.jp/representation/pdf/120518premiums_1.pdf.

D'Angelo, William. 2013. Top 10 in sales—First quarter 2013. *VGChartz*, April 9. http://www.vgchartz.com/article/250881/top-10-in-sales-first-quarter-2013/.

Demon's Souls. 2013. *VGChartz*, August 31. http://www.vgchartz.com/game/31689/demons-souls/.

Department of Defense. 2011. Active duty military personnel strengths by regional area and by country (309A). http://www.globalsecurity.org/military/library/report/2011/hst1103.pdf.

deWinter, Jennifer. 2004. Multi-media narratives: The videogame in an emerging mega-literacy. *Works and Days* 22 (43/44):73–91.

deWinter, Jennifer. 2009. Aesthetic reproduction in Japanese computer culture: The dialectical histories of manga, anime, and computer games. In *Computer Culture Reader*, ed. Judd Ethan Ruggill and Ken S. McAllister, and Joseph R. Chaney, 108–124. Cambridge: Cambridge Scholars Publishing.

Donovan, Tristan. 2010. *Replay: The History of Videogames*. East Sussex: Yellow Ant Media.

E3 06: Revolution renamed Wii. 2006. *Gamespot*, April 27. http://www.gamespot.com/news/e3-06-revolution-renamed-wii-6148462.

Final Fantasy VII. 2013. *VGChartz*, August 31. http://www.vgchartz.com/game/756/final-fantasy-vii/.

Finn, M. 2002. Console games in the age of convergence. In *Proceedings of Computer Games and Digital Cultures Conference*, ed. Frans Mäyrä, 45–58. http://researchbank.swinburne.edu.au/vital/access/manager/Repository/swin:2227.

Forster, Winnie. 2011. *Game Machines 1972–2012: The Encyclopedia of Consoles, Handhelds, and Home Computers.* Utting: GAMEplan Books.

"Game Platforms." 2012. VGChartz. http://www.vgchartz.com/platforms/

Gaudiosi, J., 2013. Fizziology exec says social media buzz points to strong demand for Sony PlayStation 4. *Forbes*, March 10. http://www.forbes.com/sites/johngaudiosi/2013/03/10/fizziology-exec-says-social-media-buzz-points-to-strong-demand-for-sony-playstation-4/.

Gibbs, E. 2008. A slump in Japan's video arcade industry spurs calls for innovation. *New York Times*, May 18. http://www.nytimes.com/2008/05/18/business/worldbusiness/18iht-arcades.1.12985638.html?_r=0.

Gilbert, B. 2011. Tokyo Game Show 2011 draws largest crowd ever as show floor continues to shrink. *Joystiq.com*, September 22. http://www.joystiq.com/2011/09/22/tokyo-game-show-2011-draws-largest-crowd-ever-as-show-floor-cont/.

GMOCloud. 2012. Japanese video gaming trends: Past, present, and future. http://www.slideshare.net/GMOCloud/japanese-gaming-market-2012.

Gran Turismo series software title list. 2013. *Polyphony Digital: Sony Computer Entertainment Group.* http://www.polyphony.co.jp/english/list.html.

Herman, Leonard. 2007. The later generation home video game systems. In *The Video Game Explosion: A History from PONG to PlayStation and Beyond*, ed. Mark J. P. Wolf, 161–172. Westport, CT: Greenwood Press.

Hori, M. 2012. 依存症"ならば自己責任論は成立しない規制なきまま社会と 共存していけるのか—ソーシャルゲームの何が問題か. *Diamond IT & Business.* http://diamond.jp/articles/-/17159.

Ishihara, T. 2013. What does the Pokémon company do? *The Pokémon Company.* http://www.pokemon.co.jp/corporate/en/interview/.

Ivan, Tom. 2009. Guinness ranks top 50 games of all time. *Computers and Video Games*, February 28. http://www.computerandvideogames.com/209385/guinness-ranks-top-50-games-of-all-time/.

Izushi, H., and Y. Aoyama. 2006. Industry evolution and cross-sectorial skill transfers: A comparative analysis of the video game industry in Japan, the United States, and the United Kingdom. *Environment and Planning* 38:1843–1861.

Iwata, S. 2011. Iwata asks: E3 2011 special edition. Nintendo.com. http://iwataasks.nintendo.com/interviews/#/e32011/newhw/0/6.

Japan Social Game Association (JASGA). 2014. About JASGA. https://jasga.or.jp/en/about/section3.html.

Japan video game industry. 2007. Japan External Trade Organization (JETRO) industrial reports. http://www.jetro.go.jp/australia/market/index.html/japanesevideo.pdf.

Johnston, E. 2007. From rackets to real estate, yakuza multifaceted. *Japan Times*, February 14. http://www.japantimes.co.jp/news/2007/02/14/reference/from-rackets-to-real-estate-yakuza-multifaceted/#.UbKXM4L3hJs.

Kelley, W. H. 2010. Censoring violence in virtual dystopia: Issues in the rating of video games in Japan and of Japanese video games outside Japan. In *Utopic Dreams and Apocalyptic Fantasies: Critical Approaches to Researching Video Game Play*, ed. J. T. Wright, D. G. Embrick, and A. Lukacs, 143–160. Lanham, MD: Lexington Books.

Kierkegaard, Alex. 2006. Arcade IC cards. *Insomnia*, September 22. http://insomnia.ac/japan/ic_cards/.

Ko, E., M. Moylan, J. Zubkavich, C. Butcher, S. King, A. Paulsen, and B. Oates, eds. 2012. *The History of Sonic the Hedgehog*. Ontario: UDON Entertainment Corp.

Kohler, Chris. 2004. *Power-up: How Japanese Video Games Gave the World an Extra Life*. Indianapolis, IN: BradyGames.

Kohler, Chris. 2012. Q&A: *Mega Man* creator wants Japan to admit failure. *Wired*, April 12. http://www.wired.com/gamelife/2012/04/keiji-inafune-qa/.

Koshi, R. (writer), and S. Watanabe (director). 1999. *Excel Saga* (Television series). Tokyo: J. C. Staff.

Kusahara, M. 2006. Japan's electronic game culture. In *Contemporary Youth Culture: An International Encyclopedia*, vol. 1, ed. S. Steinberg, P. Parmar, and B. Richard, 166–187. Westport, CT: Greenwood Press.

Lah, K. 2012. Japan's older generation turns gamers. *CNN*, February 8. http://www.cnn.com/2012/02/08/world/asia/japan-older-gamers.

Leyton, Chris. 2006. Doctor Kawashima's brain training: How old is your brain review. *Total Video Games*, June 12. http://www.totalvideogames.com/Dr-Kawashimas-Brain-Training/review-9608.html.

Makuch, Eddie. 2013. GameStop seeing "strong demand" for PlayStation 4. *Gamespot*, March 12. http://www.gamespot.com/news/gamestop-seeing-strong-demand-for-playstation-4-6405147.

McDonald, K. 2013. How successful was the Wii U launch? Now that we have the numbers, let's look at how Nintendo's latest launch went. *IGN*, January 11. http://www.ign.com/articles/2013/01/11/how-successful-was-the-wii-u-launch.

McGray, D. 2002. Japan's gross national cool. *Foreign Policy*, May 1. http://www.foreignpolicy.com/articles/2002/05/01/japans_gross_national_cool.

Ministry of Foreign Affairs of Japan. 1999. Situation of Japanese women. Article 6. http://www.mofa.go.jp/policy/human/women_rep4/part2_5.html.

Nakamura, Ichiya. 2003. Japanese pop industry. Stanford Japan Center White Paper.

Nakamura, Toshi. 2012. Portable gaming rules in Japan. *Kotaku*, October 25. http://kotaku.com/5954757/portable-gaming-rules-in-japan.

National Institute of Population and Social Security Research in Japan. 2012. Population projections for Japan (January 2012): 2011 to 2060. http://www.ipss.go.jp/site-ad/index_english/esuikei/gh2401e.asp.

Nintendo. 2000. Corporate release. http://www.nintendo.co.jp/corporate/release/2000/001227.html.

Nintendo. 2012. Consolidated sales transition by region. http://www.nintendo.co.jp/.

Nintendo company statistics. 2012. *Statistics Brain*, July 31. http://www.statisticbrain.com/nintendo-company-statistics/.

No laughing matter: Are the new crop of game consoles more—or, indeed, a good deal less—than they are made out to be? 2001. *Economist*, December 6. http://www.economist.com/node/884973.

Nouchi, R., Y. Taki, H. Takeuchi, H. Hashizume, Y. Akitsuki, Y. Shigemune, A. Sekiguchi, Y. Kotozaki, T. Tsukiura, Y. Yomogida, and R. Kawashima. 2012. Brain training game improves executive functions and processing speed in the elderly: A Randomized Controlled Trial. *PLoS ONE* 7(1), January 11. http://www.plosone.org/article/info%3Adoi%2F10.1371%2Fjournal.pone.0029676.

Okabe, D., J. Chipchase, I. Mizuko, and A. Shimizu. 2006. The social uses of Purikura: Photographing, modding, archiving, and sharing. PICS Workshop, Ubicomp.

Ono, K. 2013. A history of control in Japanese game industry. *IGDA Perspectives Newsletter*, February 26. http://newsletter.igda.org/2013/02/26/a-history-of-control-in-japanese-game-industry/.

Peterson, S. 2012. Gree, DeNA stocks plunge as Japanese government cracks down. *Games Industry International*, May 7. http://www.gamesindustry.biz/articles/2012-05-07-gree-dena-stocks-plunge-as-japanese-government-cracks-down.

Pettus, S. 2012. *Service Games: The Rise and Fall of SEGA*. CreateSpace Independent Publishing Platform.

Platform totals. 2013. *VGChartz*, August 31. http://www.vgchartz.com/analysis/platform_totals/.

PlayStation 2 breaks record as the fastest computer entertainment platform to reach cumulative shipment of 100 million units. 2005. *Games Industry International*, November 30. http://www.gamesindustry.biz/articles/playstation2-breaks-record-as-the-fastest-computer-entertainment-platform-to-reach-cumulative-shipment-of-100-million-units.

Pokémon adventures. 2002. http://web.archive.org/web/20021215144736/www.vizkids.com/pokemon/adventures/characters.html.

Purchese, R. 2009. Hirai: We're the "official" industry leader. *Eurogamer.net*, January 20. http://www.eurogamer.net/articles/hirai-were-the-official-industry-leader.

Ramsay, Morgan. 2012. *Gamers at Work: Stories behind the Games People Play*. New York: Apress.

Robinson, M. 2012. The truth about Japan: A postcard from the Japanese game industry. *Eurogamer.net*, October 10. http://www.eurogamer.net/articles/2012-10-10-tokyo-story-a-postcard-from-the-japanese-games-industry.

Rose, M. 2012. Japanese video game market revenues declined 8% in 2011. *Gamasutra*, January 5. http://gamasutra.com/view/news/39493/Japanese_video_game_market_revenues_declined_8_in_2011.php.

Ruggill, Judd Ethan, and Ken S. McAllister. 2011. *Gaming Matters: Art, Science, Magic, and the Computer Game Medium*. Tuscaloosa, AL: University of Alabama Press.

Ryan, J. 2011. *Super Mario: How Nintendo Conquered America*. New York: Penguin.

Ryckert, Dan. 2011. Tokyo Game Show 2011 report: Big booths, small impact. *Game Informer*, September 18. http://www.gameinformer.com/b/news/archive/2011/09/18/tgs2011-report.aspx.

SEGA SAMMY Holdings. 2012. Fiscal 2012 business results report. http://www.segasammy.co.jp/english/ir/ar2012/message/index.html.

SEGA SAMMY Holdings. 2013. Business summary. http://www.segasammy.co.jp/english/pr/business/.

SEGAtastic. 2009. Mega Drive sales figures: An update. *SEGAtastic*, September 1. http://segatastic.blogspot.com/2009/12/mega-drive-sales-figures-update.html.

Sheff, David. 1994. *Game Over: How Nintendo Conquered the World*. New York: Vintage Books.

Son, M. 1998. *Pachinko Industry Report*. Japan: Banseisha.

Statistical Handbook of Japan 2012. 2013. Ministry of Internal Affairs and Communications. http://www.stat.go.jp/english/data/handbook/c0117.htm#c02.

Stone, Sabbi. 2012. Arcade IC Cards? Oh, I see. http://newbreview.com/arcade-ic-cards-oh-i-see/.

Tajiri, S. 1999. The ultimate game freak. *Time Magazine*, November 22. http://www.time.com/time/magazine/article/0,9171,2040095,00.html.

Tobin, Joseph. 2004. Introduction. In *Pikachu's Global Adventure: The Rise and Fall of Pokémon*, ed. J. Tobin, 1–11. Durham, NC: Duke University Press.

Toto, Serkan. 2011. Who's actually playing social games in Japan? *Dr. Serkan Toto—Japan Mobile and Social Games Consulting*, December 29. http://www.serkantoto.com/2011/12/29/who-social-games-japan/.

Toto, Serkan. 2012. Nomura: Japanese social gaming market to double in size by 2016. *Dr. Serkan Toto—Japan Mobile and Social Games Consulting*, January 30. http://www.serkantoto.com/2012/01/30/nomura-social-gaming-market-size/.

Toto, Serkan. 2013. Japan's video game market grows to US$4.6 billion in fiscal 2012 (but social games not too far off). *Dr. Serkan Toto—Japan Mobile and Social Games Consulting*, April 7. http://www.serkantoto.com/2013/04/07/japan-video-game-market-growth/.

Video arcades. 2010. *Statistics Japan: Prefecture Comparisons*, September 3. http://stats-japan.com/t/kiji/12028.

Wong, Nicole C. 2000. Hip-hop music sweeps arcades with "Dance Dance Revolution." *SiliconValley.com*. http://www.ddrfreak.com/newpress/San%20Jose%20Mercury%20News.htm.

Y-N. Ken. 2006. Gaming in Japan. *What Japan thinks: From Kimono to Keitai: Research Japanese Facts and Figures through Translated Opinion Polls and Surveys*, May 24. http://whatjapanthinks.com/2006/05/24/gaming-in-japan/.

MEXICO

Humberto Cervera and Jacinto Quesnel

In this chapter we will explore the history of video games in Mexico. Given the lack of previously published information, this will be an exploration of the word-of-mouth side of this story. We will explore the retail industry, game journalism, and game development; there is little to no written record of Mexico's video game history, so this chapter is based on interviews with a few key players in the industry. There is still a lot missing in this tale; sadly, some key players couldn't be found or did not agree to be interviewed by the authors. This small piece is only comparable to the first relaxed steps at the bottom of a mountain before getting to the climbing. The authors hope you find this chapter as enjoyable as it is informative. The interviewees are:

Gonzalo "Phill" Sanchez, a game development teacher at the SAE Institute (School of Audio Engineering), UVM (Universidad del Valle de México), and other universities. He is also editor in chief at *Motor de Juegos* (*MDJ*) (http://www.motordejuegos.net). He has almost ten years in the video game market, and at *Motor de Juegos* he has created a community of devoted Mexican developers. He is well known for the support he offers to aspiring developers and is the Mexican authority on the state of the video game industry. *MDJ* publishes an annual report on the state of the industry.

Adrian "Carqui" Carbajal, editor in chief at *OXM* (*Official Xbox Magazine México*). Carqui has been a journalist in the industry since its birth. He has worked at the major local publications, at *Club Nintendo*, and was the editor in chief at *Atomix* and later for *EGM en Español* (*Electronic Gaming Monthly*). He has seen firsthand how the industry began and how it evolved to what it is now.

Jose M. Saucedo has almost twenty years of experience as a Mexican video game journalist; he was the founder of *Contacto PSX* (the first indie game magazine in Mexico) after he became the editor in chief of *Atomix* magazine. He now works at Team One as a sales representative for several publishers for the Latam region, and he is responsible for community management and support in marketing campaigns.

Gabriel Palacios has participated in nationwide education projects such with the SEP (Secretaría de Educación Pública), CONACyT (Consejo Nacional de Ciencia y Tecnología), among many other government institutions. He creates educational content, educational video games, and transmedial gamification strategies for education. He is also a digital artist whose work has been exhibited in several art galleries in the United States and MK&Gon Germany.

Ivan Chapela has much experience in the Mexican video game industry. He established Radical Studios in the late 1990s, one of the pioneer developing studios in Latam; in 2005 he directed *Atomix* magazine, the biggest video game publication in Latam at the time; and in 2007 he founded 3nMedia, a consulting agency for the Latam video game industry. He has positioned himself as a key player within the industry and as a leader in PR and marketing, experience which gives him unique and broad insight into game development and the retail industry.

The Market

Video games are deeply rooted in Mexican pop culture. As the southern neighbor of the United States, it is easy to understand why our consumption habits have grown in parallel with our northern neighbor. Only a handful of people had heard about video games at the beginning, and even with piracy and contraband, there is much evidence of the Mexican upper class importing consoles, such as the Magnavox Odyssey or the Atari, for personal use as soon as they were released in the United States. Almost any gamer over the age of thirty-five will remember that period. Up to this day, you can still find such consoles in thrift shops all over the country. Due to a lack of a formal retail industry in Mexico at the time, gamers in the country could only buy their original copies of the games they wanted on gray markets that sold smuggled goods. Gamers had the money to buy their games, but a lack of availability forced the consumers to acquire them through said markets.

One cannot speak of video game consumption in Mexico without addressing our northern border; at least half of the video game consumption in this country was illegal back then (there is no data about this and we can only estimate through observation). Whether it was through smuggled imports (called "fayuca" in the local slang) or piracy, a lot of gamers got hooked into video game culture through the informal market. Without recorded data, game consumption in our country is hard to put into numbers, but according to *El Universal*, a nationwide newspaper, the growth of video game consumption is three times larger than the growth of any other industry ("Videojugadores Invaden El País" 2013).

The black market (piracy) is different from the gray market (smuggled imports) and the formal (legitimate) market, and we can only infer that the growth in the formal market might not have been as much as it is now without these other markets, since there is a migration of a significant percentage of consumers from black or gray market games to the consumption of legitimate copies. This has been fueled by tighter control of game piracy, availability, and the status quo derived from the ownership of formal market games, as opposed to the social stigma created by owning pirated, black market, illegal copies of games. Again, there is no official data, and probably our next step should be to start quantifying it.

"Las Maquinitas"

To make a long story short, Mexicans, given a low enough price, thrive on playing video games; the widespread success of arcade machines on the market is proof of this. It was on arcade machines, called "chispas" (sparks), "maquinitas" (little machines), or "electros" in the local slang, that a huge number of young consumers had regular access to video games. Beginning with arcades, video games started to gain ground as the preferred form of entertainment for some people, but video games still had a long way to go to achieve critical mass. Some popular arcade games at the time were *Pac-Man* (1980), *Centipede* (1981), and *Frogger* (1981); popular games in Mexico were the same as in any other country, just not as popular in the mainstream.

People thought of video games as things that were meant for children, at least until the 1990s, when avid gamers started to include adults in the demography. When arcade games began their decline in North America, arcades were still big in Mexico, but as soon as consoles became more accessible in the mid-1990s, arcades in Mexico followed the same fate as in the United States.

Countrymen in Michoacan, a state famous for its continuous illegal migrations to the United States, still tell an unverifiable story about them being the first to—illegally—import *PONG* (1972) and hundreds of arcade machines back in the 1970s. Some gamers used these machines for gambling, a common sight in the many years to come, regardless of the game so long as it was competitive. To this day, one can see dozens of arcade machines among small populations, places where there is still no Internet or cell phone communication, and yet, video games prevail. Perhaps the reason for this is that many gamers in said areas still don't own a console. Many corner shops in poor city neighborhoods and rural areas still operate arcade machines. A common sight is either an old 2-D fighting arcade game or a newer one that holds a plethora of games running over an internal emulator (called "maquinitas multijuegos"). Over the last couple years, some of these machines have been replaced by "Perla de Oriente" (Orient Pearl), soccer-themed, illegal, gambling arcades, since they obviously provide a better business.

Chuck E. Cheese restaurants never came to Mexico, but their main competitor, Show Biz Pizza Place, opened up a Mexican branch, rebranded as Show Biz Pizza Fiesta. It was located in a high-income urban area of Mexico City, but many people remember their first experience with video games happening at Show Biz Pizza Fiesta, despite its affluent location, since their price range was still accessible. Two other arcade parlor chains, Coney Island and Recorcholis (their main differentiating factor being that neither of them was a restaurant), followed. Coney Island closed in the early 2000s, but Recorcholis, a Mexican-owned company, still operates (as of 2013) and has extended its operations to Spain.

Parallel to the growth of the arcade parlors, one could find arcade machines at the entrance of corner shops and drugstores. Of course, most of these cabinets were illegal and unlike those in the retail industry, there was never a trade association regulating the arcade business. It was because of these arcade cabinets that people could start experiencing games such as *Centipede* (1981) or *Space Invaders* (1978) right in the corner of their neighborhood; little by little, video games were taking over other forms of entertainment.

The arcade business in Mexico was a vibrant market even during the late 1990s. Near Metro Insurgentes in Mexico City, as many as ten arcade parlors were next to each other, constantly competing for customers and trying to offer the newer games. Some even paid the best players to hang out at their places to attract hard-core fans of certain games. Even with piracy, there was formal importation of Neo•Geo MVS boards, and official launch events were held in Mexico City each time a new *King of Fighters* game was released (with launch dates very close to those in Japan). As noted before, *El Universal*, reporting on video game consumption, stated that gamers in rural areas were excluded. With 52.26 million people who play video games in the country, versus 40.6 million with Internet access, we are a gaming country; there is no doubt about that ("Videojugadores Invaden El País" 2013).

The only companies that had official representation for their arcade business in Mexico were Nintendo, SEGA, Capcom, and SNK, which was because the Japanese arcade games of these companies were more popular than Atari consoles. This could be the reason why Atari never opened up an office in the country, even though it was geographically much closer, in California. A troubled market forced Nintendo and SEGA to retire at the end of the 1980s, probably because of the North American video game industry crash of 1983, and this fate would soon be followed by Capcom and SNK in the late 1990s; they managed to survive longer because of the popularity of their fighting games, but they closed their offices in the 1990s when home consoles became more popular than arcade games in the Mexican market.

In the 1980s and 1990s, the main player base for video games in Mexico existed within the arcade space; there were many more video game players in the arcades than players who owned or had access to a home video game console (sadly, there are no numbers to back this up, but this information comes to us from our interviewees' experience and what we can infer from our own experience). This was mostly because of the high price of home console systems and because of the fact that since the effort to bring the Atari 2600 back in the 1980s, there hasn´t been another large-scale effort to legally bring consoles to Mexican consumers. Most if not all of the consoles in Mexico were imported by individuals who had the financial capacity to travel to the United States and buy one for their relatives.

In the Mexican arcade scene, SNK's *The King of Fighters* (1994) has remained present since its release. One could say that in Mexico the national sport is lucha libre, and if one were to define a video game genre as the national video game genre of Mexico, one could easily say that fighting games are this genre, for in Mexico they have a whole subculture of their own. To this day, one can always find a couple Mexicans in the final round of international tournaments such as Evolution Championship (EVO). Some of the Mexican champions are Cesar Garcia (also known as TA Frutsy, 2012, fifth place on *Ultimate Marvel vs. Capcom 3* [2011], Antonio Medrano [a.k.a. Kusanagi], 2010, third place on *Melty Blood: Actress Again* [2008]), and Armando Velazquez [a.k.a. IGL Bala], 2012, second place on *The King of Fighters XIII* [2010]). *The King of Fighters* is currently played mostly on illegal emulation-enabled arcade machines. *KoF* remains one of the most-played video game series in the country.

As of 2013, one can still find arcade cabinets in grocery stores and drugstores, but the only chain of arcade parlors that survives is Recorcholis. There have been continuous efforts to bring back the arcade parlor, with franchises such as Dave & Busters or Gameworks, but they keep failing; these franchises always open their arcade parlors in the country's high-income zones, so one could infer that the reason for this constant failure is the fact that gamers in these areas have access to consoles; arcades were successful in Mexico because the people without access to consoles sustained the parlors.

The Retail Titans

As early as 1973, a local engineer, Morris Behar, produced his own *PONG*-like console. The NESA-Pong (NESA standing for "Novedades Electronicas, S.A.") was marketed and distributed all across Latin America and achieved considerable success in the market. Even though there are no sales figures available, the NESA-Pong console was still a fair success for it had a great impact within Mexican video game pop culture, and even the original *PONG* (1972) was referred to as the "NESA-Pong" by Mexicans. The NESA-Pong helped create awareness of the existence of video games, but it was still a luxury item that many lower-class Mexican families could not afford. The lack of TV or radio publicity proved crucial in NESA-Pong's failure to appeal to a mass market and not just a niche.

It was not until 1980 that the first shipment of the Atari VCS 2600 came to Mexico as the first console to legally appear in the Mexican market. Curiously, it wasn't an electronics company that chose to bring the Atari to Mexico; it was a meat packaging company (no one seems to remember the name of said company). They held the right permits to import the Atari, and they decided to try bringing this funny new electronic entertainment product to Mexico. From a business perspective, it made sense; the company was exporting meat to the US, but the trucks were empty when returning to Mexico. In order to maximize the use of the trucks and make more money, importing something was the logical thing to do. Liverpool, a department store targeting high-income individuals, attained the exclusive rights to sell the Atari 2600, and importation of the consoles began.

The Atari 2600 did not enjoy the levels of success it had in the US, but in the 1980s, Atari opened up an office in the country with Mexican engineers dedicated exclusively to the production of Atari video games; like most people working in the game industry at that time, these engineers had little to no interest in video games. It was because of Liverpool's high-income target market that the Atari did not achieve deeper penetration into the Mexican market, but as it happens with any form of media in Mexico, video games would soon be pirated and illegally imported. Due to the extended piracy and contraband in Mexico, more people had the chance to get their hands on this form of entertainment.

During the early 1980s, there was a void within the console market, a void that would not be filled until the Nintendo Entertainment System appeared in 1985. A lot of illegal importation of Famicom (the Japanese name for the console) from Asia occurred during the 1980s. The console was sold alongside an electrical adapter for the NES and multigame cartridges. You could buy the Japanese console on gray markets such as Pericoapa or Lomas Verdes around the country, and this smuggled console was cheaper than the legal one, for it avoided import taxes. The retail price for the NES in Mexico was around USD $250, while the pirated version was USD $199.

During the late 1980s, Mexican bazaars, like the nationally known "Bazar de Lomas Verdes" and "Bazar de Pericoapa," were the focus of attention for gamers hungry for new releases. Smuggled imports were desired, and even though we have no sales figures, many remember that copies of new releases sold out after only a few days.

Some years later, Teruhide Kikuchi opened up the first "Mundo Nintendo" store in Mexico, the first legal retail shop dedicated exclusively to selling video games. Kikuchi worked at C. Itoh, the official distributor for Nintendo products in Mexico. He was charged by his company with opening official Nintendo retail stores, and so "Mundo Nintendo" opened up in four locations within Mexico City. Seito opened up a Mexican subsidiary in Mexico under the name Gamela, which would operate successfully until 2002. There was no formal structure for the legal taxing and distribution of video games in Mexico; Kikuchi was the one who did all the necessary lobbying to create the legal structure for the sale of commercial video games in the country, and this would make Abraham Bautista's job much easier when he decided to become legitimate and expand his business.

Bautista, a businessman, noticed that there was a high demand for video games and that no one was supplying that market, so in 1987 he started bringing video games as personal imports from the United States to sell on his bazaar stand in Pericoapa. As a result, he was able to sell original copies of games at a more accessible price, eventually absorbing all of his local competition. He would start business relations with the first-party distributors, and, eventually, every non-Nintendo game in the country was brought in by Bautista's operations. Bautista would soon be importing millions of copies.

In 1995, Bautista founded Game Express, a store selling legally imported original games. With the foundation of Game Express, Bautista successfully moved the core of his business from personal imports into successful full retail stores that are now present in most malls across the country.

Kikuchi and Bautista both looked at video games as a business opportunity, and their goal was to make video games a viable business. These two men would come to define the retail industry as we know it today.

Media and Marketing

Meanwhile, Kikuchi, in an effort to get the industry to be taken more seriously, commissioned the creation of a Nintendo-dedicated magazine, *Club Nintendo*, in 1991. His idea was that a monthly publication would give an image of formality and seriousness while also functioning as a marketing device for Nintendo products; better yet, the magazine would turn a profit. The founders of the magazine were Gustavo "Gus" Rodriguez and José "Pepe" Sierr, and for years to come, Rodriguez would be the public face of Nintendo in Mexico (and some Latin American markets that get Mexican TV). The market penetration of Nintendo was unlike what anyone had ever seen in the video game industry in Mexico, and even today, one can hear casual gamers refer to any game console generically as "El Nintendo."

Club Nintendo had humble beginnings, but in time the magazine changed its style and format. It was top-notch and marks the turning point for the video game retail industry in Mexico. *Club Nintendo* created a sense of community between gamers in Mexico, one that you couldn't find before. This was the only magazine available for the industry and gamers. Even if you couldn't afford a console, once a month you had access to insider information about video games. Many people who didn't even own a game console would buy a copy of the magazine, and of course it was because of this magazine that Nintendo became the leader in the Mexican market, a position it would hold until Sony's release of the PlayStation 2 in 2000.

The "Family" consoles (local slang for the illegal Famicom consoles) were a big issue for Nintendo at the time, but it was because of these pirated consoles that people had such deep brand loyalty to Nintendo. This situation would repeat for the PSX, the PS2, and the Xbox, since pirated CD-ROMs flooded the gray market. Many would buy a console while owning only a pair of legal games and dozens of illegal ones, and this created a very strong user base and market for the PS3 and the Xbox 360 when they were released.

In Mexico, piracy often works as a marketing device for companies, even though it does not help the market grow. It makes certain products accessible to a segment of the public that otherwise would never buy them. Even if the people acquire the products illegally, they will still develop a deep sense of brand loyalty, and if possible, in the future, they will strive to become legitimate customers for their brand.

By the mid-1990s, video games were already a legitimate hobby, and the Mexican consumer was starting to look for more mature experiences. The problem was that Nintendo didn't grow up with the consumer, and their marketing was still aimed at children and teenagers. *Club Nintendo*, while still a household name for the Mexican gamer, failed to offer what consumers were looking for, information about other consoles apart from Nintendo, and more mature content. Some entrepreneurs saw the opportunity and started publishing their own magazines, with the sole objective of offering better and different content than one could find in *Club Nintendo*. The magazines that actually achieved this were *Atomix* in 1997 and *Contacto PSX* in 1999.

Atomix started as an Internet forum for gamers and became the first serious competitor for *Club Nintendo* when it launched as the first multi-console magazine. It would become the most successful magazine in the Latin American market. *Atomix* retired from print media in 2009 but continues to operate successfully online. *Contacto PSX*, on the other hand, was a print magazine directed by José Saucedo, specializing in PlayStation content.

There were many other magazines, but none of them experienced the levels of success achieved by *Atomix* and *Contacto PSX*. As José Saucedo pointed out, "The main problem of video game journalism is that the journalists are mainly video game fans. They often lack objectivity and professionalism." Saucedo believes that this is still a problem in the Latin American media, and this opinion is also shared by Adrian Carbajal, who told us, "Video game journalism in Mexico is produced by fans, not by professionals." This is one of the reasons why not a single Latin American magazine has achieved the level of success of say *IGN* or *Kotaku*. English-speaking gamers would rather go to these international magazines than read the locally produced ones, mostly because of the quality and the objectivity of the content. There were other efforts to license some European magazines for the Mexican market; the problem with these magazines was that the information was completely localized for the European market and totally out of date when it arrived in Mexico.

Grupo Televisa, a leading mass multimedia company, had acquired the rights to distribute a Spanish version of the US magazine *Electronic Gaming Monthly*, and Carqui was chosen as editor in chief for this initiative. Most of the content was a translation of its US counterpart, but *EGM en Español* still had local and original articles, and the magazine was in circulation until January 2009, when the US edition stopped circulation. In April 2010, when the US edition resumed circulation, the Mexican edition did not. Grupo Televisa also held the rights for *Club Nintendo* and would later buy the rights for *OXM* (*Official Xbox Magazine*) in the mid-2000s, when the rise of online magazines rendered the video game print magazine business unprofitable. The magazines that do continue are part of big publishing companies.

As of 2013, there are many print magazines on the Mexican market, most of them offering single-platform content like *OXM* has since 2011. The only multi-console magazine left is *GameMaster* (founded in 2010 by Eduardo Aké, Hugo Juárez, and Edgar Alarcón), which also happens to be the only printed magazine that covers national game development.

The second significant factor in the consumer maturation process was the PSX. This console offered a more grown-up experience and was more accessible than any Nintendo console; better yet (for the consumer), the game CD-ROMs were much cheaper and easier to copy illegally than the Nintendo cartridges. One could buy a PSX game for MXN $10 (less than USD $1.00), so if you were a really hard-core gamer, you could buy more than a game per week and your personal finances would not be affected. For hard-core gamers, the main appeal of the Super Nintendo was the JRPGs (Japanese role-playing games), but when PlayStation obtained the exclusive rights to *Final Fantasy VII* (1997), the hard-core gamer segment of the market migrated permanently to PlayStation consoles.

Today, many gamers have retired from buying pirated games, as is evident from the sales spike that the retail industry has seen in recent years; there are not only more gamers, as noted earlier, but gamers are changing over from buying pirated copies of games to buying original copies.

In terms of console games, the most-played game series in the country is *FIFA*. The desire to play it is rooted in Mexican soccer culture and the inner need to see Mexico win the World Cup at least once. Today, there are people who buy the console and only the yearly installments of *FIFA* games to play on it.

The PlayStation would continue to dominate the Latin American market until the release of the Xbox 360 (2005). One of the decisive factors in this market domination was that when the PlayStation 2 was released, Sony Computer Entertainment of America (SCEA) formalized relationships with Mexican distributors and kept a very close relationship with the members of the press. Neither Nintendo nor SEGA had attempted to do so. Following Sony's example, Microsoft did this from day one for the release of their Xbox console. Of course, it was much easier for them because they were just entering the market, whereas PlayStation had to deal with hardware modifications of their consoles and the pirated versions of their games in the Mexican market. Microsoft was able to avoid these issues.

The Dormant Giant

Apart from the NESA-Pong console mentioned earlier, there was little video game development in the country during the 1980s and 1990s—as far as we know, not even hobbyists programming home computers. But a few museum edugames were being made at Universidad Nacional Autónoma de México (UNAM) for their museum Universum, open since 1992, and as mentioned earlier, there were Mexican engineers working for Atari, but somehow the industry never managed to grow and mature as it did in the United States. It was up to the fans to make the video game development industry a reality, and to this day we continue to struggle to make this happen. The people working in the industry in those days were just businessmen or engineers who saw in video games an opportunity to make a prosperous business, and only through the importation of foreign systems and games rather than indigenous productions.

Those children who stayed up late playing video games or who went every Sunday to Show Biz Pizza Fiesta, however, would soon grow up with a dream: the dream of making video games. As noted by Gonzalo "Phill" Sanchez, people realized that they could actually produce video games themselves when *Club Nintendo*'s forty-fourth issue featured a special report focusing on the DigiPen Institute of Technology, a Canadian school for video game design. This was the first time that Mexican fans noticed that there were schools that could teach them how to make their hobby a professional reality. Some Mexicans tried to go to DigiPen, but the expensive tuition left many more with only a dream and almost no tangible way of making it happen.

Three companies, Evoga, Aztec Tech Games, and Radical Studios, were the pioneers of Mexican game development. As Chapela points out, "When the pioneer developers started in the early 2000s, the local industry was not established yet (which is to say, no retail video games, formal distribution channels, or local publisher representation). There was no support from the government. The worldwide industry didn't look at Latin America, and academically, the closest thing to video games was computer science. Therefore, most of the funding had to come from personal relations, and the workforce had to learn through experience, which in most cases was not enough." Each company had a very different business model; Evoga (2000–2004) did art outsourcing work for SNK. Considering the very high status that *The King of Fighters* games had in Mexican culture, this gave the studio a great sense of pride.

Initiating what is still struggling to become a formal industry, Aztec Tech Games (established in 1998) started the development of *War Masters* and *Hellcopters*, although neither game was ever released. This was due to inexperience and production problems within their development; nevertheless, the efforts that Aztec Tech started would soon help the industry in another way with the foundation of the Aztec Tech Institute. The institute was founded with one core principle: to be a learning center for video game industry professionals. The Aztec Tech Institute was the only school in Mexico where one could go if one was interested in the creation of video games and didn't have the resources to go to DigiPen. The school operated for five years (2003–2008), but the directors of the project wanted it to be a full-fledged university with international recognition. They didn't have the means to make this happen, and soon enough, the project was out of control and spiraling downward. Of course, it wasn't a complete failure; some people managed to take courses at the institute, and those people would become the start of the third generation of developers within the industry. Also, Aztec Tech proved to the Mexican video game fan that making video games at a professional level was an actual possibility.

Last but not least in this group of pioneers is Radical Studios (2000–2004). At the time of the studio's founding in September of 2000, massively multiplayer online games (MMOGs) were starting to become popular with PC gamers, so Radical chose to develop a simple 2-D MMOG as its premiere game. This turned out to be *Eranor* (2003), an MMOG tailored for the Latin American market and completely in Spanish. Radical Studios had a very talented team, but with little experience creating an MMOG; it ended up being much more complicated and expensive than what they had originally thought. After two years of intense learning and financial struggles (no one in the team earned more than USD $200 per month), the investment the studio had available for the game ran dry. They did achieve some respectable things for the time, such as having some players as beta testers and getting one server up and running while also engaging with a small, devoted community of fans that was heavily invested in the release of the game. Considering the inexperience of the team and the funds available to them, this was something great. Sadly, *Eranor* never became a reality, but it is still highly regarded as the first large-scale attempt at developing a commercial video game in Mexico.

The directors and ex-employees of Evoga, Aztec Tech Games, and Radical Studios moved on to other ventures, but they still have a great presence within the video game development community in Mexico. If it were not for them, the video game landscape would not be what it is today. As they say, the hardest step to take is the first one, and this is what they did.

Many little studios emerged from these first initiatives, most of them amateur studios or student groups that were finally realizing they could make their passion a reality. There were no other big developments until the foundation of Sabarasa México in 2009. Sabarasa was a very big international company from Argentina that worked mainly with licenses or did outsourcing for triple-A studios. The man behind the bringing of Sabarasa to Mexico was none other than Abraham Bautista, the retail mogul, who finally decided to take a shot at development.

Sabarasa worked as an outsourcing company, but soon enough, Bautista decided that they wanted to start producing triple-A console titles, and so Sabarasa became the first Mexican triple-A development studio and was renamed Slang. (Full disclosure: the authors could not get interviews with ex-employees of Slang/Sabarasa. They are still bound by nondisclosure agreements, and so our knowledge of what happened at the company comes from other sources and casual conversations between the authors and the parties involved.)

In the short years that Sabarasa was operating in Mexico as Sabarasa-Slang (from June 2009 to April 2012), the company acquired all the licenses and permits needed to develop games for consoles and the licenses to create games based on the AAA (a Mexican wrestling federation), *Atrévete a Soñar* (*Dare to Dream*, 2009–2010, a high school–oriented "telenovela" with musical features, aimed at children and teens), and *El Chavo del Ocho* (1971–1980, one of the most influential TV sitcoms in Mexico since its appearance). These three intellectual properties were the most popular at the time in the Latin American market, and it was difficult to get licenses as popular as these. *Atrévete a Soñar* (2011) was released for the Nintendo Wii and seemed influenced by the

High School Musical game franchise. It found mild success in the market. *El Chavo del 8* (2012) was a *Mario Party* (1998) clone and was not that well received; it was also released for the Wii console. *Lucha Libre: Heroes del Ring* (Xbox and PS3, 2010; Wii, 2011) was the game with the biggest IP and was expected to be very successful. Millions of dollars were spent on the game marketing campaign. Adrian Carbajal remembers that the launch party for that game was the biggest he had ever seen. Everyone was there; they had live music, lots of drinks, and lots of food. This comes from a seasoned journalist, someone who has been at many release events, parties from international studios and well-established franchises; he still remembers this party as the biggest one. But after years of development, the team delivered a product that didn't sell as well as expected; even after all the effort, it was not able to compete with the existing WWE franchise games.

Even today in Mexican game development forums, there is much talk about what happened at Slang. Some say the studio had a lot of internal communication problems, others that there were toxic employees in positions of leadership who were not cut out for the job and were only looking out for their own interests. Perhaps what happened was a combination of all these issues, and the inexperienced employees were unable to develop for consoles as a result. The truth is that until one of the ex-employees decides to speak up, we will never know what truly happened there. Nevertheless, one thing we know for sure is that the endeavor failed, and after two years of operation, Slang Studio closed its doors while Slang Publisher still operates.

As noted in the annual report published by *Motor de Juegos* (Sanchez 2012), today we have around seventy-five video game studios operating in the country. Gonzalo Sanchez suggests that probably only fifty of these studios actually operate or have video games in development. Most of these studios are run as a side business of a parent company or by students with no financial responsibilities. There is not a single game studio that operates with video games as its main line of business, and the people at the head of these studios struggle day to day to keep them alive. As such, the Mexican video game industry knows how to make video games but is still learning how to make money off of the games it makes. We have yet to see the first Mexican studio that can operate with video games as its main business.

Triple-A development might still be far from a Mexican reality, but the growth of the mobile market has opened a huge window of opportunity for most. There is a thriving indie player community among IT and video game making programs in most universities.

Advergaming is a very popular business model for Mexican studios. A great example of this is Artifacto Studio (see http://www.artefactoestudio.com), and having started in 2003, it might be the oldest video game studio operating in Mexico. Its main line of business is advergaming, and it has worked with companies such as Nissan, and Huevocartoon (a very popular Mexican comedy animation show). Sometimes it produces original content of its own, but taking care of the business side of the company always comes first. Currently the studio is the largest advergaming company in Mexico. Some of its games include *Kunana Island* (2013, Windows Phone and Blackberry), *Takis Air Challenge* (2013, iOS), and *Toxic Balls* (2008, iOS). Some advertising companies, such as Squad, have taken the leap from advergaming to the creation of original, intellectual properties; *Kerbal Space Program* (2012, PC), its first original title for gamers, has achieved worldwide success and was a top seller during the 2013 Steam "Summer Sale" and has since received widespread attention.

There are studios that have government support and have advergaming business as the core income for the company. Larva Game Studio, with more than twenty-five employees, is recognized as one of the biggest and most successful game studios in Mexico (see http://www.larvagamestudios.com). Larva is located in Guadalajara and has five years of experience in the industry. As its website states, the studio is developing *Last Day on Earth* (TBA, intended for Xbox Live Arcade and PlayStation Network), *Night Vigilante* (TBA, for iOS), and *Red Bull Crashed Ice Kinect* (2012, Xbox).

Educational games are also a very recognizable line of work. As a wonderful example, Enova, a company whose main line of business is operating communal learning centers, produces dozens of learning games. The company distributes the games for free on its "Chispale" portal. Some of its games are *Cibers* (2012, PC), *El Circo* (2012, PC), and *Topo-Tops* (2012, PC).

There is also much out-of-the-box thinking among game developers in Mexico. A particular example is KaraoKulta Games's approach to video game development (http://karaokulta.com/). It has around ten teams in different cities, and ideally, each team produces about one small casual game every month. It creates more quantity than quality, but as the company's founder Jorge Suarez told us, "At this moment, our main goal is to produce as many video games as we can, so we can learn and grow as a community of developers. In the future, we can augment our quality, but first comes the learning." Some of its games are *Run Zombie Run* (2013, IOS), *Bee Rush* (2013, iOS) and *Popcorn Adventure* (2013, iOS).

Waking the Giant

What does the future hold for the Mexican industry? Many agree that we have the talent and the passion. So why can't we just start an industry?

First of all, some point out a lack of cooperation within developers: there are many associations that claim to be the "official video game developer association of Mexico." If we want the industry to grow, we will have to learn how to work together, how to achieve common goals, and communicate with each other. As of this moment, the industry is very small. If any of the studios would stop developing their current video game projects in Mexico, no one outside the country would notice. There would be no actual difference.

We might be fighting for a cake that hasn't yet been baked. We need to understand that we have to start with a plan that grows from small to big. Each studio has to focus on their own goals, as the biggest and most successful studios do. Many Mexican developers are driven by nostalgia; they want to recreate the games that they loved when they were young, but we need to create a shift. We need to wake up from that dream and change it to a drive to innovate. Maybe it's better to stop asking, "How can I recreate the good games of the old days?" and start asking "How can I move the medium forward? What can I do for video games as a whole? What can I explore that hasn't been explored yet?"

If we are to wake up the dormant industry, we also need professionalization. A significant quantity of studio teams are composed mainly of animators and programmers who have the passion to make games but lack a lot of other abilities and the production knowledge it takes to actually make a game. They do not know how to make a game project budget, and they do not know how to organize or lead game-making teams. We have the technical talent, but we need good project managers and game industry understanding to lead the team in a successful direction.

Last but not least, we need businesspeople who believe in the Mexican video game industry; we have yet to see game studios able to survive solely by making video games. To create an industry, we need entrepreneurs who are willing to take risks, people able to do what Bautista and Kikuchi did for the retail industry, business-driven individuals with a passion.

Let's hope that in a decade we can reread this and see how the industry has grown, but as of this moment (summer of 2013), the only thing that we can do is work hard and keep moving forward, together. Only then will we have a Mexican industry relevant on an international level. Ten years ago, none would have thought this was even possible, so perhaps we are on our way to waking a giant.

References

Sanchez, Gonzalo, ed. 2012. *Motordejuegos.net Reporte 2012*. Rep. no. 2. MDJ ed. vol. 2. Mexico DF: Motor De Juego, 2012. Print. Reporte De La Industria.

Videojugadores Invaden El País. 2013. Web log post. *El Universal*. Ed. Juan Luis Ramos Mendoza. El Universal Online, SA De CV, August 2012. March 2013. http://www.eluniversal.com.mx/graficos/pdf12/tecno/invaden.pdf.

THE NETHERLANDS

Christel van Grinsven and Joost Raessens

Gaming is a hot topic in the Netherlands. Dutch consumers make up the most active online gaming market in Europe. The Dutch games industry is a young and dynamic sector that has a lot of potential. While there is a clear focus on entertainment gaming worldwide, strikingly, the Dutch industry shows an almost fifty-fifty split between entertainment and serious gaming. In the varied Dutch market, small independent (indie) developers, innovative serious (or applied) gaming developers, and developers of entertainment games are all represented.[1] Digital distribution of games is a big focus of Dutch businesses, and a large number of companies are active in mobile games.

In this chapter, we would like to offer an overview of the state of affairs in the Dutch games industry.[2] The chapter is divided into the following sections: In the introduction, we refer to the themes we want to explore in more detail in the rest of the chapter, such as the Dutch gaming ecosystem, research and education, financing, and the Dutch games industry as seen from an international perspective. The second section deals with the Dutch games industry in numbers, focusing on the number of game companies, their business size, the importance of serious gaming, the many team startups, the kind of games that are being developed, and turnover and profits. Finally, the third section describes the ecosystem of the Dutch games industry, paying attention to seven aspects of the Dutch ecosystem that are of special importance.

Introduction

There are roughly 330 game companies in the Netherlands that together account for 3,000 jobs. The turnover of the game industry in 2011 was an estimated 150 million to 225 million euros. These figures show considerable growth compared to earlier research into the Dutch game industry (see iMMovator 2010). The game industry in the Netherlands began developing in the 1990s. Engine Software, from Doetinchem, is generally considered to be the oldest Dutch game developer still in business. This company focuses on games for entertainment on handheld (for example, Game Boy and PlayStation Portable) and digital platforms. The Netherlands has played a small role in the development of game consoles. Early in the 1990s, the Philips CD-i

(Compact Disc Interactive) was released as an interactive multimedia CD player and was sold until 1998. In addition to games, educational and multimedia reference titles were produced, such as interactive encyclopedias, museum tours, and so on, which were popular before public Internet access was widespread. Although Philips is a Dutch company, the development of the CD-i didn't have a large impact on the Dutch video game industry. Only a small number of the published games for the platform were created by Dutch game companies (for example, Radarsoft and SPC Vision/The Vision Factory), and these are not active in the game industry anymore. The number of Dutch game companies grew steadily during the second half of the 1990s and the first half of the 2000s. With easier access to self-publishing, for example, through Apple's App Store, the number of companies has increased significantly.

This rise of the Dutch games industry has led to initiatives and partnerships to support the growing sector. Platforms such as industry associations (for example, the Dutch Games Association [DGA]), specialized events (such as Games for Health Europe, Festival of Games, and the Dutch Game Awards), and media (including *Control Magazine*, *Power Unlimited*, gamekings.tv, and gamer.nl) have firmly established themselves over the last few years. The industry holds them in high esteem, and overall, they are seen as necessary for continuing growth. Game entrepreneurs also realize that business development is impossible without professional services. For the successful development and exploitation of games, thorough knowledge of the field, intellectual property management, proper financing, business models, and market expertise are essential. The Dutch gaming ecosystem is still in its developing stages. The system does have the potential to grow into maturity and deliver companies that are both nationally and internationally successful. In order to reach this next step, support for small and medium-sized businesses in their second stage is essential. Network meetings, workshops, pitch events, and incubator/accelerator programs could offer the necessary support (such as the Dutch Games Association, Dutch Game Garden, and CLICKNL).[3]

Collaborations with centers of knowledge such as universities and research institutes are already being sought. These collaborations pave the way for further research into the effects of games, which will be valuable for the development of the industry. Up until now, Dutch researchers have mostly concerned themselves with serious gaming, while for businesses in the field of entertainment games, research into the effects of their games and user experience would also be very useful. This kind of information could aid in the development of better business models, games, tools, and engines. Close ties do exist between game companies and institutes of higher education. Half of the businesses regularly work together with a university or other institution, interns being an important link between the two. The Netherlands now has forty-four game-related study programs on offer. These, at this time, have over 8,000 registered students jointly.

Of all the game companies, more than 90% have growth ambitions. The vast majority intends to achieve this goal by using their own funds. The amounts available to be invested back into companies are on average not very high, meaning that for many game businesses, matching external funds will remain a necessity. Furthermore, there are chances to arrange clever matches between entrepreneurs who are skilled game developers and others who have a feel for (new) markets and organization.

A quick international comparison shows that in the Netherlands, the video game industry is doing comparatively well.[4] The largest video game industry is in the United States, which directly employs more than 32,000 people. With indirect employment included, the numbers add up to more than 120,000 (Siwek 2010). Canada, in part due to an extensive stimulation program, also has a sizable sector of almost 16,000 game developers. Several countries have a video game industry roughly the same size as that of the Netherlands. The United Kingdom has about the same number of businesses, but there the number of jobs is considerably smaller. The French industry is also similar in size to the Dutch, with around 3,000 jobs. The Scandinavian countries have smaller industries in absolute numbers, but, interestingly, in Sweden, Finland, and Iceland, the number of employees in gaming per 100,000 inhabitants is significantly higher. What is also remarkable is that there are relatively (that is, per 100,000 inhabitants), as many game developers in the Netherlands as there are in Japan, while the latter has a much longer standing in games. What also stands out is that the Dutch industry is not as centered on producing the "classic" games in boxes, but rather focusing on digital distribution and serious games. Because of that, the Dutch sector seems less sensitive to the effects of sales and publishing trends in games. There is also great international potential for upscaling and rollout to new and broader markets.

All in all, the Dutch video game industry is in a good position for further growth. The rise of (mobile) online gaming and the trend of gamification is a clear advantage to the Dutch industry, which is characterized by smaller developers. Internationally, Dutch games are widely praised for their mix of creativity and gameplay. In recent years, investments into research on the field of serious gaming have helped create a video game industry that can operate in both the entertainment and serious games markets. However, serious games are still often bound to a single client and a single market. This means that there are chances for international expansion. Finally, new investments into interdisciplinary research on the effects of games and gameplay could make future commercial propositions stronger with more theoretical backing. This validation would improve serious games propositions and could aid in the development of business models for entertainment games.

The Dutch Video Game Industry in Numbers

In the past five years, the worldwide games industry has seen some considerable growth (an average of 11.2% a year). This strong growth over the last couple years has made gaming an important domain within the content and media industry. It now ranks in size between the music industry (± 39 billion euros) and the film industry (± 66 billion euros) (PwC 2012).

The Netherlands stands out because it has the most active online gaming market in Europe. Gaming is seen as an interesting sector with a lot of potential for growth. The increasing number of students at university games programs is a rich breeding ground for developers and artists. Policy makers have also discovered the sector and its potential. In 2011, the Dutch government started a new policy focusing on nine Top Sectors that were selected for their strong market and export position, knowledge base, intensive collaboration between entrepreneurs and educational institutions, and the potential of offering innovative answers to challenges in society. One of the Top Sectors selected was the creative industry, and within that sector, gaming received specific attention as an area with much innovation potential.[5]

Until recently, much remained unclear, however, about the economic scope of the Dutch games industry, mostly because we are dealing with a dynamic young sector with fast-growing companies. A moving target is hard to aim at. There is also a statistical reason. Gaming, and game development especially, is not clearly defined in the standard classification of economic activities (in the SBI, the Dutch version of the European NACE classification).[6] Because of this lack of distinctiveness, in Roso (2013), the Dutch games industry was approached in a bottom-up manner using company lists owned by the game trade journal *Control* and the Dutch Games Association (DGA).

In order to set game companies apart and to be able to measure their economic scope, we first need a proper definition of what exactly a game company is. The OECD (2011) defines the content and media industry as follows: "content and media industries are engaged in the production, publishing and/or electronic distribution of content products." Gaming clearly fits this description and can, consequently, be seen as a specific branch within the broader media industry, a branch consisting of companies that occupy themselves with the development, production, distribution, and facilitation of electronic games. This definition, then, means that "games industry" includes all companies that have as one of their core activities the development, production, publication, facilitation, and/or distribution of electronic games (see fig. 1). Based upon this definition of the games industry, 330 companies and independent professionals can be identified.

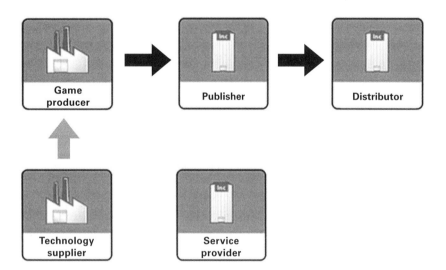

Figure 1
This is the core definition of the games industry. In this chapter, we try to develop a clear definition of the industry. Whenever the chapter mentions businesses in the games industry, we mean companies that develop games that appear in digital form. The parties involved in the production chain are: game producer, game technology supplier, service provider, publisher, and distributor.

Two remarks are of importance here:

- When we refer to games, we are referring to electronic games alone. Companies that are involved in the development and publication of physical-based games such as board games and card games are excluded from this definition.
- The development of games, by this definition, has to be one of the company's core activities. A significant part of the company's turnover (at least a third of the total earnings) should come from the development, production, publication, facilitation, and/or distribution of electronic games. This has consequences for many parties (such as clients and educational institutions) involved in serious gaming. Clients in this field can range from advertising agencies and the Ministry of Defense, to several other public authorities and training agencies. For these companies, gaming is not a core activity, but they do often employ a department or a group of people primarily occupied with serious gaming. The latter has been excluded from the core definition of the games industry.

We would like to emphasize five characteristics of the Dutch games industry here: its business size, the importance of serious gaming, the many team startups, the kind of games that are being developed, and turnover and profits.

Business Size

By 2012, the Dutch games industry consisted of approximately 330 companies (see table 1). These companies together account for 3,000 jobs. The industry's turnover for the year 2011 was estimated at 150 million to 225 million euros. Three quarters of the companies and a little over half the number of jobs fall into the "game

Table 1

Key economic figures, Games Sector, 2011.

Value chain	Turnover (in million euros)	Jobs	Establishments	Jobs per Establishment
Game Producer	80–100	1,590	250	6.4
Game Producer/Publisher	20–30	360	5	89.5
Publisher	20–30	380	30	12.7
Distributor	10–20	200	10	16.8
Technology Supplier	15–30	340	10	38.0
Service Provider	5–15	130	25	5.1
Total	150–225	3,000	330	9.1

Source: TNO, data from *Control Magazine*, LISA, CBS research. Both the number of jobs and establishments are rounded figures, and the column "Jobs per Establishment" is based on absolute figures (see note 2).

producer" or "game developer" category. The average size of a game production company is relatively small compared to businesses in the other categories such as distributors and technology suppliers. Specialized companies in the field of arts and sound design have also been included under game producers.

In terms of employment, publishers and technology suppliers are important categories for the Dutch games industry, supplying 380 and 340 jobs respectively. Examples of such companies are Ubisoft, Playlogic, and UnitedGames. Technology suppliers are relatively large companies, which in 2011 together filled 340 positions. Parties developing technology and tools for the production of games were also included in this category, for example, Vtech, Xsens, and Ex Machina. Distributors and service providers are relatively small in the Netherlands. The group of distributors consists mainly of the Dutch branches of console games publishers such as SEGA and Nintendo. By service providers, we mean those offering specialized services for the production and publication of games (such as localization). This category also includes the specialized press, events, and the Dutch trade organization DGA. Together they employ 130 people, and the companies are relatively small. In game development, there are four Dutch cities that matter: Amsterdam, Hilversum, Utrecht, and Rotterdam.

The Serious Gaming Sector and Entertainment Gaming Sector Are Equal in Size

While entertainment gaming is dominant worldwide, strikingly, the Dutch industry shows an almost fifty-fifty split between entertainment gaming and serious gaming. When looking at the number of companies, serious gaming is slightly ahead. Looking at the number of jobs, entertainment gaming takes the lead due to bigger players such as Spil Games and Guerilla Games. What should be noted is that in the serious games sector, figures concerning the clients and knowledge institutes involved have been left out of the results. They are often large organizations such as the Ministry of Defense and research institute TNO, which have separate departments working on serious games, but the development of games is not a core activity of these companies. Including these clients would give Dutch serious gaming an even larger market share than is presented here.

Few Independent Professionals, Many Team Startups

Looking at the number of companies, the small scale of the Dutch game industry is clearly visible. Figure 2 shows the 330 game companies organized by size. Out of all the companies for which the size is known, 70% have 5 employees or fewer (107 independent professionals and 98 companies of 2–5 employees). Fewer than 100 companies employ 6 people or more, with most of them falling into the medium-sized businesses category of 6–10 or 11–25 employees. Fewer than 10 companies have over 50 employees. Compared to other creative industries, the number of companies with 2–5 employees is large. In other creative industries there is a large number of people who are self-employed (business category 0–1 employees). Due to the multidisciplinary character of making games, game companies often start out with 2–3 people.

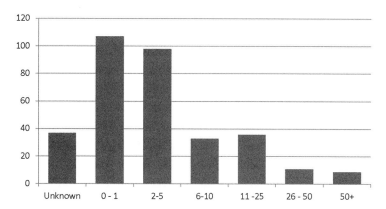

Figure 2
Number of companies in the Dutch games industry in 2011 by business size (number of employees). Source: TNO, data from *Control Magazine*, LISA (see note 2).

Games That Are Developed

Boxed games are becoming "outdated." Only one in five of Dutch gaming companies still produces games sold in boxes. Downloadable games, by contrast, are highly popular, with 60% of gaming companies developing downloadable games for PC and/or consoles. Steam and the Mac App Store are entertainment studios' most popular channels. Serious game developers often use their own websites for the distribution of their games. Out of the other platforms, iOS is by far the most popular: more than 60% of the surveyed companies develop games for it, putting the platform on par with Microsoft Windows. Android follows not far behind, but among developers, Google's mobile operating system is still not nearly as popular as the number of devices sold would suggest: worldwide, in fact, more Android devices than iOS devices were sold. The developers' preference for iOS is attributed to Android's greater diversity of machines and versions (meaning less standardization), Android's users, who are less willing to pay for features than iOS users, and the Android app store, which does not function as well as Apple's.

Developing games is an in-house matter for most studios. A staggering 95% of serious game companies, 87% of entertainment companies, and 74% of mixed companies produce their own games from start to finish. They do make use of freelance professionals, especially for art, audio, music, and animation. What is remarkable is the large percentage of studios also hiring external programmers (42%). A possible explanation for this high percentage could be the small number of programmers available and—partially for exactly that reason—the fact that they are too costly to create full-time positions. Even though there have been many shifts and changes in the industry over the past years, the number of women in the games industry is still very low. On average, 13.5% of game companies' employees (permanent and freelance) are female. For now, the games industry is still very much a man's world.

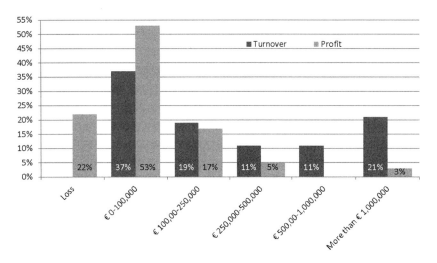

Figure 3
The turnover and profit of Dutch gaming companies in 2011. Source: *Control Magazine* (see note 2).

Turnover and Profits

A fifth of the Dutch gaming companies have an annual turnover of more than a million euros (see fig. 3). Regarding profits, the percentage is much lower: only 3.4% of all the studios make a profit of one million euros or more. Over 90% of the game companies want to grow; most intend to achieve this by their own means. Due to the small number of businesses making a large profit, the addition of external funds remains necessary.

The Dutch Games Industry Ecosystem

The ecosystem surrounding the Dutch games industry is clearly still in its developing stages: the industry is not yet fully grown, but there is great potential. There are many people within the sector who have a strong intrinsic drive; they want to create beautiful products, games that work well and look good. Less prevalent are real entrepreneurs—people from the games industry itself emerging as talented and successful managers and entrepreneurs, and people who see opportunities and build bridges between (new) markets for games and their applications, who can create a stable company structure, arrange financing, and be the "heroes" startups can look up to.

Big differences exist between Dutch game companies. The Netherlands is home to many indie developers, innovative serious game developers, and some larger developers of triple-A entertainment games. The Netherlands accommodates few publishers. Because of this, Dutch game developers have had less chance to connect to the traditional money flows from larger media groups. The rise of casual games for smartphones and tablets has therefore benefited the Dutch market, since their production does not require as much prior investment and lowers the bar for market entrance (Juul 2009).

Internationally, the entertainment games segment has some very strong competition. In the past, Dutch companies delivered a number of successful entertainment games (such as *Overlord I* [2007] and *Overlord II* [2009] from Triumph Studios, as well as *Awesomenauts* [2012] from Ronimo Games), while others developed successful international entertainment game platforms, which have become major online traffic hubs.[7] However, companies themselves say that for more durable success internationally, further growth of the Dutch gaming sector is desirable. Most serious game developers have, so far, exclusively aimed their work at the domestic market, their games and simulations usually being developed one-on-one with a client. Still, developers do see international potential through upscaling and rollout to new and broader markets.

The expansion of the games industry has led to initiatives and partnerships arising to support the growing sector (see fig. 4). Platforms such as industry associations, specialized events, and media have firmly

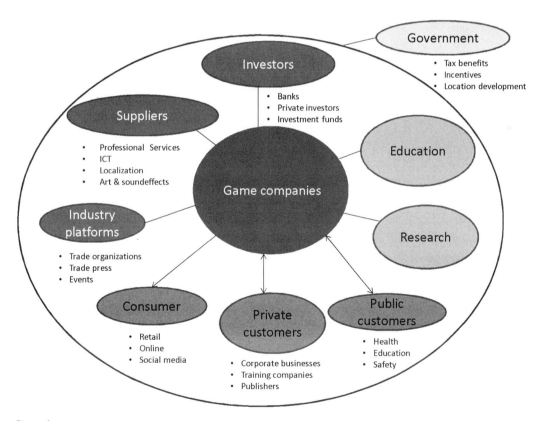

Figure 4
The Dutch games industry ecosystem.

established themselves over the last few years. The industry holds them in high esteem, and overall, they are seen as necessary for continuing growth. The "circle of clients" within the ecosystem is made up of consumers and public and private organizations. Consumers make more and more use of the Internet to find their games, and, as such, traditional retailers are disappearing from the chain; publisher-owned and developer-owned platforms, app stores, game platforms, and social media are the go-to places. Serious games often have public and private organizations as clients. Public parties are also influential in shaping the ecosystem—for example, through fiscal incentives or investing in incubators. Finally, higher education and research institutions also play an important role in the games industry ecosystem.

Companies in the games industry cooperate almost automatically because the process of development, production, and marketing of games requires it. Developing and exploiting games demands a tailor-made combination of design, technology, and commercial insight. Networks are of vital importance to a successful game company. They are built up during one's university years, and through job changes, events, collaborations, and seminars. What stands out is that there is a lot of international collaboration. Foreign partners and networks are especially important for developers and publishers in the entertainment gaming segment, indicating that entertainment gaming is a business that crosses borders. Working together with partners and customers abroad is essential.

Within the Dutch ecosystem, seven aspects are of special importance: the characteristics of Dutch game companies, the process of publishing games, the game companies' degree of organization, financing and crossovers, government influence, education, and research.

Characteristics of Dutch Game Companies

There is one large game developer in the Netherlands, and that is Guerilla Games, owned by Sony. Another big player, Spil Games from Hilversum, is one of the largest online casual game platforms in the world. The Netherlands has a relatively large number of small, independent game studios. This group especially has benefited from the rise of mobile phones as a game and download platform. The games these studios produce are relatively small projects: games (depending on the platform) in a price range of 0.79 to 15 euros. Take the iOS App Store as an example; platform owner Apple takes a 30% cut of sales, and any publisher involved takes a share of whatever amount is left. In a situation like that, with the low prices involved, it is essential to sell large amounts. That is not an easy task, considering the iOS App Store had more than 181,000 games available by the end of 2013 (Jones 2013). Still, several Dutch developers were able, over the last couple years, to produce games that ranked in the top ten of the best-selling games in the store. Recent examples include Game Oven's *Fingle* (2011), Vlambeer's *Ridiculous Fishing* (2013; chosen by Apple as the best game for iPhone in 2013), and Gamistry's *Munch Time* (2013).

Dutch studios are good at creating original games with striking designs. Another chance for success lies in exploiting concepts as part of a follow-up strategy. The biggest mobile game success story in recent history has without a doubt been *Angry Birds* (2009), developed by Finnish company Rovio. *Angry Birds*, as a series, is constantly being added to with new special editions.

Another striking point is that Dutch studios rarely make use of other capitalization strategies. A widely popular way of making money from games is through free-to-play games. Such games are offered to the consumer for free, and he or she can keep on playing them for free or pay small amounts to increase development speed, build up a character, get new content, or reach more levels. An increasing number of games are offered this way. For Dutch studios, this could become a useful area for future development.

The Netherlands has a strong indie scene. Dutch independent games are popular with the international indie game scene, a separate movement within gaming that has been very much on the rise in recent years with an ever-growing group of followers, producers, and players. The indie scene for a long time existed next to the commercial games market, dominated by large publishers and (console) releases with big budgets, and was fiercely against the world of the big bucks and constant sequels. Indies were well aware of the "big" games that appeared, but the average "commercial" gamer hardly knew indie games existed. That situation has changed: consoles are no longer the exclusive terrain of large publishers, and developers and online channels (such as PSN, XBLA, WiiWare, and DsiWare) gave indie developers a place to market their games.

In the Netherlands, serious games are on the rise. The idea that games and game technology can be instrumental in teaching or informing people is becoming more widely accepted. The Netherlands is well represented in the segment; about half of all the Dutch studios develop serious games. In other countries, the emphasis lies much more on entertainment games. Serious games can be used with several aims: training, simulation, healthcare, recruitment, marketing, and education have all been tried and tested. Dutch studios work on all terrains. VSTEP and E-Semble, for instance, create training and simulation games, both companies specializing in disaster and calamity training for emergency services. VSTEP has also taken a step toward consumer games with their *Ship Simulator* (2005) (an extended version was released in 2007, and *Ship Simulator Extremes* in 2010). Advergames are another area of expertise for Dutch studios. Little Chicken Game Company (http://littlechicken.nl) and Sticky Studios (http://www.stickystudios.com) have a strong international reputation in this area; Little Chicken's games include *Aviation Empire* (2013) for KLM, *Road to the Final* (2012) for Heineken, and *Aquafresh* (2012) (about brushing your teeth). Sticky Studios has developed games for brands including Disney, Warner Bros. (for example, *Sherlock Holmes 2: Checkmate* [2011]), McDonald's, DreamWorks (for example, *Madagascar 3: Race Across Europe* [2012]), and Red Bull.

Health care, recruitment, and education are the fields of Ranj Serious Games, IJsfontein, and Grendel Games, among others. Ranj (http://www.ranj.com) is strong in educational games, for example, *Houthoff Buruma—The Game* (2010) for a law firm; *Sharkworld* (2009), about project management; and *Juf-in-a-Box/Teacher-in-a-Box* (2011), which has writing exercises. Games made by IJsfontein (http://www.ijsfontein.nl) include *Peace of Cake?* (2013), about the ethical issues involved in questions of peace and conflict; and health games about ADHD and dementia. Grendel Games (http://www.grendel-games.com) has made health games (such as *Gryphon Rider*, 2013) and games used to train surgeons (for example, *Virtual Endosuite*, 2013). It has also expanded to the field of sustainability with a game about cradle-to-cradle development (the *Cradle to Cradle: Islands* Facebook game, 2013).

Publishing

The process of publishing games has changed dramatically over the last couple years. New distribution channels, the fading influence of retail, and the rise of mobile, social, and free-to-play games force publishers to adapt quickly or perish. The old-fashioned type of publisher, the one making "games in boxes" (the big triple-A games, now starting to become available as downloads), never really existed in the Netherlands. The first and last to make an attempt on a large scale was the Amsterdam publisher Playlogic, and indirectly, Japanese Sony Computer Entertainment, which is present in the Dutch market as the owner and publisher of Guerilla Games, creators of the *Killzone* series. The influence of these large publishers on the Dutch games industry is therefore very small in comparison. The traditional model, which has a publisher take care of the (pre-)financing, distribution, and marketing of the product, has been supplanted by alternative ways of publishing (such as self-publishing) over the last couple years. In the Netherlands, there are about ten small- to medium-sized publishers left that are still more or less traditional in their approach, though even they went through the changes necessary to stay relevant in an ever-developing market with physical sales of games going ever down as the market moves to digital distribution. Digital distribution channels (such as the App Store and Steam) have moved one of the main tasks publishers had—bringing a game to the consumers—firmly into the hands of developers.

Degree of Organization

In the Netherlands, developers and publishers often have the same interests. For this reason, there is just one comprehensive trade organization, called the DGA (Dutch Games Association). The mission of the DGA is to create a healthy ecosystem for the Dutch games industry. The DGA tries to achieve this mission by stimulating entrepreneurship and collaborations, both national and international. The DGA, for instance, facilitates Holland Pavilions at international fairs and expositions. The DGA also has a directing role as a lobby organization. Many (Dutch) game developers and publishers are members of the DGA. On the other hand, local branches and distributors of international publishers are more often connected to NVPI Interactive (the Dutch Association of Producers and Importers of image and sound carriers). NVPI Interactive is concerned with the entire entertainment industry and represents most of the Dutch record companies, video distributors, and game software distributors. Therefore, the NVPI is more focused on the physical sale of games via Dutch retail chains, while the DGA acts as a Dutch trade organization for the entire games industry. Next to the DGA and NVPI Interactive, there are smaller groups that represent a variety of game-related companies and organizations in the Netherlands. They are niche organizations such as DSSH (the Dutch Society for Simulation in Healthcare).

Financing and Crossovers

A challenge for the video game industry in the coming years is to develop a strategic approach to growing. Entrepreneurs prefer to focus on finishing their games; once a studio has managed to actually produce a very successful game, they should be able to make a good follow-up. Being aware of this is a first step toward a solution. It demands the arranging of clever matches between entrepreneurs who are skilled game developers and others who have a feel for (new) markets and organization, but also, importantly, for the process of developing creative products and services. So-called game jams, which match game design students and students from commercially oriented curricula, are already being organized, creating mixed teams able to approach questions from a broader perspective (see http://www.globalgamejam.org). This appears to be a successful approach: game developers get a reality check on the spot, and commercial students develop a feel for the creative process and the possible applications for games.

As of March 2014, a number of steps have been taken to improve the access to finance for the development of games. Several parties are joining efforts to enhance the growth of the Dutch games industry, for example, through the program Growing Games (http://www.growinggames.nl) and the Center for Applied Games (http://www.centerforappliedgames.com).

Government Influence

Governments can support their national games industries in various ways. Subsidizing (cultural) game projects, supporting infrastructure, and creating tax benefits for the games industry are all valid options. In the Netherlands, several public authorities support the games industry in different ways. On a provincial level, Utrecht has been active in giving financial contributions to congresses and toward the organization of the incubator Dutch Game Garden. On a national level, many starting game developers, until 2012, made use of the benefit program for graduates from art academies. With the rise of game curricula outside of art academies and the growing number of games graduates who have no background in the arts, the benefit programs came to be used less for initial startup capital. Moreover, many studios also made use of the benefit from law-enabling income tax reductions for employees involved in research and development. In 2013, the reduction was 38% for turnovers, up to 200,000 euros, and even 50% for starting developers.

What stands out about most countries' contributions to the games industry is that the development of entertainment games is a focal point. In the Netherlands, the relatively large serious games sector makes an even larger contribution to the applied side of the industry. An example would be the Dutch M&ICT (Maatschappelijke Sectoren and ICT, see http://www.m-ict.nl) program, which ended in 2009, working toward an answer to mobility, education, safety, and health care issues in society through the better use of IT. In many other countries, these types of subsidy programs contributing to the development of a serious game on such a scale and so directly are very rare. Game research programs, in many shapes and sizes, are more common, yet there, the Netherlands also stands out with the GATE (Game Research for Training and Entertainment) scheme, which ended in April 2012 (see http://www.gameresearch.nl), and its successor U-GATE, the Utrecht Center for Game Research and Technology (http://www.u-gate.nl). U-GATE is closely linked to Utrecht University's Game Research focus area. The current Dutch subsidy arrangements—through the Top Sector policy—are all focused on (compulsory) collaborations between game studios and research institutes (see note 5).

Education

There are close ties between game companies and educational institutions. Interns are an important link between the two. Many companies and independent entrepreneurs also offer courses and guest lectures, with the aim of making study programs match the reality of the sector. The Netherlands now has forty-four game-related study programs available (full-time vocational training, applied science and university-level programs, plus some minors and extra courses that other curricula offer) in the areas of interaction design, game design, game development, game art design, media technology, and games and play studies (http://www.gamesandplay.nl). These, as of late summer 2013, had over 8,000 registered students between them. The number of game programs and interns has reached a point where companies can pick only the best of the bunch. They do, however, voice their concerns about the size of their field—only 330 companies—and whether it will have the capacity to take on the upcoming waves of game students and graduates. For an increase in absorbency, the growth of existing companies and a reinforcement of entrepreneurial competencies is essential.

Research

In serious gaming, collaborations with centers of knowledge such as universities and research institutes are well established. Government stimulation programs, too, are usually aimed at serious games. Collaborations between serious game studios and research institutes usually center on artificial intelligence, group behavior, behavioral changes, and story-based, persuasive, and learning scenarios. There are possibilities for broader applications of games and simulations in public sectors such as health care, education, and security. Until now, knowledge networks, locally and internationally, were mainly developed around the serious games field. Opportunities, however, also exist for partnerships between entertainment game companies and research institutes. Research into game behavior and the effects of gaming are still in the early stages. One example is the Persuasive Gaming in Context project, a four-year project (2013–2017) funded by the Creative Industry program of NWO, the Netherlands Organisation for Scientific Research, that is concerned with the characteristics, design principles, and effectiveness of persuasive gaming (http://www.persuasivegaming.nl). The results of such projects will help the Dutch game industry in the development of better business models, games, tools, and engines. Further research into the effects of games seems a necessary condition for the development of the industry so that potential clients can better estimate the costs and benefits of investing in gamification and game applications. The same goes for applications with educational and training purposes: the effectiveness of game applications needs to be properly supported to allow a broader acceptance of gaming solutions. Research into the effects of gaming as a communication tool could be a great help in updating business models, both in the fields of serious and entertainment gaming.

Conclusion

The Dutch video game industry is developing very quickly: more companies, more start-ups, more jobs, more successful games for different devices, and more (international) acclaim through awards and so forth. The Dutch gaming ecosystem has the potential to grow into maturity and deliver companies that are both nationally and internationally successful in entertainment gaming and serious gaming. In order to take the next step toward maturity and fulfillment of potential in the sector, a number of necessary tasks need to be accomplished. For the short and medium term, we can think of second-stage support for small- and medium-sized businesses, better access to financing, and more research into the effects of games; in study programs, more attention should go to entrepreneurial competencies and crossovers, and more government support.

Two elements of the ecosystem seem to be worth mentioning separately. First, we have already referred to the fact that Dutch gaming companies are well represented in the segment of serious gaming, and that, especially in this segment, collaborations with centers of knowledge such as universities and research institutes are already well established. In April 2013, the Dutch Organization for Scientific Research (NWO) announced that twenty-one research applications were approved (with an investment of 7.7 million euros) that involved cooperation between universities and creative industries companies (see http://www.nwo.nl). What stood out was that in nine of the twenty-one research applications, games or gamification were the objects of research, and seven of the nine game-related research applications focused on serious gaming. We can, therefore, conclude that (research in the area of) serious gaming plays a crucial role in the Dutch gaming landscape.[8]

The second element is the trend of gamification, or the use of game-design elements in nongame contexts (Deterding, Dixon, Khaled, and Nacke 2011; Fuchs, Fizek, Ruffino, and Schrape 2014). Gamification is spreading, both as a design paradigm in systems for human engagement and as a (buzz-abundant) field with its own evangelists, conferences, and revenue streams. Gamification allows existing businesses to incorporate game design elements into their overall strategy for tasks as varied as improving customer satisfaction and maintaining high in-company productivity. Many Dutch gaming companies find themselves in the unique position of being able to use their know-how in combining the fun and playful elements of games with purposeful content for the benefit of gamification, rather than full-fledged development. Analysts predict that by 2014, 70% of the top 2,000 global companies will have used at least one gamified process; by 2015, 25% of global businesses will employ gamification strategies, a trend that will result in a USD $2.8 billion industry by 2016. Both gamification and the use of serious games can be seen as innovators for different societal challenges. Meanwhile, the consumer market for entertainment games continues to grow as more people buy smartphones and tablets. Dutch game companies, because of their focus and experience, are in a unique position to seize these opportunities and fulfill their potential.

Acknowledgments

In 2012, in cooperation with the Task Force Innovation (TFI) Utrecht region, Dutch Games Association, Dutch Game Garden, iMMovator, *Control Magazine*, TNO, and Monpellier Venture, valuable insights into the Dutch game industry were collected through different means of research (interviews, a survey, quantitative research, desk research, and roundtable discussions). All these insights were gathered first in 2012 in the

Dutch publication *Gamesmonitor 2012* and in 2013 in the English publication *The Dutch Games Industry: Facts and Figures* (Roso 2013). Without this joint effort, this chapter could not have been written. The authors would like to thank the contributors to these publications: Olaf Koops and Thomas Bachet (TNO), Matthijs Dierckx and Eric Bartelson (*Control Magazine*), Monique Roso and Evelien Boshove (TFI), and Jurriaan van Rijswijk and Alex Gekker (Monpellier Venture).

Notes

1. The terms "serious games" and "applied games" are both used for games that have a purpose that supersedes entertainment. In the Netherlands, the term "applied games" is more commonly used. In this chapter we have chosen to use the term "serious games," which is more common internationally (see Raessens 2009, 2010).

2. In the second half of 2012, a number of Dutch partners combined efforts and several methods of research to attain an overview of the Dutch games industry. This chapter is composed, for the most part, of the different sections of the overview *The Dutch Games Industry: Facts and Figures* (Roso 2013). In the appendix of that publication, the sources, approach, and research methods are explained in detail (see pp. 30–31).

3. See: http://www.dutchgamesassociation.nl; http://www.gamesforhealtheurope.org; http://www.festivalofgames.nl; http://control-online.nl; http://www.dutchgamegarden.nl; and http://www.clicknl.nl/games.

4. In *The Dutch Games Industry: Facts and Figures* (Roso 2013), two chapters focus more in detail on such an international comparison (chaps. 5–6, pp. 20–30). In this chapter, we only refer to other countries to better understand the Dutch situation.

5. The Dutch Creative Industries knowledge and innovation network is called CLICKNL (see http://www.clicknl.nl). CLICKNL GAMES is the gaming part of this network (see http://www.clicknl.nl/games).

6. NACE refers to the statistical classification of economic activities in the European community (the term NACE is derived from the French), and SBI is the Dutch version of it (*Standaard Bedrijfsindeling*).

7. This chapter is not a summary of all the quality games developed by Dutch companies; for a broader overview, see *Dutch Games Go Global*, 2012 edition (http://www.dutchgamegarden.nl/blog/detail/2012/08/07/dutch-games-go-global-english, and Control Magazine 2012). For a selection of games by Dutch independent developers, see http://www.dutchgamegarden.nl/indigo/archief. And for some recent quality games, we direct you to the Dutch Game Awards (http://www.dutchgameawards.nl).

8. See the *Control Magazine* special issue on serious games (edition 2013–2014). For an overview of these twenty-one projects, see http://www.nwo.nl/en/research-and-results/programmes/creative+industry/Funding+allocation.

References

Control Magazine. 2012. Dutch games go global 2012. *ISSUU*. http://issuu.com/controlmagazine/docs/dggg_2012_def_screen.

Control Magazine. 2013. *Serious Games—Information for businesses and governmental organisations, edition 2013-2014*. *ISSUU*. http://issuu.com/controlmagazine/docs/serious_games_2013_by_control_magaz.

Deterding, Sebastian, Dan Dixon, Rilla Khaled, and Lennart Nacke. 2011. From Game design elements to gamefulness: Defining "gamification." In *Proceedings of the 15th International Academic MindTrek Conference: Envisioning Future Media Environments*, 9–15. New York: ACM Press.

Fuchs, Mathias, Sonia Fizek, Paolo Ruffino, and Niklas Schrape, eds. 2014. *Rethinking Gamification*. Lüneburg: Meson Press.

iMMovator. 2010. *Game industrie in de Noordvleugel en Nederland (Speciale bijlage bij de Cross Media Monitor 2010)*. Hilversum: iMMovator Cross Media Network, http://files.goc.nl/files/pdf/Gaming/immovator_game_monitor_2010.pdf.

Jones, Chuck. 2013. Apple's App store about to hit 1 million apps. *Forbes*, December 11. http://www.forbes.com/sites/chuckjones/2013/12/11/apples-app-store-about-to-hit-1-million-apps/.

Juul, Jesper. 2009. *A Casual Revolution: Reinventing Video Games and Their Players*. Cambridge, MA: MIT Press.

OECD. 2011. Guide to measuring the information society. *Organisation for Economic Co-operation and Development*. http://www.oecd.org/internet/ieconomy/oecdguidetomeasuringtheinformationsociety2011.htm.

PwC. 2012. Global entertainment and media outlook: 2012–2016. http://www.careercatalysts.com/pdf/PwCOutlook2012-Industry%20overview%20(3).pdf.

Raessens, Joost. 2009. The gaming *dispositif*: An analysis of serious games from a humanities perspective. In *Serious Games: Mechanisms and Effects*, ed. Ute Ritterfeld, Michael Cody, and Peter Vorderer, 486–512. New York: Routledge.

Raessens, Joost. 2010. A taste of life as a refugee: How serious games frame refugee issues. In *Changes in Museum Practice*, ed. Hanne-Lovise Skartveit and Katherine Goodnow, 94–105. New York: Berghahn Books.

Roso, Monique, ed. 2013. *The Dutch Games Industry: Facts and Figures*. Utrecht: Taskforce Innovation Utrecht Region. http://www.economicboardutrecht.nl/over-ebu/downloads/gamesmonitor.

Siwek, Stephen E. 2010. Video games in the 21st century: The 2010 report. *Entertainment Software Association (ESA)*. http://www.theesa.com/facts/pdfs/videogames21stcentury_2010.pdf.

NEW ZEALAND

Melanie Swalwell

This chapter focuses on the history of video games in New Zealand during the "long" 1980s. Beginning in the late 1970s and extending well into the 1980s, New Zealand had a booming digital games industry across the spectrum of arcade, console, handheld, and microcomputer games. This is a largely unknown history. Apart from my own research on this topic, written accounts have largely been compiled by private collectors, who continue to deepen understanding of the specificities of particular games software and hardware (Davidson n.d.; Lord n.d.; "SegaSC-3000 Survivors" n.d.; Stephen 2012). Much of the chapter focuses on the domestic production of digital games. Also covered are the local reception of imported computers and games, the articulations between New Zealand history and policy around video games, and some of the anomalies of New Zealand's digital game consumption in the 1990s. Since this early era—the focus of this chapter—New Zealand game development has continued to grow with the changing technologies and expectations of developing games in a global environment. The New Zealand Game Developers Association was founded in 2001 (NZGDA n.d.), and during the 2000s, game development studios worked on contracts for major game companies overseas. The nation's largest development studio, Sidhe Interactive, undertook an impressive suite of contract work as well as multiple games based on the NRL Rugby League franchise. Much of the contract development work dried up when the global financial crisis hit. Following this, the industry has refocused on games for mobile devices.

This chapter draws on in-depth research into the local history of digital games in 1980s New Zealand. A range of methods has been used to inquire into the beginnings of the digital games era in this specific locale, including: consulting archival records (ephemera, companies' databases, government reports, and archives); textual research (locating primary sources such as newspaper and magazine accounts, and documentary photographs); and interviews with key informants (those involved with game production and distribution, arcade operators, players, home computer users, journalists, game authors, and collectors). I have published a number of pieces on this history, including the interactive scholarly essay "Cast-offs from the Golden Age" (Swalwell and Loyer 2006), several articles and chapters (Swalwell 2007, 2009), and an exhibition of historic photographs, "More Than a Craze" (Swalwell and Bayly 2010). Despite researching this project over a number of years, gaps in knowledge remain, and some questions are unanswered: records might not have been kept or knowledge might not have been shared. Nevertheless, the outlines of a local history of games can be discerned from the fragments.

Local and Global

New Zealand game culture was a complex meld of local and international influences during the 1980s. Though it can be tempting to look back to a "golden age" of local production, such a perspective can mask more complex intersections between local and nonlocal—or global—factors. Mixtures between local and nonlocal influences, products, and experiences are scattered throughout this history as, despite the country's geographic distance (and policies of the era, which had the effect of isolating New Zealand from much of the rest of the world), there is clear evidence of ongoing connections with the "outside."

Arcade Games — Or "Spacies" — Production

On the basis of my research, as many as fifteen companies were involved—at one time or another and to a greater or lesser extent—in arcade game manufacture in New Zealand (Aaron's Amusement Games, Advance Automatics, Century 21, Century Electronics, Chastronics, Coin Cascade, Kitset Electronics [Kitronix], Rait Electronics, Silverdale Investments, SpaceTec, Taito [or Taito-tronics], Video Vend Automatics, Videotronics, Videotech, Videoworld). Chastronics and Kitronix are two important early companies. The former was involved very early in arcade manufacturing, from the late 1970s, as I understand it, as well as having an interest in several Christchurch arcades, a model of vertical integration that a number of other companies would later adopt. Unlike its competitor, Kitronix didn't run arcades but did write its own games. In fact, Kitronix produced fully locally made arcade games. The Kitronix games *Malzak* (1981) and *Panix* (not yet dated) initially appear to be "clones" of the well-known games *Scramble* (1981) and *Space Panic* (1980) respectively. However, when the ROMs of these games were dumped to MAME (Multiple Arcade Machine Emulator) and analyzed, it was found that their code is unlike any other known game, suggesting that while visual design and gameplay elements may have been borrowed from these games, they were in fact written locally from scratch. *Malzak* is notable because everything about it is local: the coding, artwork, buttons, and joysticks were all produced locally. Some Kitronix machines are believed to have been exported to Australia. This unique piece of gaming history is profiled in detail elsewhere (Swalwell and Davidson forthcoming).

The involvement of so many companies in the early arcade manufacturing business does suggest the "why" question: why were so many companies in such a small country eager to make arcade game machines, and what caused such a number of local entrepreneurs to try their hand in the games industry?

Import Licensing

New Zealand's system of import licensing seems to be an important piece of the puzzle. Import control regulations were initially introduced in 1938. These regulations, which were then intended to control expenditure and help build up secondary industries, made it illegal to import any goods except under the authority of a license granted by the customs department. They were suspended for a time before being reintroduced in 1958 as part of the Labour Government's "Black Budget." The late New Zealand historian, Michael King, writes that it was the "temporary drop in overseas reserves" that triggered this decision and brought the "highly unpopular reimposition of severe import controls" (King 2004, 433). The difficult economic situation New Zealand faced in the 1980s brought trade barriers down very suddenly, and the import licensing system was dismantled progressively but quite quickly during the middle of the decade. It's difficult to find information on the dismantling of the scheme, but arcade operators have told me that they were able to bring in second-hand games by the latter part of the decade (1987 or 1988). By contrast, computers could be imported without containing any local content. It's not clear to me why this was, though a 1993 paper reflecting on New Zealand's IT policy suggests that computers might have been subject to different regulations, specifically, tariffs:

> Until the mid-1980s, economic policy was inward oriented, marked by high trade barriers and heavy government regulation. In 1984, the Labour government responded to a balance of payments crisis with a radical program of economic liberalization. By the end of the decade, this process had fundamentally altered the New Zealand economy through deregulation, privatization, and public sector reform. IT policy likewise moved from protectionism and centralized control to almost pure laissez faire. Tariffs on computer hardware were lowered from 40% to 10%. (Kraemer and Dedrick 1993, "Executive Summary")

In the arcade industry, the import licensing system meant that only those who had an import license were able to bring game machines into the country (legally, at least). These licenses were difficult to come by, which meant that those who had them were in a privileged position, often making a lot of money. As well as selling machines to other arcade operators, a number of import license holders seem to have run arcades themselves (they would have had a significant advantage over their competitors, being able to secure machines for their own arcades at cost).

Importers often had to show, as a condition of their license, that they were doing a certain amount of the manufacturing locally. This requirement, coupled with the large size of arcade game machines, would have motivated importers to begin building some of the cabinets locally, as shipping preassembled cabinets to New Zealand would have been very expensive. Importing the boards and building cabinets in which to house them, locally, would have been far more economical. Tax concessions were also available for goods that were (at least partly) manufactured locally. Collector of New Zealand games, Michael Davidson, attributes the number of businesses involved in arcade machine manufacture to "the cost of importing them. Can we do it cheaper ourselves? I think it all changed when they got cheaper to import" (private communication). Some arcade owners no doubt became involved in making their own games in an attempt to cut out such "middlemen."

What this means is that a number of companies (Chastronics, Coin Cascade, and many more) imported, produced and/or operated arcade machines that—at least until the end of the licensing regime in the mid-1980s—were locally manufactured, at least to some extent. This localization means that these games often look quite different from international releases. Figure 1, of a Chastronics *Black Dragon* (release year unknown) cabinet, provides an example. Not only does the marquee header feature the Chastronics logo and use artwork that was locally sourced, the area behind the screen also sports a quite remarkable advertisement for Chastronics:

> This machine is manufactured in Ch.Ch., by the Chaston Trading Co. Ltd under the brand name Chastronics. For further information Ph. (03) 50303. The best 20c of fun you'll ever get!

Chastronics' graphic panels were printed by the Christchurch company, Screen Sign Arts, which used to come up with the visual concepts around the games and screen print them onto Perspex. They reported that Chris Chaston, the owner of Chastronics, would typically deal in volumes of between 200 and 300 of each game title; these were shipped around the country and also, it is believed, to Australia. Chaston would change a game's graphics if its "take" was slowing down. At one point there were two printers working full time on the Chastronics account.

Game titles were sometimes licensed from well-known international game distributors; however, bootleg boards appear to have been reasonably common. A number of my informants recall seeing warning screens that indicated that region specificity was being ignored. For instance, according to Stephen McGregor, "It's quite common to see an arcade game saying 'Stop! If you are playing this game outside of Japan, you are breaking the law.' That is quite a regular thing, you still see that now on video games" (pers. comm.). Another reason that accounts for the degree of business interest in arcade games is the fact that arcade games were hugely popular in 1980s New Zealand (Rihari 2013; van Rooyen-McLeay 1985). Arcade machines were supposed to have been extraordinarily profitable, provided (I am reliably informed) the location and signage were right. That ex-operators for the most part remain tight-lipped about the profits they made means these are still the stuff of myth. I have heard stories of an operator's vehicle that literally groaned under the weight of coinage.

In Christchurch, readers of the *Press* newspaper got glimpses of what was at stake as the operators of two rival arcades—Wizards and The Doghouse—fought a very public legal battle over market share when Wizards sought to open an arcade in town. The plaintiff was Patrick David Sloan, operator of the fast food bar and arcade The Doghouse and of an unspecified number of "rounds" of amusement machines, originally through Chaston Electronics, one of several companies in which Christopher Chaston had an interest. In 1979, Taito's Australian subsidiary became interested in Chaston Engineering, allegedly because it wanted to obtain the skills and knowledge of Andrew Nobbs, a shareholder in the company. Chaston Engineering's takeover by Taito brought with it a change of name, first to Electronic Amusement Services Ltd in 1980, then to Taito (NZ) Ltd. Both Chaston and Sloan had concerns about Taito's possible movement into operating arcade machines in Christchurch, but they claimed to have been reassured that it was not Taito's intention to "mov[e] into the operational field" ("'Space invaders war' reaches High Court" 1982). Sloan remained a substantial customer of

Figure 1
A Chastronics *Black Dragon* arcade game cabinet, found in Greytown, New Zealand, in 2005. (Photo courtesy of Andreas Lange)

Taito's but in mid-1981 had a "liquidity problem" and found himself unable to pay Taito all that he owed (he paid NZD $30,000 of $90,000). During previous negotiations with Taito in 1980 to buy his round of machines, Sloan had shared commercial information with Taito:

> that was important commercial information because it revealed where the machines were installed and which sites returned the most revenue. Because of that concern Mr Sloan obtained an undertaking from Mr Nobbs that if the negotiations fell through Taito would not use the information which would enable it to 'pick the eyes out' of his business. (ibid.)

When Sloan could not pay Taito what he owed in 1981, he suggested they take over some of his machines in payment for the debt. "The evidence would be that Taito did not obtain the sites from Mr Sloan but took two of them over behind his back and attempted to get the third. To top it all off Taito issued a writ against Mr Sloan for $60,000" ("'Space invaders war' reaches High Court" 1982). In his new premises, Sloan had space for fourteen machines but lacked capital to purchase them, so a revenue-sharing arrangement was allegedly made with Taito where machines would be installed at Farmer Johns and "all other Taito machines [would] be moved to Auckland where Taito was involved on a substantial scale." When Sloan learned of Taito's plans to open a "large amusement emporium ... contain[ing] 60 machines, dodgem cars, and a fast food operation," he initiated proceedings against the Taito group of companies ("'Space invaders war' reaches High Court" 1982). Sloan alleged that the "defendants had breached a contractual promise they made not to work in Christchurch in competition with the plaintiff" ("Games centre delay," 1982). He further alleged "conspiracy, negligence and unlawful interference with his business" ("'Space invaders war' reaches High Court" 1982). Wizards was subject to a High Court injunction while the case was heard ("Games centre delay" 1982). Sloan ultimately failed in his attempt to prevent Wizards from opening and was expected to face a substantial damages claim. The interim injunction—granted on March 24, 1982—was lifted on July 23, 1982, meaning the legal wrangling had delayed the center's opening by four months ("Wizards may get large damages" 1982).

Arcade Culture

Writer and pop culture commentator Kerry Buchanan writes:

> In my home city of Auckland, as in many other areas, video games had a huge impact on youth culture. The spectacle and aura of the experience often influenced the way we saw life. They brought a sense of sophisticated knowledge and an acceptance of technology. Playing these games somehow linked us to a global youth culture, much in the same way that rock 'n' roll and movies formed a conduit to alternative cultures in previous generations. (Buchanan 2004, 142)

Photographic images have been an important part of my research into this local history of games. Many of the images can be viewed in the online exhibition that Janet Bayly and I curated, "More Than a Craze: Photographs of New Zealand's Early Digital Games Scene" (Swalwell and Bayly 2010). A majority of the images are taken inside arcades, allowing us to revisit this space and to remember what it was like. A number of photographs record the spatial layouts of early arcades, demonstrating the great breadth of environments in which people played. Clearly, some were more salubrious than others. The photographs show us what some of the environments were like—from the futuristic-themed to the grimy—and the other items games shared space with (cigarette machines, pool tables, etc.). Of course, "Spacies" were not only found in arcades but also in burger bars, swimming pools, and dairies (corner stores), with the sites usually receiving a commission of the money spent on the machines (40% is mentioned in one reference ["Games cause 'war'" 1981]). There were often ritualistic elements involved in these shared use spaces, such as spending the change from the fish and chips on the Spacies machines.

Beyond their appearance, the arcade was an important social space where kids could hang out, meaning it was a venue where kids usually went without adult accompaniment. In many of Ans Westra's photographs, for instance, we see groups of people we assume to be friends, simply hanging out. Others are portraits of individuals playing, flirting, or simply relaxing. Showing some of these photographs to McGregor, he commented that game arcades were like a kind of kids' club or bar:

> [The photograph has] really captured ... the social space that developed around [the games]. You'd see guys ... 14-year-olds would hang out with their 13-year-old girlfriends, and it was just a space and you could just go and do stuff. Today ... there's not for young kids of this age especially, and older kids, there is not really modes to like cruise, where you can just hang out with other kids. You can go to school, but this is a good place where you can just hang out. You can just ... yeah, it's just a good place to hang out, it is just a good place to cruise, I think.

He added, "This is nice, the combination with pool, it's like a kid's bar. The combination of pool and these video games" (pers. comm.).

Another informant recalls that in smaller suburban settings, the proprietors of the dairy where the games were located got to know all the kids who would go after school. The kids used to get in trouble for leaving their bikes in disarray out front, but there was also a sense that this was a kind of extended community space, with the owners playing the role of informal youth counselors (pers. comm.).

Attempts at Regulation

Arcades frequently attracted complaints from neighboring businesses, and there were concerns about "delinquents" and students who should have been at school. Disputes over these literal space invaders were fought in a number of locales. In Dunedin, the owners and operators of the Knightrider video parlor on Lower Stuart Street defended themselves, saying, "They [were] not to blame for the bad behavior of young people gathering in the area." They said it was "poor family life and the lack of an adequate central city drop-in centre which [was] causing the problems." They claimed to "[work] closely with the police and youth workers, and a social lounge had recently been opened upstairs for young people to use instead of hanging around on the footpath" ("Video parlour 'not to blame'" 1985). Colin and Beverley Holden, proprietors of a Riccarton dairy who had installed two arcade machines in a lean-to, had a long-running dispute with the Riccarton Borough Council over whether or not they were on dairy premises ("Games cause 'war'" 1981). Similar cases kept councils animated across the country (as in "Legal action stopped" 1983 and "Te Atatu parlour man's challenge" 1983). In Auckland, the city amended its Code of Ordinances in 1983, effectively making "Amusement Galleries" a conditional use in a number of zoning classifications so it could impose conditions:

> 3.37.2 Amusement Galleries: Amusement galleries are commonly permitted in commercial zones. They have characteristics which in a suburban context can generate special problems—noise, increased traffic, heavy parking demand, late hours and obtrusive advertising. Therefore they are classed as conditional uses in most suburban commercial zones so that conditions can be imposed.

Ellerslie Borough Council in Auckland went further, preparing a bylaw and licensing scheme whereby amusement galleries would need to be licensed, complying with a range of minimum standards on sanitation, building safety, sound-proofing, space (three square meters per "coin-in-the-slot machine"), bike stands (one for every three "coin-in-the-slot machines"), and a range of other requirements. Inspectors were to be appointed, and managers made personally responsible for the conduct of patrons: "the manager shall ... himself see that all the provisions or requirements of this Bylaw are duly carried out and observed otherwise he shall be guilty of an offence against this Bylaw" (Ellerslie Borough Council n.d.).

Consoles and Handhelds

A number of consoles and handheld game systems were manufactured locally in New Zealand in the 1980s. These include a range of early *PONG* clones (including the Sportronic [Brown 2003], and the David Reid Electronics' Vidplay), as well as consoles with cartridges such as the Fountain TV Video Game System (rereleased as the Fountain Force 2), and the Tunix. The Tunix console appears to have been the product of an early licensing deal. I have a cartridge for the game *Space Attack* (n.d.) (in a *Jump Bug* [n.d.] box), which clearly has "Made in New Zealand" on the box and "Printed in New Zealand" on the instructions. Nevertheless, this system is known in Canada, for instance, as the Leisure Vision (CongoBongo n.d.). Images of a number of these systems and their game titles are available at Davidson's *Obscure Pixels* website (Davidson n.d.). For some of the companies involved in the manufacture and sale of console (and arcade games), video games were just one of the various electronics design and manufacturing enterprises in which they were involved. Fountain was a well-known electronics company, producing record players, stereo systems, clock radios, and even microwave ovens. David Reid was the name of an electronics store. As far as arcade machines went, Rait made audio amplifiers while Kitronix's products included heat sinks, guitar amplifiers, and printed circuit boards.

Apart from the locally made consoles, fully imported systems such as the Atari VCS 2600 were also available. A sales catalog I obtained shows the Atari and Fountain systems for sale on the same page (the Atari sold for $399, the Fountain for $249) (see fig. 2). One can just imagine kids arguing with their parents about the perceived benefits of the imported system over the local, as parents (perhaps) tried to understand and justify

Figure 2
Advertisements for the Atari VCS 2600 and the Fountain Force 2 home video game systems (left), and an advertisement for Grandstand's *Invaders from Space* and *Astro Wars*. (The image of Atari VCS/2600 and the Fountain Force 2 is used with permission from Smith's City, image courtesy of the Christchurch City Library, Ephemera collection. The image of Grandstand's *Invaders from Space* and *Astro Wars* is used with permission from Toyworld NZ, image courtesy of the Alexander Turnbull Library, Wellington, NZ. Ephemera collection reference Eph-A-TOYS-1982-01-048.)

the extra outlay. Global distribution anomalies appear to explain the fact that some widely available consoles such as the ColecoVision, Intellivison, and the Vectrex did not make it to New Zealand. The collector Michael Davidson writes that he has never seen or found any of these systems in New Zealand (Davidson, *Obscure Pixels*, "Finds & Brags" n.d.). Reflecting further, he says,

> Coleco and Intellivision were both for sale in Australia. They may have been picked up by an Australian company that never expanded to New Zealand. The gaming industry is full of that sort of thing—vapourware. I wonder also how much of it was to do with how much of a margin you could make on [these consoles]. These ['local'] consoles where you went to Taiwan and got them to knock them up real cheap … and bring them back here and make double your money, sort of thing, they probably were a bit more cost-effective than trying to sign a deal with some multi-huge corporation. (pers. comm.)

Grandstand was a significant player in the local handheld (and later, the computer) market. The company was set up in Auckland by Grandstand UK's émigré cofounder, Les Kenyon. Grandstand NZ initially worked with the New Zealand toy importer R. R. (Bill) Fenton to sell the *Astro Wars*, *Firefox*, *Dracula*, *Scramble* (all circa 1982), and other tabletop games (see fig. 2). Fenton had an import license, and they decided to join forces using Kenyon's contacts in Hong Kong to broaden Fenton's toy business and bring in consumer electronics. However, at the time, they did not fully realize that New Zealand's import licensing regulations would mean they would be doing a lot more than just sending to Hong Kong for shipments. They were able to import the chips but then had to set up a production line to make the power supplies, solder components together, and assemble the units. This was done in a warehouse at the bottom of the Great South Road in Auckland. The plastics were all shot locally in plastics factories. These arrangements brought the local content in these games up to the required threshold of (roughly) 30%. Jaro Gielens's book *Electronic Plastic* (2000) has a map depicting the sites of handheld game manufacture. According to this reference, New Zealand is just another blue country on the edge of the map in which no handheld game development allegedly took place. This is incorrect. While there are no accurate numbers available, the handheld games were very, very successful for the size of the country, which at the time had a population of about 3 million people. As well as shipping large numbers of these portable games around New Zealand, Grandstand exported the handhelds to Australia through Fenton's toy industry contacts, supplying Toyworld and Kmart stores (private communication).

Microcomputers

The reception of (mostly foreign-made) 8-bit microcomputers was strong, and a vibrant local culture of game programming and production existed. Games were a key reason for people purchasing a microcomputer (Swalwell 2012a). Most of the big name systems associated with early games made it to New Zealand. These include Atari (2600, 7800, 400/800, XE); Apple; BBC; Sinclair (ZX81, Z88, ZX, Spectrum); Commodore (including Vic 20, C64, and Amiga); Radio Shack TRS-80 and its Australian clone, the System 80; SEGA SC3000; and, of course, IBM PCs. Lesser-known locally made microcomputers are the Aamber Pegasus (1981) and the Poly, both apparently made for the emerging school market. According to New Zealand vintage computer collector Philip Lord, the Aamber Pegasus was NZ's first kit computer (Lord n.d.). People were also known to assemble their own computers out of bits and pieces (see, for instance, Breen and Arrow cited in Swalwell 2012a). Some brought their computers back from overseas trips, such as Harvey Kong Tin, who brought his Atari 800 back to New Zealand when he returned from a stint living in the United Kingdom (Kong Tin 1985).

The Question of Lag

One of the questions I have looked into concerns whether there was a lag in games and associated products reaching New Zealand. There is a reasonably widespread assumption that New Zealand lags behind the rest of the world when it comes to technology. This is slightly paradoxical, given that New Zealanders also have a reputation for being "early adopters." There is, however, some truth to the perception of lag in terms of anomalies in the way that early games arrived in New Zealand. While it is hard to determine with any accuracy the dates that these systems were released locally, the general sense I have gotten from conducting archival research is that New Zealand releases were usually a year or two later than in major markets (shipping and import restrictions were two factors that may have caused delays in the early years).[1] The ZX81 provides an example here: originally released in the UK in 1981, it is advertised in the November 1983 issue of the New Zealand computer magazine *Computer Input* for $199; by Christmas 1984, David Reid Electronics was selling them for $99, "under half price." (Of course, this doesn't mean it wasn't for sale in 1981; I just haven't been able to find evidence that it was.) Other anomalies relate to which games systems actually made it into the country. As noted above, there were uneven patterns of distribution of early consoles. As far as arcade machines go, Cher Reynolds, ex-sales and marketing manager for Taito in Christchurch, commented,

> New Zealand, in my experience, has always tried to be at the forefront of new technology. Whether it's readily adopted or not is another story, but people are interested in these things. There were game shows that would regularly happen. You can find out about anything that's happening now just by getting on the Internet, but at that time you had to really go along to some kind of distributor's show to find out what was going. So we were picking up things just as fast, sometimes faster, in my opinion, than the Americans were. I can't comment on the UK, but I can comment on the west coast of the United States, because I would regularly go back there. (pers. comm.)

Microcomputers and Games

Software was written both by professionals and hobbyists, and there was a strong culture of "having a go," supported by indigenous and imported magazines, books, computer fairs, and local user groups. I have researched the user group and magazine culture in some detail and published the findings elsewhere (Swalwell 2010). The early SEGA computers, particularly the SC-3000, developed a wide and loyal following in New Zealand. There were many user groups as well as locally published magazines: *Sega Computer* was devoted entirely to the system; others, such as *Bits and Bytes*, had regular columns for different computers including the SEGA SC-3000 (Swalwell 2010). Many user groups existed and published newsletters, which have been scanned and assembled by collector Aaron Wheeler. New Zealand-written software for the system also proliferated, with a good proportion being games (Wheeler and Davidson 2008). Some of this software was published by SEGA's New Zealand distributor, Grandstand, to ensure that customers would have some software to use on this system in order to maximize sales. But a significant amount was written by users and other software developers, some of whom were professionals. The Sega Survivors collective has tracked down and interviewed some of these programmers ("SegaSC-3000 Survivors" n.d.).

While many hobbyists wrote games for fun (Swalwell 2008), some in this homebrew scene had their titles published commercially. Some were written by schoolchildren: John Perry's game *City Lander* (1984), for instance, was published by Grandstand when Perry was thirteen years old. Mark Sibly wrote and marketed his game *Dinky Kong* (1984) for the Vic-20 while he was still at school, under the name Perspective Software. As the name suggests, these titles are often "clones" of well-known games. (Sibly would go on to publish a number of

Amiga games in the late 1980s and into the 1990s and develop the *Blitz* series of games software.) Kong Tin and Andrew Bradfield developed two games together for the Atari, *Laserhawk* (1986) and *Hawkquest* (1989). These, like some of Sibly and Simon Armstrong's Amiga games (*Skidmarks* [1993], *Super Skidmarks* [1995], and *Gloom* [1995]), were published in the UK. Armstrong also published a number of games in New Zealand under the label Acid Software. While the barriers to entry in the software market for 8-bit computers were low, there was also little in the way of centralized support. As Kraemer and Dedrick observed in 1993, "The government has refused to provide any significant incentives or subsidies to the fledgling software industry, feeling the industry should succeed or fail on its own" (Kraemer and Dedrick 1993).

The End of an Era

The end of import restrictions, coupled with the demise or takeover of several influential game companies in the stock market crash of 1987, changed the game production business from one involving significant amounts of local production using some imported components to a more thoroughly global one, with offshore production facilities turning out fully assembled items. By the late 1980s, the era of local hardware production was over. The 1980s was the era when New Zealand had a home-grown hardware and software production scene. Once microcomputers took off, the low barriers to entry in making games for micros meant that there was a significant amount of locally written game software, persisting at least into the early 1990s. While there are, of course, local histories to be written after this point, global distribution models were in the ascendancy.

There are some noteworthy anomalies of reception of imported products in the ensuing years. New Zealand game consumers bought SEGA consoles in much greater numbers than Nintendo consoles in the early 1990s. The SEGA Master System enjoyed outstanding success, well and truly outselling the Nintendo Entertainment System (NES), defying sales trends elsewhere, including in neighboring Australia (Johnson 1993). SEGA's New Zealand representative, John Peppiatt, was quoted saying, "Nintendo is not a rival here" (D. Wilson 1993). According to Allen McCormick, then marketing manager at Roadshow Entertainment (NZ) and in charge of Nintendo's marketing in New Zealand,

> [Nintendo] was very much the number two brand after SEGA, and SEGA had done a very good job of establishing themselves. My job was essentially to try to catch up and replicate what had happened in Australia, where, in fact, Nintendo was the market leader and Sega was chasing hard to catch up. (pers. comm.)

It didn't go as well as McCormick had planned, as he explained: "Essentially, we had a flawed marketing strategy. We were number two and we were pricing up, with the only rationale being a reputed technical advantage, which really didn't justify the price premium" (private communication). While McCormick puts this down to SEGA's earlier release and superior marketing campaigns, one wonders whether the brands' respective 1980s introductions to the New Zealand market might also have partially shaped the reception of their consoles in the following decade.[2]

Research is currently being undertaken into the history and preservation needs of locally written computer games software from 1980s New Zealand and Australia. The "Play It Again" project is working with a range of institutional partners, including Ngā Taonga Sound & Vision, to document, collect, and preserve the remnants of the early history of digital games for microcomputers. To date, we have compiled a spreadsheet of more than two hundred game titles that were either written in New Zealand or written by New Zealanders. This can be viewed on the "Play It Again" research blog (Play It Again 2012). We are also collecting people's memories of playing games and associated artefacts from the era, via an online "Popular Memory Archive" (de Vries, Ndalianis, Stuckey, and Swalwell 2013; Stuckey, Swalwell, Ndalianis, and de Vries 2013).

Notes

1. Wilson and Swalwell's timeline plots international release dates against likely New Zealand dates, based on evidence of the system being on sale in New Zealand gleaned from ephemera such as store catalogs, magazine articles, and the like (see Wilson and Swalwell 2004).

2. SEGA also reportedly bought up redundant console games from around the world and resold them in the New Zealand market "for under $30," Wilson reported. "SEGA still makes a 50% profit on that, while fostering brand loyalty. Satisfied children will stay with the product and go on to more expensive machines. The marketing strategies are not so much about today but about next year too" (D. Wilson 1993).

References

Brown, R. 2003. Blast from our past. *Unlimited Magazine New Zealand*. http://unlimited.co.nz/unlimited.nsf/opinion/blast-from-our-past.

Buchanan, K. 2004. Video Games: Ritual or Rebellion? In *Media Studies in Aotearoa/New Zealand*, ed. L. Goode and N. Zuberi, 135–145. Auckland: Pearson.

CongoBongo. n.d. Leisure-Vision—A Canadian Console. *CGCC*. Retrieved June 17, 2011, from http://www.cgcc.ca/articles/view.php?article=Leisure_Vision.

Davidson, Michael. n.d. *Obscure Pixels: Retrogaming with a New Zealand Slant*. http://homepages.ihug.co.nz/~pinwhiz. Retrieved from www.retrogames.co.nz.

De Vries, D., A. Ndalianis, H. Stuckey, and M. Swalwell. 2013. Popular memory archive. http://www.playitagainproject.org.

Ellerslie Borough Council. n.d. Bylaw on Amusement Galleries (S8/1/25). Auckland.

Games centre delay. 1982. *Press*, April 3, 6. Christchurch.

Games cause "war." 1981. *Press*, December 23, 1. Christchurch.

Gielens, J. 2000. *Electronic Plastic*. Berlin: Verlag.

Johnson, V. 1993. Sega and Nintendo shoot it out for NZ. *Marketing Magazine*, August, 20.

King, M. 2004. *The Penguin History of New Zealand*. Albany, Auckland: Penguin.

Kong Tin, H. A. 1985. Home computers—Yesterday, today, and tomorrow: A personal view. *Hailing Frequencies 33 (fanzine)*, June.

Kraemer, K. L., and J. Dedrick. 1993. Turning loose the invisible hand: New Zealand's information technology policy. *Information Society* 9(4). http://escholarship.org/uc/item/6tx9v7g5.

Legal action stopped. 1983. *Press*, May 31, 8. Christchurch.

Lord, P. n.d. Technosys Aamber Pegasus. Retrieved January 17, 2013, from http://www.neoncluster.com/aamber_pegasus/aamber_pegasus.html.

NZGDA. (n.d.). New Zealand Game Developers Association. Retrieved September 10, 2013, from http://www.nzgda.com/about-us/what-is-the-nzgda/.

Obscure Pixels—Finds & Brags. n.d. Retrieved January 17, 2013, from http://homepages.ihug.co.nz/~pinwhiz/brag.htm.

Play It Again. 2012. New Zealand games software list online for review. http://blogs.flinders.edu.au/play-it-again/2012/10/29/1980s-new-zealand-games-software-list-online-for-review/.

Rihari, M. 2013. Old school gaming with game masters. *Te Papa Blog*, January 3. http://blog.tepapa.govt.nz/2013/01/03/old-school-gaming-with-game-masters/.

SegaSC-3000 Survivors. n.d. Retrieved from http://www.sc-3000.com/index.php/Table/Interviews/.

"Space invaders war" reaches High Court. 1982. *Press*, June 22. Christchurch.

Stephen, A. 2012. Saving the Hot Copter—Archiving our digital heritage. *Vintage 8 Bit*, March 18. http://www.vintage8bit.com/content/saving-hot-copter-archiving-our-digital-heritage.

Stuckey, H., M. Swalwell, A. Ndalianis, and D. de Vries. 2013. Remembrance of games past: the popular memory archive. In *Proceedings of the Ninth Australasian Conference on Interactive Entertainment: Matters of Life and Death*. New York: ACM.

Swalwell, M. 2007. The remembering and the forgetting of early digital games: From novelty to detritus and back again. *Journal of Visual Culture* 6 (2): 255–273. doi:10.1177/1470412907078568.

Swalwell, M. 2008. 1980s home coding: The art of amateur programming. In *Aotearoa Digital Arts Reader*, ed. S. Brennan and S. Ballard, 193–201. Auckland: Clouds/ADA.

Swalwell, M. 2009. Towards the preservation of local computer game software: Challenges, strategies, reflections. *Convergence: The International Journal of Research into New Media Technologies* 15 (3):263–279. doi:10.1177/1354856509105107.

Swalwell, M. 2010. Hobbyist computing in 1980s New Zealand: Games and the popular reception of microcomputers. In *Return to Tomorrow: 50 years of Computing in New Zealand*, ed. J. Toland, 157–169. Wellington: NZ Computing Society.

Swalwell, M. 2012a. Questions about the usefulness of microcomputers in 1980s Australia. *Media International Australia* 143:63–77.

Swalwell, M. 2012b. The early micro user: Games writing, hardware hacking, and the will to mod. *DiGRA Nordic 2012—Local and Global: Games in Culture and Society*. Tampere, Finland. http://www.digra.org/dl/db/12168.37411.pdf.

Swalwell, M., and E. Loyer. 2006. Cast-offs from the Golden Age. *Vectors: Journal of Culture and Technology in a Dynamic Vernacular* 3. http://vectorsjournal.org/projects/index.php?project=66.

Swalwell, M., and J. Bayly. 2010. More than a craze: Photographs of New Zealand's early digital games scene (exhibition). http://www.maharagallery.org.nz/MoreThanACraze/index.php.

Swalwell, M., and M. Davidson. Forthcoming. Game history and the case of "Malzak": Theorising the manufacture of "local product" in 1980s New Zealand. *Locating Emerging Media*, ed. B. Aslinger and G. Halegoua. London: Routledge.

Te Atatu parlour man's challenge. 1983. *Western Leader*, July. Auckland.

Van Rooyen-McLeay, K. 1985. Adolescent video-game playing. Master's thesis, Victoria University of Wellington.

Video parlour "not to blame." 1985. *Dunedin Star Midweek*, August 21, 1. Dunedin.

Wheeler, A., and M. Davidson. 2008. SC3K tape software list. http://homepages.ihug.co.nz/~atari/SC3K08.html.

Wilson, D. 1993. How this craze captured your kids. *Press*, October 16. Christchurch.

Wilson, J., and M. Swalwell. 2004. International—New Zealand game history timeline. http://www.academia.edu/4447512/International_NZ_Game_History_and_Popular_Culture_Timeline.

Wizards may get large damages. 1982. *Press*, July 23, 5. Christchurch.

PERU

Arturo Nakasone

The adoption and expansion of video games in Peru has been relatively slow, mainly due to the hard economic situation the country was going through during much of the 1980s and the first half of the 1990s. Video game history in Peru basically starts with the introduction of arcade machines during the beginning of the 1980s. At that time, a small number of businesses appeared, ranging from medium-sized arcade game centers, which deployed tens of machines, to small stores that had just a handful of them. The majority of arcade machines was provided by Japanese manufacturers such as Namco, Konami, and so on, and their games proved to be very successful among high school students and young adults. Among the most popular ones were Namco's *Pac-Man* (1980), Konami's *Super Cobra* (1981), Taito's *Phoenix* (1980), Irem's *Moon Patrol* (1982), and Williams Electronics's *Defender* (1980).

By the mid-1980s, home computers and game consoles started to appear, with the Atari VCS 2600 being the first to gain popularity among gamers. These devices were not always affordable to everybody or easy to buy inside the country, since there were not many import businesses interested in that kind of hardware at the time. Commonly, people would just purchase the console/computer and its cartridges/software when they traveled abroad, either for their own personal use or for sale to others. Examples of systems from this time are the Magnavox Odyssey, the Atari VCS 2600 console, the Coleco Gemini, and the Sinclair ZX81 (including their upgrades and variants such as the Sinclair ZX Spectrum, and so on). When the first systems with cartridges began to appear, several businesses offered their customers the opportunity to rent those cartridges, much like they would rent a movie in Betamax or on VHS. The most popular were the cartridges made for the Atari 2600, whose influence endured well until the end of the 1980s. The end of that decade was mainly characterized by the appearance and subsequent popularity of personal computers for gaming, with the Commodore 64/128 being its key representative. Although computer hardware stores were increasingly interested in offering this affordable computing alternative to average users, the software aspect was not covered in the same way. Games (which were the main reason people bought these computers) could only be purchased either through the same store where the hardware was bought or through "second-hand" dealers—in other words, people who would just copy the games for a price. Despite these limitations, devices such as the Commodore 64/128 started to increasingly appeal to home users because of their capability to offer a better gaming experience in terms of graphics and sound. The first Intel-based PCs started to gain acceptance in the market, but their main focus was not oriented to gamers. Soon, that would change dramatically.

The beginning of the 1990s was characterized by a revolution in terms of PC and game console development. Since the introduction of the Intel 286 processor in Peru in 1990, computers were made much more affordable for home use because of the appearance of small businesses focused on assembling computers from generic hardware. Due to the fact that there was still no software culture in Peru, most businesses would just offer unauthorized copies of software (including games), contributing to the quick adoption of the technology. In addition, this decade was also characterized by the renaissance of the game console hardware led mainly by Nintendo and its NES (Nintendo Entertainment System) family of consoles. Much like what happened with Atari, cartridges became one of the main limiting factors for the expansion of this technology, again because of economic issues. Although game-related imports had increased substantially by this time, people would still find it prohibitive to pay quite a lot of money to purchase one single game cartridge.

With Sony's PlayStation coming on the scene by the end of the 1990s, interest in game consoles started to flourish again, mainly due to the fact that PlayStation games could be obtained not only legally, but also through unauthorized copy dealers, which might have constituted the bulk of games sold in the country. By the beginning of the 2000s, however, the government began to encourage a software culture that would mainly try to deal with software piracy, motivating people and businesses to purchase software (and games) legally. These efforts have been met with encouraging success, leading to a huge economic increase in terms of imports for gamers. For example, in 2006, the value of imported game-related goods (including consoles, games, gamepads, etc.) was almost USD $2 million, but in 2010, this value increased to more than USD $10 million.[1] This trend is explained by the fact that the Peruvian economy has been enjoying a period of solid growth since approximately 2007 (up through the time of this writing [2013]), and, if this increase continues, it is expected that the growing trend of game-related product acquisition will also continue.

Domestic Video Game Production

Although video games enjoyed a rather early start in Peruvian computing history, video game production has not seen the same kind of success. Since early adopters were usually people with solid economic status, this would represent a very small number of users that could potentially be interested in the construction of games themselves. However, with the exponential increase of PC users and a growing interest in game software development for its economic benefits, computer programmers have started to polish their skills in this area. The easy availability of game construction tools, going from Adobe Flash kits to commercial 3-D engines, has greatly contributed to the minimization of the learning curve, and good examples of game initiatives have begun to appear, coming from for-profit companies, academic institutions, and individuals eager for public recognition. One of the first games that was developed in the country was a 3-D race simulation game called *Full Speed* (2009) from ArtiGames.[2] *Full Speed*'s development began in 2006, and it took almost three years before it was deployed in its earliest version. Although its concept was very simple for a racing game, it was the first attempt for a small Peruvian company to develop a full-fledged game that would try to mimic the characteristics of established commercial racing games.

The promotion of activities or projects dealing with historical settings and events has always attracted the attention and funding of governmental organizations which, in the case of Peru, are the only institutions that can actually finance these kinds of research activities. One of the first academic institutions that showed a strong interest in combining game development with historical topics is the Pontificia Universidad Católica del Perú (PUCP), a leading Peruvian university. Formed by a group of students, professors, and professionals from PUCP, the Grupo Avatar (Avatar Group)[3] started, in 2011, the development of the first strategy game based on a Peruvian historical event known as the "1814 Rebellion in Cusco."[4] The first version of the game was completed in March 2013 and was based on *Age of Empires* (1997), a strategy game from Microsoft. By the end of 2012, the game had been tested by high school students, and the results of those studies were utilized as feedback to improve game playability. This project will be sponsored in its final version by the Peruvian Ministry of Education for its final release to all interested academic institutions eager to implement this learning tool in their curricula.

Indigenous Video Game Culture

Recently, there has been growing interest in game creation as a means to express the designer's own views and beliefs, particularly in the subject area of politicians and showbiz celebrities. Most of these games are constructed using very simple Flash-based sprite techniques and are meant to present these well-known public figures in a sarcastic way rather than provide a realistic game experience. For instance, in March 2013, during the electoral period when citizens were recalling the mayor of Lima, a game called *Revoca-game* (*Recall-game*) was developed as part of the political campaign favoring the incumbent.[5] Its premise was very simple: the player controls the mayor's character, who is inside a car, while there are two types of balloons falling down the road. The "yes" type (meaning "yes to the recall") and the "no" type. The player must catch the "no" balloons while avoiding the "yes" ones. The game, despite its simplicity, became quite popular with young people. (An increasing number of these games can be found on sites such as that of Inka Games (http://www.inkagames.com).

Other initiatives aim to encourage people to show their skills either in game programming or competitive game playing. The Lima Game Jam 2013 (see http://www.limagamejam.com), implemented as part of the Global Game Jam, included as many as sixty participants, among them programmers and artists who developed sixteen games in total. Although this event has only been held twice, it is expected that for future events the number of participants will increase based on the current interest in video games in general. Recent years have seen Peru's continued participation in the World Cyber Games (WCG), with the country occupying the thirty-ninth position overall and fourth in South America. In the 2012 WCG events, the two Peruvian representatives (professional gamer Jian Carlo Joan Morayra, aka "Fénix," and Andrés Gutiérrez, aka "Andrucas," a lawyer and five-time *FIFA 2012* Peruvian champion) achieved meritorious positions in the *StarCraft II* (2010) and *FIFA 2012* (2011) games.[6]

Video Game Companies

The rapid advancement of social networking tools and the increasing user base of smartphones in Peru has made several people consider establishing their own companies to develop video games, through which it's possible to reach a considerable number of potential customers/users without a huge economic investment. Among the most established companies in Peru, we could mention the following:

- 3S Games (see http://3sgames.com). Based in Lima, 3S Games is a company whose objective is to design and implement video games for entertainment and media communication purposes. It emphasizes conceptual art, animation, and user interface design in its works. The most representative game from this company is *Corre Chasqui Corre* (*Run Chasqui Run*) (2012). In this game, the player takes control of a messenger (Chasqui, in the Quechua language), whose objective is to take messages to and from different regions in Peru. This game was published as part of the new platform of Smart TVs from LG.
- ArtiGames (see http://www.artigames.com). ArtiGames has offices in Lima as well as Barcelona, Spain. It is mainly dedicated to the implementation of applications for the entertainment business, particularly in the areas of marketing and advertising. According to its profile, it specializes in "programming and game design, 3-D visualization and real-time applications, digital animation, virtual reality simulations, and augmented reality." Recently, the majority of its projects are essentially focused on animation for marketing. Thus, they have very few projects dedicated to video games. The only known video game developed by this company is the *Full Speed* racing game mentioned earlier.
- Pariwana Studios (see http://www.magiadigital.com/principal/categoria/pariwana-studios/167/c-167). A subsidiary company of Media Digital, Pariwana Studios is in charge of video game development for mobile devices. It was established in 2012 and focuses on the development of "creative digital industries using Peruvian cultural heritage as their main resource, designing world-class video games." So far it has released two video games:
  ◦ *Inka Madness* (2012), which is the first video game developed for mobile platforms in Peru (specifically for Windows Phone 7). The player controls Atuq, a great Inca warrior whose mission is to save his people from chaos and dispel a curse cast by an evil sorcerer over the heir of the Incan empire. The development of this game took five months, and it is expected to be released for other platforms as well (including the Xbox, iOS, and Android).
  ◦ *Tadeo in the Lost Inca Temple* (2013), which is an endless-running type of game. Here, the player tries to help Tadeo, an explorer, leave a lost temple of the Incas before it collapses. To accomplish this, Tadeo must

sort out a series of obstacles and collect as many coins as possible to purchase extra lives, which help him reach his goal. This game was released for iOS, Windows Phone, and Android.

- Bamtang (see http://www.bamtang.com). This company is exclusively dedicated to the development of video games, not only for mobile platforms (including Android, iOS, and Blackberry) but also for PC and PlayStation platforms. A great deal of its games (such as *Ben10 Forever Defense* [2011], *Blockade Blitz* [2011], and *Teenage Warriors* [2011]) were developed in Flash and distributed through the Cartoon Network's Web site. Bamtang was founded in 2002, and one of the first projects it developed was *Boxing Box*, a boxing game prototype with a haptic interface. This game was never released commercially, but, thanks to it, the studio was recognized by publishers such as Novint and the Cartoon Network for its creativity, quality, and skills. A complete list of games released by the company can be found at http://www.bamtang.com/games.
- Inventarte (see http://www.inventarte.net). This company is dedicated to the development of video games for the Web and social media sites (such as Facebook). Its most well-known game is *Crazy Combi* (2009), which was initially developed for Facebook. In this game, you are a combi (minivan) driver who must avoid other vehicles on the road (either by dodging them or jumping over them) and gain points by making it to the next bus stop before time runs out. The success of this game caught its developers by surprise since it was only expected to reach around 50,000 people in total, but five days after its debut, the number of players was around 120,000. The majority of its currently developed games is available on Facebook. For a complete list of games, visit http://www.inventarte.net/quienes-somos/chicha-games/.
- Toy Catz (see https://www.facebook.com/toycatzoficial?filter=1). This company develops video games mainly for Facebook, iOS, and Android platforms. One of its most representative games is *Cuy* (*Guinea Pig*) (2010), released for iOS. In this game, the player must help Cuy, the protagonist, climb Machu Picchu collecting the ancient treasure of the Incan empire in the form of Golden Tumis (ceremonial knives).

The academic study of video games in Peru is almost nonexistent. The few essays available deal mostly with the effect that overplaying has on children's lives. CEDRO (Centro de Información y Educación para la Prevención del Abuso de Drogas) (the Information and Education Center for Prevention of Drug Abuse) has been the first to release such a study, raising tangible concerns over the possibility that online gaming is starting to show symptoms of addiction in more than 30% of young players.[7]

The Future of Video Games in Peru

Due to the great progress in 3-D graphics technology and its increasing usage in the entertainment business, people in Peru are starting to take more seriously the idea of developing business strategies in which delivered goods are not physical but digital in nature. In this context, educational institutions, mainly led by private universities, have tried to fulfill the need for digital content creators by implementing courses and programs specializing in video game development,[8] digital art,[9] and even virtual worlds.[10] Although the economic impact of video game production is still small compared to more traditional areas such as mining and manufacturing, its steady evolution will aid in the diversification of the economy and boost the technological drive necessary for the development of the country as a whole.

Notes

1. See http://blogs.peru21.pe/inserte1ficha/2010/11/la-importacion-de-productos-pa.html.

2. See http://videojuegos-peru.com/2009/07/primer-juego-3d-creado-peru/.

3. See http://avatar.inf.pucp.edu.pe/.

4. See http://elcomercio.pe/actualidad/1548172/noticia-1814-primer-videojuego-estrategia-sobre-historia-peru.

5. See http://elcomercio.pe/actualidad/1543189/noticia-revocagame-revocacion-susana-villaran-llego-videojuegos.

6. See http://www.localstrike.net/peru/noticias-313/fenix_y_andrucas_en_wcg_grand_final_2012-3575.

7. See http://publimetro.pe/actualidad/2486/noticia-32-gamers-juega-mas-4-horas-diarias.

8. See http://avatar.inf.pucp.edu.pe/diplomatura-en-desarrollo-de-videojuegos-siguen-abiertas-las-inscripciones/.

9. See http://www.eadperu.com/.

10. See http://agenda.pucp.edu.pe/educacion-virtual/mundos-virtuales-para-la-educacion/.

POLAND

P. Konrad Budziszewski

The West is the best.
Get here, and we'll do the rest.

—Jim Morrison

"You need to appreciate how much our team had to rule for guys in America to trust us with their precious baby—their biggest money-maker ever. It's as though some Polish company took its most successful product and farmed it out to Kazakhstan" (Chmielarz 2007, 126). Thus spoke Adrian Chmielarz, cofounder and creative director of People Can Fly (PFC), reflecting on the latter's relationship with Epic Games—a relationship that netted the studio a contract for the PC version of *Gears of War* (2007) and led to partial acquisition by Epic. In a peculiar combination of self-aggrandizement and self-effacement, Chmielarz extols the merits of the PFC team, while simultaneously situating Poland as the Kazakhstan of game development. To be sure, the basis of comparison here is not so much the Kazakhstan of Abay Qunanbayuli and Mukhtar Auezov, as that of *Borat* (2006): an imaginary space located in some remote and geographically vague "out there," as distant culturally as it is spatially. It is a locus of a very particular sort of otherness: a bit strange, a bit scary, very much backward. Thus, even as he strives to express the magnitude of the studio's accomplishment, Chmielarz's rhetorical choices convey a sense of skepticism (or even bitterness) about the overall condition of the Polish game industry and its relative position in the global hierarchy.[1] The measure of PFC's success effectively becomes the measure of the presumed gap separating Polish developers from the "guys in America."

It is a striking sentiment, coming from one of the more accomplished figures among Polish game creators, a veteran with fifteen years of experience, whose design credits include People Can Fly's internationally-acclaimed *Painkiller* (2004). While Chmielarz's statement does not really say much about the actual state of Poland's development community, his ambivalence reveals a great deal about said community's self-image, especially in relation to its Western counterparts. On the one hand, his words speak to a sense of technocultural distance and associated dread of inadequacy, fuelling strong determination to measure up to the elusive "Western standards"; an implicit assumption of the superiority of these standards; a desire for recognition and approval from Western developers; aspirations to compete in the global arena on equal terms; and a yearning for international success. On the other, there is considerable pride in one's accomplishments (all the stronger for the belief in the disadvantage inherent in one's position), self-respect, high regard for independence, and a firm belief in one's ability to make valuable contributions to the development of the medium.

What are we to make of this peculiar amalgam of pride, ambition, anxiety, and doubt? In order to understand the present condition—economic, creative, but especially psychological—of the Polish game industry, it is necessary to examine its past, a past shaped by the political, economic, and social realities of the final years of the Communist regime, the turbulent transitional period, and the fledgling Third Republic. Toward this end, this chapter offers a topical/historical overview of Poland's electronic gaming sector, foregrounding some of the key factors that determined the specific trajectory of its development.

The account begins with a brief analysis of the political and economic circumstances in which the Polish "computer revolution" of the 1980s took place, and their influence on the nascent gaming culture. The subsequent section, spanning the years 1984–1996, focuses on games for 8-bit and 16-bit home computers, from hobbyist roots through the emergence of profit-oriented development and publishing initiatives. The period from the late 1980s to the mid-2000s is examined in two parts. The first gives an account of the early forays into the world of IBM PC-compatibles, as well as of the industry's gradual turn toward foreign markets. The second details the parallel efforts to combat software piracy and boost domestic sales to viable levels. By way of conclusion, the final section profiles Poland's leading game developers—the studios that achieved financial success and earned critical recognition on the international stage.

The Iron Curtain and the Silicon Wave: Home Computers in the People's Republic of Poland

As fate would have it, the home computing boom of the 1980s coincided with one of the most politically turbulent and economically fragile periods in Polish post–World War II history. Edward Gierek's tenure as First Secretary of the Polish United Workers' Party (1970–1980) was characterized by a dramatic increase in prosperity (bought for the price of vast foreign loans), followed by an equally rapid decline in the second half of the decade. As a result, Poland entered the 1980s on the verge of a massive economic crisis, saddled with foreign debt in excess of US$20 billion. The situation was soon exacerbated by the growing budget deficit and economic sanctions imposed by the United States in response to the introduction of martial law in December 1981. Gross national income dropped at an alarming rate, and inflation rose even faster.

At the time, no "official" avenues for the import of computer technology existed, due to the Coordinating Committee for Multilateral Export Controls (CoCom) embargo on sales of "sensitive" technologies to the countries of the Eastern Bloc. This included devices with potential military applications (so-called "dual-use" items), such as telecommunications equipment and computers, of which only very limited exports were allowed under exception provisions, with much red tape involved. In 1984, revised guidelines relaxed the restrictions on computer hardware—in particular, the mass-produced "home computers"—but another crucial obstacle remained in place: the lack of suitable funds in state coffers. Polish złoty (PLZ) was not convertible, and all hard currency was channeled toward the repayment of the substantial—and growing—foreign debt. Although the Party wholeheartedly embraced the idea of "computerization," with its perceived manifold educational, scientific, and economic benefits, it was simply not in a position to facilitate the acquisition of necessary technology.[2]

Accordingly, the "silicon wave" that swept the country was driven for the most part by semi-legal private initiatives. Chief among them was so-called trade tourism—the practice of bringing in small quantities of computer equipment for private use or sale, most commonly from West Germany and the United Kingdom. Although private import of goods was not prohibited in itself, severe restrictions on foreign travel and possession of hard currency often forced "suitcase importers" to rely on various under-the-counter "arrangements." Still, the government did little to curb the practice and, in 1985, exempted personal computers from import duty—a convenient way to boost the inflow of technology from beyond the Iron Curtain and usher the society into the Information Age at no expense to the state. As a result, computers were soon arriving in the country at an approximate rate of 30,000 units per year (Szperkowicz 1987, 32–33), then spreading throughout the population via classified ads, consignment shops, and specialized flea markets. Also in 1985, the first official import and distribution channel was established thanks to Lucjan Wencel, a Polish emigrant to the US. His California-based company, Logical Design Works, supplied Atari 8-bit computers and peripherals to Pewex, a chain of state-owned hard currency stores, while LDW's Polish subsidiary, Przedsiębiorstwo Zagraniczne "Karen,"[3] provided technical support and service. The following year, Portuguese-made Timex microcomputers began to appear—albeit erratically and in limited quantities—in the stores of the Central Scouting Depository (Centralna Składnica Harcerska).

Pricing remained the primary determinant of the availability of technology. Because of the economic chasm separating the country from the "capitalist West," the only feasible option, in most cases, was buying behind the curve. Of course, even older, less expensive hardware could not be considered affordable by any stretch of imagination. To illustrate, in 1986, the average monthly net salary was approximately 24,000 PLZ ("Przeciętne miesięczne wynagrodzenie w gospodarce narodowej w latach 1950–2012" 2013). A second-hand Sinclair ZX Spectrum cost 75,000–85,000 PLZ. For approximately the same price, one could become a happy owner of an Atari 800XL, although, in that case, the need for a dedicated tape drive would ultimately raise the tally by a further 30,000–35,000 PLZ. A Commodore 64 could be had for a meager 150,000–160,000 PLZ.[4] Newer, 16-bit machines would remain out of reach of all but the most affluent for several more years.

Even with equipment costs ranging from merely extravagant to outright prohibitive, home computing was rapidly gaining popularity. By 1987, the Central Customs Office (Główny Urząd Ceł) gauged the number of microcomputers in the country at around 100,000; less conservative estimates suggested the figure could well be doubled (Szperkowicz 1987, 33). The willingness to adopt the new technology was certainly stimulated by exceedingly low software prices, which helped offset the high initial investment in hardware. The key was the outdated copyright law, which did not (explicitly) extend to computer programs. Thus, software brought in from the West was stripped of copy protection, modified as necessary, and sold by cassettefuls, often at prices not much higher than the cost of blank media.⁵ Though morally questionable, the practice was technically legal and, in the absence of practicable alternatives, accepted as a necessary evil.

Cheap and plentiful software notwithstanding, electronic gaming in 1980s Poland was often a tricky proposition for a number of reasons. Without guarantees or support, any purchase carried inherent risks. The reliability of duplication methods left much to be desired, and the various creative interventions into the code, necessary to rid it of irksome restrictions, could, every so often, make a game impossible to complete. Because the overwhelming majority of software arrived from abroad, language barriers posed a major problem, adding whole new levels of challenge to genres such as simulation, strategy, and role-playing, and rendering text adventures completely unplayable. Finally, lack of proper (or, more often than not, any) documentation removed gameplay from whatever narrative context the supplementary materials might provide and left hapless players guessing as to the rules, objectives, and even basic controls. The state of affairs improved marginally when the early computing magazines stepped into the gap, dedicating a few pages per issue to game coverage.⁶ Shaped as they were by the shortcomings of the software market, the reviews covered both recent and older titles and combined the evaluative function with abbreviated instructions (plot summary, overview of objectives and controls), rudimentary localization (translation of key terms and phrases), and a more traditional game guide (detailed maps, tips, or even complete walkthroughs).

The experiences of those formative years determined the shape of both game development and consumption for decades to follow. One of the most obvious repercussions of the political turmoil and economic scarcity of the 1980s was a pronounced gap between Poland and the First World. Relative to their adoption in the West, computer technologies (as well as the associated know-how) were reaching the country with a delay of three to five years. Immediate consequences of this lag were apparent well into the second half of the 1990s, and some of its less direct implications can still be felt today. Chief among them are the compulsive need to "close the distance" and resolve to follow the course set by Western developers—a legacy of the long years when software arriving from Western Europe and the United States quite literally represented the future of the industry.

Another crucial product of the era was a decade of unchecked, practically institutionalized piracy. The undetermined status of computer software with respect to intellectual property laws provided a solid foundation for a thriving gray market, which, until the early 1990s, accounted for the vast majority of all sales. Even after the introduction of the revised copyright regulations in 1994, attitudes and behaviors inherited from the earlier period persisted. We may never fully understand the extent to which the lack of legal protection hindered the growth of the Polish software industry. One of its more evident consequences, however, was the gradual shift of developers' attention toward foreign markets in an effort to reclaim the profits lost to "alternative distribution channels." This, in turn, resulted in the relative dearth of games purposefully drawing on the idiosyncrasies of Poland's history, cultural tradition, and socio-political milieu. Sandwiched between the massive influx of software from abroad in the early-to-late 1980s and the industry's Western turn in the second half of the 1990s, the few uniquely Polish titles amounted to no more than a noteworthy but brief interlude.

Finally, an interesting characteristic of the first decade was the relative rarity of game consoles. (A pricy electronic "toy" was an expense much harder to justify than an "educative" computer.) Among the handful of exceptions, the most prominent were the Soviet-manufactured derivatives of the Nintendo Game & Watch handhelds. The situation changed in the early 1990s, with the introduction of the Pegasus—an affordable Chinese clone of the Nintendo Entertainment System. Several hundred thousand units were sold over the years, giving the system unchallenged rule over the console niche until the introduction of the Sony PlayStation in 1996. The increasing popularity of consoles has not, however, eroded the primacy of computer-based gaming: as recently as 2010, PCs still accounted for an estimated 60% to 70% of the market (enki 2010, 87).[7]

Opening Pandora's Box: The First Ten Years

It is not surprising, then, that game development in Poland originated with home computing enthusiasts. The earliest known efforts—*Ships* (J. Cichorski), *OiX* (O&X, Jarosław Cichocki), and *Królewna Jebaczka* (*Princess Fuckerella*, X-Prog) for the ZX Spectrum; *Invers*, *Minefield*, and *Slider15* (Jerzy Wałaszek) for the ZX81—date back to 1984. The programs were simple, written in BASIC, with rudimentary, text-mode graphics and straightforward, usually puzzle-based gameplay. As Kluska and Rozwadowski point out, many of the surviving titles from that period bear more than passing similarities to Western counterparts, with sources of "inspiration" ranging from commercially released games to type-in listings from computer magazines. For example, *Barman* (anon. ca. 1984) for the Meritum computers, was a text-mode clone of Marvin Glass and Associates' *Tapper* (1983), while J. Rajzer's *Tajpan* (1984) was a minimally modified and translated version of Jaysoft's British-developed *Taipan* (1983) (Kluska and Rozwadowski 2011, 35–39). The most immediately evident characteristic of that period, compared to countries such as Great Britain, Germany, or the United States, was a pronounced difference in quality—a direct consequence of the technological distance separating Poland from "the West." The delay with which computer technology entered the country was obvious with regard to hardware, but no less significant was its crippling impact on the formation of expert knowledge, where access problems were further exacerbated by the language barrier. Issues of computing magazines and photocopies of programming handbooks from the other side of the Iron Curtain (circulated through the same "informal" distribution networks that provided access to software) were comprehensible only to a select few, and it was not until 1986 that their Polish equivalents began to appear in appreciable numbers. In short, while in many Western countries game development had already become a booming industry, Poland was firmly stuck in the "bedroom coder" stage.

The first breakthrough came in 1986, with Marcin Borkowski's *Puszka Pandory* (*Pandora's Box*), a text adventure for the ZX Spectrum, created entirely in assembly language. Borrowing the basic premise from Marek Baraniecki's post-apocalyptic novella, "Głowa Kasandry" ("Cassandra's Head," 1985), the game is set several decades after the devastation of World War 3 and tasks the player with scouring a remote island in the Pacific in order to find and disarm the titular Pandora's Box—an automated missile system, capable of destroying whatever human life is left on the planet. The presentation is austere, with bare-bones descriptions enlivened only by a scrolling strip of minimalist graphics depicting one of the few basic terrain types. There are no sound effects or music, save for the mournful tune signaling the hero's premature demise. The interaction is fairly limited as well, as the parser only recognizes about 20 commands. Still, what *Puszka Pandory* lacks in looks and complexity, it makes up for in sheer size. The challenge of the game lies just as much in successfully navigating the 1,600 locations that make up the island and its immediate vicinity as in dealing with the perils—from minefields to magnetic anomalies to deadly currents—contained therein. Borkowski's oeuvre certainly did not revolutionize the genre, but, unlike any of its contemporaries, it did not require knowledge of foreign languages. ("Having trouble with *The Hobbit*? Don't know English well enough? Buy *Puszka Pandory*!" encouraged the cover.) The first Polish adventure game was also, by all accounts, the first Polish computer game in legal commercial circulation, courtesy of Enter Computing, a game rental business with publishing aspirations, which acquired the distribution rights for a respectable sum of 40,000 PLZ.[8]

The year 1987 brought further text adventures for the "Speccy" (ZX Spectrum). Bogusław Woźniak's relatively obscure *Hibernatus* (also distributed by Enter Computing) finds its hero awakened from cryosleep—prematurely and in dire circumstances—in the year 2057. Tomasz Kopiejć and Zbigniew Wymysłowski released two fantasy-themed freeware titles: the brief *Conan: Spotkanie w krypcie* (*Conan: An Encounter in the Crypt*)—closely modeled after "The Thing in the Crypt" (1967), a Conan the Cimmerian short story by L. Sprague de Camp and Lin Carter—and a wholly original work titled *Metropolis 1*. The latter was developed as the first part of a planned trilogy of games revolving around the quest for three mythical crystals of power. Unfortunately, the remaining two installments never materialized. By far the most successful of the crop was Piotr Kucharski and Krzysztof Piwowarczyk's *Smok Wawelski* (*The Dragon of Wawel*), inspired by a popular Polish legend of a dragon terrorizing the city of Kraków and its demise at the hands of a young shoemaker's apprentice. The story may not have been novel, but the game distinguished itself with detailed illustrations accompanying all the locations, as well as with its sophisticated parser, capable not only of processing full sentences, but also of coping with the flexibility of Polish syntax. *Smok Wawelski* was picked up for distribution by Krajowe Wydawnictwo Czasopism (National Periodical Publishing) and, priced at an attractive 600 PLZ, reportedly sold 2,500 copies (Kucharski 1990, 4).

Krajowe Wydawnictwo Czasopism, along with Krajowa Agencja Wydawnicza (National Publishing Agency), played a key role in the early efforts to establish a legal software market. As divisions of Robotnicza Spółdzielnia Wydawnicza (Workers' Publishing Cooperative) "Prasa Książka Ruch," a state-owned and operated publishing syndicate, they attracted aspiring programmers with something no one else could offer: access to a nationwide distribution network. Although both focused primarily on educational software, their respective catalogs did include a number of games. Thus, in addition to *Smok Wawelski*, Krajowe Wydawnictwo Czasopism released logic games *Magiczne krzyże* (*Magic Crosses*; Jarosław Cichocki, 1986) and *TIM* (Janusz Gajewicz, 1986), as well as a simple turn-based strategy game, *Gwiezdne imperium* (*Star Empire*; Arkadiusz Lisiecki, 1988). KWC also became the first Polish distributor of foreign-developed titles, with its 1987 releases of *The Trap Door* (Don

Priestley, 1986) and *Oddział Cobra*, a localized version of *Strike Force Cobra* (Five Ways Software, Ltd., 1986). Krajowa Agencja Wydawnicza's offer included *Nim 2* and *Tixo* (Cezary Waśniewski, 1986), *Kółko i Krzyżyk* (*Naughts and Crosses*; Jarosław Cichocki, 1987)—an update of the 1984 *OiX*, *Młynek* (*Mill*; Stefan Życzkowski, 1984/1987), as well as a trio of titles by Poland's first female game creator, Wiesława Waśniewska: stock market sim *Czarny poniedziałek* (*Black Monday*), resource management game *Handel zagraniczny* (*Foreign Trade*), and 3-D labyrinth *Hexan* (all 1987).

Meanwhile, encouraged by the relative success of *Smok Wawelski*, Kucharski and Piwowarczyk, joined by Wiesław Florek, started their own development business, Computer Adventure Studio, and in 1989 released what was without a doubt the most advanced Polish text adventure of the time. *Mózgprocesor* (*Brainprocessor*) puts the player in the role of a special agent of the United Nations on a mission to recover the eponymous artifact: a biological computer capable of mimicking the workings of the human brain. It is revealed that the creator of the miraculous artifact—one Professor Thompson—has suffered severe injuries in an airplane crash. Now his only chance is to be implanted with his own invention. Naturally, as befits a true genius, the professor is a proper recluse, and so the device is stashed away somewhere in his abode, located on—where else?— a remote island in the Pacific. Despite some superficial similarities to *Smok Wawelski*, *Mózgprocesor* improves on its antecedent in every way. The upgraded text parser accepts a wider vocabulary set, puzzles are more challenging, and gameplay more complex. In addition to fully illustrated locations, the game features

Figure 1
You are at a dock. What now? (*Mózgprocesor* [1989] for the ZX Spectrum).

a functional compass and radiation meter, as well as a clock, relentlessly counting down the time left before the mission ends in failure.

At the time of its release, *Mózgprocesor* represented a new standard in game development. In a 1990 interview, Kucharski spoke of the studio's future in very optimistic terms, mentioning preparations to release the game outside of Poland, ongoing work on a specialized scripting language and tool suite for game creators, and a desire to venture into the publishing business (Kucharski 1990, 4). Alas, the ambitious plans never came to fruition; CAS managed to port *Mózgprocesor* to Atari computers and then quietly disappeared off the radar. There are no records of what exactly transpired, but, with only one title in its catalog, waning popularity of the ZX Spectrum, and unbridled piracy, the studio was most likely unable to generate sufficient profits.

The end of the 1980s also marks the beginning of a renaissance for the 8-bit Atari machines. By this point, the platform—dominant on the Polish market—was obsolete and increasingly irrelevant in the West.[9] As the flow of new software steadily slowed down to a mere trickle, local developers capitalized on the situation. In 1989, Tomasz Pazdan and Janusz Pelc, joined shortly thereafter by Mirosław Liminowicz, founded Laboratorium Komputerowe Avalon (Avalon Computer Labs). The studio debuted with Pelc's *Robbo* (1989), a puzzle game with arcade elements. The player's task is to guide Robbo the robot through fifty-six planets/levels. In each level, Robbo must locate a specified number of spare parts and reach the escape pod, all the while avoiding hostile creatures and other hazards such as automated turrets and giant electromagnets. With challenging but rewarding gameplay, reminiscent of such classics as *Sokoban* (Thinking Rabbit, 1982) and *Boulder Dash* (First Star Software, 1984), the game was a great success. However, it was the following year's release of *Robbo Konstruktor*—a user-friendly level construction utility—that catapulted it to cult status. *Robbo* has routinely been counted among the best and most popular Polish games of the 8-bit era, and new collections of user-created levels continue to appear to this day.[10]

Avalon followed up on the initial success with hit platformers *Misja* and *Fred*,[11] as well as another puzzle game, *Lasermania* (all 1990). *Lasermania* was the last game developed in-house; from that point on, the studio shifted focus to publishing and distribution, aggressively expanding its software catalog with a number of titles from independent programmers such as Henryk Cygert's minesweeping puzzler *Saper* (1991), Roland Pantoła's trio of point-and-click graphic adventures, *A.D. 2044* (1991), *Klątwa* (*The Curse*; 1992), and *Władcy ciemości* (*Lords of Darkness*; 1993), and a number of flip-screen maze platformers, including Rafał Wójcik's *Adax* (1992), Dariusz Żołna's *Hans Kloss* (1992), and its follow-up, *Spy Master* (1993).

Two more notable publishers were associated with the Atari platform: Mirage Software (est. 1989) and A.S.F. (est. 1991). The latter, an outgrowth of a prominent demoscene group of the time, Atari Star Force, gained recognition for arcade adventures programmed by Henryk Cygert (*Miecze Valdgira* [*Valdgir's Swords*, 1991], *Artefakt przodków* [*Artifact of the Ancestors*, 1992], *Miecze Valdgira II: Władca gór* [*Valdgir's Swords II: Mountain King*, 1993]) and Piotr Kulikiewicz (*Magia kryształu* [*Magic Crystal*, 1992], *Dwie wieże* [*Two Towers*, 1993], *Incydent* [*Incident*, 1993]). All in all, between 1989 and 1993, the three studios released more than 140 games for the Atari 8-bit line. Still, despite the influx of new titles, the popularity of the rapidly aging computers soon began to wane. By 1994, LK Avalon moved away from "little" Atari, while A.S.F. shut down altogether. Mirage

Software continued to support the platform until 1995. The more noteworthy of its later releases included Adrian Galiński's arcade platformer *Rockman* (1995) and a joystick-driven RPG/text adventure hybrid *Inny Świat* (*Other World*; Andrzej Wilewski, Adam Mateja, and Przemysław Piotrowicz, 1995).

Parallel to the gradual decline of the Atari brand, the early 1990s saw Commodore 8-bit computers rise to a more prominent position. The years 1993–1996, in particular, marked a brief surge in commercial releases for the Commodore 64. As Kluska and Rozwadowski note, this closely paralleled the pattern of the earlier boom in Atari development, wherein domestic production was jump-started by the sharp drop-off in the availability of new software from Western Europe and the US. In the same vein, quality Polish games for the C64 began to appear in noticeable quantity only after the platform had lost commercial viability in the West (Kluska and Rozwadowski 2011, 127). LK Avalon ported over some of its Atari best-sellers—including *Robbo*, *Hans Kloss*, *Klątwa*, and *Władcy ciemności*—and released a number of original games such as arcade puzzler *Arktyczne polowanie* (*Arctic Hunt*, 1994), graphical adventure *Wyspa* (*The Island*, 1996), and trading simulation *Gwiezdny kupiec* (*Star Merchant*, 1996). The smaller but very prolific TimSoft (est. 1989) also published several interesting titles, most notably Krzysztof Augustyn's arcade-adventure platformers *Lazarus* (1994) and *Eternal* (1995).

The year 1996 saw the effective end of commercial development not only for 8-bit micros but also for the Amiga. (Amiga's direct competitor, the Atari ST line, never gained much of a foothold in Poland, despite the great popularity of its younger siblings.) In the latter's case, serious efforts did not begin in earnest until 1994, with a delay similar to that affecting the earlier generation. However, the Polish 8-bit market was, by and large, closed and self-sustaining. The Amiga, in contrast, while slowly reaching the end of its life, was still very much an active platform with a steady stream of new software. Thus, Polish releases had to coexist—and compete—with titles developed in the West. The inevitable comparisons were rarely favorable to local developers.

The top-down shooter *Rooster* (TSA, 1994), for instance, while drawing obvious inspiration from *Alien Breed* (Team17, 1991), failed to match the sophisticated level design and dynamic action of the original. Similar criticisms were levied against *DOOM* (1993) clones *Za żelazną bramą* (*Behind the Iron Gate,* EGO, 1995), *Project Battlefield* (TSA, 1995), and its sequel, *Project Intercalaris* (TSA, 1996). Even *Cytadela* (*Citadel*, Arrakis Software, 1995), by far the most accomplished of the first-person shooter crop, was quickly outclassed by *Gloom* (Black Magic Software, 1995) and *Alien Breed 3D* (Team17, 1995). Platform games did not fare any better. Titles such as *Magiczna księga* (*The Grimoire*, 1994) and *Jurajski sen* (*Jurassic Dream*, 1994), both by Midnight Computer Games, and *Forest Dumb* (Eagle Software, 1995) were proficiently coded but marred by uninspired design and poor playability. The most common gripe, though, was overall lack of polish (no pun intended). Considering the small (often relatively inexperienced) teams, minuscule budgets, and publishers' willingness to release just about anything in hopes of recouping at least a fraction of the sales lost to the black market, this was hardly surprising. Even generally well-received titles were, more often than not, found wanting when measured against the coveted "Western standards"—as set by such powerhouses as the Bitmap Brothers, DMA Design, Psygnosis, or Sensible Software. A review of strategy RPG *Legion* (Gobi, 1996), for example, asserted the game was a "decidedly professional product, by any measure" and proof positive that the "Made in Poland" label should no longer be a reason for concern but ultimately ranked it "slightly above average."

Even so, game players appreciated the efforts of domestic developers, especially if the end-product could offer something uniquely Polish. The runaway success of *Franko: The Crazy Revenge* (World Software, 1994) is a prime example of "local color" trumping the painfully evident quality gap. The side-scrolling beat-'em-up action, driven by the story of the protagonist's bloody vengeance for the murder of his friend, is set against the surprising backdrop of the developers' native Szczecin, circa 1992. The real-life locations, recreated with keen attention to detail, may not have been widely recognizable, but the bleak atmosphere of a rundown urban neighborhood of the early post-socialist era definitely was. Crude presentation and grotesque violence notwithstanding, *Franko* managed to evoke an unprecedented sense of familiarity. Immensely popular, the game received a number of accolades—including *Top Secret* magazine's 1994 Game of the Year award—and quickly became a cult classic.[12] Another representative of the genre, Myys Art's *Prawo krwi* (*Blood Law*, 1995), placed the player in the role of an undercover police operative punching and kicking his way through the ranks of the mob. Despite its digitized graphics, the game did not quite turn out to be the *Mortal Kombat 3*-killer the advertising campaign had made it out to be (the poorly implemented first-person shooter and top-down driving levels did not help, either), but its distinctly Polish vibe earned it a warm place in many a teenage heart.

Figure 2

Neighborhood cleanup, Polish style (*Franko: The Crazy Revenge* [1994] for the Amiga).

Specifically Polish themes were also successfully explored in a number of point-and-click adventures. *Kajko i Kokosz* (*Kajko and Kokosz*, Seven Stars, 1994), based on a popular comic book series by Janusz Christa, followed the eponymous pair of Slavic warriors on a quest to protect their village against the malicious schemes of the Teutonic-esque Banditknights. *Skaut Kwatermaster* (*Kwatermaster the Scout*, LK Avalon, 1995) employed art style and interface design reminiscent of LucasArts' *Day of the Tentacle* (1993) to deliver a delightfully wacky take on a scouting adventure set in the Polish countryside. The small-town setting appeared again in *Alfabet śmierci* (*Death Alphabet*, Arte, 1996), this time as a stage for a tale of supernatural horror. The reviewers praised the story and the genuinely spooky atmosphere but unanimously pointed out that the very concept of a game based entirely on static, digitized photographs was woefully archaic.

By that point, the popularity of the Amiga was dwindling and—coincidentally—so was the willingness to "encourage" local developers by turning a blind eye to bugs, dated graphics, and questionable design choices. For instance, *Rajd przez Polskę* (*Race across Poland*, Digital Munition, 1996) managed to generate some (short-lived) interest among players, but the sheer novelty of racing the iconic products of the domestic automotive industry toward familiar skylines was not enough to win the increasingly critical press.[13] Somewhat symbolically, later the same year, the PC port of *Franko*—once the most popular Polish game for the Amiga computers—was dismissed by one of the leading gaming magazines as a "morbid joke" that "should be given away for free" (Gulash 1996, 31). The 16-bit era was over.

Westward Ho! PC Developers on a Quest for Profits

The beginnings of PC game development in Poland can be traced to Lucjan Wencel and the software division of P.Z. "Karen"—arguably the country's first true software development studio. Established as a porting house, the branch took advantage of economic discrepancies between Poland and the US to offer services at highly competitive prices. Wencel ensured the talented young programmers he recruited had access to everything their contemporaries sorely lacked: capital, up-to-date equipment, know-how, and—perhaps most importantly—industry connections. Thanks to the latter, the company soon secured contracts from such big leaguers as Strategic Simulations, Inc. (for which it created the Commodore 64 ports of *Phantasie* and *Panzer Grenadier*, as well as the IBM PC version of *Wizard's Crown* [all 1985]) and Electronic Arts (the Apple II conversion of *Mike Edwards' Realm of Impossibility* [1986]).[14] In 1987, under the banner California Dreams, it began to release original titles, starting with a simple collection of casino games titled *Vegas Gambler*. Significantly, as Wencel deemed domestic sales unfeasible due to rampant piracy (Wencel 1991, 9), the studio developed for export only, with an eye primarily toward the US market.

The next release would become California Dreams' best-known—if not most financially successful. *Blockout* (1989) was a three-dimensional take on *Tetris*, retaining the principle of clearing out blocks by building complete layers but transposing it into a 3-D well, presented in a top-down view. Despite its seeming complexity, the game was received very enthusiastically—to the surprise of the publisher. With the demand exceeding the relatively meager supply, the game quickly became a hot commodity in illegal circulation, effectively falling victim to its own popularity. The same year, the studio also released a sci-fi actioner, *Tunnels of Armageddon*, as well as its second most acclaimed title, *Street Rod*, a street racing game with heavy emphasis on vehicle modification and tuning, set in 1960s California. Although not as popular as *Blockout*, the loving homage to the Golden Age of American car culture was successful enough to warrant a sequel, *Street Rod 2: The Next Generation* (1991).

By that point, however, rising development costs were severely undercutting the profitability of the enterprise. In 1991, Wencel decided to shut down California Dreams—shortly before the premiere of the studio's most ambitious product yet, the political strategy sim *Solidarność* (*Solidarity*). Created as a potential tie-in to the planned Hollywood movie about opposition leader Lech Wałęsa, the game challenged the player to coordinate the actions of local chapters of the Solidarity movement in the struggle against the Communist government. Development was completed, but the title never saw official release. Fortunately, the final build leaked out onto the gray market, preserving the record of Poland's first openly political game.

In 1992, Janusz Pelc (of *Robbo* fame) left LK Avalon, and, along with Maciej Miąsik, joined the newly formed xLand. The same year, the studio released side-scrolling platformer *Electro Body*, and, in 1993, arcade puzzlers *Robbo* and *Heartlight PC*—both based on Pelc's earlier creations for Atari 8-bit computers. All three titles were picked up for distribution by Epic MegaGames (now Epic Games)—the first two as *Electro Man* and *Adventures of Robbo* respectively—making xLand the first Polish developer to succeed in releasing a game domestically as well as internationally. While neither became a big seller, the mere fact that *Heartlight* made its way onto the cover disk (a floppy disk attached to the front cover of a magazine) of the British *PC Format* magazine was reported by the Polish gaming press as a cause for celebration ("Polacy nie gęsi..." 1994, 74; "X-Land Rulez" 1994, 4). In 1995, Pelc and Miąsik left xLand to start their own development house, Chaos Works. The company put out three titles, *Fire Fight* (1996), *Excessive Speed* (1999), and *Akimbo: Kung-Fu Hero* (2001), and disbanded in 2001, after which Pelc retired from the game business.

Meanwhile, Miąsik, who departed from Chaos Works shortly after the release of *Fire Fight*, joined the very studio Pelc helped to establish just a few years earlier. At the time, the Avalon team was putting finishing touches on *A.D. 2044* (1996), an enhanced remake of Roland Pantoła's 1992 graphical adventure game for Atari computers. Like its earlier incarnation, the game is loosely based on Juliusz Machulski's 1984 cult comedy *Seksmisja*. The hero—a test subject of a cutting-edge experiment in suspended animation—is awakened half a century into the future to find himself the last surviving male in a quasi-totalitarian, matriarchal society. The unprecedentedly extensive advertising campaign proclaimed boldly, "The 21st century begins on 9/9/96"—referring as much to the adventure's temporal setting as to the new era in game development it ostensibly ushered in. Indeed, with two CDs' worth of 3-D graphics and professional voice work (courtesy of a local puppet theater troupe), *A.D. 2044* did raise the bar for the Polish games industry. Although reviewers were not enthused about the character models and animations and pointed out that—between the static scenes and the pre-rendered movies simulating movement between locations—opportunities for exploration were rather limited, the write-ups were very favorable. Even the harshest critics conceded that, uninspired plot and some logical lapses in puzzle design aside, the game "approached world standards" (Jor's Tymoteoo de wrajter 1997, 96).

Much like Machulski's movie, *A.D. 2044* was steeped in the realities of life under the communist regime. It was to these idiosyncrasies of the source material that the studio attributed the game's less than stellar performance outside the country. As Tomasz Pazdan, one of Avalon's executives, would later note, the experience effectively "cured" the studio of any notions to base scripts on "local folklore" (Banzai 1998, 14). Accordingly, its next product, *Reah: Zmierz się z nieznanym* (*Reah: Face the Unknown*, 1998), adopted a more straightforward science fiction narrative revolving around the mystery of the planet Reah and its alternate dimension. With a more expansive world, significantly more detailed high-resolution graphics, ambient animations, and twenty-five minutes of live action scenes with professional actors, the game took up six CDs or a double-sided DVD. Since software piracy was still a pervasive problem, there was little hope a project of such scope could generate enough domestic sales to ensure profits for the studio. *Reah* was therefore envisioned as a product targeted primarily at Western markets.

That philosophy found its ultimate expression in Avalon's next and final release, *Schizm: Prawdziwe wyzwanie* (released as *Schizm: Mysterious Journey* and *Mysterious Journey: Schizm*, 2001). At the time the most expensive Polish game ever, it was created entirely in English from a script written by Australian sci-fi writer Terry Dowling. The story follows the two-person crew of a supply vessel en route to a science station on Argilus, a planet once inhabited by a highly advanced civilization, now seemingly abandoned. Following a strange malfunction, the pair finds themselves stranded on the alien world, the research team gone without a trace. The player takes control of both characters, coordinating their actions to solve the puzzles. The press, while impressed with the overall quality, was somewhat bitter about the fact that a Polish-developed product had to be localized for domestic release, bemoaning the "wooden" quality of the translated dialogues and obvious mismatch between the dubbed lines and characters' lip movements (Hazard 2001, 50; Micz 2001, 51).

Schizm turned out to be Avalon's last game; shortly after its release, the company changed focus to educational and hobby software. In the meantime, the six-person development team (Roland Pantoła, Maciej Miąsik, Danuta Sienkowska, Robert Ożóg, Łukasz Pisarek, and Krzysztof Bar) went on to form Detalion. Under the new banner, the team continued its collaboration with Dowling, who provided scripts for *Schizm II: Kameleon* (released as *Schizm II: Chameleon* and *Mysterious Journey II*, 2003) and *Sentinel: Strażnik grobowca* (released as *Sentinel: Descendants in Time* and *Realms of Illusion*, 2004), the latter an adaptation of the author's short story, "The Ichneumon and the Dormeuse." The games utilized an entirely new engine, eschewing the pre-rendered panoramas of *Reah* and *Schizm* in favor of a fully explorable 3-D environment. Both titles were warmly received by genre enthusiasts. Alas, by the early 2000s, the genre had been consigned to niche status. The studio kept afloat for a few more years by taking on outsourced work (under the name Detalion Art). In 2007, it was merged into City Interactive.

Another casualty of the late 2000s market shakedown was Metropolis Software. Founded in 1993 by Grzegorz Miechowski, Andrzej Michalak, and Adrian Chmielarz, Metropolis debuted with *Tajemnica statuetki*

(*The Mystery of the Statuette*, 1993), a crime mystery putting the player in the role of an Interpol agent investigating the disappearance of ancient religious artifacts. The game was well received, though mostly on the strength of its script. The production values were Spartan, with game locations represented through digitized photographs, lack of animations or music, and minimal sound effects. Still, it was the first Polish adventure for the PC platform, and for that reason alone the reviewers were willing to overlook its shortcomings. As one of them reminded the readers, any attempts to draw direct comparisons between *Tajemnica* and the latest offerings from such giants as Sierra On-Line, LucasArts, Infogrames, or Delphine were obviously misguided. The distance separating the studios—with regard to experience, financial and technological resources, infrastructure, and legal protection—was tremendous, he pointed out, and would not be closed overnight (Zgliński 1993, 15). In any case, the draw of a Polish-language adventure (coupled with Metropolis's aggressive advertising efforts) was enough to move approximately 5,000 copies of the game over the following two years (Alex 1996, 58)—exceptional market performance, by then current standards.

By comparison, the studio's next release represented a tremendous step forward. *Teenagent* (1995) featured colorful, hand-drawn backgrounds and rotoscoped, fluidly animated character graphics. With its quick-witted and sharp-tongued teenager-cum-secret-agent protagonist, a dastardly world domination plot to foil, and a healthy dose of oddball humor, the game became an instant success, with universal praise culminating in the 1995 Game of the Year award. An enhanced, fully voiced[15] version was released the following year as *Nowy Teenagent* (*New Teenagent*, 1996)—the first Polish game on compact disk. Metropolis's subsequent releases included the side-scrolling shooter *Katharsis* (1997) and another graphic adventure, *Książe i tchórz* (*The Prince and the Coward*, 1998). Increasingly disillusioned with domestic sales figures, the studio sought access to foreign markets, finding a suitable partner in TopWare Interactive. As a result, the next two titles, sci-fi real-time strategy *Reflux* (released as *Robo Rumble*, 1998) and horror-themed, turn-based *Gorky 17* (released in North America as *Odium*, 1999), went on to achieve moderate success outside of Poland. Following the poorly received action RPG *Archangel* (2002), Metropolis revisited the *Gorky* universe with *Gorky Zero: Fabryka niewolników* (literally "Slave Factory" but released as *Gorky Zero: Beyond Honor*, 2003) and *Gorky 02* (released in Europe as *Aurora Watching* and in North America as *Soldier Elite*, 2004). Narratively a prequel and a sequel to the original game, the follow-ups abandoned the strategy/role-playing roots in favor of third-person stealth action—a move neither reviewers nor players appreciated. Despite the overwhelmingly negative reactions, the studio continued to tinker with the third-person action-adventure format, the efforts culminating in the 2007 supernatural shooter *Infernal*. Not without some faults (such as linear level design and questionable AI), the game nevertheless seemed to bode well for the developer's future. Unfortunately, the next project, a post-apocalyptic first-person shooter under the working title *They*, never saw the light of day. Acquired by CD Projekt in 2008, Metropolis closed the following year, by some accounts due to the difficult market situation, by others, due to internal tensions. The work on *They* was suspended indefinitely. A large part of the former Metropolis team, led by Grzegorz Miechowski, left to found 11 bit studios (*Anomaly: Warzone Earth*, 2011); the rest were integrated into the ranks of CD Projekt RED, where they assisted with the development of *Wiedźmin 2: Zabójcy królów* (*The Witcher 2: Assassins of Kings*, 2011).

Hearts, Minds, and Wallets: Software Distributors and the Battle for the Market

At the same time, the game market was undergoing significant transformations. In the wake of the 1989 fall of the Communist regime, rapid and far-reaching changes swept through virtually every aspect of the country's life, electronic entertainment business being no exception. With the newly opened borders, exchangeability of the złoty, and liberalization of the economy, official importation and distribution of software from the West were at long last possible. One problem remained, however. The outdated copyright regulations—hitherto an effective counterbalance to access restrictions—were now a major stumbling block on the way toward a legal software market. That such a market began to emerge in spite of complete lack of legal protection is a testament to the valiant early efforts of publisher-distributors such as LK Avalon and Mirage Software. Arguably, their greatest contribution to the development of the Polish games industry lay not in the specific titles for which they are now remembered, but in their efforts to promote legal software and educate the consumers. The latter, for their part, did not prove unreceptive. After years of shoddily duplicated compilation cassettes and disks, a properly published game—with its full color cover, instruction manual, warranty, and reasonable price to boot—was something of a treat. The "original" quickly became a marker of exclusivity and many gamers accepted buying legitimate copies from authorized distributors as good form—not so much *despite* as precisely *because of* the absence of proper regulation.

The combination of relatively low prices and good quality worked in favor of the domestic games industry. Still, while many people were willing to support local developers, most had no qualms at all about copying software of foreign origin. Entrepreneurs attempting to set up official importation and distribution channels were facing a seemingly Sisyphean task. The Western 8-bit markets had all but dried up, and most of the titles finally receiving official release in Poland had already been in "informal" circulation for years. In the case of 16-bit and 32-bit software, the most pressing problem—for years to come—was the price tag. To be sure, prices were still substantially lower than in the West, as Polish distributors, attuned to the peculiarities of local economy, negotiated cuts as high as 80%. Even operating at minimum profit margins, however, they were hard-pressed to withstand the competition from computer markets and hundreds of semi-legal "computer studios," where as many as thirty games could be procured for a price equivalent to that of a single title from an authorized retailer. The most logical course of action was to offer something the gray market for the most part could not: Polish-language instruction manuals.

Such was the strategy adopted by IPS Computer Group (established 1991). Under its "be original—buy original" slogan, the company distributed games and utilities for the Amiga, Atari ST, and PC platforms. An unfortunate drawback of the approach it adopted was the narrowing of the range of games that entered authorized circulation in those early years, as IPS focused particularly on genres that benefited the most from proper documentation, such as simulation, strategy, role-playing, and (to some extent) adventure. In 1992, LK Avalon expanded its catalog with a small selection of Zeppelin Games releases for the Atari 8-bit computers but soon thereafter channeled most of its efforts into development and publishing. Mirage Software took note and, from 1993 on, established distribution agreements with a large roster of Western publishers—including

Codemasters, Ocean, US Gold, Infogrames, Core Design, Krisalis, Millenium, Silmarils, and others—quickly becoming Poland's second largest software distributor.

Meanwhile, a heated debate was taking place regarding the forthcoming revisions of copyright regulations. Gaming magazines, in particular, became a key forum for discussion over the necessity of such reform, its potential pitfalls, and possible consequences. While developers, publishers, and distributors treated software piracy as a moral and (after 1994) legal problem,[16] the majority of gamers expressing their opinions on the issue approached it from a purely economic perspective. Despite all the drawbacks associated with unauthorized copies (the lack of documentation, warranty, and tech support, as well as a very real risk of ending up with cut down or altogether nonfunctional games), they simply represented better value for the money, especially for the allowance-dependent youth. The economic reality of the transition years was certainly a key factor: with exchange rates unfavorable to the złoty and the population's spending power still very low, the prices of high-profile new releases appeared exorbitant.[17] Equally significant were the attitudes toward the value (or lack thereof) of software, inherited from a decade of gray market practices. While the high costs of hardware had always been—and continued to be—regarded as a matter of fact, the enduring, widespread view assumed that they would be offset by low costs of software, which could be procured cheaply, if not for free. The power of this belief was so strong that some seriously considered the possibility that the introduction of the new copyright law would lead to a market crash and thoroughly destabilize Polish computer culture, or, at worst, bring its development to a screeching halt. Understandably, then, publishers and distributors' immediate goal was not *eradicating* software piracy so much as *unsettling* it as the dominant market model. Thus, up until the late 1990s, the fight for legal software placed heavy emphasis on efforts to change consumer attitudes toward a more expensive but authorized product.

The Copyright and Related Rights Act of 1994 at long last extended legal protections to computer software. However, since matters of copyright fell under the Civil Code, violations could only be prosecuted on complaint. This effectively limited the protections to software with an official distributor in the country. The years 1994–1998 saw a series of raids on Warsaw computer markets, organized by IPS in cooperation with the local police department. The initiatives generated publicity for the cause of legal software but had relatively little quantifiable effect, as most cases were ultimately dismissed due to "low civil harm." As late as in early 1998, Stowarzyszenie Producentów i Dystrybutorów Oprogramowania (the Association of Software Publishers and Distributors) complained in an open letter to the Minister of Justice Hanna Suchocka about the minimal conviction rate in copyright infringement cases, colossal losses to both the software industry and the state, and, most alarmingly, the mounting evidence of increasing involvement of organized crime in the software black market. Finally, in September of that year, amendments to the Criminal Code made copyright infringement a criminal offense, prosecutable *ex officio*.

The act may not have brought about instant and dramatic changes, but the market situation did nevertheless begin to improve. According to Grzegorz Onichimowski, the CEO of IPS Computer Group, by early 1993, sales averaged 2,000 copies per game (Onichimowski 1993, 4); by 1995, that estimate increased to 4,000–6,000 copies (for the better-selling games) (Onichimowski 1995, 60). Of course, Onichimowski pointed out, the numbers were still a far cry from the 30,000–40,000 unit sales necessary for a game of "reasonable" quality (by "Western standards") to be profitable. With the domestic market unable to support requisite sales levels, he concluded, it was only natural for Polish developers to seek access to foreign markets instead (Onichimowski 1995, 61).

More significant changes took place in the late 1990s and early 2000s, as sales surged upward in response to new pricing policies. A key role in these developments was played by TopWare Interactive Poland. Established circa 1995 as a branch of TopWare Interactive AG—at the time, the largest German software publisher—the studio debuted in 1997 with two titles created by its internal development team, Reality Pump: *Jack Orlando: A Cinematic Adventure*, a point-and-click murder mystery set in the 1930s US; and *Earth 2140*, a post-apocalyptic sci-fi RTS. With the financial backing and access to Western European markets provided by its parent company, TopWare was able to set the retail price for its offerings at an unprecedentedly low 39.95 PLN[18]—just over a quarter of the average cost of a new release. As a result, each game went on to sell approximately 20,000 copies (Wawrzeszkiewicz 1999, 80). (*Earth 2140* also managed to achieve moderate success outside of Poland—with 120,000 units sold in Germany [Hassinger 1997, 89], for instance—and spawned two sequels: the imaginatively titled *Earth 2150* [Reality Pump, 2000] and *Earth 2160* [Reality Pump, 2005].) TopWare's distribution efforts focused on "affordable over revolutionary,"[19] with inexpensive releases such as the German-developed *Knights and Merchants: The Shattered Kingdom* (Joymania, 1998) and *Emergency: Fighters for Life* (Sixteen Tons Entertainment, 1999). Their hands forced by TopWare's rock-bottom pricing, other publishers and distributors responded—some reluctantly—with budget lines of their own, contributing to further normalization of the market.[20]

Some credit is also due to gaming magazines, which, in the late 1990s, began to include older games on their cover CDs—much to the chagrin of the industry, whose representatives claimed this devalued software in the eyes of the consumers. In 1999, a consortium of publishers and distributors attempted to block the practice under threat of sanctions. The effort was unsuccessful, and cover disks with full versions of games became an important part of the Polish gaming landscape.

By the early 2000s, as the *per capita* income gradually climbed and prices continued to drop, social acceptance of piracy declined considerably. According to the data collected by *CD Action* magazine, in 1995, the average (gross) monthly salary could buy 6.4 games; by 2002, that number rose to 29.4 (Jasiński 2002, 91). As a result, between 1994 and 2002, software piracy rates in Poland dropped by nearly 30% (Moores 2005, 11). The sales of the most successful games (like *The Sims* [Maxis, 2000] and *Heroes of Might and Magic IV* [New World Computing, 2002]) began to approach 50,000 copies (Jasiński 2002, 92). With budgets rising even faster, however, Polish developers eyed the Western markets more intently than ever before.

Are We There Yet? The Polish Game Industry in the Twenty-First Century

The closures, acquisitions, and mergers of the 2000s left indelible marks on the industry landscape. From the settling dust, four major studios rose to prominence. The oldest among them, Techland, started out in 1991 as a software publisher and distributor. Soon, the company expanded its scope to localization work and, in the late 1990s, moved into game development. Techland's first notable product was the vehicular combat actioner *Crime Cities* (2000), whose design (inspired by *Blade Runner* [1982] and *The Fifth Element* [1997]) earned it the Excellence in Visual Art award at the 1999 Independent Games Festival. The studio's breakthrough release, however, was the 2003 *Chrome*—a sci-fi first-person shooter with mild RPG trappings. In spite of some problems (particularly in the level design and AI departments), *Chrome* was praised for its original story and impressive visuals. The game was powered by Techland's proprietary 3-D engine, to which it lent its name. Subsequent iterations of the Chrome Engine have been used in all of the developer's major products since, beginning with the highly rated racing game *Xpand Rally* (2004). After revisiting the futuristic *Chrome* universe in *Chrome: SpecForce* (2005), the studio set its sights on the American Old West with *Call of Juarez* (2006). The game alternates between stealth-based and gunplay-based segments as the player takes on the dual role of a fugitive falsely accused of murder and a pursuing gunfighter. With a largely negative response to the stealth missions, the 2009 prequel *Call of Juarez: Bound in Blood* focused primarily on combat—close or ranged, depending on the player's character choice. In 2010, Techland delivered the ATV and dirt bike racer *nail'd*, and the following year saw the release of two high-profile titles. *Call of Juarez: The Cartel* (2011) took the franchise into the present day, pitting the protagonists against Mexican drug lords. The game was poorly received, as reviewers bemoaned the loss of the Wild West charm, shallow and repetitive gameplay, as well as insensitive treatment of drug-related violence. *Dead Island* (2011), a survival horror action RPG set in a tropical resort overrun by zombies, generated more positive reactions, as well as brisk sales.[21] The story of the zombie outbreak continues in *Dead Island: Riptide*, while the *Juarez* series returned to its Old West roots in *Call of Juarez: Gunslinger* (both 2013).

CD Projekt, the brainchild of Marcin Iwiński and Michał Kiciński, was founded in 1994 as a distributor specializing in CD-ROM-based games. The company began with low-volume imports but soon obtained formal distribution rights from several major US publishers. In 1996, it began partial localization of foreign titles, providing Polish-language packaging and instruction manuals. In the same year, deeming the market sufficiently strong for Polish games to be profitable, CD Projekt established exclusive distribution agreements with developers Seven Stars (*Kajko i Kokosz, Wacki: Kosmiczna rozgrywka* [*Cosmic Caper*, 1998]) and Leryx-LongSoft (*Lew Leon* [*Leo the Lion*, 1997], *Clash* [1998], *Golem* [2003]). Beginning in 1997, selected titles received full localization, often involving high-profile actors; particularly significant was the 1999 publication of the Polish version of BioWare's *Baldur's Gate* (1998)—the first game to sell more than 100,000 copies ("Nasza historia" 2013). In 2002, CD Projekt established a development branch: CD Projekt RED. The studio debuted in 2007 with *Wiedźmin* (*The Witcher*), an action RPG featuring itinerant monster hunter Geralt of Rivia, the protagonist of a cycle of works by fantasy writer Andrzej Sapkowski. The game presented an original narrative, but retained the dark tone and moral complexity of the source material, forcing the player to make difficult choices with far-reaching consequences. A runaway hit, *Wiedźmin* sold more than 35,000 copies in the first three days—the most successful premiere in the history of the Polish game industry ("Nasza historia" 2013). *Wiedźmin 2: Zabójcy królów* followed in 2011 and quickly matched its predecessor's performance.[22] By 2013, the combined worldwide sales of both titles surpassed 5 million ("CD Projekt zakończył 2012 rok z najwyższym zyskiem w historii" 2013), making *Wiedźmin* Poland's most popular game franchise to date. The final chapter of the planned trilogy, *Wiedźmin 3: Dziki gon* (*The Witcher 3: The Wild Hunt*) is currently under development, along with *Cyberpunk 2077*, a new futuristic RPG based on the *Cyberpunk* pen-and-paper system. In 2008, CD Projekt launched Good Old Games, a digital distribution platform for DRM-free "classic" games. In 2012, the service changed its name to GOG.com and expanded the catalog to include modern (primarily independent) titles. Soon thereafter, the publishing and distribution branch was renamed cdp.pl.

Cofounded in 2002 by Adrian Chmielarz (who left Metropolis in 1999), Andrzej Poznański, and Michał Kosieradzki, People Can Fly was catapulted into the spotlight by *Painkiller*. The frenetic first-person shooter follows its tragically deceased hero's quest for redemption as he fights his way through the nightmarish realms of purgatory and hell. Enthusiastically received, the game was lauded for its fast-paced, violent combat, harkening back to the genre's roots, as well as haunting visual design. The studio's bid to license the Unreal Engine from Epic Games unexpectedly led to a commission to create additional content for *Gears of War* and, subsequently, to develop the PC version of the game (2007). In the wake of the success of the port, Epic acquired a controlling interest in the company. People Can Fly's next oeuvre, sci-fi FPS *Bulletstorm* (2011), refined the *Painkiller* formula, focusing on inventive and gory kills ("skillshots") strung into elaborate combos. While its over-the-top violence and coarse language stirred a minor controversy in the US media, *Bulletstorm* received very strong notes from the gaming press and went on to sell close to 1.5 million copies. In August 2012, PCF—then at work on *Gears of War: Judgment* (2013)—was bought out by Epic Games. At that time, the three original founders departed from the company to start a new enterprise, the independent development studio The Astronauts. Their first title was a horror adventure, *The Vanishing of Ethan Carter* (2013).

Also a comparatively young company, City Interactive came into existence in 2002 as a result of a merger of the development branch of Lemon Interactive with two smaller studios, We Open Eyes and Tatanka. A developer, publisher, and distributor, the company made a name for itself as a creator of budget first-person shooters (among others, the *Terrorist Takedown* series [2004–2008] and *Code of Honor* series [2007–2009], *Sniper: Art of Victory* [2008], *Mortyr III: Akcje dywersyjne* [*Battlestrike: Force of Resistance*, 2008], *Mortyr: Operacja Sztorm* [*Operation Thunderstorm*, 2008]) and historical arcade flight simulators (in particular, *Wings of Honor* [2003], its sequel *Wings of Honor: Battles of the Red Baron* [2006], and the *Combat Wings* series [2005–]). In 2007, City merged with its sister studio, Oni Games (also established in 2002, as a separate publishing and distribution branch for titles not developed in-house), and acquired Detalion. Capitalizing on the experience of the former Detalion team, the company branched out into adventure games with the *Art of Murder* series and *Chronicles of Mystery* series (both 2008–2011). Its most successful product thus far, however, has been the stealth-based FPS *Sniper: Ghost Warrior* (2010). Although reviewers were not enthused about its linearity and extreme difficulty, the game sold approximately 3 million copies. A sequel, *Sniper: Ghost Warrior 2*, was released in 2013.[23]

Naturally, while the aforementioned studios stand out in terms of international visibility and market performance (each with at least one title to have sold over a million units), they represent only a portion of

Figure 3
Geralt of Rivia: mutant, outcast, national treasure (concept art from *Wiedźmin 2: Zabójcy królów* [*The Witcher 2*, 2011]).

Poland's diverse and prolific game industry. Following in their footsteps are several up-and-coming mid-sized teams, most notably Reality Pump Studios[24] (*Two Worlds* [2007], *Two Worlds II* [2010]), The Farm 51 (*NecroVisioN* [2009], *Painkiller: Hell and Damnation* [2012]), Flying Wild Hog (*Hard Reset* [2011]), and 11 bit studios. The past decade has also seen rapid expansion of the casual sector, driven by such developers as One2Tribe, Ganymede, Fabryka Gier (Game Factory), Vivid Games, Bloober Team, and Artifex Mundi, to mention just a few ("Video Games Industry in Poland: Information Booklet" 2013).

From 1980s guerilla computing through the sweeping changes of the 1990s and the gradual normalization of the 2000s, the Polish gaming industry has traveled a long way indeed. Where is it now? Projections in 2012 estimated the value of the Polish game market at USD $350 million to $450 million. In itself, the number may not be especially impressive—less than 1% of the global total, even by the most optimistic accounts. Even so, as of 2013, Poland ranked among the fastest-growing electronic entertainment markets in Europe (PricewaterhouseCoopers 2013). With no shortage of skill, enthusiasm, and resources, the outlook is better than ever before. Established and new IPs are developed for all current platforms, including a variety of mobile devices and eight-generation consoles. Critically and financially successful franchises have earned the industry international recognition and numerous accolades. It seems quite ironic, then, that one of the most pressing challenges lurks, as it were, in its own backyard. According to a recently published report, despite increasingly enthusiastic media coverage, Polish consumers, by and large, expect domestically produced titles to be mediocre and derivative (Monday PR and SW Research 2012). In other words, the Polish game industry seems to have an image problem. How will it negotiate this hurdle? Only time can tell, but if Chmielarz's remarks are to be taken seriously, it might have to face some of its own demons first.

Notes

1. As well as a certain deficit of cultural sensitivity, which, in light of some recent missteps, appears to be a troublingly common malady among Polish game developers.

2. Homegrown efforts to create computers for the masses—the Mera-Elzab Meritum, the Elwro 800 Junior, and the Unipolbrit Komputer 2086 (clones of, respectively, the TRS-80, the ZX Spectrum, and the Timex Sinclair 2068)—soon proved to be textbook examples of "too little, too late."

3. "Przedsiębiorstwo zagraniczne" (literally, foreign enterprise) was a legal status designator, indicating a company under foreign ownership.

4. "Giełda" 1986a, 25; "Giełda" 1986b, 48. Unsurprisingly, the relatively low-priced ZX Spectrum was the most popular choice in the early years, accounting for a guesstimated 70% of all home computers in Polish homes (see Majewski 1986, 3).

5. At a price equal to or exceeding that of the computer itself, a floppy disk drive was a relatively rare item until the 1990s.

6. In particular, *Bajtek*, since 1985, and *Komputer*, since 1986. Dedicated gaming magazines did not begin to appear until 1990.

7. Understandably, this also had an impact on the industry. A few isolated exceptions aside (such as the *Kao the Kangaroo* series from Tate Interactive [formerly X-Ray Interactive, 2000–2005] and *Painkiller: Hell Wars* [People Can Fly, 2006]), console game development did not begin in earnest until the seventh generation.

8. As Borkowski recalls, the sum was more or less equivalent to three months' worth of the wages he collected as a lab assistant at the Industrial Chemistry Research Institute, arguably making him the first Polish software developer to make significant profit off a computer game—no small feat at a time when any attempts to monetize the hobby were snuffed out promptly and definitively by thriving software piracy. (See Marcin Borkowski at the speccy.pl forum, November 17, 2011, http://speccy.pl/forum/index.php/topic.29.msg683.html#msg683.)

9. According to *Bajtek*'s poll results, 60% of the magazine readers were users of Atari 8-bit computers ("W waszych oczach" 1989, 4).

10. A particularly noteworthy example is GR8 Software and AtariOnline.pl's *Robbo Forever* (2009), a collection of fifty-nine extra-challenging levels released to celebrate the game's twentieth anniversary.

11. Although both games were very well received in Poland, the history of their distribution outside the country poignantly illustrates the problem of which Kucharski spoke only a few months earlier. In 1991, *Misja* (renamed *Mission*, or, in some territories, *Mission Shark*) and *Fred* were licensed out to Zeppelin Games. With the Western European 8-bit market rapidly declining, the latter earned the dubious distinction as the worst-selling title in the company's history (see "Quick Facts," Eutechnyx Ltd., available at http://press.eutechnyx.com/sites/default/files/article/eutechnyx-quick-facts/eutechnyxquickfactssept09.pdf).

12. World Software followed up with *Doman: Grzechy Ardana* (*Doman: Ardan's Sins*, 1995), this time inserting the impossible physiques and over-the-top brutality into a heroic fantasy setting.

13. Even the most positive reviews could not help noting that, despite superficial similarities to *Lotus Esprit Turbo Challenge* (Magnetic Fields, 1990), the game had more in common with the age-old *The Great American Cross-Country Road Race* (Activision, 1985).

14. Interestingly, Apple computers were virtually unknown in Poland at the time.

15. As one of the selling points, the voices were provided—with varying results—by the staff of *Secret Service* magazine.

16. Of course, the issue was complicated by the fact that most individuals involved in the software publishing and distribution business at the time hailed from the infamous gray market background themselves—a detail those less enthusiastic about the legal reform were always quick to point out.

17. Toward the end of 1993, as much as 20% of average gross monthly income.

18. In response to the hyperinflation of 1989–1990, the złoty was redenominated in 1995 at the ratio of 10,000 PLZ to 1 PLN.

19. TopWare's marketing director, Jarosław Owczarek, quoted in "TopWare wchodzi!" (1997, 15).

20. Ironically, budget titles turned out to be TopWare's undoing. In 2001, TopWare was bought out by Zuxxez Entertainment, a German development and publishing house founded by a group of employees of the recently bankrupt TopWare Interactive AG. Under new management, the company turned almost exclusively to budget titles, with increasing focus on vehicle simulation games (most notably the *18 Wheels of Steel* and *Euro Truck Simulator* series from Czech developer SCS Software). Despite some successes—chief among them the surprising performance of the internally developed arcade shooter franchise *Kurka Wodna* (*Chicken Shoot*)—its situation declined. After years of financial troubles and personnel changes, Zuxxez finally closed the subsidiary in 2010.

21. With over 5 million copies sold, *Dead Island* currently holds the title of the best-selling Polish game in history ("O grze" 2013).

22. The game was also a major critical success, earning the top spot in six categories, including Best Game, Best Game Design, and Best Gameworld, at the 2012 European Games Award competition ("Winners 2012," *European Games Award*, accessed June 14, 2013, http://www.european-games-award.com/).

23. Since late 2012, the development branch of City Interactive operates under the name CI Games—a move widely interpreted as an indication of the company's desire to break away from its image as a bargain bin developer.

24. Formerly an internal development team at TopWare, Reality Pump split off as an independent entity in 2003.

References

Alex [pseud.]. 1996. Z wizytą w Metropolis. *Gambler*, February.

Banzai [pseud.]. 1998. Mała, wielka firma: Z wizytą w Rzeszowskim LK Avalon. *Secret Service*, September.

Borkowski, Marcin. 2011. Message to speccy.pl forum. November 17. http://speccy.pl/forum/index.php/topic.29.msg683.html#msg683.

CD Projekt zakończył 2012 rok z najwyższym zyskiem w historii. 2013. *CD Projekt*, March 21. http://cdprojekt.com/Relacje_inwestorskie/Wydarzenia,news_id,1858.

Chmielarz, Adrian. 2007. Niech szlag trafi PC-gry! Interview by Hut Sędzimir [pseud.]. *CD-Action*, October 2007.

enki [pseud.]. 2010. Jak powstają gry: Dystrybucja. *CD-Action*, February.

Giełda. 1986a. *Bajtek*, July.

Giełda. 1986b. *Komputer*, July.

Gulash [pseud.]. 1996. Review of *Franko*. *Secret Service*, November.

Hassinger, Dirk. 1997. Interview, *Secret Service*, November.

Hazard [pseud.]. 2001. Review of *Schizm: Prawdziwe wyzwanie*. *Świat Gier Komputerowych*, October.

Jasiński, Maciej. 2002. Ceny gier pod mikroskopem. *Świat Gier Komputerowych*, November.

Jor's Tymoteoo de wrajter [pseud.]. 1997. Review of *A.D. 2044*. *Gambler*, May.

Kluska, Bartłomiej, and Mariusz Rozwadowski. 2011. *Bajty polskie*. Łódź: Samizdat Orka.

Kucharski, Piotr. 1990. Myśli zaprogramowane. Interview by Maja Agata Wójcik, *Top Secret*, December.

Majewski, Władysław. 1986. Wokół komputera. *Komputer*, March.

Micz [pseud.]. 2001. Review of *Schizm: Prawdziwe wyzwanie*. *Secret Service*, October.

Monday PR and SW Research. 2012. Polska branża gier komputerowych: Analiza wizerunku medialnego i świadomości marek polskich producentów gier. *Poznań Game Arena*. http://www.gamearena.pl/pl/pga_w_pigulce/raport_gamingowy.pdf.

Moores, Trevor T. 2005. An analysis of the impact of culture and wealth on declining software piracy rates: A nine-year study. In *Proceedings of the Ninth Pacific-Asia Conference on Information Systems* (PACIS 2005). Bangkok.

Nasza historia. 2013. *CD Projekt*. http://www.cdprojekt.com/Grupa_kapitalowa/Historia_korporacyjna.

O grze. 2013. *Dead Island: Riptide*. http://www.deadislandgame.pl/riptide/ogrze.html.

Onichimowski, Grzegorz. 1993. Powiedz mi w co grasz... Interview by Marcin Przasnyski. *Secret Service*, January.

Onichimowski, Grzegorz. 1995. Nie jestem donkiszotem. Interview by Aleksy Uchański and Wojciech Setlak. *Gambler*, September.

Polacy nie gęsi ... 1994. *Gambler*, April.

PricewaterhouseCoopers. 2013. Global entertainment and media outlook: 2013–2017. *PwC.com*. http://www.pwc.com/gx/en/global-entertainment-media-outlook/segment-insights/video-games.jhtml.

Przeciętne miesięczne wynagrodzenie w gospodarce narodowej w latach 1950–2012. 2013. *Główny Urząd Statystyczny*. http://www.stat.gov.pl/gus/5840_1630_PLK_HTML.htm.

Quick Facts. 2009. *Eutechnyx Ltd*. http://press.eutechnyx.com/sites/default/files/article/eutechnyx-quick-facts/eutechnyxquickfactssept09.pdf.

Szperkowicz, Jerzy. 1987. Skąd się biorą komputery? *Horyzonty Techniki. Suplement: 64 strony o komputerach.*

TopWare wchodzi! 1997. *Gambler.* November, 15.

Video games industry in Poland: Information booklet. 2013. *Let's PLay!* http://lets-play.com.pl/Download/33.

Wawrzeszkiewicz, Barbara. 1999. Top Ware: Dobrze, tanio i po polsku. Interview by Piotr Orcholski. *Świat Gier Komputerowych,* April.

Wencel, Lucjan. 1991. Nie tylko Atari. Interview by Jarosław Młodzki and Wojciech Zientara. *Moje Atari,* March–April.

Winners 2012. 2012. *European Games Award.* http://www.european-games-award.com/.

W waszych oczach. 1989. *Bajtek,* June.

X-Land Rulez. 1994. *Top Secret,* February.

Zgliński, Dariusz. 1993. Review of *Tajemnica statuetki. Computer Studio,* July.

PORTUGAL

Nelson Zagalo

It took ten years after PONG (1972) and the arrival of Sinclair microcomputers (Adamson and Kannedy 1986) for Portuguese game development history to begin. The first video game we have been able to trace was José Oliveira's *Laser* (1982), created on a Sinclair ZX81. During the last decades, the Portuguese video game community has been able to create and implement for almost any game platform, creating around 350 video games; however, the main platform has always been personal computers, rather than proprietary consoles, and the latter figure very little into the country's video game history.

Portuguese video game history is mainly characterized by the production of various highly interesting games and game technologies, but ones without international consequence. This means that from 1982 to 2012, we can find highlights in every five-year period; however, these hardly relate to each other, thus failing to grow and empower a Portuguese industry in the domain of game development. This is not unlike the situation faced by the other arts in Portugal, namely the film industry. Therefore, we'll spend this chapter listing and explaining the importance of each of these highlights in the history. We believe that looking back, putting together our story, and sharing it will contribute to the avoidance of repeating errors from the past, and so contribute to a stronger Portuguese game industry.

The First Games

In the 1980s, a revolution arrived from England with Sinclair microcomputers, the ZX81, and ZX Spectrum. José Oliveira, a college computer science student, received his first ZX81 during Christmas of 1982. Since he had already played with programmable calculators, he started programming the ZX81 right away and developed more than fifty little programs in a couple of days. After having implemented a program for drawing mathematical functions (such as $y = f(x)$), Oliveira went further and decided to add a goal to the graphics being drawn on the screen. The addition of a goal, and points for each successful goal, inspired the ideas for his first two games, *Laser* (1982) and *Bala* (1982).

The two games had very similar mechanics, which involved changing the angle of a line drawn on the screen. In *Laser*, the line would be drawn linearly, imitating a laser beam, while in *Bala* the line included a variable for gravity that would add the curvature typical of a cannonball trajectory. In both games, the goal was for the player to set the angle at which the computer would draw the line and attempt to reach the blinking pixel. The game featured a points system so that every time it hit the pixel, the player not only received a congratulatory message, but also a point was added to the player's score. We can define these games as basic shooters, and in the case of *Bala* (taking into account its particular use of physics), we have, to some extent, the basic mechanics used in a game like *Angry Birds* (2009). Oliveira never thought about selling his games, but he managed to distribute them at the town coffeehouse, which made him quite popular and suggested that games could become commercial products.

The First Commercial Games

In 1983, the ZX Spectrum arrived in Portugal and game fever began, not only for playing, but also for creating new titles. These were promising times for microcomputing, not only in Portugal but also abroad (Metzstein 2009). Most of the titles were limited to sharing among friends, but some people started to think that maybe they could do something more with their creations. Thus, two friends, Marco Carrasco and Rui Tito, from Portimão in the Algarve, after having created their first game in BASIC, *Galaxy Patrol* (1983), signed a contract with the British company Wizard Software for distribution in the UK of three new games written in Assembly language: *Mr. Gulp*, *Defenders*, and *Moon Megatron* (all 1984). The games were not a success, but their creators referred to the importance of the surrounding culture for maintaining the motivation that kept them creating games. According to Marco Carrasco,

> For us, the appearance of Portuguese magazines was very important (*Cérebro* and later the magazine bi-monthly *Mini-Micro*). They were a stimulus to programming games for publication. In 1985, *Mini-Micro* launched a game programming competition to encourage the creation of programs; we competed and won a dot matrix printer for the ZX Spectrum, a luxury at the time. Thereafter, the editor of *Mini-Micro* invited us to start writing regularly for the magazine and challenged us to develop a book on machine code programming. The invitation resulted in a book dedicated to the programming of games, *Super Programas em BASIC e Código Maquina* (1986).[1]

Super Programas em BASIC e Código Maquina (*Super Programs in BASIC and Machine Code*) (1986) would be the first Portuguese book completely dedicated to game development. As a result of their rising interest and achievements, Mario Carrasco and Rui Tito tried a second venture in the UK, this time working with Gremlin Graphics. They developed their biggest commercial success, *Alien Evolution* (1987) (see fig. 1), which returned more than £20 thousand. Gremlin's producer invited both friends to Sheffield, UK, to get to know them personally and show them the studios. As Marco told us, they couldn't believe that two eighteen-year-old guys could have done all the work required to load the machine assembler code with a tape recorder.

Figure 1
Paradise Café (1985) (top, left); *Alien Evolution* (1987) (top, right); *Portugal 1111* (2004) (bottom, left); and *Under Siege* (2011) (bottom, right).

The First National Hit

The year 1985 represented one of the higher moments for pop culture in Portugal in music, films, clothes, and also games, which included one of the biggest successes of the Portuguese game industry, *Paradise Café* (1985) (see fig. 1). It would be hard to find a person who had a ZX Spectrum in Portugal in the 1980s and did not know the game, but even those who did not will certainly have heard of it. However, despite several unsuccessful attempts by magazines and websites over the last decades to identify the game's author, we only know his nickname, Damatta. Analyzing the content of the game, we can easily understand why the author wants to remain unknown.

Paradise Café became famous not only for its adult themes, despite the poor quality of its gameplay, but because one of the elements that contributed most to its success was the fact that it was based on the so-called "Reinaldo story," which was a big topic of conversation in coffeehouses throughout the country. People were talking about a famous football player named Reinaldo, who had allegedly sent a pop singer to the hospital after they had intercourse. The singer belonged to the first Portuguese pop girl-band, Doce, a big success in those years. The story was never confirmed, no hospital records were ever found, and it is now classified as an urban myth. But these elements were appropriated by the game, creating a very direct relationship with the myth and triggering an intense spread via word-of-mouth, leading to big success for a very basic game.

Game Technology Made in Portugal

Timex, the American watch manufacturer, which was established in the nineteenth century, came to Portugal in the 1970s and settled in Costa da Caparica, where it still resides today. Its history is relevant in the world of video games because in 1980 its CEO, Fred Olsen, made a great transformation in all its manufacturing processes, closing factories and introducing new technologies into the remaining ones. More importantly, he met Clive Sinclair in 1981 and signed an agreement with him; afterward, Timex Corporation would manufacture and distribute Sinclair computers. In Europe, the product would be called the Sinclair and would be manufactured at Timex Scotland. For the US market, Latin America, and Eastern Europe, the product would be called the Timex Sinclair and would be manufactured in Portugal. Although the factory in Portugal was initially only an assembly line, the hiring of Alvaro Oliveira, an electrical engineering PhD, gave the factory an important role in terms of research and development (Beira 2004; Beira and Heitor 2004).

Timex Portugal began manufacturing the Timex Sinclair TS1000, which was the ZX81, to sell in the United States, but it completely failed in the US market. Commodore seized the market through a highly aggressive pricing policy, offering discount coupons to those with used consoles or machines. Soon the TS1000 was out of the American market, and in Portugal, design work began on the new Timex Sinclair for the American market, the TS1500, which would be an intermediate version between the ZX81 and ZX Spectrum, with 16KB RAM. However, the effort was not rewarded as it arrived late, when the ZX Spectrum was also coming to market.

Although the American market was not buying the expected numbers, production at the Portuguese factory was still quite significant, quickly reaching a thousand employees and the production of ten thousand machines per day. The factory was not just a simple assembly line, but included research and development in areas such as electronic integrated circuits, the molds for plastic boxes, the membranes for keyboards, and even innovative metallic inks as insulators. With the arrival of the ZX Spectrum, Timex Portugal created the TC 2048, which turned out to be a machine sold mostly in Portugal and Poland. In Portugal it had strong penetration, and most people bought it without even knowing it was made in Portugal.

Apart from microcomputers, the Portuguese plant, together with INESC (Instituto de Engenharia de Sistemas e Computadores), developed an add-on for the TC 2048, the 3000 FDD (itself almost a complete computer), which was connected to the TC 2048 to expand processing (via a Z80 at 4MHz) and memory (with 64KB RAM and three-inch floppy drives). It even had an operating system based on the CP/M by INESC. According to Professor António Dias Figueiredo, "The role of Timex Portugal in awakening the country's interest in computers was particularly important, as the company put a very high number of personal computers, Spectrum and

Timex, in the Portuguese market at a very low price, which in mid-1985 reached an internal volume exceeding 150,000 units, one of the highest penetration rates in Europe of computers per household" (Figueiredo 2004).

After the entire venture and achievement of success, the home computer market shifted to the IBM PC, and Timex Corporation (as well as Timex Portugal) left the computer market and returned to the world of watches. It is interesting to note that despite the fact that Timex was not a Portuguese brand, its impact on the country's electronics industry was tremendous, creating new companies and perfecting others. In this sense, the history of Timex is one of the greatest national success stories of the video game industry. Despite not having directly contributed to the creation of new games, it contributed greatly to the creation of a national culture of adherence to a whole new technological world, and with it, the acceptance of a new art form, the video game.

The Age of the CD-ROM

By the middle of the 1990s, domestically produced games were using the IBM PC platform but leaving behind diskettes and adopting the CD-ROM. This was not merely the result of technological change but a necessity, since the games were using bitmap images, raster graphics, and CD-quality music instead of polyphonic sound. In the case of the moving image, the CD-ROM permitted the appearance of the first full-motion video scenes, although in the early years programmers had to use very high compression, reducing quality, including the reduction of the number of frames per second. Portuguese game development in the time of the CD-ROM was divided into three main types: discoveries, education, and entertainment. Of the three, education is the least interesting for our account here because it mainly involved the copying of game mechanics from previous games or just imitations of imported games. Thus, we'll only present here the most interesting games of the "discoveries" and "entertainment" types.

During the 1990s, Portugal commemorated the 500-year anniversary of the discoveries of the Admiral Vasco da Gama and others, which would translate into a lot of available funding for the creation of artwork related to the theme. Also in 1998, the Lisbon World Exposition (Expo '98) was held, which would also fund projects related to the oceans and national endeavors. New companies emerged in those years, as a Grupo Forum that began activities in the area of multimedia production. However, we must note that of the over twenty CD-ROMs that Grupo Forum developed, only two had the characteristics of a video game and were not merely databases of digital illustrations, videos, and text. This is evident in the words of Marco Morais, coordinator of the group's first video game, *Aventuras da Peregrinação* (*Adventures of Pilgrimage*) (1997): "Since I started programming, I always had the ambition to make a game set in Portugal, but it was not an easy task to convince people to support the idea of making a game at the Group; one of the ways that I did this was to give the game educational attributes" (Zagalo 2013, 82).

Aventuras da Peregrinação was dedicated to the Portuguese explorer Fernão Mendes Pinto and presented a dramatization of his pilgrimage. It was divided into six narrative frames or levels in which the player had to solve some of the problems the explorer faced on his trip. In the process, the video game put players in the place of the explorer, creating a strong sense of identification and thus facilitating learning.

Then, in 1998, the Grupo Forum dedicated its second video game to Vasco da Gama, with *Vasco da Gama: A Grande Viagem* (*Vasco da Gama: The Great Journey*) (1998). *Vasco da Gama* was a great project in terms of multimedia experience. The video game came on a CD-ROM, but beyond this, it had a logbook (with activities stimulating the player's interaction with the events of the game) that could be used in classrooms. The game also came with a second CD, an audio CD with the game's entire soundtrack. The experiment was then complemented by a website with specific information for teachers and parents. In the online component, one could download more games, including those corresponding to the return voyage of Vasco da Gama.

The game, created by Alice Alcobia and Luís Paulo Pinho, presented a collectivist perspective of the trip, diminishing the leading role of Vasco da Gama and giving it to the storyteller. The idea was to avoid the creation of the hero figure and make the player understand that piloting a boat on the high seas was impossible to do alone.

Despite these interesting efforts, it was never easy to convince investors in Portugal to support games that only had entertainment as the goal. Thus, the only two commercial games made for the CD-ROM format between 1995 and 2004 were *Gambys* (1995) and *Portugal 1111* (2004). In 1993, one of the members of the team who had created the first commercial games for ZX Spectrum, Rui Tito, came back to start a new foray into the world of video games. This time he decided to join illustrator and animator Luis Peres, programmer Carlos Leote, and game designer Nelson Russa, plus Jorge Monica and Rui Rosa, under the name Portidata. The idea was to create a graphically advanced video game, far beyond the limitations of the Spectrum games of the 1980s.

In *Gambys*, gameplay was designed with a hundred levels that led through dozens of puzzles, guaranteeing a sense of progression through the increasingly difficult levels, ensuring more than eight hours of entertainment. The theme of the game revolves around an impending ecological disaster on planet Earth and the salvage effort carried out by the player, who controls one of the oldest species on the planet, the Gambys. The game was a milestone in the history of Portuguese game development on several fronts, being the first game to make use of three-dimensional computer animation, the first national game to be sold in a box, and the first game to be made by a team working professionally full time, developing the game for two years.

Gambys was very well received in Spain and Portugal, but Portidata wasn't able to distribute the game outside these two countries, despite having CD-ROM versions in Portuguese, Spanish, English, French, and German. There were advanced negotiations between Portidata and Psygnosis for worldwide release, but then the unexpected happened. The game had been made to run on MS-DOS, but in 1995, Windows 95 appeared, and Psygnosis wanted to launch only those games that were compatible with it. With such a small team, Portidata could not possibly create a conversion in time, and so the release was limited to Portugal and Spain. Because of that, after the game's launch, Portidata left game developing and dedicated itself to the development of management software.

The CD-ROM market was very strong in the second half of the 1990s, but the early 2000s saw this format decline very quickly with the emergence of technologies such as Flash, which could create online content that previously could only be created offline. Thus, 2004 was the year of the last commercial game released in Portugal. *Portugal 1111* would be launched on April 22, 2004, and distributed with *Visão*, a national news weekly magazine.

Portugal 1111 (see fig. 1) was a recreation of the Moors's territorial conquest in Portugal, made as a turn-based strategy game. Ciberbit created the game, with investment from the municipality of Soure and with scientific advisement from academic historians at the University of Coimbra. The project was conceived and coordinated with conviction by Professor Joaquim Carvalho and was designed to be a game, not merely educational software, since its inception. In terms of gameplay, *Portugal 1111* is similar to games such as *Age of Empires* (1997) and does not lag behind in anything, showing remarkable quality both in graphics and in artificial intelligence. All these attributes make *Portugal 1111* one of the most important games in Portuguese game development history.

Mobile Innovation

In 1999, António Câmara returned from the United States, where he spent a year as a professor at the Massachusetts Institute of Technology (MIT), and decided to create a company that turned academic research ideas into innovative products. He joined Eduardo Dias, Edmundo Nobre, Miguel Medicine, and Nuno Correia, who had worked together at the Faculty of Science and Technology at the New University of Lisbon. YDreams was founded in June 2000. Their goal was to move forward in the field of new communication technologies, a heavily hybrid area, and so they built the company with people from very different backgrounds, from engineering to the arts.

In 2002, this group, working together, delivered a demonstration of a mobile game made in Java. The demo triggered a green light for Eurico Moita, Ricardo Andrade, and Tiago Carita, who developed the first YDreams mobile game, *RockStar* (2002), which was released by Vodafone in November of that year and would become the most-played game of Vodafone Live.

Game development teams at YDreams were small, and projects did not last more than two to four weeks. Development was carried out on a proprietary YDreams platform initially developed by Eurico Moita and then improved over various games. Games were running in a 2-D environment, yet Tiago Carita developed some scenarios and 3-D animations which were then converted into 2-D sprites.

In 2003, Blast Theory, in conjunction with the Mixed Reality Lab at the University of Nottingham, presented the experimental game *Can You See Me Now?* (2003) at Ars Electronica (Benford et al. 2006). This was a new type of game involving "urban hunting" using GPS systems integrated into PDAs. *Can You See Me Now?* triggered enormous interest in location-based content development, which would bring together the scientific and artistic communities, and to which YDreams was not indifferent.

In that same year, YDreams entered the history of international video game development by creating the first commercial location-based game for mobile platforms, *Undercover* (2003), a game with limited multiplayer ability. The game's story goes something like this: after Russian documents with data on biological weapons are stolen by a group of terrorists, the player assumes the role of an agent of the TIA (Terrorism Intelligence Agency) with a mission to save the world from imminent war. To do this, players need to use their location in order to act, to take shelter in predefined locations, to prevent terrorist attacks, and to use the power of "teleportation" to travel between places on the map.

With in-house know-how, YDreams created the sequel *Undercover 2: Merc Wars* (2003). The bar was raised and the whole system was made more complex, leading YDreams to create what is probably the first commercial MMO mobile location-based video game. Tiago Carita, one of the game designers, described the complexity and challenges behind the game:

> *Undercover 2* was undoubtedly a pioneering milestone in my entire career, the game had complicated technology, it had to be a LBG (Location Based Game) and have actual maps as backgrounds. We created an RPG (Role-Playing Game) with various characters, we had a multitude of weapons and equipment, we could equip the character and that could change the points of combat and movement. The characters had to move in real time on the screens of Nokia S60s with explosions, power shots, and even flamethrowers. We had a highly-evolved system with multilevel missions in forking structures; depending on the choices, the player's mission adapted. The game was fully managed via GPRS and we had close to 500 players playing simultaneously, at a time when accessing the Internet on a phone was very expensive, slow, and required players to load the actual maps. The game was created as a global game that ran on all carriers and everyone played against each other.[2]

This was one of the highlights of the history of Portuguese video game development; however, and despite the greatness of the feat, many problems were present, which meant that the game failed to inspire a wave of innovation in Portugal. It is true that this lack of interest in innovation was due more to external factors than to YDreams itself. In 2003, mobile technologies were still far from the iPhone or Android. Furthermore, developing a game that ran globally ran contrary to the vested interests of the operators, who saw no value in sponsoring a platform where their customers would share equal content. On top of all this, there was still the problem of the lack of standardization: for each family of devices, games had to be reprogrammed and redesigned, sometimes even within the same family; and so every game could be redone almost ten times to cover a number of handsets that operators considered relevant. At that time, game companies thought it would be easier to move to online platforms and leave mobile behind.

From Downloadable to Online

In the early 2000s, the Internet began to take the place of the CD-ROM format, as the emergence of technologies like Shockwave and Flash made possible the development of graphically dynamic applications and complex environments online. The national educational games on CD-ROM started to move onto the Web, and portals for kids appeared. Online games and MMOs began to appear, and in the second half of the 2000s, downloadable casual games arrived, completely changing the face of national game production.

In 2003, Portugal gave birth to its first MMO in the area of space adventures, *Orion's Belt* (2003), created by Pedro Santos and Nuno Silva on the Microsoft platform .NET with ASP.NET and C#. The game had its beginnings as a student project, as the result of a joint project for two courses—Information Technology and Software Engineering—at ISEL (Instituto Superior de Engenharia de Lisboa). *Orion's Belt* is based on games such as *Master of Orion* (1993) and presented as an innovation of the battle system board game, depicting a naval battle in space. The game was published as open source on SourceForge.net and was seen by the media as a promising work in the field of national games, having been featured in several television programs on AXN and RTP, as well as in several magazines. In 2006, it won the awards for Best Technology and Best Internet Game at GAMES 2006. But after finishing their degrees, and without having found a way of financing and sustaining the game, the creators were obliged to look for companies willing to support the game. So it was in 2008 that Pedro Santos and Nuno Silva founded PDMFC (*Projectos, Desenvolvimentos, Manuteção, Formação, e Consultoria*), a company capable of promoting *Orion's Belt* according to the creators' wishes, launching *Orion's Belt 2.0* in 2009 and turning a simple student game into a professional production.

The year 2008 was a historic one for Portuguese downloadable games. After several games were developed and launched for international game portals, RTS reached a production agreement with the portal BigFishGames.com. This resulted in *Farmer Jane* (2008), the first Portuguese video game to be featured in an international top ten of downloadable games. Being in fourth place at BigFishGames.com and the fifteen days that it stayed on the list allowed *Farmer Jane* to reach half a million downloads, legitimizing the Portuguese video game development industry and giving it worldwide exposure.

Farmer Jane used a proprietary 3-D game engine developed internally by RTS. Unusual for games of that time, the entire game had three-dimensional graphics that were rendered in real time. The game's design, which was praised in reviews, was largely responsible for the game's success. *Farmer Jane* was an agricultural time management game, a theme that was in vogue at that time in casual games; one year later, Zynga's *FarmVille* (2009) appeared on Facebook and became its most successful game. As for features, *Farmer Jane* allowed players to customize their avatars, something unusual for a casual game, as were the three-dimensional graphics that dazzle the player with game production values well beyond ordinary casual games.

Finally, in 2010, the Portuguese company Biodroid produced the most successful game to date for the iPad, *Billabong Surf Trip*, which was then distributed by Chillingo worldwide in the App Store. The game was designed with a simulator of real beaches, heaving beach-specific types of waves and permitting different kinds of maneuvers. The interface, controls, and game mechanics were designed to take into account feedback from professional surfers and Billabong. The game's designers tried to create a virtual experience of surfing that would mirror real experiences that surfers felt in the water.

Since its launch, *Billabong* has earned more than 1.5 million downloads worldwide, and the game became such a success that Biodroid even created an internal group just for the production of sports games, which was responsible for launching *Cristiano Ronaldo Freestyle* (2011) and *MegaRamp* (2012).

PS3, Under Siege

Since its inception, the Portuguese video game industry has tried to create a console game, but most of the projects created in the 1990s and 2000s remained incomplete and were never published. At last, in 2008, the first console video game fully created in Portugal was published: *Toy Shop* (2008) for the Nintendo DS. It was developed by the Portuguese Seed Studios, produced by Gameinvest, and then distributed by the American company Majesco Entertainment.

Following this first step, other companies have been able to publish their own games. Biodroid's very interesting game, *Miffy's World* (2010), was made from a very important IP. The Miffy license belongs to Mercis BV, which represents a brand that has published more than 120 books and sold more 85 million copies worldwide since its debut in 1955. *Miffy's World* was designed and produced as a simulation game world for children that featured the rabbit Miffy and her cast of animal friends. It was launched in WiiWare in summer of 2010.

After these first games, Portugal was able to produce a successful console game, this time for the PS3, with the game *Under Siege* (2011) (see fig. 1). Completely created and produced by Seed Studios, a company based in Porto, it became one of the ten best-selling indie games in the PS3 network in 2011. *Under Siege* was innovative in the sense that it presented an RTS game to be played on a console and with a game pad. The art of the game was brilliant, in 3-D and 2-D, making it feel more like a professional game sold in a box rather than an indie game for the PS3 online network.

Producing a game with the production values of *Under Siege* was something completely new in Portugal, and it marked a historical achievement in the country's history of game development. The game received funding from the government and business, allowing a production cost higher than most of the feature-length motion pictures in Portugal. The team behind this creation had worked together previously in game engines as well as the production of demos and games in order to gain community respect so as to be able to attempt production of a console game. It took them ten years to see their dream come true. Pride in their achievement made them give everything to the game, and it explains why they produced such a high-value production (which could easily have cost more than 5 million euros but only ended up costing 1.4 million euros). The game launch was accompanied by strong production value trailers and comic books that helped to generate interest in the game and its story.

Portuguese Video Game Research

Portuguese research on video games was born from the interdisciplinary overlap between two sciences: computation and communication. The first Portuguese university degrees in both fields appeared during the 1970s. Then, in the 1980s, the association of the Portuguese Group for Computer Graphics (GPCG) was founded, and in the 1990s, the Association of Communication Sciences (SOPCOM) appeared. Both associations promoted regular annual events in each scientific branch, creating spaces for the discussion of multimedia in the 1990s and for video games in the early 2000s. Consequently, by the end of the 2000s, people from both associations wanted to create a specific association for video game studies, and thus the Portuguese Society of Videogames Sciences was born in 2009.

Prior to the appearance of this association, research on both sides had been growing. In 2000, the first Portuguese book on video games appeared, grounded in the master thesis of Jorge Martins Rosa at the New University of Lisbon, entitled *No Reino da Ilusão, a Experiência Lúdica das Novas Tecnologias* (*In the Kingdom of Illusion, the Playful Experience of New Technologies*) (2000). Then, in 2003 came the fourth edition of the magazine *Caleidoscópio—Revista de Comunicação e Cultura*, with the first special issue on the *Cultura de Jogos* (*Game Culture*) (2003), organized by Luís Filipe Teixeira and with texts by Espen Aarseth, Luís Filipe Teixeira, Patrícia Gouveia, Jorge Martins Rosa, and others. The next year, 2004, Luís Filipe Teixeira published another book, *Hermes ou a Experiência da Mediação* (*Hermes or the Mediation Experience*) (2004) with a full section dedicated to ludology.

Also in 2004, the Portuguese Group for Computer Graphics launched the national conference on Person-Computer Interaction, which held the first national scientific workshop dedicated solely to video games, the "Games 2004" in Lisbon. Although it was organized by the computation group, the workshop's participants came in equal numbers from the computation and communication sectors. In 2006, the first International Digital Games Conference was held in Portalegre, with plenty of international keynote speakers. This academic conference was part of a bigger event dedicated to video games, GAMES 2006, which declared bankruptcy the following year, resulting in no event in 2007.

In 2008, however, a new academic event was held in Portugal, Digital Games 2008, which was full of transdisciplinary elements, from education to psychology, design to arts, and computer science to narrative. It was a solid event, with good presentations, and papers passing though peer review with a rejection rate of around 50%. It was this event that inspired the creation of the Portuguese Society of Videogames Sciences in 2009. Since then, every year in Portugal, the academic conference *Videojogos* (*Videogames*) has been held, which is organized by the Portuguese Society of Videogames Sciences together with the university, which elected to host the conference that year. This national conference has served as the point of encounter for people interested in video games who come from other academic disciplines. It has also established partnerships with its Brazilian companion conference, SBGames.

Since then, scholarly publication has flourished in Portugal, with the organizers of the *Videojogos* conference publishing the annual proceedings and making them freely available online for anyone to access. In 2009, I published the monograph *Emoções Interactivas, do Cinema para os Videojogos* (*Interactive Emotions: From Film to Videogames*) (2009) with the Portuguese publisher Grácio Editor and CECS/UM. Over the next two years, the Edições Universitárias Lusófonas published two books, *Jogos de Computador e Cinema* (*Computer Games and Film*) (2009) by Filipe Costa Luz, and *Artes e Jogos Digitais, Estética e Design da Experiência Lúdica* (*Digital Arts and Games, Aesthetic and Design of the Ludic Experience*) (2010) by Patrícia Gouveia. More books are sure to appear as more and more master's theses and PhD dissertations are being written in Portugal on the subject of video games.

Conclusion

The history of Portuguese game development began in 1982, and the country has produced more than 350 games since then. Most of them are short games done within independent frameworks and with no industry to support greater developments. Game development in Portugal has always been highly fragmented in time, and in the early years we can find various important historical landmarks, moments that could have changed the development industry but did not. Looking at the history, we can see that many of the more qualified artists and programmers in the field left and went to work in advertising or telecommunications. Others just abandoned Portugal, emigrating to countries where the field was better recognized. From 2009 to 2012, we've been researching in-depth the history of game development in Portugal, and we've found dozens of great stories and strong motivation in its people. Our goal in this study is to publish it as a book and make game development communities aware of our history. We believe that giving names, listing timelines, and presenting the good quality work that has been done will serve to remind people that they are not alone in this venture. More than that, though, we believe that knowing history helps in the avoidance of errors and will help people know each other better, creating possibilities for collaborations. All this said, we strongly believe that knowledge of the past helps to create the future.

Acknowledgments

This chapter is a summary of the results of a three-year project done by the Portuguese Society of Videogames Sciences to trace the first history ever done in our country on national video game development. Complete results were published in the book *Videojogos em Portugal: História, Tecnologia e Arte* (*Videogames in Portugal: History, Technology, and Arts*) (2013).

Notes

1. Marco Carrasco, in an interview with Nelson Zagalo, 2013.
2. Tiago Carita, in an interview with Nelson Zagalo, 2013.

References

Adamson, Ian, and Richard Kennedy. 1986. *Sinclair and the "Sunrise" Technology*. London: Penguin Books.

Beira, Eduardo. 2004. *Protagonistas das Tecnologias de Informação em Portugal*. Braga: Associação Industrial do Minho.

Beira, Eduardo, and Manuel Heitor. 2004. *Memórias das tecnologias e dos sistemas de informação em Portugal*. Braga: Associação Industrial do Minho

Benford, Steve, Andy Crabtree, Martin Flintham, Adam Drozd, Rob Anastasi, Mark Paxton, Nick Tandavanitj, Matt Adams, and Ju Row-Farr. 2006. Can you see me now? *ACM Transactions on Computer-Human Interaction (TOCHI)* 13 (1): 100–133.

Carrasco, P., and R. Tito. 1986. *Super Programas em Basic e Código Maquina*. Lisbon: Edicoes Socedite.

Figueiredo, António Dias. 2004. Engenharia Informática, Informação e Comunicações. In *Momentos de Inovação e Engenharia em Portugal no século XX*, vol. 3, ed. M. Heitor, J. M. Brandão de Brito, and M. F. Rolo, 551–573. Lisbon: Dom Quixote.

Gouveia, Patrícia. 2010. *Arts and Digital Games, Aesthetics and Design of the Ludic Experience (Artes e Jogos Digitais)*. Lisbon: Edições Universitárias Lusófonas.

Luís, Filipe B. Teixeira. 2004. *Hermes ou a experiência da mediação (Comunicação, Cultura e Tecnologias)*. Lisbon: Pedra de Roseta.

Luís, Filipe B. Teixeira (org.). 2003. *Cultura de Jogos, Caleidoscópio*, no. 4. Lisbon: Edições Universitárias Lusófonas.

Luz, Filipe Costa. 2009. *Jogos de computador e cinema*. Lisbon: Edições Universitárias Lusófonas.

Metzstein, Saul. 2009. *Micro Men*. London: BBC Four.

Zagalo, Nelson. 2009. *Emoções Interactivas, do Cinema para os Videojogos (Interactive Emotions: From Film to Videogames)*. Coimbra: Grácio Editor & CECS/UM.

Zagalo, Nelson. 2013. *Videojogos em Portugal: História, Tecnologia e Arte. (Videogames in Portugal: History, Technology and Arts)*. Lisbon: FCA Editora.

RUSSIA

Alexander Fedorov

The history of video games in Russia goes back to the early 1980s. In the beginning of the computer era, Soviet arcade games, very primitive compared to today's standard, included a large number of slot machines or electromechanical arcade games such as *Morskoi Boi* (*Sea Battle*, 1981), *Tankodrom* (1981), *Rally-M* (1981), *Sniper* (1981), and *Safari* (1982). In *Sea Battle*, the player had to shell enemy ships, while *Rally-M* was a racing game, and in *Sniper*, the player had to shell the target (see http://www.15kop.ru/en/). These games were coin-operated; it was necessary to insert fifteen kopecks into the machine, which gave you about three minutes of play. These kinds of gaming machines were usually installed in urban parks and cinema lobbies, and they were very popular among Soviet children and adults.

From the late 1970s to the early 1990s, Soviet military factories produced some seventy different arcade game models like these, but "production of the games ceased with the collapse of communism, and as Nintendo consoles and PCs flooded the former Soviet states, the old arcade games were either destroyed or disappeared into warehouses and basements" (Zaitchik 2007).

From 1984 to 1985, Alexéy Pazhitnov developed the first Russian video game, *Tetris*, which was based on tetrominoes falling on the screen and disappearing as players filled each row. The game soon became very popular not only in Russia but also abroad. *Tetris* was not copyrighted at first, allowing it to spread even faster than it would have otherwise.

In 1989, Nikita Skripkin created (for the developer Locis) what was a very opportunistic computer game for the time, *Perestroika*. Named in honor of the Perestroika movement, the splash screen shows Mikhail Gorbachev against the Kremlin wall. The game comes with a soundtrack and a gramophone recording of the song "Dubinushka" by Feodor Chaliapin (a sort of comparison with the PC computer's sound system, the speakers). The game's objective was to use the keyboard's arrow keys to send a frog "democrat" through the swamp by jumping from lily to lily, while trying not to drown or be eaten by other frogs, which were "bureaucrats." Lilies periodically reduced in size, disappearing and reappearing in other places, all at different rates (overall speed increased with each level). On some lilies were blue pills that increased the frog's "welfare" (increasing the score), and when they were eaten, the frog produced a characteristic "yum" sound. Red pills, on the other hand, had a negative effect. At higher levels ("milestones"), different colored computer-controlled opponent frogs appeared—the "bureaucrats" who would try to catch the player's frog "democrat." The round would end when the lily pad that the frog was sitting on disappeared, the frog jumped into the water, or it "ate" a bureaucrat. The game was a success with consumers.

Also established in 1991, the Russian company Gamos released successful logic-based video game *7 Colors* (1991) and *Color Lines* (1992). Gamos managed to enter the foreign market and beginning in 1999 made the switch to online video games while working with the TV channel TNT. Another company, Nikita (renamed Nikita Online in 2007), was launched in 1991 by Skripkin and Stepan Zotov and produced educational video games including *Wunderkind* (1995), *Happy Birthday* (1995), *Anatomik* (1996), *Journey through Europe* (1996), *Twigger* (1996), *Magic Dream* (1997), *Parkan* (1997), the arcade game *Hunter on the Road* (1997), and more. Though Gamos successfully started in the 1990s, when it later tried to enter the online gaming market, it could not withstand the competition, and in 2005 it ceased operations.

In 1993, Buka Entertainment entered the video game market and became not only a distributor of video game consoles from SEGA, Nintendo, and Sony, but also a manufacturer of its own games (since 2010, for the iPhone). Some of Buka's games include *Education of Neznaika* (1999), *Magic Chest* (1999), and *Magic Game* (2012).

The rapid development of the video game industry in Russia, since 1990, contributed to the massive expansion of the Internet in Russia. The early twenty-first century was a turning point for Russia in the intensive development of multiplayer browser video games on social platforms (such as the social network Facebook), including *Mafia* (2006), *Virtual Russia* (2007), and *Soul* (2013), and was also the period when Russia began to develop video games for mobile phones.

It should be noted that almost the entire history of video games in Russia is inextricably linked to video game piracy. Pirated copies of Western and Russian video games (especially in 1990) successfully took away profits from legitimate companies. The fight against video game piracy has been conducted and carried out in Russia through legislation; however, the pirate audiovisual market continues its activity.

Today, social video games are dominant in Russia. The games are created and distributed by companies Mail.ru Group, Crazy Panda, Plarium, and Social Quantum, which control 50% of the Russian market for video games ("Review of the game market in Russia in 2011" 2012, 18).

The Reception of Foreign Imports

By the 1990s, the video game market in Russia was basically a pirate one, with many Western novelty games coming into Russia illegally, and although they probably enjoyed great popularity among the people, it is almost impossible to find exact figures as to the income they brought in. Even today, a significant percentage of the video game market in twenty-first century Russia is imported products (almost all well-known foreign video games, especially hits, can be found in the Russian market), while Russian manufacturers seek to keep up with the domestic market developments. In 2012, the total turnover of the Russian video game market reached USD $1.3 billion, almost one-and-a-half times as much as in 2010. This turnover exceeds the income received from the cinema box office in Russia, which in 2012 was equivalent USD $1.2 billion. Despite all these achievements, the worldwide market share of video games for Russia is low, at only 2.2% ("Game Market in Russia" 2012, 4–6).

The growth in profit of the video game market in the country is largely dependent on increasing the influence of online games (the market share of the online segment increased in 2012 by 64% to USD $0.9 billion). In general, their volume has grown from 2010 to 2012 to 2.4 times what it was before, while the market for offline games dropped by about 12%. ("Game market in Russia" 2012, 4–6).

Significant increases occurred in the spread of video games in Russian social networks, and by the end of 2012, they reached a monthly audience of 52.6 million people. For example, playing video games occupies about one-third of social networks' daily audience (which generally includes 9 million users). The average social network users who play are willing to spend about USD $15 per month to play online games ("Games market in Russia" 2012, 7).

Massively multiplayer online games (MMOGs) are popular in Russia, in particular because they are free-to-play games (with no mandatory payments, although gamers can purchase additional in-game objects or features). Profit from casual games in 2012 dropped significantly; however, due to the fact that these games are only 1% of the total video game market in Russia, this decline was unobtrusive.

The Russian console game market in 2012 grew by only 2%, mainly due to the Sony PlayStation and PSP. In the area of video games designed for mobile phones and tablets, Android and Apple's iOS are the leading platforms ("Game market in Russia" 2012, 8–10), and due to the fact that the number of users of smartphones and tablets in Russia is growing, this segment of the market will expand.

The Influence of Russian National History on Video Games

During the Soviet era, in the 1970s and 1980s, the primitive nature of the games produced did not allow historical themes in video games to develop. Games such as *Sea Battle* (1981) and *Sniper* (1981) simply perfected the skills of shooting at moving targets. The communist regimes did not suppress video game production, but before the 1990s, all Soviet video games were primitive. The Soviet regime did not buy Western video games so as to save money and not compete with domestic game products. In general, the communist regime rarely bought Western audiovisual products (movies, TV shows, etc.). The change came after the fall of the communist regime, when technology had become more sophisticated, and it became possible to develop more complex plots and themes in games, including historical ones.

Despite the opportunities that Russian history can give for the plots of video games, game developers have focused mainly on military issues and fighting the battles of the Second World War (Belyantsev and Gerstein 2010, 282). However, there are exceptions. In 2008, 1C Company released a military strategy game, *XIII Century: Rusich*, in which the player can enter into the role of the Prince of Pskov. Employees of the Chelyabinsk

Regional Juvenile Library created a video game, *How the Urals Saved the Battle of Borodino* (2012), modeled on the events of the War of 1812. The main characters of this video game, Ural teenagers, go to war in 1812 to fight the French. This game attracts lovers of history, adventure, and logic puzzles. One of the most popular video games about Russian history is the strategy game *European Wars: Cossacks XVI–XVIII Centuries* (2001) (see http://www.cossacks.ru).

It is worth noting that in recent years, Russian schoolteachers of history have tried to use video games in the learning process, as they provide an opportunity in an interactive way to "survive" the historical events, which boosts secondary students' interest in historical facts. The best of these video games not only impart knowledge of history, geography, ethnography, and culture, but they also help students to understand the causes and consequences of certain events and learn about what life was like for people of various eras. As such, games in historical settings need a certain knowledge base derived from history (Chernov et al. 2009, 46).

There are, in Russia, strategy-based video games that address recent history. For example, *The Truth about the Ninth Company* (2010) is a virtual reconstruction of the events of the war in Afghanistan in 1988. Another example of a game on the theme of contemporary history, *Confrontation: Peace Enforcement* (2008), opportunistically plays on the military events in Georgia and South Ossetia in August 2008, mixing history with fantasy (Russobit M 2008).

Unfortunately, historical video games in Russian schools are not extensively used because teachers are not well trained for such an activity, which reiterates the need for the development of media education that aims at raising the level of information literacy, media literacy, and media competence of people of all ages.

Domestic Video Game Production and Exports

One of the first specialized companies in Russia in 1991 was Gamos, created by E. Sotnikov. Gamos's *7 Colors* (1991) and *Color Lines* (1992) found success with what was then a fairly narrow domestic audience, those who had computers. Gamos is best known for such video games as *Sobor* (1991), *Sky Cat* (1991), *Columbus Discovery* (1992), *Balda* (1993), *Tank Destroyer* (1993), *Corners* (1993), *Kalah* (1993), *Magnetic Labyrinth* (1993), *Regatta* (1993), *Wild Snake* (1994), *Flip Flop* (1997), *Snake Battle* (1995), *Pilot Brothers: On the Trail of the Striped Elephant* (1998), and *Pilot Brothers: The Case of the Serial Maniac Sumo* (1998).

Gamos's main competitor was the Russian company Nikita (renamed Nikita Online in 2007; http://www.nikitaonline.ru), organized by Nikita Skripkin in 1991. From 1992 to 1997, her company released educational video games *Wunderkind* (1995), *Happy Birthday* (1995), *Twigger* (1996), *Magic Dream* (1997), *Parkan* (1997), and others. During this time she was able to enter into agreements with a number of Scandinavian countries to obtain data on the educational games used on three thousand computers at kindergartens and schools. Since 1999, Nikita began producing browser-based minigames and entered the German video game market. From

2002 to 2003, Nikita, along with 1C Company, developed the first Russian multiplayer online game, *Sphere* (2004).

In 2006, Nikita released three important projects for the Russian market: the browser-based game *WebRacing*, the economic strategy game *Truckers: Transport Company*, and together with the television channel TNT, the online simulation game *House 3*, playing on the success of the reality TV show for youth (later, the project was called *Avatarika*). In 2012, Nikita Online relaunched the entertainment portal *GameXP* (http://www.gamexp.ru) with dozens of online games and social networking games. By 2013, the number of *Nikita Online* users had reached 10 million, and the company had developed more than 100 video games in a variety of genres.

In 1993, two years after the establishment of Gamos and Nikita, another Russian company, the distributor and video game producer Buka Entertainment, was created (see http://www.buka.ru/). In 1996, Buka created its first video game, *Russian Roulette*. Two years later, the Buka collection added the quest *Petka and Vasily Ivanovich Save the Galaxy* (1998). Since 2000, the company has developed new video games for children under the name "Bukashka." During the twenty-first century, Buka expanded its offerings, with *Pacific Storm* (2004) and *Metro: Last Light* (2013). Since 2010, the company has produced video games for the iPhone, such as *Adventures of the Hunter* (2010) (see http://www.youtube.com/watch?v=FHgYX3YZAEY). Another well-known Russian company, IT Territory, was founded in 2004 and is engaged in developing and publishing massively multiplayer online games such as *Legend: Legacy of the Dragons* (2006) and casual games. In December 2007, IT Territory became a part of the holding company Astrum Online Entertainment, along with Nival Online, Time Zero, and Nikita Online. Finally, the company Gameland is the leading publisher of game magazines including *Land Games* and *PC Games*, and the owner of the Internet portal http://www.gameland.ru.

Without downplaying the role of the "old" Russian companies that produced video games, it should be recognized that today Russia is dominated by relatively "new" companies for the development and creation of social online video games: Mail.ru Group (http://corp.mail.ru), Crazy Panda (http://crazypanda.ru), Plarium (http://plarium.com/ru/), and Social Quantum (http://www.socialquantum.ru), which control 50% of the Russian video game market ("Review: Gaming market in Russia in 2011" 2012, 18). Mail.Ru Group is the most popular Russian free e-mail service, and also—as the operator of two leading Russian social networks, MoiMir@Mail.Ru and Classmates—owns a significant stake in the social network Vkontakte. Mail.RuGroup is actively engaged in browser-based video games (including games in social networks and mobile devices), and owns the rights to seventy online games in Russia, including the foreign *Perfect World* (2005), *The Lord of the Rings Online* (2007), *Warface* (2013), and their own developments such as *Legend: Legacy of the Dragons* (2006), *Allods Online* (2010), and more.

The young company Crazy Panda is another developer of social and mobile video games, such as *Zaporozhye* (2011), and its online games have (as of summer 2013) more than 50 million registered users. Among Plarium's hits are the popular video games *Stormfall: Age of War* (2012), *War of Thrones* (2011), *MarketCity* (2013), and *Poker Shark* (2012).

Indigenous Video Game Culture

Today, video games are not only used for entertainment, they are used extensively in teaching languages, history, geography, art, and science. But they still attract a broad audience, especially games that are fun and entertaining, with a fascinating story. In the spirit of postmodern trends, modern video games have absorbed almost the entire arsenal of entertaining tales and myths, comic books (with their brutal one-dimensional characters), and film genres (action, science fiction, thriller, detective, comedy, romance, erotica, etc.). According to Savitskaja, for players, the "official role" of mythology in modern computer games is

> as a hidden language of the unconscious in common with the style of their dreams, hence the increasing popularity of psychoanalytic interpretations of games based on the identification of natural, virtual dreams with mass-market versions of virtualization awareness, one of the most popular formats of high-tech global mass culture (special effects in blockbuster movies, computer games, amusement parks, computerized laser shows, etc.). (Savitskaja 2012)

Social communication skills are an important aspect related to video games, which include the ability to play in a band, the sharing of information about video games in networks, forums, chat rooms, and the use of mobile communications. For video game players, communication is characterized by the diversity of the virtual world (which is patterned after the real world or immerses players in a world of fantasy that encourages the development of creative thinking activities); the reversibility of acts done in the virtual environment (except, perhaps, in MMORPGs); and the anonymity of people entering into voluntary gamer communication in social networks, which is observed as far as it is acceptable for them (Yugay 2008, 22).The main properties of the virtual culture of video games (as a product of globalization) include bringing people together in new subcultures as a form of communication; the formation of new types of relationships that characterize geographic, democratic, and broad social and cultural differentiation; psychological manifestations of human creative freedom in virtual environments; and active use of the opportunities that are not available to a person in real life (Yugay 2008,12).

Video games can develop certain abilities, including skills involving working with three-dimensional and two-dimensional spaces, attention (selectivity and distribution), working memory, logical and strategic thinking (in certain game genres), and spatial reasoning. Typically, gamers make informed, deliberate decisions, but are also willing to take risks. According to some authors, a gamer's willingness to take risks can be useful in business (Voiskunsky and Bogacheva 2013, 4–5, 12–13). Negative effects are also possible, such as emotional enthusiasm that develops into addiction or causes a full withdrawal into the virtual world, resulting in irreparable damage to the health of the gamer. In addition, many video games that involve the user interactively in acts of bloody violence can negatively affect the psychological state of players, especially underage ones. I conducted research regarding the gaming audience in Taganrog, which showed that Russian students tend to choose exactly this type of game, with virtual worlds that allow one to kill with impunity and beat or fight opponents (Fedorov 2005, 88–96).

Boys are especially emotionally reactive and aggressive while playing video games, enthusiastically recounting the bloody scenes and weapons list. "My favorite game is about worms," says Peter W. (seven years old). "They have 'wet' worms, so that the blood is sprinkled on all sides, for it gives life!" (Brevnova 2012, 22). Of course, video games clearly meet the need to discharge and release aggression in a safe direction for society, but the child "still very often confuses fiction with reality, especially since the game involves everyday activities" (ibid.).

A sociological study in Russia (the survey was conducted in December 2012 in cities with populations of more than 100,000 people; 2,033 people over the age of thirteen were surveyed) showed that motivation and behavior in video games is expressed as follows: achievement of goals in a game (76%), the training of intelligence and skills development (73%), immersion in a story and its atmosphere (64%), rest from everyday life (62%), entertainment (45%), obtaining an aesthetic pleasure from the game's story/characters, etc. (33%), and playing with friends (19%) ("Game market in Russia" 2012, 22). At first glance, it seems paradoxical that the motive of entertainment gained only 45% of gamers' votes. However, it should be noted that achieving the game's objective (76%), immersion in the story and its atmosphere (64%), and rest from one's daily routine (62%) in video games are also related to their main function, which is entertainment.

The average age of video game players among the urban population of Russia is thirty-three years (54% of them women and 46% men), 45% of them are married, and 58% have children ("Game Market in Russia" 2012, 13, 29), which proves that games are interesting to not only teenagers, but also adults. Furthermore, 87% of the Russian Internet audience plays video games more often than once per month, and 50% of them play every day. Due to the intense proliferation of tablet computers and smartphones, the number of Russian gamers who play video games on these devices has increased, to about 40% to 50% of the surveyed Internet users. And about 60% of Russian Internet users play video games online in cities with populations of more than 100,000 people. Most Russian players spend about 30% of their leisure time playing video games, both on weekdays and weekends. As much as 75% of Russian gamers are paying for the use of video games, and their spending on this hobby is approximately 19% of their total spending on leisure. Comparatively, the other major expenses of Russian gamers are restaurants and cafés (24% of total spending), sports (20%), and cinema (16%) ("Game Market in Russia" 2012, 24).

Thus, Russian video game players are said to form a subcultural association; "gamers share a certain view of the world, members of the gaming community have a similar status in the real world (a single age group, income level, and education). This community is characterized by its own symbolic level (attributes, symbols, jargon, and subcultural folklore). Gamers are aware of themselves as the elite community *Homo ludens*, and are a significant part of the network society, which has already become an integral part of many cultures of everyday life" (Vasilyeva, Efimov, and Zolotov 2009, 208).

In recent years, Russia has increasingly created special training courses on video games, in which university students learn basic approaches and concepts of cultural and anthropological analyses of video games, the history and theory of media and video games, and the structural and generic features of video games and computer games, and they learn to competently discuss the problems of the uses of video games in culture and everyday life. Dr. Dmitry Galkin from Tomsk State University, Russia, developed the following content for the training course on video games:

- Introduction to the study of the phenomenon of culture studies problems of the game
- Historical and cultural analyses of the development of video games
- Genre structure and variety of video games
- Aesthetic features of computer games
- The narrative and visual structure of video games
- Cognitive effects: video games and the development of age-related problems
- Social effects: the proliferation of video games and violence
- Therapeutic effects: video games as medical instruments
- Gaming experience in a multimedia environment: trends and technologies (Galkin 2008, 2)

Video Game Players in Russia

Russian researchers of the twenty-first century have repeatedly addressed the topic of video games (Fedorov 2005; Tkacheva 2006; Savitskaya 2012; and others). According to I. V. Anisimova, 78.1% of video game players are thirty years old or younger, and 90.3% are male. At the same time, young men prefer games with three-dimensional graphics, role-playing games, strategy games, and puzzle games, while girls prefer adventure games and card games (Anisimova 2004, 20). These studies prove that video games have aroused people's aggression, anger, addiction to the scenes of virtual violence, emotional alienation (Anisimova 2004, 20), and are addictive (Lipkov 2008; Piljugin 2010). Similar phenomena were identified in my own study of underage gamers in Taganrog (Fedorov 2005).

Comparing the responses of children from different years, we find an increase of interest in the virtual world not only in adults, university students, and grade-school children, but also in preschool children. In 2007, 80% of Russian preschoolers said that they have a home computer; by 2008, this figure had increased to 92%, and by 2009, it had risen to 98%. The percentage who were able to play and enjoyed playing computer games rose from 58% of preschool children in 2007, to 82% in 2008 and 94% in 2009. An even more rapidly growing number of children play computer games on their own without the help of adults; from 28% in 2007 to 62% in 2009. The most popular games among preschoolers are various video game simulators that allow the player to control cars, planes, and helicopters. Boys are usually more passionate about these games, whereas girls seem to prefer games requiring the care of virtual animals. (Brevnova 2012, 20–21).

While the sociological study of the urban population of Russia (with a sample of 2,033 respondents in cities with a population of over 100,000 residents) showed that 87% of the urban Internet audience plays video games more often than once a month, and 50% play every day ("Game Market in Russia" 2012, 11, 20–21), in the whole of Russia, these figures are much more modest, according to the research company GfK-Rus, which conducted a survey in April 2010 in fifty-two regions, territories, and republics of the Russian

Federation with a sample of 2,205 respondents (including small towns and rural populations, for whom Internet access is often difficult). According to GfK-Rus, the number of Russians who play video games totaled 28.4 million people (that is, not more than 24% of the adult respondents), 34% of whom play video games each day, which is 16% less than in cities with a population of over 100,000 residents. The number of video gamers in rural areas (17.7%) is significantly lower than the urban average in Russia (Davydov and Nemudrova 2011, 110–111).

In general, the share of gamers among Russian men is 32.6% (including 12% who are active), and among women these figures are much lower (16.5% and 4.9%, respectively). The sixteen- to nineteen-year-old age group accounts for the peak gaming activity (62.1% of these, including 30.3% with a very high level of activity). Among Russians ages forty to forty-nine, 15.1% are active players, and with a further increase in age, the figure is sharply reduced (Davydov and Nemudrova 2011, 111). If 45% urban gamers (in Russian cities with a population of more than 100) are married ("Game market in Russia" 2012, 13, 29), in general, only 19% of Russian married respondents play video games.

According to sociological studies, the typical domestic employment of video game players are school or university students (66.6%), employees with higher education (33.1%), or unemployed (31.5%), with the time spent by Russian gamers on video games at 126 minutes per day, on average (Davydov and Nemudrova 2011, 111). The largest subgroup, 34.2% of the total number of Russian players, make up the so-called conservatives, for whom games are an insignificant part of their lives. The average age of its members is 34.6 years, and it is the only subgroup in which the majority (56.2%) were women. Only 14.5% of "conservatives" play video games daily. But gamers who are "fans" (9.1% of Russian video game players) are young men with a mean age of 25 years. The average playing time for this group (often spent on online games and social network games) is the highest of all groups at 195 minutes per day. "These people tend to collect media about their favorite games, are interested in software development (a little less than half of them also want to become a developer). The most active fans get their information about new games from the media" (Davydov and Nemudrova 2011, 113–115).

Moreover, among the entire gaming audience, the most popular genres of games were puzzle games and jigsaw puzzles. They attracted 47.5% of respondents, whereas 30.8% prefer games like everyone else. In second place were racing games (41% of gamers' preferences), and in third, shooting games (27.1%) (Davydov and Nemudrova 2011, 117).

The Future of Video Games in Russia

As mentioned earlier, in recent years video games have become more and more popular in Russia, which led to the fact that at the end of 2012, profits from sales of various types of video games for the first time exceeded the profits of film distribution. Due to the rapid growth in sales of new generation TVs, a significant increase in games is possible, especially those capable of full HD and 3-D.

With the development and further expansion of the Internet in Russia (including niche rural areas), we can expect significant increases in the online game market, and with the increase in sales of smartphones and tablets will come increased profits from mobile games and cross-platform games.

References

Anisimova, I. V. 2004. Features of computer culture of students in Modern Russia: A sociological analysis. PhD diss., Ekaterinburg: Ural State University.

Belyantsev, A. E., and I. Z. Gerstein. 2010. The image of the country through a computer game: The historical and political aspects. *Bulletin of the Nizhny Novgorod University* 6:279–283.

Brevnova, Y. A. 2012. Computer games in the modern subculture of childhood (socio-cultural aspects). PhD diss., Moscow: State Academy of Slavonic Culture.

Chernov, A. I., A. U. Morozov, P. A. Puchkov, and E. N. Abdullaev. 2009. *Computer Lessons for History and Social Science: A Guide for Teachers.* Moscow: Education.

Davydov, S. G., and T. A. Nemudrova. 2011. The experience of the Russian audience segmentation gamers. *Sociology* 32:104–123.

Fedorov, A. V. 2005. School students and computer games with screen violence. *Russian Education & Society* 47 (11): 88–96.

Galkin, D. V. 2008. *The Work Program of the Discipline:" Computer Games as a Cultural Phenomenon."* Tomsk: Tomsk State University.

Game market in Russia. 2012. Final Report. Moscow: Mail.ruGroup.

Gulyaeva, E. V. and Y. A. Soloviev. 2012. Computer Games in the Lives of Preschoolers. *Psychological Science and Education* 2:5–12.

Lipkov, A. I. 2008. *Pandora's Box: The Phenomenon of Computer Games in the World and in Russia.* Moscow: LKI.

Piljugin, A. E. 2010. Dependence on video games as a consequence of the deficit experienced by adolescent subjectivity. *Herald TSPU* 5:115–118.

Review of the game market in Russia in 2011. 2012. Moscow: Mail.ruGroup.

Russobit, M. 2008. Confrontation: Peace Enforcement. http://www.russobit-m.ru/catalogue/item/protivostoyanie-prinuzhdenie_k_miru/.

Savitskaya, T. E. 2012. Computer games: A step to the culture of the future? *Culture in the Modern World* 4. http://infoculture.rsl.ru.

Tkacheva, N. 2006. Russian users of computer games. *Regular research "Russian Target Group Index"—TGI-Russia* 1:14.

Vasilyeva, N. I., P. I. Efimov, and T. A. Zolotov. 2009. "The man who plays": A picture of the world in the subculture of gamers. In *Internet and Folklore*, 202–208. Moscow: Russian Folklore Center.

Voiskunsky, A. E., and N. V. Bogachyova. 2013. The learning potential of computer games. In *Third International Scientific and Practical Conference, "Psychological assistance to vulnerable persons using remote technology."* Moscow City Psychological and Pedagogical University.

Yugay, I. I. 2008. The computer game as a genre of art at the turn of the twentieth–twenty-first centuries. PhD diss., St. Petersburg: St. Petersburg University of Humanities.

Zaitchik, A. 2007. Soviet-era arcade games crawl out of their Cold War graves. *Wired*, June. http://archive.wired.com/gaming/hardware/news/2007/06/soviet_games.

SCANDINAVIA

Lars Konzack

Scandinavian countries (Sweden, Denmark, Norway, and Iceland) are not known for their video game consoles. Neither is Scandinavia known for big video game distribution companies. However, there have been a number of significant video game production companies throughout Scandinavia, and major video game hits have been launched. Furthermore, Scandinavia has been deeply involved in demo parties and the development of video game theory. But let us first go back to the beginning.

Early Scandinavian Computer History

Computers were a new invention in the 1950s. The first Swedish computer, BARK[1] (Binär Automatisk Relä-Kalkylator), was from 1950, and the Swedes already had their second computer in 1953, BESK[2] (Binär Elektronisk Sekvens-Kalkylator). Norway had its computer NUSSE[3] (Norsk Universell Siffermaskin, Sekvensstyrt, Elektronisk) in 1954. Denmark had to wait until 1958 before it had DASK[4] (Dansk Aritmetisk Sekvens-Kalkulator), and Iceland's first computer, an IBM 1620 model II, was donated in 1964 by the Icelandic Development Bank to the Icelandic University in celebration of the bank's tenth anniversary (Magnússon 2003). Needless to say, none of these computers really had any impact on Scandinavian video game design. But even so, it meant that the new field of computer science and the education of programmers took off, which would later turn out to be useful for video game design in Scandinavia.

In the early days of Scandinavian computer history, none of the computerized games turned into a real commercial success. The earliest documented public display of a computerized game in Scandinavia was in Blindern, Norway, in 1954 on the newly invented Norwegian computer NUSSE, a name suggested by docent Ole Amble (1913–1996). Three years earlier, the mathematical strategy game *Nim* had been presented at the summer Festival of Britain on the NIMROD computer, and Ole Amble took on the task of programming *Nim* for NUSSE. On a spring day, NUSSE was put on display as an Open University event. Whenever NUSSE played the *Nim* game, she would almost always win, in which case she wrote "Hurra!" on the screen; on the rare occasion the opponent won, she would write "Bravo!" The machine got an overwhelmingly positive reception and the

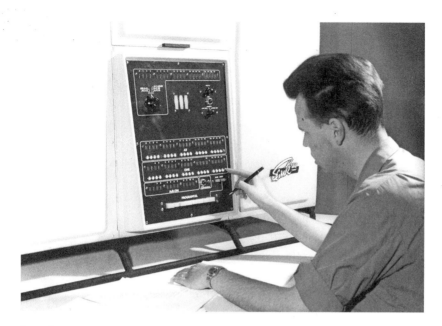

Figure 1
Göran Sundqvist worked with the navigation calculations for the video game *SANK* on the Saab D2.

mission to gain public acceptance of computers was a success (Haraldsen 1999). Two years later, in 1956, the Swedes presented *Nim* on their newly developed computer SMIL (Siffer-Maskinen I Lund) (Sandquist 2010).[5]

In 1960, Göran K. Sundqvist (born 1937) developed a missile game to impress officers of the Swedish Air Force (Flygvapnet). The game was designed for the SAAB D2 computer, aka SANK (SAAB's Automatiska Navigerings-Kalkylator) (see fig. 1).[6] In the game, the player launched a missile and followed the course on an oscilloscope. When the missile hit the target, a simulated explosion occurred. It was the first Scandinavian video game with movable graphics (Ernkvist 2007; Sandquist 2010).

In the early 1960s, the Danish poet and designer Piet Hein (1905–1996) met with Norbert Wiener (1894–1964) in Copenhagen and told him about his idea for a game called *Nimbi*, an auxiliary development of the *Nim* game. Impressed by the game, Norbert Wiener helped him to make contact with Danish computer scientists, and together with the newly employed programmer Søren Lauesen (born 1942), Piet Hein created a computerized version of *Nimbi* on the GIER (Geodætisk Instituts Elektroniske Regnemaskine) computer.[7] The game was ready to be played in 1963 but never turned into a commercial success (Jørgensen 2009).

It is uncertain if the original *PONG* (1972) from Atari made it to Scandinavia, but similar arcade games such as *Winner* (1973), *Hockey TV* (1973), and *Paddle Battle* (1973) have been found in Sweden. Later arcade games such as *Datsun 280 Zzzap* (1976), *Sprint 2* (1976), *Boot Hill* (1977), *Canyon Bomber* (1977), and *Circus* (1977) were more common in Denmark and Sweden.

The Birth of the Scandinavian Video Game Industry

Inspired by Will Crowther's *Colossal Cave Adventure* (1975), the Swedish brothers Kimmo and Viggo Eriksson, together with their friend Olle E. Johansson, developed their own text adventure *Stugan* (English title: *The Cottage*) in 1977 and 1978 (Sandquist 2010). Considering they were only ten, twelve, and fourteen years old, respectively, when they started writing the game, it is an impressive achievement. Written in BASIC on the mainframe computer Oden at Stockholm's Datamaskinscentral, the game involved an adventure down nine levels below a Swedish cottage in Småland, with the player moving in cardinal directions (north, south, east, and west) and collecting items. In the late 1970s and early 1980s, the game had cult-like status among the people that had access to Oden. Later, the game was developed for home computers, and in 1986 they launched Swedish, Danish, Norwegian, and English versions of the game.

One attempt at making a Danish commercial video game was *Kaptajn Kaper i Kattegat* (English title: *Privateer*). The game was developed by IBM employee Peter Ole Frederiksen (1940–2011) for the IBM PC in 1981 and was a privateer simulation game, a combination of strategy and shooting, which took place during the Napoleonic Wars when Danish captains were permitted to become privateers against the English aggressor that had sunk or stolen the Danish fleet after the bombardment of Copenhagen in 1807. Peter Ole Frederiksen tried to convince IBM that it was a sellable product, but IBM did not believe in the value of video games at that time. *Kaptajn Kaper i Kattegat*, however, did become a minor success on IBM PCs in Denmark because Peter Ole Frederiksen made it into shareware through Danadata, receiving donations, though not enough to make a living as a full-time game designer (Jensen 2011).

During the 1980s, several attempts at making video game businesses were made, mostly for the popular home computers of the 1980s, such as the ZX Spectrum, Amstrad, Commodore 64/128, and Commodore AMIGA. The first commercial Scandinavian video game success, *Space Action* (1983), a side-scrolling shooter, was made by the Swede Arne Fernlund for the Commodore 64 (Salman 2008). Distributed by Handic, it sold 10,000 copies; of these, 7,000 copies were sold in Italy, and the game was a minor success in Australia as well.

Greve Graphics in Sweden was the first Scandinavian video game developer. Founded in 1984 by Lars Hård in cooperation with his brother Nils Hård and programmer Bengt Caroli, the three university students managed to create several games within a few years (Nissen 1993). Their first game was *Soldier One* (1986), which sold 75,000 copies in Sweden, and was quickly followed up by *Supercan* (1986), *Captured* (1986), *Blood and Guts* (1986), *1943: One Year After* (1987), and *Gorminium* (1987). Likewise, in Denmark, a video game business venture, Kele Line, was launched in 1986 by Keld Jensen, and had twenty employees at its peak (Stokholm 2003). In a short while, Kele Line created four titles for the Commodore 64: *The Cyborg* (1986), *The Vikings* (1986), *Zyron's Escape* (1986), and *Tiger Mission* (1987). But as was the case for many game companies of the time, it proved difficult to uphold a long-lasting business.

Discovery Software International was another video game development company that, for a couple of years, made a few game productions. Most notable was their game *Sword of Sodan* (1988), developed for the Commodore Amiga by the Danes Søren "Sodan" Grønbech and Torben Larsen. The hack 'n' slash gameplay wasn't that remarkable since they used all of their energy on sounds and visuals, designing a more film-like experience compared to other games at the time that only had small figures and "bip-bip"-style sounds. In that respect, *Sword of Sodan* worked as an inspiration for the video game market, receiving the "Software Oscar 1988" for "Arcade Game of the Year" ("Oscars 1988—The megastars" 1989).

Commercial Success

While the 1980s were a playground for inexperienced game designers to try and sell a few games, the 1990s became the decade for professional game development companies, with bigger teams and more ambitious aspirations. The domestication of computing took hold, and by the end of the decade, most families were on the Internet from home. The influx of money during the dot-com wave meant that investors were more willing than before to invest in video games, even though they knew almost nothing about the industry, giving rise to some interesting video game experiments.

One game franchise made for television stands out. It was *Skærmtrolden Hugo* (English title: *Hugo the Troll*) from 1990, published by ITE Media (Konzack 2004; Moesgaard 2010). The game was originally produced for Danish television TV2 as a video game that one viewer could play while the rest of the viewers watched him or her succeed or fail. A television host would select a randomly chosen viewer and phone him or her and have a little talk before playing the game, using the telephone buttons as input (it did not work with old rotary dial phones). The other viewers could watch if the player on the phone would succeed or fail. If the player succeeded, she or he would win merchandise. In the Danish version, it was presented by game show hostess Nina Klinker Jørgensen. Later, the concept was adapted for a number of television broadcast stations worldwide and did sell video game copies as well. *Hugo* aired in more than forty countries, localized with native tongues—even the names of the game's characters were changed as part of the localization. Hugo only spoke Danish in Denmark, and consequently many people in other countries mistakenly believed *Hugo the Troll* to be a local invention. At first it was mostly popular in European countries, but as of 2012, the troll has grown to be admired in Vietnam and China. In 1996, *Hugo the Troll* was even turned into a theatrical musical, entitled *The Magical Kingdom of Hugo*, in Tel Aviv, Israel. The *Hugo* franchise has spawned more than thirty video games for multiple platforms including PCs, consoles, handheld devices, and mobile phones. Today the *Hugo* video game franchise is owned by Krea Medie, which also acquired other popular Danish children's video game brands from the 1990s such as *Pixeline*, the first title launched in 1994, and *Magnus og Myggen* (English title: *Skipper and Skeeto*) launched in 1996.

Another Danish contribution was *DikuMUD*, an online multi-user dungeon that was based on *Dungeons & Dragons* (1974) role-playing game rules and written by Sebastian Hammer, Katja Nyboe, Tom Madsen, Michael Seifert, and Hans Henrik Stærfeldt from Datalogisk Institut Københavns Universitet (Computer Science at Copenhagen University). *DikuMUD* opened to the public in 1991, and although it was a successful textual game world, they did not manage to make it into a commercial success. Richard Bartle stated that *DikuMUD* was one of the five major codebases used for designing virtual worlds (Bartle 2003). Furthermore, *DikuMUD* worked as inspiration for later MMORPGs such as *Ultima Online* (1997), *EverQuest* (1999), and *World of Warcraft* (2004) (Donovan 2010).

Norwegian video game production took off as well. Scangames Norway was a Norwegian game development company that in the early 1990s had some success with the video games *Fuzzball* (1991) and *Combat Cars* (1994). But what really put Norway on the map of video game development was Funcom. Founded in 1993 by Erik Gloersen, Ian Neil, Andre Backen, Gaute Godager, and Olav Mørkrid, Funcom first released a 2-D fighting game entitled *Samurai Shodown* (1993), also known as *Samurai Spirit*. *Samurai Shodown* was named game of the year by *Electronic Gaming Monthly* and Best Game of 1993 in the Seventh Annual Gamest Grand Prize in the February 1994 issue of *Gamest* magazine in Japan. Another great game from Funcom was *Den Lengste Reisen* (English Title: *The Longest Journey*) from 1999, a single-player adventure game about upholding the balance between two dimensions, one magical and one technological (Liestøl 2001). *The Longest Journey* was followed up by the sequel *Dreamfall: The Longest Journey* (2006).

Born out of the demoscene from a group named The Silents, the Swedish video game development company DICE (Digital Illusions Creative Entertainment) was founded in 1992 by Ulf Mandorff, Olof Gustafsson, Fredrik Liliegren, and Andreas Axelsson (Lemon 2007; Elofsson 2011). Its first games were pinball simulators for the Amiga, but by the end of the 1990s, DICE changed its focus toward racing simulators. Later in the 2000s, DICE would become one of the leading game companies in Scandinavia.

A different kind of Swedish game, *Backpacker* (1995), was a travel simulator developed by Stefan Gadnell and Jens Thorsen (Bergin 2003). It was a success in Scandinavia but did not get much attention outside of Scandinavia, maybe because it presents a particular kind of Scandinavian travel style in which you put on a backpack and travel the world. In the game, the player travels the world, soon discovers money problems, and may have to work and answer questions in order to travel further. Trains are fairly cheap, while airplane tickets cost more. *Backpacker* turned into a series with *Backpacker 2* (1997), *Backpacker 3* (2003), and the concept was followed up by the Danish game development company Deadline Games, creating *Globetrotter* (2000) and *Globetrotter 2* (2001).

Experimenting with the video game medium, Deadline Games launched a series of adventure games in the late 1990s, putting a lot of effort into the narratives. The first of these games was *Blackout* (1997) by author Michael Valeur and game director Simon Jon Andreasen (Dahlager 1998). *Blackout* was a nonlinear narrative set in a film noir-like setting, with players becoming an amnesic character with four different personalities. The character, personality, and story changed in the game, depending on what the player chose to do.

Blackout led to the development of the *Crosstown* trilogy, which, as it turned out, only resulted in two adventure games instead of three as originally scheduled. The concept was built around a little globe-shaped game world that one could circumnavigate in a few hours. The whole globe was a city that got its name from two crossing roads. The first game in the series was called *Giften* (*The Poison*) (1998) by author Merete Pryds Helle and game director Simon Jon Andreasen. Made as an interactive narrative, the story centered on a toxic spill, which the main character was about to discover (Larsen 1998). The next *Crosstown* game was *Englen* (*The*

Angel) (1999) by author Michael Valeur and game director Simon Løvind. As a thought-provoking, interactive, poetic, emotional thriller, the game has the player controlling an angel that may influence the destiny, thoughts, courage, or future of the characters in the story (Jensen 1999). The last game, originally meant to be the first game of the trilogy, *Manipulator*, was delayed and never published due to low sales of the first two games.

Triple-A Dreams

The 2000s began with the crash of the dot-com wave. This, however, did not discourage game designers from living out their dreams. They wanted to go for the big score, creating triple-A games, and some of them actually succeeded. It was the decade in which massively multiplayer online games really caught on, and some Scandinavian MMOGs became known in the global game market. One could argue that indeed, the video game industry of Scandinavia only really began in the twenty-first century.

The Swedish game *Europa Universalis* and the Danish game *Hitman: Codename 47* both came out in 2000, forever changing the Scandinavian attitude toward video games, proving that indeed, successful, good quality, immensely popular video game production wasn't something unique to the United States and Japan but could just as well be done in Scandinavian countries if one had a globally oriented mindset. That same year, Norwegian Funcom published the English version of *Den Lengste Reisen* (*The Longest Journey*), and another Norwegian game developer, Innerloop, published *Project I.G.I.: I'm Going In* (2000), which was followed up three years later with *I.G.I. 2: Covert Strike* (2003).

Paradox Interactive's *Europa Universalis*, created by lead programmer Johan Anderson and based on a French board game by the same name, became the start of a series including *Europa Universalis II* (2001), *Europa Universalis III* (2007), and *Europa Universalis: Rome* (2008). This sudden success led to a landslide of more than seventy games released during the decade. Paradox Interactive had become a player in the video game industry.

The stealth game *Hitman: Codename 47* by Janos Flösser's game development company IO Interactive is without a doubt the most successful Danish game, and the game's main character, the hit man known as Agent 47, is the most well-known Scandinavian video game character (Cancel 2004). Using stealth, skill, and disguises, the player is rewarded for sophisticated sneaky assassinations. The instant success of *Hitman* led to the *Hitman* series with titles such as *Hitman 2: Silent Assassin* (2002), *Hitman: Contracts* (2004), *Hitman: Blood Money* (2006), and *Hitman: Absolution* (2013). Award-winning and critically acclaimed as one of the best video game composers, Jesper Kyd composed the score for all of the first four *Hitman* games (Schollert 2008). The *Hitman* series even spawned a Hollywood movie, *Hitman* (2007), directed by Xavier Gens, but the film did not receive a positive reception from critics, becoming number four on *Time* magazine's Top Ten Worst Video Game Adaptations Ever (*Time* 2008). Apart from *Hitman*, IO Interactive has also produced the *Kane & Lynch* series, third-person shooters in which two escaped convicts, Kane and Lynch, have to shoot their way out of trouble.

In 2002, the Swedish game development company DICE had its first major hit with *Battlefield 1942*, from an original concept developed by lead programmer Johan Persson. The game is a first-person shooter, but the center of attention is more on joint effort and vehicle combat than in previous first-person shooters. The player either takes the role of a scout, an assault, a medic, an anti-tank, or an engineer, and joins in famous WWII battles presented with large maps. *Battlefield 1942* was crowned the best Swedish game of the decade (Prage 2010), and the *Battlefield* series, as it came to be known, went beyond WWII with games involving the Vietnam War, fictional modern warfare, and even futuristic combat situations. DICE also had artistic success with the parkour-inspired running game *Mirror's Edge* (2008) by game writer Rhianna Pratchett (Collins 2011), daughter of fantasy writer Terry Pratchett.

During the 2000s many video game companies in Scandinavia had ambitions toward making triple-A video games, a goal aided by foreign investments and a number of international acquisitions. In 2004, British video game publisher Eidos Interactive acquired IO Interactive, and in 2006, the American video game publisher Electronic Arts fully acquired DICE, changing its name to EA Digital Illusions CE. Furthermore, in 2009 the Japanese video game publisher Square Enix acquired Eidos Interactive, changing its name to Square Enix Europe. As a consequence of these acquisitions, the Scandinavian video game industry became more professional.

Other companies did not have the same luck as IO Interactive and EA Digital Illusions CE. One such company was Danish Deadline Games, which had come into the video game business during the 1990s; in the 2000s it succeeded in developing the triple-A game *Total Overdose* (2005), a GTA-like game set in Mexico. A later remake of the game for the PlayStation Portable was titled *Chili con Carnage* (2007). Two years later, Deadline Games released part one and two of *Watchmen: The End is Nigh* (2009) as a prequel to the film adaption of the comic book *Watchmen* by Alan Moore and Dave Gibbons. While it may have seemed from the outside as if everything was going in the right direction, the company was caught up by the financial crisis and applied for bankruptcy in 2009 (Elkær 2009).

Another strand of video game development in the 2000s included massive multiplayer online games (MMOGs). Norwegian Funcom was the first Scandinavian country to take up this challenge with its science fiction MMORPG *Anarchy Online* (2001) by game designer Gaute Godager and writer Ragnar Tørnquist, a game in which players take on the role of new colonists in Rubi-Ka or the Shadowlands (Schiesel 2003). The game development began as early as 1995, and accordingly, it was an all-or-nothing launch that almost failed. Funcom had trouble with stability problems, registrations, and controlling payments, leading to an overall impression of a failed attempt; however, it did manage to make a comeback by improving the game and offering free trial subscriptions, a marketing strategy that has been adopted by other MMOGs. *Anarchy Online* was also the first MMORPG to have in-game advertising, fitting well with its science fiction theme. Since 2001, *Anarchy Online* has had five expansions, including *Shadowlands* (2003) and *Legacy of the Xan* (2009).

In 2008, Funcom released a new MMORPG titled *Age of Conan: Hyborian Adventures* by game director Gaute Godager. Only a few months after its release, the game was turned over to Craig Morrison when Godager left Funcom due to dissatisfaction with elements in the game's development since its release (Fahey 2008). Based on Robert E. Howard's fantasy universe, *Age of Conan* allows the player to create a virtual avatar, choosing between human races Aquilonian, Cimmerian, Stygian, or Khitan, and the professions rogue, priest, mage, or soldier, inspired by the classic *Dungeons & Dragons* concept; overall, the game is based on the good old hack 'n' slash Gygaxian-style RPG.

Contrary to other fantasy MMORPGs, naked female breasts were visible to players in *Age of Conan*. In June 2008, shortly after the release, gamers became furious when the voluptuous sizes of breasts in the game were suddenly reduced. It turned out that it was a mistake due to an unintended change in data that was introduced in an earlier patch, data which control the so-called morph values associated with character models and the size of their respective body parts. They quickly fixed the problem (Moore 2008), and later, in 2009, Funcom introduced a veteran system in which loyal players would gain point benefits, one of which was a breast-enlarging potion for female characters. Whatever one might think of the use of cleavage in the MMORPG, it essentially stays true to the stories written by Robert E. Howard (Schiesel 2008a); it is interesting that a Scandinavian video game developer has remained true to the nature of the American author Robert E. Howard's stories in a way that an American video game company probably wouldn't dare to do.

Following in the footsteps of Robert E. Howard, it was natural to make a game based on the Cthulhu Mythos created by Howard's friend H. P. Lovecraft; yet his was not the original idea behind Funcom's next MMORPG, *The Secret World* (2012) ("Funcom and EA partners unleash *The Secret World* worldwide on PC today" 2012). Originally, Funcom wanted to develop a game set in the contemporary world filled with secrets and mysteries as the working titles *Cabal* and *The World Online* suggest. However, the final result created by the game designers Ragnar Tørnquist and Martin Bruusgaard was Lovecraftian Cthulhu Mythos combined with Indiana Jones.

Funcom wasn't the only Scandinavian video game developer to succeed with MMOGs. Icelandic CCP Games developed *EVE Online* (2003), a space science fiction MMORPG. Approximately half of the original workforce of twenty-one came from the (at the time) recently ruined company OZ Virtual, which, up until then, had created only a 3-D world viewer. Having a skilled staff with 3-D programming experience, CCP Games was able to turn *EVE Online* into a commercial success, putting Iceland on the map of video game production in Scandinavia (Schiesel 2008b).

The early twenty-first century has also seen the rise of independent video game developers, especially those taking advantage of online distribution possibilities. An example of a successful artistic independent video game developer is the Danish company Playdead, makers of the game *Limbo* (2010), a puzzle-platform game with a stylish, eerie black-and-white noir feel to it, like that of the films of directors Fritz Lang and Tim Burton (Brophy-Warren 2010). Mood and style are the primary focus of game designer Jeppe Carlsen and game director Arnt Jensen, along with Martin Stig Andersen, who constructed the soundscape of the game.

Another example of an independent video game developer is the Swedish company Mojang, which found success with the online video game *Minecraft* (2011), a sandbox game in which the player can build anything he sets his or her mind to (Lynley 2011). Designed by Markus "Notch" Persson and Jens "Jeb" Bergensten, and heavily inspired by Zachary Barth's *Infiniminer* (2009), *Minecraft* has grown into a space for crafting and building copies of such places as the Hogwarts School of Wizardry, the *USS Enterprise* of *Star Trek*, or Minas Tirith from *The Lord of the Rings*.

Finally, there is a growing market in casual mobile gaming, including tablet computers and smartphones, which Norwegian game maker Håkon Bertheussen has exploited with his *Scrabble*-like word game *Wordfeud* (2010), which has two million users per day (Halleraker 2011). Likewise, Danish video game developers Sybo Games and Kiloo Games have found success with a game called *Subway Surfers* (2012) by Sybo's Sylvester Rishøj and Bodie Jahn-Mulliner. In *Subway Surfers*, the player-character is a graffiti artist surfing away and avoiding apprehension by the authorities. During the free game, the player earns gold coins, but it is possible to buy 7,500 gold coins for a dollar, which has turned the game into a financial success (Nielsen 2012).

Overall, the early twenty-first century has seen Scandinavian video game development become an industry with retail sales in 2009 exceeding more than 500 million EUR, and a total of more than 2,700 employed in the industry (Norden 2011).

Scandinavian Geek Culture

While Americans rightfully feel that "geek culture" comes from their own backyard, with writers such as H. P. Lovecraft, Robert E. Howard, Robert A. Heinlein, and Ursula K. LeGuin, comic book artists such as Jerry Siegel and Joe Shuster, Stan Lee, and Jack Kirby, and role-playing game designers such as Gary Gygax, Greg Stafford, and Mark Rein·Hagen, the Scandinavian experience with geek culture was that of cultural colonization. One reason for this was that Scandinavian modernist, minimalist art, and social realism in literature had almost wiped out competing fantastic art and literature at the time, and consequently Scandinavia lay prone to influences from the outside since Scandinavians did not create anything of importance in this field. Nature abhors a vacuum, and for this reason, it became common for Scandinavians to read American science fiction, fantasy, and horror novels since they couldn't find what they wanted in their local Scandinavian tongue.

In the 1970s, video game consoles and arcade machines were imported to Scandinavia mostly from the United States and Japan. Only the wooden cabinets were made locally, except from some wall-mounted *Yatzy* (*Yahtzee*) machines made by the Danish company CompuGame around 1980. According to Rune Keller from Spilmuseet (the Danish game museum), coin-operated video game arcade machines were widely distributed in Denmark and Sweden, especially during the 1980s, while Norway didn't have as many due to the long distances within the country, which might explain why the Norwegian video game industry took longer to get started.

The Scandinavian board game and role-playing scene began with the Swedish game convention KON-VENT77 (renamed Goth Con two years later) in 1977 at Chalmers Tekniska Högskola in Gothenburg (Löf 2000). There, twenty-five people learned different war games, *Dungeons & Dragons*, and video games on a Commodore PET. In Denmark five years later, inspired by Goth Con, Copenhagen Gamecon (renamed Viking-Con the following year) was founded (Johanneson 2006), and in Trondheim, Norway HexCon was started in 1984. Game culture has, of course, worked as a breeding ground for game design and led to the first Scandinavian role-playing game *Drakar och Demoner* (1982) based on Chaosium's *RuneQuest* (1978). Later in 1997, the first Nordic live-action role-playing (LARP) conference Knutepunkt became a reality in Oslo, Norway. The conference shifts location each year between Norway, Sweden, Denmark, and Finland, changing its name from Knutepunkt, Knutpunkt, Knudepunkt, and Solmukohta, respectively. The conference is known outside Scandinavia as highly artistic and thought provoking (Stark 2012).

In Scandinavia, the 1990s saw the rise of the demoscene, where lots of computer enthusiasts gathered together, making computer demos in a positive competitive atmosphere, playing and copying video games, and not getting much sleep. Some of the most well-known demo parties of the 1990s were Hackerence (Sweden), DreamHack (Sweden), The Party (Denmark), and The Gathering (Norway). Scandinavian computer enthusiasts, of course, went to these parties and to gatherings in neighboring countries as well (such as Assembly and Alternative Party in Finland, and SymMek in Germany), creating a feeling of community beyond borders.

The Swedish demo party Hackerence began as early as 1989 and was held twice a year until 2000. Another Swedish demo party, DreamHack, began in 1994 and still exists (as of 2012), and has the world's largest LAN party. The Danish demo party, The Party, took place for the first time in 1991 and was held each year up until 2002. The release of id Software's *DOOM* (1993) a few weeks before The Party in 1993 turned the event into the first major LAN party. The Norwegian demo party, The Gathering, began in 1992 and still exists (as of 2012). While the demo parties are a subject to be researched in their own right, it is important to stress that they had immense influence on video game development in Scandinavian countries, creating friendships within the community of computer culture and having educational value in regard to what could be done with computers. Many programmers, graphic designers, and sound artists came out of the demo scene; for example, in Denmark, video game music composer Jesper Kyd often performed at The Party long before he became famous (Dyar 2012).

Another cultural phenomenon directly related to video games is e-sports. Scandinavians have been active in this discipline. The Swedish e-sport group [9] won the Electronic Sports 2003 World Cup in the *Counter-Strike* tournament, while Alborz "HeMaN" Haidarian won the *Warcraft III* tournament. In the *Counter-Strike* female tournament that same year, the Swedish group Femina Bellica came in second, followed in third place by the Danish group Denmark Girls. In 2004, the Danish group Titans won the *Counter-Strike* tournament and the female *Counter-Strike* group, Team All 4 One, won the *Counter-Strike* female tournament, while Sweden won the *Quake III* tournament. In the Electronic Sports World Cup 2008, the Swede Kalle "Frostbeule" Moertlund Videkull won the *Trackmania Nations* tournament. In the 2008 Master of Paris, the Danish group Mortal Teamwork won the *Counter-Strike* tournament, and Swedish group Fnatic came in second, while the Swedish group Les Seules won *Counter-Strike* female, and the Swedish group SK Gaming won the *Defense of the Ancients* tournament. At the 2009 Master of Cheonan, the *Counter-Strike* tournament was won by the Swedish group Fnatic, and the Swedish group SK Gaming came in second. At the Electronic Sports World Cup 2010, the Norwegian group Bergie won the *Trackmania* tournament, and in 2011, Swedish SK Gaming won the *Counter-Strike* tournament. As of Electronic Sports World Cup 2010, the medal tally was as follows for the Scandinavians: Sweden came in fourth overall (with the highest number of silver medals), Denmark in eighth, and Norway in fifteenth. The top three nations were France, the United States, and South Korea. Scandinavia, combined however, would reach the number one spot.

In 2002, Denmark's first video game museum opened. At first it only had fifty machines, but ten years later, Spilmuseet had more than 800 arcade machines and 3,000 arcade games, institutionalizing video game culture. Since 2000, Sweden has held the Swedish Video Game Awards, and Norway has held the Gullstikka (Golden Joystick) Awards since 2002, while Denmark had to wait until 2010 for the Danish Game Awards. The rise of the geek culture in Scandinavia was a central ingredient in establishing the video game production in Scandinavia because a lot of the ideas and networking emanated from this culture.

Scandinavian Anxiety

Whenever a new and vibrant culture appears, there will always be reactionary anxiety. The first real game scare in Denmark wasn't focused on video games but role-playing games. On November 17, 1987, the monopoly Danish television station Danmarks Radio presented *Dungeons & Dragons* as dangerous on *Søndagsavisen* (*The Sunday News*). A state-authorized Freudian psychologist, Verner Regli, said it was satanic and led to suicide, although he did not have any real evidence for his claims. This ten-minute news episode was the Danish public's first exposure to role-playing games, and even though it wasn't a video game as such, it still gave a bad image to game culture in general because this was not merely an attack on role-playing games but on the whole geek culture of which role-playing and fantasy were inherent components.

In 1995, Jan Esmann wrote an article saying that indeed the video game *DOOM* was so violent, realistic, and at the same time tempting that it would lead to what he called cyber-psychosis (Esmann 1995). This view, however, was challenged by Laila Ingrid Rasmussen, who defended video game players (Rasmussen 1995). In Norway in 1996, Stein A. J. Møllerhaug published a government report guiding insecure parents on how to control children's video game behavior, based on a suspicious notion of violent video games (Liestøl 2001). A report from *Statens Filmtilsyn* (the Norwegian Board of Film Classification) in Norway pointed out that some violent video games were of a nature that was not suitable for children, suggesting an age classification of video game content (Asbjørnsen, Borgnes, and Ulrichsen 1998). In Denmark, Professor Albert Gjedde argued that video game players became dependent on the substance dopamine (which plays a major role in reward-driven learning) produced in their brains while they played, arguing the players could not distinguish between fantasy and reality. These claims were, however, easily dismissed by game researcher Carsten Jessen, since Professor Gjedde couldn't prove his point as it was based on wild assumptions; while dopamine production occurred, that did not mean the players could not distinguish between fantasy and reality. In fact, one criticism of the research Albert Gjedde referred to was that the players received £7 for each level they accomplished, which in itself could account for the high level of dopamine in the brain (Rubin 2001). In Sweden, Christer Sturmark, chairman of the Swedish Humanist Association, Humanisterna, claimed that since there were all of these violent games, it would only be a matter of time before rape games came on the scene, which would then lead to lack of norms and trust in humanity (Sturmark 2002). As a direct answer to this criticism, entrepreneur Mikael Pawlo declared video games were art, and consequently, they could only be understood in an art context of free speech (Pawlo 2003).

In February 2001, the Swedish EU presidency suggested harmonized protection of minors against unsuitable content in video games, and in 2003 the Pan European Game Information (PEGI) age rating system was introduced in many European countries and among them the Scandinavian countries. In Denmark, Norway, and Sweden, this was done with official support and representation in the council, but without any real legislative power; while in Iceland, which had representation in the council as well, it also became legislative by law.

The video game violence debate in Scandinavia re-emerged in the wake of right-wing terrorist Anders Behring Breivik's bombing of government buildings in Oslo and his mass murder of sixty-nine people on the island of Utøya on July 22, 2011. Since he was known to have played *World of Warcraft* (2004) and *Call of Duty: Modern Warfare 2* (2009), the question was raised whether his video game enthusiasm had led to this violent act. As a result of this, the Scandinavian super-market chain Coop in Norway removed fifty-one violent video games from its stores (Nunneley 2011). However, when it became apparent that he had planned this for ten years and that he wasn't motivated by video games, the Norwegians focused on how they could fight racism, by singing the one song Anders Breivik couldn't tolerate: *Barn av regnbuen* (English title: *My Rainbow Race*) (Peters 2012; Holmberg 2012).

Another issue regarding video game anxiety is that of gender in video games. In 2012, the trailer for *Hitman: Absolution*, in which Agent 47 kills eight heavily armed, high-heeled, latex-clad "nuns," suddenly became an issue on how the male-dominated video game industry treated women, and *Hitman* director Tore Blystad had to tell the public that he meant no harm (Sheffield 2012). One should think this would be discussed eagerly in Denmark, but there was no major outcry. To understand this, one has to know that Denmark was the first country in the world to legalize pornography in 1969, and to a Dane, it is actually much more provocative to see a person with a real gun in his hand than to see a naked woman. Since *Hitman* already was about shooting and killing people in the first place, considerations regarding the fetish nuns being killed were seen as a minor issue in that context. It was not because Tore Blystad was insensitive to women or nuns as a male game director; the Danish culture surrounding this issue has different priorities compared to that of, for example, American culture, in which gun ownership is less threatening to some than female nudity.

Game Research and Education

In the twenty-first century, the reaction to video games changed as the generation that had played video games as kids became adults. They began researching video games, educating video game developers, and handing out awards.

Actually, Scandinavian research into computers had already influenced the video game industry long before the turn of the millennium. Norwegian computer scientists Kristen Nygaard (1926–2002) and Ole-Johan Dahl

(1931–2002) had invented object-oriented programming; Danish computer scientists Bjarne Stroustrup (born 1950) invented C++; and Anders Hejlsberg (born 1960) invented Turbo Pascal and C#. A humanistic approach to video games came later.

In 2001, the IT University of Copenhagen held the conference Computer Games & Digital Textualities. Aki Järvinen, Susana Pajares Tosca, Anja Rau, Jesper Juul, Markku Eskelinen, Selmer Bringsfjord, Ulf Wilhelmsson, Ragnhild Tronstad, Frank Schaap, Marie-Laure Ryan, Raine Koskimaa, Lisbeth Klastrup, and Seppo Kuivakari all presented papers about issues relating games to narratives (IT University of Copenhagen 2001). This conference led to the first DiGRA (Digital Game Research Association) conference in Tampere, Finland, the following year, and in 2012 the first Nordic DiGRA conference was held. During the 2000s, the IT University of Copenhagen became one of the leading universities in video game research, welcoming young game researchers from all over the world to come and study games together.

In Scandinavia, famous game researchers such as Espen Aarseth (Norway) and Jesper Juul (Denmark) began to discuss the nature of games (Newman 2004). But they were not the only ones; a whole new generation of game researches followed suit. In Denmark, Simon Egenfeldt-Nielsen, Jonas Heide Smith, Carsten Jessen, Bo Kampmann Walther, and Lisbeth Klastrup came along, together with Norwegian game researchers such as Rune Klevjer, Hilde Corneliussen, Ragnhild Tronstad, Torill Elvira Mortensen, and Kristine Jørgensen. In Sweden, Jonas Linderoth, Staffan Björk, Annika Wærn, and Mikael Jakobsson took the challenge, and all of them together created a Scandinavian atmosphere of vision and ingenuity.

At the same time, it became clear that Scandinavia had to educate people about the growing video game industry, and in 2000 the Swedish education PlaygroundSquad was established in Falun, focusing primarily on programmers and visual artists, but game designers as well. In April 2012, PlaygroundSquad entered a partnership with Sony concerning designing games for Sony's video game platforms, such as the PlayStation Portable (PSP).

In 2005, the Danish DADIU (Det Danske Akademi for Digital, Interaktiv Underholdning) launched a whole new ambitious concept of cooperation between design and production schools and universities in close contact with the video game industry. The coordination between education and industry consists of joint productions and a joint curriculum. In 2011, EUCROMA (the European Cross Media Academy) was established as a Pan-European extension of DADIU, and during the mid-2000s, different colleges and universities in Norway likewise formed educational video game programs, though without any coordination with the industry (Bryne 2005). Following up on these educational investments, the Nordic Game Program was launched in 2006 and has since supported more than twenty games, most notably *Limbo*.

The Saga Continues

One might think that the Scandinavian video game industry would make heavy use of Scandinavian folklore and mythology, but that is not the case; it has been used only to a minor extent. So is there any kind of special Scandinavian video game style at all? That is difficult to say; most big Scandinavian game companies focus on the global market rather than using their unique cultural Scandinavian heritage. On the other hand, some games have a tendency to be gloomy and dark, such as *Hitman*, *EVE Online*, *Battlefield*, *Limbo*, and *The Secret World*. The video game saga has just begun, and time will tell if this gloomy and dark style becomes a trademark of Scandinavian video games.

Notes

1. "Bark" is the same as the English *bark*, meaning the outermost layers of stems and roots of woody plants.
2. "Besk" in Swedish means *bitter taste* or a kind of bitter alcoholic drink.
3. "Nusse" in Norwegian means an *innocent kiss*.
4. "Dask" in Danish means *to slap*, and is sometimes slang for money.
5. "Smil" simply means *smile*.
6. "Sank" in Swedish means *sinking*, as in a sinking ship.
7. "Gier" in Danish means *hoist device*. The word, however, is rarely used.

References

Asbjørnsen, D., M. Borgnes, and C. Ulrichsen. 1998. *Regulering av Dataspill*. Oslo: Statens Filmtilsyn.

Bartle, Richard. 2003. *Designing Virtual Worlds*. Indianapolis, IN: New Riders.

Bergin, Erik. 2003. Lärorik fotvandring är inne på åttonde året. *SvD Näringsliv*, September 27. http://www.svd.se/naringsliv/larorik-fotvandring-ar-inne-pa-attonde-aret_110992.svd.

Brophy-Warren, Jamin. 2010. Limbo: A videogame that blends art, puzzles, and dreams. *Wall Street Journal*, August 4. http://blogs.wsj.com/speakeasy/2010/08/04/limbo-a-videogame-that-blends-art-puzzles-and-dreams/.

Bryne, Snorre. 2005. Vil du lage dataspill?: Stadig flere læresteder i Norge har spillstudier. Søknadsfristen er i dag! *Dagbladet.no*, April 14. http://www.dagbladet.no/kultur/2005/04/14/428880.html.

Cancel, C. 2004. En lejemorder bliver til. *Politiken.dk*, March 11.

Collins, Karen. 2011. Making gamers cry: Mirror neurons and embodied interaction with game sound. In *I AM '11 Proceedings of the 6th Audio Mostly Conference: A Conference on Interaction with Sound*, 39–46. New York: ACM Press.

Dahlager, L. 1998. Et blandet menneske. *Politiken.dk*, December 17.

Donovan, Tristan. 2010. *Replay: The History of Video Games*. East Sussex: Yellow Ant Media Limited.

Dyar, Amanda. 2012. Exclusive interview with Darksiders II composer Jesper Kyd. *BioGamerGirl.com*, August 13. http://www.biogamergirl.com/2012/08/exclusive-interview-with-darksiders-ii.html.

Elkær, M. 2009. Deadline Games går konkurs i dag. *Computerworld*, May 29. http://www.computerworld.dk/art/51716/deadline-games-gaar-konkurs-i-dag.

Elofsson, J. 2011. Svenskarna som erövrar spelvärlden. *Affärs Världen*, November 1. http://www.affarsvarlden.se/tidningen/article3305793.ece.

Ernkvist, M. 2007. *Svensk Dataspelsutveckling, 1960–1995*. Stockholm: Kungl, Tekniska högskolan.

Esmann, J. 1995. Cyberpsykose. *Politiken.dk*, January 1.

Fahey, Mike. 2008. Age of Conan game director quits dissatisfied. *Kotaku*, September 17. http://kotaku.com/5051102/age-of-conan-game-director-quits-dissatisfied.

Funcom and EA partners unleash *The Secret World* worldwide on PC today. 2012. *Businesswire.com*, July 3. http://www.businesswire.com/news/home/20120703005124/en/Funcom-EA-Partners-Unleash-Secret-World-Worldwide.

Halleraker, Tormod. 2011. Wordfeud har gjort Håkon (28) til millionær. *E24*, November 9. http://e24.no/media/wordfeud-har-gjort-haakon-28-til-millionaer/20119002.

Haraldsen, A. 1999. *Den Forunderlige Reisen Gjennom Datahistorien*. Oslo: Tano Aschehoug.

Holmberg, K. 2012. Norrmän sjunger sången Breivik hatar. *DN.se*, April 25. http://www.dn.se/nyheter/varlden/norrman-sjunger-sangen-breivik-hatar/.

IT University of Copenhagen. 2001. *Computer Games and Digital Textualities: A Collection of Papers*. Copenhagen: IT University of Copenhagen.

Jensen, K. D. 1999. Gør-det-selv-mirakel. *Berlingske Tidende*, December 21.

Jensen, K. V. 2011. Kaptajn Kapers far er død. *Computerworld*, April 12. http://www.computerworld.dk/art/133432/kaptajn-kapers-far-er-doed.

Johanneson, M. B. 2006. Viking-Con gennem 25 år. In *Spilkultur gennem 25 år*, ed. I. S. Flamant, 12–23. Copenhagen: Viking-Con.

Jørgensen, Anker H. 2009. Context and driving forces of the early computer game Nimbi. *IEEE Annals of the History of Computing* 31 (3): 44–53.

Konzack, Lars. 2004. Interaktivt TV og TV-spil. In *Interaktivt tv ... vent venligst ...*, ed. Lars Holmgaard Christensen, 187–193. Aalborg: Aalborg Universitetsforlag.

Larsen, Erik K. 1998. Interaktiv øko-gyser. *Information.dk*, November 28. http://www.information.dk/25390.

Lemon, K. 2007. Amiga interview—Andreas Axelsson at Digital Illusions. *Lemon Amiga*, April 18. http://www.lemonamiga.com/interviews/andreas_axelsson/.

Liestøl, E. 2001. *Dataspill: Innføring og analyse*. Oslo: Universitetsforlaget.

Löf, P. 2000. Personligt med en GothCon-general. *Rollspelstidningen Runan*, May 15. http://www.runan.info/personligt-med-en-gothcon-general/.

Lynley, Matthew. 2011. Indie Darling Minecraft "incompatible" with Steam Digital distribution service. *VentureBeat.com*, August 30. http://venturebeat.com/2011/08/30/minecraft-steam-incompatible/.

Magnússon, Magnús. 2003. The advent of the first general-purpose computer in Iceland and its impact on science and engineering. *CiteSeerX*, August 28. http://citeseerx.ist.psu.edu/viewdoc/summary?doi=10.1.1.109.7486.

Moesgaard, J. 2010. Computerspil er kulturarv. *Magasin fra Det kongelige Bibliotek* 23(2): 35–43.

Moore, Matthew. 2008. Age of Conan: Gamer fury as breasts reduced. *Telegraph*, June 17. http://www.telegraph.co.uk/news/newstopics/howaboutthat/2144975/Age-of-Conan-Gamer-fury-as-breasts-reduced.html.

Newman, James. 2004. *Videogames*. London: Routledge.

Nielsen, M. K. 2012. Dansk gratis-app tjener 100.000 kroner—om dagen. *Berlingske Business*, June 21.

Nissen, J. 1993. *Pojkarna vid Datorn: Unga Entusiaster i Datateknikens Värld*. Stockholm: Symposion Graduale.

Norden. 2011. The Nordic Game facts. Norden, February 15. http://www.nordicgameprogram.org/docs/TheNordicGameFacts.pdf.

Nunneley, S. 2011. Norway store cans 51 brands after Oslo, Utøya murders. *VG24/7.com*, July 29. http://www.vg247.com/2011/07/29/coop-norway-to-remove-51-titles-after-oslo-and-ut%C3%B8ya-murders/.

Oscars 1988—The megastars. 1989. *Commodore Magazine*, March.

Pawlo, M. 2003. Datorspel är konst. *IDG*, January 7.

Peters, Tim. 2012. Spill-eksperter: Breivik snakker tull om dataspill. *VG.no*, April 19. http://www.vg.no/nyheter/innenriks/22-juli/rettssaken/artikkel.php?artid=10065496.

Prage, N. 2010. "Battlefield" tog storslam. *Aftonbladet.se*, March 19. http://www.aftonbladet.se/nojesbladet/spela/article12726861.ab.

Rasmussen, Laila I. 1995. Forsvarstale for DOOM. *Politiken.dk*, January 29. https://bibliotek.kk.dk/ting/object/870971%3A82339671.

Rubin, M. 2001. Videospil som kokain. *Politiken.dk*, June 16. http://politiken.dk/kultur/ECE9082/videospil-som-kokain/.

Salman, S. 2008. Fynd från dataspelens stenålder. *Norrländska Socialdemokraten*, April 17. http://www.nsd.se/nyheter/artikel.aspx?ArticleId=3623147.

Sandquist, U. 2010. *Digitala drömmar och industriell utveckling: En studie av den svenska dator- och tv-spelsindustrin 1980–2010*. Umeå: Umeå Universitet.

Schiesel, Seth. 2003. Voyager to a strange planet. *New York Times*, June 12. http://www.nytimes.com/2003/06/12/technology/voyager-to-a-strange-planet.html?pagewanted=all&src=pm.

Schiesel, Seth. 2008a. At play in a world of savagery, but not this one. *New York Times*, June 4. http://www.nytimes.com/2008/06/04/arts/television/04conan.html?_r=0.

Schiesel, Seth. 2008b. Face to face: A council of online gamers. *New York Times*, June 28. http://www.nytimes.com/2008/06/28/arts/television/28eve.html?_r=1&oref=slogin.

Schollert, Peter. 2008. Succes med musik til spil. *Jyllands-Posten*, February 3. http://jyllands-posten.dk/kultur/musik/ECE3921335/succes-med-musik-til-spil/.

Sheffield, Brandon. 2012. Opinion: Video games and male gaze—are we men or boys? *Gamasutra*, June 29. http://www.gamasutra.com/view/news/173227/.

Stark, L. 2012. *Leaving Mundania: Inside the Transformative World of Live Action Role-Playing Games*. Chicago: Chicago Press Review.

Stokholm, F. 2003. Hvordan Forhandler du? *Berlingske Nyhedsmagasin* 19 (May 26): 56–58.

Sturmark, C. 2002. Snart våldtar vi kvinnor i dataspelen. Aftonbladet.se, December 14. http://www.aftonbladet.se/debatt/article10327934.ab.

Top 10 worst video game movies. 2008. *Time*, October 20. http://entertainment.time.com/2008/10/20/top-10-worst-video-game-movies/.

SINGAPORE

Peichi Chung

Digital entertainment has become one of the world's most profitable industries, with global annual sales that now surpass sales generated by traditional media industries. Since 2007, the rapid growth of this new industry developed most conspicuously in the area of video games. In 2011, software and hardware constituted 60% and 24% of global video game sales, earning USD $44.7 billion and USD $17.8 billion respectively, while online game sales took 16% of the market share, with earnings of USD $11.9 billion. Major video game titles have already achieved higher market success than even blockbuster Hollywood movies. Major video game titles such as *Halo 3* (2007), *Grand Theft Auto IV* (2008), *Call of Duty: Modern Warfare 2* (2009), and *Call of Duty: Black Ops* (2010) have demonstrated strong market power in comparison to contemporaneous Hollywood blockbusters such as *Harry Potter and the Order of the Phoenix* (2007), *The Dark Knight* (2008), *Avatar* (2009), and *Iron Man 2* (2010). The four video games mentioned above collected more than double the market revenue during their first weekend release than the four Hollywood blockbusters that were released in the same years.[1] This rapid growth shows no sign of slowing either. According to the 2011 ITU Technology Watch Report, global sales in the digital entertainment industry will reportedly increase from USD $74.4 billion in 2011 to USD $112 billion in 2015.

This expansion of market scale has created a whole new scenario for the global video game industry. As easy distribution networks for digital download allow the industry to successfully divert consumer spending and advertising dollars online (Wager 2007), emerging industry players explore new business models that allow them to enter a previously concentrated global video game market (Kline, Dyer-Witheford, and De Peuter 2003). By 2010, these new distribution networks had changed the global video game market, from one dominated solely by the console to one diversified among five categories: console (39%), arcade (27%), online (18%), mobile (10%), and personal computer (4%) (PricewaterhouseCoopers 2011). The data on the 2010 market structure also indicates that preferences in gameplay now vary geographically, while at the same time, the choice of cross-platform gaming is on the rise.[2] Analysts expect that networked gaming will become the major trend for future gaming, as Internet penetration has already expanded to reach one-third of the world's population (2.4 billion) in 2012 (Internet World Stats 2012).

This essay studies Singapore's game industry and the evolving process by which it integrated with the global video game industry. Like many other Asian nations, Singapore began to recognize the economic potential of the interactive digital entertainment industry in the early 2000s.[3] The country has actively promoted game development through the formation of alliances between game companies in Singapore and the region (Tan 2012). For a small state, Singapore enjoys a large reputation for maintaining close links with transnational game companies and making efforts to attract foreign capital and encourage the creation of regional offices in the city-state (Chung 2009). Over the past decade, the ecosystem of the country's game industry has formed within parameters set by government leadership. A stable technological environment enhances the industry's infrastructure. Singapore's IT achievements have gradually transformed the city into a "global city of digital media." As a result, Singapore ranked in 2012 as the world's second most network-ready country (World Economic Forum 2012). The local game market also reportedly reached more than 3.7 million online gamers in 2012.[4] In particular, the government's creative industry policies, such as Media 21, the Singapore Media Fusion Plan, and the iN2015 Masterplan, paved the way for stable industry growth as game companies from Singapore began to produce cultural content for the global video game market, beginning with the release of the country's first produced-in-Singapore title, *Romance of Three Kingdoms On-line* (2008).[5]

This chapter introduces the history of video games in Singapore from the perspective of industry development. It is based on fieldwork interviews that the author conducted with forty-nine Singapore game industry professionals from 2006 to 2012, and includes historical analysis of the policy influence on video game industry development in Singapore.[6] Singapore's IT infrastructure, media regulatory environment, and particular political structure are also taken into consideration in the discussion of Singapore's stable business network for game distribution and publishing. It describes the popularity of foreign games that prevail among Singapore gamers and examines the market they create in Singapore by analyzing the popularity of games, be they Western or Japanese. The chapter concludes with a discussion about the future of regional game development and its connection to Singapore's rapid transformation into a creative hub of digital content in Asia.

History of Video Games in Singapore

Literature in the field of the political economy of communication has debated the problem of unequal resource allocation in the concentrated global media industries (Flew 2011; Garnham 2000; Wasko, Murdock, and Sausa 2011). It is argued that transnational capital (channeled mostly through American and Japanese companies) often enters different markets to compete with small- and medium-sized companies that are supported by their national governments. Because of the lively competition that exists in these various geographic locations, international companies encounter resistance from established national and regional companies. However, Miege (2011) argues that this competitive model, which sees transnational and national capital locked in rivalry, indicates a simplistic view of the transnational-national relationship. A media industry entering into the global market needs to consider complicating factors, including protective regulations, the development of business networks, and the erosion of national cultures by transnational standards. All these factors shape the complexity and diversity of the transnational-national relationship in the global media industry.

A review of game consumption before 1995 shows the influence of foreign imports in establishing demand for video games while also forming audience expectations in Singapore (Ng 2001; Wolf 2007). Among all the Southeast Asian nations, Singapore is the most receptive country of Western popular cultural content (White and Winn 1995). The penetration of English versions of games in Singapore is the highest in Asia (Wolf 2007, 212). Due to government regulation, imported video games tend to target the entire family as an audience. The earliest foreign video games entered Singapore during the late 1970s and early 1980s. These were games such as *PONG* (1975), *Breakout* (1976), *Space Invaders* (1978), *Galaxian* (1979), and *Pac-Man* (1980), all on coin-operated machines (Ng 2001). American and Japanese companies were the main source of influence in setting up the popular video game culture at this time. For instance, Atari and Midway imported video games with software designed by Japanese companies Namco and Taito. In the 1980s, as console games became the main form of home entertainment systems, Nintendo dominated the game market of Singapore with its popular games including *Mario Brothers* (1983), *Donkey Kong* (1983), and *Contra* (1987). According to Ng, unauthorized compatible Nintendo Entertainment System machines and pirated Nintendo cartridges also contributed to the success of Nintendo products in this period. In the 1990s, the Singapore market became more diverse as SEGA and Sony entered the competition with improved technology that presented better graphics and colors. For instance, the SEGA Genesis captured about 50% of the video game market share in the early 1990s. In the mid-1990s, the Sony PlayStation arrived in Singapore and became the top video game system in the country, with games including the *Tekken* series (1994), *Ridge Racer* (1983), *Parasite Eve* (1988), and *Time Crisis* (1997).

Since 1995, Singapore's progression into global game industry development indicates a collaborative relationship between transnational and national capital. This chapter evaluates the development pattern that formed between transnational companies, local firms, and the nation-state. The Singaporean state has played a major role in facilitating the outcome of efforts to structure the current industry ecosystem. The industry's value chain reflects an ongoing effort between transnational and local game companies, while at the state level three government offices, the Economic Development Board (EDB), the Media Development Authority (MDA), and the Infocomm Development Authority (IDA), play a supportive role by cultivating a supportive environment for game development and publishing.[7]

Next, this chapter discusses the dynamics that emerged during the industry formation process. The history of this formation is divided into three waves that arose between the mid-1990s and the present.

The First Wave: Initial Trial

The first wave of industry development is characterized by trial and error between the Singapore state, international companies, and local companies. It began in 1995 when the government decided to explore the potential of game development (Moss 1995). Development occurred sporadically during this first wave; however, the main structure of the industry did not take shape during this time. The major actors involved during this period were limited mainly to government-related offices, in particular the Economic Development

Board (EDB). For instance, the EDB collaborated with Electronic Arts in order to introduce startups to the local scene.[8] As Electronic Arts worked with educational institutions to set up the Innovation Lab and create game curricula at Ngee Ann Polytechnic, EDB also worked with Sembawan Corporation[9] to invite foreign programmers to develop games in Singapore. Games developed during this period include *Games On-line*, *Cyberpunk 2020* and *iPower* (Moss 1995).[10] These games were never released, but their development showed a focus on online game components; *iPower* explored 3-D elements, and *Cyberpunk 2020* focused on the fantasy role-playing component in game design.

In game research, projects undertaken by academic institutions came mainly from the Institute of System Science lab (ISS) of the National University of Singapore. The ISS lab, later renamed Kent Ridge Digital Lab (KRDL), played a significant role in transforming research projects into start-ups in the early days of game research and development.[11] By 1998, KRDL had grown to include about 292 employees. Programmers in the lab worked on virtual world projects even as the first-generation Internet, Singapore One, was being introduced. Projects that contain historical value include *HistoryCity* (1997), which presented a local story based on the then popular television series of the same name.[12] Figure 1 shows the virtual community of Singapore in 1870, as portrayed in the game. The project was intended to be a massively multiplayer online game (MMOG)

Figure 1
Historical presentation of virtual Singapore, circa 1870, in *HistoryCity*. Source: http://www.psenthil.8m.com/papers/html/vrst97/vrst97.html.

used for educational purposes. As a project under the auspices of the National Library Board and National Heritage Board, the game was designed for children between the ages of seven and eleven to learn about the history, legends, and folktales of Singapore. In contrast to *HistoryCity*, *Pau Chu Kang* is an online game project that allows 500 simultaneous users and is Singapore's first CD-ROM-based game. The game was intended for commercial release, but the project did not make it that far in the process.

The Second Wave: Shaping the Major Industry Ecosystem

The Transnational

Singapore's game industry development has been shaped by the strong presence of international game companies that form the development portion of the country's industry value chain. The early stage of its second wave appeared when Japanese international game companies began to explore Singapore as a possible location for their offshore subsidiaries in Southeast Asia. Since 2002, the government has established training programs (such as the KOEI-EDB Training Attachment Programme) that send local Singaporeans to Japan for talent training. Among the four Japanese companies KOEI, SEGA, Genki, and Capcom, KOEI officially established its biggest development studio outside Japan in Singapore in 2004.[13] KOEI Singapore has invested heavily in training its local workforce in Singapore. Its Singapore studio grew while creating content for the markets of Japan, Korea, China, Taiwan, Hong Kong, and Macau.[14] As one of KOEI's subsidiaries around the world, KOEI Singapore mainly focuses on offering internal support to projects initiated by the main office in Japan. The studio has also developed a niche within KOEI by producing titles for new platforms and emerging markets. As a result, in 2008, KOEI Entertainment Singapore released the first online game produced in Singapore, *Romance of Three Kingdoms On-line*, to the Japanese market. The studio also released KOEI's iPhone version of the top-selling PC game, *Nobunaga's Ambition* (2010), in 2010. KOEI Singapore released the company's first Facebook game, *Jollywood*, to the Southeast Asian market in 2011.

A direct connection between these transnational capitals and the local game industry appears in the level of government that expects transnational media companies to increase local employment and game development capabilities. This can be seen in the following interview, which discusses the long-term influence that the government expects the game industry to exert. It considers these companies contributors to their long-term plan to encourage the specialization of the talent pool within Singapore's video game industry, which will eventually move out of the big multinational corporations and start their own businesses in Singapore:

> Big multinational companies coming in provide an opportunity for people to work in triple A [high-end] title[s]. It helps the local industry because this way the local industry can be seen with three industry groups. One group is for the mobile game industry. The other group is a very high-end industry group. My personal opinion is that Singapore can target the mid-level—which means handhelds or even casual games—industry group. Big companies need a 200-man team to do console games now. Singapore does not

have these experienced people. But big companies help us with the transition to move to mid-level industry. (Ms. T., Media Development Authority, interview with the author, August 30, 2006)

American and European developers, including Lucasfilm, Electronic Arts, Ubisoft, Emerging Entertainment, and Real U, are among the international companies that moved to Singapore during the latter part of the second wave (2006 to 2009). Electronic Arts has been active in the local game industry since it relocated its headquarter in Asia from Hong Kong to Singapore. Strong government support and a friendly living environment that helps to bring in foreign talent have been the elements that have attracted these companies to Singapore to set up development studios there.[15] The EA Singapore studio focuses on localizing the company's popular content for Asian markets. By comparison, other international game companies such as Lucasfilm and Ubisoft have focused on employing locals to build a studio that incorporates Southeast Asian talent in order to produce new titles for their existing global market.

EA Singapore, for example, set up a localization studio in Singapore as part of the company's entry into emerging markets in places like Taiwan, India, and Vietnam.[16] Transnational-national connections, as in the case of EA Singapore, happen whenever a company needs to outsource projects to small and medium (SME) game developers in Singapore and the surrounding region. Similarly, Ubisoft Singapore has connected with the local game industry by developing its Southeast Asian team, through which it can identify artists and programmers from Malaysia, Indonesia, India, and Singapore for its talent pool.[17] Since 2008, Ubisoft's major contribution to Singapore's industry includes its active involvement in assisting local industry in the development of educational curricula in game programming, game art, and game design. It also followed the government's initiative to collaborate with DigiPen, an international game school, in hiring graduates from the school's program after they completed their curriculum on the Ubisoft-DigiPen Singapore campus.

In the publishing arena, Singapore's game industry connects to global game publishing through international companies such as Atari, Namco Bandai,[18] Boonty Asia, Nexway, Microsoft, and Asiasoft. Most of these companies became more actively involved in Singapore's industry during the second wave. Local publishing began when Asiasoft, a regional online game publisher from Thailand, entered the Singapore market in 2005. Regional publishing has always been the main concern as international publishers selected Singapore to be their gateway into the Southeast Asian market for their PC and console game products. Atari Singapore and Microsoft Singapore are the international publishers that reflect this regional function in game publishing at an earlier time. Atari Singapore later became Namco Bandai's office in 2009, as a consequence of the global financial crisis. Likewise, Boonty Asia is a European e-content aggregator that sees the potential of electronic publishing in the region. The company's Singapore office opened in 2007. Two years later, it merged with Nexway, a European publisher, and became Nexway Boonty Asia. Generally speaking, these international publishers selected Singapore thanks to the support of both local talent and the government, which allowed Western game companies like them to engage in a regional market that is highly diverse, sophisticated, and unregulated.

In all, international console and PC game publishing companies see Southeast Asia as a market of limited scale.[19] The problem of rampant piracy compels console and PC game publishers to work with regional distributors in order to obtain reliable product sales.[20] On the one hand, international game publishers promote newly released titles by shipping out copyrighted games from Singapore to nearby countries. On the other hand, they also engage in third-party publishing so that they can fully utilize their understanding of the Southeast Asian market in order to increase business growth. For example, Namco Bandai Singapore not only publishes the company's global titles, such as *One Piece* (2013) and *Naruto* (2010), but also engages in third-party publishing by releasing products such as Codemasters' *F1* (2009) and Square Enix's Xbox version of *Final Fantasy* (2011) to the Southeast Asian market (Toyad 2011). Similarly, during the second wave, Microsoft Singapore formed a promotion team, the Developer Platform Evangelist Group, in order to advocate game development for Windows and Kinect.[21] In the best example, the Singapore government agency MDA in 2008 announced the MDA-Microsoft XNA Development Initiative in order to encourage budding developers to produce games for the Microsoft platform. As a result, in 2008, the company negotiated a contract to distribute the first Singapore game, *Carneyvale: Showtime* (2008), through the Xbox LIVE Indie Game Channel.

By contrast, e-content publishers see Southeast Asia as an emerging market. These international publishers consider Singapore to be a gateway into the rapidly expanding regional online market. For instance, the European e-content aggregator Boonty Asia distributes its 6,000 games through various electronic systems located within the region. The company is stationed in Singapore, but collects gamers' payments through regional Asian electronic channels such as Money On-line, Asia Pay, and Nets set up within the region. Singapore's efficient Internet connection provides companies with technological convenience by allowing them to expand their sale of online content into electronic markets nearby. This transnational-local interconnection is evident whenever a publisher utilizes a convenient server in Singapore in order to undertake regional publishing.[22] One manager at an international game publisher spoke about the company's direct connection to the local game industry through incorporation in the following interview:

> We are a company incorporated in Singapore. We have links with Singaporean companies, government, agencies such as IE Singapore, MDA, IDA. ... We know people inside. We also maintain contracts with Singapore local companies to bring their games to our distribution system. ... Singapore's market is our smallest market in Southeast Asia, but political stability and the government's support in [the] gaming business affected our decision to stay in Singapore. Singapore has efficient technology as we bring European games to bigger markets in Malaysia, Philippines, Indonesia, and Thailand. (Mr. N., interview with the author, August 12, 2009)

The National

From the vantage point of national capitals, local game developers avoid direct competition with transnational companies in Singapore. These developers explore outward into the international casual game market. The ecosystem that formed during the second wave allowed local small- and medium-sized game companies to emerge as independent game developers in the global industry. Local companies that started to do this during the second wave include Boomzap, Envisage Reality, Magma Studios, Mikoishi, Nabi Studios, NexGen Studio,

Protégé Production, and Orange Gum. While these companies concentrate on online and mobile games, some of them also strategize as outsourcing studios, contracting to work for international game companies from abroad.

In the mid-2000s, local start-ups began to emerge with the government's support. In 2003, the Media Development Authority also started to fund a few projects in order to assist the start-up of local game companies. For instance, NexGen began its business as an independent game developer, releasing PC and mobile games through a European distributor.[23] The company received a grant from the MDA for its start-up but maintains that it is economically self-managed, without funding, through the selling of games to the international market. Similarly, Envisage Reality also received a grant from the MDA to develop the real-time strategy game *Immortals: The Heavenly Sage* (2004). The company later received projects from Singapore's Ministry of Defense to develop a real-time interactive 3-D simulation game.

The following interview describes Singapore's industry ecosystem as one that provides a proactive environment for local game companies. The CEO of a local company considers the presence of multinational national companies (MNCs) a boon to local industry:

> Singapore has a pretty proactive business environment. We have a stable economy, stable government and a very efficient e-portal. We can find a lot of information about grants. ... Now with the government encouraging digital creative industry, with a lot of MNCs coming to help, this definitely helps the local industry. In the short run, the local companies will have to compete to hire talent, because they are also considering MNCs. But so far people who know how the environment really works, they also sign up for us. That's a plus for local game companies. (Mr. Y., interview with the author, June 12, 2007)

The contribution of multinational game companies to local industry can be seen in the increase of foreign capital invested in start-up companies in Singapore. Mikoishi was once Singapore's largest game studio, producing two mobile games, *Dropcast* (2008) and *Steam Iron* (2008). Unlike NexGen and Envisage Reality, which chose a conservative approach to growth, Mikoishi rapidly expanded from a small- to mid-sized company because it received knowledge transfer from MNC and support from the government. At the senior management level, Mikoishi's top management came from Electronic Arts. The company eagerly ventured into the global mobile game industry. However, due to its expansive growth, which resulted in too many projects to handle, the company closed down in 2010 after losing talent during the financial crisis.

Singapore's industry ecosystem encourages an environment that nurtures small development companies. New developers such as Boomzap and Magma Studio emerged in 2005. Boomzap is a virtual company based in Singapore with employees working online in many countries. The company targets the US market. It benefits from Singapore's tax-exemption scheme and plans to gradually grow into a mid-sized virtual company in the casual game market. Another company, Nabi Studios, started in 2006 after receiving a grant from the MDA. The company produced its first game, *Toribash*, in 2007. This game is a turn-based fighting game that runs on a free-to-play business model. The game received an international award at the 2007 Independent Games Festival. It has also become a successful MMOG, which has helped it to enter the emerging markets of Russia and Brazil. In all, local game companies have adopted various development approaches to entry into specialized

Figure 2
Concept game art from *Immortals: The Heavenly Sage* (2004).

game genres in independent game production. These game companies have formed a particular fusion game development concept. They present a particular fusion style of game art that is based upon the multicultural and multiethnic background found among game developers in Singapore.

Figure 2 demonstrates such fusion style in the game *Immortals: The Heavenly Sage.* The game presents the story of a folk character, the Monkey King, in a Western art style. The Monkey King theme is widely popular among gamers in China. This East-West connection reveals a collaborative industry ecosystem between the transnational and the local in Singapore's industry value chain.

The local publisher Infocomm Asia Holdings (IAH) is the only Singapore-based publisher managing with both local and Southeast Asian markets. The company was originally a subsidiary of a Taiwanese publisher, GigaMedia, but later nationalized to become a Singaporean-owned company during the global financial crisis. Since 2007, IAH has managed to distribute eight online games in seven Asian countries. It has gradually grown to reach 27 million registered users in Southeast Asia and a total of 140 staff in its regional offices in 2012. The company promoted competitive gaming in regional e-sports tournaments. Dynamic growth of the company in the region within Singapore's current ecosystem is discussed in the following interview:

> Since [*our*] establishment, we expanded with investment from Gigamedia, Softbank, The9, etc. However, as Southeast Asia is a very lucrative market, harder than China, in order to break the number of gamers and

the critical mass in our operation, we have to overcome our problems in earning profit through cross-border trading, maintaining margin in currency exchange, and overcoming cultural expense by managing offices in different countries in Southeast Asia. (Mr. O., interview with the author, May 20, 2012)

It is noted that game industry policy in Singapore is implemented within a neoliberal marketization framework, so that government offices such as the IDA, EDB, and MDA promote the game industry by proposing various funding schemes to support it. The government does not interfere with company performance within the organization in order to maintain the direction of development within Singapore's ecosystem. As a result, the second wave ended in 2009 and 2010, when a local company, Mikoishi, and international game companies such as Electronic Arts, Second Life, and a few other European firms closed their studios in Singapore during the global financial crisis.[24] The move of some MNCs away from Singapore continued as Lucasfilm Singapore closed its operations due to a global acquisition deal of Lucasfilm by the Walt Disney company in 2012. In addition, the German company Real U also closed down its Singapore office in 2013. These closings led to the government's notice on the emergence of local social media games that then began to enter the local industry and achieve publication across various new media platforms.

The Third Wave: The Emergence of Local Social Games

Singapore's game industry after 2010 was marked by the rise of made-by-Singapore content in new genres. As international game companies remained concentrated on developing games for their existing global market, small-scale local game companies emerged as a new force of content creators in Singapore. New international game companies include NHN Singapore and Dragonfly from Japan and South Korea, and Namco Bandai game studio from Japan (Tan 2013). These international and local small-scale game companies not only utilize Singapore's geographic potential in the Southeast Asian market, they also use new media technologies to develop Web-based games (for sites such as Facebook) and games for mobile devices (such as the iPhone and iPad). At the same time, they also publish games in Android and iOS systems in order to maximize their reach to potential gamers within the new social network market. Local games released in global publishing platforms include *CarneyVale: Showtime* (2008), *Armor Valley* (2010), *Rocketbirds: Hardboiled Chicken* (2010), and *Autumn Dynasty* (2012). These games are either student projects or independent projects that won competitions at the Independent Game Festival or International Game Developer Conference. The innovation presented in these games collectively characterizes a literary cinematic art style that indicates a new dimension of game development for young entrepreneurs in Singapore's game industry.

As new social games continue to emerge during the third wave, local developers have started to connect to the global game market by publishing their games in the emerging mobile network. Along with social media game content from other local game companies, this global connection reveals a wider level of game ecosystem that extended from transnational capitals during the second wave to local small-game companies in the third wave.

In addition, the government is also active in promoting a positive gaming culture through serious game initiatives. In 2009, the MDA facilitated cyber wellness initiatives to prevent cyberbullying of youths (MDA 2011). In 2010, the government also set up a think-tank office, called the Game Solution Center, to bridge collaboration between a local educational institution (Nanyang Polytechnic) and an international company (Sony Entertainment) with its Unreal Technology Lab. In 2012, Singapore's Cybersports & On-line Gaming Association (SCOGA) also worked to further promote made-by-Singapore content in the Chinese market. The organization also works with the IDA and MDA to promote made-by-Singapore game content in an increasingly popular e-sport community in Singapore.

Table 1 is a list of selected game companies in Singapore. As of 2013, new companies and start-ups are still entering Singapore due to the country's strategic location as a gaming hub in Asia. The diversity of game development is seen in the "sector" column. This list reveals Singapore's game industry after more than ten years of development since the first wave.

Influence of National History on Video Games

The influence of national history on video games shows a development pattern wherein the state used technology to advance its nation-building agenda. The political structure of Singapore over the past four decades has revealed the priority that the single-party state gave to promoting economic development through policy concentration on political and managerial powers (Chua and Kwok 2001). Video game industry development reflects the state's intention to construct nationhood through market-making in global capitalism. Since the 1990s, the government has produced specific developments in order to transform the country into a technological hub and international community. Since then, Singapore has been regarded as an "intelligent island" (Lim 2002). This technological trend relates to the country's historical past as a trading-port city under British rule. It also shows the state's effort to expand Singapore's economic performance into the creative economy by encouraging the city's rapid rising as a base for global key players in the technology business and creative industries (Kong 2012).

Game industry development exemplifies the government's intention to turn national development from traditional industries to artistic and cultural fields. Since the 1990s, the government has been active in exercising power to create a sense of local place in its arts and cultural policy (Kong and Yeoh 2003; Lee and Willart 2006). Particularly in game development, the Media Development Authority provides schemes such as the Start-Up Enterprise Development Scheme and the Productivity and Innovation Credit Scheme to encourage the growth of local companies. The government has accomplished the goals of several infrastructure projects as indicated in table 2. It is evident that the government focuses on building both infrastructure and also university and research centers in order to promote new growth in terms of local talent and to boost Singapore-based innovation. Specific developments also indicate the government's achievements in creating a local industry scene under the policy of Media 21.

Table 1

List of selected game companies in Singapore in 2013. Source: https://docs.google.com/spreadsheet/ccc?key=0Aio1Iu3YinPxdHNUbkYxczlUV3FMckFBaVdxMGZNYmc#gid=0.

Name of Company	Sector	Year Established	Links
Activate	Facebook games	1997	http://www.activateplay.com
Asiasoft On-line Pte Ltd.	Online games and portal	2001	http://www.asiasoftsea.net
Boomzap Entertainment	Multiple (Help games get published by small developers)	2005	http://www.boomzap.com
Cabal Entertainment Software	iPhone	2005	http://www.cabalsoft.com
Cherry Credits	Online payment/gaming	2007	http://www.cherrycredits.com
Corous	Cloud service for games	2005	http://www.corous360.com
Daylight Studios	Mobile games	2011	http://www.daylightstudios.com
Double Negative Singapore	Special effects for film	2009	http://www.dneg.com
Electronic Arts	EA's IPs	2005	http://www.ea.com/asia
Forever Young Studio	Some mobile	2008	http://www.foreveryoungcreative.com
Game Exchange Alliance	Help game companies get their products into Asia	1999	http://www.gxa.org.sg
Garena On-line Pte Ltd.	Online games	2002	http://www.garena.com.sg
Gumi Asia	Online and mobile games	2012	http://www.gumi.sg
h.a.n.d. Global Singapore	Online and mobile games	2013	http://www.hand.co.jp/singapore/
Immersive Play Spaces Pte Ltd.	Game content for iOS and Web	2008	http://www.ipspaces.com/
Konami Digital Entertainment	Online and mobile games	2012	http://www.konami-digital-entertainment.com.sg
L.A.I. Singapore Pte Ltd.	Coin-operated machines	1999	http://www.laigames.com
Magma Studio Pte Ltd.	Online and mobile	2005	http://www.magma-studios.com
Mikoishi	Mobile	2001	http://www.mikoishi.com
Nabi Studios Pte Ltd.	Physics fighting games	2006	http://www.nabistudios.com
NexGen Studio	Mobile	2003	http://www.nexgenstudio.com
NHN Singapore	Online and mobile games	2012	http://www.nhncorp.sg/singapore/index
Nordasia Partners Pte Ltd.	Connecting Nordic regions with Asia	2011	http://www.nordasiapartners.com
Personae Studio	Mobile	2007	http://www.personaestudios.com

Table 1
(continued)

Name of Company	Sector	Year Established	Links
Planet Arkadia Pte Ltd.	MMORPG	2010	http://www.planetarcadia.com
Playware Studios	Educational learning games	2004	http://www.playwarestudios.com
Rainbow Media Pte Ltd.	Online games	2007	http://www.rbmedia.com
Ratloop Asia	Console games (PS3 and PS Vita)	2008	http://www.ratloop.com
Razer Pte Ltd.	Gaming hardware	1998	http://www.razerzone.com
Redsteam Gameloft	Gameloft subsidiary	2009	http://www.redsteam.net
Rock Nano Private Ltd.	Digital advertising and entertainment	2009	http://www.rocknano.com
SEA Gaming Pte Ltd.	Helping publishers and developers to set up in Southeast Asia	2011	http://www.seagaming.net
Tecmo KOEI Games Singapore	Console games	2004	http://www.koei.com.sg
TheMobileGamer Pte Ltd.	Mobile games	2008	http://www.tmgamer.co
Think! Studio	Mobile games	2008	http://www.thinkstudio.com.sg
Tyler Projects Pte Ltd.	Social and mobile games	2007	http://tylerprojects.com/index.php
Ubisoft Singapore	Console games	2008	http://www.ubi.com
Van der Veer Games	Board and mobile games	1998	http://www.vanderveergames.com
Vinova Pte Ltd.	Mobile games (iOS, Android and Windows)	2010	http://vinova.sg
Vocanic	Marketing and advertising	2005	http://www.vocanic.com
WooWorld Pte. Ltd.	Marketing and distribution for mobile games	2003	http://www.wooworld.net
Zealot Digital Pte Ltd.	MMORPG	2007	http://www.zealotdigital.com
ZekRealm Interactive	Localizing social games	2008	http://www.zelrealm.com
Zingmobile Pte. Ltd.	Mobile games	2002	http://www.zingmobile.com

Table 2
Establishment of the gaming industry under Media 21.

Goal	Details of the Establishment
Position Singapore as a hub for digital media exchange	Mediapolis was built to increase transnational collaboration. It allows small enterprises and game enterprises to benefit from the clustering effect in collaboration with international players.
Establish a flagship media university program	In 2008, DigiPen collaborated with the Economic Development Board to set up the first international branch campus in Singapore. In 2009, the new Art, Design, and Media building at Nanyang Technological University was established. In 2012, a new university, the Singapore University of Technology and Design, began operation.
Establish a media lab in Singapore	In 2006, the interagency Interactive Digital Media Programme Office was created to support the long-term vision of developing Singapore into a global interactive and digital media capital. The Singapore-MIT GAMBIT Game Lab is a five-year research initiative established in 2006. It has a core focus on identifying research problems using a multidisciplinary approach that can be applied by Singapore's game industry.

A representative from the Media Development Authority commented on the government's pragmatic approach to facilitating an active industry environment in the following interview:

> The MDA works with the local companies to actually produce games for the international market. It is not because we want it that way. Our interest has been to generate jobs, build up the GDP and build up the economy. We help these companies to produce made-by-Singapore content that actually falls on the international market. Local companies are mostly small companies. But when the MDA matches them as a group, the group becomes large in government-to-government talk. ... Similarly, most MNCs are much bigger. They come in with big projects that can generate expertise. So it does not matter if we matchmake the local companies or the MNCs. We want both to prosper. (Mr. C., Media Development Authority, interview with the author, July 4, 2007)

In all, Singapore's use of technology, art, and culture reveals an ideology of pragmatism (Kong 2000). This pragmatism adopts an optimistic view toward cultural governance. Because of its top-down approach to governance, the state has held an elitist perspective on how cultural policy should lead national development. The perspective assumes that economic drive will ultimately create cultural regeneration (Kwok 1995). However, cultural policy scholars have disputed this view, arguing that Singapore's recent creative industry policy does not encourage authentic Singaporean art and culture (Chong 2012; Kong 2012; Lim 2012; Ooi 2010).

Reception of Foreign Imports

All in all, the political and economic industry structure results in the Singaporean game market being receptive to foreign imports. Factors that contribute to the easy reception of game imports include players' preference for foreign content, the government's control over piracy, and the industry's convenient retail environment.

Singaporean gamers are selective consumers who are knowledgeable about the news of the international gaming world. They often gain information about foreign titles as a result of previous viewing experience in the local media. Sometimes, gamers go online to read animé or manga versions of the same story, especially when it comes to Japanese content.[25] Mostly, international console game publishers see Singapore as a profitable market.[26] Game imports face limited regulation. Registered companies can apply for auto-approval permits through the MDA's online declaration system before they distribute original titles for Xbox, PlayStation 3, and Wii to the local market. Similarly, foreign MMORPG and social media content are also popular among online gamers in Singapore. As Singapore gamers play games in a highly interconnected network environment, they often access international titles by directly entering the company's servers as subscribers. They can also access foreign titles by connecting to local publishers in order to play their published games.

In general, Singapore gamers mainly prefer blockbuster titles from the United States, Europe, and Japan. However, Singapore gamers also play games that originate in Korea and China. Western games such as *Diablo III* (2012), *World of Warcraft* (2004), and *StarCraft* (1998) are popular, but gamers also seek out Korean and Chinese games through computers in Internet cafés or at home. Local online game publishers provide East Asian game content such as *Ragnarok Online* (2002), *MapleStory* (2003), *Tianlongbabu* (2003), *Counter-Strike* (1999), *FIFA Online* (2010), *Rohan* (2012), *Superstar On-line* (2010), and *Grandchase Chaos* (2011). As Wi-Fi has grown pervasive in public places, casual games on Facebook such as *Bejewelled Blitz* (2010), *CityVille* (2012), and *Pet Society* (2008) have also met with approval from Singaporean gamers ("On-line poll series unveils insights into social gaming in Singapore" 2011). Dynamics that shape the popularity of imported content are discussed in the following interview with Blizzard Singapore, which describes the success of *StarCraft II* (2012) and *World of Warcraft* in Singapore and Southeast Asia:

> Singapore is a mature game market. Its rating follows the US rating. The market is easy, effective, and efficient. When we first enter, we work with a local publisher. We host e-sport events to promote the games. ... When we launch new games, we are widely successful in getting people to promote and people to buy the game. So here in Southeast Asia we plan to lower prices [for] three days and seven days [to] 1.5 and 3 SGD. This is only for gamers who cannot pay [a] fee to play our game. We hope that by doing this we will attract more players in the region to play games as often as our gamers in Singapore. (Mr. O, Blizzard Singapore, interview with the author, August 5, 2010)

Domestic Video Game Production and Exports

Games that are produced in Singapore range across various platforms. However, they are mostly produced for the foreign rather than domestic market. One platform, MMOG, illustrates Singapore's outward-looking production strategy. These games were created in response to the development trend that emerged in the global game industry during the second wave. Products such as KOEI's *Romance of Three Kingdoms On-line* and Nabi Studios' *Toribash* now sell on the international market in places such as Japan, Russia, and Brazil. KOEI developed the first game, *Three Kingdoms* (1995), as an online version of its previously successful war strategy game, which it originally published for Nintendo Entertainment System, and later extended it to the PC, PlayStation, and Wii. The company's Singapore studio explores the online dimensions of the game by focusing on the narrative of the ancient Chinese tale, *Chronicles of the Three Kingdoms*. The company published the game in Japan in 2008; however, the service ended in 2011. The other MMOG, *Toribash*, emphasizes multiplayer martial arts. The game was created by a Swedish developer who set up Nabi Studio with the support of MDA. In 2007, *Toribash* won an award at the Independent Developers Festival. The online service became popular in emerging markets such as Russia and Brazil and by 2007 reached 700,000 members (MDA 2010). Later, the game also became available for the Wii system. The game's features include 3-D fighting and in-game virtual trading of microtransaction items. Gamers in the overseas market have expanded, and the game still maintains active service to this day through e-distribution channels such as Steam.

Table 3 shows the list of selected games made in Singapore since 2008, their developers, and their publishing platforms. Because of the release of these titles, Singapore became more visible in the global game scene. Most of the titles listed are console and PC productions released through international publishing companies. The aims of corporation branding ensure that little local cultural content related to Singapore is reflected. Successful titles that demonstrate the capability of Singaporean companies to produce global blockbuster titles include *The Clone Wars* (2008), *Jedi Alliance* (2008), and *The Force Unleased II* (2010) from Lucasfilm. Selected

Table 3

List of games made in Singapore and their developers and publishing platforms.

Year	Title	Developer	Publishing Platform
2008	Romance of Three Kingdoms On-line	KOEI SG	MMORPG (Japan)
2008	The Clone Wars: Jedi Alliance	LucasArts SG	Nintendo DS
2008	The Force Unleashed	LucasArts SG	Xbox 360, PS3
2008	CarneyVale: Showtime	Gambit	Xbox Live Community Game
2009	Armor Valley	Protégé Production	Xbox Live Indie Game
2009	The Clone Wars: Republic Heroes	LucasArts SG	Nintendo DS
2009	Teenage Mutant Ninja Turtles: Turtles in Time Re-Shelled	Ubisoft SG	Xbox 360, PlayStation Network
2010	Star Wars: The Force Unleashed II	LucasArts SG	Nintendo DS
2010	Assassin's Creed 2: Brotherhood	Ubisoft SG	PS3, Xbox 360, PC
2010	Nobunaga's Ambition	KOEI SG	iPhone
2011	Assassin's Creed Revelations (Partial level)	Ubisoft SG	PS3, Xbox 360, PC
2011	Rocketbirds: Hardboiled Chicken	Ratloop Asia	PSN, Steam
2012	Autumn Dynasty	Touch Dimensions	iPad

titles from Ubisoft include *Teenage Mutant Ninja Turtles* (2009) and *Assassin's Creed* (2010). *Nobunaga's Ambition* is a PC game from KOEI. Its story focuses on the cat version of KOEI's previous successful strategy game series, *Nobunaga's Ambition*. This game was released in the iPhone channels and has won popularity among gamers in Taiwan.

For small- and medium-sized companies, networked games across multiple platforms represent another type of production global release. These platforms allow niched social games to enter the global market by means of Web-based and mobile distribution. For instance, casual games created by a medium-sized local company named Boomzap show how this niche production strategy developed among local game developers. Games such as *Otherworld* (2012), *Awakening* (2011), *Death Upon an Austrian Sonata* (2010), and *Botanica* (2007) reflect the international content that companies in Singapore create in order to meet the popular tastes of older female gamers in the US market.[27] The visual narrative style derives from a computer graphic art that is familiar to readers of romance novels in the United States. In Singapore, small- and medium- sized companies have turned their appeal to internationally niched gamers in the United States and Europe into a common production and exporting practice. Their strategy involves a particularly market-oriented production process that draws talent to their base in Singapore from multicultural backgrounds as diverse as Singapore, Malaysia, the Philippines, Indonesia, Japan, and other parts of the world.

Indigenous Video Game Culture

Singapore is a "techno" nation, where an indigenous video game culture has grown to flourish in an environment where gamers can access content from a range of different genres and platforms through a variety of channels. The Singaporean government tends to be more lax in screening video game content imported from the United States, Europe, and Japan, in contrast to its approach to traditional media, where the local government applies censorship in order to restrict commentary that may jeopardize the political sovereignty of the dominant People's Action Party (PAP). The government's support in facilitating faster growth for the domestic game industry has led to the emergence of a dynamic video game culture. This, and the fact that gamers play in a largely English-language environment, means that Singapore's indigenous game culture reflects a style of gameplay that is closer to Western gaming styles, in contrast to countries in East Asia such as China, Taiwan, and Korea, where hardcore MMORPG players shape the main form of collective video game culture. Moreover, local gaming trends have evolved toward casual gaming, an indigenous gaming culture that reflects a mix of hybrid popular cultures imported from Japan, Europe, and the United States (STGCC 2012).

The rise of gaming culture in Singapore has grown in tandem with the introduction of information and communication technology into local society. In 2011, Singapore reportedly reached the highest rate of Internet penetration in Southeast Asia; the penetration rate among its youth reached 97% among fifteen- to nineteen-year-olds going online (Kwek 2011). As of 2013, the country also has the world's fourth-highest smartphone penetration, reaching 78%, following South Korea, Hong Kong, and Norway (Anjum 2013). The city-state's continuous IT development allows gamers a variety of sources through which to access different genres of content. Publishers such as Asiasoft, for example, provide their service to customers through forty-two Internet cafés with whom they contract (Asiasoft 2013). Besides, gamers can also access these games by connecting to the company's servers through their personal computers at home. Singapore's high penetration of mobile phones also encourages the popularity of mobile and Web-based games. Playing app games has become a naturalized part of young people's daily lives as an important part of their leisure and entertainment.

As parents have come to accept video gaming as a form of media entertainment for their children, competitive video gaming has grown to become a common activity among youth. In order to avoid the potential social problems that arise with game addiction among teenagers, the government has set up a serious game initiative directed at the promotion of healthy gaming. This policy initiative receives support from local non-profit organizations. For example, in 2009, a group of avid gamers who are passionate about competitive gaming formed Esport Association Singapore. Interregional tournaments offer opportunities for Singapore gamers to participate in competitions with gamers from nearby nations. Another organization, the Singapore Cybersports and On-line Gaming Association (SCOGA), provides a seed fund of 20,000 SGD to assist competitive gamers who wish to represent Singapore at international tournaments (SCOGA 2013). Indigenous video game culture is therefore growing in a number of directions that move computer game play beyond the traditional stereotype. This provides further means to give game developers and game players new influxes of capital that will raise the social status of people who are involved in game production and game competition in Singapore's business-driven society.

As stated throughout this chapter, local game companies in Singapore concentrate on independent game development in the global game industry. After a decade since the second wave, Singapore has formed an industry that sees both transnational and local game companies growing game development into various platforms. It is estimated that there were more than sixty-two game companies in Singapore in 2010 (Media Development Authority 2012). Through 2013, companies that were active in Singapore collectively formed a dynamic globalization network that shows the effort of transnational and national integration in the making of Singapore's digital gaming hub. These companies include Asiasoft Online (Thailand), Boonty Asia (France), DF Interactive (Korea), Electronic Arts (US), Infocomm Asia Holdings (Singapore), and Namco Bandai (Japan) in the publishing sector. Small- and mid-sized developer firms include Activate Interactive, Boomzap, Envisage Reality, Gevo Entertainment, Lionstork Studios, Magma Studios, MatchMove Games, Nabi Studios, NexGen Studio, Orange Gum, Playware Studios, Protege Production, Ratloop Asia, Tecmo KOEI Singapore, Time Voyage, Touch Dimension Interactive, and Tyler Projects.

Video Game Content Description

Since the end of the second wave, games made in Singapore have begun to win awards at international game festivals. These games include *CarneyVale: Showtime* (2008), *Armor Valley* (2009), *Rocketbirds: Hardboiled Chicken*

(2011), and *Autumn Dynasty* (2012). These games have their own particular art style as part of a hybrid production strategy. For instance, *CarneyVale* is a game that evolved from a student project at the Singapore-MIT GAMBIT Game Lab. The game narrative shifts away from commercial popular tastes by focusing on strong visual artistic presentation in coloration and storytelling. *CarneyVale: Showtime* has a story about rising to stardom in the world of the circus. The game requires little localization because the theme speaks to a niche audience that already favors the aesthetic of French animation and the circus story (Chia and Wong 2012).

Armor Valley, *Rocketbirds: Hardboiled Chicken*, and *Autumn Dynasty* are also games that tell universal rather than nationally oriented stories. Both *Armor Valley* and *Rocketbirds: Hardboiled Chicken* focus on the use of music in storytelling. In *Armor Valley*, players enjoy real-time war game action in a 3-D environment. In *Rocketbirds*, players engage in a setting that displays a cinematic 2-D view that provides the setting for chicken chases. The game *Autumn Dynasty* has its own innovation in the form of calligraphic drawing through a touch-based medium. This game takes Chinese themes in its retelling of a famed Chinese story about war games and popularizes it among global gamers. Regardless of artistic value, these games have experienced a gradual rise in commercial value as they've begun to be consumed and reviewed by gamers in the international market.

World of Temasek (2011), in contrast, is one local game that aims specifically at Singapore gamers. The game tells the early history of Singapore. With support from the National Heritage Board, Magma Studio produced this 3-D nonfiction game for educational purposes. The National Heritage Board is the government office that oversees cultural heritage under the Ministry of Culture, Community, and Youth. The game has an element that allows users, preferably Singapore children, to log in so that they can interact with each other and experience life in the land of Temasek during the fourteenth century. This game has built a virtual world using advanced 3-D technology in order to present ancient Singapore, set within the parameters of the city landscape as it was in 2010. The game concentrates on presenting Singapore's multicultural historical past, featuring the lives of the territory's different races—Malays, Chinese, Indians, and indigenous Orang Laut. Players also learn about Singapore's struggle for autonomy through negotiating with neighboring political powers, the Javanese Empire, and the Siamese Empire. With research support from academics at local universities, this game reconstructs a digital landscape of nationalism by re-representing heritage as players are immersed in the game and develop their own stories.

Video Game Studies in Singapore

Institutions that provide game education curricula range from universities, to community colleges, to professional art schools. The major academic institutions are the DigiPen Institute of Technology Singapore, Lasalle College of the Arts, Nanyang Polytechnic School of Interactive and Digital Media, Nanyang Technological University, the National University of Singapore, and the Ngee Ann Polytechnic School of Infocomm Technology. Some of these are pioneer academic institutions, including Nanyang Poly and the National University of

Singapore, executing the planning of game education during the first wave. Some of the institutions, such as DigiPen, are the outcome of game policy from the second wave. Game research is carried out at the university level between the National University of Singapore and Nanyang Technology University. The Singapore MIT-GAMBIT Game Lab, under the National University of Singapore, represents one of the achievements in Singapore's game education. The MIT-GAMBIT Game Lab provides student exchange opportunities through research collaboration between MIT and Singapore. This five-year initiative, along with other game education training in Singapore, has contributed to the rise of start-ups in the third wave. Companies such as Ratloop Asia, Time Voyager, and Touch Dimensions demonstrate the capacity of students to benefit from the training of game education in the local academic system.

The Future of Video Games in Singapore

An overview of the history of the development of the video game industry in Singapore shows the progression of the local game industry from a location of no significance to one that enjoys key geographic influence in game development and publishing in the Southeast Asian region. The industry demonstrates a collaborative pattern of industry development that integrates local state, transnational, and national capitals.

Developments that took place during each of the three waves illustrate a gradual industry transition from transnational game companies to local SMEs. The lessons learned in the Singapore case are three-fold: First, game industry development for a small state in Asia requires a flexible policy framework that skillfully makes use of the geographic resources that transnational and national capital can provide. The implementation of a collaborative model in the transnational-national relationship proves to work more effectively than a competitive one. Second, the Singapore case reveals the potential social outcome of an economy-driven cultural policy. The growth of Singapore's video game industry reflects the social change that is underway in the Internet of the nation. The emergence of award-winning local content shows the diversity of content creation that Singapore is able to provide to the existing video game world. Third, developing an interlinked industry ecosystem in Asian video game development requires very active state intervention in policy-making so that the local industry can receive more support to strategize ways to free themselves of the constraints set up by the globally dominant players in a highly concentrated video game market.

Despite the achievement described above, Singaporean video game developers need to look further into the transformation of creative capital found at local levels into economic capital, where dividend policy should be addressed. As local SMEs adjust their ways to secure niches for their independent game development, industrial development policy should address the flexibility in opening up space for local SMEs to gain, profit, and prosper. In addition, opportunities should exist to tap game development and game publishing at the local level in order to break the myth that Singapore is a small market. Local video game content can become a national resource for local consumers to enjoy. Asian game developers still have room to consider what innovation means in designing popular commercial game content for their national citizens.

Notes

1. The four video game titles collected USD $300 million (*Halo 3*), USD $500 million (*Grand Theft Auto IV*), USD $550 million (*Call of Duty: Modern Warfare 2*) and USD $650 million (*Call of Duty: Black Ops*) during the first weekend of their releases, while the four Hollywood blockbuster titles reached total sales of USD $140 million (*Harry Potter and the Order of the Phoenix*), USD $150 million (*The Dark Knight*), USD $230 million (*Avatar*) and USD $120 million (*Iron Man 2*).

2. In 2011, global gamer demographics show various market sectors in the world. For instance, there were 211 million video gamers in the United States, China had 324 million online game users, and there were 70 million mobile gamers in the European Union (Snider 2012; China Internet Network Information Center 2012; "Mobile games trend report" 2012).

3. The other Asian countries that also actively developed game industries include China, South Korea, and some Southeast Asian countries such as Malaysia, Thailand, and Vietnam.

4. Industry data is provided by the online game publisher Infocomm Asia Holdings (IAH).

5. Email interview with Ms. C at the Ministry of Information and Communication and Arts on June 11, 2012.

6. The interviewees include government officials in game policy divisions within the Economic Development Board and Media Development Authority. The interview list also includes company CEOs, game designers, and animation artists in major international game companies and local game companies. Anonymity is used to protect the privacy of the interviewees in this research.

7. Personal interview with the game division at the Economic Development Board, Ms. P, on May 11, 2007.

8. Institutions that resulted from this EDB-EA cooperation include Silicon Illusions and Kent Ridge Digital Lab.

9. A multifaceted engineering and technology company undertaking projects for the Singapore government.

10. *iPower* was a real-time strategy game involving tank combat and was meant to be rendered in 3-D rather than sprite-animated. *Cyberpunk 2020* was based on a fantasy role-playing game, which was a near-future science fiction game. The game was code-named "Year of the Rat," which was Chinese historical fantasy. Both of those games were to be based on the same engine and were role-playing games.

11. KRDL is now named A*STAR (Agency for Science, Technology, and Research). The office is, at present, one of Singapore's most prestigious research labs that conducts the world's advanced projects in the fields of biomedical science, engineering, and technology.

12. Personal interview with game programmer at Kent Ridge Digital Lab, Mr. M, on June 12, 2012.

13. KOEI later merged with Tecmo to form Tecmo KOEI Holdings, Co., in 2009.

14. Personal interview with Mr. N., from KOEI Singapore, on June 12, 2012.

15. Personal interview with Mr. T., from Electronic Arts, on November 7, 2007.

16. Personal interview with Mr. T., from Electronic Arts, on November 7, 2007.

17. Personal interview with Mr. R., from Ubisoft Singapore, in October 2008.

18. In 2009, Namco Bandai acquired a 34% share of Atari Europe. Atari's Asian offices in Korea, Taiwan, and Singapore were taken over by Namco Bandai. In January 2013, Atari's US subsidiary filed for chapter 11 bankruptcy protection in the United States.

19. It is estimated by one international game publisher that Singapore occupies 20% of the total market revenue of the company in Southeast Asia, while its European market appears to be twenty times larger than its market in Southeast Asia.

20. Personal interview with Mr. S., from Namco Bandai Singapore, on June 12, 2012.

21. Personal interview with Ms. T., from Microsoft Singapore, on August 16, 2007.

22. Personal interview with E-Club Malaysia on October 22, 2012.

23. Personal interview with Mr. Y., from NexGen, on May 15, 2007.

24. E-mail correspondence with Ms. T., from the Singapore-MIT GAMBIT Game Lab, on November 21, 2011. The industry, however, continues to evolve with transnational game company Namco Bandai setting up a game development studio in Singapore in 2013.

25. Personal interview with Mr. S., from Namco Bandai Singapore, on June 12, 2012.

26. Personal interview with Mr. T., from Electronic Arts, on November 7, 2007.

27. Personal interview with Mr. S., from Boomzap, on June 20, 2006.

References

Anjum, Zafar. 2013. South Korea and Hong Kong beat Singapore in smartphone penetration, *Ericsson*, June 18. http://www.cio-asia.com/tech/industries/south-korea-and-hong-kong-beat-singapore-in-smartphone-penetration-ericsson/.

Asiasoft. 2013. Internet cafe locations. http://acafe.asiasoftsea.com/acafe_member/SG/main/locations.aspx.

Chia, Bruce, and Desmond Wong. 2012. Postmortem: Singapore-MIT GAMBIT's CarneyVale: Showtime. *Gamasutra: The Arts & Businesses of Making Games*. http://www.gamasutra.com/view/feature/132331/postmortem_singaporemit_gambits_.php?print=1.

China Internet Network Information Center. 2012. Statistic report on Internet development in China. http://www1.cnnic.cn/IDR/ReportDownloads/201209/P020120904421720687608.pdf.

Chong, Terence. 2012. Manufacturing authenticity: The cultural production of national identities in Singapore. *International Journal of Cultural Policy* 45 (4): 877–897.

Chua, Beng Huat, and Kian Woon Kwok. 2001. Social Pluralism in Singapore. In *The Politics of Multiculturalism: Pluralism and Citizenship in Malaysia, Singapore and Indonesia*, ed. Robert W. Hefner, 86–118. Honolulu, HI: University of Hawaii Press.

Chung, Peichi. 2009. The dynamics of new media globalization in Asia: A comparative study of the on-line gaming industries in South Korea and Singapore. In *Gaming Cultures and Place in Asia-Pacific*, ed. L. Hjorth and D. Chan, 58–82. New York: Routledge.

Flew, Terry. 2011. Media as creative industries: Conglomeration and globalization as accumulation strategies in an age of digital media. In *The Political Economies of Media: The Transformation of the Global Media Industries*, ed. D. Dynseck and Y. Dal, 84–100. London: Bloomsbury Professional.

Garnham, Nicholas. 2000. *Emancipation, the Media, and Modernity: Arguments about the Media and Social Theory*, 39–62. New York: Oxford University Press.

International Telecommunication Union. 2011. Trends in video games and gaming. *ITU-T Technology Watch Report*. http://www.itu.int/dms_pub/itu-t/oth/23/01/T23010000140002PDFE.pdf.

Internet World Stats. 2012. World Internet users and population stats. http://www.internetworldstats.com/stats.htm.

Klein, Stephen, Nick Dyer-Witheford, and Greig de Peuter. 2003. *Digital Play: The Interaction of Culture, Technology, and Marketing*. Montreal: McGill-Queen's University Press.

Kong, Lily. 2000. Cultural policy in Singapore: Negotiating economic and socio-cultural agendas. *Geoforum* 31:409–424.

Kong, Lily, and Brenda Yeoh. 2003. *The Politics of Landscapes in Singapore: Constructions of "Nation."* Syracuse, NY: Syracuse University Press.

Kwek, Carmelita Miki. 2011. Singapore has the highest rate of Internet penetration in Southeast Asia. *Jakarta Globe*, July 11. http://www.thejakartaglobe.com/tech/singapore-has-the-highest-rate-of-internet-usage-in-southeast-asia/452152.

Kwok, Kian Woon. 1995. Singapore: Consolidating the new political economy. *Southeast Asian Affairs*: 291–308.

Lee, Terence, and Lars Willnat. 2006. Media research and political communication in Singapore. Working Paper No. 130. http://wwwarc.murdoch.edu.au/publications/wp/wp130.pdf.

Lim, Alwyn. 2002. The culture of technology of Singapore. *Asian Journal of Social Science* 30 (2): 271–286.

Lim, Lorraine. 2012. Constructing habitus: Promoting an international arts trend at the Singapore Arts Festival. *International Journal of Cultural Policy* 18 (3):308–322.

Media Development Authority. 2008. MDA collaborates with Microsoft to fuel Made-in-Singapore game content through Xbox LIVE Community Games. http://www.idm.sg/mda-collaborates-with-microsoft/.

Media Development Authority. 2014. Game brochure. http://www.mda.gov.sg/IndustryDevelopment/IndustrySectors/Documents/Brochures/MDA%20Games%20E-Brochure.pdf.

Media Development Authority. 2014. *Cyber Wellness.* http://www.cyberwellness.org.sg/SitePages/PublicHome.aspx.

Media Development Authority. 2012. Singapore's video game industry. http://www.mda.gov.sg/Industry/Video/IndustryOverview/Pages/Overview.aspx.

Miege, Bernard. 2011. Theorizing the cultural industries: Persistent specificities and reconsiderations. In *The Handbook of Political Economy of Communications*, ed. J. Wasko, G. Murdock, and H. Sousa, 83–108. Malden, MA: Wiley-Blackwell.

Mobile games trend report. 2012. *Newzoo.* http://www.newzoo.com/trend-reports/mobile-games-trend-report/.

Moss, Will. 1995. Report from Singapore. http://imagethief.com/report-from-singapore/report-from-singapore-part-1/.

Ng, Wai-Ming. 2001. Japanese video games in Singapore: History, culture, and industry. *Asian Journal of Social Science* 29 (1):139–162.

On-line poll series unveils insights into social gaming in Singapore. *Money On-line.* 2011. http://www.molglobal.net/online-poll-series-unveils-insights-into-social-gaming-in-singapore/.

Ooi, Can Seng. 2010. Political pragmatism and the creative economy: Singapore as a city for the arts. *International Journal of Cultural Policy* 16 (4): 403–417.

Singapore Cybersports and On-line Gaming Association. 2013. Gamer Assistance Program. http://scoga.org/gamer-assistance-program/.

Singapore Toy Games and Comic Convention. 2012. STGCC strikes with competitive gaming content. August 15. http://www.singaporetgcc.com/media/15-Aug-2012.

Snider, Mike. 2012. NPD: Total number of US video game players drops. *USA Today*, September 5. http://content.usatoday.com/communities/gamehunters/post/2012/09/npd-total-number-of-us-video-game-players-drops/1#.UeXdPaX-tUQ.

Tan, Elizabeth. 2013. Japanese game developer Namco Bandai to open studio in Singapore. April 17. http://e27.co/2013/04/17/japanese-game-developer-namco-bandai-to-open-studio-in-singapore/.

Tan, Margret. 2012. Promises and threats: iN2015 Masterplan to pervasive computing in Singapore. *Journal of Science, Technology Society* 17 (1): 37–56.

Toyad, Jonathan Leo. 2011. Namco Bandai to distribute Xbox 360 Final Fantasy XIII-2 in Asia. *Gamespot*, October 12. http://www.gamespot.com/news/namco-bandai-to-distribute-xbox-360-final-fantasy-xiii-2-in-asia-6339843.

Wager, Skiff. 2007. Media and entertainment—Use of digital assets and digital management: Industry outlook 2006–2010. *Journal of Digital Asset Management* 3:5–9. http://www.palgrave-journals.com/dam/journal/v3/n1/full/3650055a.html.

Wasko, Janet, Graham Murdock, and Helena Sousa. 2011. *The Handbook of Political Economy of Communications*. Malden, MA: Wiley-Blackwell.

White, Timothy, and J. E. Winn. 1995. Islam, animation, and money. *Kinema: A Journal for Film and Audiovisual Media*. http://www.kinema.uwaterloo.ca/article.php?id=349&feature.

Wolf, Mark J. P., ed. 2007. *The Video Game Explosion: A History from PONG to PlayStation and Beyond*. Westport, CT: Greenwood Press.

World Economic Forum. 2012. The Global Information Technology Report 2012: Living in a hyperconnected world. http://www3.weforum.org/docs/GITR/2012/GITR_OverallRankings_2012.pdf.

Yu, Eileen. 2005. KOEI unveils US$1.8 million game plan for Singapore. *Business Times*, February 15.

Yue, Audrey. 2006. The regional culture of New Asia: Cultural governance and creative industries in Singapore. *International Journal of Cultural Policy* 12 (1): 17–33.

SOUTH KOREA

Peichi Chung

This chapter introduces the developmental history of South Korea's video game industry. It first traces the political and social factors that led to the birth of Asia's online game industry. It then examines the history of video gaming in South Korea since the 1970s, highlighting the types of video games and processes of production and consumption that developed during particular stages of the industry's growth. This provides an opportunity to document foreign game imports and analyze their influence on South Korea's domestic industry, particularly the country's development as a cultural exporter between 2002 and 2012. The indigenous video game culture that the industry spawned is derived from gamer socialization patterns that are particular to South Korean society. Next, this chapter breaks down the industry's pyramid structure in order to demonstrate the competitive nature of the industry in South Korea. It profiles important large-scale game companies in addition to a selection of innovative, middle-scale firms. The chapter ends by acknowledging the contributions that the Korean game industry has made to the development of game industries, markets, and users throughout Asia.

Background

Korean video games have shaped new types of game production and consumption in the conventional video game world. In Korea, as game design moved from consoles to online platforms, developers created a market niche apart from the greater domain of global game corporations. Unlike the content created by Microsoft, Sony, and Nintendo, which focuses on console games for individual users, Korean online game design focuses on innovations that enable large-scale socialization activities, such as massively multiplayer online role-playing games (MMORPGs).

For example, Korean companies have produced MMORPGs such as NCSOFT's *Lineage* (1998), GRAVITY Co.'s *Ragnarok Online* (2002), and Wizet's *MapleStory* (2003), which have succeeded in both domestic and international markets. The Korean companies that published them, GRAVITY and Nexon, respectively, first released them to Korean markets before going on to create international success in the markets of China, Japan, North America, and Southeast Asia. Even today, after more than ten years in operation, *Lineage* and *MapleStory* remain among the top five grossing online games in Korea.¹ The third game, *Ragnarok Online*, was published in many Asian markets before it was acquisitioned by the Japanese company Softbank in 2005. Consequently, *Ragnarok Online* remains one of the most popular online games in Southeast Asia, especially so in Thailand and the Philippines.²

In the area of consumption, Korean video games have earned their reputation thanks to the country's advanced development in the online game market. Through their products, Korean companies have successfully aggregated a group of sophisticated gamers that has in turn vastly increased the number of consumers. Since the 2000s, the development of low-cost, high-speed broadband connections has allowed Korean users to enjoy quick access to games whenever they play online.³ As a consequence, the maturity of these gamers has made Korea a preferred market for multinational game companies such as Electronic Arts and Blizzard, which now consider Korea a test bed for new products. Take, for example, *StarCraft II: Wings of Liberty*, which Blizzard released in 2010. Thanks to the popularity of its games in Korea, Blizzard incorporated features into the content design of *StarCraft II* that appealed to Korean gamers (Steed 2009).

Dyer-Witheford and de Peuter (2009) point out that since the 1960s, the predominant features of game content have developed within the framework of a global military-corporation that has grown as a result of imperialist expansion. Indeed, the military-industrial complex in the United States has utilized games for a long time (Nichols 2010). However, the Korean video game industry apparently has developed a structure that transcends the discourse of empire expansion. This has encouraged a different type of innovation process to develop in Asia. The structure of this current process tends to begin with the imitation of game content imported from abroad. A creative stage in the domestic market then follows once the local game market has matured. This has produced a pattern of innovation that highlights the complex government-corporation relationship that propels Korea's local game industry from behind and encourages new content creation going forward.

This chapter includes eight sections that examine the complex government-corporation relationship that underlies the development of the online video game industry in South Korea. It aims to document this development by first discussing the formation of Korea's game market and its online gaming industry in particular. These sections include: (1) a history of video games in Korea, (2) the reception of foreign imports, (3) domestic video game production and export, (4) indigenous video game culture, (5) video game company profiles, (6) video game content description, (7) video game studies in Korea, and (8) the future of Korean video games. The study's data are derived from industry fieldwork interviews conducted in Korea, China, Singapore, and elsewhere in Asia, as well as interviews with Korean government officials, game company managers, and gamers, collected between 2006 and 2011.

The History of Video Games in Korea

Korea's online game industry has grown to become an emerging production center in Asia (Ahonen and O'Reilly 2007). Korean game developers, in general, are competitive players who actively internationalize their business operations in order to integrate with game powerhouses in the United States, Japan, and China. By 2010, they transformed the Korean game industry into the world's second largest online game market. Following China, Korea's revenue from online games reached USD $4.123 billion, which represented about 26% of the world's total online game market. As 18% of the global market in 2010 was made up of online games, this means that Asia assumes about 70% of the world's online game market (Korea Creative Content Agency 2011).

A detailed breakdown of Korea's video game market illustrates that online games occupy 64% of the total video game market, making them the dominant sector in the market. Console and mobile games followed, although each of these sectors occupied only 10% of the video game market in 2010. By comparison, sales from PC games, computer game rooms, and arcade game rooms are in rapid decline.

The 1970s to the Mid-1990s

Two historical stages shaped the current government-corporate relationship in the Korean game industry. The first stage began in the late 1970s, when illegal underground slot machine operators imported arcade machines into the local market (Deerbo 2010). Due to the wide circulation of pirated game materials, arcades and PC games became the dominant mode of game playing, shaping popular tastes in Korea.[4] The Japanese content of these games gradually shaped the popular tastes of game consumers in Korea, and so this period reflects the influence of Japanese cultural content. Although the experience of Japanese colonization prompted the Korean government to ban Japanese cultural content (Jin and Chee 2008), Japanese influence persisted nonetheless. Many Korean professionals, for example, continued to enjoy playing arcade or pachinko games in electronic entertainment rooms in order to relieve the stress of their daily lives.[5] This game room culture cultivated a market that spawned the PC Bang culture that emerged during the era of online game development (Huhh 2008). PC Bang culture is prevalent in Korea because people visit the computer game room together with large groups of friends. The computer game rooms function like extended social space for Koreans because they offer friendly, social settings where people can play online games, check e-mail, and chat with others online (Chee 2006).

Politically, the Korean government, in particular the Ministry of Health and Welfare, maintained strict control over electronic entertainment rooms due to concerns about gambling addiction. Although they were illegal during the 1970s, many of these rooms were built in public places, such as theaters and amusement parks.[6] Between 1978 and 1980, some companies imported game and machine merchandise from Japan and reassembled it in pachinko or casino game rooms for adults to enjoy. Electronic entertainment rooms remained popular, although they only became legal in the 1990s.

The government gradually relaxed video game regulation beginning in the 1980s, when it decided to transform Seoul into an electronics city. The PC diffusion project exemplifies this policy. It influenced the growth of Korean gaming because the government requested industry support from major Korean corporations, including Samsung, Hyundai, and LG.[7] The active participation of domestic mega multinational companies led the PC industry to grow into a nationwide phenomenon.[8] Samsung, Hyundai, and Daewoo started to produce

Japanese-style personal computers (Deerbo 2014). At the same time, they maintained their impact on game industry development and hosted competitions to discover competitive game developers during an early stage of video game industry development.

As the government increased its support for the corporate profiling of game talent, competitors who won the software award at the Samsung tournament became influential figures in the Korean game industry. Hong Dong Hee, Kim Bum Wong, and Nam In-Hwan all won awards for their software and later produced important products. Hong, for example, wrote software for IBM, and Nam developed the first commercial PC game, *Dream Traveler* (1987), which used Korean Hangeul scripts (see fig. 1). The game *Liar: Legend of Sword* (1987) was exported overseas to countries including the United States and Russia ("Liar: Legend of Sword: Korean action RPG for the Apple II, first title in the Liar's franchise" 2012). Additionally, Kim Bum Wong imitated the Japanese arcade game *Hexa* (1986) and produced Korea's local arcade game *Turtleship* (1988). The game's narrative is based upon a traditional Korean folk story that tells how the military hero Admiral Yi used turtle ships to

Figure 1
The first Korean role-playing game, *Dream Traveller* (1987).

successfully defend Korea from Japanese invasion during the *Imjin War* of 1592–1598. The game engages users to experience Korean cultural context through the play of a space shooter game. Korean cultural elements are also seen in the design of the game.

Economically, the game market formed in a relaxed political and regulatory environment during its early stages. Before 1987, when the government enacted its first strong copyright regulation, game companies tended to import illegal game materials from Japan (Deerbo 2014). A strong cloning culture, which first emerged in the 1980s, has existed among game companies. These companies consider the government's support of the PC industry to be an indicator of potential game business in Korea. In response to the government's policy, many companies started production with imitations of foreign games in order to connect to local game consumers. For instance, the software company Zemina produced several imitation titles, *Brother Adventure* (1987), *Super Boy* (1989), *Super Bubble Bubble* (1989), and *Street Master* (1992). These games show a strong resemblance to Japanese console games such as *Mario Bros.* (1983), *Super Mario Bros.* (1985), *Bubble Bobble* (1986), and *Street Fighter* (1987).

Until the mid-1990s, independent studios such as Makkoya, Mirinae Software, Sonnori, and Planet Entertainment changed their focus to the localization of PC game content for market purposes. The situation became more obvious after Korea joined the World Trade Organization (WTO). In particular, the East-West Game Channel began official operation in 1990 as megacompanies such as Samsung, Hyundai, and SK Telecom saw the potential of PC game distribution. Korea's game industry filled with competitors, and companies such as Softmax (http://softmax.co.kr/) and Ntreev Soft (http://www.ntreev.com/) extended their business into online games where they continued to operate.

The Mid-1990s to the Present

The second historical stage moves Korea from game market formation to game industry creation. During this period, game development changed from console game imitation to PC game innovation. Nam (2005) argues that Korea's online game development can be further divided into two subperiods: (1) 1996–1999 and (2) 2000–present. This first subperiod saw the beginning of radical innovation in Korea's creative content industry. This explosive growth happened in part due to the emergence of several young talents who graduated from the Korean Advanced Institute of Science and Technology (KAIST) and Seoul National University (SNU).[9] For example, the president of Korea's largest online game company, Song Jae-Kyung, who set up NCSOFT in 1997, graduated from KAIST. Kim Jeng Joo and Lee Hye Jing, who in 1994 founded Korea's most successful game development company, Nexon, also graduated from KAIST, as well as SNU.

Table 1 shows the growth of the Korean online game market from 2003 to 2010. Specifically, the stage of initiation for the online game industry began when Nexon released Korea's first online game title, *Kingdom of Winds* (1996). In 1998, NCSOFT completed Korea's first and foremost marketable online game, *Lineage*. Two years later, major Internet companies such as Hanbitsoft, Naver, and Hangame also started business operations. The government played a role, too, when it established the office of the Korean Game Promotion Center

Table 1

Korean online game market expansion from 2003 to 2010 (in trillions of Korean won). (Source: "The 2011 White Paper on Korean Games," Korea Creative Content Agency 2011, 571.)

2003	2004	2005	2006	2007	2008	2009	2010
39,387	43,156	86,798	74,489	51,436	56,047	65,806	74,312

in 1999. This office acted as an agent to promote Korean game industry and culture. The office was later upgraded to the Korean Game Development Institute (KGDI) in 2001, then restructured in 2009 under Article 31 of the *Framework Act on Cultural Industry Promotion* as it merged with other offices to form the Korean Creative Content Agency (KOCCA).[10]

The second subperiod, which began in 2000, reveals a change in the government-corporation relationship that favors a market-driven approach to achieve internationalization. During this period, the Korean government adopted a proactive approach to promoting the industry (Jin 2010). Policies that support IT-related industries reflect the government's intention to position Korea as Northeast Asia's IT hub. For example, the Ministry of Information and Communication implemented policies such as *Dynamic U-Korea* in order to facilitate a competitive market structure favorable to IT industry growth.[11] The Korean Game Development Institute (KGDI), the main government body overseeing industry development, also proposed plans (such as Strategic 2010) in order to ensure quality game education, labor training, and a healthy game market.[12] In addition, the Korea IT Industry Promotion Agency (KIPA) proposed the Global Service Platform (GSP), which focuses on the international promotion of game content produced by small- and medium-sized companies. One of KIPA's major achievements in this sub-period is its installment of servers in the geographic locations of Britain, Germany, Russia, Singapore, Japan, and the United States that allowed gamers in these regions to access Korean games.[13]

The government's support of the industry through policy can be seen in two aspects. First, the government adopted a neoliberal approach seeking to manage a structural change that encouraged a market-driven industry environment in Korea. After the 1997 financial crisis, the Korean government instituted the Venture Registration Rule, which sought to ensure the free flow of foreign investment capital into Korea. This regulation encouraged Korean markets to open up and sought to ensure investment opportunities for new technology-based venture firms (Nam 2009). The Korean government's policy-making can be summarized by the strategy of "catch-up policies."[14] That is, the government supported healthy industry growth by adopting policies to break monopolies in the telecommunications industry, promote the proliferation of high-speed communication networks, and ensure effective policies such as the Broadband Policy and the Special Military Exemption Program (Wi 2009). Other relevant policies include the PC Bang Healthy Culture Promotion Project promoted by the KGDI to ensure that a healthy gaming network is established through the PC Bang system.

The second aspect of policy influence involves game consumption, where the government applied a restricted approach to regulating game ratings in order to prevent gambling and addiction. In 2000, the government began its classification based upon the Sound Records, Video and Game Product Act (Korea Game Development & Promotion Institute 2003). This act defines the area of government control in its monitoring of the country's game industry. In 2006, the government enacted the Relevant Implementation Order and Implementation Rules for the Game Industry Promotion Law (Jin 2010). In it, the government set strict laws intended to prevent people from developing addictions to gambling. Similar concerns shaped the creation of the Deliberation of Video Products regulation, which seeks to prevent the side effects of online gaming. For example, the Korea Game Rating Board monitors online game playing among Korean youth. In 2012, the Ministry of Culture, Tourism, and Sports adopted a "Shutdown Law," which forbids children under the age of sixteen from playing online games between midnight and 6:00 a.m. (Lee 2011).

After 2000, the government-corporate relationship changed. For one, Korean online game companies had grown to become the main stakeholders in the country's game market. Table 2 presents a historical development of the online game industry by showing the early years of various government offices and major game companies. Korean online game companies were in a phase of swift expansion during the 2000s, in both domestic and international markets. In the domestic market, a trend of consolidation has driven the industry's expansion. For example, in 2000, about thirty to fifty game developers moved to online games due to the

Table 2

The founding years of major game companies and government offices.

Year	Game Firm	Government Office
1994	Chung Media Nexon	
1997	NCSOFT Joymax	
1999	Hanbitsoft Naver Hangame	Korean Game Promotion Center
2000	JCE (previously Chung Media) Wemade Netmarble Webzen	
2001	NHN (previously Hangame and Naver)	Korean Game Development Institute (KGDI)
2003	Neowiz Playnus	
2004	CJ Internet (Previously Playnus)	
2005	Shanda (Previously Actoz)	G-Star Game Conference
2009		Korean Creative Content Agency (KOCCA)

Table 3

The growth of game companies and workforce employment in Korea's game industry, 2000 to 2010.

Year	Number of Game Companies	Number of Workers
2000	952	13,520
2001	1,381	23,594
2002	1,774	33,878
2003	2,059	39,104
2004	2,567	47,051
2005	2,839	60,669
2006	2,786	32,714
2007	2,792	36,828
2008	3,317	42,730
2009	N/A	43,365
2010	N/A	48,585

competition they faced from the products of Electronic Arts and Blizzard Entertainment.[15] These companies followed the successful model of *Lineage* in Asia's MMORPG market, producing about 100 MMORPGs in 2000 alone (Wi 2009). As indicated in table 3, the number of game companies grew from 952 in 2000 to 3,317 in 2010. While the majority of game companies in Korea is small in size (Jin 2010), table 3 also shows that the industry's workforce has grown, despite market concentration. In 2000, the industry had about 13,500 workers. This number grew to 60,669 in 2005 and then fell back to 32,714 workers the following year. By 2010, the number of workers in the game industry had reached 48,585.

Table 3 reveals an expanded industry structure based upon market consolidation led by five large-scale companies. Due to fierce competition, Korea's online market began to approach saturation in the mid-2000s. By late 2005, five companies—NCSOFT, NHN, CJ Internet, Neowiz, and Nexon—controlled 65% of the total market share in Korea (Jin and Chee 2008), and the Korean game industry had created an industry value chain that only allowed small- and medium-sized game companies to develop and publish games within the corporate network of these five major companies.[16]

A market share analysis of the industry shows that NCSOFT and Nexon operated as vertical portals, with concentration on game-related businesses in development and publishing. NHN, Neowiz, and CJ Internet functioned as portals focused on game publishing.[17] In 2001, NHN started the first portal business after a merger between Hangame and Naver. Because of its ownership of Naver, NHN emerged a strong player in the game market. The company successfully entered Korea's competitive MMORPG market by releasing new genres to differentiate itself from NCSOFT and Nexon. Similar approaches also applied in the case of Neowiz's market entry in 2003. As it started its own portal service, pmang.com, Neowiz rose to top performer status based upon its previous success in managing the community dating website Say Club. The company claimed its niche market with an innovative business model in virtual avatar and sports game development.[18] Their sports game *FIFA Online* (2010) (codeveloped with Electronic Arts) and the first-person shooting game *Special Force* (2004) both have become top-selling games. As CJ Internet started its game portal business, Netmarble—in the wake of its acquisition of Playnus—rose to become the third largest game portal, publishing both Korean and international titles.[19] With strong conglomerate support from the CJ Group, the company applied aggressive publishing strategies, releasing games such as *Perfect World* (2005) from China. Games that enjoyed market success include *Sudden Attack* (2005) and *MaguMagu* (2010), the first of which became the second best-selling game in the Korean market for more than forty weeks in 2009.[20]

In the international market, the transnationalization of Korean video games took off when NCSOFT first exported *Lineage* to Taiwan. Wemade and Webzen were companies that followed in later years and achieved market success in China with such games as *Legend of Mir* (2001) and *MU Online* (2003). In 2004, as the Chinese market began to rise with the help of strong government protection, Nexon switched its focus to the Japanese market and began to set up portal service from its Nexon Japan office.[21] Collectively, Korean game companies created a regional corporate network that allowed Korean games to be exported to Japan, China, Taiwan, Hong Kong, Singapore, Thailand, Vietnam, Malaysia, and the Philippines, among other places. The globalization of Korean games promoted a marketable Korean MMORPG culture in both East Asia and Southeast Asia.

There are three general distribution approaches that Korean game companies use to form the collective regionalization network in Asia. The first seeks direct entry into major markets by setting up regional offices. Nexon, NCSOFT, and NHN adopted this strategy and set up offices in Tokyo and Shanghai in the early 2000s. The second internationalization game distribution approach sought to publish games with local partners. Such collaborations are particularly important because Asian markets were heterogeneous and fragmented. In Southeast Asia, Korean game companies often needed to rely on local partners in order to localize their game content. In the case of China, support from local partners became even more important because the Chinese government imposed quota limits, which aimed to restrict the number of foreign games in its review of license approvals in China. Finally, the third distribution strategy called for a direct service, commonly used by small- and medium-sized companies. One medium-sized game company, Joymax, has been the greatest beneficiary, totaling 88% of the company market revenue in global publishing with its hit game *Silkroad Online* (2004) in 2009.[22]

In all, the regional penetration of Korean game companies illustrates the increasing ease of Korean game distribution throughout the Asian market. However, an examination of the relationships between government and corporations reveals the obstacles to internationalization by Korean companies, in particular to the Chinese market. Despite the localization effort, Korean game companies lost their market advantage in China due to the strong protection of the Chinese game market by the Chinese government. Figure 2 shows the

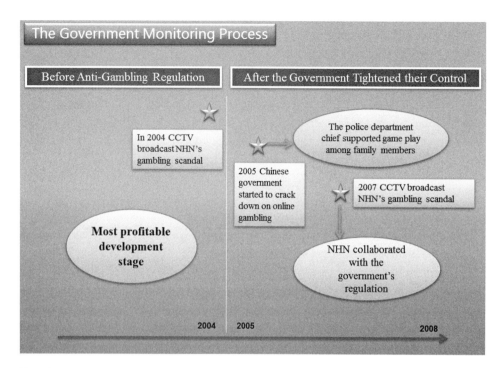

Figure 2
The influence of the Chinese government's game monitoring policy on the market performance of NHN Games in China. (Source: Internal corporate data from fieldwork interviews.)

Chinese government's implementation of anti-gambling regulations and its influence on market performance by NHN in China. The illustration reveals a rapid decline in NHN's market share between 2004 and 2008. Such an event corresponds to the overall timing of the Korean game industry's loss of overseas market share in 2005. However, the Korean game industry as a whole continued to grow, thanks to the internationalization of Korean firms. Nexon, for example, took 50% of the total revenue produced by the overseas market, while CJ Internet and Neowiz accumulated only about 30% and 10% of the total overseas market revenue, respectively.[23]

In 2011, Nexon, Neowiz, NHN, NCSOFT, and CJ Internet collectively claimed ownership of more than 88% of the online video game market.[24] An analysis of the industry value chain in the late 2000s illustrates that most small-scale and middle-scale game companies look for major game companies for resource support in game development and publishing. Independent game developers NeoPle and Dragonfly both released their top-selling games, *Dungeon Fighter Online* (2005) and *Special Force* (2004), under the auspices of Hangame and Neowiz. In 2011, the oligopoly this created reached a new level after Nexon surpassed NCSOFT as the leader of the Korean game market. Further consolidation occurred when Nexon purchased game developers NeoPle and JCE; this resulted in Nexon's ownership of six top-selling game titles, *MapleStory*, *Dungeon Fighter Online*

(2005), *Sudden Attack*, *FreeStyle* (2004), *FreeStyle Football* (2010), and *FreeStyle 2* (2011). The earnings from these six games earned Nexon 1.3 trillion won (USD $1.2 billion), 35% of the market's total share in 2011.

The Influence of National History on Video Game Development

Two historical factors contributed to the development of the Korean video game industry. The first factor relates to the formation of a video game market in the 1970s and 1980s. The memory of Japan's colonial occupation of Korea between 1910 and 1945 has cast a long shadow over the Korean government's policy-making. During the periods of military authoritarian regimes (from 1962 to 1992), Korean governments enforced restrictive regulation of broadcasting media in order to preserve the domestic market from the influx of overseas programming (Kim 2011). This stringent regulation also applied to the gaming market, where illegal imports of Japanese games offered the means for Korean gamers to access Japanese game content. To some degree, this circulation of illegal gaming content cultivated a taste for Japanese styles among Korean gamers. On the one hand, restrictive regulation between 1962 and 1992 also prevented Korea from developing console games and arcade games. On the other hand, piracy culture encouraged Korean game designers to focus on Japanese graphic art patterns. Iwabuchi (2002) argues that a positive perception of Japanese media content among East Asian audiences reveals the influence of Japan's colonial legacy in the region. The cultural proximity that links Japan and Korea together explains the possibility of Japanese games influencing the Korean game market. Chan (2006) considers such growing appreciation about emerging networked games in East Asia as a regional phenomenon that benefitted from cross-media literacy that existed among gamers and developers in Japan, Korea, and China.

A second historical event, the Asian financial crisis of 1997, also helped spark the growth of Korea's online game industry (Chan 2006; Jin 2010; Nam 2009; Wi 2009). After the era of military regimes, South Korean governments adopted a neoliberal policy-making approach to strengthening economic growth and industrial transformation (Lim and Jang 2006).[25] The change that the government brought to Korean society after the financial crisis with its promotion of IT innovation led to the country's rapid economic recovery.[26] This recuperation is evident in online games, which during this era grew to diversify and expand in all aspects. That is, workers who lost their jobs after 1997 began to use information technology—some of them in order to create new PC Bang businesses, others to patronize them (Huhh 2008). After fifteen years of development, the industry diversified and expanded in all aspects ranging from genre, firm size, business model, gamer scale, e-sports, and platform extension, from online to mobile (Volz 2012).

Table 4 illustrates the stable distribution network that PC Bangs created for the online game market. By 1999, the growth rate of PC Bangs reached 405% (Nam 2009), and by 2000 their number reached an average of 20,000 or above. This made PC Bangs the major distribution outlets for game companies.[27] In addition, PC Bangs became learning centers where beginner gamers gathered to exchange information and learn special skills needed to achieve gaming success (Nam 2009).

Table 4

The number of PC Bangs from 1998 to 2010. (Source: Korea Creative Content Agency 2011, 142.)

	2001	2002	2003	2004	2005	2006	2007	2008	2009	2010
Number of PC Bangs	22,548	21,123	20,846	20,893	22,171	20,986	20,607	21,496	21,547	19,014

Domestic Video Game Production and Exports

Analysis of the Korean video game industry reveals a growing sector of game development. Converging MMORPG culture in East Asia has established a standardization process that began with Korean game development and then influenced content creation in the Chinese online game industry (Chung 2010). In the early evolution of Asian game culture during the 1980s, Japanese game development dominated game creation and consumption in East and Southeast Asia (Ng, 2001, 2006). In the 2000s, this dominance gave way to Korean firms through online games, pitting Korea against Japan as the epicenter of game production in Asia. Consequently, game developers in East and Southeast Asia now tend to imitate Korean titles in order to enter their own local markets.

For example, *The World of Legend* (2003), *QQ Audition* (2008), and *Fist of Fu* (2010) are Chinese games that imitate Korean titles such as *Legend of Mir*, *Audition Online*, and *Grand Chase* (2003). In Southeast Asia, the 2-D online local game in Thailand called *Asura Online* (2008) imitates *Ragnarok Online*. All of these imitated games have entered local markets by appropriating the brand identities of pirated Korean games. *Asura Online* benefits from cloning *Ragnarok Online* because of the latter game's existing popularity in Thailand. In 2010, *Asura Online* was exported to Taiwan under a similar Chinese title. Images in figure 3 show the similarities in game content between Korean games and their imitators.

In 2003, 383 out of 1,500 games produced domestically were rated and approved for operation in the Korean game market. While MMORPGs, represented by *Lineage*, *Lineage II* (2003), and *MU Online*, constituted the leading growth force behind the industry when it first started, new genres also developed, as exemplified in first-person shooting games and online sports. Up to the present time, Korea's game industry export revenue grew from USD $173 million in 2003 to USD $671 million in 2006 and reached a total of USD $1,606 million in 2010 (Korea Creative Content Agency 2011). In 2006, online games alone accounted for 89% of the total volume of exports. The largest export market was Japan (32%), followed by China (23%), the United States (20%), Taiwan (9%), Southeast Asia (7%) and other countries (1%). Annual growth rates in the Korean game market remained stable, as imports of foreign games lessened. In 2010, the proportion of exporting countries changed slightly from 2006, with China making up 37% of Korea's total exports, followed by Japan (27%), Southeast Asia (15%), North America (10%), Europe (8%), and other countries (3%). However, the total sales in Asia remained high as the continent consumed 80% of Korea's total sales in the international market.

Figure 3

Korean online games and their foreign imitators from China and Thailand. T3 Entertainment's *Audition Online* from Korea (top, left) and Tencent's *QQ Audition* from China (top, right); Nexon's *Grand Chase* from Korea (center, left) and Outspark's *Fist of Fu* from China (center, right); and GRAVITY's *Ragnarok Online* from Korea (bottom, left) and *Asura Online* from Thailand (bottom, right).

In 2011, Korean online game exports earned revenue of USD $2.262 billion, comparatively higher than revenues from movies and popular music, which totaled USD $80 million and $181 million respectively (Korea Creative Content Agency 2011).

Reception of Foreign Imports

Games from foreign sources such as the United States have made their presence felt in the domestic game market. However, their influence on Korean society cannot compare to games such as *StarCraft* (1998). In the late 1990s, when Blizzard released the first *StarCraft*, PC Bang operators faced the challenge of looking for quality content to fill their distribution networks. High-speed Internet connections allowed gamers to enjoy playing *StarCraft* at PC Bangs (Huhh 2008). As for the game itself, *StarCraft* is a real-time strategy game that meets the gameplaying tastes of most Koreans. Its continuous popularity demonstrates the strength of Blizzard's ability to develop products that can hold the attention of gamers for long periods of time, in this case, over fifteen years. Release timing has also played a role in the game's success. As a result, *StarCraft* played a key role in the later development of e-sport culture in Korea. Two television broadcast channels, Ongamenent and MBC Games, provide all-day broadcast services that provide updates about game competitions in the pro-gamers' StarCraft League. Samsung demonstrates its support by hosting the World of Cyber Games (WCG), an annual game competition held in cities in the US, Singapore, Italy, Germany, and China since 2000. The popularity of *StarCraft* has formed a culture among pro gamers, with top competitive *StarCraft* pro gamers such as Lim Yo-Hwan receiving rock star celebrity status (Jin 2010). In Korea, an e-sport event can draw a crowd of 120,000 spectators to a stadium to watch *StarCraft*.

In addition to *StarCraft*, foreign games also appear as competition titles in the World of Cyber Games. These include *Counter-Strike* (1999), *World of Warcraft* (2004), the *FIFA Soccer* series, *Tekken 6* (2007), and *Asphalt* (2004). In the late 2000s, Korean game publishers began to import MMORPGs such as *Perfect World* (2005) and *League of Legends* (2009). However, because popular foreign games tend to be mostly packaged games rather than online games, another major foreign title entered the Korean market, *Diablo* (1996), and another global title entered the PC Bangs.[28]

Indigenous Video Game Culture

Indigenous video game culture creates a game space that Korean gamers use to construct their digital national identity. To Korean gamers, the concept of a digital Korea represents an imaginary space of Korean community where people play games together. Unlike gamers in the United States and Japan, whose gaming experience tends to be individualized, Korea's indigenous video game culture represents a new form of youth culture that allows young gamers to engage in social interaction through gaming with friends at PC Bangs. In this culture, entertainment happens at the moment when gamers are able to "shout and play games together."[29] This experience of social gaming creates a particular taste of gameplay that also leads to further immersion in a gaming narrative particular to most Korean gamers (Ok 2011). It is said that Korea is a mad gaming nation (Ahonen and O'Reilly 2007), and the country has the highest penetration rate for a single online game; 12 million South Koreans have driven a car in the Nexon online game *Crazyracing Kartrider* (2004). In addition to social gameplay within the Korean community, nation-building sentiments also arise in the context of Korean player-killing, where Korean gamers engage in social gaming on the international servers of an online game.[30] Thomas (2008) describes such gameplay as a cultural location that reflects existing racial tensions between Korean and American gamers. Similarly, political tension also appeared in a game massacre event, when Chinese gamers hacked into a Korean server and sparked mass killing between Chinese and Korean gamers in *Legend of Mir II* (2001).

On a macro level, gaming as a national pastime can be seen in the rapid spread of e-sports in all aspects of Korean society, and this e-sport culture is an indigenous gaming culture that receives support from the government, media institutions, and passionate gamers. E-sports have become recognized as an international sports phenomenon with their origins in Korea. With their emphasis on professional gamers, they have also become an emerging new media phenomenon, an international spectacle in video games (Jin 2010).

Video Game Company Profiles

Since 2005, five game companies, NCSOFT, Nexon, NHN, Neowiz, and CJ Internet, rose to dominate the Korean market. Detailed overviews of company expansion strategies are listed below.

NCSOFT

The NCSOFT company earned a standing in Korean game history for its development of the *Lineage* series. Ever since its early success with *Lineage* achieved the first mover advantage, NCSOFT's main corporate strategy has focused on developing blockbuster titles for the global market. The company has diversified its publishing strategy into three main genres: casual Web games, MMORPGs, and Web-and-board games. Main MMORPG titles such as *Lineage, Lineage II, Guild Wars* (2005), and *AION* (2008) are the primary source of revenue for the company, although its challenge remains at the level of achieving market success outside existing titles. In 2008, for example, revenues from *Lineage* and *Lineage II* accounted for 80% of the company's total sales. The game series has contributed the most to NCSOFT's leading position in the international market. At the same time, NCSOFT receives criticism for its overreliance on a single game series. A revenue breakdown shows that NCSOFT in 2012 collected 35% of its revenue from *Lineage I*, 15% from *Lineage II*, 2% from *City of Heroes* (2004), 42% from *AION*, 1% from *Guild Wars*, and 5% from other games ("NCSoft investor report" 2012).

NCSOFT's early global strategy has contributed greatly to its advancement as a competitive global game company. NCSOFT's first worked with local publishers to achieve market penetration, then it established local subsidiaries in Japan, Taiwan, the UK, and the United States. Compared with other Korean game companies, NCSOFT has a more aggressive globalization strategy. The company earned 634 billion Korean won (USD $490.5 million) in total sales in 2009; in 2011, it reached 117 billion won (USD $109 million). The company's global network has benefitted from the first mover advantage that *Lineage* created. In general, NCSOFT's innovation thrives on the continuous efforts the company makes to produce a global market for content and to explore localization strategies in markets in the United States and Europe.

Nexon

Nexon's strength also derives from its focus on development. Its leading status comes from pioneer work following the microtransaction business model and its shift in focus from hardcore to casual gamers. The company has achieved this transition thanks to several internationally successful titles, including *MapleStory*, *Mabinogi* (2004), *Dungeon Fighter Online*, *Crazyracing Kartrider*, *Counter-Strike Online* (2008), and *Sudden Attack*. In 2012, for example, the eight-year-old game *Kartrider* continued to rise in popularity, as its number of registered users reached 270 million worldwide. Another game, *Dungeon and Fighter*, set a new record when the number of its concurrent users in China surpassed 3 million. The company attained 1.3 billion registered users and total revenue of USD $1.1 billion in 2010 (Gaudiosi 2012). In 2012, its market was divided among Japan (12%), China (47%), Korea (27%), North America (7%), and other regions (7%) (Nexon 2012).

Unlike NCSOFT, Nexon adopted a late globalization strategy to enter the international market. The company maintains a conservative expansion strategy by concentrating on Korean markets as its primary focus of development. With headquarters established in Japan, Nexon expanded beyond Korea using two offices, Nexon Japan and Nexon America, but did not begin to enter the US market until 2007. Like NCSOFT, Nexon relies on two titles, *MapleStory* and *Kartrider*, for 50% of its total income. As Nexon's cautious growth has contributed to its competitiveness, it achieved a 46% growth in sales in 2011. Only recently, in 2012, has Nexon adopted an aggressive global expansion strategy. The company plans to enter the mobile games market and new emerging markets by purchasing Japanese mobile game developer Gloof and the Taiwan-based Asian game publisher Gamania. Internally, Nexon has merged with several top-performing independent game developers and purchased 14% of the limited shares offered by its major game competitor, NCSOFT (Lee 2012).

NHN Corporation

NHN appeared in the online game business after its previous success as a search engine service. The company owns Korea's largest search engine, Naver, which occupies 75% of the country's market. Unlike NCSOFT and Nexon, NHN is an Internet company that provides a variety of businesses, including games, a search engine, microblogging, and a donation portal. The company embarked on internationalization in 2002 with the creation of NHN Japan and expanded into China two years later with a joint venture it formed with the Sea Rainbow Holding Corporation. It set up its US office in 2007. As NHN reached saturation in its domestic sales, it considered online games the most effective approach to internationalization. However, its global strategy in China and the US did not earn satisfactory-enough profits to maintain game publishing operations in two locations overseas. NHN operates its domestic game publishing business through the Hangame portal.

Neowiz

Neowiz entered the Korean game industry in 2003 and created a game portal to publish games. Thanks to its successful experience working with Electronic Arts to develop *FIFA Online* (2010), Neowiz established its niche as a game portal that also excels at in-house development. The company works with independent game developers such as Dragonfly, Smilegate, Red Duck, and others to publish reputable titles. Genres that Neowiz targets include first-person shooting games, racing games, and role-playing games. Major titles include *FIFA Online*, *Slugger* (2009), *CrossFire* (2007), and *Special Force*. As the company has consistently earned 10% of the domestic market in Korea,[31] Neowiz adopted a global strategy in 2010 when it began looking into overseas markets and publishing games through the global gaming portal on Neowiz's website.

CJ Internet

CJ Internet belongs to Korea's largest multimedia conglomerate, CJ Groups. The company shares the same parent company with CJ Media, CJ Entertainment, and others. Major titles of CJ Internet include *Sudden Attack*, *MaguMagu*, and *Prius Online* (2008). The company concentrates on game publishing through its online portal, Netmarble. Published genres mainly include online, mobile, and card games as well as Web board games. In 2009, 60% of the company's market revenue came from global markets, while 40% came from local markets. The company diversified its publishing strategies by dividing its partnership into three main channels.[32] First, CJ Internet works with its development team to create games. It also publishes games from other independent developers, primarily Anipark, which produces the successful baseball game *MaguMagu*. Sometimes CJ Internet also shares titles with NHN's Hangame. International publishing is executed mainly through local partnerships with companies in China, Japan, Taiwan, Singapore, and Thailand. CJ Internet developed its global strategy in the late 2000s when it launched a direct service that provided online feedback for international gamers through Global Service.

Video Game Content

The Korean game industry can be seen as having developed through genre expansion pushed by innovative game content that came to the market at various stages. Four types of game genres developed in the online game industry in Korea: MMORPGs, casual games, sports games, and first-person shooting games. Among the four categories, MMORPGs developed the earliest when *Lineage* came onto the market in 1998 and met with great success. Other MMORPG blockbuster titles, such as *MU Online* and *Legend of Mir*, quickly filled the demand for this genre. Other successful titles followed, such as *Guild Wars* and *AION* from NCSOFT. Among these MMORPGs, *Lineage* achieved the best market results due to its first mover advantage. The game's story took place in the Kingdom of Aden at the time of the European Middle Ages. The games *Guild Wars* and *AION* share similar historical references. Graphic design improved as visual representation moved toward three-dimensional graphics technology. The difference, however, between *Legend of Mir*, *MU Online*, and these other titles created by NCSOFT lies in the styles of narrative that structure them. *Legend of Mir* has become popular in China, while *MU Online* has become a top-selling game in Vietnam. These two games contain much Chinese-style storytelling in the structure of their narratives, which allows users to experience gameplay in the cultural context of Chinese mythology and Wushu martial arts. The game *Legend of Mir* establishes the tension in the game narrative between races of warriors, sorcerers, and Taoists. It reinforces this narrative with cinematic graphic representations of an Eastern fantasy story during medieval times.

Casual games make up the second genre, which came onto the market with Nexon's *MapleStory* and *Kartrider*, two casual MMORPGs. Nexon's contribution, its innovation of the free-to-play business model, has marked a new extension of MMORPG genres in Korean video game history. The company focuses on cute designs in MMORPGs. The basic storyline for *MapleStory*, for example, prompts gamers to answer quests and join any one of five factions to fight the villain, Black Magic, in the world of Maple. Similarly, *Kartrider* presents cute visual aesthetics. The main storyline of *Kartrider* provides gamers with various maps for casual racing. Gamers play with different features, such as drifting, that engage users driving in different modes. They can also compete in races using a variety of fictitious vehicles. Money transactions take place whenever gamers purchase virtual items at in-game shops.

The sports game genre also follows the developmental track of the Korean market. In contrast to the previous two genres, JCE's *FreeStyle* was Korea's first MMO basketball game. The company followed a free-to-play model to design a street-style basketball game. Here, the visual narrative moves away from medieval times and focuses on a modern urban environment where gamers play street basketball to the audio background of hip-hop music. Thanks to Viviendi's investment, the graphics tend to emphasize Western styles. Gamers can maximize their moves through a simple interface. Another game that demonstrates the rise of the sports genre is Neowiz's *FIFA Online*, an online extension of the *FIFA Soccer* series from Electronic Arts. In addition, the sports genre further extended to baseball; CJ Internet's *MaguMagu* and Neowiz's *Slugger* appeal to sports genre gamers. The narratives of the two games are presented with very different graphic representations. Both games, however, use updated information from Korea's baseball leagues to provide a sense of timeliness.

The fourth genre that emerged is the first-person shooter. The continuous success of various titles, starting with NHN's *Special Force* (2004), Neowiz's *CrossFire* (2005), and Nexon's *Sudden Attack* (2009), highlights the popularity of the first-person shooter genre among Korean gamers. These games follow the typical narrative structure of an FPS game; however, specific narratives, action, and choice of play modes vary. The genre now extends to third-person shooters, where visual aesthetics and cinematic effects become new elements in the shooter game genre.

Video Game Studies in Korea

Video game studies has become an increasingly popular subject in the higher education of Korea, and the government has begun to build institutions that encourage game education. The Game Promotion Center Academy has played a significant role as one of the few official institutions established for the promotion of game education. This institution later became a role model for other countries, showing how government can successfully promote game study education.

At the university level, there are several dozen university institutions offering a game study degree (see table 5). While some universities place game studies under the subjects of cultural and media studies, most universities offer a curriculum within the department of game technology, game design, game software, and computer game and graphic studies. Educational institutions that offer a game curriculum include Dong Eui University, Hongik University, Ho Seo University, Chunnam Technology College, Seogang University Game Education Center, and Yeosei Digital Game Institute.

The Future of Korean Video Games

From the 1970s to the present, the Korean game industry transformed itself from an industry of imitation to one of innovation. Between the 1970s and late 1990s, the Korean game industry helped to create a market of cultural tastes. An analysis of the political and economic relationships between government and the corporate sector shows the importance of local factors shaping every stage of development in the industry as well. The government-corporate relationship that developed from the late 1990s to the present reveals a new globalization context that shaped Korea's industry following its entry into the World Trade Organization (WTO). Analyzing the historical factors that shaped the online game industry's development, it becomes clear that local factors played an important role in shaping the globalization of Korea's online game industry.

Future Korean video game studies should connect two main lines of literature and further engage the interconnectivity of arguments raised in the existing research. On the one hand, literature that perceives game research from the perspective of innovation emphasizes the study of an effective entry mode for a game company's global strategy. On the other hand, Korean video game studies that propose a political economic framework consider Korea's current transnationalization a form of empire expansion from Korea. This chapter brings together both bodies of literature. It proposes a discussion for future research that evaluates the complexity of this development, as Korea continues to see online games as a way to represent the country's innovation and demonstrate the power of ubiquitous computing at all societal levels. As the industry becomes more oligopolistic, with even fewer Korean global game companies concentrating the ownership of the market, there is a real need to carefully consider the value of the "local factors" that help to make Korea an empowered online game nation. A more critical rhetoric should be engaged and formulated in order to

Table 5

A list of colleges and universities that offer a game studies curriculum in Korea.

No.	Name of University	Department	Location
1	Dong Eui University	Game Technology	Pusan
2	Pai Chai University	Game Technology	Daejeon
3	Ho Seo University	Game Technology	Chungnam
4	Tongmyong University	Game Technology	Destrict
5	Korea Polytechnic	Game Technology	Pusan
6	University, KPU	Game Multimedia	Gyeonggi
7	WooSong University	Game Multimedia	Destrict
8	Yongsan University	Game Contents	Pusan
9	Jeonju University	Game Contents	Jeonju-si
10	Joongbu University	Game Contents	Seoul
11	Hongik University	Game Software	Seoul
12	Hallym University	Ubiquitous Game Technology	Chuncheon
13	Ho Won University	Computer Game	Gunsan
14	Ye Won Arts University	Game Animation	Jeollabuk
15	Kongu University	Game Design	Seoul
16	Korea National College of Rehabilitation and Welfare	Computer Game Design	Gyeonggi
17	Dongguk University	Game Multimedia	Gyeongju
18	Kyungwon	Game Software Administration	Gyeonggi
19	Kaywon School of Arts and Design	Gameware	Kyunggi-do
20	Korea Polytechnics	Computer Games	Gyeonggi-do
21	Dong Seoul College	Computer Software	Seoul
22	Jangan University	Department of Computer Games	Seoul
23	Yong In Song Dam College	Computer Game Information	Yongin
24	Chunnam Technology College	Game Production	Gokseong County
25	Seoul Soongeui Woman's College	Computer Games	Seoul
26	Doowon Technical University	Computer Games	Anseong
27	Dong Ah Institute of Media and Arts	Game Animation	Anseong

Table 5
(continued)

No.	Name of University	Department	Location
28	Yeoju Institute of Technology	Game Entertainment	Yeoju County
29	Sangji Youngseo College	Game Production	Wonju
30	Dong—Pusan College	Game Consulting	Busan
31	Daegu Mirae College	Game Creation	Gyeongsan
32	Daekyeung University	Department of Internet Games	Gyeongsan
33	Byuksung College	VR Game Development	Jimje
34	Seogang University	Game Education Center	Seoul
35	Yonsei University	The Game Research Center	Seoul
36	Hangyang University	Cultural Content/Animation/Characters/Film	Seoul
37	Koomin University	Game Education Institute	Seoul
38	Sejong Cyber University	Game Animation	Seoul
39	Sang Myung University	Game Production	Seoul
40	Daeduk University	Game Animation	Daejeon
41	Sungduk University	Game Video/Image Production	Yongcheon
42	Keimyung College	Computer Games	Daegu
43	Kangwon Tourism College	Computer Games	Taebaek
44	Far East University	Game Software Studies	Gamgok-myeon
45	Ajou University	Information Communication—Media Department	Suwon
46	Tamna University	Information Studies, Game Development Studies Major	Jeju City
47	Dong A Injae University	Computer Games	Busan
48	Hyechon College	Computer Games and Graphics	Sejong City
49	Seoul Digital University	Multimedia Department	Seoul

safeguard any democratic value that may have been overlooked as the Korean online game industry rises to represent another form of corporate globalization in the new media world.

Notes

1. The lifetime sales of NCSOFT's three biggest games, *Lineage*, *Lineage II*, and *AION*, reached USD $2.7 billion in 2011 (Elliot, 2011). The lifetime revenue of Nexon's two games, *MapleStory* and *Kartrider*, was USD $1.2 billion in 2009.

2. Personal interview with Filipino game publisher Mr. Ramo on May 4, 2011.

3. In 2005, Korea had the lowest-cost broadband among twelve IT-advanced countries. Following South Korea were Singapore, Japan, Sweden, the United States, France, Hong Kong, the UK, Canada, Belgium, Germany, and the Netherlands (Ahonen and O'Reilly 2007).

4. Personal interview with game magazine reporter Mr. Lee on June 20, 2006.

5. Personal interview with Korean game expert Mr. Jiang on December 20, 2011.

6. Personal interview with Korean game expert Mr. Jiang on December 20, 2011.

7. Personal interview with game magazine reporter Mr. Lee on June 20, 2006.

8. Personal interview with game magazine reporter Mr. Lee on June 20, 2006.

9. Among all the universities in Korea during the 1990s, only KAIST and Seoul National University enjoyed the privilege of access to fast-speed Internet broadband service (Wi 2009).

10. The KDGI became a model to many other countries as government support for game industries. The Korean Creative Content Agency includes five organizations: the Korean Broadcasting Institute, the Korean Culture and Content Agency, the Korean Game Development and Promotion Institute, the Culture and Content Center, and the Digital Content Business Group of the Korea Software Industry Promotion Agency.

11. In 2004, Korea's foreign direct investment increased 98%. In particular, investment in ICT-related parts increased 233%.

12. Personal interview with game policy-maker Mr. Kim, from KGDI, on June 22, 2006.

13. Personal interview with game policy-maker Mr. Chung, conducted at the Ministry of Information and Communication, on June 18, 2006.

14. Personal interview with game policy-maker in KGDI, Ms. Hung, on July 5, 2007.

15. Personal interview with game reporter Mr. Lee on June 20, 2006.

16. Personal interview with game company manager Ms. Kim, from Nexon, on November 18, 2009.

17. Personal interview with game company manager Mr. Park, from Neowiz, on November 17, 2009.

18. Personal interview with game company manager Mr. Kim on July 3, 2006.

19. Personal interview with game company manager Mr. Myong, from CJ Internet, on November 20, 2009.

20. Personal interview with game company manager Mr. Myong, from CJ Internet, on November 20, 2009.

21. Personal interview with game industry consultant Mr. Kim on July 3, 2006.

22. Personal interview with game company CEO Mr. Park, from Joymax, on November 18, 2009.

23. Personal interview with game company manager Mr. Park, from Neowiz, on November 17, 2009.

24. The ownership of the top five companies shows that Nexon owns 35% of the market share. It is followed by Neowiz with 16%, NHN with 16%, NCSOFT with 16%, and CJ Internet with 5% of the domestic game market in 2011.

25. The presidents since 1992 are Kim Young-Sam (1993–1998), Kim Dae-Jung (1998–2003), Roh Moo-Hyun (1998–2003), and Lee Myun-Bac (2008–present).

26. From 1992 to 1996, South Korea achieved an average of 7.9% in GDP growth. The growth rate dropped to −1.3% during the economic crisis from 1997 to 1998. However, the GDP increased to 7% from 1999 to 2002 during the post-crisis period.

27. For example, in 2002, NCSOFT acquired 40% of the Korean game market due to its access to PC Bangs.

28. Personal interview with PC Bang owner in Seoul, November 22, 2011. At that time, Korean titles appearing through the PC Bang network were *Lineage*, *Free Style*, *MapleStory*, *Hero* (2006), *Freestyle Football* (2009), *Mabinogi*, *Grand Chase*, *Tera* (2009), *Crazy Arcade* (2002), *Mu Continent of Legend* (2003), *AION*, *Continent of the Ninth* (2012), and *Slugger*.

29. Personal interview with competitive gamer Mr. Park on June 18, 2006.

30. Player killing is a particular type of gameplay that is popular among Asian gamers. It is usually referred to as PK, which means the type of gamer who continues nonstop killing in the gaming environment.

31. Personal interview with game company manager Mr. Park, from Neowiz, on November 17, 2009.

32. Personal interview with game company manager Mr. Myong, from CJ Internet, on November 20, 2009.

References

Ahonen, Tomi, and Jim O'Reilly. 2007. *Digital Korea: Convergence of Broadband Internet, 3G, Cell Phones, Multiplayer Gaming, Digital TV, Virtual Reality, Electronic Cash, Telematics, Robotics, E-Government and the Intelligent Home*. London: Futuretext Limited.

Chan, Dean. 2006. Negotiating intra-Asian games networks: On cultural proximity, East Asian games design, and Chinese farmers. *Fibreculture Journal* 8. http://eight.fibreculturejournal.org/fcj-049-negotiating-intra-asian-games-networks-on-cultural-proximity-east-asian-games-design-and-chinese-farmers/.

Chee, Florence. 2006. The games we play online and offline: Making Wang-tta in Korea. *Popular Communication* 4 (3): 225–239.

Chung, Peichi. 2010. The MMORPG culture of Korea and China: Cultural standardization and the shaping of regional convergence culture in East Asia. *Communications & Convergence Review* 2 (2): 27–39.

Derboo, Sam. 2014. A history of Korean gaming. http://www.hardcoregaming101.net/korea/korea.htm.

Dyer-Witheford, Nick, and Greig de Peuter. 2009. *Games of Empires: Global Capitalism and Video Games*. Minneapolis: University of Minnesota Press.

Elliot, Phil. 2011. MMO giant marks new revenue record for veteran title Lineage. *Gamesindustry International*, February 11. http://www.gamesindustry.biz/articles/2011-02-11-ncsoft-sales-up-2-percent-but-profits-down-6-percent.

Gaudiosi, John. 2012. Nexon celebrates seventh anniversary of MapleStory game with continued success. *Forbes*, May 23. http://www.forbes.com/sites/johngaudiosi/2012/05/23/nexon-celebrates-seventh-anniversary-of-maplestory-game-with-continued-success/.

Huhh, Jun-Sok. 2008. Culture and business of PC bangs in Korea. *Journal of Games and Culture* 3 (1): 26–37.

Iwabuchi, Koichi. 2002. *Recentering Globalization: Popular Culture and Japanese Transnationalism*. Durham: Duke University Press.

Jin, Dal Yong. 2005. Socio-economic implications of broadband services: Information economy in Korea. *Information Communication and Society* 8 (4): 503–523.

Jin, Dal Yong. 2010. *Korea's Online Gaming Empire*. Cambridge, MA: MIT Press.

Jin, Dal Yong, and Florence Chee. 2008. Age of new media empires: A critical interpretation of the Korean online game industry. *Journal of Games and Culture* 3 (1): 38–58.

Kim, Milim. 2011. The role of the government in cultural industry: Some observations from Korea's experience. *Koei Communication Review* 33:163–182.

Klein, Stephen, Nick Dyer-Witheford, and Greig de Peuter. 2003. *Digital Play: The Interaction of Culture, Technology, and Marketing*. Montreal: McGill-Queen's University Press.

Korea Creative Content Agency. 2011. 2011 White Paper on Korean games. Seoul, South Korea: Ministry of Culture, Sports, and Tourism.

Korea Creative Content Agency. 2014. Main statistics of Korea's content markets. http://eng.kocca.kr/en/contents.do?menuNo=201450.

Korea Game Development & Promotion Institute. 2003. *The Rise of Korean Games: Guide to Korean Game Industry and Culture*. Seoul: Ministry of Culture & Tourism.

Lee, Jiyeon. 2011. South Korea pulls plug on late-night adolescent online gamers. *CNN Digital Biz*, November 22. http://edition.cnn.com/2011/11/22/world/asia/south-korea-gaming/index.html.

Lee, Mark. 2012. Nexon buys $685 million stake in NCSoft, becomes biggest holder. *Bloomberg*, June 8. http://www.bloomberg.com/news/2012-06-08/nexon-buys-685-million-stake-in-ncsoft-becomes-biggest-holder.html.

Liar: Legend of Sword: Korean action RPG for the Apple II, first title in the Liar's franchise. 2012. *Giantbomb*. http://www.giantbomb.com/liar-legend-of-the-sword/3030-38103/.

Lim, Hyun-Chin, and Jin-Ho Jang. 2006. Between neoliberalism and democracy: The transformation of the developmental state in South Korea. *Development and Society* 35 (1): 1–28.

Nam, Youngho. 2009. Service innovation in digital contents industry: A case of Korean online games. *Journalism* 17 (1): 119–148.

NCSOFT investor report. 2012. NCSOFT. http://global.ncsoft.com/global/board/attachmentfile.aspx?BID=ir_pr&BNo=24&No=0.

Nexon. 2012. Nexon investor presentation. http://pdf.irpocket.com/C3659/JA1b/CJUy/HlTD.pdf.

Ng, Benjamin. 2001. Japanese video game in Singapore: History, culture, and industry. *Asian Journal of Social Science* 29 (1): 139–162.

Ng, Benjamin. 2006. Street Fighter and the King of Fighters in Hong Kong: A study of cultural consumption and localization of Japanese games in an Asian context. *International Journal of Computer Game Research* 6 (1). http://gamestudies.org/0601/articles/ng.

Ng, Wai-Ming. 2007. Video games in Asia. In *The Video Game Explosion: A History from PONG to PlayStation and Beyond*, ed. Mark J. P. Wolf, 211–222. Westport, CT: Greenwood Press.

Nichols, Randy. 2010. Target acquired: America's army and the video game industry. In *Joystick Soldiers: The Politics of Play in Military Video Games*, ed. Nina Huntemann and Matthew Thomas Payne, 39–52. New York: Routledge.

Ok, Hyeryoung. 2011. New media practices in Korea. *International Journal of Communication* 5:320–348.

Steed, Adam. 2009. Starcraft: A national pastime: How Korea stands to shape Blizzard Entertainment's future. *Suite101.com*.

Thomas, Douglas. 2008. KPK, Inc.: Race, nation, and emergent culture in online games. In *Learning Race and Ethnicity: Youth and Digital Media*, ed. Anna Everett, 155–174. Cambridge, MA: MIT Press.

Volz, Pat. 2012. The original Hallyu: Korean video game industry. *10mag.com*, September 23. http://10mag.com/korean-video-game-201209/.

Wi, Jong H. 2009. *Innovation and Strategy of Online Games*. London: Imperial College Press.

Yoo, Byungjoon, and Hyunmyung Do. 2011. MMORPG development company's successful market entry strategy: A case study of NHN games. *Journalism* 2:159–168.

SPAIN

Manuel Garin and Víctor Manuel Martínez

Let us begin with a cliché: just like Spain has been considered, historically and aesthetically, a place of and for contrasts, full of cultural crossbreeding, the history of Spanish video games is one of imbalances, changing from mystified golden ages to looping crises and—maybe too—great expectations. If world-renowned critics such as Erich Auerbach (2003, 357) and Harold Bloom (1995, 124) have stated that *the world as play* constitutes the greatest creative contribution of Cervantes' *El Quijote* (1605), a sort of literary *gameplay*, we are forced to admit that four centuries later the—still young—history of Spanish game design is shaped by that very "Quixotic" interplay between fantasy (the adventures of talent) and realism (the misadventures of business).

Those are the two main conflicting poles throughout the history of video games in Spain: talent and business. As far as the former is concerned, a variety of works within different periods and contexts has proven the creative strength of indigenous games, from *La Abadía del Crimen* (1987) to *Radikal Bikers* (1998), and between *Commandos: Behind Enemy Lines* (1998) and *Lords of Shadow* (2010). Even so, ironically enough, the imagination and creativity of Spanish game designers have continuously crashed against the lack of a proper industry, one that transforms talent into business. So the contrast remains still uncertain—in spite of the many wonders and letdowns that we would like to approach in this text: a popular and critical history of Spanish game design.

The Good Old Pioneers of the 1980s: From Popular Wit to Mystified Origin

While other media such as cinema were forced to struggle with the restrictions of Franco's dictatorship between 1939 and 1975 and therefore were marked by social, political, and economic deficiencies (the films of Berlanga and Azcona prove it), the video game explosion matched the most promising and active years of recent Spanish history after the democratic elections of 1977 and the Constitution of 1978. It was a period of excitement and unleashed creativity, exemplified by the countercultural movement of "La Movida" and epitomized in Pedro Almodóvar's movies. As such, a dynamic atmosphere in the country's social fabric accompanied the early years of video games in Spain.

Throughout the 1970s, the popular "futbolín" machines, a national version of table football, were gradually replaced by pinball and slot machines as early forms of electronic gameplay. But the greatest change took place between 1972 and 1973, when the first *PONG* (1972) arcade games hit the market and reshaped the context of electronic machines forever: indigenous newspapers even blamed those early foreign imports for a decline in the flourishing—and mostly national—slot machine industry (Ortega 1981, 25). In any case, during the following years, the consumption of video games kept rising, and a solid gaming culture was established. In 1977, a national company called Electrónica Ripollés started to distribute the popular Palson CX consoles (similar to the German model Interton 2000), which included different home versions of *PONG*, known as "tennis" by most Spanish gamers. The social success of video games soon gained public resonance, and we can grasp a certain kind of pre-crash triumphalism in the way national newspapers portrayed the promising industry of their time: "From fourteen manufacturers in 1977 we have increased to 100 in 1980; from 800 jobs directly related to video games to 5,000 jobs, adding an important auxiliary industry which now feeds up to 50,000 families" (Ortega 1981, 25).

In the early 1980s, newspapers' classified advertising sections often included one or two positions in video game distribution companies, proving the emergence of a demanding industry. Video games and consoles soon became standard in many homes—on a national scale, the Atari 2600 being one of the most popular choices. In 1982, a national *Pac-Man* (1980)—locally known as "comecocos"—competition was held in a famous department store group, El Corte Inglés, the winner of which qualified for the international final that took place in Montecarlo (ABC 1982). One of the most successful foreign imports was a pack of games for the Atari 2600 called *Los Cuatro Magníficos* (*The Magnificent Four*), which included *Phoenix* (1980), *Vanguard* (1981), *Centipede* (1981), and *Galaxian* (1979), and sold for 29.500 pesetas (ABC 1983). By that time, thanks to the increase in foreign imports, video games had become a reality at the national level, and important distributors such as Erbe Software soon lowered the games' prices in order to expand sales and fight piracy. Between 1981 and 1983, the total number of video game consoles in the country increased to 200,000, and the world classic *Donkey Kong* (1981) became the biggest hit of its time, praised by the press as a "different" kind of gaming experience that went beyond standardized space battles (Pineda 1983).

The year 1983 is canonically considered the first "important" year in Spanish video game history: the first national gaming magazine was published (*ZX-Microhobby*), and the first indigenous game, *La Pulga* by Paco Suárez, was developed for the Sinclair ZX Spectrum. Although *La Pulga* is popularly regarded as the earliest title of Spanish design, there were a couple companies developing arcade games before it: the pioneers of Cidelsa in Madrid, maker of *Destroyer* (1980), *Altair* (1980), and *Draco* (1981), and Tecfri in Barcelona, maker of *Ambush* (1983) and *Hole Land* (1984), whose main partners left the company to establish Gaelco in 1985. Considering that, during those same years, the North American video game industry crash was already taking place. It seems obvious that game development arrived a bit late to Spain, but it is also true that the early 1980s marked a stage of intense national production known as "the Golden Age of Spanish Software" (Flores 2012), a triumphalist assumption that we would like to discuss, in retrospect.

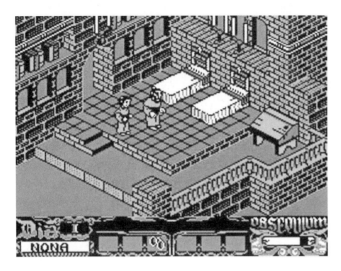

Figure 1
Screenshot from *La Abadía del Crimen* (1987).

Between 1983 and 1992, a considerable number of interesting games was produced, essentially for Spectrum, Amstrad, and Commodore 64 home computers. Paco Suárez conceived *La Pulga* as an experiment to teach his brother how to program a parabola in a ZX 81, but in a few months, the experiment grew bigger, and the game was soon exported to the United Kingdom under the name *Bugaboo (The Flea)*, achieving great success. Spanish artists, programmers, and musicians gradually became interested in game design, and a young and thrilling scene evolved. Companies such as Dinamic Software, Topo Soft, and Opera Soft were born during those years, developing pioneering games such as Opera Soft's *Livingstone, Supongo* (1986), and *La Abadía del Crimen* (1987); Dinamic Software's *Abu Simbel Profanation* (1985) and *Army Moves* (1986); and Erbe Software's *Las 3 Luces de Glaurung* (1986), all of which became hits in the 8-bit European market.

At the same time, traces of Spanish cultural identity were introduced in national productions, and popular comic books such as *Mortadelo y Filemón* (1958–) and TV shows such as *Un, dos, tres ... responda otra vez* (1972–1988) were adapted into video games, while sports celebrities such as Emilio Butragueño, Perico Delgado, and Fernando Martín were licensed by several games, furnishing the social puissance of the new medium. A distinctive trait of the age was the suggestive cover artwork of Alfonso Azpiri for up to two hundred games, which became a sort of national trademark in the late 1980s (Azpiri 2009), an emblem of local thriving times and great expectations.

Critics and consumers favored Spanish productions, therefore creating an atmosphere of praise and benevolence toward "national flavor" (Pérez 2007, 10), which was very clear in the reviews, interviews, and advertising reports of gamer magazines such as *Microhobby* and *MicroManía*. Indigenous games were more appreciated and vindicated than foreign imports, as well as cheaper to develop, publish, and buy—a unique, pro-national situation that never again occurred throughout the history of video games in the country. Completely different from present-day, multimillion-dollar projects that require hundreds of people to work on a game for several years, the Spanish designers of the 1980s worked in small but incredibly dynamic groups, developing several low-budget games per year that recovered costs quickly, as they were rapidly consumed by national audiences. On the other hand, while other countries were already celebrating at conferences and trade fairs, establishing the foundations of a stable industry (having learned from their mistakes during the crash), the international reach of video games in Spain was a bit more limited, and sales outlets were more rudimentary (El Rastro, a second-hand market in Madrid, was one of the most popular). Those are the traditional factors that make authors talk about "the Golden Age of Spanish Software," a sort of lost arcadia that we would like to humbly discuss.

If the period between 1983 and 1992 is regarded as such a fruitful stage in the history of Spanish game design—no less than a golden age—it is, above all, because of the deep crisis and decline that hit the national industry in the following years (discussed below). The period after that first decade was so tragic for the hopes of designers and creators that they automatically considered the past a lost paradise, thus transforming the passion and intensity of early Spanish software into a mystified idea, a maybe-not-so glorious past. "Golden age" is quite an important term in Spain, since it recalls the unparalleled splendor of Spanish arts between the sixteenth and seventeenth centuries, from the literature of Cervantes, Lope de Vega, and Quevedo, to the paintings of Velázquez, El Greco, and Murillo. We believe that comparing such internationally undisputable masterworks with early Spanish game design and using the expression "golden age" when referring to that stage, may work as some kind of *boutade* or inside joke, but definitely not as a historical and aesthetic reality. Amazing games were created and successfully exported to Europe during those years (for example, *Bugaboo*), and young programmers earned a well-deserved reputation (for example, the very Wertherian genius, Paco Menéndez); but the true value of that self-proclaimed "Golden Age of Spanish Software," in terms of the art of video games and its international scope, is still a subject for further research, largely mystified by national discourses.

Film historian Jean-Louis Comolli used the French concept *fétichisation de la première fois* (fetishization of the first time) to refer to a mystifying tendency within film history (Comolli 1971), which sings the praises of early and silent film simply for "being there first," extolling pioneers in a false and idealized way, automatically, regardless of its true artistic merits. We find that something similar has happened with the early ages of Spanish game design, heavily praised inside Spain's borders by way of nostalgia because the period that followed it—the 1990s—became a time of uncertainty and despair, causing some to yearn for the "good old times."

In any case, there were very specific reasons for the decline of homeland production; analysts have listed at least three key points, mostly related to the difficulties involved in transforming local talent into international business. First of all, Spanish companies couldn't match the shift from an almost "artisan" industry during the 1980s to the standardized production criteria of the 1990s (De La Fuente and López 2008). Secondly, that process was aggravated by the change from 8-bit computers to game consoles; young programmers needed a series of permits, licenses, and design tools owned by Japanese and American companies, and thus became unable to adapt to such a changing scenario (Pérez 2007, 8). Finally, the lack of a proper distribution system or marketing structure to commercialize Spanish titles added to the absence of state public funding to game development and put the power of the industry in the hands of big foreign corporations, which focused on distributing games produced elsewhere (Checa-Godoy 2009, 181). Many indigenous companies and game studios went into bankruptcy, and while other countries were already adapting to the aesthetic and technological challenges of a new generation of consoles and computers, moving from 8-bit to 16-bit machines, the game industry in Spain seemed to enter a dark age. But, as usually happens with dark ages, Spanish game designers were going to learn an invaluable lesson: how to bridge talent and business, sailing the wild seas of a global video game market.

Learning the Game: Five Significant Exceptions within the Decline of the 1990s

The general atmosphere of crisis lasted through the whole decade of the 1990s. Very promising companies and creators were forced to cease their activities, and that period is widely regarded as an evolutionary burden that diminished the expectations of Spanish designers. Yet, within that pessimistic context, a small group of solid studios managed to survive. Their importance in the history of game design in Spain is major, since they were able to keep up the good work during hard times and therefore establish a proper basis for the industry's recovery in the 2000s. We have chosen to focus on five key examples—Gaelco, Bit Managers, Dinamic, Pendulo, and Pyro—which not only defined indigenous production in the 1990s but also shaped its future in the decades that followed. Therefore, we will examine how those companies struggled during the last decade of the twentieth century and how they reemerged (or collapsed) in the early-to-mid-2000s.

For many years, the most internationally renowned Spanish studio was probably Gaelco, the creator of the worldwide hit *Radikal Bikers*. It should be noted, however, that the studio's talent came from an earlier company, the aforementioned Tecfri, which developed excellent arcade games such as *Ambush*, which was even distributed in Japan by Nippon Amuse. In 1985, Xavier Valero, Josep Quingles, and Luis Jonama, three Tecfri employees, decided to establish their own company in Barcelona and develop the first Gaelco game: the quiz-show arcade *Master Boy* (1985). In the early 1990s, the hack 'n' slash *Big Karnak* (1991) and the action-platform *Thunder Hoop* (1992) achieved great success, foretelling the European mass success of the racing game *World Rally* (1993).

Interestingly, Gaelco was one of the few companies that not only developed software but also patented its own hardware, producing graphic engines and chips for 2-D and 3-D environments (Pérez 2007, 13). The company's biggest success came in 1998 with *Radikal Bikers*, a fascinating arcade game based on delivering pizzas on a rare Italian scooter. As one of its creators explained, "No one expected the success of that game, we were doing things never done before, so we had no references. The gameplay combined platforms concepts with

Figure 2
Screenshot from *Radikal Bikers* (1998).

straightforward and intuitive driving. Hundreds of stunts were developed so that fun things happened along the way, all the time, to make the player advance no matter what. It was one of the first arcade hits that [was] massively played by girls. It wasn't elitist, everybody was able to play" (Xavier 2009). Arguably, *Radikal Bikers* inspired SEGA's designers when they developed *Crazy Taxi* (1999), which remains a true milestone in Spanish video game history.

In spite of having fewer resources than its foreign competitors, Gaelco was able to find its own place in the international market during the late 1990s: it went beyond the localism of other Spanish companies and aimed for global audiences. The proof of that strategy is *Tokyo Cop*, an ambitious and sophisticated driving arcade game that hit the Japanese market in 2003. Unfortunately, the consolidation of home consoles and PCs as the main platforms for games diminished the expectations of arcade game developers such as Gaelco. In a few years, its benefits were drastically reduced and the popular culture of arcades was overtaken by the mainstream culture of online games. The studio couldn't cope with the change and, although some of its games were ported into console versions (such as *Radikal Bikers*) and they gained the control of important indigenous franchises, such as *PC Fútbol*, their benefits took a nosedive, and Gaelco went bankrupt between 2007 and 2009. In any case, the studio remains part of the symbolic identity of Spanish game design, rooted in the gamer memories of several generations.

Another Catalonian company, one of the few able to overcome the shift from Spectrum and Amstrad in the 1980s to consoles in the 1990s, was New Frontier, later renamed Bit Managers. A business trip to Lyon stands as a crucial point in the studio's history, when one of their executives visited the quarters of Infogrames to sell an 8-bit conversion of the game *Hostages* (1990). That trip involved the beginning of a fruitful collaboration between the French company and Bit Managers, which remained one of the few indigenous companies developing globally distributed games for consoles, such as *Asterix* (1993) and *Tintin* (1996), thanks to their "French connection" (Pérez 2007, 15). The studio became famous for producing conversions of games that Infogrames had already developed for other platforms; a good example is *The Smurfs* (1994), for which Infogrames was contracted for the Super Nintendo Entertainment System (SNES) and Sega Mega Drive versions, while Bit Managers was in charge of the Nintendo Entertainment System (NES), SEGA Master System, SEGA Game Gear, and Nintendo Game Boy versions. Bit Managers' creative peak was reached in 1998, when it was among the first studios to develop games for Game Boy Color's launch, being appointed by Acclaim to make the portable version of *Turok 2: Seeds of Evil* (1998) and by Infogrames for the licensed adventures of *Sylvester & Tweety* (1998). Precisely during that time, Gaelco bought Bit Managers and took over the PlayStation version of *Radikal Bikers*. In 2001, the original partners bought the company back from Gaelco and returned to their previous collaboration with Infogrames, developing ports of existing games for the Game Boy Advance until, in 2005, Bit Managers was bought and renamed Virtual Toys Barcelona by the Spanish group Virtual Toys.

Dinamic Software, one of the pioneering companies in the 1980s and responsible for great Spectrum, Amstrad, MSX, Amiga, Commodore 64, and Atari 2600/ST titles such as *Abu Simbel Profanation* and *Navy Moves*, stands as another interesting example of how national studios dealt with the change of paradigm toward consoles and VGA computers. Unlike other studios, who insisted on developing new games and finally went into bankruptcy (unable to adapt to the new scenario), Dinamic changed its business scope by focusing on distribution instead of production, and founded Dinamic Multimedia, a stable company that distributed national and international products until its closure in 2001. Today, the name "Dinamic Multimedia" still prevails as a distinctive cultural phenomenon in Spain, since millions of gamers link it with the *PC Fútbol* franchise, one of the most renowned sport simulators in Europe and a true mass success in the Spanish context. In 1992, during the studio's most difficult financial stage, a first version of what later became *PC Fútbol* was developed: the economic reliability of Dinamic was so precarious that it had to put the game's distribution in alien hands, those of Editorial Jackson, a company devoted to serialized publications and installments. The strategy was a desperate one, and the product was far from being "top of the line"; while games including *Alone in the Dark* (1992) and *Wolfenstein 3D* (1992) were launched internationally, *PC Fútbol 1.0* looked like an out-of-fashion, late 1980s game. But, as history later proved, the monumental appeal of soccer in Spain and the product's low price turned into huge—and unexpected—sales all around the country.

The game's success became a strategic standpoint for Dinamic, which immediately changed its name from "Software" to "Multimedia," thus broadening its commercial activity to include distribution. HobbyPress, the publishing corporation that owned the two most important national gamer magazines, *Micromania* and *Microhobby*, played an important role in that process—a distinctively Spanish trait of video game culture, we might add. *PC Fútbol* games started to be sold in newsstands, taking advantage of the cohabitation between entertainment software and leisure magazines, openly advertised in the HobbyPress's periodicals. So the strategy was a success, empowered by the critical mass of soccer fans in Spain who were eager to manage virtual versions of their favorite teams, therefore transforming *PC Fútbol* into both a sales hit and a social phenomenon. In an industry that has remained reluctant to explore "national themes" (unlike Spanish cinema, indigenous games don't portray cultural symbols and clichés such as bullfighting, macho-melodrama, or temperamental *pathos*), the *PC Fútbol* case remains the main example of a truly picturesque product, reenacting the myths and trades of "el deporte rey" (the popular term referring to soccer in Spain). The formula was simple: a soccer management simulator for PCs, straightforward and easy to control, which was gradually updated year by year, with player transfers, new stadiums, and emerging clubs (a sort of renewed approach to Kevin Tom's *Football Manager* [1982]). Sales were extremely positive during its first seven versions: *PC Fútbol 7* sold 400,000 copies in 1998 and became the best-selling title ever produced by a national studio within Spanish borders.

While other franchises, such as Electronic Arts' *FIFA*, focused on in-match gameplay, *PC Fútbol* exploited all the elements associated with soccer: the social, economic, and cultural *aura* of European football, the ancestral rivalries between historical clubs such as Real Madrid and F. C. Barcelona—identities that remain outside the sport itself but reach deep into every fan's soul. That's why the game achieved considerable success in other countries when it was launched as *PC Calcio* (1994–2006) in Italy, *Premier Manager* (1996–1998) in the UK, and *PC Fútbol Apertura/Clausura* (1994–1999) in Argentina. Dinamic also expanded the formula to other sports such as basketball and cycling (*PC Basket* [1993] and *PC Ciclismo* [2001]). But ironically enough, just as great soccer teams rise and fall in cycles, Dinamic Multimedia went into crisis after its main partners and founders, the brothers Ruíz, abandoned the studio in 1999 and established their own company, FX Interactive. The head of Dinamic at the time, José Ignacio Gómez-Centurión, then tried to expand the company by taking advantage of the dot-com bubble, but the meager sales of the studio's games in the following three years and the cooling down of the Internet boom took Dinamic to bankruptcy in 2001. In spite of a number of isolated efforts to relaunch the *PC Fútbol* franchise (Gaelco developed three new games from 2005 to 2007), the series has never regained the social significance it used to enjoy in the late 1990s and early 2000s.

Among the several games distributed but not produced by Dinamic Multimedia, we would like to highlight three graphical adventure games designed by Pendulo Studios: *Igor: Objetivo Uikokahonia* (1994), *Hollywood Monsters* (1997), and *Runaway: A Road Adventure* (2001). Drawing upon the aesthetics and the success of LucasArts games such as the canonical *Monkey Island* series (Tim Schafer's work is admired and studied by Pendulo's designers), *Igor* became the first great achievement of the genre fully produced in Spain. Instead of trying its luck in different genres like other indigenous studios, Pendulo insisted on the creation of graphical adventure games with their second game, *Hollywood Monsters*, and eventually acquired a style of their own, fully crystallized in the success of the *Runaway* franchise (first distributed by Dinamic, later by FX Interactive). The studio's continued use of the point-and-click aesthetic remains one of the few distinctive game design styles in the Spanish context; and although the popularity of graphic adventures has gradually diminished, Pendulo has gained a modest but strong reputation in the European market. Since its games tend to have more success in countries such as Germany or Italy than inside the Spanish borders, it has become customary for Pendulo to launch its products in foreign countries before hitting the indigenous market. Without making too much noise or boasting about it, this humble studio is one of the few national companies that has remained working in a genre and a gaming idea for almost two decades, faithful to the ways of graphical adventure games, but also trying to renew its aesthetics, as proven in *New York Crimes* (2012), a darker and more mature approach to its own particular way of understanding the genre (Martínez 2012).

The most distinguished Spanish game of the 1990s, however, was Pyro Studios' *Commandos: Behind Enemy Lines*, a single-player real-time tactical computer game published by Eidos Interactive. The greatest achievement of studio head Gonzo Suárez and his team was to expand the possibilities of a genre that hadn't been fully exploited at the time, focusing strategy on small but detailed maps and specialized tactical squads like a sort of immanent puzzle, where the members of each commando were deployed as pieces in an intricate operational campaign. Audiences were fully delighted by the game, and international gaming magazines praised its wonders: "If you're into strategy games, own an attention span, and are looking for something that's a lot different than anything we've seen since *Cannon Fodder*, you owe it to yourself to give this game a try" (Ward 1998). The consistent sales of that first game allowed Pyro to publish an expansion disc, *Commandos: Beyond the Call of Duty* (1999), which added to the good fortune of its predecessor and granted the development of further franchise installments. In 2001, after canceling a hybrid project called *Heart of Stone* (half graphical adventure game, half arcade game), Pyro launched *Commandos 2: Men of Courage*, achieving great commercial and critical success, as the high praises of *The Electric Playground* prove: "It was a long wait for a sequel, but Pyro spent the time innovating a game that was already innovative to begin with" ("Commandos 2: Men of Courage—Review" 2001). Originally designed for PC, the Xbox and PlayStation 2 conversions of the game weren't as successful: gameplay felt a bit unnatural on the consoles.

The series' next installment, *Commandos 3: Destination Berlin* (2003), and the traditional 3-D strategy game *Praetorians* (2003) were Pyro's following projects, though they were not as well received as their predecessors. In the following years, the company reorganized its activity (just like one of *Commandos*' tactical teams) into three key branches: Pyro, in charge of video game design; Ilion Animation Studios, focused on 3-D computer-generated cinema; and Zed, devoted to mobile/Internet products and services. It should be noted that the critical weight of Pyro Studios inside that corporate triangle seems to have declined: after developing another two games that didn't sell as well as expected (the real-time tactical game *Imperial Glory* [2005] and *Commandos: Strike Force* [2006], a translation of the *Commandos* series into the first-person shooter genre), the studio focused on creating a subsidiary game for the movie *Planet 51* (2009), produced by the company's animation branch and released as a cross-media franchise in 2009. Since then, the company has concentrated on social networking-based games.

While these five studios aren't the only ones, they do give a fair and overall idea of the state of affairs in the Spanish game design industry between the 1990s and 2000s. Gaelco, Bit Managers, Dinamic, Pendulo, and Pyro produced games that achieved greater repercussions in the international market, but they also represent the different tendencies found within indigenous game creation during those years. They function as fundamental attempts to bridge the talent/business gap that condemned other national studios during the "golden age" crash, a series of contrasting yet effective ways of surviving in the global market of video games, which emerged at the turn of the millennium. Gaelco represents the past glory of arcades and the constant struggle toward new ways of understanding gameplay, which are not always successful. Bit Managers embodies the paradigm of a great "assignment studio" that fulfilled the requirements of foreign companies in faithful and consistent conversions, mainly for the Game Boy. Dinamic Multimedia exemplifies the synergies between the production of solid franchises, epitomized in *PC Fútbol*, and the distribution of other studios' games. Pendulo and Pyro both started as creative-focused teams, concerned with achieving quality through a distinctive style, and they evolved in different ways: the former continues to believe in the possibilities of graphical adventure games, in a modest yet firm way; the latter has evolved into a media conglomerate that expanded from game design to film production and mobile development, making sure business takes proper care of talent. So these five contrasting ways of learning the game, which shaped the present and future of Spanish video game design, were both the echoes of a mystified past and the solid bases of a promising future.

Global Hopes: The Foundations of an Emerging Game Industry

Throughout the 2000s, Spanish game design has been strengthened due to a number of solid foundations that foretell a promising horizon: a process that encompasses social, cultural, economic, and even educational factors. First of all, the consumption of video games in the country surpassed that of any other entertainment medium, including music and cinema, and Spain remains the fifth biggest European sales market, very close to Italy. As of 2012, a home console system can be found in three out of four Spanish homes (Checa-Godoy 2009, 180), adding to an estimated gamer population of fifteen million players, and in 2007, the sales volume reached a peak of €1.454 billion, which stabilized around €1 billion in the following years (ADESE 2012, 30). Good sales numbers tend to confirm the existence of a solid consumer base, but, more importantly, the social acceptance of game culture has recently found the support of institutional authorities; in 2009, video games were declared "Bien de Interés Cultural" (Constenla 2009), a state recognition long vindicated by indigenous developers.

Along with that distinction, public funding policies have empowered national production, and although the percentage of state support doesn't reach the level it does in France (up to 20%; see Checa-Godoy 2009), it is now perceived that Spanish studios—at least the big ones—count, with institutional and social support never achieved in the past. Game conferences and trade fairs, such as the GameLab (which originated in 2004 in Gijón, now celebrated every year in Barcelona), have contributed to the expansion of indigenous works; little by little, GameLab has gained more and more public support inside the country while sharpening the relations of Spanish designers with key foreign creators such as Peter Molyneux, Hideo Kojima, Cliff Bleszinski, Trip Hawkins, and Ian Livingstone, who have participated in the event's global strategies. We would like to trace the *renaissance* of Spanish game design by focusing on two specific studios that represent the national industry's present and future: Novarama, solid proof of the links between game creation and higher university education, and MercurySteam, a successful case of local talent evolving into international business.

The links between game development and higher education have become an essential standpoint, as proven by world-acclaimed initiatives such as the EA Game Innovation Lab at the University of Southern California. In Spain, Daniel Sánchez Crespo, the driving force behind Universitat Pompeu Fabra's master's in video game design, has pioneered that fundamental liaison between industry and university since the program's creation in 2002 (the first one in the country). Year after year, other national universities have tightened links with game developers: for instance, Universidad de Alcalá has been establishing a research project linked with the Spanish branch of Electronic Arts in order to bring video games to school classrooms as an educational tool for children. In addition, the Universidad Complutense de Madrid has developed a collaboration program with Nintendo España, fostering meetings between designers, academics, and students. But what was interesting about Crespo's pioneering initiative at the Universitat Pompeu Fabra was how it combined the talent of the master's students with his successful design studio, Novarama, bridging education and industry. After achieving international recognition with *Invizimals* (2009), an augmented reality (AR) game for PlayStation Portable (PSP), Novarama established a solid relationship with Sony and became one of the Japanese company's official third-party developers in 2011, expanding the game into a sound, ongoing franchise. As far as game aesthetics are concerned, the bestiary-driven logic of *Invizimals* allowed players to summon legendary monsters into quotidian spaces, bringing the fighting genre closer to everyday realism.

In 2001, a studio called Rebel Act launched *Blade: The Edge of Darkness.* Developed in San Sebastián de los Reyes and distributed worldwide by Codemasters, it achieved an unexpected cult following. It had state-of-the-art graphics and proved what a young design team with deep gameplay knowledge could do. In spite of the solid reputation that the game has won over the years, a failed marketing campaign and a number of distribution issues (for legal reasons it had to be renamed *Severance* in the United States) turned promising hopes into commercial failure, and the studio closed. Some of the company's partners then established MercurySteam, probably the most exciting design group in the current indigenous context. Its first project was American McGee's *Scrapland* (2005), a third-person action-adventure game set in a futuristic world, which achieved considerable success in its PC and Xbox versions and, more importantly, accomplished a high quality standard through world-building visuals and challenging gameplay, described by IGN as "a refreshing sense of personality" (McNamara and Perry 2005). Its next main project, *Jericho* (2007), followed the same strategy by taking advantage of Clive Barker's premise for the game's storytelling arc; but in spite of its popularity among traditional gamers, it didn't achieve critical success and was considered a hard-core horror product. Even so, by that time, MercurySteam had already been established internationally as a solid studio, capable of creating games with punch and style, ready for the global market.

While MercurySteam was working on its next game, *Lords of Shadow,* a happy turn of events led to Konami's interest in the project: it asked the Spanish studio to turn it into a game in the legendary *Castlevania* franchise. Hideo Kojima, on his way to being Konami's vice president, aided the team with powerful ideas during the game's development. The game's critical and popular success in late 2010 has widely confirmed a

Figure 3
Screenshot from *Castlevania: Lords of Shadow* (2010).

traditional assumption in the indigenous game design trade: once a project gains enough financial backing and the designers are able to work with the same tools owned by other major studios overseas (both granted through Konami's collaboration), a Spanish design team can truly concentrate its efforts on creating the best possible game. It may sound obvious, but for an industry accustomed to struggling with bad finances and low production standards, finding "the proper way to work" is a big deal, since it grants the required business channels for national talent, just like the compelling gameplay of *Castlevania: Lords of Shadow* (2010) demonstrates. As the international game scene is becoming more global, MercurySteam's game exemplifies an ideal future path for indigenous productions. It is not only a big blockbuster that can compete with American and Japanese titles, but also a demonstration of aesthetic talent and proof of strong business relations in the international market, capable of resurrecting a franchise such as the *Castlevania* series with the sequel *Lords of Shadow 2*, released in 2014.

The relatively good health of the Spanish game design industry shines bright in other markets and channels as well, from online downloads to mobile apps. For instance, the Catalonian studio Digital Legends has completed a series of solid games for mobile phones over the last decade, while Virtual Toys—the company that bought Bit Managers in 2005—established a relationship with UbiSoft by developing a number of games for the Nintendo DS, specifically the vocational simulator *Imagina Ser* (2007) and other childhood-oriented projects. More *indie*, innovative-oriented studios, such as Over the Top Games, have also established interesting collaborations with foreign companies, for example, the console versions of Brad Borne's *Fancy Pants Adventures* (2011). At the same time, other indigenous studios such as EnjoyUp and Akaoni have focused on developing small yet solid downloadable games for Nintendo's platforms, very carefully aimed at the Japanese market; a retro arcade game such as *99 Bullets* (2011) and a cross-genre extravaganza such as *Zombie Panic in Wonderland* (2010) are good examples. More recent cases are Deconstructeam, Teku Studios, and Asthree Works, small studios that are building a considerable reputation within the global indie scene, bridging local talent with Kickstarter initiatives. And, as far as the new generation of consoles is concerned, indigenous companies such as Tequila Works have already started developing interesting and international projects (such as *RiME*, a PS4 exclusive).

All things considered, with economic crises and national uncertainty spreading all over the European Union, the Spanish game industry remains one of the very few national trades (besides tourism) that appear to be considerably optimistic about the future. Of course, there has been a sales decline in recent years (around 15%), just as in other European countries, but hopes remain high because the loss is much smaller than the one suffered by other entertainment industries such as film and music (ADESE 2012, 33). As for the two conceptual poles considered in this chapter, talent and business, it seems now that the mistakes of the 1990s have been learned from; indigenous studios are finding ways to connect their gameplay ideas with global marketing strategies, sustaining talent through the uncertainties of business. Good signs are everywhere: the institutional and social prestige of the video game industry was officially confirmed in 2012, with Shigeru Miyamoto winning a Príncipe De Asturias prize, the highest cultural and symbolic award in the country. Meanwhile, the GameLab conference seems to have found institutional support and global projection in its new site in Barcelona, bringing together the best indigenous design with the promised land of international markets. It seems that, between golden ages and industry crashes, from pioneering code to certified third parties, we are finally enjoying a period of modest pride and international licenses. After all, we may very well be facing the best years of Spanish game design, whether we consider the glass to be half full or half empty—so, once again, (too) great expectations?

References

ADESE (Anuario de la Industria del Videojuego). 2012. Spanish Association of Entertainment Software's Distributors and Editors, May, Madrid, Spain. http://www.adese.es/docs/documentacion/el-anuario-del-videojuego.

Atari Presenta Los 4 Magníficos. 1983. *Diario ABC*, June 26. http://hemeroteca.abc.es/.

Auerbach, Erich. 2003. Mimesis. Princeton, NJ: Princeton University Press.

Azpiri, Alfonso. 2009. Spectrum: El Arte para Videojuegos de Azpiri. Barcelona: Norma Editorial.

Bloom, Harold. 1995. The Western Canon. New York: Riverhead Books.

Checa-Godoy, Antonio. 2009. Hacia una Industria Española del Videojuego. *Revista Comunicación* 7, Sevilla. http://www.revistacomunicacion.org/comunicacion_numero_7.htm.

Commandos 2: Men of Courage—Review. 2001. *Electric Playground.* http://www.metacritic.com/game/pc/commandos-2-men-of-courage/critic-reviews.

Comolli, Jean-Louis. 1971. Technique et Idéologie, la profondeur de champ primitive. *Cahiers du Cinéma* 233 (Nov.): 40–45.

Constenla, Tereixa. 2009. Un Respeto para el Videojuego. *El País*, March 26. http://elpais.com/diario/2009/03/26/cultura/1238022002_850215.html.

De La Fuente, Manuel, and Guillermo López. 2008. Historia, Mercados y Culturas del Videojuego. In *Industrias de la Comunicación Audiovisual*, ed. J. Duran and L. Sánchez, 221–256. Barcelona: Universitat de Barcelona.

Flores, Alberto Del Rio. 2012. Spain. In *Encyclopedia of Video Games: The Culture, Technology, and Art of Gaming*, ed. Mark J. P. Wolf, 613–614. Westport, CT: Greenwood.

Gran Campeonato Comecocos de Atari en El Corte Inglés. 1982. *Diario ABC*, June 20. http://hemeroteca.abc.es/.

Martínez, Víctor Manuel. 2012. Pendulo Studios: La Aventura Gráfica como Dogma. *Anait Games*, March 28. http://www.anaitgames.com/articulos/entrevista-pendulo-studios-new-york-crimes/.

McNamara, Tom, and Douglass C. Perry. 2005. American McGee Presents Scrapland. *IGN*, February 23. http://www.ign.com/articles/2005/02/23/american-mcgee-presents-scrapland.

Ortega, Juan Manuel. 1981. Tres Juegos para tres Décadas: Futbolín, Pinball, Video. *Diario ABC*, November 8. http://hemeroteca.abc.es/.

Pérez, Óliver. 2007. La producción de videojuegos en Cataluña. *OPA Observatori de la Producció Audiovisual*, Barcelona. http://www.upf.edu/depeca/opa/dossier0/info4_esp.htm.

Pineda, A. Vicente. 1983. Otros Juegos: Videojuegos. *Diario ABC*, November 5. http://hemeroteca.abc.es/.

Ward, C. Trent. 1998. Commandos: Behind Enemy Lines. *IGN*, September 1. http://www.ign.com/articles/1998/09/01/commandos-behind-enemy-lines.

Xavier A. G. (Gaelco). 2009. *Retrovicio* 4077. http://www.retrovicio.org/miscelanea/entrevista-xavier-ag-gaelco.

SWITZERLAND

Matthieu Pellet and David Javet

In order to consider the status of video games in Switzerland, one must adopt a cultural, historic, and geopolitical approach regarding the place of this country on the international stage: four national languages and five bordering countries are unique elements that define Swiss culture, with the idea of diversity at its core. First, its geographical position—at the center of Western Europe—places Switzerland in the position of a complex cluster for market strategies and legal practices in and out of its bordering countries. In addition, while being at the heart of the European continent, Switzerland is not part of the European Union. Thus its legal system as well as its distribution policies can be influenced or inspired by French or German domestic decisions, for example.

Diversity: A Double-Edged Sword

Switzerland's diversity also has negative aspects: one good example can be seen in the process of localization between the French- and German-speaking regions, where the distribution is subject to the French or German market. This problem, enhancing the differences of practice and experience between the Swiss linguistic regions, hinders the potential for the production of video games with national characteristics that represent Swiss culture. In other words, this multilingual and multicultural specificity builds a barrier to the creation of a video game corresponding to the whole of the Swiss public and that would thus be the mirror of a national identity. Furthermore, in spite of a stable and powerful economy that has made Switzerland a country of important cultural offerings, this multilingual barrier, as well as a persistent retrograde vision of video games, has pushed Swiss participants in the gaming scene outside of its borders. Only recently is creativity coming back and with it, a new appreciation of this medium.

Video Games in Switzerland: A Brief Market History

Up until recently, in Switzerland, as in most European countries, gaming platforms (such as computers or home consoles) were introduced around six months to one year after the production and distribution of new systems in the United States or Japan. In the 1970s, the arrival of video games would be heralded by the Magnavox Odyssey (1974) home console that was followed by the so-called first generation of video game consoles (unless specified, we will be referring to the Swiss launch dates, not the original ones). In the beginning of the 1980s, while the Swiss people could also acquire second-generation home consoles (including the Atari 2600, the Mattel Intellivision, and the GCE/Milton Bradley Vectrex), it was only with the boom following the launch in 1981 of Nintendo's Game & Watch portable devices that video games would reach a wide audience, bringing down the generational barrier between kids and adults for a few years.[1] Still, this barrier would be maintained at the level of home consoles and most prominently in video arcade centers, hubs for young people attracted by such technologically advanced products. Finally, it was in 1986 that the general public of Switzerland accepted video games as a new medium, as they achieved popularity with the arrival of the Nintendo Entertainment System (NES).[2]

The 1990s, with the arrival of the major 16-bit systems (for example, the SEGA Genesis in 1990 or the Super Nintendo Entertainment System [SNES] in 1992), saw the establishment of a new phenomenon among video game aficionados: the semi-legal importation of foreign market hardware and software. Since the Swiss market depends on the delayed European distribution schedule and localization, most fans of video games turned toward small shops specializing in the importation of Japanese or US products. This practice continued throughout the 32-bit generation and up until the present day, even though it has gradually been confronted with new promotional campaigns targeting adults, including new networks of supermarket distribution.[3] Furthermore, the global economic success of video games that enables distributors to schedule international releases will be a second blow to this practice of importation in Switzerland. Now, without a doubt, video games are an economic force that the Swiss people can no longer ignore.[4]

Consequently, it was no surprise that in 2003 the Swiss Interactive Entertainment Association (SIEA) was founded. The SIEA, composed of a board of veteran distributors, has as its main concerns the respect and understanding of age ratings and the fight against piracy. The association also began observing and studying the Swiss market. Looking at their data, we can clearly see how video games sales have resulted in a very lucrative industry. In 2005, sales of hardware and software generated 271 million CHF (USD $216,800,000).[5] This figure rose significantly in the following years with annual sales of 300 million CHF (USD $239,479,500) in 2006, and 420 million CHF (USD $350,387,940) in 2007. The highest point was reached in 2008 with revenue of 438 million CHF (USD $405, 499,962). But, starting in the year 2009, these impressive figures for Switzerland would gradually start to regress toward the 2005 figure. In 2009, video games generated 382 million CHF (USD $352,550,856), while 2010 and 2011 brought in 342 million CHF (USD $328,741,686) and 297 million CHF (USD $335,929,869) respectively. There are many explanations for this significant drop; without a doubt, the subprime mortgage crisis in 2007, the following global financial crisis of 2008, and the resultant feelings of fear and caution on the part of consumers all have a large part in it.[6] Switzerland, at the center of Europe but using its own strong and stable currency, watched as the depreciation of the euro forced a decrease in the prices of hardware and software. So, in that sense, these annual figures do not simply testify to consumer restraints in spending, but also to cheaper products that bring lower revenues.

Swiss Game Design: A Complex Start

The geographical position of Switzerland, along with its cultural and linguistic diversity combined with a stable economy and strong academic infrastructures, makes it a land of promise and potential in terms of game design. If, on the one hand, linguistic and cultural barriers hinder the production of narrative-based games, on the other hand, they invite game designers to break them down, emphasizing interactivity and visuals. Still, a lack of communication and networking persists not only between the different linguistic regions but also, more basically, between the different participants of the Swiss video game scene. But policy-makers are reacting: from September 2010 to November 2012, the Swiss Arts Council Pro Helvetia, a foundation funded entirely by the Swiss government, faced this problem with the help of a nationwide promotional program called *GameCulture*. The main goal of the foundation was first and foremost to establish connections between every facet of the Swiss gaming scene. Before they could launch the program, they had to get a general idea of the current state of the Swiss video game industry. Therefore, they asked Beat Suter from the Zurich University of Arts to conduct a survey in 2009 and 2010. The survey was sent to around 150 game designers and developers living in Switzerland or abroad. The goal was to assess their needs and their preferences.[7] For the first time, independent Swiss game designers and their companies and products would be at the center of a cultural promotional program that supported their work, conceptually and substantially, rather than one that put the focus on the market-driven aspects of video games for international trade. Because of this, the program undoubtedly represents a key moment in the history of Swiss video games.[8]

Technological advances and the creation and development of new platforms, conceived not only for games but allowing, nonetheless, a wide variety of people to play them, helped Swiss game creators access a greater range of possibilities. The Android, the iPhone, and personal computers are therefore chosen as the favorite platforms to work on. The reasoning behind this has to do with the current state of most Swiss video game studios: the wide majority of them do not have more than fifteen employees.[9] Indeed, in order to develop a small but professional game for Steam or an app for iOS, for example, one does not need important resources, in terms of manpower and money. As Swiss developers said in Suter's surveys, these platforms allow them to display their originality and innovative side without taking heavy risks. Thus, there is an almost complete lack of Swiss games for the most recent home consoles. An exception would be Codebox's *The Path of Go* (2010), the first Swiss game available on the Xbox Live Arcade (see http://www.codebox.ch/de/web/index.php?p=1&a=7).

This situation has led to the production of original work that was noticed outside of the Swiss scene. Inside the small family of Swiss games that have achieved international recognition, GIANTS Software's *Farming Simulator* series proved that a company based in Switzerland (Zurich, to be precise) can achieve high enough sales in order to see a release on a contemporary portable system such as the Nintendo 3DS. Every game made by GIANTS Software would have its own full-featured game engine, the *GIANTS Engine*—the creation of game engines being, as we will see, one of Switzerland's most promising fields (see http://www.giants-software.com/index.php). Geneva-based EverdreamSoft's online trading card game *Moonga* (2010) was another work that achieved international renown. Considering the success of trading card games in Asia and especially in Japan, it is remarkable that a Swiss production could reach the top spot on Apple's Japanese App Store in the role-playing game category against such fierce competition. The hundreds of thousands of downloads since 2010 allowed EverdreamSoft to expand into other Asian markets, and a distribution contract with Chinese distributor The9 was signed. As the game was first developed on iOS and then became accessible on Facebook, the creators themselves did not expect such success in this geographical area. This story proves how Swiss games generate unexpected perspectives and validated the company's distribution choices (see http://www.everdreamsoft.com).

If financial success is possible for Swiss productions, artistic acknowledgment is also possible. For example, independent game creator Mario von Rickenbach is recognized internationally for his artistic, innovative, and surprising games. Initiated as his bachelor's thesis project at the Zurich University of Arts, *Mirage* (in development as of February 2014) was nominated at the 2012 Independent Games Festival (IGF) in San Francisco for "Excellence in Visual Art" and received the Marseille European Indie Game Days "Grand Prize" and "Originality Award" (see von Rickenbach's website at http://www.mariov.ch). In addition, with a team composed of Michael Burgdorfer and Phil McCammon, *Krautscape* (in development as of February 2014) was exhibited in Paris at the digital arts center La Gaîté Lyrique, showing that his games are appreciated for their artistic qualities (see http://www.krautscape.net). Yet another example of a critical success is Florian Faller and Adrian Stutz's independently developed game *Feist* (in development as of February 2014). The game was a finalist at the 2009 IGF for the "Excellence in Visual Art" prize and among the winners of the "Student Showcase" prize (see http://www.playfeist.net). In both cases, we can see that two of their important assets are the quality and originality they present on a visual level.

Despite such examples, however, more efforts are needed in order for the Swiss video game scene to mature. For example, the government could invite major game publishers to establish studios or offices in the Swiss territory. This would allow young Swiss game designers to have access more directly to training courses and jobs and would prevent them from leaving the country. Switzerland has the advantage of a stable economy, a central position geographically, and a high standard of living, which could potentially interest any studios looking for an ideal productive environment. But, only the canton of Geneva actually has such a promotional policy, with Electronic Arts and Take-Two Interactive Software installing branches there in 2006, though these were not development studios but administrative offices. Electronic Arts transferred its European publishing headquarters to Geneva, while Take-Two Interactive established its international publishing headquarters there for only two years. In addition, in 2011, the French publisher Ubisoft, interested in the research and academic environment of the Zurich area, established a development studio specializing in free-to-play titles for PC and tablet devices. Finally, Miniclip, the international digital entertainment and games distribution platform, chose as a place for its headquarters the Swiss city of Neuchâtel (see http://corporate.miniclip.com). As for other publishers in the Swiss territory, they only set up distribution and marketing offices. This is, without a doubt, still one of the weaknesses of the Swiss video game scene.

At the same time, the formative and researching structures provided by various Swiss universities are becoming a solid basis for the growth of a mature creative environment.[10] Starting with a bachelor's degree in game design in 2004 at the Zurich University of the Arts (ZHdK)—a program that the aforementioned Mario von Rickenbach, Florian Faller, and Adrian Stutz followed—the establishment of academic courses all around Switzerland related to video games will help to train the new generation of Swiss game designers.[11] This forward-thinking act by the ZHdK, combined with special programs from the Swiss Federal Institute of Technology Zurich (ETHZ), gave a significant advantage to the Swiss German area and explains why most of the activities in the Swiss video game scene are located around Zurich. While not benefiting from as many years of experience, the Swiss French region shows promise, notably with the creation, in 2009, of a master's degree program in media design at the Geneva University of Art and Design (HEAD) (see http://head.hesge.ch/made/media-design), within which game design occupies an important place.[12] Located in Saxon, the Academy of Contemporary Arts (EPAC) started in 2011 a two-year program in game art (see http://www.epac.ch/index.php/en/). Furthermore, in 2013, Ceruleum, the Lausanne-based School of Visual Arts, began offering a new master's degree in concept art (see http://www.ceruleum.ch/). This led to the creation, inside their walls, of a new Swiss video game studio founded by former students, Sunnyside Games (http://www.sunnysidegames.ch/).

This polarity between the Lake Geneva region and Zurich can also be found in the field of digital technological research. But the advantage that Zurich possesses over the other Swiss urban centers lies in how the Swiss Federal Institute of Technology Zurich has taken the initiative of creating research projects in the field of video game development and conception in collaboration with renowned companies. For example, the Disney Research Lab, a collaboration between the ETHZ Visual Graphics Lab and the North American Walt Disney Company, opened in 2010 (see http://www.disneyresearch.com/research-labs/disney-research-zurich/). Without a doubt, their work will influence the digital technology used in the future. In Geneva, where the relative lack of collaboration between public and private enterprises is unfortunate, the company Pixelux Entertainment, founded in 2003, developed the "Digital Molecular Matter" technology, a physics engine that LucasArts used in its *Star Wars: The Force Unleashed* series (see http://www.pixelux.com/DMMengine.html). In the future, there will hopefully be more new enterprises specializing in the development of new technologies serving the worldwide video game design industry. The field of technological research could be the backbone of the slowly emerging Swiss video game industry.

Politics and Video Games in Switzerland: An Enduring Vision

On the cultural side, one can see that video games, due to their image, still have a long way to go before they are accepted by the general public as an important medium in Switzerland. As mentioned earlier, this image results from a persistent retrograde vision judging these products as mind-numbing, childish, and breeding violence. From arcade centers to recent political debates, this discourse has hindered the blossoming of this medium. If arcade centers were gathering points for the young generation, they are also considered by nongamers as meeting spots for juvenile delinquents and thugs;[13] following a vigorous social debate in France during the 1990s around violence in imported Japanese cultural products such as anime or video games, the problem was also discussed in Switzerland owing to the neighboring nature of both countries. This debate surrounding the intrinsically violent aspects of video games continues in the twenty-first century.

A climax to this debate was reached in March 2010 when the Swiss Federal Council, the highest authority in the Swiss government, was mandated by the Council of States, the Assembly's upper house, to write a new law leading to a possible banning of the importation and production of "violent" video games.[14] More precisely, the two motions proposed by the Swiss left wing's members of the cantons of Bern, St. Gallen, and Fribourg were as follows: the first was an enforcement of the interdiction to sell over-sixteen age-rated games to underage people,[15] and the second a complete prohibition of the distribution, importation, and advertisement of all "violent" games by strengthening Article 135 of the Swiss Criminal Code. Finally, what motivated the Swiss Federal Council to reject both motions was the difficulty of defining and assessing the concept of violence, a concept that is at the core of the aforementioned Article 135. The article deals with the representation of acts of violence and says in its first paragraph that a game could be banned if it has no cultural or scientific value. It tackles the difficult question of the definition of violence, describing it as an act of cruelty against human beings or animals that "offends basic human dignity." As it applies to the production of video games, the problematic question here for Swiss policy-makers is whether the said "violent" games can be considered cultural works or not; and to answer this question, one should be able to define clearly what a cultural work is. Faced with such difficulties, the current state of Article 135 of the Swiss Criminal Code was judged unsatisfactory by those proposing these motions.[16] Still, the Swiss Federal Council considered the current law to be sufficient and, referring to scientific surveys, added that there was no proven relationship between violence in games and violent behaviors.[17] Even though the German law already allows the interdiction of selected titles, Switzerland would have been the first European country to reach such a strict decision. This discussion was reviewed by international news agencies and also triggered a worldwide debate on the Internet, concerning not only video games but also the Swiss international image among gamers.[18] It also shows how Switzerland, with its strong legal and political independence, functions as a laboratory for new measures, notably in the field of video games.

Strengthening the Idea of Video Games as Cultural Expressions in the Eyes of the Swiss Public

The two motions proposed in 2010 showed one side of the Swiss vision of video games, a persistent backwardness considering them mind-numbing and not cultural. Therefore, the concerns regarding violence would be balanced with a need to align the Swiss art and academic scene to the European scene already in motion. There is no question whatsoever that there exists in Switzerland a strong tradition of artistic and cultural support: this can be seen in the founding of festivals, academic conferences, and museums. But, up until the end of the first decade of the twenty-first century, efforts in the field of video games were mainly directed toward the support of exhibition platforms rather than exchange and creation. This has, of course, a lot to do with the necessity of first changing the perception of video games in the eyes of the general public. This changed, along with the *GameCulture* program, as more emphasis was put on showing the public what Swiss creators were capable of. A more pragmatic way of considering this sudden emphasis on video games as art or culture would be to notice how, as they became an economic force that could not be ignored, they were finally considered cultural expressions. In this sense, Switzerland did not differ from the rest of the world: economic success (in sales, not in development here) had to be reached before the general public would be forced to accept the new media as worthy of interest—the state-funded *GameCulture* program was put in motion in 2009, one year after Switzerland's best year in terms of annual sales. Consequently, the Swiss public now faces two seemingly divergent and conflicting images of video games: one presenting mainly the ultrarealistic action-oriented first-person or third-person shooters, and one underscoring the independent artistically engaged works of a few creators.

Thankfully, to put these confusing images in order, actions were undertaken through the establishment of exhibitions and museums devoted to digital culture/art and the history of video games. Located in Basel, one of Europe's prominent artistic capitals, the House of Electronic Arts, a space dedicated to new media and technologies, officially opened in May 2011. One of their activities includes the organization of the Shift—Electronic Arts Festival, an event offering concerts, exhibitions, and scientific lectures for an ever-growing audience.[19] From March to December 2012 in Yverdon-les-Bains, the Maison d'Ailleurs (Museum of Science Fiction, Utopia and Extraordinary Journeys) had an exhibition called "Playtime—Videogame Mythologies," which pondered the link between "play, the various manners of gaming, and technology" (for more information on the exhibition, see http://www.ailleurs.ch/en/expositions/archives/playtime-videogame-mythologies/).

As mentioned earlier, while courses in game design and development were gradually being established, the creation of classes and conferences to discuss this medium as a newcomer in the field of human sciences demonstrates how Swiss researchers are now expanding it and constructing discourses around it. Switzerland's first international academic conference, "Powers of Video Games: From Practices to Discourses" (our translation) was held in June 2012. It was the fruit of the collaboration between the University of Lausanne (UNIL), the Swiss Federal Institute of Technology Lausanne (EPFL), the Geneva University of Art and Design (HEAD), and the Maison d'Ailleurs museum.[20] The logical follow-up in the future will be the establishment of video game departments in the human science faculties of some universities. Furthermore, gatherings such as the annual Imaging the Future (ITF) symposiums organized during the Neuchatel International Fantastic Film Festival (NIFFF) have allowed, since 2006, video game discourse to evolve and reach new territories (see the official website of the ITF symposium at http://www.imagingthefuture.ch).

As this brief chapter on Switzerland shows, the Swiss video game scene, while having followed in the footsteps of the European video game market, is still in an infant state in terms of game design and development. But, during the first decade of the twenty-first century, new promotion policies and cultural changes established the auspicious roots of a future evolution in the relationship between the Swiss people and video games.

Notes

1. This portable device boom is easily comparable to the introduction in 2005 of the Nintendo DS portable system in Switzerland, where people of all ages could be seen acquiring it. Still, compared with Japanese practices, only young people would play in public, since video games were still considered childish.

2. Even though it only began to be distributed in 1990, the importance of Nintendo's Game Boy portable system should also be taken into account in order to assess this popularization.

3. Here we refer mainly to the Sony PlayStation (1994) advertising strategy that was soon followed by the more adult-oriented PlayStation 2 (2000) and Microsoft Xbox (2001) advertising campaigns.

4. In terms of numbers, Swiss annual sales of DVD/Blu-ray for 2011 were around 29 million Swiss francs less than the 297 million generated by video games sales (SIEA 2012).

5. To simplify, we decided to round off the figures in million CHF. For more information about the SIEA and the quoted numbers, see the official website of the association, http://www.siea.ch. Also, USD figures are based on the exchange rates during the years in question and are calculated from the rounded CHF figures given, making them only rough approximations.

6. The SIEA made a similar analysis of the reasons behind this fluctuation: "After several years under the sign of growth, the video game and PC market suffered a decrease last year which corresponds to the unfavorable economic conditions" (SIEA 2010, paragraph 1, our translation).

7. Beat Suter's enlightening surveys were the main basis for the writing of this part of the chapter. His work can be found on the *GameCulture* program official website, at http://www.gameculture.ch/developers-technology/.

8. Since Beat Suter's 2009 survey demonstrated the lack of communication networks between the participants in the Swiss video game industry, actions have been undertaken to correct this: the International Game Developers Association's (IGDA) Switzerland chapter was reactivated in 2009, and a LinkedIn group called "GameCulture Switzerland" would quickly come to function as a hub for Swiss game developers.

9. Beat Suter's 2010 survey implies that the average number is around five employees, which shows that the Swiss industry is one that is constructed around very small development groups.

10. It should be emphasized here that every university or graduate institute in Switzerland is a public establishment.

11. The ZHdk bachelor's degree program was followed in 2010 by a master's degree program. For more information, see http://gamedesign.zhdk.ch.

12. Of course, there exists an expanding number of courses oriented toward game design, both in the Swiss German and Swiss French areas. For more information on them, see Beat Suter's 2010 surveys or the *GameCulture* "directory" page at http://www.gameculture.ch/directory.

13. The authors would like to note that since the Swiss press has no electronic archives for the period before 1994, no articles were found concerning this precise moment in the history of games in Switzerland. If paper archives exist, the persons responsible for their filings were not able to direct us to relevant articles.

14. A motion had already been passed by the National Council, the Assembly's lower house, in June 2009. For more information on the debate that took place in the Swiss-French region, see Haltiner (2010) and RTS.ch (2010).

15. While the PEGI system offers age-rating suggestions, this new motion would imply a strictly respected banning (see http://www.pegi.info/en/index).

16. For more details, see the complete English version of Article 135 at http://www.admin.ch/ch/e/rs/311_0/a135.html.

17. For more information, see the official Web page of the Swiss parliament about the rejection, at http://www.parlament.ch/f/suche/pages/geschaefte.aspx?gesch_id=20093422.

18. For selected examples of the international reactions to the episode, see http://www.ecrans.fr/La-Suisse-vers-l-interdiction,9510.html, http://www.mcvuk.com/news/read/switzerland-passes-violent-games-ban/09233, http://gamepolitics.com/2010/04/05/swiss-game-ban-may-feature-only-little-censorship, and http://www.gamesindustry.biz/articles/switzerland-votes-to-ban-violent-games.

19. See the museum's website at http://www.haus-ek.org/en/home and the festival's website at http://www.shiftfestival.ch/en.

20. For the keynotes of the international conference, see Atallah et al. 2014.

References

Atallah, Marc, Christian Indermuhle, Nicolas Nova, and Matthieu Pellet. 2014. *Pouvoirs des jeux vidéo: Des pratiques aux discours.* Gollion: Infolio.

Haltiner, Nadine. 2010. Jeux vidéo violents bientôt bannis. *24heures.ch*, March 19. http://archives.24heures.ch/actu/suisse/jeux-video-violents-bientot-bannis-2010-03-18.

RTS.ch. 2010. Une motion pour interdire les jeux vidéo violents. *RTS.ch*, June 28. http://www.rts.ch/info/suisse/1733146-une-motion-pour-interdire-les-jeux-video-violents.html.

SIEA. 2010. Contenu du bulletin SIEA—Chiffres du marché 2009 et T4/2009. *Swiss Interactive Entertainment Association (SIEA),* March 1. http://www.siea.ch/wordpress/wp-content/files/100301_pm-siea_chiffres-2009-et-T4-2009_F1.pdf.

SIEA. 2012. SIEA chiffres du marché 2011. *Swiss Interactive Entertainment Association (SIEA)*, June 11. http://siea.ch/wordpress/wp-content/files/110612_pm-siea_ChiffresMarche_2011_FR.pdf and http://www.svv-video.ch/downloads/Presse/Communique_KJ_2011_franz_final.pdf.

THAILAND

Songsri Soranastaporn

Video games (sometimes called "VDO games" in Thailand) are popular and attract massive numbers of players around the world. Thai players engage console games, computer games, online games, and handheld games, and many organizations and companies are involved in video games in Thailand, yet there are few studies on the subject. Therefore, this chapter will describe the history of video games, the current situation of video games, and explain the behavior of video game players in Thailand. The population in this study included four groups who are involved in video games, including e-learning, animation and computer graphics, movie production companies, and software and digital content engineers. Along with other documentation, interviews were conducted with the chief officers of education programs, presidents of various associations, owners and managers of companies, and game players. Data were analyzed (in categories) and provided interesting results.

Informants

Four informants were selected and interviewed: the president of the Thai Game Software Industry Association (TGA), the manager of a game provider, a teacher and trainer, and the founder of a game company.

Sittichai Theppaitoon (BankDebuz) is the president of the Thai Game Software Industry Association (since 2011). He studied computer science at Kasetsart University, one of the famous computer science programs in Thailand. He has been involved in games since 2001 and has owned three companies in Thailand: Debuz (established in 2001) and TumGame Company Limited (established in 2010), which both developed online games, and GAMEINDY (established in 2006), a company that supports customers of the other two companies. Each company functions differently: Debuz developed *Asura Online* (2007), and TumGame developed three games: *GodsWar* (2010), *PetsWar* (2010), and *Avatar* (2010). Theppaitoon teaches game production at Sripathum University.

Rittirong Keawvichien is CEO of the Goldensoft Company. He graduated with a bachelor's degree in civil engineering and a master's degree in business from Chulalongkorn University, the oldest and most famous university in Thailand. He worked at Asiasoft, a game company, before becoming a cofounder of Goldensoft Company in 2005. His work involves finding and evaluating new online games from abroad to be sold in Thailand.

Dendej Sawanthat was a computer science teacher at Dhurakit Bundit University and is now a trainer in the area of Web mobile applications and Web security for organizations in both the government and the private sector. He and Keawvichien have been friends since they were students at Chulalongkorn University. Sawanthat received a bachelor's degree in computer engineering from Chulalongkorn University and a master's degree in computer science from the University of Louisiana. He is also a member on the committee of the e-Learning Association of Thailand.

Santi Lothong is the founder of Compgamer Company (http://www.compgamer.com), established in 2000. Compgamer promotes the game industry and creation of quality media, making the company a center for players, providers, and developers. Compgamer acts as a news office and publishing office, collecting and advertising news and information regarding games and the information, computer, and technology (ICT) sector, such as game competitions and organizations. The company publishes its products as hard copies in stores as well as downloadable copies.

Since no history of video games in Thailand has been written, I interviewed these informants and cross-checked their data. Also, Lothong, the president of Compgamer Company, and his team worked hard to write the history of video games in Thailand (Thossapol 2012a). This team agreed that in the past, some twenty to thirty years ago, people did not pay attention to video games, and video games were new to Thailand, so there were no agencies that documented information about them. At that time, sellers just flew to Hong Kong, Japan, or Singapore and bought video games to sell in Bangkok. The history of video games in Thailand, then, is summarized below.

Early Video Games in Thailand

Video Games in Thailand can be categorized into four types: arcade games, console-based games, PC and LAN games, and online games. Among these games, arcade games were the first type in Thailand, followed by console-based games. Then, when computers and Internet technology progressed far enough, PC and LAN games and online games began to play significant roles and today are very popular with Thai gamers.

Arcade Games

Thailand never produced any arcade games because of the cost of development. The Galaxy Group Thailand Company has been importing arcade games into Thailand from Japan since 1993. As in other countries, these arcade games were mostly offered in shopping centers, and popular games included *Space Invaders*, *Phoenix*, *Daytona USA*, *Street Fighter II* (all 1993), and *The King of Fighters* (1994). Fighting games were the most popular genre. As of 2012, rhythm-and-dance games such as *Dance Dance Revolution* (1998) and *DrumMania* (2000) were popular among Thai teenagers, who played them in groups. The prices of new imported games are varied, ranging from USD $10,000–$17,000. The highest-priced arcade game is *Outrunner Special* (2012), which cost 10 million Baht (USD $333,333).

Recently, as the cost of old arcade games has decreased and the Thai economy has improved, some Thais bought old arcade games for playing in their homes as well as in big restaurants. Typical prices for these games are USD $400–$500 for fighting games, USD $1,500–$2,000 for shooting games, and USD $3,300–$3,500 for racing games. Galaxy Group Thailand (http://galaxy.co.th) is still the leader in the arcade game market in Thailand, and from 2003 to 2006, the company tried to build its own arcade games with the same capacity and ability as imported arcade games such as *Initial D3* (2003) and *Outrun 2* (2003). The cost of development, however, was very high because there was no factory in Thailand that supplied parts and devices, so the company had to order pieces that were handmade. As a result, the cost of development was very high, and in 2005, free trade law was instituted in Thailand, so the company gave up constructing arcade games. Consequently, all arcade games are imported.

Console-Based Games

The earliest video games in Thailand date back to 1977, when Atari Inc., USA, launched the Atari VCS 2600. This was the first home console system in Thailand, and it cost about 5,000 Baht (USD $199.99). Though Atari produced and sold many new games, they were not popular in Thailand because of their high prices. Three years later, in 1980, Nintendo launched a series of handheld games in Japan, namely the *Game & Watch* series, and Thais called it a "Press Game" because players pressed the buttons while playing. These games cost 5,800 yen. Later, Nintendo produced more games and their costs were varied, from USD $4–$400. Thais loved these games so much that Thai gamers kept them in their pockets and played them constantly. In 1983, Nintendo launched a console system, the 8-bit "Family Computer" or "Famicom," which cost 14,800 yen. Four years later, in 1987, the NEC Company of Japan produced its new console system, the PC-Engine, which cost 24,800 yen. Though the sophistication of this game was higher than the Famicom, it did not attract Thai gamers because it was too expensive and there weren't many games offered for it. Unfortunately, the foreign companies that produced these systems did not have offices in Thailand, so there was no one to promote and repair their games, which made it difficult to interest consumers in them.

Later, in 1988, SEGA released the SEGA Mega Drive (renamed the SEGA Genesis for its North American release in 1989), but it was not sold in Thailand. An updated version, the SEGA Mega Drive II (also known as the SEGA Genesis 2), came out a year later and was distributed in Thailand by Galaxy Company, but this console was not popular in Thailand either.

To compete with SEGA and NEC, Nintendo released a new console system in Japan in 1990, the Super Nintendo Entertainment System (SNES) (also known as the Super Famicom), which was a 16-bit system that cost 25,000 yen. The SNES was very popular in Thailand for three main reasons. First, Nintendo provided additional devices such as the Nintendo Super Scope and the Power Glove (see fig. 1). The Super Scope functioned as a light gun and had an infrared transmitter, while the Power Glove was able to input the player's movement.

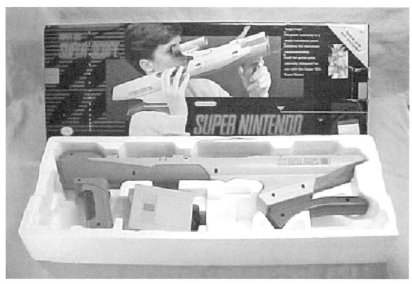

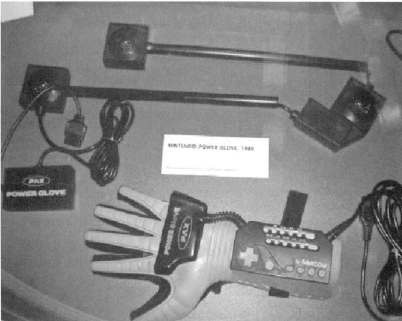

Figure 1
The Nintendo Super Scope (top) and the Power Glove (bottom). Source: http://www.online-station.net/feature/feature/15213.

Figure 2
Pro machines (left) and the Super Famicom with games on 3.5-inch diskettes (right). Source: http://www.online-station.net/feature/feature/15213.

Second, there was a wide range of games available, including such games as *Final Fantasy II* (1988), *Final Fight* (1989), *Super Mario World* (1990), *NBA All-Star Challenge* (1992), and *Dragon Ball Z: Super Butōden* (1993). Finally, players could play pirated copies of these games on 3.5-inch diskettes by using "Pro" devices (see fig. 2). Most games for the Super Famicom could be replicated on 3.5-inch diskette and were much cheaper than legitimate cartridges, making them more affordable for Thai players.

In 1990, SNK (Shin Nihon Kikaku) Playmore Company of Japan also released a new console system, the Neo Geo, but it was not popular because of its price and limited number of games, so Thai gamers rented it and played its games at game shops instead. Unlike SEGA, however, SNK did not have an office in Thailand.

In 1994, Sony Computer Entertainment released the PlayStation in Japan and also launched it in Thailand the same year. This system was very popular because of its graphics and the pirated CD-ROMs of the systems' games, which were sold in Thailand. Two Thai companies, Gamefreak and Uptron Multimedia (Thailand) Co., Ltd., legally imported and distributed the PlayStation, but had to sell it at a high price because of copyright, so the system did not sell well in Thailand.

The 2000s could be called a "Next Generation" and were a time of warring console systems because the three major console-producing companies launched their systems around the same time. Sony launched the PlayStation 2 on March 4, 2000, Microsoft launched its first console system, the Microsoft Xbox, in the US on November 15, 2001, and Nintendo returned to compete in the console game market by launching the Nintendo GameCube on September 14, 2001, which was not popular in Thailand, unlike the PlayStation 2 and the Xbox. The prices of the PlayStation 2 and the Xbox were about the same, but the former was much more popular than the latter. The reasons were as follows: the PlayStation 2 was backward-compatible with the first PlayStation, allowing PS1 CD-ROM games to be played on the PS2, and most of these games were developed by famous Japanese companies such as Enix, Namco, and Capcom, which could design and develop games that corresponded to the style and culture of Thai gamers. The Xbox, meanwhile, had a problem with its electric circuitry, which Thai gamers referred to as "red light," and which caused the machine to break down. Moreover, these broken machines were not under warranty at the shops where they were sold in Thailand, so Thai gamers were afraid to buy them. The price of the CD-ROM games for both the PlayStation 2 and the Xbox were similarly high, so Thai gamers preferred to buy pirated CD-ROM games.

Four years later, in 2005, Microsoft released another new console system, the Xbox 360, but it was not popular in Thailand even though the price was lower and the electrical problem was resolved. Nintendo launched the Nintendo Wii on November 19, 2006, and it was distributed in Thailand the same year. The Nintendo Wii was designed and developed differently from both PlayStation 2 and Xbox; the machine was thinner and smaller, and the wireless Wiimote and nunchuk allowed players to control and play more conveniently and remotely, with new experiences that challenged players. Also, the Nintendo Wii was designed to be used by a group, so more family members played together. Furthermore, the Nintendo Wii was popular with Thai adults, many of whom exercised or played sports using the Nintendo Wii. Around the same time, November 11, 2006, Sony released the PlayStation 3, which was also a multimedia computer and Blu-ray player, so both the quality and the price of the PlayStation 3 were high. However, since Thai gamers still enjoyed using the PlayStation and PlayStation 2, many Thai gamers did not buy the PlayStation 3.

During the 2000s, some Thai companies became legal agencies of foreign game companies, so Thai gamers could buy games at a reasonable price and these games were promoted. For example, the company Original Recreation is an official partner shop of Sony Entertainment Hong Kong, meaning it sells PlayStation 3 merchandise officially. SICOM Amusement bought copyrighted games from Capcom, and Saluzi (now defunct) was an official agency of EA (Electronic Arts) Sports. The companies CD Gamar and Bangkok CD Entertainment Co., Ltd. have been official agencies of Namco Bandai and Square Enix, respectively. Thus, this decade saw more Thai game companies and Thai gamers selling and playing games legally.

PC and LAN Games

In 1963, IBM 1620s (costing about 2 million Baht or USD $200,000) became the first computers to be imported into Thailand for the purposes of education and research, which would spur development in the country; for example, an IBM 1620 was used for statistical analysis at the Secretary of National Statistics Office, and by Professor Bundit Kantabut, chair of the department of statistics at Chulalongkorn University. Another computer, installed at the faculty of commerce and accounting, Chulalongkorn University, was placed on display in 2012 at the Science Center for Education and Bangkok Planetarium (asoftBiz 2013). One year later, in 1964, IBM 401 mainframe computers were imported for the National Statistical Office, the Siam Cement Group, and Bangkok Bank. In 1974, minicomputers were imported for the Stock Exchange of Thailand, and by 1979, small business enterprises were using them. Later, in 1982, PCs were imported and mostly used in universities and schools (asoftBiz 2013). In 1986, the Macintosh Plus (USD $2,495) was imported by SVOA (established 1982) for King Bhumipol (ITeXcite 2010). As of 2012, iStudio (http://www.istudio.in.th) is the current importer of Apple computers.

Some rich and educated Thai gamers played their first PC games using MS-DOS, but it was not popular because computers were too expensive to use for playing games. Finally, in 1995, the launch of Microsoft's Windows 1995 allowed more people to use PCs, as well as play games on them. PC games were distributed on 1.44 MB floppy diskettes and run on the A drives of PCs. (Even though the 1.44 MB diskette came out in 1987, games on CD-ROM drives and diskettes were not launched in Thailand until 1997.) With more sophisticated computer technology, Thai gamers were able to enjoy games such as *Warcraft: Orcs & Humans* (1994), *Command & Conquer* (1995), *Diablo* (1996), and *StarCraft* (1998). At first, Thai gamers played these games in single-player mode but began playing in groups (multiplayer) when LANs were established in Thailand in the late 1990s. Thai gamers enjoyed playing *Counter-Strike* (1999) and other first-person shooters (FPS), which were so popular that many game shops were opened in Bangkok so that Thai gamers could play them. Most were located near schools and universities, such as Kasetsart University, where there were many large game shops, and from there, game shops spread to provinces all over Thailand; this was the starting point of competitive game businesses in Thailand. B. M. Media (Thailand) Co., Ltd. licensed games such as Valve's *Half-Life* (1988) and Sierra Entertainment's *Pharaoh* (1998), and licensed *StarCraft*, *Warcraft*, and *Diablo II* (2000) from Blizzard Entertainment. Foreign companies also established branches in Thailand; for example, Electronic Arts (EA) set up EA Thailand as its agency, and the New ERA Company, established in 1997, was another agency for selling many license games.

Online Games

The first online game in Thailand, launched in 2003, was *King of Kings*, which was developed by Lager, a Taiwanese company (Aimsupasit 2012), and sold in Thailand by Just Sunday Cybernation, but it was not popular. On the other hand, GRAVITY's *Ragnarok Online* (2002) from Korea, which was sold by Asiasoft Thailand, was very popular. *Ragnarok Online* was launched September 16, 2002 with three servers; then AsiaSoft provided an Open Beta test (OBT) service on October 25, 2002, with six servers. Finally, AsiaSoft started to charge gamers on March 3, 2003. Though Thai gamers had to pay, there were enough gamers playing *Ragnarok Online* that AsiaSoft needed to establish twenty-three servers. The same year, Warax imported AMPED's *N-Age Online* (2002) from the Philippines, and New Era imported Webzen's *MU Online* (2001) from Korea. Since then, online games in Thailand have become a growing business and many of them have been provided free of charge.

In the early days of online games in Thailand, game companies provided their services by charging "air time" either daily or monthly. The average air-time charge was USD $10–$15 per month for one account, so game companies bought foreign games for release in Thailand. Around this time, computers and the Internet had become more efficient through the use of Asymmetric Digital Subscriber Lines (ADSLs) with speeds of 128–256K, which had replaced connections using 56K modems. After this, around 2006–2007, the online game industry changed to a free-to-play model, no longer charging for air time, meaning companies now earned their income and profit by selling in-world merchandise. Thai gamers played for free and bought items to help them play games more easily and efficiently. As of 2012, most games are imported, with the most popular ones coming from South Korea and Japan, and Thai companies now import Chinese online games as well.

Nowadays, due to the progress of computer technology, the Internet, and Thai education, some Thai game players have become game developers. In the past (2003–2006), Galaxy Group Thailand Company tried to develop arcade game machinery, but the cost was very high so the company gave up. Presently, game developers tend to develop console games because the cost of making console games is much lower than the cost of making arcade game machinery (Keawvichien 2012; Sawanthat 2012). Later, Debuz (established in 2001) and TumGame Company Limited (established in 2010), for example, developed online games as their business. As of early 2013, there were about 150 online games and thirty game providers in Thailand. One indicator showing that online games are popular in Thailand is that an educational program of game development is offered in some Thai universities, such as Sripathum University.

The Current State of Video Games in Thailand

Currently, many video game companies in Thailand are responding to this big-business market, and the Thai government supports these companies so that they can export their games. One prominent project, "The Workshop of Thailand's Digital Content International Market Development Program" (2009), was launched to help these companies learn and gain experience as part of the international video game industry with the hope that domestic video games will be exported and currency will flow into Thailand. To describe the current situation of video games in Thailand, I will next examine domestic video game production and exports, discuss Thai game associations and the Software Industry Promotion Agency (Public Organization) (SIPA), consider the hiring of experts for the development of both animation and video games for export, and summarize video game industry trends.

Domestic Video Game Production and Exports

Many Thai small and medium enterprises (SMEs) develop video games, and a few Thai companies integrate some Thai culture and tradition into their games. Examples of these domestic video game companies include the Samart Corporation, Play Thai, and AIM Advance.

The Samart Corporation developed the online game *Kankluay: The Last Battle* (2006), a two-player fighting game based on the first animated movie in Thailand, *Kankluay* (2006), which was developed by the Kantana Animation Company and was very successful. The Samart Corporation licensed the movie's characters and operates this game as an online business, sharing profits with the Kantana Animation Company (http://www.kantana.com).

Play Thai, a branch of Asiasoft co-operated with Promit Production and PromptNow from the FEC Company, developed the action MMORPG *King Narasuan Online* (*KNO*) (2012), which opened for beta testing on February 16, 2012 (GD Game-DED 2012). *KNO* underwent more local development than any other domestic game. It was designed as a three-dimensional action MMORPG and focused on fighting performance. The game's story was based on Thai history in the age of King Naresuan the Great (AD 1555–1605), a famous fighter, and its graphics were meticulously detailed because Queen Sirikit asked the producers to make new costumes for the game's characters. Players control characters and choose their styles of living and fighting, while taking part in events from Thai history. The experts (Frank Holz, Jurgen Reusswing, and Marc Kuepper) from the project "Workshop of Thailand's Digital Content International Market Development Program" (2010, see below) suggested that this game might be exportable because the game is interesting and graphics are very good. However, they also felt the game must be adapted, that some parts of the story change, and that it must not be too limited to Thai history so that the content could be more flexible for players. In order that the new version sold in the United States and Europe be more attractive, they recommended that the game be renamed so Americans and Europeans would understand what it was about as soon as they read the title. Characters should be able to fight each other, and various weapons should be sold as in-game merchandise (SIPA 2010, 22, 24, 50, 67, 76, 146, 161).

The Thai developer AIM (Advance Intelligence Modernity) developed *Gang Xa* (2011), which was distributed by Asiasoft. *Gang Xa* is an action 2.5-D side-scrolling MMORPG. Players use joysticks to control characters, which move on a horizontal plane (flat) or along the z-axis (deep), and there are various attack modes, such as quick attack (quick but soft), power attack (slow but strong or hard), and combo (continuous and able to choose special attack). Players can play this game via social networks such as Facebook, hi5, and Twitter (http://www.asiasoft.co.th/ourproducts/gangza.html).

Though there are some Thai games that integrate Thai history or culture into their stories, only a few have been made into English versions, such as *The Spirit of Khon* (2010), a 3-D game developed by Novaleaf Company for PC and the Xbox platform (see fig. 3). The developers were inspired by Khon (traditional Thai dance), murals of the Emerald temple, and the story of Ramayana, from which the game's simulations were made. In the game, Rama, the protagonist, walks his troop to the city of Longka to help Sida (his wife). Along his journey, he has to fight to survive through various obstacles, challenges, and quick time events (QTE), in which cinematic scene techniques are used (Tai 2010). *The Spirit of Khon* was funded by the Thai government and advertised on *Thailand Planet* (a virtual world that simulated Bangkok Metropolitan) (Compgamer 2014a, 2014b).

My interviews revealed that most Thai games are made using the Thai language, making them unlikely to be exported, especially to Europe or the United States. Moreover, companies developing games alone may not be successful in the globalized market; they need to seek help and support. In order to gain more power through networking, they have gathered together and established the Thai Game Software Industry Association.

Figure 3
Screenshots from *The Spirit of Khon* (2010).

The Thai Game Software Industry Association

Though the video game industry in Thailand is young, some Thai game developers are already visionary leaders, and in 2006 they formed the Thai Game Software Industry Association (TGA) to encourage Thai game companies to work together to share knowledge and experience and gain negotiating power. In 2012, the TGA proposed some interesting policies to the Thai government:

1. Help the Thai people to understand the pros and cons of playing video games. Thai people should be informed how to live and practice on-line life safely and smartly. They should learn to be critical thinkers, so they will search for and use information logically before they believe what they read or follow it.
2. Improve and correct Thai laws concerning movies and video so that they correspond to the current state of technology and society. Moreover, laws concerning video games should be separate from laws concerning movies and karaoke because they are different.
3. Provide a regional app store so that Thai games can be sold more globally instead of through Apple's App Store or the Android Market.
4. Finally, promote cooperation between private sectors and educational institutes for the development of Thai games. Provide e-learning resources related to the development of games on operating systems such as Windows, iOS, Android, and social network platforms such as Facebook and Google+, as well as resources that describe how to administrate and establish the security of online games (source: http://blog.tga.in.th/)

What the TGA proposes will greatly affect the Thai game industry because video games constitute a bigger market than those of both film and music. The Thai government needs to help and support it, and the Software Industry Promotion Agency (Public Organization) has done so.

The Software Industry Promotion Agency (Public Organization) (SIPA)

The Software Industry Promotion Agency (Public Organization) (SIPA) was established in 2003 to promote and develop software and digital content developers and entrepreneurs, to promote and support the marketing of software and digital content, and to develop organizations that promote software and digital content. SIPA realizes that the digital content industries, animation, and games, in Thailand have a big market; SIPA's 2008 study estimated that the digital content industry, in terms of animation, film, and games, cost USD $560 million (17,000 million Baht) and that its growth is 12% per year (SIPA 2009a, 7). Additionally, this industry is young and new. It needs to be nurtured and supported by the Thai government so both animation and game markets and their revenue will expand. Research has revealed that the enormous growth and market of the animation and game industry came from foreign products and the Thai people; the youth especially preferred them over domestic products, despite the fact that a few Thai companies developed quality animation and game products (SIPA 2009a, 18–24; SIPA 2009b, 7-1-7-71). As a result, SIPA tried to help these companies by hiring international consultants to help them improve their video games and marketing (SIPA 2010).

The Hiring of Experts to Aid in Developing Animation and Video Games for Export

SIPA launched the project "The Workshop of Thailand's Digital Content International Market Development Program" (2010) to help the game industry in Thailand. (Though SIPA hired experts to aid the development of both animation and video games for export, only the information regarding video games will be discussed here.) The main purpose of this project was to help Thai video game companies to think globally in terms of product and business development. The process of training Thai video game companies to be competitive with foreign companies was as follows: companies submitted their products, and consultants selected companies whose products were of high quality and had potential for export. Next, an "Intensive Development Program" was arranged for each company, and consultants prepared presale packages for each company and evaluated them. Though seven companies submitted, only five were selected by consultants (see table 1), for which they developed marketing plans, sold or published their products abroad, and then reported on their operation, sales, and any other interesting issues that emerged.

Several criteria were used to select the consultants. They had to have experienced development teams and strategic marketing teams, and had to be able to provide business development consultation. They had to offer agency services, a world-class publishing network, and international media partners. Finally, they had to have a good track record. In the end, the IEM Consulting Company from Germany was selected, with key trainers including Frank Holz, Jurgen Reusswing, and Marc Kuepper. Finally, only five companies and one game from each company were evaluated (see table 1).

The results of the consultants' evaluations were summarized in five areas: game design, gameplay, technology application, presentation, and marketing strategies.

Game Design and Play

If games are developed strictly based on Thai history, flexibility will decrease because imagination and creativity will be more difficult to integrate, even without language barriers or cultural barriers. A game's

Table 1
Companies Selected for Evaluation (SIPA 2010, 14–22, 48).

No.	Company	Product Title	Type of Game	Technology
1	Novaleaf Software Co., Ltd.	*CASIE: Combat*	Console Game	All of these companies developed their own game engines, and Novaleaf used C#
2	Gamesquare Interactive Co., Ltd. (Public)	*UCube*	Handheld Game	
3	AIM (Advance Intelligent Modernity) Advance Co., Ltd.	*Dark Silence*	Handheld Game, PC Online Game	
4	PromptNow Ltd.	*King Naresuan Online Game*	PC Online Game	
5	CyberPlanet Interactive (PCL) Ltd. (Public)	*Celebrity Salon*	Console Game for Nintendo Wii	

story must be planned, designed, and plotted well so that gameplay will correspond to the whole story. Gameplay must be balanced between being too easy and too hard, and the games themselves must not to be too simple or too complex. Game manuals that describe a game in detail should be provided in the game's packaging.

To evaluate games, a competitive analysis was done by experts and compared with other games of the same type so that gameplay and game balance could be adjusted accordingly (SIPA 2010, 53). Developers must design their games in such a way as to distinguish them from existing games and be able to make them into MMORPGs to allow for large numbers of players.

The Application of Technology

New technology needs to be used to develop games so that they can serve future markets. While small- and medium-sized companies can develop their own game engines, it is much more difficult and requires large investments for Thai companies to develop game engines for others to use and develop their new games. As of 2010, Thai companies lacked new technology to develop their games, so these companies needed to look for new technology to serve the future game market. For example, UCube speakers were developed for playing on the PSP and could also be developed for multiple platforms. Data backup and security also needed to be managed.

Presentation

Thai game companies had difficulties in presenting their games to experts, so experts provided suggestions as follows: First, all documents had to be complete and well organized, especially the game design document. The priority of each topic had to be set and balanced. Second, a game's vision needed to be established and explicit. Third, a game's unique selling points had to be distinguished in a manner that is concise and attractive. Fourth, Thai companies had to present their games in a simple, direct, and concise way. This is very necessary when the story or contents of games are complicated. Enough of a game must be presented so buyers and publishers can understand the concepts behind the games when previewing them. Finally, the track record of each Thai game company had to be provided.

Marketing Strategies

Although marketing is essential in business, only two Thai game companies planned to market in Southeast Asia, the United States, Europe, and China, through their websites and community networks. Marketing plans had to be set and specify target markets such as Southeast Asia, the United States, and Europe, as well as multiple channels in which advertisements would be placed. Advertising should run eight to twelve months before a game was launched. Data encryption should be used to fight piracy. Tie-in sponsors must be matched with the targeted game players.

The game companies had to present their games to buyers or publishers. To sell games, a sales package had to be made, which should include the game's unique selling points, a synopsis of the game, a character bible, guidelines for playing the game, examples of game animation (not more than three minutes' worth), a work plan that includes a production timeline and budget, a list of licensees for the given platform, and a company biography and track record. This business plan was proposed at Game Connection in San Francisco in March 2010, where a consultant network is established.

Training

To educate Thai game companies that joined the workshop, experts presented information regarding the game industry and game businesses in the global market. Globally, in 2006, this business was valued at USD $31 billion (SIPA 2006), and in Thailand, it was valued at USD $2 billion (SIPA 2006), larger than movie and music markets, and its growth was estimated to be 5% to 7% in 2011. The future of the video game market is the digital distribution system, in which game developers will gain more than 50% in profits. Moreover, four factors are keys to success: games must be unique in design and high quality, they must be available in the online marketplace, demo versions should be provided for users to experiment with, and finally, global advertisement strategies should be used and make it possible for game players to meet each other. Details regarding these factors are explained in the four sections below.

Business Development and Environment

A new generation of games will be moved to online digital distribution systems. The most commercially competitive games today are MMORPGs (54%), first-person shooters (26%), and casual games (15%) (Asiasoft 2012b, 12). The growth of casual games will increase in the future because they are easy to play and require less time to play. Casual games will be marketed to all groups, especially office workers.

As the game business environment changes, developers must adapt quickly. Social network games, mobile games, and MMORPGs have important roles within the game industry. To be competitive in this business, companies must focus more on quality and innovation than on technology because game engines such as Unreal Engine 3, Gamebryo, CryEngine 3, idTech 5, and Unity are developed and accepted by many publishers, decreasing the developer's risk while increasing reliability.

Game Development Strategies

To succeed in developing games, strategies are essential. These include the analysis of products, competitors, opportunities and profit, trends of new markets, target population, and resources. Above all, the quality of games must be focused on. To summarize, there are many Thai SMEs that develop video games and struggle because of novices and investment; fortunately for the industry, there are some Thai companies such as Asiasoft that earn about USD $14 million (420 million Baht) per year (Asiasoft 2012a).

Video Game Company Profiles

The number of video game companies has been increasing. In 2008, there were 23 companies, and together they published 65 games. In that same year, five new companies were established and 25 new games were published. As of 2011, there were 33 companies publishing 127 games, and during that year, five new companies were established and 39 new games were published (Asiasoft 2012b, 5). In what follows, only the leading companies will be presented, including Asiasoft, True Digital Plus Co., Ltd., Winner Online Co., Ltd., Galaxy Group, Goldensoft, Novaleaf, GAMEINDY, and Compgamer.

Asiasoft, established in September 2001, is Thai-owned, located in Thailand, and a leading online game operator with a dominant market share in Thailand, Singapore, Malaysia, and Vietnam, making it the online entertainment leader in Southeast Asia. Asiasoft's market share by genre in Thailand was 38% in 2011 and 40% in the first quarter of 2012. In 2011, Asiasoft's market share by publisher was 46% in Singapore and 17% in Malaysia, and in 2012, Asiasoft's market share by genre in both Singapore and Malaysia was 26%. Asiasoft Corporation Public Company Limited, the company's branch in Thailand, was established in 2003, and publishes three kinds of online games: MMORPGs such as Eyedentity Games' *Dragon Nest* (2010, released in Thailand in 2011) and T3 Entertainment's *Camon Hero* (2012); casual games such as *R2Beat* (released in Thailand in 2012, developed by playpark.com) and GNI Soft's *Gigaslave* (2011); and shooting games such as Pentavision's *S4 League* (2011) and *Alliance of Valiant Arms SEA* (*AVASEA*) (2011). Its first imported MMO game was *Ragnarok Online* (2002), imported from South Korea in 2002. The game's language was translated from Korean to Thai, and there were 110,600 players during the first year of release. Overall, the company earned about 1.54 billion Baht (US $49.9 million), 1.52 billion Baht (USD $51.3 million), and 1.684 billion Baht (US$56.13 million) during the 2009, 2010, and 2011 fiscal years, respectively (Asiasoft 2012a).

True Digital Plus Co., Ltd. (previously True Digital Entertainment Company Limited), founded in 2006, is a subsidiary of the True Corporation, which has five branches: Truemove, TrueOnline, TrueVision, TrueMoney, and TrueLife+ (http://www.truedigitalplus.com). True Digital Plus Co., Ltd.'s market share by genre in Thailand was 24% in 2011. The company provides games such as Dragonfly GF's *Special Force* (2004, released in Thailand in 2006), Nexon's *Crazyracing Kartrider* (2004, released in Thailand in 2008), EA Sports' *FIFA Online 2* (2006), *HipStreet* (released in 2008 but shut down in May 2012 due to lack of popularity), *Karma Revo* (2010), and *Battery Online* (2012). All of True Digital Plus's games are consolidated under the GG, which stands for "Good Game" portal site at http://www.gg.in.th. Beginning with *Special Force*, True Digital Plus has been a pioneer in e-sport competition, hosting several world-class championships in Thailand since 2008, when the competitions began. The first round of the *Special Force* Thailand Championship 2012 started on June 9, 2012, and the final round was on November 10, 2012. The venue was IMAX Siam Paragon Department Store, Bangkok (http://www.gamegigs.com/game-event/6614.html). The fourth GG ECL for e-sport competition was August 25–26, 2012, at the Fortune Department Store, Bangkok (http://www.compgamer.com/home/229596/gg-ecl). The competition was presented live on cable television on TrueVison channel 67 (http://www.truelife.com).

Winner Online Company Limited (WOCL), established in 2003, publishes both casual games and MMOs for both gamers in Thailand and abroad. WOCL imported twenty-three online games, and as of 2012, still serviced nineteen of them. Recent games include *Devil* (2012, developed by LineKong), *Emperor* (2012, developed by Chinese Gamer International Corp.), *Born to Fire* (2012, developed by Funtree and NHN), and *Eden Online* (2012, developed by X-Legend). But two games, Kingsoft's *Xshort* (2009) and *Glory Destiny Online* (2012, also known as *Spirit Tales*, created by X-Legend and published by Koramgame), are especially popular among Thai gamers. WOCL attracts its customers by using a membership club and provides benefits for them. It has organized a two-day e-sport competition for gamers every month since 2003. The competition is regularly held on Saturdays (for FPSes) and Sundays (for MMORPGs). WOCL's market share by genre in Thailand was 7% in 2011.

Galaxy Group, established in 1985, is the oldest company and the leader in the arcade game industry in Thailand. Amorn Apithanakoon, the founder and owner, was appointed as the first (and only) video game distributor of SEGA Enterprises. He established his first game arcades with twenty game machines in 1985, and he expanded his enterprise to five locations in Bangkok one year later. As of early 2013, there are more than 5,000 arcade game machines located all over Thailand (at 121 locations), including games by SEGA, Namco, Capcom, Pentavision, Saintfun, Gamecon, Hypaa, Merit, and Stern Pinball. Galaxy Edu Hitech, a subsidiary of Galaxy Group, functions as an educational resource, providing information for players, especially kids, to manage their time, money, and selves when playing games. Galaxy Edu Hitech selected games that are safe and fun for kids, so the Ministry of Culture and the Censorship Agency gave them a certificate of Videotape Materials for Home and Public Viewing (meaning that the content is suitable for children and adults). One marketing strategy of the Galaxy Group is to send an expert gamer to play games in each location. This gamer plays games as if he is a movie star, and customers and game players can come to watch him playing (see the YouTube video "The Greatest Awesome Gamer in Thailand," http://www.youtube.com/watch?v=mxX7SRuU0ho, 2012).

Goldensoft was established in 2006 and publishes only MMORPGs from Taiwan, Korea, and China, such as Aeria Games and Entertainment's *Lime Odyssey: The Chronicles of Orta* (open beta testing began January 5, 2012, and closed on September 6, 2012), Dovogame's *WarFlow* (*WF*) (2011), and *Weapons of the Gods Online* (2011). To promote games and gain customers, the company operates game competitions in each region of Thailand. Moreover, the company arranges regular seminars in Internet cafés to introduce new products, at the same time offering promotions such as reduced prices for in-game items and accessories. Goldensoft's market share by genre in Thailand was 6% in 2011.

Novaleaf, established in 2007, a Thai game developer, is supported by both the Board of Investment (BOI) and the Software Industry Promotion Agency (SIPA). Novaleaf develops games for the Xbox 360 (specifically XBLA) and the Windows PC using Microsoft C# and XNA technologies, including *The Spirit of Khon* (2010), *AARRR! Ninjas Beware* (prototype, 2009), *Biology Battle Xbox* (2008), *Biology Battle PC* (2010), *C.A.S. Combat Armor Suit* (prototype, 2010), *Brawler* (2011), and *Thailand Planet* (2011). Additionally, Novaleaf helps to train students to design and develop games. Programmers teach students at Sripathum University, and students from the university are trained at the company. Novaleaf administrators said that there was a lack of game developers, so training these students would help grow the game industry.

Debuz and TumGame were both established in 2006 by Sittichai Theppaitoon (BankDebuz), the president of the Thai Game Software Industry Association. Both companies develop Thai games specifically. Debuz produced *Asura Online* (2007), the first Thai MMORPG produced by Thai developers. TumGame developed three games: *PetsWar* (2010), *GodsWar* (2010), and *Avatar* (2010). *Asura Online* was created with PlayBrowser technology. *PetsWar* is the first Thai online game that can be played on Facebook. All customers who play these games are supported by the company GAMEINDY.

Compgamer was established in 2000 to support the game industry, to produce quality game media, and to be a center for all gaming stakeholders such as gamers, providers, and developers. Compgamer, therefore, serves as a news center and publisher to collect and advertise information about game exhibitions and events. Compgamer also acts as an organizer of game and information technology competitions. Now Compgamer has produced an e-book called *Gamer News*.

In conclusion, video games play a significant role in Thailand. The business is growing, the market value is huge, and the Thai government promotes and supports game companies, enabling them to produce and publish their games worldwide.

Finally, we turn to the players themselves and statistical data regarding them.

Behavior of Video Game Players in Thailand

Research results regarding the behavior of video game players in Thailand showed the following: There were twice as many male players as female players (Chokevisanchai and Juwarahawong 2006; Trongtosak 2006; Duangrat 2010). Younger gamers (age twelve and older) played games between one and four times per week for one to three hours each time (6:00 p.m. to 9:00 p.m.) (Chokevisanchai and Juwarahawong 2006), whereas older players (ages twelve to twenty-five) played games one or two days per week for one to three hours each time (from 9:00 p.m. to midnight) (Trongtosak 2006). Those gamers who were considered addicted to games played five to six days per week for eight to ten hours per day (Sennan 2006). These gamers like to play MMORPGs, such as *Ragnarok Online* or T3 Entertainment's *Audition Online* (2004), more than others, and they knew and played these games because they were introduced to them at Internet cafés (Trongtosak 2006). They learned the details of these games from websites and game magazines (Chokevisanchai and Juwarahawong 2006) as well as from their friends (Sennan 2006). These gamers were attracted by game design, game specification (quantities) (Sennan 2006), game content, background, characters and their occupations, graphics, game soundtracks, the ability to see 360 degrees, virtual images that looked like real humans, customizable avatars, and activities such as quests (Duangrat 2010). As a consequence, a gamer's social status depends on the games that gamer plays (Hutarom 2007). Parents encourage their children to become gamers because they want their children to keep up with technology and so their children will have technological skills that they can apply to their studies and work (as opposed to children who do not play) (Hutarom 2007). Gamers also provided reasons for playing games: entertainment and peer pressure to play (Hutarom 2007; Sawakjun

2009), poor relationships with their family (Sennan 2006), and as a result of advertisements for games seen on television, in magazines, and on the Internet (Hutarom 2007).

Gamers who are addicted to games will likely face problems with their health, studies, finances, and relationships (Lerdsupavaree 2008; Mahajarean 2004). Health problems include pain in their eyes and back, and stiff fingers and hands as a result of playing for long stretches; they did not eat meals on time, so they had stomach problems; and finally, they went to bed too late and so were sleep-deprived. Study problems included failing courses or poorer academic performance as a result of not spending enough time on their studies; some did not attend their classes and did not do their homework assignments. Financial problems included gamers spending more on games than what their income allowed; paying for games and game items also increased their expenses. Relationship problems included slightly increased trouble with both family and friends.

Conclusion

Video games appeared in Thailand relatively late, as compared to the United States, and the video game industry in Thailand grew mainly due to the development of the Internet and personal computers. The Thai government sees this opportunity and supports companies by providing workshops for training Thai developers in order to export their games internationally. However, some Thai companies have imported video games to Thailand and expanded their businesses to neighboring countries. As elsewhere, video games have found great popularity in Thailand, but also abuses, and players need to have self-discipline, be able to manage their time, and take care of their health and finances.

References

Aimsupasit, P. 2012. Talk about games. DMA: Digital Media Animation Television. http://www.dmaonline.in.th/index.php?modules=article&f=writer&writerID=5&typeID=5.

Asiasoft. 2012a. Asiasoft Corporation PLC, (AS) 1Q12 Opportunity Day. http://www.set.or.th/th/company/files/opp_day/1_2012/as.pdf.

Asiasoft. 2012b. Asiasoft Corporation PLC, (AS) 4Q11 Opportunity Day. http://www.dcs-digital.com/setweb/downloads/2554q4/20120314_as.pdf.

Compgamer. 2014a. Spirit of Khon drawing attention from both Thais and foreigners. http://www.compgamer.com/home/14286.

Compgamer. 2014b. Novaleaf developed Thai game emphasized on Thai culture. http://www.compgamer.com/home/13219.

Chokevisanchai, N., and S. Juwarahawong. 2006. Behavior and factors affected to play Ragnarok online game in Bangkok. Master's thesis, Naresuan University.

Duangrat, A. 2010. Online game playing behavior of young people in Bangkok metropolis. Master's thesis, Ramkhamhaeng University.

Game Gigs. 2012. Special Force SFTH Championship 2012 for OPEN Round. http://www.gamegigs.com/game-event/6614.html.

GD Game-DED. 2012. A list of online games in Thailand. http://www.game-ded.com/listgameonline.

Hutarom, D. 2007. The reduction of juvenile online game addiction: A case study of the participants of the Center of Game Addict Prevention, Child, and Adolescent Mental Health, Rajanagarindra Institute. Master's thesis, Chulalongkorn University.

ITeXcite. 2010. Followed the first computer of King Bhumipol. http://www.itexcite.com/article/.

Leesa-nguansuk, Suchit. 2010. Virtual kingdom pushes country's potential: Tourist spots, business promoted in 3D world. *Bangkok Post*, October 27. http://www.bangkokpost.com/tech/computer/203391/virtual-kingdom-pushes-country-potential.

Lerdsupavaree, N. 2008. A study of utilization behavior and affect of playing online game according to junior secondary school students opinions of SuanKulab Nontaburi School. Master's thesis, Durakit Pandit University.

Mahajarean, J. 2004. The study of behaviors and effects of online game exposure among high school students in Muang District, Nakornpathom Province. Master's thesis, Thamasrt University.

N-Age Online International Edition! 2007. *RPG.net*, March 19. http://forum.rpg.net/showthread.php?317373-N-Age-Online-international-edition!%28Korean-MMORPG%29.

New Era. 2012. NETH: Enjoy the games with New Era. http://www.neweragames.co.th/home.php.

Project One and Beginning of Asura. 2012. http://wiki.asura.in.th/index.php.

Puru, Z. 2012. GAME: I'm real gamer. Additional devices for consoles in the past. *Online Station*, October 17. http://www.online-station.net/feature/feature/15213.

raytaker555. 2011. The greatest awesome gamer in Thailand. *YouTube*, September 17. http://www.youtube.com/watch?v=mxX7SRuU0ho.

S! Game. 2013. Introducing new games. *S! Game*. http://game.sanook.com/preview/00350.php.

Sawakjun, T. 2009. The study of behaviors and effects of online game exposure among high school students in Amphoe Mueang, Surat Thani Province. Master's thesis, King Mongkut University of Technology Thonburi.

School of Digital Media. 2013. Sripatum University website. http://web.spu.ac.th/sdm/project/novaleaf.

Sennan, T. 2006. Behavior and impact of online game addiction: Case study of student. Master's thesis, Chulalongkorn University.

SIPA. 2006. Thai Digital Content White Paper: Animation & Games 2006. Bangkok, Thailand: Knowledge Development Center Co., Ltd.

SIPA. 2009a. *Thailand's Digital Content Industry 2008: Animation & Game.* Bangkok: Software Industry Promotion Agency.

SIPA. 2009b. A Study of Thailand's Digital Content Industry 2009: Animation & Game. Bangkok, Thailand: International Institute for Trade and Development. http://khonkaen.sipa.or.th/ewt_dl_link.php?nid=54.

SIPA. 2010. *The Workshop of Thailand's Digital Content International Market Development Program, Software Industry Promotion Agency.* Bangkok: International Institute for Trade and Development.

Softbizplus. 2013. History and pioneer of computer in Thailand. *Amazon Softbizplus.* http://www.softbizplus.com/computer/1478-computer-history-thailand.

Tai. 2010. Spirit of Khon. *Compgamer.com*, November 29. http://www.compgamer.com/home/13871/the-spirit-of-khon.

Tai. 2012. The 4th GG ECL and E-Sports Competition. *Compgamer.com*, August 14. http://www.compgamer.com/home/229596/gg-ecl.

TGA. 2012. Thai Game Software Industry Association (TGA). http://www.tga.in.th/post/11565334673/plan2012.

Thossapol. 2010. Karma REVO Online. *Compgamer.com*, December 15. http://www.compgamer.com/home/22158/C-karma-revo.

Thossapol. 2012a. Origination of Electronic Games. *Compgamer.com*, July 5. http://www.compgamer.com/home/221470/.

Thossapol. 2012b. Types of video games. *Compgamer.com*, July 6. http://www.compgamer.com/home/?p=221815.

Thossapol. 2012c. Video game and Thai market. *Compgamer.com*, July 7. http://www.compgamer.com/home/?p=222045.

Trongtosak, R. 2006. Bangkok youth's behavior in online game playing. Master's thesis, Kasem Bundit University.

Winner Online. 2010. *Winner Online.* http://www.winner.co.th/about.asp.

TURKEY

Cetin Tuker, Erdal Yılmaz, and Kursat Cagiltay

Turkish people may have encountered computer games later than many Westerners, but they have wasted no time catching up. The modern Turkish game industry is one of the most rapidly growing markets in the world (Newzoo 2012). Thus, almost all game hardware producers and major game development companies have been paying special attention to Turkey. The average game playing durations, habits, and preferences of Turkish survey respondents are similar to those of developed countries (Karakus, İnal, and Cagiltay 2008; Durdu, Tüfekçi, and Cagiltay 2005). In the area of game development, however, Turkey remains far behind; no game hardware development activity in Turkey warranted mentioning at the time of the writing of this chapter. In the following sections, we will discuss digital game development activities, beginning with the late 1980s. In order to better convey this uncovered history, significant Turkish game development activities have been summarized through the present. Lack of academic studies in this field required the use of popular resources, personal observations, and anecdotal comments. A discussion of game development history is followed by a current view of the Turkish game market. The future of Turkish digital game development is addressed in the conclusion.

The 1980s: Level 1, Beginner

Until the mid-1980s, most of the Turkish population could not afford personal computers or game console systems. Thus, no game development activity occurred in the late 1970s or early 1980s, excluding the basic work of computer science or electronics researchers and students. Some programmers had developed simple games such as tic-tac-toe on mainframe computers prior to 1980. In the early 1980s, famous electronics journal *Elektor* presented a hardware-based (AY-3-8500 chip set) circuit diagram for a *PONG*-like game. Based on this diagram, some small companies tried to build and sell locally produced *PONG*-clone consoles. However, no extant documentation is available about these basic projects.

The Sinclair Spectrum, Sinclair ZX81, Commodore VIC-20, Commodore 64, Amstrad, and MSX series were the first computers that allowed Turkish people to develop software. Among them, the Commodore 64 was the most popular, offering many games. The first Turkish computer magazines, such as *Commodore, Sizin Amstrad, 64ler,* and *Elo Elektronik,* played an important role in delivering the printed source codes of simple games. Many people had to learn game programming from similar written documents because it was difficult to find appropriate resources in Turkish. In 1984, a programmer named Özden Kılıçay, PhD, published a book in Turkish, *Applied BASIC Programming Language Book for Sinclair Spectrum, Sinclair ZX81, Commodore VIC-20, Commodore 64, BBC A/B, Electron, Oric-Atmos, Texas Instruments 99/4a, Apple IIe, and Sharp MZ80K,* which contained a special section on writing games, including code examples. His next book, *Commodore 64, Peek-Poke & Machine Code,* which was published in 1986, focused mainly on graphics, sprites, and other game-related capabilities of the Commodore 64; it included twenty example games.

One of the first commercial development attempts in this period was *Keloğlan,* a game about a hero from a famous Turkish fable. The game was released in 1989 as a group effort by Ahmet Ergen, Ilgaz Akbaş, and Derya Yalazkan from the Byte Computer Company. *Keloğlan* was developed using *Adventure Game Maker,* a third-party software program. *Keloğlan* was not a commercial success, and only a few people played it. However, it was the first attempt targeting Turkish gamers with a story built upon a national folk hero.

Toward the end of the 1980s, the AMIGA 500 series, which was the main platform for early Turkish game titles, became popular. Although no local or global commercial titles emerged from Turkey in the 1980s, Turkish game developer Mevlüt Dinç, who was living in the United Kingdom, released or adapted several games on new platforms, such as Firebird Software's *Gerry the Germ Goes Body Poppin'* (1985), System 3 Software's *Last Ninja 2: Back with a Vengeance* (1988), Electric Dreams Software's *Prodigy* (1986), and Activision's *Hammerfist* (1990). Mevlüt Dinç continued to produce successful titles in the 1990s in the UK, returning to Turkey in 2000. His work from the 1990s and 2000s will be mentioned in the following sections.

The 1990s: Level 2, Amateur

With its superb multimedia capabilities, the AMIGA encouraged Turkish developers to release text-based adventure games, card games, and puzzle games such as *Çarkıfelek* (*Wheel of Fortune,* 1992), *Para ve Güç* (*Money and Power,* 1992), and *Pis Yedili* (a card game, 1992). However, some developers planned to use the AMIGA's power to release more professional games. Sedat Çöloğlu was a main game producer of this period. In 1992, he led Digital Dreams during the publication of its first boxed game, *Hançer* (*The Dagger*) (see fig. 1).

The following summarizes the story of the game's development based on experiences shared by Sedat Çöloğlu: in 1990, Çöloğlu organized the Amiga Professional Users Club (APUC), a group with 450 members. One of his most outstanding project ideas was producing a Turkish strategy game. In November 1990, he finished the game scenario of *The Dagger,* a game about the rise of the Ottoman Empire, and shared the idea with club members. A variety of people ranging from professional graphic artists to historians supported the project, reviewing details such as important events and accurate accessories of the medieval age. Some academics voluntarily advised about historical facts of the era. Before the release of the game, Sedat Çöloğlu and

Figure 1
Cover illustration and screenshots from Hançer (*The Dagger*, 1992). (Courtesy of Sedat Çöloğlu.)

his development team, Digital Dreams, advertised it in a local AMIGA magazine. One of the most interesting parts of this story took place during the diskette preparation pipeline. Each set included three diskettes, and copying one diskette required approximately five minutes, including verification. Volunteer APUC members with external floppy drives helped speed up this production problem. After a year of hard work, the game was ready for release in 1992. In the early 1990s, software piracy was quite rampant in Turkey. Original boxed game sales for very popular AMIGA titles were only around 300 units. *The Dagger* sold more than 2,200 original copies in two years. The price was USD $15, while the development cost of one set was USD $7. These figures were clearly not encouraging for Turkish game entrepreneurs in the 1990s.

Siliconworx comprised several high-school student game veterans and was another important player of this era. It released AMIGA games such as *Umut Tarlaları* (*Fields of Hope*, 1993), *İstanbul Efsaneleri: Lale Savaşçıları* (*Legends of Istanbul: Tulip Warriors*, 1994), and *Muhtar* (the term for an official village headman, 1995). *Fields of Hope* was Siliconworx's first commercial game, a farm management simulation. This game had almost all of the properties of a professional game, such as retail packaging, a user manual, an original musical score, and copy protection. Such professional properties were handled by assigning multiple roles to every member of the team. For example, lead programmer Özgür Özol also worked on game music, design, graphics, and the user manual. Further, each team member reportedly visited potential sellers with backpacks full of game

Figure 2
Umut Tarlaları (*Fields of Hope*, 1993); front cover of the game's box (left) and a screenshot from the game (right). (Courtesy of Siliconworx.)

boxes. The game was offered on consignment, not much different than goods in arts and crafts markets. They managed to sell 1,000 copies via this method. Total revenue was barely enough to recover box, diskette, and labeling costs. No matter the number of sales, this game is still remembered as one of the most successful and enjoyable Turkish digital games. Siliconworx released its second game, *Legends of Istanbul: Tulip Warriors*, in 1994 (see fig. 2). Development and delivery phases show many parallels to its first game. *Tulip Warriors* was the first Turkish role-playing game (RPG), remembered for its original music, graphics, and easy controls. This game also sold fewer than 2,000 copies.

In the mid-1990s, personal computers (PCs) began to overtake the AMIGA in Turkey. In 1996, Cartoon Studios released its comedy adventure game *Dedektif Fırtına* (*Detective Storm*), and later, a movie adventure game, *Gerçeğin Ötesinde* (*Beyond the Truth*, 1998).

Considering the increasing role of PCs, Siliconworx and Compuphiliacs (another young game development team) decided to cooperate on a port of *Legends of Istanbul* for the PC. Compuphiliacs rewrote the game several times to keep up with changing technology and overcome technical challenges. Eventually, they began to consider delivering the game on CD-ROMs. One of the leading music industry companies of Turkey offered to distribute the game. In 1996, after another period of development in a basement office provided by the publisher,

Compuphiliacs completed the game. The publisher set the price at USD $35, which was quite expensive for a single game in 1996. Total retail sales were around 2,000. One year later, 45,000 copies of the CD-ROM were bundled with popular game magazines. Distribution by a professional company was new for the Turkish game market, and total revenue of the developers was only USD $10,000 including bundle sale shares. The same multiple-role approach was used in this game; it was possible for a programmer to work as a cameraman or composer as well.

Piri the Explorer Ship (1997) was an outstanding edutainment application produced by SEBİT Education and Information Technologies. It was a point-and-click adventure game published on CD-ROM in which all scenes were pre-rendered 3-D images from a first-person perspective. The player takes on the role of the captain of a sixteenth-century ship following routes on the first world map as prepared in 1513 by Piri Reis, a Turkish admiral. During gameplay, the player solves puzzles to advance, learning cultural, social, historical, technological, and artistic facts about the sixteenth century. This game was released in Turkey, the United States, and some Asian countries.

The 1990s were a transition period from amateur game development to professional game development in Turkey. Mevlüt Dinç continued his game development work in the UK, developing several well-known games such as *First Samurai* (1991), *Street Racer* (1994), and *S.C.A.R.S.* (1998). The amateur groups that had developed

Figure 3
Screenshot from *Piri the Explorer Ship* (1997). (Courtesy of SEBİT.)

games used ties forged in their high school years to create team spirit. Although no specific gaming company arose and game sales recovered only material expenses, the developers insisted on developing these games in dark basement rooms in the face of adverse conditions, motivated only by their desire to produce local games for Turkish people.

The 2000s: Level 3, Professional

An important event in 2000 for the Turkish game industry was the return of Mevlüt Dinç after twenty years. Dinç, having produced many titles in the UK, decided to continue with local Turkish talent. One year after he started his business in Turkey, Dinç Interactive (later Sobee Studios), released the first Turkish game for the international market, which was completely written by Turkish university students. *Dual Blades* (2002), a clever mix of fighting games featuring Turkish heroes, was developed for the Game Boy Advance and published by Metro 3D in the United States and Japan. The next release was *Football Manager* (2003), a real-time, competitive multiplayer management simulation. The next Sobee game was *Semih Saygıner ile Magic Bilardo* (*Magic Billiards with Semih Saygıner*, 2004), an online competitive 3-D billiards game developed with the help and endorsement of billiards world champion Semih Saygıner.

After this game, Sobee started working on Turkey's first MMORPG, *İstanbul Kıyamet Vakti* (*IKV, Istanbul Apocalypse Time*), which was released in 2006. This game became very popular, with more than 700,000 registered players. Sobee released four more noteworthy games in the 2000s: a third-person action game, *Kabus 22* (*Nightmare 22*), which was developed with Son Işık Game Studios and published in 2006 in several European countries; *Citroën C4 Robot* (2008), a high-quality racing game also developed with Son Işık Game Studios; *I Can Football* (2009), the world's first eleven-by-eleven online soccer game; and *SuperCan* (2011), which introduced Turkey's first child hero. Within a short time, the game, featuring Marvel superheroes, reached almost two million players. A second version of this game was published in 2012.

Sobee was joined by other professional game houses in Turkey in 2005. Yoğurt Technologies, Cinemedia, and 3TE Games published their first game, *Pusu: Uyanış* (*The Ambush: Awakening*), in May 2005 (see fig. 4). The game was a 3-D third-person shooter/action-adventure game in which a Turkish army officer fights criminals who have kidnapped his scientist father. This game was not a straightforward shoot-'em-up game; the storyline and character backgrounds were well developed, and actions and events were interrelated. Both the protagonist and antagonist had supporting characters that were present during gameplay and cut-scenes with their own subplots. The game was voiced by famous actors and actresses in a professional recording studio. *The Ambush* was a success for the Turkish game market, and 10,000 boxed copies were sold.

The studios 3TE and Yoğurt completed another project in 2006, a 3-D third-person action-adventure game for the new product (YAMI) of a famous biscuit brand, using an updated version of *The Ambush*'s game engine. This game, *YAMI: Mechanical Invasion*, was based on the TV commercial for the product, which was completed almost eight months before the game was released. The animator and character modeler of *The Ambush*, who

Figure 4

Screenshots from *Pusu: Uyanış* (*The Ambush: Awakening*, 2005): a scene in the garage from around the middle of the game (top, left); the house of Fatih, the game's main character (top, right). Below them are screenshots from the Istiklal Street scene in the game *YAMİ: Mechanical Invasion* (2006) (bottom, left) and on top of a historic tram on Istiklal Street in the YAMİ television commercial created before the game (bottom, right). One of the authors, Cetin Tuker, worked on both of these projects. (Courtesy of 3TE Games.)

is also one of the authors of this chapter, Cetin Tuker, left the game development team after completing animation for *The Ambush* and started working for a studio specializing in TV commercials. When the advertisement agency came up with the idea of producing an animated film with video game aesthetics, he suggested the idea of creating a complementary video game. The agency agreed and decided to work with the producers of *The Ambush*. The game's interface, world architecture, and overall look were very similar to the commercial and can be considered its successor.

During the intermediate stage period, some developers found a chance to publish games in the international market. In the early years of this period, Turkish design studios gained experience developing games for the Turkish market; later they used their experience to develop games that met international standards and satisfied international publishers' needs. Five titles were released by international publishers during this period: *Dual Blades* (2002), *Kabus 22* (2006), *Darkness Within: In Pursuit of Loath Nolder* (2007), *Culpa Innata* (2007), and the most successful title of all, *Mount & Blade* (2008). *Darkness Within: In Pursuit of Loath Nolder*, a point-and-click Lovecraftian horror-thriller-adventure game, was developed by Zoetrope Interactive and published by Lighthouse Interactive in 2007. The protagonist is Howard E. Loreid, a British cop looking for Loath Nolder, a private investigator and the main suspect in the murder of Clark Field, an amateur archeologist who has just escaped from a mental hospital. The storyline is constructed around Loreid tracking down Nolder while trying to determine if he is the killer.

Culpa Innata, an adventure game inspired by Alev Alatlı's novel *Schrödinger's Cat* (1999), was developed by Momentum DTM and published by Strategy First in 2007. The events take place in 2047 in an Orwellian society in the city of Adrianopolis, part of the idealized government World Union. The protagonist is World Union Peace Officer Phoenix Wallis, who is investigating the murder of a friend in a neighboring "Rogue State" in Odessa, Russia. *Culpa Innata*'s game space is a fully 3-D, real-time rendered environment with a diurnal cycle that is dynamically related to the appearances of characters and events. The main technological advance in this game is its nonlinear narrative; the player's actions affect how the game will end. The same player can experience a different storyline and various endings in repeated plays based on decisions made during gameplay. Changes in players' actions result in dynamic changes in the sequence of locations visited. The logic engine of the game can allow 10,000 different outcomes, offering unique games for nearly every player.

With sales of more than 1 million units, the *Mount & Blade* series is probably the most popular Turkish game in the international market. It is a medieval, nonlinear action RPG with sandbox-style play from the Turkish company TaleWorlds Entertainment. The most outstanding feature of the game is its artificial intelligence algorithm, which creates outstanding combat scenes. The first *Mount & Blade* game was released in the fourth quarter of 2008 in the United States and Europe; a sequel, *Mount & Blade: Warband*, was released in 2010, and an expansion pack, *Mount & Blade: With Fire and Sword*, was released in 2011 (see fig. 5).

Turkish game development efforts in the 2000s were not limited to studios. Some young game designers had enough courage to start projects on their own. One reason for this phenomenon is the considerable circulation of monthly game magazines. Critics' articles about how games were developed sparked their enthusiasm. Another reason for independent development was the affordability of the hardware. Finally, with the

Figure 5
Screenshot from *Mount & Blade: Warband* (2010). (Courtesy of Taleworlds.)

help of the Internet, programming tutorials and tools also became more accessible. Although many of these projects ended in failure, three of them were completed successfully: a first-person shooter, *Koridor* (*Corridor*, 2004) by Hakan Nehir; *İstila* (*Invasion*, 2008), a turn-based strategy game by Ozan Gümüş; and a horror-adventure game, *Lanetin Hikayesi* (*The Story of the Curse*, 2003), by İdris Çelik. The first two of these games were distributed as freeware by several game magazines, and it is still possible to download them for free. According to the developers, around 75,000 to 100,000 copies of these games have been distributed by magazines or downloaded from the Internet.

Another important advancement in the 2000s was the increasing popularity of mobile gaming. This new platform has attracted the attention of several Turkish game developers. In the early 2000s, Turkish mobile games such as *Asmalı Konak* (*The Mansion with Vines*, 2003), based on the story of a famous TV serial, and *Kötü Kedi Şerafettin* (*Şerafettin the Bad Cat*, 2004), based on the story of a famous cartoon cat, by Başarı Mobile and Pixofun companies respectively, have achieved success in the local market. Shortly after the first release of the iPhone in 2007, Turkish developers began to follow smartphone trends. One of the first outstanding examples, with its 3-D graphics and creative game controls, is *Mission Deep Sea* (2009), published by Reo-Tek. The multitouch screens of mobile devices provided game designers with new potential for interfaces. Designers who successfully use these controls produce games that can only be played on mobile devices. In *Mission Deep Sea*, the player controls a sea turtle that dives into the deep sea. The head and arms of the turtle are visible in the lower part of the screen, as it carries a small camera. To control the diving speed and direction of the turtle, the player swipes his fingers on the mobile device's screen, activating the front arms of the turtle. The game's objective is to find and remove objects that are harmful to natural sea life, such as barrels full of chemicals.

The Turkish Game Industry as of 2013: Facts and Advancements

Commercial Advancements

Although game development in Turkey has increased substantially since the 1980s, Turkey still has a fairly small industry compared to the size of its economy and population. Currently, the Turkish game industry comprises small-scale 3-D action game development studios, mobile/social/casual game studios, and game localization companies, all on small, medium, and large scales. A few companies work on world-class 3-D games such as Taleworlds, Sobee Studios, Ceidot, and MadByte. None of these companies employs more than fifty workers. TaleWorlds Entertainment, creator of the *Mount & Blade* series, and Sobee Studios are the largest. In 2009, Turk Telekom, a leading technology company in Turkey, acquired Sobee Studios. This merger can be seen as the first economic success in the Turkish game industry. Currently, there are nearly fifty mobile, casual, and social game studios. Among them, Peak Games offers several casual/card games that reach millions of players. The success of Peak Games has attracted investments from various global venture capitals. Peak Games is the biggest social gaming company in Turkey and MENA (Middle East and North Africa), and it operates several offices around the world as well. It is the largest game company in Turkey, with more than 200 employees. Several mobile and casual game companies have also attracted attention, such as Picnic Hippo (known for *Bucketz* [2012]), Duello Games (known for *iSlash* [2010] and *Bellyfish* [2013]), Zibumi (*Wedding Snake* [2012]), Pixofun (*FootboCity* [2012]), and NoWhere Studios (*Monochroma* [2013]).

Game development is not the only way to make money in the Turkish game market. Joy Game, a pioneer in localization and publishing in Turkey that is expanding into other countries, has gained the attention of South Korean gaming company CJ Games, which now owns 50% of the company. Several other companies focus on localization and publishing, and Peak Games is also investing in this field. With Riot Games and Crytek opening offices in Turkey, 2013 was quite an interesting year for the Turkish game market. Such news bolsters its existing potential and promising future. Such advancements also help to accelerate the growth of the Turkish digital game development market. Currently, nearly 1,000 professionals work in this industry in Turkey. Although there are more in Western cities, this growth is hopeful, given the fact that there were only ten professionals twenty years ago and around 100 professionals just ten years ago.

Academic Activities Involving Computer Games

During the last few years, digital game development has garnered attention from academia and government-supported technology development centers. The following section summarizes related advancements.

METUTECH-ATOM

The METUTECH Animation Technologies and Game Development Center (METUTECH-ATOM), the first and only sector-specific preincubation center in Turkey, was established in 2008 as part of Middle East Technical

University-Technopolis (METUTECH, http://www.metutech-atom.com/). The main goal of METUTECH-ATOM is to prevent the loss of creative ideas regarding digital games and animation, converting these ideas into marketable values. It is aimed at providing entrepreneurial and financial support, leading to commercial value. The goals of METUTECH-ATOM are to

- create a highly professional cluster of digital game and animation sectors
- create a platform for academics, firms, and students to work together
- create a knowledge pool for and awareness of digital arts
- make university-industry cooperation more active and sustainable
- strengthen activities of the sector and increase the reputation of ATOM and its groups on both the national and international level
- develop sector-specific international cooperation

Beneficiaries are selected through two processes. In the first step, projects are evaluated according to their content and framework. Next, project plans are assessed. At ATOM, groups are supported for one year. After that, the new firms are brought to an incubation center, where they are supported for two or three more years; successful companies then become METUTECH firms.

Since 2008, eight digital game enterprises have been established out of twelve projects through ATOM services. Currently, twelve groups are working at METUTECH-ATOM. Turkey has a small number of companies in the game sector, and as of summer 2013, approximately 20% are ATOM Companies. ATOM also organizes national and international events to engage the digital sector with the public space, such as the International Workshop on Mobile Games, Global Game Jam, and university conferences.

METU Game Technologies Program

METU Game Technologies (GATE) is a postgraduate program established in 2009 within the Informatics Institute at Middle East Technical University, one of the world's top-ranked universities. The primary aim of the program is to educate the next generation of game developers and designers. The program has a strong research component with a prominent interdisciplinary edge reflected not only in the academic backgrounds of the faculty but also in the diverse backgrounds of enrolled students. Courses include computer science and engineering, electronic engineering, industrial design, music, and architecture. Students select one of two tracks: game development or game design. These tracks consist of related courses for either the more or less technical development aspects of games. The work carried out by the students reflects novel aspects of cutting-edge research in game technologies. Research infrastructure is at the disposal of GATE students and staff, including the GameLab, an audio research lab, a motion-capture lab, and a virtual reality research lab. The program has strong collaboration and research links with other game technology labs and within the game technology industry in Turkey, other countries in Europe, and overseas countries. The program is also supported by game companies in METU-Technopolis, such as METUTECH-ATOM. METU-GATE's strong links with game technology companies have resulted in the formation of an industry board: a group of experts who help steer the program by fine-tuning the curriculum with timely knowledge about the actual needs of the industry.

Hacettepe University, Computer Animation and Game Technologies

Hacettepe University's computer animation and game technologies programs was established in 2011. It is an interdisciplinary program that gathers disciplines such as computer science, art, physics, biomechanics, and anatomy with a wide range of output from motion pictures and games to medicine and scientific simulation. For this reason, computer animation and game technologies uses algorithms based on techniques of scientific calculation, statistics, signal processing, linear algebra, control theory, geometry, and artificial intelligence. The nonthesis master's degree program requires education in computer animation, including the creation of three-dimensional models lit with realistic illumination, and the training necessary to use tools and technologies to create computer games. Within this program, students explore basic design concepts, the sensitivity of human perception, interaction design, the modeling of objects, camera control, and character equipment for graphic animation, particle systems usage, and surface rendering techniques. Students create characters, scenery, and 3-D graphics animations, and they program behaviors with artificial intelligence, applying new skills to the making of short films and computer games. The program is based on technical and aesthetic principles for students who want a career in computer animation (see http://www.bil-grafik.hacettepe.edu.tr/animasyonIng.html).

SIMGE Research Group

Middle East Technical University's Simulations and Games in Education (SIMGE) research group was established in 2004 by Professor Kursat Cagiltay. The main aim of this group is to conduct research on the use of simulations and games in education. In addition to pedagogical and cognitive issues, SIMGE researchers also focus on the social and technical issues of simulations and games. According to SIMGE, electronic games are becoming more and more popular in Turkey. As the game industry develops, its effects can clearly be seen. However, not enough research has been conducted about the potential and problems of this new medium. A limited number of studies have investigated the effects of gaming on Turkish people; SIMGE aims to close this gap (see http://simge.metu.edu.tr/).

BUG

Founded by Dr. Güven Çatak (communication design department, Bahçeşehir University), BUG focuses on video games with playful interaction design. As a game lab, BUG aims to be a communication and education platform in the İstanbul region for the young enthusiast who wants to develop games. BUG started out in 2011 with the shooting of the documentary film *Game.Doc*, which covered the game scene in Turkey. BUG's first R&D project was realized in May 2012 by the workshop Playful Interactions in collaboration with the MIT Media Lab; this event introduced BUG to the public. A three-week boot camp titled Game Workshops followed in summer 2012. BUG has produced seven teams that have designed seven games, and it has become a meeting point for indie game studios and developers. In addition to the game lab activities, BUG plans to develop academic programs on game design and support game studies (see http://bug.bahcesehir.edu.tr/).

Conclusion

It is very difficult to find academic resources regarding Turkish game development activities. Excluding a few research papers (such as Yılmaz and Cagiltay 2005; or Binark and Bayraktutan-Sütcü 2012), popular resources such as gaming forums or blogs and personal connections were the most useful resources for this study. Considering these facts, it is possible that some information was missed, especially for the period before 1995. However, this essay covers all major game development activities in Turkey that are worth mentioning.

Many of the games listed from the mid-1980s to the present use national figures and local cultural features. Even games for the international market have contained Turkish heroes. *The Dagger* features historical figures from the Ottoman Empire, while *Tulip Warriors* and the *İstanbul Kıyamet Vakti* MMOG use many figures from İstanbul. Authentic Turkish folklore and national history will likely be incorporated by game developers in the future, too.

Considering the game development activities presented in this chapter, the blossoming Turkish game industry will have more opportunities to produce titles in the future with growing support from universities and technology development centers. If the current game houses start to make more money, investments will also continue to grow, employing a talented population of people who want to be involved in the digital game development business. The future of the Turkish game development industry seems very bright; current trends predict important developments in the near future.

References

Binark, M., and G. Bayraktutan-Sütcü. 2012. A critical interpretation of a new "creative industry" in Turkey: Game studios and the production of value chain. In *Computer Games and New Media Cultures: A Handbook of Digital Games Studies*, ed. J. Fromme and A. Unger, 371–391. Heidelberg: Springer.

Durdu, P. O., A. Tüfekçi, and K. Cagiltay. 2005. Üniversite ogrencilerinin bilgisayar oyunu oynama aliskanlikları ve oyun tercihleri. *Euroasian Journal of Educational Research* 19:66–76.

Karakus, T., Y. İnal, and K. Cagiltay. 2008. A descriptive study of Turkish high school students' game-playing characteristics and their considerations concerning the effects of games. *Computers in Human Behavior* 24 (6): 2520–2529.

Newzoo. 2012. Infographic 2012—Turkey. http://www.newzoo.com/infographics/infographic-the-turkish-games-market/.

Yılmaz, E., and K. Cagiltay. 2005. History of digital games in Turkey. In *Proceedings of DiGRA Conference 2005: Changing Views—Worlds in Play.* http://www.digra.org/dl/db/06276.58368.pdf.

UNITED KINGDOM

Tristan Donovan

The United Kingdom is one of the world's foremost producers and consumers of video games. Based on revenue, the country is the largest consumer of games in the world after the United States and Japan (Padilla and Swift 2012). A third of its 63 million people identify themselves as game players, and half of its households own at least one game console (UKIE 2011). Britain also has some of the highest levels of smartphone ownership, social media use, and broadband Internet access in Europe (ComScore Data Mine 2013; Woollaston 2013; Jackson 2013).

Alongside its voracious appetite for playing games, the UK is also a major producer of them. Measured on revenue, the UK stands as the world's fourth-largest producer of video games, outperformed only by the United States, Japan, and Canada; although there are concerns that South Korea may have pushed it into fifth place by now (Skillset 2011).

Furthermore, many British games have had a significant impact on the medium's history and development.

Early Stirrings: Games in the UK Prior to the Formation of the Domestic Industry, 1945–1979

The earliest British games were created in computer labs during the late 1940s to 1960s and were, primarily, re-creations of traditional games such as chess or nim, which were the fruit of artificial intelligence research or military simulations rather than products of a desire to entertain. Most were never seen outside the computer labs of universities or the British military, but one did give the British public an early glimpse of what was to come when NIMROD, a nim-playing computer built by the Manchester computer firm Ferranti, went on display in London during the 1951 Festival of Britain (Donovan 2010). It would, however, take another two decades before video games became commercially available in the UK.

It is unclear exactly when commercial video games first arrived in the UK, following their creation in the United States. However, they were certainly on sale from 1973 onward—the year when Atari Inc. opened a short-lived sales arm in the country and Magnavox began exporting its Odyssey games console to Europe (Goldberg and Vendel 2012; Baer 2005).

By the following year, Britain's first homegrown responses to the emergence of video games as a consumer product arrived with the launch of two consoles that played *PONG*-type games: the Videomaster Home T.V. Game and the Videosport MK2, the latter being the creation of hi-fi retail chain Henry's (Wolf 2007).

It is unknown how successful these consoles were, but the Videosport MK2 was still on sale in 1977, and Videomaster was still releasing new single-game consoles as late as 1979, suggesting that demand was high enough for games to remain an attractive market for both companies for several years (Winter n.d.; Old-computers.com n.d.).

While these and later British-made consoles of the 1970s were largely derivative of what had come out of the United States, the UK did have a major global influence on the home *PONG* boom of that decade thanks to the AY-3-8500 chip. Developed in 1976 by General Instruments in Scotland, this integrated circuit ran several bat-and-ball games and, in doing so, offered manufacturers a cheap, off-the-shelf means of building their own game console. The AY-3-8500 became the basis for hundreds of *PONG*-like consoles released worldwide, including ones produced by Coleco, Magnavox, and Sears as well as the British manufacturers Binatone and Grandstand.

In contrast, the UK had far less success in the coin-operated video game business, which was the other major way of accessing video games during the 1970s. Britain's first arcade video game company, Subelectro, formed around 1977, and while a few more followed—most notably Century Electronics, the creator of the popular game *Hunchback* (1983)—none were able to achieve success comparable to their American and Japanese rivals. While the UK's own coin-ops floundered, US and Japanese games were widely available by the late 1970s and could be found not just in dedicated arcades but also in cafés, pubs, and fish-and-chips shops.

Arcade video games would remain a significant part of British video game culture right through to the late 1990s, when the arrival of more advanced home game systems, the Internet, and mobile phones led to players losing interest in the arcades. Today, most British arcades are dominated by redemption games and gambling machines, usually fruit machines.

Beyond the commercial sphere, the UK produced two significant computer games during the 1970s: John Conway's zero-player cellular automaton *Game of Life* (1970) and *MUD* (late 1979, early 1980).

Whether Conway's creation is truly a video game is disputable given that player involvement is limited to defining the initial conditions of the "game," but what is not disputable is the influence it has had on games. Many game designers cite it as an inspiration, including *SimCity* (1989) creator Will Wright and *Braid* (2008) developer Jonathan Blow. Some have also incorporated its concepts into their games; for example, *Populous II: Trials of the Olympian Gods* (1991) lets players "plant" fungi that then grow across the game's landscape in accordance with the rules of Conway's work.

MUD, short for *Multi-User Dungeon*, has been even more influential. Developed by two University of Essex students, Roy Trubshaw and Richard Bartle, *MUD* was a text-only multiplayer role-playing game set within a semi-persistent fantasy realm. Its audience was limited to the few who could access computer networks and afford the phone bills that resulted from long periods of play, but *MUD* and its tight-knit player community laid the foundations for the massively multiplayer online games (MMOGs) of today.

MUD introduced design ideas that are now MMOG conventions, such as a persistent world and the chaining together of player quests. Its players also had an important role in shaping the culture of online gaming, spawning slang terms such as "noob" and fostering an acceptance of people playing characters of a gender different than their own. Indeed, as Richard Bartle has argued, not unreasonably, this latter trait of online game communities would have been less likely to have taken root if *MUD* originated outside the UK, where cultural acceptance and tolerance of homosexuality was (and is) more widespread than in most of the world (Donovan 2010).

Significantly, *MUD*'s creators shared the source code for the game with other computer users and allowed them to use it however they wished. This openness resulted in such a proliferation of *MUD*-based games during the 1980s and 1990s that "MUD" shifted from being the name of a single game to a catch-all term for an entire genre of text-based virtual worlds. These MUDs, most notably the Danish variant *DikuMUD* (1991), greatly informed the development and design of commercial MMOGs such as *Ultima Online* (1997), *RuneScape* (2001), and *World of Warcraft* (2004), as well as inspiring the virtual world *Second Life* (2003). As such, *MUD* probably remains the UK's single most important contribution to video games.

Cottage Industry: The Birth of the British Games Industry, 1979–1987

By the time work on *MUD* began in 1979, mass-produced home computers from the United States—such as the Commodore PET and TRS-80—were on sale in the UK, but the country's economic problems and the high cost of these imported systems restricted ownership to organizations and the most affluent hobbyists. This situation would, however, change rapidly in 1980 when British inventor Clive Sinclair's technology company Science of Cambridge (later renamed Sinclair Research) launched the ZX80 home computer.

Although less powerful than its American rivals, the ZX80 was the cheapest computer on the market by far, costing as little as £79.95, a price point achieved in part by it being sold in kit form rather than preassembled. This low price not only undercut other home computers but also cartridge-based game consoles; a TRS-80, for example, retailed for around £431.25 at this time, while an Atari VCS 2600 cost £95.45 (Tandy [advertisement] 1979; Silica Shop [advertisement] 1981). This gave the ZX80 a major advantage at a time when the average salary in the UK was around £6,000 a year (Swaine 2009).

The ZX80 sold around 50,000 units, an unprecedented volume for a home computer at this time, and Sinclair's 1981 successor, the ZX81, achieved sales of more than 1.5 million units. Sinclair's third computer, 1982's ZX Spectrum, enjoyed even greater success, selling in the region of 5 million units by the time it was discontinued in 1992, by which time Russian clones of the computer running the TR-DOS operating system were playing an important role in the development of that country's games industry (Anderson and Levene 2012; Donovan 2010; Stewart 2013). Sinclair's computers, together with the other budget systems that followed his lead, made computing affordable in the UK.

The adoption of home computers was further encouraged by the government of Prime Minister Margaret Thatcher, which launched a major drive to get schools to teach computer programming—a skill that the government and many others at the time believed would be as vital as math and English in the future—using BBC-branded computers developed by the UK company Acorn Computers. The result was a boom in computer ownership in the UK. By 1985, home computers were more common in British households than in American ones, present in 12.6% of British homes, compared to 8.9% of US homes—a lead the UK maintained until 1995 (Schmitt and Wadsworth 2002).

In retrospect, Britain's decision to buy computers rather than game consoles appears to have been highly significant in the emergence of its games industry. First, the adoption of British-made computers such as the Amstrad CPC, BBC Micro, Dragon 32, and ZX Spectrum created a barrier to entry to the UK market for the more developed game businesses of the US and Japan, which were focused on other home game platforms. This isolation gave the UK's industry a chance to build itself up with relatively little competitive threat from larger foreign game publishers and developers (the crash of the United States' game industry around the same time helped, too, by curtailing American companies' plans to expand into Europe).

Second, home computers, unlike consoles, came with a version of the programming language BASIC standardly preinstalled, and so when people bought a computer, they were not just getting something they could play games on, but also something that equipped them with the tools they needed to make games. And make games they did.

Many computer users learned the principles of computer programming from typing in command lists for BASIC games that were published in computer magazines and then began creating games of their own. In keeping with the Thatcher government's rhetoric about entrepreneurialism, many of these amateur game makers—including numerous schoolchildren—began trying to sell them.

Initially, these homebrew games were sold via mail-order ads in magazines and newsletters, but as the boom in computer ownership fuelled demand for games, high-street shops began to stock them on their shelves. Many of these so-called bedroom programmers were further encouraged by media reports of teenage computer whiz kids striking it rich with a hit game. As the momentum built, the UK went from producing almost no commercial games in 1979 to boasting a thriving cottage industry that released more than 1,180 games for the ZX Spectrum alone in 1983 (from the author's own analysis of data on the World of Spectrum website).

These programmers found much of their inspiration in the American and Japanese games in arcades, but the influence of British culture was evident too. Games such as *Football Manager* (1981) and *Graham Gooch Test Cricket* (1986) reflected the popularity of soccer and cricket in the UK, the British children's comic *The Beano* sparked the idea for the naughty schoolboy sim *Skool Daze* (1984), and the surreal and very popular platform game *Jet Set Willy* (1984) featured unmistakable nods to *Monty Python's Flying Circus* (1969–1974) and the cover of the Pink Floyd album *Animals* (1977). Some even drew on British current affairs. The platform game *Wanted! Monty Mole* (1984) found inspiration in the Miners' Strike, a nation-defining clash between the Thatcher government and the National Union of Mineworkers, and was set within a fictional secret mine owned by union boss Arthur Scargill.

British developers of the time also explored new forms of play as well. These were often a byproduct of attempts to push the technology of home computers as far as possible. The landmark space trading and combat sim *Elite* (1984) offered not just fast wireframe 3-D visuals, but used procedural generation to create an open-world galaxy with 256 planets within the BBC Micro's 32KB of RAM, while Incentive Software's *Driller* (1986) was one of the first games to give players freedom of movement within a 3-D world built from polygons. Another couple of games, *Alien* (1984) and *The Rats!* (1985), were among the very first to explore the potential of horror as a genre (as in games that seek to frighten players, as opposed to games with a horror setting such as Atari's VCS title *Haunted House* [1982]).

Isometric arcade adventures became an especially popular type of game in the UK. These games coupled the isometric visual approach first used in games such as SEGA's *Zaxxon* (1982) with a mixture of platforming, arcade action, and puzzle-solving. The first step toward the creation of this genre was *Ant Attack* (1983), a British game that challenged players to save their girlfriend or boyfriend from a city infested with giant ants, but it would be Ultimate Play the Game's *Knight Lore* (1984) that became the template for the genre and established it as a mainstay of British gaming during the 1980s.

Many British game developers reveled in creating oddball, surrealistic games, too. Jeff Minter became a cult figure in the UK's industry by conjuring up trippy shoot-'em-ups such as *Attack of the Mutant Camels* (1983), which were defined by trigger-happy shooting, psychedelic visuals, and plenty of ruminants. Matthew Smith, the creator of the aforementioned *Jet Set Willy*, has suggested that the bizarre enemies the player encounters within that game were the result of taking hallucinogenic mushrooms (Edge 2012). Oddest of all is Mel Croucher's *Deus Ex Machina* (1984), a ZX Spectrum game designed to be played while listening to a synchronized audio cassette soundtrack that was so bizarre that one review concluded: "It's hard to decide whether this is an extension of the computer video game by music, or an extension of the 'concept album' by the addition of games playing" (Candy and Kean 1984). That the game's storyline involved playing a mouse dropping that disrupts a human-rearing factory in a dystopian world similar to Aldous Huxley's sci-fi novel *Brave New World* (1932) did little to help ease the confusion.

Few of these games made it out of the UK and even fewer beyond Western Europe. Eventually, *Elite* did get converted to the Nintendo Entertainment System (NES) and the isometric arcade adventure *The Last Ninja* (1987) became a global hit with sales of more than 1.25 million outside Europe, but most British games were unknown outside their home country (Spitzley n.d.).

Foreign Influence: Non-domestic Games and Consoles in the UK Market During the 1980s

While British-made games dominated the UK home games market, some US and Japanese titles were available, as were a few games from other Western European nations. This was primarily due to the efforts of British publishers such as Ocean Software and U.S. Gold, the latter of which became one of the nation's biggest publishers by releasing versions of foreign games on computer formats that were popular in the UK. However, since most computer owners in the UK were still using cassette tapes rather than floppy disks for storage during the 1980s, the bulk of the games that crossed the Atlantic were simpler, arcade-orientated efforts such as Epyx's *Impossible Mission* (1984) than more complex creations such as Sierra's adventure games or Origin Systems' *Ultima* series.

Game consoles, meanwhile, would continue to struggle throughout the 1980s in the UK. Nintendo's NES and the Atari 7800 had little impact on the UK, and the consoles created by British companies were even less successful. The Amstrad GX4000, a repackaging of its CPC computer, flopped, and the much-hyped Konix Multisystem never made it out of development. SEGA's Master System did better, establishing a sizeable—if far from dominant—foothold in the UK market, and the Nintendo Game Boy became a huge success, but the high price of console games and a lack of support from UK publishers (which were wary of the restrictions that console manufacturers demanded in return for access to their systems) conspired against the adoption of consoles.

Another factor in the relative disinterest in consoles was technology. By the late 1980s, British game players were abandoning their ZX Spectrums and Commodore 64s for the new breed of 16-bit processor-based computers: the Atari ST, Commodore AMIGA, and—a bit later—IBM PC compatibles. These systems offered significantly better audio-visual capabilities and processing power than the 8-bit consoles of the time, and this made the idea of opting for a relatively underpowered console less attractive to UK consumers. The UK's general disinterest in consoles only really began to thaw with the arrival of the SEGA Mega Drive in 1990 and the Super Nintendo in 1992.

Growing Up: The Evolution of the UK Industry During the Late 1980s and Early 1990s

The UK market's late 1980s, early 1990s transition to these new 16-bit computers and consoles also brought the era of British-made computers to an end. Amstrad, which had absorbed Sinclair's computer business in 1986, moved into making generic PCs instead of developing a new computer format of its own design, while Acorn's Archimedes struggled to win over game players.

By this time, the UK's games industry had professionalized considerably from its formative days. A small group of large publishers dominated the market, and developers had begun to club together into teams rather than working solo. As a result, there was a greater focus on producing games that would attract sales rather than producing off-the-wall experiences, although the zaniness of those early years of British games was still evident in the crazed surrealism of *Wizkid* (1992) and Team17's enduring *Worms* series.

More developers were, however, producing games that were not instantly identifiable as British. The only real giveaway that The Bitmap Brothers' popular shoot-'em-up *Xenon 2: Megablast* (1989) originated in the UK came from its acid house theme tune and the developers' marketing—which presented them as shades-and-leather-jacket-clad pop stars—spoke of a new commercial awareness within the country's game industry. Revolution Software's point-and-click adventure *Lure of the Temptress* (1992) aped the work of America's LucasArts and Sierra, while Magnetic Fields' *Lotus Esprit Turbo Challenge* (1990) and the role-playing game *Legend* (1991) could have come from anywhere.

But if British games shed some of their cultural distinctiveness around this time, the country's developers continued to push the technological boundaries, particularly when it came to 3-D visuals. The late 1980s and early 1990s saw a rush of British games based around polygonal 3-D visuals years before they became the industry norm. Among them were games such as *Zarch* (1987), a 3-D twist on the coin-op *Defender* (1980), the real-time strategy game *Carrier Command* (1988), Argonaut Software's *Starglider 2* (1988), the open-world adventure *Midwinter* (1989), and the first-person shooter *Corporation* (1990).

Again, most of these games only made waves within the UK, but the nation's 3-D programming talent went on to play an important role in the wider, global adoption of polygonal 3-D graphics within games. Argonaut's expertise resulted in a deal with Nintendo that saw the British developer helping the Japanese console giant learn how to make 3-D games and designing the Super FX graphics accelerator chip that would power Super Nintendo games such as *Star Fox* (1993) (Anderson and Levene 2012).

Argonaut's experience reflected how the UK's industry had begun to expand beyond British shores as the country's domestic audience transitioned to the new 16-bit computers. Rare, the developer that evolved out of Ultimate Play the Game, was among the first to see the potential. In 1985, just a year after rising to the top of the British industry with *Knight Lore*, the company turned its back on its British audience in order to concentrate on the more profitable task of developing NES games for American and Japanese audiences. In 1994, Rare's reputation within the console market and its own 3-D graphics technology prompted Nintendo to buy 49% of the business and give it the opportunity to create the hugely popular *Donkey Kong Country* (1994).

Codemasters, a Warwickshire-based publisher that started out by releasing inexpensive games, also made a lucrative move into the United States when it created the Game Genie, a device that let NES owners enter codes that would allow them to cheat in games, following in the footsteps of British devices such as Datel's Action Replay, which let players hack games on their computers. Others began expanding overseas, including Virgin Games (which established a development studio in Irvine, California), and Ocean (which opened offices in the United States and France). This international expansion also saw more British games become successful overseas. Bullfrog Productions' pioneering god game *Populous* (1989) became a global hit and turned its designer Peter Molyneux into one of the world's best-known game developers, while DMA Design's slapstick *Lemmings* (1991), where players had to guide dozens of tiny creatures through levels filled with traps and dangers, achieved international sales of more than 15 million copies across all formats.

This opening up of the British industry was a two-way process. Just as more British games were reaching foreign markets, more foreign games were entering the UK. The late 1980s and early 1990s not only saw Japanese consoles growing in popularity, it also saw numerous American computer game publishers open their European headquarters in Britain, including Electronic Arts and MicroProse. The rise of the AMIGA, Atari ST, and IBM PC as the most popular gaming platforms also gave British players more access to disk-based titles such as those created by Cinemaware, LucasArts, and Sierra On-Line.

Globalization: The UK Games Industry from the Mid-1990s to the Present Day

This process of globalization accelerated as the 1990s continued. By the late 1990s, the game platform barriers that once kept the UK semi-isolated from foreign games (and kept British games within the country) were gone. The PC had come to dominate the computer games market and, like players in North America and Japan, British audiences had embraced the Sony PlayStation as their favorite game console. The UK's platform choices have remained largely in line with those of the US ever since.

Game publishing within the UK also consolidated significantly during this period. Leading publishers Core Design, Domark, and U.S. Gold fused together to create Eidos Interactive. Ocean Software and Gremlin Interactive were swallowed up by the French publisher Infogrames, and Psygnosis was bought by Sony to form the basis of Sony Computer Entertainment Europe. By the end of the first decade of the 2000s, most of the major UK publishers founded during the 1980s had disappeared.

While British publishers were becoming a thing of the past, however, British developers were continuing to reach a global audience with titles such as *Theme Park* (1994), *Tomb Raider* (1996), *Grand Theft Auto III* (2001), *Fable* (2004), *Moshi Monsters* (2008), and *Batman: Arkham Asylum* (2009), demonstrating the country's ongoing ability to continue producing hit games. However, while UK developers have remained competitive, the focus on the international audience has meant that most British-made games are now indistinguishable from the games being produced in the United States, Canada, France, and beyond. *LEGO Pirates of the Caribbean* (2011), *DmC: Devil May Cry* (2013), and *F1 2011* (2011) might all have originated in the UK, but there is little sign of their British links. Yet while the more overt allusions to Britishness are gone, British games still contain subtler hints about their nationality, such as Lara Croft's aristocratic English accent, the toilet humor of *Fable*, and the undercurrent of satire within the *Grand Theft Auto* games.

British games are also continuing to have a sizeable influence on video games globally, from *RuneScape*'s role in popularizing the free-to-play business model and *Grand Theft Auto III*'s groundbreaking realization of an open-world city to the pioneering motion control device EyeToy (2003) and the game-creating aspects of *LittleBigPlanet* (2008). *Tomb Raider*'s use of a female lead character and subsequent success can also be seen as an important moment in the medium's history, as it made the business case against the use of female lead characters weaker and, through Lara Croft's appeal to mainstream media, played an important role in improving the image of games as an acceptable adult pastime during the late 1990s.

Another important, if unintended, contribution made by the UK's games industry has been how *Grand Theft Auto: San Andreas* (2004) led to the United States Supreme Court decision that video games are covered by the first amendment of the US Constitution, which occurred after a legal battle prompted by a Californian assemblyman's attempt to criminalize the sale of mature-rated games to minors following the "Hot Coffee" scandal surrounding that game.[1]

Regarding game consumption, as mentioned earlier, coin-operated video games are now rare in the UK and most games are consumed on consoles, PCs, smartphones, and tablets. In terms of consumer tastes, the

British audience appears to be most closely aligned with France, despite the greater popularity of the Xbox 360 in Britain. In 2010, eight of the ten best-selling games in France also appeared in the UK's top ten. In comparison, only five of the UK's top titles made the top ten list in the United States and Germany (Mazel 2011). That said, the similarities are slightly skewed by the presence of *Madden NFL* games on the US top ten lists and *FIFA* games on the European top ten lists, which is more a reflection of national differences in sports culture than video game tastes.

Tax Breaks and Game Studies: The Health of the Present-Day UK Games Industry

While the UK still has a strong presence and influence on the international games industry, the country's importance has waned slightly in recent years. In recent years, some of the country's larger development studios, such as Bizarre Creations and Eurocom, have closed down, and the country has slipped from being the world's third-largest producer of games to fourth and possibly fifth as Canada and South Korea's industries have grown in strength.

The global economic downturn that began in 2007 and 2008 also resulted in a shrinking of the UK's industry, although the latest figures released by the country's game developer trade association TIGA (The Independent Game Developers Association) suggest that a recovery is underway, with the number of people employed in game development rising 4% during 2012 to bring the total number of employees to 9,224 and the number of studios from 329 to 448 (TIGA 2013). TIGA attributes part of this growth to the rise of smartphones and tablets plus the formation of new start-ups, most of which are focusing on smartphone/tablet apps or indie games, in the wake of larger studios closing down.

The UK's games industry cites a lack of tax breaks similar to those on offer to developers in Canada and elsewhere as a key reason why the country has been falling behind in recent years. British governments have been slow to introduce these incentives, and TIGA claims that this was responsible for a 7% decline in the size of Britain's games industry between 2008 and 2012 and why 41% of the jobs lost within the UK industry have been caused by development jobs moving overseas. The UK government finally introduced tax relief for video game companies in August 2014 offering developers the opportunity to claim back up to 25% of their production costs (Rose 2014). Although the impact of the tax relief, which the UK government says could provide game studios with support worth $58.2 million a year (Rose 2014), is yet to be seen, TIGA says that the government's decision to adopt the policy has already encouraged more investment in the British games industry with new studios being started by companies such as Activision Blizzard, GREE, and Microsoft during 2012.

The country has also seen significant improvements in the quality of game courses during recent years, although evidence of dissatisfaction among employers still exists, with 44% of games industry employers in the UK indicating in one survey that they did not value formal qualifications (Skillset 2011). According to the website of the Universities and Colleges Admissions Service (UCAS), the agency that oversees university course applications within the UK, there are around 270 higher education courses on offer in the UK from 97 institutions. University courses are also doing more to build links with the games industry. Bournemouth University, for example, has appointed an advisory board made up of industry representatives to help inform the content of its courses (Edge 2013). However, Skillset, a British government-industry agency that seeks to improve skills within the creative sectors, has only accredited eighteen games courses in the UK, in terms of their meeting the needs of the country's games industry to date (Skillset 2013).

In addition to university courses, there is also Train2Game, a program founded in 2008 and endorsed by TIGA, which seeks to provide a fast-track route into the industry for the best graduates from computer and video game degree courses. There is also a renewed push by the government for schools to teach children computer science and, complementary to that, the Raspberry Pi, a computer designed by the Raspberry Pi Foundation, which helps schools teach children how to program. As yet, however, it is too early to know whether these initiatives and the introduction of tax breaks will have much impact on the future of the UK's games industry.

Note

1. "Hot Coffee" was a modification or "mod" created by a hacker that gave players access to an otherwise inaccessible sex game within *Grand Theft Auto: San Andreas*. The sex game had been abandoned during the game's development but remained hidden in the computer code. When the Hot Coffee mod was made available, the presence of the sex game provoked political outcry even though it was never intended to be accessible to players of the game (Donovan 2010).

References

Anderson, Magnus, and Rebecca Levene. 2012. *Grand Thieves and Tomb Raiders: How British Video Games Conquered the World*. London: Arum Entertainment.

Baer, Ralph H. 2005. *Videogames in the Beginning: The Invention of Home Videogames*. Springfield, NJ: Rolenta Press.

Candy, Robin, and Roger Kean. 1984. Deus ex machina. *Crash* 10 (November): 53.

ComScore Data Mine. 2013. Smartphones reach majority in all EU5 countries. *ComScore Data Mine*, March 15. http://www.comscoredatamine.com/2013/03/smartphones-reach-majority-in-all-eu5-countries/.

Donovan, Tristan. 2010. *Replay: The History of Video Games*. Lewes: Yellow Ant Media.

Edge 2012. The making of: Jet Set Willy. *Edge*, August 5. http://www.edge-online.com/features/making-jet-set-willy.

Edge. 2013. University Profile: Bournemouth University. *Edge* 255 (July): 188.

Goldberg, Marty, and Curt Vendel. 2012. *Atari Inc.: Business is Fun*. Carmel, NY: Syzygy Computer Press.

Jackson, Mark. 2013. EU scoreboard shows UK superfast broadband penetration improving. *ISPreview*, June 12. http://www.ispreview.co.uk/index.php/2013/06/eu-scoreboard-shows-uk-superfast-broadband-penetration-improving.html.

Mazel, Jacob. 2011. The top 20 games of 2010 in the USA, UK, France, Germany, and Spain. *VGChartz*, January 28. http://www.vgchartz.com/article/83788/the-top-20-games-of-2010-in-the-usa-uk-france-germany-amp-spain.

Old-Computers.com. n.d. Videomaster Star Chess. http://www.old-computers.com/museum/computer.asp?st=2&c=1116.

Padilla, Ana, and Hannah Swift. 2012. Impacts of video games. Houses of Parliament Office of Science and Technology, March. http://www.parliament.uk/Templates/BriefingPapers/Pages/BPPdfDownload.aspx?bp-id=post-pn-405.

Rose, Mike. 2014. UK video game tax breaks now finally available. http://www.gamasutra.com/view/news/223698/UK_video_game_tax_breaks_are_now_finally_available.php.

Schmitt, John, and Jonathan Wadsworth. 2002. Give PC's a chance: Personal computer ownership and the digital divide in the United States and Great Britain. Centre for Economic Performance, London School of Economics and Political Science, April. http://cep.lse.ac.uk/pubs/download/dp0526.pdf.

Silica Shop [advertisement]. 1981. *Computer & Video Games* 1 (Nov.): 7.

Skillset. 2011. Computer Games Sector—Labour Market Intelligence Digest. http://www.creativeskillset.org/uploads/pdf/asset_16891.pdf?4.

Skillset. 2013. Computer games industry accredited courses. http://www.creativeskillset.org/games/accreditation.

Spitzley, Kai. n.d. Greetings, traveler. *The Last Ninja Archives*. http://lastninja.lemon64.com/intro.php.

Stewart, Graham. 2013. *Bang! A History of Britain in the 1980s*. London: Atlantic Books.

Swaine, Jon. 2009. UK recession: Life in Britain in 1980. *Telegraph*, January 23. http://www.telegraph.co.uk/finance/recession/4323171/UK-recession-Life-in-Britain-in-1980.html.

Tandy [advertisement]. 1979. *Personal Computer World*, August, 71.

TIGA. 2013. UK games industry returns to growth. Press release. March 18. http://www.tiga.org/news/press-releases/uk-games-industry-returns-to-growth.

UKIE. 2011. UKIE Fact Sheet Q1 2011. http://ukie.info/sites/default/files/documents/UKIE_Fact_Sheet.pdf.

Winter, David. n.d. VideoSport MK2. http://www.pong-story.com/vsmk2.htm.

Wolf, Mark J. P., ed. 2007. *The Video Game Explosion: A History from PONG to PlayStation and Beyond.* Westport, CT: Greenwood Press.

Woollaston, Victoria. 2013. The meteoric rise of social networking in the UK: Britons are the second most prolific Facebook and Twitter users in Europe with a fifth of over 65s now using these sites. *Mail Online*, June 13. http://www.dailymail.co.uk/sciencetech/article-2340893/Britons-second-prolific-Facebook-Twitter-users-EUROPE-fifth-aged-65.html.

UNITED STATES OF AMERICA

Mark J. P. Wolf

As the birthplace of video games and a major producer of them, the United States of America is the location of much of video game history, which is often covered in detail when the history of video games is recounted. Since this history is too extensive to cover adequately in a few thousand words and has already appeared a number of times (many book-length histories of video games center on the United States), this chapter will be different from the others in this collection in that it will focus specifically on the way US history shaped and influenced its video game industry, beginning with the precursors and situations that gave rise to video games and the various forms they took, and continuing through each decade, examining how industry and culture affected the developing medium. The chapter will also consider whether or not there is a national style or character to games produced in the United States, and it will look at the influence of the growing number of imports from other countries, since many large video game producing companies from other countries, such as Nintendo, SEGA, and Sony, also have divisions or subsidiaries headquartered in the United States.

Since its beginning, the large domestic audience of the United States has provided enough support for the existence and growth of a video game industry, which can be divided into arcade games, console-based games, computer games, handheld games, and online video games. Although video game production companies can be found across the country, they tend to be more numerous in larger cities such as Los Angeles, New York, and Chicago, as well as other areas such as Silicon Valley in California, and Austin, Texas, where game production has grown into a subculture. But while a large domestic audience may explain how video games quickly became a viable industry, other initial conditions must be examined to explain how video games came about in the first place.

A Convergence of Conditions

The 1880s through the early 1900s saw the rise of arcades as places of entertainment as a result of the introduction of coin-operated amusement devices, including strength testers, slot machines, card machines, racing games, and other "trade stimulators" as well as the coin-operated Mutoscopes and Kinetoscopes that paved the way for cinema. Like these machines, early arcade games were simple, visual, based on action, and inexpensive to play, making them similar to much early film, which appealed to a wide mass audience of limited means and education that made up a significant percentage of the US public around the turn of the century. As a nation of immigrants and many languages, visual media (including film, comics, television, and video games) have usually found widespread popularity that crosses language barriers. One of the most popular types of games to emerge from the arcade was pinball, which became electrified and grew in popularity from the 1930s to the 1950s. Other electromechanical games, housed in upright cabinets, would provide the format that would be adopted by arcade video games, and frequently they were shooting games or racing games, genres that would become mainstays for video game makers. So the arcade provided a ready venue for video games to enter once they could fit into arcade cabinets and were cheap enough to operate.

The actual invention of video games required another set of conditions to appear. From the late 1800s onward, the do-it-yourself mentality, an outlook that began with the immigrant homesteading pioneers, was increasingly applied to the rapidly spreading electrical technology. Inspired by entrepreneurs and inventors such as Thomas Edison, many boys and young men experimented with telegraph and radio technologies, resulting in the formation of technological elites and the culture of hacking, which involved playing with technology and finding new uses for it. The boom in technological expansion that came with the Second World War—particularly in the area of computer technology—along with other resources such as the G. I. Bill, which increased college enrollments, encouraged the development of university laboratories (such as those at MIT and the University of Utah). These labs became homes to mainframe computers and the hacker subculture, and it was in such places that hacking led to the first mainframe games, such as *Spacewar!* (1962), inspired as much by the technology as the US-Soviet Space Race and a renewed cultural interest in science fiction. While concurrent movements in the US art world such as interactive installation art, abstract art, electronic music, and experiments that combined art and technology may not have had a direct influence on the rise of video games, they at least helped contribute to an environment that helped validate the minimalistic appearance of early video games.

During the postwar period, television was making inroads into US homes, which were now moving away from the city into the suburbs, where residents had greater leisure time to enjoy family life and activities in the home, including watching television. As a cultural force that helped immigrants become US citizens, television quickly became a ubiquitous technology, leading inventor Ralph Baer to consider new uses for it. In 1948, Baer earned a degree in the new field of television engineering, and by the 1960s he was working at Sanders Associates, where he was a division manager supervising around 500 engineers, technicians, and support staff, giving him the means to pursue new ideas. As Baer described it,

> Along the way I wandered off the straight and narrow into interactive video areas that initially had nothing at all to do with the normal work going on in my division at Sanders. The invention of videogames in 1966, the concept of using TV sets for something other than watching network fare, came first; building the early feasibility models came next. The activity started as a skunk works operation but it didn't stay that way for long. (Baer 2005, 3)

Baer's work led to the production of several "Brown Box" prototypes, which would eventually lead to the production of the first home video game console system in 1972, the Magnavox Odyssey, which was made by Magnavox, a maker of television sets.

Throughout the 1960s, projects of the US Government Space Program and Department of Defense provided the motivation and funding for the integrated circuit industry to integrate ever-greater numbers of circuits on individual microchips, and the increased production also brought prices down low enough that by the 1970s, integrated circuits could start appearing in consumer products. Integrated circuits led to the miniaturization of electronic components, leading to pocket calculators, digital watches, and other new gadgets. Entrepreneur Nolan Bushnell, having seen a demonstration of the Odyssey and familiar with *Spacewar!*, used the newly available technology to develop a smaller, arcade-cabinet-sized version of *Spacewar!* that could be placed in arcades, resulting in *Computer Space* (1971), the first mass-produced video game (but not the first coin-operated one; another game based on *Spacewar!* named *Galaxy Game* [1971], built by Bill Pitts and Hugh Tuck, appeared a month earlier in the Stanford student union, but was too big to be mass produced). Bushnell and his partner Ted Dabney began the first company devoted to the production of video games, Atari.

With the 1972 appearances of both the Magnavox Odyssey and its arcade imitator, Atari's *PONG*, the first arcade video game hit, video game businesses were up and running, and the large generation of US baby boomers, the prime consumers of a burgeoning youth culture that hungered for new experiences, helped push video games from a novelty into a rapidly expanding industry and subculture.

The Video Game Industry Begins

The invention of large-scale integrated (LSI) circuits in the mid-1970s led to an outpouring of new electronic products, including home video game consoles and home computer technology. Arcade video games began displacing electromechanical arcade games, eventually becoming the dominant type of game at the arcade. The Odyssey and *PONG* inspired many copycat games and systems, as other companies began producing ball-and-paddle games for the home and arcade. Home computers appeared around this time as well, with Micro Instrumentation and Telemetry System's Altair 8800 selling as a mail-order kit for hobbyists beginning in 1975, and Steve Wozniak's handmade Apple I computers in 1976.

In 1976, General Instruments released the AY-3-8500 chip, which included all the components needed to make a ball-and-paddle game, encouraging their production. Coleco was the first company to order the chip, and a rush of other companies followed suit. The rush to cash in on the home video game craze made dozens of companies produce second-rate products to try to beat their opponents to market. Some systems, such as General Home Products' Wonder Wizard, did not even bother to have a copyright notice on their boxes. Even systems that have long since been forgotten by the public found success initially; National Semiconductor's Adversary home game system, for example, sold over 200,000 units in 1976 (Eimbinder and Eimbinder 1980, 55). Around seventy companies were among the competitors, and one of the most successful, Atari, was bought by Warner Communications for $28 million. Many of these systems made their way overseas, introducing video games to countries around the globe. The Magnavox Odyssey, for example, was exported to Argentina, Australia, Belgium, Canada, Egypt, Germany, Great Britain, Greece, Hong Kong, Israel, Italy, Japan, Mexico, Switzerland, and Venezuela (Wolf 2012b, 372). The Odyssey's international success inspired imitators and also encouraged other companies to join in the new market; for example, Nintendo, which had been around since 1889, entered the video game industry in 1974 by distributing the Odyssey in Japan.

The market flooded with these games, both in the United States and Europe, while their novelty faded, eventually leading the video game industry to the crash of 1977. Other sectors of the US technology market, like the pocket calculator industry, had seen similar crashes, hinting at the possibility of the home video game industry's crash. In both cases, there was no lack of consumer demand, which was great enough to lead to chip shortages; it was the large number of competitors coupled with severe price drops and the lack of profits that followed, especially for smaller companies that could not withstand huge losses, that was the cause of these crashes.

The industry recovered quickly, with the appearance of the second generation of home video game technology: programmable video game consoles that used cartridges, as opposed to the dedicated consoles with built-in games. The first such system appeared in 1976, from Fairchild Semiconductor, the Fairchild Channel F (originally named the Video Entertainment System), which sold for $169.95 and did fairly well during the 1976 Christmas season. In 1979, Zircon International bought the rights to the Channel F, and in all, twenty-six "videocarts" (cartridges) were released for the system.

The Channel F was quickly overshadowed by another cartridge-based system released the following year, the Atari Video Computer System (VCS), which was later renamed the Atari 2600. Not only were its name and products already well-known, Atari was the only company (and would remain the only one) to produce arcade video games, home console systems, and home computers, allowing it to port its popular titles from one platform to another. The game, however, that really helped to popularize the system was actually a port of a foreign import: Atari's home version of *Space Invaders* (1978), the Japanese arcade game that was so popular in Japan that the country suffered a shortage of the 100-yen coins needed to play the game and had to mint more of them (Kent 2001, 116). *Space Invaders* became the first game licensed to another company, and its success led to Atari licensing other hits such as Williams's *Defender* (1980) for its home system.

The popularity of *Space Invaders* in the United States is also important to consider because it can be seen as a turning point in US video game history; it was the first foreign import to find a mass audience in the United States, foreshadowing the dominance that Japanese companies such as Nintendo and Sony would have after the great North American Video Game Industry Crash of 1983. It is rather ironic, then, that the United States' powerhouse video game company of the pre-crash era, and the first company devoted solely to the production of video games, was given a Japanese name (taken from the game Go, "atari" means a situation in which pieces are in danger of being captured on the next turn; a foreshadowing of the taking over of the market when Nintendo eclipsed Atari in the mid-1980s).

Atari's success helped motivate the appearance of other new home console systems including the Coleco Telstar in 1978, the Mattel Intellivision in 1979, the ColecoVision in 1982, and Atari's own Atari 5200 in 1982. Like silent films in their earliest days, video games did not include credits, and designers at Atari saw little of the profits from their popular games since they worked on salary. The lack of recognition led one programmer, Warren Robinett, to include his name as an Easter egg in *Adventure* (1979), and four other Atari programmers, David Crane, Larry Kaplan, Ed Miller, and Bob Whitehead, even left Atari in 1979 to start their own company, Activision, which made cartridges for the Atari 2600, becoming the first independent developer and distributor of video games for home console systems. Activision actively publicized the names of their programmers and found success with their games. In the following years, many other companies made cartridges for the Atari 2600, including Imagic, Data East, Mystique (which was taken to court by Atari for producing X-rated games that Atari felt had tarnished its name), and other companies including Parker Brothers, 20th Century Fox, and even Quaker Oats.

Over the years, more than a thousand different cartridges would be produced for the Atari 2600, making it the leading home console system of the late 1970s and early 1980s. Home consoles, however, were not just competing with each other, but with the other sectors of the industry, including arcade games and the new market sectors of handheld games, home computer games, and online games.

New Technologies and Growing Competition

In 1976, a new type of toy appeared that would begin an entire industry: the handheld electronic game. That year, Mattel released *Auto Racing*, the next year it released *Football* (1977), and in 1978, other sports games such as *Baseball* and *Basketball* would appear. Other companies followed suit; Coleco had its own "Head-to-Head" series of electronic handheld sports games, and Milton Bradley's Microvision, the first cartridge-based handheld system, appeared in 1979. Soon foreign competition joined in; Nintendo entered the market in 1980 with its *Game & Watch* series, and Bandai and Mega Corp. each had their own series of pocket-sized handheld games.

Electronic handheld games also pushed the boundaries of what were popularly called "video games." The screens used by handheld games were either LED-based (for example, Mattel's *Football*) or LCD-based (such as Mega Corp.'s *Fireman, Fireman* [1980] and Bandai's solar-powered *Invaders of the Mummy's Tomb* [1982]). Most games using LCD-based screens had elements that simply turned on or off instead of pixel-based displays, though Milton Bradley's Microvision system did have a 16×16 pixel-based display, allowing it to produce low-resolution imagery. While they were technically not video games since they did not use video, the structure of the games, gameplay, and the activities they involved were similar to video games, and some video games and foreign franchises (such as *Pac-Man* and Nintendo's Mario) would be ported over to handheld games. Due to all these similarities, the games were often seen as part of the video game craze of the time.

The popular use of the term "video game" was also expanded through the use of another new technology that appeared in the arcade with the arrival of Cinematronics's *Space Wars* (1977): vector graphics. Unlike raster graphic monitors that scanned the entire screen for every frame of imagery, vector graphic monitors drew their images on screen line-by-line, drawing only what was needed and leaving the rest of the screen dark. The resulting images were sharper and faster to draw, but they were mostly limited to wireframe renderings. The technology was not video, but the games used cathode ray tubes and produced moving images. To the general public of the time, they didn't seem all that different from video.

Vector games were produced during the late 1970s and early 1980s by companies including Cinematronics, Vectorbeam, Atari, Exidy, and SEGA, and one vector-based home console system appeared in 1982, the GCE/Milton Bradley Vectrex. Notable vector games include *Speed Freak* (1978), *Warrior* (1979), *Asteroids* (1979), *BattleZone* (1980), *Tempest* (1981), and *Star Wars* (1983). Because their graphics were based on coordinates and lines, vector games were the first to present computationally true three-dimensional graphics to video games, beginning with the car crash scene in *Speed Freak* and perhaps most notably in the three-dimensional environment where *BattleZone* took place. Vector games would be produced until around the mid-1980s, eventually falling out of favor when three-dimensional filled-polygon graphics became the main method used for three-dimensional graphics by the end of the 1980s.

The most revolutionary new technology to appear in the mid- to late 1970s, however, was the home computer, which quickly became a platform for video games. Even though older technologies such as the pocket calculator, typewriter, rolodex, and card catalog were still around, US citizens loved gadgets and the new technologies that were appearing, and video games provided a reason to buy a home computer in many households. The Texas Instruments TI99/4a home computer even had a cartridge slot built into it, and home computers in general also allowed users to write their own games or type in the code for games that appeared in hobbyist magazines, carrying on the pioneer "do-it-yourself" (DIY) tradition. Atari joined in, producing its own line of home computers, including the Atari 400, Atari 800, and the Atari ST series. Home computers also helped video games gain respectability since they could help teach programming and the procedural, logical thinking that it required (even the Atari 2600 had a cartridge named *Basic Programming* [1979]).

The United States had one of the most advanced telephony infrastructures in the world, so with the addition of modems, home computers could use telephone lines and network with each other, resulting in new possibilities for gameplay. Online games had been around since networked games on mainframe computers, but these were mainly at universities and not available to the general public. In 1978, the first publicly available bulletin board system (BBS) came online in Chicago (around the same time that the first multiuser domain [MUD] came online in Essex, England). BBSs allowed users to post messages and trade files with other members of the online community, whereas MUDs allowed real-time interaction between users who were logged in simultaneously. Both BBSs and MUDs would eventually become venues for online gaming in the early 1980s, as user interaction was structured into role-playing games. Some early online games, such as *Sceptre of Goth* (1983), allowed as many as sixteen users to dial in at once and play. The first online console gaming came shortly after, with the release of Mattel's PlayCable service in 1981 and the CVC Gameline service in 1983. Mattel's PlayCable delivered games for its Intellivision system, while the CVC Gameline allowed Atari 2600 users to receive games online. The CVC Gameline Master Module plugged into the Atari 2600's cartridge port and had an internal 1200-baud modem through which games could be loaded into the unit's 8 KB of RAM, allowing games to be downloaded and played but not saved.

During the late 1970s and early 1980s, home video game console systems and home computers were also being exported to Asia, Europe, Australia, New Zealand, and South America, starting new industries, first for their importation and later for the indigenous production of consoles and computers (see the other chapters in this collection). By the end of the 1970s, video games had become a powerful force in popular culture and were discussed in the media. Stores sold out of them during the holiday season, and many companies rushed to join in the frenzy, producing third-party games, console systems, computer games, handheld games, and arcade games. New arcades opened across North America, and by 1982, arcade video games had reached their peak, with about 10,000 arcades in operation in the United States alone (Mehrabian and Wixen 1983, 72; Alexander 1983, 64). It looked like the craze would continue for some time.

The home market, however, was growing saturated with cheap imitations of successful games, and despite prices being slashed, consumer interest could only be sustained for so long. After a decade of constant innovations, consumers came to expect technological advances, waiting for the next generation of them, and video games had been around long enough that their novelty began to wane. Soon it was clear that another industry crash was imminent.

The Great North American Video Game Industry Crash and Recovery

The 1977 video game industry crash was certainly a warning of what could happen, and other indicators hinted that consumer interest was dropping and that the boom would not continue. In October of 1981, Atari held its Coin-Op $50,000 World Championship in Chicago, expecting around 10,000 players to attend, but only around 250 came (*Game Informer* 15 [2005]: 150). In the last quarter of 1982, arcade profits fell unexpectedly. The number of arcades had nearly doubled from 1980 to 1982, but in 1983, over 2,000 of them would close (Mehrabian and Wixen 1983, 72; Alexander 1983, 64). The market's oversaturated nature became especially apparent in 1983, when industry-wide profits amounted to only $2.9 billion, down about 35% from 1982 ("The trend is back to pinball machines" 1984, 37).

The crash worsened into 1983 and lasted more than two years. Companies that were testing the waters of the video game market withdrew, and even major players such as Mattel, which had once been the third largest video game maker, earning $250 million from its Intellivision system in 1981, left the market altogether in 1983. Even Atari was hit hard, losing over half a billion dollars in 1983, despite being the market's major player. Only one new home video game system appeared in 1984, Rick Dyer's Halcyon system, and it was a failure. The home video game market had sunk to a new low. Video games' novelty had worn off, the golden age of arcade games was over, and no one knew how long it would be until the industry bounced back, or if it even would.

Dyer's Halcyon system was based on laserdisc technology, which some hoped would pull the industry out of its slump. Laserdisc games had first appeared with Electro Sport's *Quarter Horse* (1981), a game that had one screen for game graphics and one for horse race video clips used by the game. Most games, however, used their video clips directly in their gameplay, either as a background the player flew over, as in Mylstar's *M.A.C.H. 3* (1983), or as animated segments representing a player's actions during the game, as in Cinematronics's *Dragon's Lair* (1983). While the games had better graphics as a result, the fact that the video clips could not be changed, only played in different sequences, greatly limited the interactivity that was possible. In *Dragon's Lair*, for example, there were only a few choices a player could make in any given situation, all of them animated scenes that would play once the player had made a choice. Laserdisc games typically cost twice as much as standard arcade games, charging players 50 cents per play. In the end, laserdisc games failed to catch on due to the lack of interesting gameplay and higher expenses for both players and arcade operators.

The crash finally ended in 1985 when a foreign import became a hit in the North American market, assuring US game makers that home consoles could still be popular. The system that brought the industry out of the crash had been selling in Japan since 1983 as the Nintendo Famicom, which was renamed the Nintendo Entertainment System (NES) for its North American release. The NES introduced the third generation of home video game consoles and was an 8-bit machine, making it the most advanced home system yet to appear, and Nintendo allayed retailers' hesitance to stock another video game console by promising to buy back unsold systems. The system was a success, the NES soon had a large library of games, and Nintendo was more careful than Atari had been when it came to third-party licensing. Hundreds of cartridges were released for the NES, and Nintendo dominated the US video game market even after the release of other third-generation systems such as the SEGA Master System in 1985 and the Atari 7800 Prosystem in 1986.

Video games also gained strength through association with other media and transmedial franchises. The rise of blockbuster cinema during this time, along with the development of the franchising and merchandising of popular film and television shows, led to video games that were adaptations of content from other media, and in the 1980s, video games would themselves become the source of character-based franchises (such as those surrounding Pac-Man, Mario, and Zelda) which would spread to other media.

The Growing Home Computer Market

Although the video game industry had been saved by a foreign console system (the NES), the United States would continue to lead in the area of home computer gaming thanks to domination of IBM, Microsoft, and Apple in the computer industry, though these, too, would find growing international competition. While the arcade and home console markets had been undergoing consumer doubt and crises, home computers were steadily growing stronger as a market sector, with the release of such machines as the Apple II, Commodore

PET, and Radio Shack TRS-80 in 1977; the Atari 400, Atari 800, and Texas Instruments TI99/4a in 1979; the Tandy Color Computer in 1980; the IBM PC, Commodore VIC-20, and the Xerox Star in 1981; the Commodore 64 in 1982; the Apple IIe, Apple Lisa, and Coleco ADAM in 1983; the Apple Macintosh and Tandy 1000 in 1984; and the Commodore AMIGA in 1985. All these computers could play games, and some, such as the Texas Instruments TI99/4a and Commodore VIC-20, even had a slot for game cartridges.

Home computers, in addition to their ability to offer software other than just games, had some advantages when it came to games: better storage capacity because of floppy disk drives, the ability to save games (although the Nintendo cartridge for *The Legend of Zelda* [1986] had a battery in it that allowed games to be saved), online capabilities (thanks to modems) allowing access to a growing number of BBSs and MUDs, and the potential for hobbyists to be able to write and exchange their own homebrew games (some programmers, for example Richard Garriott, creator of the *Ultima* series, went on to sell thousands of games and start their own companies). These advantages encouraged certain genres, such as adventure games, which had longer stories and play times, relying more on thoughtful consideration than fast action, and sports games and simulations that required more complicated interfaces and more detailed graphics and gameplay. After the Apple Macintosh brought the graphical user interface (GUI) to a wide audience, and Microsoft changed over from DOS to Windows, home computers were better equipped to offer games with good graphics and would try to compete with console-based games in all genres.

Computer memory was another area of advancement during the 1980s. Just as the cartridges for home consoles were approaching the storage capacity of floppy disks, the home computer industry responded with new media that could hold even more data, for example the 3.5-inch diskette, and in the early 1980s, two foreign companies, Philips and Sony, developed the CD-ROM, an optical disc that could store 650 MB of data. Cyan's *The Manhole* (1987) is generally considered to be the first game released on CD-ROM, leading the way for CD-ROM-based games that could contain even larger game worlds due to the greater memory capacity, such as Cyan's own *Myst* (1993), a best-selling game that encouraged the sales of computers with CD-ROM drives. More memory also allowed room for full-motion video (FMV) to be incorporated into games, essentially picking up where laserdisc games had left off.

US video game companies, however, were slow to incorporate CD-ROM drives into their game systems. Japanese companies did so much sooner; a CD-ROM add-on peripheral was released for the NEC PC-Engine/Turbogrfx-16 in 1989, and for the SEGA Mega Drive/SEGA Genesis in 1992; and in 1991, Fujitsu's FM Towns Marty (released in Japan), became the first 32-bit system and the first console system with a built-in CD-ROM drive. Finally, in 1993, a US system, Trip Hawkins's 3DO Interactive Multiplayer, came with a CD-ROM drive, but the console was a commercial failure and ceased production only three years later. Consequent systems, however, would use CD-ROMs as they became the industry standard, and by 1996, only the Nintendo 64 was still cartridge-based.

During the late 1980s and into the 1990s, arcade games were struggling to stay ahead of home game systems. Before the great industry crash, arcade video games were the cutting edge, and arcades were where the best new games were found; they had faster processing speeds, better graphics and sound, and were often the place where new advances in video game technology debuted. Although computationally true three-dimensional graphics had first appeared in mainframe games, *Speed Freak* (1978) was the first game to bring them to the public in the arcade, and *BattleZone* (1980) was the first publicly available game with a three-dimensional environment. Both were vector games with wireframe graphics, but another arcade game, *I, Robot* (1983), became the first arcade game to feature filled-polygon three-dimensional graphics. The demands of such graphics resulted in simpler game worlds than what two-dimensional graphics could offer using hundreds of sprites (as in games such as *Pole Position* [1982] and *Space Harrier* [1985]), and so three-dimensional graphics remained on the sideline a few years until the late 1980s, when computing power and speed improved.

The Rise of Three-Dimensional Graphics

By the early 1990s, advances in computer technology led to, and were encouraged by, the use of computer graphics in all areas of visual media. Computer graphics had been used in Hollywood feature films in the 1970s and 1980s, and television commercials also provided the high budgets and hunger for novelty that computer graphics needed in order to advance. Film and television made computer graphics companies possible, and those companies' developments would find their way into video games. Video games, however, would always lag behind cutting-edge computer graphics due to games' need for graphics that were rendered interactively in real-time (some games used pre-rendered imagery but were less interactive as a result). But even with lower polygonal resolution, flat lighting, and limited textures, three-dimensionally rendered computer-generated imagery (CGI) brought video games closer to other media such as film and television, which were coming to rely on it more as well.

At the end of the 1980s, three-dimensional filled-polygon graphics returned to the arcade in such games as *Hard Drivin'* (1989) and *S.T.U.N. Runner* (1989), this time with more representational imagery, and during the 1990s they would become the new standard in arcade games. Many genres had switched over from two-dimensional graphics to three-dimensional graphics, especially racing games, fighting games, sports games, and shooting games. But arcade games' leading edge was diminishing, and home computers and home console systems were not far behind them. During the 1990s, home games began to feature three-dimensional graphics as well, both as pre-rendered graphics (such as those of *Myst* [1993] and *Riven* [1997]) and those rendered in real time (such as those of *Tomb Raider* [1996]).

Arcade video games fought back by introducing new features that would be more difficult for home games to offer. The number of three-player and four-player arcade games increased dramatically, and there were even some six-player games such as Atari Games' *T-Mek* (1994). Many arcade games also featured innovative interfaces and cabinet designs, which players could sit on, ride in, or balance on (for instance, games with skis, skateboards, or snowboards). Two games using virtual reality equipment even appeared, Virtuality's *Dactyl Nightmare* (1992) and its sequel, *Dactyl Nightmare 2: Race for the Eggs!* (1994). Sports games remained a popular genre in the United States and featured novel interfaces with fishing rods, soccer balls, bowling balls, pool cues, boxing gloves, and more. They also found competition from a new genre of games imported from Japan: rhythm-and-dance games. These games introduced new interface devices such as the dance pads for Konami's *Dance Dance Revolution* series of games, full-sized guitars in their *GuitarFreaks* series, and sets of miniature drumheads, short piano keyboards, and DJ turntables for their *Beatmania* series. Eventually, the popularity of these games would lead to similar peripherals for home games, bringing them even closer to arcade games.

Although some games featured technological innovations, others were part of a cautious retreat to more tried-and-true genres, including vertically oriented shooting games, fighting games, and driving games. The number of games produced in these genres expanded, while other genres shrank, and fewer innovative game designs were produced. Following the US film industry's example, more sequels and series games were made, relying on their predecessors' successes for instant recognition and acceptance. Few, if any, arcade games from the 1990s became household names among nonplayers. Perhaps only Midway's *Mortal Kombat* (1992), a foreign game, was widely known among the general public, and infamously at that, due to the controversy surrounding the game's extreme violence. Soon, even the arcade itself would be redefined as a game center or cybercafé, as operators tried to lure players back. By the end of the 1990s, it was clear that the home was now the main venue for video gaming.

The Domination of Home Games and Foreign Systems

Home game consoles grew more powerful with each new generation of technology, and the fourth generation (1987–1999), which introduced CD-ROM technology, was dominated by Japanese companies such as Nintendo, SEGA, SNK, and NEC. The late 1980s and 1990s were a difficult time for US console manufacturers. The RDI Halcyon, the home laserdisc game system from 1984, was too expensive and had too few games to gain a user base. The 3DO Interactive Multiplayer (1993), a CD-ROM-based system, was also expensive and unsuccessful. Atari was the only US company that actively released more consoles, such as the Atari 7800 Prosystem (1986), the Atari XE (1987), the Atari Lynx (1989), and the Atari Jaguar (1993), but none of them found the success that the Atari 2600 had enjoyed. Atari's systems were also all cartridge-based, although a $199.99-CD-ROM peripheral became available in 1995—a case of too little, too late. After the failure of the Lynx and the Jaguar, Atari was sold to JTS Inc. in 1996, was acquired by Hasbro in 1998, and eventually sold to the European company Infogrames in 2008, which would change its name to Atari in 2009. By the mid-1990s, with the North American release of the Sony PlayStation in 1995, the Nintendo 64 in 1996, and the SEGA Dreamcast in 1999, US console manufacturers had fallen far behind their foreign competitors.

The home computer market, however, was still booming in the 1990s, and computer games continued to provide some of the impetus behind home computer purchases. As IBM and Apple came to standardize and dominate the US industry, their user bases came to constitute even more lucrative audiences for computer games, which had now become the main competition for home console systems. Online gaming continued, and when the World Wide Web went worldwide in 1993, making the Internet easier to access, online gaming became easier as well. Games were able to accommodate larger and larger numbers of players, eventually becoming known as massively multiplayer online role-playing games (MMORPGs), with hundreds of thousands of players around the world. The first game to be considered an MMORPG was 3DO's *Meridian 59* (1995), but by the end of the 1990s, the three main MMORPGs with the most players were Origin Systems' *Ultima Online* (1997), 989 Studio's *EverQuest* (1999), and Turbine Entertainment Software's *Asheron's Call* (1999). Online gaming gave home computers another advantage over console systems, until modems were included in the next generation of systems, starting with the SEGA Dreamcast, released in Japan in 1998 and in North America in 1999.

Since 1993, Web-based games have grown in number and popularity, both on personal computers as well as other devices used for mobile gaming such as cell phones, iPods, and iPads. The ability to play games online has also led to the production of advergames used to advertise products in a way that engages the user, and games designed for use with social media, such as *FarmVille* (2009) and *Mafia Wars* (2009). With the growth of cloud computing, an increasing number of games can be leased online instead of purchased. As digital distribution and downloadable content (DLC) and the bandwidth of home Internet connections continues to expand, online games may one day begin to supplant home console and home computer gaming, becoming the dominant form of gaming, as has already happened in other countries where consoles never had the prominence they did in the United States.

By the mid-1990s, the members of "Generation X," who had grown up with video games, were beginning to have children of their own and were also becoming nostalgic for the early games of their own youth. As home computers became more powerful, emulators for earlier game consoles began to appear, leading to the retrogaming movement and renewed interest in early games. Communities formed on the Web to collect and exchange old game information, and homebrewers even wrote games for older systems such as the Atari 2600 and the Vectrex. Over the next couple decades, not only would older games be re-released in new forms for newer systems, but the old games would prove to be good fodder for the tiny screens of cell phones and other mobile gaming devices, which had neither the power nor the screen resolution to compete with contemporary home console games and home computer games.

Handheld Games and Mobile Games

As mentioned above, rapidly growing technologies such as cell phones and handheld computing devices opened up new venues for video games. Since the post-World War II mass exodus to the suburbs and the rise of the interstate expressway system, US citizens spent more time commuting and on the move, and both handheld games and mobile devices were ideal for such users. Earlier handheld games and game systems such as the Microvision and the Atari Lynx were limited in their abilities and the number of games available for them, and their small user base made the games produced for them less profitable to make. Cell phones and mobile devices such as Apple's iPod (introduced in 2001) and iPad (introduced in 2010), however, had other uses, allowing a much larger user base to form, and since these devices also had wireless access to the Internet, games could be bought and downloaded directly as they appeared, making the distribution and marketing of the games much easier and more successful. Games played on mobile devices could be played whenever players had a few minutes of spare time, waiting for mass transit, or during rides, between meetings, and so forth. Consequently, casual games, which are typically easy to learn and take only a few minutes to play, became popular, due in part to the rise of mobile gaming. As mobile gaming platforms are also smaller and less complicated than consoles, game production became more affordable for independent developers, leading to a greater diversity of games, including more experimental ones that became commercially feasible due to lower development costs and the larger potential user base.

Reentering the Console Race, and Other Arenas of Competition

The year 2001 finally saw a US company producing a state-of-the-art console system again. Microsoft's Xbox was released that year, joining the PlayStation 2 (released in 2000) and the Nintendo GameCube (released in 2001) in the sixth generation of console system technology. The Xbox was the first console to have a built-in hard disk drive, eliminating the need for memory cards. The system was also known for its Xbox Live online gaming service, and the *Halo* series of games, which was introduced with the system's launch. The Microsoft Xbox sold 24,750,000 units and was finally discontinued in 2006 (see http://www.statisticbrain.com/xbox-statistics).

Only four years later, Microsoft released the Xbox 360, on November 22, 2005. The Xbox 360 had the jump on the other two seventh-generation consoles, the PlayStation 3 and the Nintendo Wii, both of which were released in November of 2006. All three systems had online capabilities and services, could play other media such as CDs and DVDs (and in the case of the PS3, Blu-ray as well). Due to the popularity of the motion-sensing Wiimote controller, Microsoft produced its own motion-sensing peripheral, the Kinect, which was released in November 2010 for the Xbox 360 and in February 2012 for Windows. During November 2013, the next three major systems debuted: on the 15th, Sony released the PlayStation 4; on the 18th, Nintendo released the Wii U; and on the 22nd, Microsoft released the Xbox One. Together, these three systems make up the eighth generation of home video game consoles.

Another area where US video game companies have been active is online games, including MMORPGs such as Blizzard Entertainment's *World of Warcraft* (2004), which holds the record for most subscribers, at around 7 million (as of August 2014), although in 2008 the game had over 10 million subscribers. Other U.S. MMORPGS, such as the abovementioned *Meridian 59* (1995), *Ultima Online* (1997), *EverQuest* (1999), and *Asheron's Call* (1999), also helped MMORPGs to develop technologically and attract players. Other companies developed free-to-play social media games for websites such as Facebook; for example, San Francisco–based Zynga's *Mafia Wars* (2008), *FarmVille* (2009), *Café World* (2009), *CityVille* (2010), *Mafia Wars 2* (2011), *FarmVille 2* (2012), and *CityVille 2* (2012).

Despite being ranked the country with the third-highest number of mobile phones in use (with more mobile phones [327.6 million as of April 2014] than citizens [317.9 million as of April 2014]),[1] only a few of the best-selling mobile games are from companies headquartered in the United States, including Lima Sky's *Doodle Jump* (2009) with 5 million downloads, Bolt Creative's *Pocket God* (2009) with 2 million downloads, and Freeverse Inc.'s *Flick Fishing* (2009) with 1.3 million downloads.[2]

Video games have also found a wide variety of applications beyond entertainment. Particularly during the 2000s, new uses for video games and game-like programs have appeared in higher education, where their interactive nature requires students' participation, and even in physical education, where games such as *Dance Dance Revolution* (1998) have been used for exercise. Educators have been finding new uses for games for years, and much literature has been published regarding the use of video games in education. Businesses use video games for job training, simulations, and even customer service.[3] Artists such as Jason Rohrer and Jenova Chen have used video games for artistic expression, creating games that embody ideas and lead to reflection and contemplation (see, for example, Rohrer's *Passage* [2007] and *Gravitation* [2008], and Chen's *flOw* [2006] and *Flower* [2009]). The wide variety of interactive experiences now available through game-like technologies and interfaces has expanded the range and blurred the boundaries of what we consider to be "video games," as on-screen interactivity becomes more common in everyday life.

Video Games as Objects of Study and Preservation

In the United States, video games were an object of study for hobbyists and journalists from the 1970s onward, discussed in such venues as *Popular Electronics*, which featured a multipart essay by Jerry and Eric Eimbinder on the history of video games, that appeared beginning in October of 1980 (Eimbinder and Eimbinder 1980). In the 1980s, entire magazines devoted to games appeared, such as *Electronic Games* (1981–1985), *Computer Gaming World* (1981–2006), *Atari Age* (1982–1984), *Commodore Power/Play* (1982–1985), *Amiga World* (1985–1995), *GamePro* (1989–2011), and others. During the 1980s, psychologists such as Patricia Marks Greenfield also studied video games and their effects on youth (Greenfield 1984), and in 1985, the first doctoral dissertation on video games, *Interactive Fiction: The Computer Storygame "Adventure,"* was written by Mary Ann Buckles at the University of California, San Diego. Alongside the retrogaming movement, video games were gradually acknowledged as objects worthy of preservation, with *Hot Circuits: A Video Arcade*, the first museum exhibit devoted to video games appearing from June 1989 to May 1990 at the American Museum of the Moving Image in Astoria, New York.

Writings about video games increased during the 1980s and 1990s, with books such as Chris Crawford's *The Art of Computer Game Design* (1984) and Leonard Herman's *Phoenix: The Fall and Rise of Videogames* (1994). After the turn of the millennium, video game studies became more prominent in academia in the form of a growing number of books, journals (both print and online), conferences, museum exhibits, websites, archives, and university classes and degree programs.

The study and discussion of older games is also popular in many online communities of gaming hobbyists and video game collectors who together have built extensive online archives of game-related materials, for example, the Killer List of Videogames, the Arcade Flyer Archive, and the International Arcade Museum. The retrogaming movement has also increased interest in early games, and a growing number of archives and organizations, such as the Video Arcade Preservation Society, are attempting to document and preserve as much early video game history as possible. At the same time, physical archives are collecting and preserving examples of arcade games, home games, handheld games, and gaming peripherals as well as a wide variety of the paraphernalia associated with gaming. Ironically, as television and monitor technology continue the changeover to LCD screens and flat-panel displays, "video games," by a strict, technical definition that requires the use of a cathode ray tube, will soon be a thing of the past. At the same time, the popular usage of the term has become the broadest and most inclusive it ever has been, as the ever-expanding world of video games leaves its old technological roots behind.

The Character of US Video Games

While the accumulated influences over five decades would suggest that video games made in the United States might retain some cultural resemblance stemming from the country's cultural values, two of those values, invention and innovation, have expanded the variety of those games to the point where only a few broad generalizations can be suggested. First, the highly competitive nature of US society makes for fertile ground for competitive gaming, and it is no surprise, then, that the majority of video games are competitive ones. Although it is easy to take the competitive nature of most video games for granted, this need not have been the case, since other styles of gameplay, such as cooperative play, sandbox-style play, or games in which exploration, navigation, or puzzle-solving are the sole objectives, could have become the dominant style, but did not. Even single-player arcade games are capable of becoming competitive through the use of high-score tables, which allow all players of a single game to compete and compare their performances, and for players to compete with their own performances.

Competition is often involved with the acquisition of goods, property, and position (which at various levels in US culture has resulted in consumerism, materialism, and imperialism), and equivalent activities in video games include the collecting of objects or points and the attaining of high scores or higher levels within a game. The fast pace of these games reflects, and contributes to, the pace of life in the United States, which has for the most part accelerated since the 1970s, as technologies such as the computer, microwave oven, cell phone, Internet, and so forth increase the convenience and speed of many of the tasks found in daily life.

Other oft-noted aspects of US culture, such as its relative extroversion, inventiveness, pioneer spirit, and appetite for novelty, are also apparent in many of the games developed in the United States, as well as in the way US game companies are run. Besides competition and acquisition, other themes such as cross-cultural conflict, cultural assimilation, and the importance of personal identity (found in such games as the *Grand*

Theft Auto series, the cities of which are modeled after US ones in their depiction and action), are explored in many games. Some games, such as *Red Dead Redemption* (2010) and *BioShock Infinite* (2012), have designs that are highly influenced by US history, even when their settings are entirely fictional.

As time goes on, however, most of the largest video game companies, US and otherwise, now have offices and branches on multiple continents, and their outlook is more global in scope. Cultural iconography and themes are borrowed from one country by another and mixed together, resulting in a range of styles and content that can leave one guessing as to a game's culture of origin. Still, as the video game's country of origin, the United States of America has influenced the medium of video games and has even encouraged its development as an eclectic and global phenomenon.

Notes

1. See http://en.wikipedia.org/wiki/List_of_countries_by_number_of_mobile_phones_in_use.

2. According to the figures given in the "Paid Downloads" section of "List of Best-selling Video Games," available at http://en.wikipedia.org/wiki/List_of_best-selling_video_games#Paid_downloads.

3. See Jana 2006; Paynter 2013, 35.

References

Alexander, Charles P. 1983. Video games go crunch! *Time*, October 17, 64.

Baer, Ralph H. 2005. *Videogames: In the Beginning*. Springfield, NJ: Rolenta Press.

Eimbinder, Jerry, and Eric Eimbinder. 1980. Electronic games: Space-age leisure activity. *Popular Electronics*, October, 53–59.

Game Informer. 2005. *Game Informer* 15 (145, May).

Greenfield, Patricia Marks. 1984. *Mind and Media: The Effects of Television, Video Games, and Computers*. Cambridge, MA: Harvard University Press.

Herman, Leonard. 1994. *Phoenix: The Fall and Rise of Videogames*. Springfield, NJ: Rolenta Press.

Jana, Reena. 2006. On-the-job video gaming: Interactive training tools are captivating employees and saving companies money. *Bloomberg Businessweek*, March 27 http://www.businessweek.com/magazine/content/06_13/b3977062.htm.

Kent, Steven L. 2001. *The Ultimate History of Video Games: The Story Behind the Craze that Touched Our Lives and Changed the World*. New York: Three Rivers Press.

Mehrabian, Albert, and Warren Wixen. 1983. Lights out at the arcade. *Psychology Today* 72 (December).

Paynter, Ben. 2013. Level up: Play this game, get that job. *Wired* 21 (7).

The trend is back to pinball machines. 1984. *Business Week*, May 7, 37.

Wolf, Mark J. P., ed. 2007. *The Video Game Explosion: A History from PONG to PlayStation and Beyond.* Westport, CT: Greenwood Press.

Wolf, Mark J. P., ed. 2012a. *Before the Crash: Early Video Game History.* Detroit, MI: Wayne State University Press.

Wolf, Mark J. P., ed. 2012b. *Encyclopedia of Video Games: The Culture, Technology, and Art of Gaming.* Westport, CT: Greenwood/ABC-CLIO Press.

URUGUAY

Gonzalo Frasca

Uruguay is a small country squeezed between two giants, Argentina and Brazil. Unlike its neighbors, Uruguay never produced consoles or computers but benefited from their hardware endeavors. Circa 1975, Argentina created the Telematch de Panoramic, a hacked Magnavox Odyssey clone that removed many of the superfluous original games. It even added a brand new game that was almost a cultural requirement of the region—soccer—and incorporated extra buttons to control the goalies. Little is known about this machine, but it is very likely that a few made it to Uruguay. Brazilian clones had a bigger impact in the mid-1980s, when the TK90X—a Sinclair ZX Spectrum clone—made the 8-bit computer the most popular game machine in Uruguay.

With only 3 million people, Uruguay was never an attractive market for hardware sales. This led to the importation of broken machines that would be later refurbished in the country. Such is the case for thousands of malfunctioning Coleco ADAM computers that arrived in the country and were fixed or completed with local parts. The machines were very affordable, and this led to the popularity of the platform in Uruguay, even after it was discontinued in the United States. Most of the software and games that were imported lacked manuals and, since this was before the Internet, they had to be rewritten from scratch. My first game-related job ever was quite original: playing unlabeled games and writing down their instructions. In many cases, I even had to come up with backstories and christen the main characters with made-up names.

The first professional Uruguayan video game studio was Iron Byte, a seven-member team developing 8-bit and 16-bit games during the late 1980s and early 1990s. During its brief existence, the studio managed to launch a series of relevant games in the 8-bit and 16-bit European market, including *Freddy Hardest in South Manhattan* (1989) and most notably, *Narco Police* (1990). Iron Byte was based in Uruguay's capital city, Montevideo, and worked with Spanish publisher Dinamic Software, which was a major player in the European ZX Spectrum market but had trouble adapting to the 16-bit era. *Narco Police* was a highly polished game for its time, and the game received good reviews and sold well mainly because of its third-person perspective and 3-D maze levels. Unfortunately, its publisher went out of business in 1992, which led to the disbandment of the Uruguayan studio.

The advent of the World Wide Web in the mid-1990s helped a few Uruguayans to freelance in production jobs, but it was not until 2002 that the current local industry took off. It was in that year that Powerful Robot

Games—which I cofounded with Sofía Battegazzore—began creating Web-based games for American animation studios including the Cartoon Network, Disney, Pixar, Warner Bros., and Lucasfilm.

The Powerful Robot team was also behind several serious games. One of them was the Newsgaming.com project that included *September 12th: A Toy World*, a game that went viral after its launch in 2003. While it was highly controversial at its launch, this online critique of the US-led "War on Terror" became a popular example of both political and news-based video games. The game offered a top view of a Middle Eastern town and allowed the player to control a sniper rifle target. The main goal was to kill the terrorists without hitting the civilians. However, once the player clicked the mouse, she realized that the target did not fire a bullet but a missile. The large area of impact caused huge amounts of so-called collateral damage among the civilian population. Once civilians died, their relatives would turn into terrorists. After a few shots, the town would be thriving with terrorists, hence making the game impossible to win.

The following year, the Newsgaming team launched *Madrid* (2004), a whack-a-mole-style game that was crafted in less than forty-eight hours as a response to the Madrid train bombing attacks. The game featured a candlelight vigil where the player had to click on the candles to prevent them from fading. The player would win if all the candles were shining at their maximum strength. In 2009, the Knight Foundation awarded the Newsgaming team a Lifetime Achievement Award for its contribution to the genre of newsgames.

The Powerful Robot team also codesigned and developed *The Howard Dean for Iowa Game* (2003), the first commissioned video game in the history of US presidential elections. It was launched on Christmas Day, 2003, and generated a large buzz in the media and among the candidate's supporters (Bogost and Frasca 2007, 237). The game was codesigned and published by Ian Bogost's Persuasive Games studio. That game was followed by *Cambiemos* (2004), another official online presidential campaign game for the 2004 Uruguayan elections. The three-level puzzle game was commissioned by the Encuentro Progresista/Frente Amplio political party, which ended up winning the presidency.

In 2007, Uruguay began the "Plan Ceibal," the world's largest experiment in educational online connectivity. As of 2013, Uruguay is the only country that has fully implemented the One Laptop Per Child project, providing free laptops and connectivity to all grade-school and high-school students in the public education system, resulting in 75% of all children in the country having access to Wi-Fi within 300 meters of their homes. The project closed the digital gap and the gender gap in technology among local children and teenagers. However, based on a study commissioned by the authorities in 2013, it had no relevant academic impact in mathematics, reading, and writing.

The massive penetration of the XO laptop computer (the OLPC XO-1, the laptop used for the One Laptop Per Child project) led to the development of educational video games created by studios such as Batovi Games Studio (established in 2005; see http://www.batovi.com/) and Trojan Chicken (established 2008; see http://www.trojanchicken.com). Because the laptops are underpowered and relatively unfriendly as a developing platform, the number of games available for them remains limited. However, for these same reasons, the system's video games are extremely popular among Uruguayan and international XO users. The vast majority of XO games has been either commissioned by the state or created through commercial sponsorships.

In 2011, a Flash-based tower defense Web-based game was launched on the Armor Games portal. Its name was *Kingdom Rush*, and it was developed by a three-member studio named Ironhide Games (no relation to Iron Byte). It quickly became Uruguay's first international commercial hit. The game gathered a large fan base and became a hit when it was ported to the iPad in 2012. Both the original game and its sequel, *Kingdom Rush: Frontiers* (2013), generated rave reviews. The *Kingdom Rush* series is currently being developed and published by Ironhide Games on both iOS and Android platforms. The first game in the series received a score of 89 (out of 100) in Metacritic and was called "an instant classic" by IGN (Davis 2012a) and "one of the best tower defense games released ... ever" (Davis 2012b). Its sequel was number one in sales in the App Store in over thirty-five countries, including the United States.

Although Uruguay still lacks a local chapter of the IGDA, the community is quite active. In addition to the Global Game Jam, programming jams are often organized for creating games and software for the XO platform. Since 2006, the professional community has organized the annual Uruguayan National Videogame Creation Contest (http://www.ConcursoVideojuegos.com), which encourages amateurs to submit their games and to participate in conferences and workshops.

Figure 1

The *Kingdom Rush* series for iOS and Android became the first Uruguayan international commercial hit.

In 2010, Uruguay became one of the first countries in South America to offer an undergraduate university degree in video game design. The program is offered by the Universidad ORT Uruguay (disclosure: I am currently employed there) and will release its first class of graduates in 2014.

In spite of its small size, Uruguay has historically been one of the largest developers of nonentertainment software in its region, in part because of its good educational system and computer science schools. Another Uruguayan asset—at least when compared to countries in its region—is that its tiny local market forces start-ups to aim at the international market from the very beginning.

References

Bogost, Ian, and Gonzalo Frasca. 2007. Videogames go to Washington: The story behind the Howard Dean for Iowa Game. In *Second Person: Role-Playing and Story in Games and Playable Media*, ed. Noah Wardrip-Fruin and Pat Harrigan, 233–246. Cambridge, MA: MIT Press.

Davis, Justin. 2012a. Kingdom Rush review. *IGN.com*, January 31. http://www.ign.com/articles/2012/01/31/kingdom-rush-review.

Davis, Justin. 2012b. Today's free game: Kingdom Rush. *IGN.com*, February 7. http://www.ign.com/articles/2012/02/07/todays-free-game-kingdom-rush.

Metacritic. n.d. Kingdom Rush iOS. *Metacritic*. http://www.metacritic.com/game/ios/kingdom-rush.

VENEZUELA

Thomas H. Apperley

There is also exclusion of people and territories which, from the perspective of dominant interests in global, informational capitalism, shift to a position of structural irrelevance.

—Manuel Castells (1998, 162)

The central role of information in the global networked society creates and entrenches regions of inclusion and exclusion, thus establishing classes of locations and people that are not valued by, or fully connected to, global networked society. Two worlds are emerging: "one is information as well as economically rich, the other is information and economically poor" (Yar 2008, 617). However, new forms of communication that are emerging on the same networks also offer feasible strategies for economic and social inclusion (Martin-Barbero 2011, 47). In Venezuela, digital games are one of these technologies that potentially both increase and mitigate forms of exclusion. Compared to other areas of the world, digital games in Venezuela have largely remained a form of public—or semi-public—entertainment. While popular, only the wealthier middle-class Venezuelans have been able to afford to purchase the latest, contemporary gaming consoles. Despite recent controversies that have led to somewhat draconian laws governing the publication and importation of digital games, they remain a common form of entertainment, and cybercafés are an everyday feature in the country's urban landscapes.

This essay examines the role that access to digital games has in facilitating social and economic inclusion by providing access to knowledge, literacies, and skills that are considered useful, even essential for inclusion in cultures and economies of the global networked society. Digital games' status as entertainment media often positions them as an irrelevant concern for social inclusion projects in the developing world, which are often more interested in more practical or directly didactic programs. It is important, however, to consider the undervalued and underacknowledged role that digital games have in building basic technological efficacy with computers, as well as supporting the development of key informal knowledge, literacies, and skills. They do this by providing opportunities to play as well as make participation in online gaming communities possible. They also have the potential to contribute to grassroots development and social inclusion.

Latin American communications scholarship traditionally has been focused on "alternative and participatory communication" (Barranquero 2011, 156). The influential work of Paulo Freire was pivotal in establishing a commitment to grassroots participatory praxis in Latin American communications scholarship (Barranquero 2011, 165). The influence of Latin American tradition on global scholarship of alternative media was recently acknowledged by Leah A. Leivrouw (2011, 227–229). She singles out the work of Jesús Martin-Barbero, who is particularly concerned with understanding how digital media and technologies offer marginalized communities strategies for social inclusion. He states:

> New technologies multiply, in every country, the presence of global images and intensify the globalization of the images of the local. But at the same time, democratization movements from below find in technology—of production as in the portable camera, of reception as in parabolic antennae, of postproduction as the computer, of diffusion as in cable—the possibility of multiplying the images of our society, starting from the regional or municipal level, or even from the barrio. (Martin-Barbero 2003, 100)

Digital media greatly enhances the possibility of the inclusion of local movements in the production of global culture. He also points to the ambiguous manner that digital networked media distribute power; while in many ways inequalities are amplified, these technologies also produce considerable scope for enhanced democracy.

> So although the technological revolution in communication aggravates the *breach* in terms of inequality between social sectors, cultures and countries, it also mobilizes the social imaginary of communities, strengthening their capacity for survival and association, protest and democratic participation, for defending their socio-political rights and activating their expressive creativity. (Martin-Barbero 2011, 44–45)

Through a discussion on ethnographic fieldwork in a cybercafé in Caracas, Venezuela, this chapter examines how digital gaming is situated in relation to this breach that simultaneously aggravates inequality and creates potency for inclusion. After providing a background to Venezuela and a description of gaming practices there, this essay will focus on examining how digital gaming can contribute to more inclusive forms of citizenship and literacy. The chapter constantly returns to the issue of digital piracy, which in the context of Venezuela, is thoroughly imbricated in digital game play and must be reconsidered in light of its contribution to social inclusion.

Background: Venezuela

The regulatory environment surrounding the media in Venezuela is characterized by an anti-neoliberal and antiglobalization stance. The politics of the *Revolución Bolivariana* (Bolivarian Revolution) platform of the current *Partido Socialista Unido de Venezuela* (United Socialist Party of Venezuela) government, led by Nicolás Maduro since the death of Hugo Chávez (1954–2013), are commonly portrayed as being socialist- or communist-leaning. Political analysts have characterized the government's stance as a broad anti-neoliberal agenda that posits an "alternative economic orientation" (Azicri 2009, 100; Al Attar and Miller 2010, 341). Under Chávez, Venezuela became the instigator of several notable "South-to-South" regional collaborations, spearheaded by the *Alianza Bolivariana para los Pueblos de Nuestra América* (Bolivarian Alliance for the Americas), typically referred to by its acronym, ALBA. The organization—led by Venezuela and including Antigua and Barbuda, Bolivia, Cuba, Dominica, Ecuador, Nicaragua, and Saint Vincent and the Grenadines—"puts forward a cohesive counter-vision of international law" (Al Attar and Miller 2010, 347). Two key ALBA initiatives that are considered to have significant regional impact are: the *Banco de Alba*, a regionally funded development bank; and *TeleSUR*, a regionally funded public news network built on a model similar to *Al Jazeera* (Azicri 2009, 100; Cedeño 2006). While the *Revolución Bolivariana* has met with considerable positive economic, political and social success, both locally and regionally, these achievements have been criticized for being relatively slow-paced and having little impact on the diversification of Venezuela's oil-dependent economy (Ellner 2010, 92–93).

The current government has challenged the perceived pro-United States and anti-*Revolución Bolivariana* stance of many global and local media outlets operating in Venezuela. Of particular significance for digital gaming are the steps that have been taken to regulate the telecommunications industry and the play of digital games. Plans to nationalize the telecommunications industry were first unveiled after Chávez's reelection in December 2006 (Apperley 2007, 171). This decision challenged the otherwise increasing consolidation of the Latin American mobile sector—across the region generally, by 2009, the Spanish company Telefónica and Mexican-owned Grupo Carso Telecom together controlled 64% of the telecommunications market (Mariscal 2009, 25). Overall, the typical experience of media markets in Latin America is that digital networked infrastructures have led to remarkable market concentration dominated by a few large stakeholders (Martin-Barbero 2011, 51).

The widespread ownership of mobile devices in Venezuela has made Internet-based services more accessible. However, there are still relatively few domestic connections; by the end of 2009 only 33% of the population of Venezuela had access to the Internet at home.[1] Consequently, many Venezuelans rely on privately owned, semi-public cybercafés for access to computers, software, and the Internet (Lugo and Sampson 2008, 102–118). Cybercafés are popular, social locations, not just because they provide access to digital services, but because they are also relatively cheap compared to other forms of communication and entertainment (Apperley 2010; Powell 2003). The phenomenon of cybercafés providing democratic and grassroots community access to the Internet in Venezuela and other areas of Latin America is called *cibercafezinhozación* (Horst 2011, 45–46; Lemos and Martini 2010; Powell 2003, 173–175; Uxo 2010, 12.6). The prohibitive cost of owning a dedicated gaming console or even a home computer has meant that the cybercafé has a central role in providing general access to digital gaming.

Digital games are a popular form of entertainment in Venezuela, even among marginalized and impoverished groups (Márquez 2002). The key scholarly analysis of the digital games sector in Venezuela suggests that—like many other countries in Latin America—Venezuela had more potential for the global digital games industry "as a low-cost producer and exporter" than as a "consumer market" (Lugo, Sampson, and Lossada 2002). While digital gaming did create economic opportunities, these were largely confined to the informal sector of the economy through the provision of services pertaining to digital games—particularly the distribution of pirated copies of software and access to computers and the Internet through cybercafés (Apperley 2010; Lugo and Sampson 2008). In 2002, Lugo, Sampson, and Lossada's study noted that there was potential for Venezuela to become an exporter of hardware, and that producing export quality software would be extremely difficult, primarily because of the need to pay developers competitive salaries (Lugo, Sampson, and Lossada 2002). In the years since this study, Venezuelan-based game development has not occurred extensively. Most work is done by way of cheap outsourced labor for North American companies, although one success story has emerged: the (originally) Caracas-based studio Teravision (Rodríguez 2011). In 2008, Teravision announced that it had been approved by Nintendo to develop games for the Wii and the DSi, making it the first Venezuelan-based studio to become a third-party developer for a major game publisher (Teravison Games Team 2008). Another notable accomplishment for Venezuelan-based developers was *Battle Tennis*, an XBA game designed by a team of university students led by Jose Alberto Gomez that was awarded second place in Microsoft's 2008 Dream-Build-Play competition (IGN Staff 2008).

By 2008, locally organized support for the games industry had developed. That year, Venezuelan-based independent developers and studios established INVENTAD, the *Industria Nacional del Videojuego, Entretenimiento y Artes Digitales* (National Industry of Video Games, Entertainment and Digital Arts) with the goal of fostering relationships within the industry and establishing stronger relationships with universities (Rodríguez 2011). The year 2009 saw two important developments for the Venezuelan games industry. First, the Caracas Game Jam was initiated and has continued to run annually (see http://www.caracasgamejam.com/), and second, an annual conference, Gamexpo, was held for the first time, bringing together representatives from the industry and academia (see http://gamexpo.org/). While the national film industry has enjoyed government support in recent years, through Villa du Cine, there is currently no support from the government for the Venezuelan digital games industry (Rodríguez 2011).

Subsequent to the fieldwork, there were two other—possibly related—developments that shaped consumption of digital games in Venezuela. The first erupted in 2007, when publicity announcing the development of *Mercenaries 2: World in Flames* (2008) by Pandemic Studios caused a scandal among Venezuelans and US-based Venezuelan expatriates (Apperley 2010, 115–116). The game positioned Venezuela as a rogue state, whose leader was threatening the United States by withholding oil. By capitalizing on US President George W. Bush's announcement that Venezuela (along with North Korea and Iran) constituted an "axis of evil," the game effectively dramatized the War on Terror (Riegler 2010, 54). When the game was finally released in 2008, its content had been modified to avoid close association between Chávez and the game's antagonist, but it still included a sandbox-style destructible portrayal of the cities of Maracaibo and Mérida, as well as much of the Venezuelan countryside. At the time of the release the Venezuelan government and Chávez himself denounced the game and "imperialist toys" more generally (Apperley 2010, 118–119).

The above events set the stage for the announcement of tough new laws on digital games and toys. In December 2009, the National Assembly banned the importation, distribution, sale, and use of "*bélicos*" digital games and toys in Venezuela. *Bélicos* is typically translated to "war," making the prohibition quite specific. Article 3 of the legislation, "Ley para la prohibición de videojuegos bélicos y juguetes bélicos" (Law for the prohibition of war video games and toys), focuses the ban on digital games that "que contengan informaciones o simbolien imágenes que promuevan o inciten a la voilencia o al uso de armas" (contain information or images that promote or incite violence or the use of weapons), and toys that imitate weapons used by the armed forces of Venezuela. One notable aspect of the new law was the extremely high penalties that could be applied; a sentence of three to five years of imprisonment could be imposed on people importing, producing, selling, renting, or distributing illegal games or toys. The law was enacted in early 2010 and has been in effect since then; however, as of mid-2013, there have been no high-profile prosecutions under this law. The broad definitions supplied in the legislation, however, have caused a great deal of uncertainty for Venezuelan-based studios, leading at least one studio—the celebrated Teravision—to relocate most of its headquarters to Colombia (Rodríguez 2011).

During the fieldwork period, an additional, localized issue that framed the use of digital games arose at Cybercafé Avila. In response to concerns from local parents and the administration of the nearby high school that the cybercafé was being used by truant students, the owner of the café (who we will refer to as "Xavier" to protect his identity) banned the wearing of school uniforms inside the premises. This was done through a rather toothless notice which effectively allowed students to control the issue, since, despite the heat, it was necessary to wear a t-shirt under their flimsy uniforms.

ATENCION ESTUDIANTES
Por disposición de Authoridades
NO podemos permitir el ingreso al
Local estudiantes uniformados
Amigo Estudiante: Si necesitas de los services del Cyber,
Favor ven a nuestro local sin tu uniforme.

[ATTENTION STUDENTS
By arrangement with the authorities
We cannot allow entry to
Local uniformed students
Student Friends (customers): If you need the services of the Cybercafé
Please come without your uniform.]

Even so, the stipulation not to wear the uniform in the cybercafé was often overlooked by Xavier and his employees. The ban, then, was largely of symbolic value; in essence, nothing changed.

Case Study: Cybercafé Avila

From late March through mid-July 2005, I conducted an ethnographic study of digital game play in Cybercafé Avila in the suburb of San Bernardino in the Libertador district of Caracas. The site had been selected during a previous research trip to Venezuela to scope out feasible fieldwork sites. The cybercafé was selected for two reasons: first, because it was representative of a typical small, community-oriented business model often found in the inner city and suburbs of Caracas (and indeed elsewhere in Venezuela); and second, because it was located in a—relatively—safe area of San Bernardino, the suburb in which I had arranged long-term accommodation. For my institution's research office, the potential risk faced by the investigator was a key concern.

Martin-Barbero has suggested that "the inclusion of Latin American countries in the challenges and possibilities of digital technology" involved both an appreciation of new forms of literacy and a detailed understanding of

> the ways in which local cultures, whether towns, ethnic groups or regions are making use of or appropriating virtual culture, that is, the means of interaction with information networks which communities select and develop, the transformations that their usage introduces into community life and the new resources, both technical and human, that are required in order to render those interactions socially creative and productive. (Martin-Barbero 2011, 59)

Consequently, the following discussion focuses on examining players' use of digital games in Cybercafé Avila to understand the resources that are required for a "socially creative and productive" engagement with the global network society.

The goal of the fieldwork was to gather data that supported a detailed understanding of the "cultures of use" of digital gaming technologies in Venezuela (Sassen 2006, 347–348). I had particular interest in two issues: first, the role that location and local consumers might have in shaping which games were played; and second, how informal, local arrangements shaped the skills and literacies that were developed. This chapter builds upon the second issue, in order to extend the argument beyond just literacies to also consider how such literacies can be put to use. There is something qualitatively different at stake in the strong connection between gaming and literacies in the developing world; these literacies are meaningless without a more thorough integration with the infrastructures and economy of the network. This chapter will examine what is at stake, arguing that an examination of the cultures of use demonstrates how digital games support the development of contemporary literacy practices and therefore social inclusion.

Cybercafé Avila is a grassroots assemblage of commercial and community interests run as a business by Xavier, who had begun the business in partnership with his brother after losing his job in the petroleum sector. Now the sole owner, Xavier had built up the café to an eleven- or twelve-computer business, which provided printing, typesetting, and computer repair services, as well as snacks and drinks to patrons and passersby. The café was also a crucial part of the community, with local small business owners—including a nearby McDonald's—using the computers and other services as part of their day-to-day work practices. Teachers from local schools used the café's computers to prepare classes and do their administration, university students did their coursework there, occasionally begging Xavier to stay open all night so that they could make an important deadline for an assignment. The café employed one full-time staff member and up to three part-timers, who provided customer support for the operation of any requested software. Xavier and his employees would often do maintenance on computers at peoples' homes or places of business, occasionally forgoing cash payment in favor of the rendering of a reciprocal service. While Xavier was happy when this meant maintenance for his car, the time he negotiated a week's worth of lunch from the canteen of the local elementary school for himself and his employees in return for upgrading the software on their computers was not considered to be a success.

Crucial for the business were the students from the local high school. While Cybercafé Avila offered a variety of services, people paying to use the computers were the cornerstone of the day-to-day trade, and a large proportion of these people were students. High school students—who were easily identified by their uniforms, with white (junior), blue (middle), or brown (senior) identifying their grade—attend school in shifts, some in the morning, others in the afternoon. Daily routines at Cybercafé Avila were shaped by this particular rhythm. Students from the morning shift would be the first customers in the morning, before classes started; students from the evening shift would linger at the cybercafé on their way home. In the middle of the day, Cybercafé Avila was one of several neighborhood locations where the students mingled, some on their way to, others on their way from, school.

This cybercafé was characterized by sociality. It was used by a remarkable cross-section of locals, and a large number of visitors to the district, particularly through the large number of hospitals and medical specialists, and the military bases on an adjacent street. Gaming contributed a great deal to this sociality, forming a mutual interest between many of the customers. This common interest allowed talk about games and displays of skill with games to occur between different peer groups and between locals and visitors. In turn, this engendered transfer of knowledge about different digital games and specific practices within them—for example, the use of hotkeys in *Grand Theft Auto: Vice City* (Rockstar Games, 2002) or *Counter-Strike* (Valve, 2000). It also facilitated cooperative and antagonistic group play, establishing connections between people that potentially endured beyond the café (Apperley and Leorke 2013).

Observing the—relatively—free exchange of information and knowledge about games led me to wonder: how exactly were these skills and literacies acquired? Clearly experience with digital games was a major factor. Research on the Internet, using paratexts—defined as "both texts and the surrounding materials that frame their consumption, shape the readers' experience of a text and give meaning to the act of reading" (Apperley and Beavis 2011, 133)—was also crucial to finding out about particular techniques within a game, including hotkeys, cheats, and shortcuts (Consalvo 2007). General research into digital games themselves was also a factor; patrons often found out about new games and informed the staff about them, who then downloaded and installed them. In some cases they would bring digital games to the cybercafé to install, and staff would make copies if they believed the game would generate sufficient interest. The availability of pirated copies of digital games was crucial for this flexibility.

Piracy, because it both permitted and multiplied access to digital games, was a crucial factor that shaped the experience of digital play (Apperley 2010, 50–110). Piracy is considered crucial for the sustainability (and profitability) of this type of business in Venezuela (Lugo and Sampson 2008, 109–110). Small businesses operating in the ICT (Information, Computer, and Technology) sector were relegated to the informal or gray sector of the economy. This gave operators effective permission "to overcome the technological, financial and political obstacles to the importing of this media technology and its content" (Mattelart 2010, 313). In terms of digital gaming, piracy is not a part of play but is essential to it as it establishes the possibility for play. Thus, understanding the operation and implementation of pirated software in a technical sense is an absolutely crucial element of the informal literacies circulating in cybercafés. At Cybercafé Avila, piracy provided material access to otherwise closed networks that were supported by a grassroots community of common interest, which provided a fuller experience of inclusion in global networked culture (Mattelart 2010, 309). The remainder of this chapter focuses primarily on evaluating what is at stake in that inclusion, and also—with Martin-Barbero's point in mind—considering the resources that are necessary to utilize this inclusion in productive and creative ways; first, by elaborating the notions of civics and citizenship in relation to digital gaming, and second, by examining the connection between gaming and literacy.

Play, Civics, and Citizenship

Citizenship and play have a historic connection. Huizinga's foundational cultural history of play, *Homo Ludens: A Study of the Play Element in Culture* (first published in English in 1949) maps out how the inclusion in sacred forms of play was a key indicator of citizenship status in Ancient Greece (Huizinga 1970). The close ties between play and citizenship have also been noted by scholars of digital games. Ian Bogost's work on procedural rhetoric focuses on digital games as rhetoric or persuasion, establishing a clear segue between digital play and contemporary citizenship. He states, "Procedural rhetoric is the practice of persuading through processes in general and computational processes in particular" (Bogost 2007, 3). However, he carefully traces the notion of "rhetoric" back to Plato to establish its roots in "public speaking for civic purposes" (Bogost 2007, 15). Procedural rhetoric is at its core a discussion of how games can be meaningfully deployed for civic purposes, and as Bogost points out, "Procedural rhetoric is precisely what is missing from current uses of technology for political and civic engagement" (Bogost 2007, 135). His more recent work continues to elaborate on the underdeveloped civic role of digital games, for example, how they could be deployed by the fourth estate (Bogost, Ferrari, and Schweizer 2010). Other research on digital games makes a strong case for the need to understand the role that they play in fostering civics (Gordon and Koo 2008; Raphael et al. 2010). Early ethnographies of massively multiplayer online worlds noted the way that players, both individually and collectively, contributed to the governance of the game world (Humphreys 2005; Taylor 2006). A wide-ranging study funded by the MacArthur Foundation found that

> teens who have civic gaming experiences, such as helping or guiding other players, organizing or managing guilds, playing games that simulate government processes, or playing games that deal with social or moral issues, report much higher levels of civic and political engagement than teens who do not have these kinds of experiences. (Kahne, Middaugh, and Evans 2009, 30)

Existing studies of civics gaming experiences have focused on classroom-based social learning experiences, leading some researchers to explore the question of whether social and individual play can foster the same levels of civic engagement (Raphael, Bachen, and Hernández-Ramos 2012). Certainly, a strong link between digital gaming and civics has been established, which will no doubt develop into more nuanced understandings that are able to pinpoint specific practices and texts within this sphere in years to come.

In the discussion of the connection between digital games and citizenship, the issue of consumption is a key concern. If gaming has civic potential, this potential is constrained by uneven access to digital games and gaming infrastructures. An early study of digital play by Marsha Kinder earmarked a similar issue; while she noted that the children in her study gained a sense of personal empowerment through their play of digital games, this was substantially mitigated by how this empowerment was enabled through consumption (Kinder 1991). More recently, there has been a recognition that in the convergent media environment, as access to media becomes more crucial for basic participation in public life in developed economies, the notions of the "citizen" and the "consumer" become conflated (Trentmann 2007). Henry Jenkins's model of convergence culture—where content is coproduced by the audience through creation and remixing—addresses this issue, arguing that as the audience gets more actively involved in creating content, the issue of content creation and consumption becomes increasingly politicized (Jenkins 2006).

The work of Néstor García-Canclini anticipates this issue. In *Citizens and Consumers: Globalization and Multicultural Conflicts*, originally published in Spanish in 1997, he argues for a radical reformulation of the notion of citizenship *vis-à-vis* consumption. Cultural consumption, which very much includes media consumption and production in García-Canclini's formulation, is "an ensemble of practices that shape the sphere of citizenship" (García-Canclini 2001, 22). Core to his argument is that access to information through media and other cultural channels (such as libraries, museums, etc.) is a right of citizenship because it is crucial for individuals to have access to information in order to make informed decisions, and thus "act autonomously and creatively" (García-Canclini 2001, 45, 130–131). The creation, circulation, and storage of information is not exclusively the realm of the government; it is supplemented by market forces. This requires a reevaluation of the role of consumption in civic life. García-Canclini argues that as a consequence of consumption becoming imbricated with citizenship and human rights, the *rights themselves* should be redefined in terms of contemporary consumption rather than enduring as abstract ideals (García-Canclini 2001, 5, 21).

This reframing of access to media and digital networks as a civic right mounts two important challenges to the exclusivity of global networks. First, it shifts discussion of media piracy away from one of *absolute criminality* toward its potency to provide access to networks. Second, it suggests that issues of access to networks are underpinned by key skills and literacies that allow them to be used autonomously, productively, and creatively (García-Canclini echoes Martin-Barbero in this respect). In light of convergence culture, it is essential that both skills and literacies are understood in a manner that includes access to technologies that allow citizens to *consume*, *remix*, and *create* content.

This suggests the right to access information is underpinned by an expanded notion of literacy. Martin-Barbero's reevaluation of citizenship in the contemporary media environment suggests that citizens should have "*access to information* not only as receivers, but as producers" (Martin-Barbero 2011, 57, original emphasis). Thus he has called for an examination and understanding of "*virtual literacy*," which he describes as a "*set of mental skills, operation habits and interactive spirit* without which the presence of technology among the majority of the population would go to waste" (Martin-Barbero 2011, 58, original emphasis). Virtual literacy thus points to the skills and literacies that allow people to use media productively and creatively. The rest of this chapter will explore how the skills and literacies that gaming and gaming cultures provide may contribute to Martin-Barbero's virtual literacy project. Particularly important is how these readily available engagements with popular culture contribute to informal literacies that many individuals in Venezuela, and throughout the developing world, might not have the opportunity to learn through more formal means.

Accounts of gaming cultures indicate two important ways that the literacies developed through gameplay align with what it means to be literate in convergence culture. First, the relationship between digital games and paratexts exemplifies how the convergent audience uses other media, especially the Internet, to coordinate, collaborate, and conduct research, and in the process, build communities (Banks and Potts 2010; Bruns 2008; Sotaama 2010). Second, the multiple and versatile productive practices of game cultures demonstrate the new modes of audience participation that prioritize engagement in the production, and the sharing, of user-generated content (Sotaama 2010). Play is an extremely important component of the literacies essential for convergence culture: "In a hunting culture, kids played with bows and arrows. In an information society, they *play with information*" (Jenkins 2006, 6, emphasis added).

The articulation of "convergence culture" through the work of Jenkins marks a key shift toward understanding the cultural impact of the technologies of convergence. This highlights the ongoing processes that alter "the relationship between existing technologies, industries, markets, genres, and audiences" (Jenkins 2006, 15). The impact that this shift to a more dynamic media environment has on global media industries and products is also important, and change is still underpinned by the drive for the media industries to extract profit from the audiences' emergent uses of new technology. This means that, in many respects, Venezuela and other countries in the developing world are marginal to these developments because it is often the technological and cultural practices of the "early adopters" of convergence culture that drive change (Jenkins 2006, 23). Consequently, there is more than *just inclusion* at stake in terms of consuming cultural materials, but also the opportunity to actively participate in the shaping of global culture. At Cybercafé Avila, for example, it was very common to play *Counter-Strike* (1999) using various user-generated maps, but not to make them. The strictly confined times that most people had to play simply did not accommodate such activities.

While access to technology is essential, technological rollouts cannot sustain systemic change without political and pedagogical support. It has long been a moot point to scholars of development that technological development cannot be understood as merely a physical problem to be solved by introducing technology. The implementation of technological change is not solely a technical issue (Martin-Barbero 2011, 58). It must be accompanied by political and educational approaches that are suitably revised to support the desired change. Often, it is a lack of investment and understanding in these latter areas that leads technological development to fail. Without a strong program that encourages the development of relevant skills to make use of technology among the population, economic inequality is likely to persist, despite increased access to technology (van Dijk and Hacker 2003, 322). There is also a need for an appreciation of the stakes of participation at the political level. According to Claudia Padovani and Kaarle Nordenstreng,

> Knowledge societies are supposed to be spaces in which citizens will be able to communicate, interact and participate. But this risks remaining only rhetoric if transformations, challenges and political solutions are perceived as highly technical issues removed from the public. (Padovani and Nordenstreng 2005, 270)

The informal learning that takes place in cybercafés, such as Cybercafé Avila, is accompanied by a government program to bring similar facilities to the barrios of the major cities and to poor remote areas of Venezuela (Robinson 2003, 47). However, their location in the informal or gray economy means that the future is relatively uncertain. A change in how digital copyright is implemented technologically or by how the local laws are applied by the federal government could effectively end the flexible conditions under which cybercafés in Venezuela operate.

At stake in the lack of access to digital games and gaming communities are the knowledge, literacies, and skills that they foster. For advocates of gaming literacy, as the digital sector becomes increasingly important, this lack of access will perpetuate exclusion from the mainstream economy. According to Zimmerman,

> In the coming century, the way we live and learn, work and relax, communicate and create, will more and more resemble how we play games. While we are not all going to be game designers, game design and gaming literacy offer a valuable model for what it will mean to become literate, educated, and successful in this playful world. (Zimmerman 2008, 30)

Jenkins also suggests lack of access to forms of digital play may have unfortunate consequences, as "the *skills we acquire through play* may have implications for how we learn, work, participate in the political process, and connect with other people around the world" (Jenkins 2006, 23, emphasis added). Gaming as a practice provides access to skills and literacies that contribute to use of global networks in a creative and productive manner. As these skills and literacies involve the emergent communicative, connective, and productive elements of networked technology, to lack full access to such services curtails the development of literacies and skills that are essential to function in the global economy and contribute to global culture.

Conclusion

This combination of high stakes and precarious access has grave implications for cultural diversity in the region, and equally for the diversity of the "network culture" of globalization. By examining the unevenness of access to the medium of digital games, it is apparent that digital media piracy cannot simply be understood in a proprietary manner. In a global economy based on knowledge and networks, exclusion equals poverty, and in some cases, piracy enables inclusion in the economy.

The digital games piracy that takes place in the informal economy and is actualized in the grassroots community of the cybercafé has a significant role in mitigating exclusion. However, this also raises the stakes of inclusion because total reliance on piracy for inclusion is extremely precarious. Primarily, this is because of its illegality, and the constant legal and technical challenges that are developed to prevent and limit it. In the developing world, increasingly high stakes are placed on digital game play as a practice that engenders creativity, collaboration, and computing skills. Participation in digital play and gaming cultures provides access to *unambiguous segues* into working in the knowledge economy. To apply the same stakes to the developing world may have particularly serious repercussions, considering the partial and precarious ways that people from those areas have access to digital games.

Thus, to consider digital games purely as a medium of entertainment *is suspect*. While in part, digital play is a consumer luxury, it also informally provides skills and literacies that contribute to autonomous participation in civic and economic life in the global network society. García-Canclini's reconfiguration of citizenship to include the right to consume information suggests that the absolute enforcement of copyright law is ethically ambiguous. Particularly, this is the case when laws are enforced in a way that prevents access to those who lack the economic power to be legitimately included. In the developed world, audience members have considerable power to push back against "harsh" or "unfair" intellectual property rights when, for example, they are used to shut down fan-based productions or implement unpopular digital rights management systems. However, similar strict enforcements in the developing world would not just threaten the lively creativity of fan cultures, but the access to software and networks—and therefore the livelihood of the grassroots communities—where they are cultivated.

Acknowledgments

The travel and fieldwork expenses for this project were funded by the School of Culture and Communication, School of Graduate Studies, and the Faculty of Arts at the University of Melbourne, Australia. Thanks also to Gabriel Rodríguez and Julián Rojas Millán for sharing their insights on the state of the Venezuelan games industry.

Note

1. Conatel (*Comissión Nacionale de Telecomunicaciones*), http://www.conatel.gob.ve/ (the information is reproduced in English at http://www.internetworldstats.com/sa/ve.htm).

References

Al Attar, Mohsen, and Rosalie Miller. 2010. Towards an emancipatory international law: The Bolivarian Reconstruction. *Third World Quarterly* 31 (3): 347–363.

Apperley, Thomas. 2007. Games without borders: Globalization, gaming, and mobility in Venezuela. In *Mobile Media 2007: Proceeding of an International Conference on Social and Cultural Aspects of Mobile Phones, Convergent Media, and Wireless Technologies*, ed. Gerard Goggin and Larissa Hjorth, 171–178. Sydney, Australia: University of Sydney Press.

Apperley, Thomas. 2010. *Gaming Rhythms: Play and Counterplay from the Situated to the Global.* Amsterdam: Institute of Network Cultures.

Apperley, Thomas, and Catherine Beavis. 2011. Literacy into action: Digital games as action and text in the English and literacy classroom. *Pedagogies: An International Journal* 6 (2): 130–143.

Apperley, Thomas, and Dale Leorke. 2013. From the cybercafé to the street: The right to play in the city. *First Monday* 18 (11). http://firstmonday.org/ojs/index.php/fm/article/view/4964/3794.

Azicri, Max. 2009. The Castro-Chavez alliance. *Latin American Perspectives* 36 (1): 99–110.

Banks, John, and Jason Potts. 2010. Co-creating games: An evolutionary analysis. *New Media and Society* 12 (2):253–270.

Barranquero, Alejandro. 2011. Rediscovering the Latin American roots of participatory communication for social Change. *Westminster Papers in Culture and Communication* 8 (1): 154–177.

Bogost, Ian. 2007. *Persuasive Games: The Expressive Power of Videogames.* Cambridge, MA: MIT Press.

Bogost, Ian, Simon Ferrari, and Bobby Schweizer. 2010. *Newsgames: Journalism at Play.* Cambridge, MA: MIT Press.

Bruns, Axel. 2008. *Blogs, Wikipedia, Second Life, and Beyond.* New York: Peter Lang.

Castells, Manuel. 1998. *End of Millennium: The Information Age: Economy, Society, and Culture*, vol. 3. Cambridge: Blackwell.

Cedeño, Jeffrey. 2006. Venezuela in the twenty-first century: New men, new ideals, new procedures. *Journal of Latin American Cultural Studies* 15 (1): 93–109.

Cedeño, Jeffrey. 2006. Venezuela in the twenty-first century: New men, new ideals, new procedures. *Journal of Latin American Cultural Studies* 15 (1): 93–109.

Consalvo, Mia. 2007. *Cheating: Gaining Advantage in Videogames.* Cambridge, MA: MIT Press.

Ellner, Steve. 2010. Hugo Chávez's first decade in office: Breakthroughs and shortcomings. *Latin American Perspectives* 37 (1): 77–96.

García-Canclini, Nestor. 2001. *Citizens and Consumers: Globalization and Multicultural Conflicts.* Minneapolis: University of Minnesota Press.

Gordon, Eric, and Gene Koo. 2008. Placeworlds: Using virtual worlds to foster civic engagement. *Space and Culture* 11 (3): 204–221.

Horst, Heather A. 2011. Free, social, and inclusive: The appropriation and resistance to new media technologies in Brazil. *International Journal of Communication* 5:437–462.

Huizinga, Johan. 1970. *Homo Ludens: A Study of the Play Element in Culture*. London: Paladin.

Humphreys, Sal. 2005. Productive players: Online computer games' challenge to conventional media forms. *Communication and Critical Cultural Studies* 2 (1): 37–51.

I.G.N. Staff. 2008. Microsoft reveals the Dream-Build-Play 2008 winners and makes indie game development dreams reality. *IGN Australia*, October 29. http://au.ign.com/articles/2008/10/29/microsoft-reveals-the-dream-build-play-2008-winners-and-makes-indie-game-development-dreams-reality.

Jenkins, Henry. 2006. *Convergence Culture: Where Old and New Media Collide*. New York: New York University Press.

Kahne, Joseph, Ellen Middaugh, and Chris Evans. 2009. *The Civic Potential of Video Games*. Cambridge, MA: MIT Press.

Kinder, Marsha. 1991. *Playing with Power in Movies, Television, and Video Games: From Muppet Babies to Teenage Ninja Turtles*. Berkeley, CA: University of California Press.

Lemos, Ronaldo, and Paula Martini. 2010. LAN Houses: A new wave of digital inclusion in Brazil. *Information Technologies & International Development*, 6 (special edition): 31–35.

Lievrouw, Leah A. 2011. *Alternative and Activist New Media*. Cambridge: Polity.

Lugo, Jairo, and Tony Sampson. 2008. E-informality in Venezuela: The "other path" of technology. *Bulletin of Latin American Research* 27 (1): 102–118.

Lugo, Jairo, Tony Sampson, and Merlyn Lossada. 2002. Latin America's new cultural industries still play old games: From the Banana Republic to Donkey Kong. *Game Studies: The International Journal of Computer Games Research* 2 (2). http://www.gamestudies.org/0202/lugo/.

Mariscal, Judith. 2009. Market structure and penetration in the Latin American mobile sector. *Info* 11 (2): 24–41.

Márquez, Patricia C. 2002. *The Street Is My Home: Youth and Violence in Caracas*. Stanford, CA: Stanford University Press.

Martin-Barbero, Jesús. 2003. Cultural change: The perception of media and the mediation of its images. *Television and New Media* 4 (1): 85–106.

Martin-Barbero, Jesús. 2011. From Latin America: Diversity, globalization, convergence. *Westminster Papers in Communication and Culture* 8 (1): 39–64.

Mattelart, Tristan. 2010. Audio-visual piracy: Towards a study of the underground networks of cultural production. *Global Media and Communication* 5 (3): 308–326.

Padovani, Claudia, and Kaarle Nordenstreng. 2005. From NWICO to WSIS: Another world information and communication order? *Global Media and Communication* 1 (3): 264–272.

Powell, Adam Clayton. 2003. Democracy and new media in developing nations: Opportunities and challenges. In *Democracy and New Media*, ed. Henry Jenkins and David Thorburn, 171–178. Cambridge, MA: MIT Press.

Raphael, Chad, Christine M. Bachen, Kathleen Lynn, Jessica Baldwin-Philippi, and Kirsten A. McKee. 2010. Games for civic learning: A conceptual framework and agenda for research and design. *Games and Culture* 5 (2): 199–235.

Raphael, Chad, Christine M. Bachen, and Pedro F. Hernández-Ramos. 2012. Flow and cooperative learning in civic game play. *New Media and Society* 14 (8): 1321–1338.

Riegler, Thomas. 2010. On the virtual frontlines: Videogames and the war on terror. In *Videogame Cultures and the Future of Interactive Entertainment*, ed. Daniel Riha, 53–62. Oxford: Inter-Disciplinary Press.

Robinson, Scott. 2003. Cybercafés and national elites: Constraints on community networking in Latin America. In *Community Practice in the Network Society*, ed. Peter Day and Doug Schuler, 47–58. New York: Routledge.

Rodríguez, Gabriel. 2011. Venezuela and the videogame industry: A follow up. *Lakitu's Dev Cartridge*, May 2. http://lakitusdevcartridge.wordpress.com/2011/05/02/venezuela-and-the-videogame-industry-a-follow-up-i/.

Sassen, Saskia. 2006. *Territory, Authority, Rights: Global Assemblages*. Princeton, NJ: Princeton University Press.

Sotaama, Olli. 2010. Play, create, share? Console gaming, play production, and agency. *Fibreculture Journal* 16. http://sixteen.fibreculturejournal.org/play-create-share-console-gaming-player-production-and-agency/.

Taylor, T. L. 2006. *Play between Worlds: Exploring Online Game Culture*. Cambridge, MA: MIT Press.

Teravison Games Team. 2008. Teravison Games is now Nintendo authorized DS developers. *Teravision Games—News*. http://www.teravisiongames.com/news/index.php?year=2008#Nintendo.

Trentmann, Frank. 2007. Citizenship and consumption. *Journal of Consumer Culture* 7 (2): 147–158.

Uxo, Carlos. 2010. Internet politics in Cuba. *Telecommunications Journal of Australia* 60 (1): 12.1–12.16.

van Dijk, Jan, and Kenneth Hacker. 2003. The digital divide and a complex dynamic phenomenon. *Information Society* 19:315–326.

Yar, Majid. 2008. The rhetorics and myths of anti-piracy campaigns: Criminalization, moral pedagogy, and capitalist property relations in the classroom. *New Media & Society* 10 (4): 605–623.

Zimmerman, Eric. 2008. Gaming literacy: Game design as a model for literacy in the twenty-first century. In *The Video Game Theory Reader 2*, ed. Bernard Perron and Mark J. P. Wolf, 23–32. New York: Routledge.

CONTRIBUTORS

Ahmad Ahmadi is a graduate of the MBA program at Tehran University. With more than ten years of experience in the IT business, he is Chief Business Development Officer of the Iran Computer & Video Games Foundation. He teaches business and marketing in the game industry, and is head of the team in the game development laboratory of the Department of Computer Engineering at Sharif University. He coordinates development projects and consultants' designs for video games such as *Paper War* (2014), *Mice City* (2014), *E.T. Armies* (2014), *Awakening* (2013), *Hate the Sin, Love the Sinner* (2013), and *Super Hungry Monsters* (2012). In 2013, he established DropFun, which is a knowledge-based company specializing in the conception of serious games and gamification development, and is one of the leaders of this field in the Middle East. He can be reached at ahmadieng@gmail.com.

Lynn Rosalina Gama Alves has a degree in Education from the Faculty of Education of Bahia (1985), master's (1998), and PhD (2004) in Education from the Federal University of Bahia. Her postdoctorate work was in the area of electronic games and learning from the Università degli Studi di Torino, Italy. She is currently a Professor Titular and researcher at the SENAI-CIMATEC-Regional Department of Bahia (Center for Computational Modeling) and the University of Bahia. She has experience in education and has conducted research on the following topics: gaming, interactivity, online learning, and education. She coordinates research and development projects in digital games such as *Triad* (FINEP/FAPESB/UNEB), *Buzios: Echoes of Freedom* (FAPESB), *Forest Guardians* (CNPq), *Salvador Sim* (SEC-Ba), *Institu* (SEC-Ba), *Dom* (SEC-Ba), *Janus* (SEC-Ba), and *Games Studies* (FAPESB), among others. The productions of her research group are available at http://www.comunidadesvirtuais.pro.br. She can be reached at lynnalves@gmail.com.

Thomas H. Apperley, PhD, is an ethnographer who specializes in researching digital media technologies. His previous writing has covered broadband policy, digital games, digital literacies and pedagogies, mobile media, and social inclusion. Tom is currently a Senior Lecturer at UNSW Australia (the University of New South Wales). He is the editor of the open-access peer-reviewed journal *Digital Culture & Education*. His open-access, print-on-demand book *Gaming Rhythms: Play and Counterplay from the Situated to the Global*, was published by the Institute of Network Cultures in 2010. Tom's more recent work has appeared in the journals *Digital Creativity*, *First Monday*, and *Westminster Papers in Culture and Communication*. He can be reached at thomas.apperley@gmail.com.

Dominic Arsenault is Assistant Professor of Film Studies at the Université de Montréal. He researches, presents, lectures, and publishes on video game narration and design, fictional and systemic immersion, video game history and music, and genres. His work has appeared in journals such as *Eludamos* and *Loading*, and as essays in *The Video Game Theory Reader 2*, *The Encyclopedia of Video Games*, and *The Routledge Companion to Video Game Studies*. His current research interests revolve around graphics and space in video games, the Super NES as an industry-, genre- and culture-defining platform, collaborative storytelling practices, and transmedial storytelling. He has a website, which is never up to date: http://www.le-ludophile.com. He can be reached at dominic.arsenault@umontreal.ca.

Guillermo Averbuj teaches at USAL (Salvador University) and is a game design consultant with a ludotecary technical degree (play studies), guiding creative production on starting teams, and is a regular lecturer at the biggest local event, EVA. For the last ten years he's been working in Argentina to encourage the local industry in the areas of game development and entrepreneurial events, while making games as a job. As the director of EVA (Argentine Videogames Expo) for seven years, Averbuj got to know the entire local industry. He's been working the last three years with the government to improve the cultural support of the video game industry and the indie community. As part of the Gamester group, he's been producing several entrepreneurial gatherings under the name "Game Work Jam," where anyone can learn to develop games for free with the help of known mentors from the local industry. See http://www.gamester.com.ar/guilleaverbuj/?a=v. He can be reached at guille.averbuj@gmail.com.

Tamás Beregi acquired his first computers (a ZX Spectrum and later a Commodore 64) sometime around 1983. He spent the next fifteen years in front of various TVs and monitors playing a wide range of video games and became an adventure game fanatic after playing the first text adventures released by the British company Magnetic Scrolls in the mid-1980s. Realizing that his childhood had suddenly passed, he quickly started to research game history. His richly illustrated, 450-page book *Pixel Heroes: The First Fifty Years of Computer Games* was published in 2010. It is the first book in Hungarian on the history of video games. Since the early 1990s, Beregi has regularly contributed to leading Hungarian computer and film magazines, where his main interest is the representation of the fantastic and the relationship between video games and movies. He was the editor of the "Letters to the Editor" column in the magazine *Computer Mania* in 1990–91. In 1999 he founded the first online computer game museum in Hungary. Beregi first studied biology and geography at the university level, but his interests later shifted to art history and film studies. He received his PhD in Art History from Manchester University in 2008. He is the author of three fantasy novels and wrote a highly-successful book in Hungary entitled *My One and Onlies*, which was later made into a feature film. The book combines his interest in partying and the underground scene of Budapest with scientific observations on how to find the ideal woman. (The book has a 100-page-long appendix that guides the reader through this task with the use of mathematical and geometrical formulas.) Beregi has written scripts for a number of movies, including a romantic comedy (an anarchistic burlesque about a bee-keeper's revolt against consumer society), a historical drama, and a Hungarian-Icelandic CGI animation movie. He also wrote a successful Alternate Reality Game in 2008. In 2013 he spent a half year in Japan with a postdoctoral scholarship from the Japanese Fund, researching classic Japanese video games at the Ritsumeikan University of Kyoto, and interviewing numerous classic Japanese video game developers. He is currently working for the Hungarian National Film Fund as a script advisor, and he is also adding the finishing touches to his fourth novel called *Noctambulo*. He lives with his cat and dog (both female) in the hills of Buda, where he often goes to collect his favorite mushrooms. He can be reached at tamas.beregi@gmail.com.

Alexis Blanchet is Associate Professor in Film Studies at the Université Paris III Sorbonne Nouvelle (France). Formerly associated with the French National Library, member of the Observatoire des Mondes Numériques en Sciences Humaines (Omnsh), he is currently studying the cultural, economic, and technical synergies between cinema and video games, and is interested in how the public nowadays is being offered extended and transmediatic fictional worlds. He can be reached at alexis.blanchet@univ-paris3.fr.

P. Konrad Budziszewski is a PhD candidate in the Department of Communication and Culture at Indiana University, Bloomington. His current research focuses on the technologies of electronic gaming, as reflected in and constituted by the practices, discourses, and affects that surround them. He can be reached at pbudzisz@indiana.edu.

Kursat Cagiltay is Professor in the Department of Computer Education and Instructional Technology at the Middle East Technical University (METU), Ankara, Turkey. He earned his BS in Mathematics and MS in Computer Engineering from Middle East Technical University. He holds a double PhD in Cognitive Science and Instructional Systems Technology from Indiana University. He is the founder of two research groups at METU: Simulations and Games in Education (SIMGE; simge.metu.edu.tr) and the HCI research group (hci.metu.edu.tr). He also established Turkey's first HCI research lab with an eye-tracking facility at METU. His research focuses on human-computer interaction, instructional technology, the social and cognitive issues of computer games, sociocultural aspects of technology, distance learning, and human performance technologies. He can be reached at kursat@metu.edu.tr.

Humberto Cervera is a self-taught game designer with three years of experience working in the Mexican video game industry. He was the Lead Game Designer for *Construye tu México* (2012) and *El Camino Para Vivir Mejor* (2012), both government projects that were part of the "Vivir Mejor" program. He also worked as Lead Designer for *Club VLP* (2012) by Playstop, *Banamex: My Life* (2013) by Revarts, and *Robo Jungle Rush* (2013) by

Fraktalia Studios. Cervera has also worked as a Gamification consultant for Viral.org and was also part of the "Catalysts for Change" project by the Institute for the Future, where he worked as Game Guide and data analyst for the complete duration of the project. Currently Cervera is leading his own game studio, Chaos Industries, which is working on its first original video game, *Agent Awesome*. Cervera never misses the chance to speak highly of video games and their benefits; he is convinced that video games help us develop the skills needed to succeed in the twenty-first century. He has done public speaking at TEDxDF, DevHour, Fodarte, and various universities such as the ITESM (Instituto Tecnologico de Estudios Superiores de Monterrey), UIA (Universidad Ibero Americana), and UVM (Universidad del Valle de México). He can be reached at humberto_cervera@hotmail.com.

Peichi Chung is an assistant professor in the Department of Cultural and Religious Studies at the Chinese University of Hong Kong. Her teaching interests include new media and culture, global communication, and Asian popular culture. Her research interests include the social impact of new communication technologies, film studies, and global entertainment media industry research. Her current research projects involve comparative online game industry studies across different countries in Asia. She started her first ethnographic research on South Korean online games in 2006 and has thus far expanded similar research to other countries in East and Southeast Asia. She has published book chapters and journal articles on online game industries about South Korea, China, and Singapore. Her present research project focuses on Southeast Asian game industries, in which she examines the social transformation process of global indie game development among outsourcing studios that produce games in ASEAN countries. She is currently working on her manuscript on digital gaming culture and online game industries in South Korea, China, and Southeast Asia. She can be reached at peichichung@cuhk.edu.hk.

Hikmat Darmawan is well known as a comics expert and pop culture observer and has written a number of books in the field. He has dedicated most of his time organizing many cultural activities in Indonesia. He is currently working as an editor for *Madina Magazine* and RumahFilm.org and was granted an Asian Public Intellectuals Fellowship (2010–2011) from the Nippon Foundation for the researching of the Subcultural Globalisation of Manga and Visual Identity in Japan and Thailand. He can be reached at hikmatdarmawan@gmail.com.

Jennifer deWinter is an Assistant Professor of Rhetoric and faculty in the Interactive Media and Game Development program at Worcester Polytechnic Institute. She teaches courses on game studies, game design, and game production and management. Additionally, she codirects and teaches in the Professional Writing program. She has published on the convergence of animé, manga, and computer games, both in their Japanese contexts and in global markets. Her work has appeared in numerous journals, including *Works and Days*, *The Journal of Gaming and Virtual Worlds*, *Eludamos*, *Computers and Composition*, and *Rhetoric Review*. Additionally, she is coeditor of the book *Computer Games and Technical Communication: Critical Methods and Applications at the*

Intersection (2014) with Ashgate's series in Technical Communication, and she is the editor of the textbook *Videogames* for Fountainhead as well as coeditor of *Video Game Policy: Production, Circulation, and Consumption* for Routledge's Game Studies series. In collaboration with Carly A. Kocurek, she is launching a new book series with Bloomsbury on game designers for which she is writing the inaugural book on Shigeru Miyamoto. She can be reached at jdewinter@wip.edu.

Tristan Donovan is a UK-based journalist, author, and gaming historian. He is the author of *Replay: The History of Video Games* (2010), which has also been published in Danish, French, and Russian. He has written about games for a wide range of publications including *BBC News*, *Gamasutra*, *Eurogamer*, *Stuff*, *Edge*, the *Times*, and the *Guardian*. He also speaks regularly to the media about games and was the opening keynote speaker at the 2014 Game History Annual Symposium in Montreal, Canada. He is also the author of *Fizz: How Soda Shook up the World* (2013) and *Feral Cities: Adventures with Animals in the Urban Jungle* (2015). He also edited Jamie Russell's book *Generation Xbox: How Video Games Invaded Hollywood* (2012) and the e-book *Death Threats and Dogs: Life on the Social Work Frontline* (2013). He also contributed a chapter about video games to *The Routledge Companion to British Media History* (2014). He can be reached at tristan.donovan@gmail.com.

Graciela Alicia Esnaola Horacek has a PhD in Pedagogy and an MA in Educational Research and Teaching Quality from the University of Valencia, Spain. She graduated as a specialist in e-learning from the University of Venetto, Italy. She has been a research manager in educative technology and video games and is currently exploring the "One Laptop per Child" pedagogical model. She has been invited by international universities as an expert in research on video games and learning, and is a Professor in the doctoral program at the University of Salamanca and in e-learning courses at the University of Valencia. She coordinates High Studies in Psychopedagogy and is the author of the book *The Construction of Knowledge in Today's Culture: What Do Videogames Tell Us?* (Edition Alfagrama) and of many scientific publications. She is a researcher at the Universidad Nacional de Tres de Febrero, Argentina. See https://sites.google.com/site/dragracielasnaola/. She can be reached at graesnaola@gmail.com.

Alexander Fedorov has been the president of the Russian Association for Film & Media Education since 2003 and vice rector of the Anton Chekhov Taganrog State Pedagogical Institute (Russia) since 2005. He is a member of the Russian Academy of Film Arts & Sciences (since 2002), the Russian Union of Filmmakers (since 1984), CIFEJ (International Center of Films for Children and Young People, Canada), and FIPRESCI. He holds a Master's degree from the Russian Institute of Cinematography (VGIK, 1983), a PhD (1986), and an EdD (1993) with an emphasis in media education from the Russian Academy of Education (Moscow). His postdoctoral affiliations include Visiting Senior Research Scholar (1998 and 2006) at Central European University, Budapest, Hungary; Visiting Senior Research Scholar (2000) at Kassel University, Germany, and at Humboldt University (Berlin, 2005); Visiting Senior Research Scholar (2002 and 2009) at Maison des Sciences de l'Homme (MSH, Paris, France); and Research Scholar (2003) at the Kennan

Institute (The Woodrow Wilson Center, Washington, D.C., USA). He received scientific grants/fellowships from the Russian Education Ministry (1997–2012), the Russian President Program for Leading Scientific Schools (2003–2005), the MacArthur Foundation (USA), for travel to UNESCO International Media & Media Education Conference (Paris, France, 1997) and for an individual research project (2003–2004); the Russian Foundation for Humanities (1999–2012); the Cultural Foundation of the President of the Russian Federation (2002); the program "Russian Universities" in the humanities area (2002); the Soros Foundation (USA); Research Support Scheme (RSS, 2000–2002); the program "Civil Society" (1998–1999); HESP (HESP-CDC—Course Development Competition, 1998); and Education Program for the best text of university lectures (1997). He was the speaker in many international media and media education/literacy conferences, including the UN "Alliance of Civilizations" (Madrid, 2008), UNESCO Media Literacy Conference (Paris, 2007), Council of Europe Media Literacy Conference (Graz, 2007), Association for Media and Technology in Education, Concordia University (Montréal, Canada, 2003), World Congress "Toys, Games and Media," University of London, Institute of Education (London, UK, 2002), The Council of Europe's Hearing on Internet Literacy (Strasbourg, France, March 2002), the 3rd World Summit on Media for Children (Thessaloniki, Greece, 2001), International Council for Educational Media ICEM-CIME—Conference "Pedagogy and Media" (Geneva, Switzerland, 2000), World Summit 2000: Children, Youth and the Media—Beyond the Millennium (Toronto, Canada, 2000), AGORA European Children's Television Center (Thessaloniki, Greece, 1999), Educating for the Media and the Digital Age (Vienna, Austria, UNESCO, 1999), World Media Education/Literacy Summit (São Paulo, Brazil, 1998), Media & Science (Montréal, Canada, 1997), and Youth and the Media, Tomorrow (Paris, France, UNESCO, 1997). He is the author of four hundred articles and twenty books about media, film, and media education/literacy. See http://mediaeducation.ucoz.ru/load/media_education_literacy_in_russia/8 and http://www.mediagram.ru. He can be reached at 1954alex@mail.ru.

Gonzalo Frasca, PhD, is CEO of Okidoko, an educational game company. He is also Chair of Videogames at the ORT University in Montevideo, Uruguay. As a researcher, his work focuses on game rhetoric, education, and serious games. Frasca received a Lifetime Achievement Award from the Knight Foundation for his Newsgaming.com project. He also codeveloped the first official video game for a US presidential campaign. He can be reached at furazaka@gmail.com.

Anthony Y. H. Fung is a Professor and Director in the School of Journalism and Communication at the Chinese University of Hong Kong. He is also the Pearl River Chair Professor in the School of Journalism and Communication at Jinan University of China. He received his PhD from the School of Journalism and Mass Communication at the University of Minnesota. His research interests focus on cultural industries and cultural studies, popular music, gender and youth identity, the political economy of culture, and new media technologies. Recently, he has been conducting research on Asian game industries and China's media globalization. His new books are *Global Capital, Local Culture: Transnational Media Corporations in China* (2008), *Policies for the Sustainable Development of the Hong Kong Film Industry* (2009) (coauthored with Chan and Ng), and *Asian Popular Culture: The Global (Dis)continuity* (2013). He can be reached at anthonyfung@cuhk.edu.hk.

Enrico Gandolfi is a postdoctoral research fellow at the Research Center for Educational Technology of the Kent State University in Ohio. He has a PhD in Social Theory and Research from the Department of Social and Economic Sciences of La Sapienza University of Rome. His research interests include game studies, subcultures, and qualitative methods with a specific focus on the role played by micro and macro contexts within the gaming experience. He is author of several articles and book chapters about these topics, and of *Piloti di Console* (Edizioni Paoline, 2011), *Nerd Generation* (Mimesis Edizioni, 2014) and *Independent Videogames between Culture, Communication, and Participation* (Unicopli, 2015).

Manuel Garin works at Universitat Pompeu Fabra in Barcelona, where he teaches as Assistant Professor in Film and Media Studies. In 2012, he defended his PhD thesis, which focused on visual gags, later published as a book: *El Gag Visual: De Buster Keaton a Super Mario* (Cátedra, 2014). He is currently expanding the comparative media project Gameplaygag (gameplaygag.com), developed during research stays at the Tokyo University of the Arts and the University of Southern California. He can be reached at manuel.garin@upf.edu.

Global Game Designers Guild (GGDG) is an organization whose goal is to promote learning spaces, information exchange, and business opportunities for present and future studios and gaming start-ups in Latin America. GGDG was born in Colombia in January 2012, led by Luis Ernesto Parra, Ivonne Tovar, and René Serrato, three Colombian entrepreneurs who wanted to create community around the video game industry. The first initiatives were created to inform the community through alternative media, with the support of international guests. Educational programs also emerged with a strong emphasis on game design. Several chapters were born in different regions of Colombia such as Bogota, Antioquia, Valle, Cauca, Bucaramanga, and the Caribbean, as well as international chapters in Mexico, France, and Australia. Thus, GGDG has the ability to reach thousands of members of the gaming community in Latin America. The chapters in Australia and France act as a point of reference and support for people who need information from these countries regarding the video game industry in professional and educational contexts. GGDG has alliances with major international companies and organizations, with the goal of supporting and strengthening the gaming industry in Latin America. See www.ggdg.co and http://www.ggdgcolombia.co.

Daniel Golding is a writer, researcher, and critic from Melbourne, Australia. He has a PhD from the School of Culture and Communication at the University of Melbourne, Australia. His research underlines a spatial logic of video games, drawing on a number of cross-disciplinary theories of space, architecture, and cultural consumption. Daniel also teaches in the fields of cinema, culture, video games, and digital media. As a critic, Daniel writes commentary on video games and gaming culture for ABC Arts. He also regularly writes for *Hyper Magazine* and has been published in Crikey.com.au, *Meanjin*, *Kill Your Darlings*, and *Kotaku*. He can be reached at dangoldingis@gmail.com.

Louis-Martin Guay is Assistant Professor of game and interactive design at the University of Montréal. With a Bachelor's degree in drama and a Master's degree in film studies, he has been involved in game design for more than ten years. His projects include many video games, the theatrical collective Cinplass, and the Open House concept. He can be reached at louis.martin.guay@umontreal.ca.

Bryan Hikari Hartzheim is a game designer and PhD candidate in Cinema and Media Studies at the University of California, Los Angeles (UCLA). He can be reached at bhartzheim@gmail.com.

Alejandro Iparraguirre is Senior Technical Marketing and Electronics Technician and Producer and Developer of video game events. He is also a Professor at the University of Palermo in the Multimedia Department at the Faculty of Design and Communication. He currently takes part in the video game industry, working for Gamester and Codenix, and has an active role in ADVA (Game Developers Association of Argentina) and Duval (Video Game Developers States of Latin America), and is Coordinator of Video Games for Argentina's Secretary of Culture. He is a producer of video games who coordinates a development team, and also acts as a technology consultant in the area of industry marketing. As personal and complementary enterprises can conflict, he is also a part of Work Game Jam (Entrepreneurship) and VgMap (Virtual Video Game Industry Map). He is also the coordinator of the gaming sector in the Ministry of Culture of the Nation (Videogames as a Culture Industry) and coordinates Videogame Work Jams all over the country. He can be reached at aleiparraguirre @gmail.com.

Toru Iwatani is a Game Creator and Professor in the Faculty of Arts, Department of Games at Tokyo Polytechnic University. Born in Tokyo in 1955, Iwatani joined Namco (currently Bandai Namco Games) in 1977, and created the video game *Pac-Man* (1980). Based on the theme of "eating," the game garnered worldwide critical acclaim and, in 2005, was recognized by Guinness World Records as the world's Most Successful Coin-Operated Arcade Game. In addition to *Pac-Man*, he has produced more than fifty other titles including *Pac-Land* (1984), *Ridge Racer* (1993), *Alpine Racer* (1995), and *Time Crisis* (1995). He became a professor in the faculty of Tokyo Polytechnic University in 2007. He is also a director of the Digital Games Research Organization Japan (DiGRA Japan) and a fellow at Bandai Namco Games. He is the author of *Pakkuman no Gēmu Gaku Nyūmon* (*The Pac-Man Game Methods Primer*, Enterbrain, 2005).

David Javet is an SNSF PhD student in Film Studies at the University of Lausanne. His PhD thesis explores the representations of technology in Japanese films and video games. He was a guest speaker at the first Swiss international academic conference "Powers of Video Games: From Practices to Discourses." David is also game designer and cofounder of Tchagata Games, a creative hub dedicated to the exploration of gaming in whatever form of support: video games, card games, and things in between. He can be reached at david.javet@gmail.com.

Radwan Kasmiya is a pioneering game creator in the Middle East, having created the first video game and computer-generated animation there, and many other bestsellers such as *Under Ash* (2001), the first

first-person shooter video game in the Middle East; *Quraish* (2005), a real-time strategy game; *Under Siege* (2004); *Road to Jerusalem* (2009); and others. Kasmiya was awarded the Creative Young Entrepreneur Award in Syria in 2007 as founder of AfkarMedia and has worked on many gaming projects with Disney-Pixar, THQ, Ubisoft, EA, and others on titles such as *BattleField 2* (2005), *Assassin's Creed* (2007), *World in Conflict* (2007), *Far Cry 2* (2008), *Tom Clancy's EndWar* (2008), *WALL·E* (2008), *Red Faction: Guerilla* (2009), and *Up* (2009). Kasmiya holds an Electronic/Computer Engineering degree from the University of Damascus and is passionate about creation. He is cofounder of Falafel Games and creator of *Knights of Glory* (2011). See http://ae.linkedin.com/pub/radwan-kasmiya/3/b07/829. He can be reached at kasmiya@gmail.com.

Wesley Kirinya is a tech entrepreneur based in Nairobi, Kenya. He has founded three software companies in the last eight years. His first company, Sinc Studios, was a video game development company. He released his first title, *Adventures of Nyangi*, in 2007. Because of the game, he won an award as Africa's Top ICT Youth Innovator in 2008, and made valuable contacts in the global professional game development industry. His experience in launching the game and founding his first company set the foundation for his entrepreneurship journey. His latest company, Leti Games, focuses on developing games and comic content for mobile. These are based on African mythology and folklore. Besides developing games, he consults for other up-and-coming game development companies in the African continent. He has attended two universities, Wichita State, Kansas, and University of Nairobi. However, in 2005, he deferred his studies to pursue entrepreneurship in video games development, to not only build a video game company, but also a video game industry in Africa. This has led him to speak and showcase his works at various universities and conferences, the largest being the Game Developers Conference, San Francisco, and the Mobile World Congress, Barcelona. Wesley has also worked as a Software Engineer in Genkey Africa and BRCK, developing enterprise and embedded systems. He can be reached at wesley@letigames.com.

Lars Konzack is an Associate Professor in Information Science and Cultural Communication at the Royal School of Library and Information Science (RSLIS) at the University of Copenhagen in Denmark. He has an MA in Information Science and a PhD in Multimedia. He was cofounder of the game-developing academy DADIU. He is working with subjects such as ludology, game analysis and design, geek culture, and sub-creation. Currently, he is teaching subjects such as video game analysis, Internet culture, media culture, knowledge media, culture analysis, and the various genres of the fantastic. He has, among others works, published "Computer Game Criticism: A Method for Computer Game Analysis" (2002), "Rhetorics of Computer and Video Game Research" (2007), "Video Games in Europe" (2007), and "Philosophical Game Design" (2009), "A Characterology of Tabletop Role-Playing Games" (2013), and "The Cultural History of LEGO" (2014). See https://nru.academia.edu/LarsKonzack. He can be reached at lars@konzack.dk.

Andreas Lange is director of the Computer Game Museum in Berlin. He studied Comparative Religions and Dramatics (MA) at Freie Universität Berlin. His 1994 graduation work, *The Stories of Computer Games—Analysed* *as Myth*s, was one of the first academic works in which computer games were treated as cultural artifacts. Since then, he has worked as a curator, author, consultant, and expert, among other professions, in the business of interactive digital entertainment culture for the German age-rating system USK. Since 1996, he has been the director of the Computer Game Museum in Berlin, which opened in early 1997 as the world's first permanent exhibition dedicated solely to interactive digital entertainment culture. On that basis, Lange lectures in academic and other contexts. He has held positions as the museum's project manager of the European research project KEEP and speaker of the SIG Emulation of the German competence network for digital preservation. In addition to that, he is a member of the Academy of the German Game Developers, the jury of the German Games Award Lara, and the advisory council of the Deutsche Gamestage and the Stiftung Digitale Spielekultur. Lange is co-initiator of EFGAMP (European Federation of Game Archives Museums and Preservation Projects). See www.computerspielemuseum.de. He can be reached at lange@computerspielemuseum.de.

Sara Xueting Liao is currently a PhD student in the Moody College of Communication at the University of Texas at Austin. She graduated from the Chinese University of Hong Kong with a Master's of Philosophy in Communication and was an assistant on a Hong Kong creative industries project for two years. Her research interests lie in fields of new media, cultural studies, globalization, transnational media, and modernity, with a specific focus on Asian and Chinese communities. She can be reached at saraconan881106@gmail.com.

Michael Liebe is founder and head of the International Games Week Berlin. He is a Berlin-based consultant for the games, online technology, and "new media" industries. Moreover, he is Ambassador of the industry network, games.net berlinbrandenburg, and among others, responsible for the Berlin-meets-Poland project. In the past eight-plus years he has initiated the industry network, games.net berlin-brandenburg, the indie games and art festival, A MAZE., and the Digital Games Research Centre of the University of Potsdam (DIGAREC), as well as the Computer Games Collection of European Media Studies at the University of Potsdam, the European Masters program, Ludic Interfaces, and many other events and programs surrounding the video game industry and culture. In 2013, he was Project Lead at Germany's leading game developers' event, DGT 13—Deutsche Gamestage, and also worked on the German Israel Congress as well as the International Media Convention. Until 2012 he was Advisor–New Media at the Medienboard Berlin-Brandenburg and responsible for marketing and networking for the games industry in the German capital region. See http://www.michael-liebe.de, http://www.gamesweekberlin.com, http://www.berlin-meets-poland.de, and http://www.digarec.org. He can be reached at mjw@michael-liebe.de.

Víctor Manuel Martínez is the Chief Editor of AnaitGames.com, where he writes about video games, cinema, and pop culture. He was in charge of the video games section of the Spanish edition of *VICE Magazine* and collaborates with the Spanish editions of Eurogamer and FHM as a critic. He has published fiction in *VICE Magazine* and *Mamajuana!* He can be reached at victor.martinez@anaitgames.com.

Frans Mäyrä is Professor of Interactive Media, Game Studies and Digital Culture at the University of Tampere. He has studied the relationship between culture and technology since the early 1990s. He has specialized in the cultural analysis of technology, particularly on the ambiguous, conflicting, and heterogeneous elements in this relationship, and has published on topics that range from information technologies, science fiction and fantasy, to the demonic tradition, the concept of identity, and role-playing games. He is currently teaching, researching, and heading numerous research projects in the study and development of games, new media, and digital culture. He has also served as the founding President of the Digital Games Research Association (DiGRA). Publications: *Koneihminen* (*Man-Machine*, editor, 1997), *Demonic Texts and Textual Demons* (1999), *Johdatus digitaaliseen kulttuuriin* (*Introduction to Digital Culture*, editor, 1999), *CGDC Conference Proceedings* (editor, 2002), *Lapsuus mediamaailmassa* (*Childhood in the World of Media*, editor, 2005), *The Metamorphosis of Home* (editor, 2005), and *An Introduction to Game Studies* (2008). See http://www.uta.fi/~frans.mayra/ and http://www.fransmayra.fi. He can be reached at frans.mayra@staff.uta.fi.

Deborah Mellamphy completed her PhD in the School of English, University College Cork, Ireland in 2010. Her thesis is entitled *Hollyweird: Gender Transgression in the Collaborations of Tim Burton and Johnny Depp*. She has taught undergraduate and postgraduate courses and has published journal articles in *Widescreen and Film* and *Film Culture*, with chapters in upcoming publications on film, television, and video game studies. She is coeditor of *Alphaville: Journal of Film and Screen Media*, an online journal aimed at graduates and early career researchers. She can be reached at deborahmellamphy@gmail.com.

Konstantin Mitgutsch is a thinker, game design consultant, coach, and lecturer in the field of learning, play, and personal development. At present, he works as a Research Affiliate at the MIT GAME LAB at the Massachusetts Institute of Technology in Cambridge and as a coach and consultant in Vienna. His work focuses on purposeful game design and learning processes, and in particular, on the role of perspective change and innovative forms of learning. In his approach, he combines novel insights into learning, with concepts of game creation and meaningful play experiences. Mitgutsch studied Media Education and Philosophy of Education at the University of Vienna and the Humboldt University of Berlin. He is participating as an expert member for the Austrian Federal Office for the Positive Assessment of Computer and Console Games and is on the expert council of the Pan European Game Information (PEGI). Since 2007, he has organized the annual Vienna Games Conference FROG and has published books on games, learning, and culture. He can be reached at k_mitgut@mit.edu.

Souvik Mukherjee is Assistant Professor of English Literature at Presidency University (previously Presidency College), Calcutta. Souvik has been researching video games as an emerging storytelling medium since 2002 and completed his PhD on the subject at Nottingham Trent University in 2009. He has done postdoctoral research with the Humanities faculty of De Montfort University, UK, and as a research associate at the Indian Institute of Technology in New Delhi, India, where he worked on digital media as well as narrative analysis. His research examines their relationship to canonical ideas of narrative and also how these games inform and challenge current conceptions of technicity, identity, and culture, in general. His current interests involve the representations of empire in video games, the analysis of paratexts of video games such as walkthroughs and after-action reports, as well as the concept of time and telos in video games. His research has been published in the *Journal of Gaming and Virtual Worlds* and *Writing Technologies*, as well as in various edited collections on game studies and popular culture. His other interests are digital humanities and early modern and classical literature. Souvik regularly writes about his research on his blog "Ludus ex Machina" (http://readinggamesandplayingbooks.blogspot.com/) and welcomes comments and feedback. As a gamer, he prefers strategy games, although he is a big fan of *Fallout 3* and the *STALKER* games. He can be reached at prosperoscell@gmail.com.

Arturo Nakasone is a former Research Associate Professor and Project Manager at Japan's National Institute of Informatics. His research interests include 3-D visualization, virtual storytelling, and virtual world and game interactivity. He collaborated with the *Grupo Avatar* at the Pontificia Universidad Católica del Perú (Pontifical Catholic University of Peru) to define strategies for virtual-world-based applications and game scenarios. Nakasone has a PhD in information science from the University of Tokyo, and is currently engaged in the development of visualization tools for mobile platforms and the Oculus Rift headset technology. He can be reached at arturo.nakasone@gmail.com.

Benjamin Wai-ming Ng is currently Professor of Japanese Studies and the director of the Research Centre for Comparative Japanese Studies at the Chinese University of Hong Kong, teaching and researching Japanese popular culture, Japanese intellectual history, and Japan-Hong Kong relations. He received his doctorate in East Asian Studies from Princeton University in 1996 and was an Assistant Professor in Japanese Studies at the National University of Singapore from 1996 to 2001. He is working on a research project on the interaction and collaboration between Japan and Hong Kong in the creative industry and preparing a book manuscript on Japanese popular culture in Hong Kong. He can be reached at waimingng@cuhk.edu.hk.

María Luján Oulton (@LuOulton) is a Cultural Producer focused in Art and New Media. She graduated in 2004 in PR and Communication and afterward she specialized in Arts and Cultural Management and is currently working on her Master's thesis about art games. María Luján specializes in creating new media art events and activities that promote cross-disciplinary practice. After some initial experience in a technology-focused agency, in 2007 she started an independent family project named *Objeto a: An Art Gallery* that also works as a cultural producer. The project first focused on contemporary art but then evolved to include new media art, encouraging the realization of exhibitions and events that involve art, science, and technology. Currently, *Objeto a* is recognized as one of the few institutions in Argentina that cares for new media art in the form of exhibitions, contests, and a variety of transdisciplinary activities. María Luján runs *Game on! El Arte en Juego*, the first and only exhibition in Argentina focused on art, technology, and play. It originally started in 2009 and has already had three editions, hosted and invited by several educational and cultural institutions. She also worked on the divulgation of art games, giving lectures in institutions and universities such as Universidad de Palermo (Buenos Aires), TEDxRiodelaplata (Buenos Aires), Universidad Jorge Tadeo Lozano (Bogotá), and DevHourmx (Mexico DF), among others. See http://www.gameonxp.com. She can be reached at lujan.oulton@objeto-a.com.ar.

Luis Parra is CEO and cofounder of Press Start Studios, a mobile gaming company located in Colombia and San Francisco, California, and graduate of the Academy of Art University. Mr. Parra's professional experience includes more than seven years working on casual and mid-core games for PC, mobile, and Web at the best development houses in Colombia and at Press Start Studios. He is also a Gamification professor at the Academy of Art University. His games, including those he's worked on at both Press Start and at other studios, have garnered more than a million mobile users. He is also president and cofounder of Global Game Designers Guild (GGDG), a Latin American game association with the goals of educating, uniting, and promoting the LATAM game industry. He can be reached at lparra@pressstart.co.

Matthieu Pellet has a PhD in the history of religions, specializes in Mediterranean antiquity, and has been a lecturer at various universities. Of course, he was also bathed in video games during his childhood and continues to spend a lot of time with them. Organizer and speaker at the first Swiss international academic conference "Powers of Video Games: From Practices to Discourses," Matthieu also held a class called "Video Games and Virtual Worlds" at the Swiss Federal Institute of Technology Lausanne (EPFL). Matthieu is also graphic designer and cofounder of Tchagata Games, a creative hub dedicated to the exploration of gaming, in whatever form of support. He can be reached at matthieu.pellet@unil.ch.

Jacinto Quesnel was drawn into the industry in 2008, when his passion for making art and games pushed him to attempt a video game indie studio in Mexico. Soon he realized this was not an easy task, and then, trying to help everyone else, cofounded DEVHR.MX, the largest game development event in Latin America, for which he's still the CEO. He also loves teaching and has taught classes, most remarkably at SAE Institute Mexico, Universidad del Valle de México, and Universidad Nacional Autónoma de México (UNAM). He's currently leading the Centro de Cultura Digital's Game Innovation Labs, organizing Game Jams and doing Game Studies research at UNAM. He can be reached at jacinto.quesnel@gmail.com.

Joost Raessens holds the chair of Media Theory at Utrecht University, The Netherlands. He has been Visiting Professor, both at the University of California, Los Angeles, and at the University of California, Riverside. His research concerns the "ludification of culture," focusing in particular on the playful construction of identities, on applied or serious gaming, and on the notion of play as a conceptual framework for the analysis of media use. Raessens was the conference chair of the first Digital Games Research Association (DiGRA) conference, Level Up, in Utrecht (2003), and is on the editorial board of *Games and Culture* (SAGE). He coedited *Level Up: Digital Games Research Conference* (Utrecht University, 2003), the *Handbook of Computer Game Studies* (MIT Press, 2005), and *Digital Material: Tracing New Media in Everyday Life and Technology* (AUP,

2009), and *Playful Identities: The Ludification of Digital Media Cultures* (AUP, 2015). Raessens was supervisor of the Playful Identities (NWO), the Design Rules for Learning through Simulated Worlds, and the Mobile Learning (both GATE) research programs. Currently, he supervises the research project Persuasive Gaming: From Theory-Based Design to Validation and Back as part of the Creative Industries Programme of the Netherlands Organisation for Scientific Research (NWO). He is the scientific director of GAP: The Center for the Study of Digital Games and Play, Faculty of Humanities, Utrecht University. See http://www.raessens.nl and http://www.gamesandplay.nl. He can be reached at j.raessens@uu.nl.

Inaya Rakhmani is Head of the Communication Research Centre, Universitas Indonesia, and an Associate at the Asia Research Centre, Murdoch University, Australia. Her work has appeared in the *Review for Indonesian and Malaysian Affairs and Internetworking Indonesia Journal* and the *Asian Journal of Social Science and International Communication Gazette*. Her current research interests are related to the media industry, media politics, media and identity, as well as power and cultural expression in the Indonesian context. She can be reached at inaya.r@ui.ac.id.

Herbert Rosenstingl is director of the youth department and manager of the "Bundesstelle für Positivprädikatisierung von Computer- und Konsolenspielen" (BuPP) at the Austrian Federal Ministry of Families and Youth. His main responsibilities include the national implementation of a positive assessment policy for computer and console games. Rosenstingl graduated in Computer Game Studies at the Danube University Krems and is coauthor of the book *Schauplatz Computerspiele*. He can be reached at herbert.rosenstingl@bmfj.gv.at.

Songsri Soranastaporn is an Associate Professor in English. She has a BN in nursing, an MA in Applied linguistics from *Mahidol University, Thailand*, and a PhD in Educational Administration and Foundations from the College of Education, Illinois State University. Her books include *Effective Reading and Writing English Texts*, and she has contributed book chapters to numerous edited collections. She is an expert in English for Specific Purposes. She is the Coordinating Editor of the "Association News & Notes" column of *Simulation & Gaming Journal*, a Sage publication. She is the cofounder (2008) and Secretary General of the Thai Simulation and Gaming Association (ThaiSim), which invites scholars around the world to join its international conference every year (http://www.thaisim.org). She is on the committee of the e-Learning Association of Thailand (e-LAT), which aims to strengthen, promote, and establish e-learning quality and standards (http://www.e-lat.or.th). She is an editor and reviewer of both Thai and international journals and conferences such as *Current Topics in Management*, the *Kasetsart Journal* (Social Sciences) at Kasetsart University, the *Journal of Language and Culture* at Mahidol University, Khonkan University's *International Journal of Humanities and Social Science*, *RMUSTV Research Journal*, *International Conference on Advances in Management* (ICAM), ThaiSim International Conference, and the International Conference on Computers in Education (ICCE). She can be reached at songsrisora@hotmail.com.

Melanie Swalwell is a scholar of digital media arts, cultures, and histories and Associate Professor in the Screen and Media Department at Flinders University, Adelaide. Melanie has published a suite of projects on New Zealand's unique games history and was keynote speaker at the first International History of Games conference, held in June 2013. She is Project Leader of the multidisciplinary Linkage project "Play It Again," which is concerned with the history and preservation of digital games from 1980s Australia and New Zealand. She is also an ARC Future Fellow, researching a history of "Creative Micro-Computing in Australia, 1976–1992." Together with Denise de Vries, she runs the "Australasian Heritage Software Database," http://www.ourdigitalheritage.org. She can be reached at melanie.swalwell@flinders.edu.au.

Cetin Tuker is an animator, illustrator, instructor, and architect specializing in 3-D animation, interactive multimedia, and motion graphics design. He completed his undergraduate and graduate education in the METU Department of Architecture in 1992 (BA) and 1996 (MA), and also completed his thesis for Doctor of Fine Arts (DFA) at Mimar Sinan Fine Arts University, Department of Graphic Design, in 2009. Since 1993, he has been awarded several times for his animations and designs in several fields. He continues to teach, which he has done since 1994 as an Assistant Professor in Mimar Sinan Fine Arts University, Department of Graphic Design. In 2007 he published his first book, *From Design to Application: 3D Environment Design*, and in 2007 and 2008, his educational DVDs, *Introduction to 3DS Max* (2007) and *Fluid Simulation with GLU 3D* (2008), were released by the Gnomon Workshop. He can be reached at cetin_tuker@yahoo.com.

Patrik Vacek served as a research assistant at the Department of Education (Faculty of Education, Masaryk University, Czech Republic) from 2008 to 2012, where he realized the first state-supported course related to video games in education. He has an MA in Film Studies from the same university. Currently, he is a freelance researcher focused on media and game literacy and video game history and aesthetics. He can be reached at patrikvacek@seznam.cz.

Christel van Grinsven holds an MSc and is operations manager at Dutch Game Garden, an incubator and network organization for the Dutch Games industry. She is responsible for a number of projects in which cooperation between game developers, knowledge institutions, and other industries is the essence. These projects concern both the development of the game industry as a whole (for example, by stimulating talent in game education) and stimulating the use of applied games in some specific sectors (for example, health and education). The international promotion of the Dutch Games industry, supporting demand articulation between potential clients and game companies, and disseminating knowledge about the sector, are part of her experience. Van Grinsven was one of the initiators and the director of the *Gamesmonitor* (2012 in Dutch) and *The Dutch Games Industry: Facts & Figures* (2013). See www.dutchgamegarden.nl. She can be reached at christel@dutchgamegarden.nl.

Mark J. P. Wolf is a Full Professor and Chair of the Communication Department at Concordia University Wisconsin. He has a BA (1990) in Film Production and an MA (1992) and PhD (1995) in Critical Studies from the School of Cinema/Television (now renamed the School of Cinematic Arts) at the University of Southern California. His books include *Abstracting Reality: Art, Communication, and Cognition in the Digital Age* (2000), *The Medium of the Video Game* (2001), *Virtual Morality: Morals, Ethics, and New Media* (2003), *The Video Game Theory Reader* (2003), *The World of the D'ni: Myst and Riven* (2006), *The Video Game Explosion: A History from PONG to PlayStation and Beyond* (2007), *The Video Game Theory Reader 2* (2008), *Before the Crash: An Anthology of Early Video Game History* (2012), the two-volume *Encyclopedia of Video Games: The Culture, Technology, and Art of Gaming* (2012), *Building Imaginary Worlds: The Theory and History of Subcreation* (2012), *The Routledge Companion to Video Game Studies* (2014), *LEGO Studies: Examining the Building Blocks of a Transmedial Phenomenon* (2014), the four-volume *Video Games and Game Cultures* (Routledge Major Works) (forthcoming), and two novels for which he is looking for an agent and publisher. He is also founder and coeditor of the Landmark Video Game book series from the University of Michigan Press. He has been invited to speak in North America, South America, Europe, Asia, and Second Life, and is on the advisory boards of Videotopia and the *International Journal of Gaming and Computer-Mediated Simulations*, as well as on several editorial boards including those of *Games and Culture*, *The Journal of E-media Studies*, and *Mechademia: An Annual Forum for Anime, Manga and The Fan Arts*. He lives in Wisconsin with his wife Diane and his sons Michael, Christian, and Francis. He can be reached at mark.wolf@cuw.edu.

Erdal Yilmaz has BS degrees in civil engineering (1993) and geomatics engineering (1995). He received his MSc (2003) and PhD (2010) from the Information Systems Department of Informatics Institute, METU. His main areas of interest are computer graphics, computer games, and GIS. Until 2013, he worked as an R&D Manager of Sobee Studios, where he contributed to several mobile games and research activities. In 2014, he decided to bridge the gap between video gaming and GIS industries. Currently he is the CTO of Kanava Tech, a company that specializes in City Visualization by using video game engines. He can be reached at erdal.yilmaz@kanavatech.com.

Nelson Zagalo is Professor of Interactive Media at the University of Minho, Portugal. He earned his PhD in Communication Technology working on new interaction paradigms in virtual environments. He is director of the Master of Interactive Media degree program and on the board of directors for the Master of Technology and Digital Art degree program. He cochairs the research group engageLab and was a founding member of the Portuguese Society of Videogame Sciences. He has more than seventy peer-reviewed publications in the fields of video games, film studies, interactive storytelling, and aesthetics of emotion. He is author of the books *Interactive Emotions, from Film to Videogames* (2009) and *Videogames in Portugal: History, Technology and Art* (2013) and is also coeditor of the books *Interactive Storytelling* (2009) and *Virtual Worlds and Metaverse Platforms: New Communication and Identity Paradigms* (2011). See http://nelsonzagalo.googlepages.com and http://virtual-illusion.blogspot.com. He can be reached at nzagalo@ics.uminho.pt.

INDEX

10th Art Studio (company), 310
11 bit studios (company), 412
1814 Rebellion in Cusco, 395
1869 (game), 72
1943: One Year After (game), 453
1C Company (company), 441, 443
1NSANE (game), 232
20th Century Fox (company), 595
2K Czech (company), 154. See also Illusion Softworks (company)
3000 FDD (computer), 428
324—El chofer mercenario (game), 38
3-D graphics. *See* Three-dimensional graphics
3DO, 52, 602
3DO Interactive Multiplayer, 599, 601
3DS, 312
3nMedia (company), 346
3S Games (company), 396
3TE Games (company), 570–571
4 minutos 44segundos (game), 45
44 Bico Largo Multimídia (company), 89
5th Generation (console system), 122
50 Cent: Bulletproof, 64
505 Games, 309
5-3-3 São Paulo Futebol Clube, 91
576 Kilobytes, 231
6 Grandes Juegos Vol. I (game collection), 37
6 Great Games Vol. II (game collection), 37
6 Waves (company), 42
6 ženichů a 1 navíc (*6 Grooms and 1 More*) (game), 154
64ler (magazine), 566
7 Colors (game), 440, 442
7 dní a 7 nocí (*7 Days and 7 Nights*) (game), 153
8 Segundos (game), 91
989 Studio (company), 602

AAA (Mexican Wrestling), 354
Aamber Pegasus (computer), 386
Aaron's Amusement Games, 378
AARRR! Ninjas Beware (game), 560
Aarseth, Espen, 98, 463
Abadía del Crimen, La (game), 523
Abandoned Places (game), 226
Ability to see 360 degrees, 561
A bosszú (*The Revenge*), 226
ABRAGAMES, 87, 101
Accolade (company), 108, 184
Accordi Rickards, Marco, 314
Acid Software (company), 388
ACIGAMES, 87, 101
Acorn Computers, 582
Across the Dnepr: Second Edition, 59
Act against Unjustifiable Premiums and Misleading Representations, 335
Acte Europa (game), 154
Action Games (magazine), 53
Action Replay (device), 585
Activate (company), 480

646 INDEX

Activision (company), 14, 110, 186, 196, 203, 221, 566, 595
Activision Blizzard (company), 295, 297
Activision-Blizzard Italia, 310
Actoz (company), 125
A.D. 2044 (game), 410–411
Ádám, Zoltán, 223, 225
Adams, Ernest, 235
Addiction, 100, 127–128, 152, 212, 215, 497, 500–501, 561–562
Adeline Software International, 183–184
Adrianopolis (fictional city), 572
Advance Automatics, 378
Adventure (Atari VCS 260 game), 595
Adventure Game Maker (program), 566
Adventures of Nyangi, The (game), 9, 20, 22, 26
Adventures of the Hunter (game), 443
Advergames, 31, 36, 42, 139, 355–356
Adversary (console), 593
Aeria Games and Entertainment (company), 560
AESVI (Associazione Editori Sviluppatori Videogiochi Italiani), 305, 308, 310, 311, 312, 313, 316, 317n3
AESVI4Developers, 310
Afghanistan, 442
Afkar Media, 10, 30, 32
AFL Finals Fever, 59. *See also* BlueTongue Entertainment
Africa Journal, 20
Africa, 3, 6, 17–28
African Game Conference, 25
Afrikan tähti, 159
Afroes, 20, 24, 28
Agartha (game), 185
Agate Studio (company), 253
Age of Conan: Hyborian Adventures, 457
Age of Empires (game), 395, 431
Age of Empires III: The Asian Dynasties (game), 236
Age of Pahlevanas I (game), 272
Agent 47 (game character), 456, 462
Aggression, 19, 445–446
AIM Advance, 552
AIOMI (Associazione Italiana Opere Mediali Interattive), 310, 313–314

AION, 509
Akbaş, Ilgaz, 566
Al-Alamyyeh, 30
Alan Wake (game), 167
Alatlı, Alev, 572
Alberti, Diego, 45
Alcobia, Alice, 430
Alianza Bolivariana para los Pueblos de Nuestra América (Bolivarian Alliance for the Americas) (ALBA), 615
Alien (1984 game), 583
Alien (film), 179
Alien Evolution (game), 426–427
Al-Khayal, 31
All Civilized Planets (company), 75
All4Games, 313
Alliance of Valiant Arms SEA (AVASEA), 559
Allods Online, 443
Al-Majd (television channel), 32
Al-Moosiqar, 31
Altair 8800, 593
Alternative World Games, 223
ALT-PLAY: Jason Rohrer Anthology (game collection), 40
Aluna (game), 141
Amanita (company), 154
A MAZE., 201, 205
Amazing Studio (company), 183
Amazone (arcade chain), 251
Amazônia (game), 87–89
Amble, Ole, 451
Ambulance for gaming addiction (Spielsucht Ambulanz), 204
America's Army (game), 31
American Museum of the Moving Image, 604
AMIGA. *See* Commodore Amiga
Amiga Professional Users Club (APUC), 566–567
Amiga World, 604
Amorn Apithanakoon, 560
Amstrad (computer), 180, 299, 453, 565, 584
Amstrad CPC, 582, 584
Amstrad GX4000, 584
Amusement Game Centres Ordinance, The, 208

Amutronics (company), 176
Ananse, 20, 23
Anarchy Online, 293, 457
Anastasov, Andrej, 149
Anatomik, 440
Ancel, Michel, 186–187, 189–190
Andalussoft, 31
Andersen, Martin Stig, 458
Andrade, Oscar, 142
Andrea, Pessino, 305, 309
Andreasen, Simon Jon, 455
Andreotti Law, 6
Andrés y la Ballena, 140–141
Android (company), 238
Android Market, 155
Android, 28, 33, 40, 213, 365, 396–397, 432, 441, 537, 611
Andromeda (company), 220, 221, 223
Andrucas, 396
Angry Birds (game), 11, 168, 213–214, 253, 368, 426
Animal Crossing: New Leaf, 312
Animé, 319, 320, 323, 325
Anipark (company), 511
Anisimova, I. V., 446
Ankama (company), 188
Anno 1503 (game), 72
Anno 1602 (game), 71–73
Anno Online, 72
Anno series, 72, 194, 201
Annunziata, Ettore, 225
Another World (game), 181–182
Anstoss series, 194
Antarctica, 13–15
Ant Attack, 583
Antichamber (game), 61
Anzola, Carlos, 141
Aoi Sekai no Chuushin de (game), 323
Apoku, Donald, 28
Apollo Entertainment (company), 89
Appaloosa (company), 225
Apple (company), 143, 180, 386, 441, 598, 601–602
Apple I (computer), 593

Apple II (computer), 598
Apple IIe (computer), 599
Apple Lisa (computer), 599
Apple Macintosh (computer), 599
Apple Store, 155, 360, 365, 368, 370
Applied BASIC Programming Language Book for Sinclair Spectrum, Sinclair ZX81, Commodore VIC-20, Commodore 64, BBC A/B, Electron, Oric-Atmos, Texas Instruments 99/4a, Apple IIe, and Sharp MZ80K (book), 566
Aquafresh (game), 369
Aquiris, 91–92
Arab World, 29–34
Arabian Lords, 31
Arabian Nights, 31
Arai, Akihiko, 252
Arbeitsgemeinschaft Games (SIG Game), 205
Arbister, Ariel and Enrique, 36–37
Arcade artwork, 380–381
Arcade clones, 378
Arcade Flyer Archive, 605
Arcade games, 37, 121, 127, 146, 177–178, 207–209, 213–214, 306, 378–381, 383–384, 393, 440, 505, 546, 552, 560, 580, 593, 599–601, 605, 439
Arcade operators, 380
Arcades (game centers, game parlors), ix, 5, 17, 19–20, 53, 124, 127, 176–177, 251, 321, 322, 327–328, 329, 332–34, 335, 347–349, 378–384, 393, 497–498, 525, 536, 594, 596–598
Arcania: Gothic IV, 74
Archibald's Adventure (game), 157n17
Archivio Videoludico of the Cineteca of Bologna, 314
Arc the Lad, 331
Arctic Shipwreck, 223
Arenas, Carlos, 141
Argentina, 11, 35–56, 354, 594, 609
Argentine Videogames Expo (EVA), 40, 50
Argentum (world), 39
Argentum Online (game), 38–39
Argonaut Software, 585
Ariolasoft, 223
Arkane (company), 188

Arkanoid (game), 37
Arkis (application), 141
Armanto, Taneli, 169
Armor Games portal, 611
Armor Valley, 478
Armstrong, Simon, 388
Army of Two (game), 110, 114
Ars Electronica (event), 431
Arsenal Brno (company), 147, 155
Artech Studios (company), 108
Artefacto Studio (company), 355
Artematica Entertainment (company), 308
Artex Studios (company), 232
ArtGame (company), 226
Artificial intelligence (AI), 30, 572
Artificial Mind and Movement (A2M) and Behavior Interactive (company), 109
Artificial Studios (company), 138
Artigames (company), 395–396
Art of Computer Game Design, The, 604
Art Scene (newspaper), 20
Arvin Tech (company), 272
Ascaron (company), 194
Ascione, Ciro, 315
A.S.F. (company), 406
Asheron's Call, 602–603
Asiama, Opuni, 20
Asian financial crisis of 1997, 505
Asia Pay (channel), 475
Asiasoft (company), 474, 486, 546, 553, 558, 559
Asiasoft On-line Pte Ltd. (company), 480
Asmalı Konak (*The Mansion with Vines*), 573
Asmandez (game), 272–275
Asobo (company), 188
Asociación de Desarrolladores de Videojuegos of Argentina (Argentine Videogame Developers Association, ADVA), 35–36, 39–40, 42, 48, 50
Asphalt (game), 506
Assassin's Creed (game), 110, 114, 296, 310–311, 484
Assassin's Creed II (game), 110, 232

Assassin's Creed III (game), 232, 311
Assassin's Creed: Brotherhood (game), 110
Assassin's Creed: Revelations (game), 110
Assembly (event), 161, 460
Asteroids (game), 17, 37, 293, 596
Astragon (company), 194
Astro Wars (game), 385
Astronauts, The (company), 417
Astrum Online Entertainment (company), 443
Asura Online (game), 506–507, 545, 561
Asylum (game), 41
AT 286/386 (computer), 148
Atari 400 (computer), 386, 596, 599
Atari 7800 Prosystem (console), 386, 584, 598, 601
Atari 800 (computer), 386, 596, 599
Atari 800XL (computer), 147
Atari Age (magazine), 604
Atari Electronics Company (company), 87
Atari Games (company), 600
Atari Inc. (company), ix, xii, 6, 29, 37, 87, 94, 138, 176–180, 185, 196, 203, 208–209, 233, 293, 297, 321–322, 346, 348, 349, 353, 394, 471. 579, 593–597, 601
Atari Jaguar (console), 52, 601
Atari Lynx (console), 601–602
Atari ST (console), 182, 584–585, 596
Atari VCS 2600 (console), 5, 10, 29, 53, 87, 196, 208–209, 306, 349, 384–386, 393, 536, 581, 594–597, 601–602
Atari XE (console), 386, 601
Atomix (magazine), 345–346, 351
ATP Tour Championship Tennis (game), 210
ATR Multimedia, 89
Atrévete a Soñar (game), 354
Attack of the Mutant Camels, 583
Attacks on digital entertainment, 314. See also Movimento contro la disinformazione sui videogiochi
Attic (company), 194
Auckland, New Zealand, 382, 384–386
Audition Dance Battle Online (game), 212, 215
Audition Online, 506–507, 561

Augmented Reality Lab, Digital Games and Digital TV (research group), 98
Australia, 7, 9–10, 57–70, 144, 213, 380, 386, 388–389, 594, 597
Australian Bureau of Statistics, 58
Austria, 11, 71–86, 227
Austrian Federal Ministry for Youth, 71
Austrian Game Jam, 82
Auto Racing, 595
Autumn Dynasty, 487
Avatar (game), 545, 561
Avatarika, 443
Avatars, 125, 236
Aventura na Selva, 90
Aventuras da Peregrinação (*Adventures of Pilgrimage*), 429
Averbuj, Guillermo, 36, 39–40
Aviation Empire (game), 369
Awakening (game), 273, 275, 484
Awesomenauts (game), 367
Axelson, Andreas, 455
Axeo5 (company), 42
AY-3-8500 chip, 565, 580, 593
Ayo Dance (game), 252
Ayrton Senna Pole Position (game), 91
Ayrton Senna Super Mônaco GP II (game), 89
Azpiri, Alfonso, 523
Aztec Tech Games (company), 353–354
Aztec Tech Institute, 353
Azzigotti, Luciano, 45

Bacca, Alvaro Felipe, 140
Backen, Andre, 455
Backpacker series, 455
Bad Influence (television show), 155n3
Baer, Ralph, ix, xii, 592–593
BAFTA Interactive Entertainment Award, 232
Bagnard, Le (game), 177–178
Baidu (website), 131
Bala (game), 425–426
Balda (game), 442
Baldur's Gate franchise, 109
Bally (company), 177
Balogh, Zsolt (Talent), 225
Bamtang (company), 397
Bandai (company), 595
Banderas, Antonio, 313
Bang! (game), 311
Banks, Rick, 108
Banzzaï (magazine), 189
Bao (traditional game), 18
Barbagallo, Leandro, 38
Barbie: Race and Ride (game), 331
Bard's Tale, The (game), 108
BARK (computer), 451
Barking Dog (company), 110
Barking Dog Studios (company), 108
Barn av regnbuen, 462
Barravento, 89
Barth, Zachary, 458
Bartle, Richard, 454, 580
Bartók, Béla, 219
Başarı Mobile (company), 573
Baseball (handheld game), 595
BASIC (computer language), 30, 175–176, 179–180, 219, 426, 582
Basic Programming (Atari cartridge), 596
Basketball (handheld game), 595
Batalla del Rio de la Plata (game), 45
Batalla Naval (game), 37
Batman: Arkham Asylum, 586
Batovi Games Studio (company), 610
Battegazzore, Sofía, 610
Battery Online, 559
Battlefield: Bad Company 2, 296
Battlefield 1942, 456, 457
Battlefield series, 312, 457, 464
Battle Isle, 201
Battle Tennis, 616
BattleZone, 596, 600
Bautista, Abraham, 350, 354, 357

Bayley, Janet, 382
Bazar de Lomas Verdes, 349–350
Bazar de Pericoapa, 349–350
Bazi Resaneh (company), 273, 275
Bazzoni, Giovanni, 308
BBC, 386
BBC Micro, 220, 582, 583
BBSs. *See* bulletin board systems (BBSs)
B.C.'s Quest for Tires (game), 108
Beam Software (company), 7, 58–59
Beardedbird Studio (company), 278
Beatmania series, 333, 601
Beenox (company), 110
Bee Rush (game), 356
Behar, Morris, 349
Bejewelled Blitz, 483
Belegost (game), 153
Belgium, 594
Bellyfish (game), 574
Ben10 Forever Defense, 397
Ben 10 franchise, 41
Benmergui, Daniel, 39, 43–44, 51
Bergensten, Jens "Jeb," 458
Bergie, 460
Bertheussen, Håkan, 458
Bertolino (company), 306
Besson, Luc, 185
Best, Cool & Fun Games (company), 213
Betamax, 393
Beta testing, 553
Bethesda Game Studios (company), 57, 63–64, 188
Bethesda Softworks, 195
Beyond: Two Souls (game), 185
Beyond Good & Evil (game), 187
Bhagath Singh (game), 237
Bichos (game), 37
Big Brother (game), 90
Big Brother Brasil 3 D On-line, 90
Big Fish Games, 295, 296, 298
Bigpoint (company), 194, 195, 201, 202

Billabong Surf Trip (game), 433
Billy the Kid 2: Hunted (game), 41
Binatone (company), 580
Bingo de Letras, 91
Bio-challenge (game), 182
Biodroid (company), 433
Biohazard (Resident Evil) (game), 331
Biohazard 2 (game), 210
Biology Battle PC, 560
Biology Battle Xbox, 560
Biopus (group), 45
BioShock (game), 63, 109
Bioshock 2, 60
Bioshock Infinite, 606
BioWare (company), 107, 109, 114, 116
Bitcrafters Inc. (company), 90
Bitkom (company), 204
BIT-LET, 229, 230
Bit Managers (company), 527
Bitmap Brothers, The, 584
Bittanti, Matteo, 307, 314
BIU (Bundesverband für Interaktive Unterhaltungssoftware), 201, 203–204
Bizarreh (game), 38
Björk, Staffan, 463
Blackberry, 397
Black Box Games (company), 108, 110
Black Dragon, 380–381
Black Gold (game), 272, 273
Blackish (company), 75
Black market. *See* Piracy
Black Net Bars, 125
Blackout, 455
BlackSoft (company), 20
Bladeslinger, 20–21
Blanchfield, Sean, 297
Blanco, Agustín, 45
Blast Theory (company), 431
Blitz series, 388
Blizzard Entertainment, 120–121, 212–213, 496, 502, 603

INDEX 651

Blizzard Singapore, 483
Blockade Blitz, 397
Blockout, 409–410
Blood and Guts, 453
Blow, Jonathan, 580
Bluebyte Studios, 194, 201
Bluehole Studio, 201
BlueTongue Entertainment, 7, 59
Blu-Ray, 603
Blystad, Tore, 462
BMX XXX (game), 61
BNV Entertainment (company). See Bonvicino, Matteo
Board of Investment (BOI), 560
Boarder X Battle (game), 43
Boarder Zone, 164
Bobblebrook (company), 75
Bodnár, István, 225
Bogatz, Denys, 202
Bogost, Ian, 50, 610
Bohemia Interactive, 154
Bola (game), 47
Bola de Gude (game), 90
Bolivarian Revolution, 614, 615
Bollywood, 237
Bolt Creative (company), 604
Bomberman (game), 209
Bongfish GmbH (company), 75
Bonnell, Bruno, 180, 183
Bonvicino, Matteo, 308
Boodhouse (company), 162
Boogie (game), 110
Boomzap Entertainment (company), 476, 480
Boong-Ga Boong-Ga, 333
Boonty Asia, 474, 486
Boot Hill, 452
Bootleg boards, 380
Borkowski, Marcin, 403–404
Born to Fire (game), 560
Boron Studios (company), 141
Bose, Atindriya, 238

Botanica (game), 484
Botanicula (game), 154
Bournemouth University, 588
Boxing Box, 397
Bradfield, Andrew, 388
Braid (game), 580
Brain Training (*Brain Age*), 320
Brainz Games (company), 142–143
Brak Show, The, franchise, 41
Brasfoot, 90
Brawler, 560
Braza Games (company), 92
Brazil, 3, 35, 40–41, 48, 87–104, 122, 213, 609
Brazilian Institute of Opinion and Statistics (IBOPE), 101
BreakAway Games (company), 31
Breakout (game), 176
Breakthrough (company), 138
Breivik, Anders Behring, 462
Bringsfjord, Selmer, 463
Britain. *See* United Kingdom
British Columbia, 108
British culture and games, 582, 586
Broken Rules Studios (company), 75
Brother Adventure, 499
Brown, Lee, 74
Browser Games Forum, 201
Bruton, Richard, 294
Bubble Bobble (game), 499
Bubble Up (game), 92
Buchanan, Kerry, 382
Bucketz (game), 574
Buckles, Mary Ann, 604
Budapest, 227
Budapest Micro (chiptune community), 233
Buffalo Roundup, 220, 225
BUG (Bahçeşehir University Game Laboratory), 576–577
Bugz Villa (game), 20
Buha (game), 31
Buka Entertainment, 440, 443
Bukashka, 443

Bulgaria, 146
Bulkypix (company), 188
Bulletin board systems (BBSs), 5, 596, 599
Bulletstorm, 64, 417
Bull Fighter, 335
Bully (game), 110, 114
Bunch of Heroes (game), 41
Bundesbeauftragter für Kultur und Medien (BKM, Federal governor for cultural affairs and media), 204
Bundesliga Manager (game), 194
Bundesprüfstelle für jugendgefährdende Schriften (Federal Department for Media Harmful to Minors), 203
Bunny Shooter (game), 213
BuPP. *See* Federal Office for the Positive Assessment of Computer and Console Games (Bundesstelle für die Positivprädikatisierung von Computer- und Konsolenspielen [BuPP])
Burbuja John (game), 37
Bureau: XCOM Declassified, The, 60
Burguener, Yamil, 45
Burton, Tim, 186, 458, 639
Business Software Alliance (BSA), 313
Butler, Paul, 108
Byte Computer Company, 566

C# language, 556
C2 Game Studio, 139
Cabal Entertainment Software, 480
Caesar the Cat, 220–221, 224
Café World, 603
Cage, David, 185, 189–190
Caillois, Roger, 98
Caimán Company (company), 37
Cale, Mark, 223, 233
California Dreams (company), 409–410
Call of Duty: Black Ops, 469
Call of Duty: Black Ops II, 311
Call of Duty: Modern Warfare (game), 53, 170, 311–312
Call of Duty: Modern Warfare 2 (game), 294, 302n8, 462, 469, 489n1

Call of Duty: Modern Warfare 3 (game), 236
Call of Juarez series, 416
Câmara, António, 431
Camon Hero, 559
Canada, 7, 11, 105–118, 138, 213, 353, 384, 594
Canoasoft, 91
Canyon Bomber, 452
Can You See Me Now? (game), 431
Capcom (company), 122, 188, 208–210, 213, 319, 348, 560
CAPES Bank of Theses, 94–95, 98–100
Capoeira Legends: The Path to Freedom, 10, 88, 91, 93
Captain Forever, 61
Captain Tsubasa Vol. II: Super Striker, 43
Captured (game), 453
Caracas Game Jam, 616
Carbajal, Adrian "Carqui," 345, 351, 352, 355
Carbone, Raoul, 310
Card catalog, 596
Cardona, Jesús, 138
Carita, Tiago, 432
Çarkıfelek (Wheel of Fortune) (game), 566
Carlà, Francesco, 307, 315, 316n1
Carneyvale Showtime, 475
Carrasco, Marco, 426
Carrier Command, 585
Carter, Royston, 300
Cartoon Network, 40–41, 397, 610
Cartoons on the Bay (company), 314
Cartoon Studios (company), 568
Cartridges, 51–52, 87, 121–122, 129, 147, 210–211, 384, 594–596, 598–599
Casa de Juegos (online project), 45
C.A.S. Combat Armor Suit, 560
Cash (game), 72
CASIE: Combat, 556
Casseta & Planeta em Noite Animal, 89
Castilla, Sebastian, 141
Castillo, Julian, 138
Castlevania: Lords of Shadow (game), 532
Casual Connect, 201

Casual games, 28, 31–33, 40, 54, 94, 124, 126, 141, 311, 334–335, 366, 368, 443, 475, 512, 558, 560, 574, 602
Çatak, Güven, 576
Caterpillar (company), 143
Cathode ray tubes (CRTs), ix, 596, 605
Catmoon Productions, 227, 232
Causa, Emiliano, 45
Cavagnaro, Juan, 43
Cawley, Anthony, 295
Caxy Gambá Encontra o Monstruário, 89
CCEBA Media Lab, 45
CCEC, 45
CCP Games, 458
CD Projekt, 412, 416–417
CD Projekt RED, 195, 412, 416
CD-ROMs, 37, 52, 129, 131, 210, 212, 429–430, 432, 568–569, 599, 601
CDV (company), 201
CeBit (event), 148
Ceidot (company), 574
Celebrity Salon, 556
Celestial Digital Entertainment, 215
Çelik, İdris, 573
CellFactor: Combat Training (game), 138
CellFactor: Psychokinetic Wars (game), 19, 138
CellFactor: Revolution (game), 138
Censorship, 9–11, 58, 61–66, 71–72, 76, 79, 120–121, 127–128, 132, 335–336, 485, 560
Center for Applied Games, 371
Center for the Study of Digital Games and Play (GAP), 372
Center of Studies for the Desarrollo Económico Metropolitano (CEDEM), 49
Centipede (game), 293, 347
Centro de Información y Educación para la Prevención del Abuso de Drogas (CEDRO), 397
Century 21, 378
Century Electronics, 378, 580
Cervantes, Miguel de, 521
CGPC (China Game Publication Committee), 119
Chahi, Éric, 181–183
Chaliapin, Feodor, 439

Champ Chase, 24
Champions of Regnum (game), 41
Channel 3 (in Brazil), 87
Chaos Works (company), 410
Chapela, Ivan, 346, 353
Charlie and the Chocolate Factory (film), 297
Chaston, Chris, 380
Chaston Engineering, 380
Chastronics, 378, 380–381
Chaturanga (traditional game), 236
Chávez, Hugo, 614, 615, 616
Chavo del Ocho, El (television show), 354
Chelyabinsk Regional Juvenile Library, 441–442
Chen, Jenova, 604
Cherry Credits (company), 480
Chess, 236
Chiado, Marcus, 87
Chiaroscuro (game), 91
Chile, 35, 48
Chili con Carnage (game), 457
Chilkowski, Andrés, 36–37, 39, 41
Chillingo (company), 139
Chimera Entertainment (company), 201
China, 8, 29, 31, 119–136, 139, 212, 214, 257, 485, 505–508, 510–511
China Internet Networks Information Center, 120, 133
ChinaLabs, 129
China On-line Games Copyright Protection Alliance (COGCPA), 125
Chine (game), 184
Chinese Gamer International Corp. (company), 212, 560
Chinese Hero Online (game), 212
Chinese Juggler, 220
Chmielarz, Adrian, 399, 411–412, 417
Christmas Jumper, 92
Christmas Magic, 89
Chrome, 416
Chronicles of Orta, The, 560
Chuck E. Cheese, 347
Chulalongkorn University, 545, 546
Chung Media, 501

Ciberne Software, 89
Cibers (game), 356
Cinematronics (company), 596, 598
Cinemedia (company), 570
Circo, El (game), 356
Circus, 452
Citizenship, 620–621
Citizen Zero, 59
C. Itoh (company), 350
Citroën C4 Robot (game), 570
City Games LLC, 213
City Interactive, 411, 417
City Landers, 387
City Network (website), 43
City of Heroes, 509
CityVille, 483, 603
CityVille 2, 603
Civilization III (game), 123
CJ Entertainment, 511
CJ Games (company), 574
CJ Groups, 511
CJ Internet (company), 501–502, 504, 509, 510, 511
CJ Media, 511
Clarín Group, 47
Claro (company), 94
Classification (Publications, Films and Videogames Act) 1995, 61–66
Classmates (social network), 443
C-lehti, 161
CLICKNL, 360
Cliffhanger Productions Software GmbH (company), 75
Clive Barker's Undying, 300
Clockspeed Mobile, 20
ClockStone Software (company), 75
Clones (of systems), 51, 121–122, 129, 132, 147, 176, 349, 353, 378, 386–388
Clone Wars, The, 484
Club Audition, 252
Club Nintendo (magazine), 345, 350–353
Club Transmediale, 205

Clustering, 106–107, 113–114
CNPQ, 96–98
Cobrasoft (company), 180
Coca-Cola Super Coach, 90
CoCom embargo 197–199
CODEAR (Concurso Desarrolladores de Videojuegos Argentinos), 36, 43
Codebox (company), 537
Codemasters (company), 237, 585
Codenix (company), 36, 40
CODEVISA (Cluster de Desarrolladores de Videojuegos de Santa Fe), 48
Coin Cascade (company), 378, 380
Coin-operated machines, 591
Coin-Op World Championship, 597
Coktel Vision, 182, 184
Coleco (company), 179, 580, 595
Coleco ADAM (computer), 599, 609
Coleco Gemini (console), 393
Coleco Telstar (console), 595
ColecoVision (console), 7, 51, 306, 385, 595
Collins, Dylan, 297
Collins, Stephen, 296
Çöloğlu, Sedat, 566–567
Colombia, 19, 137–144
Colombia Games (company), 143
Colonization, 17–19
Colony Wars (game), 154
Color Lines, 440, 442
Colossal Cave Adventure, 453
Columbus Discovery, 442
Combat Cars, 455
Combate a Dengue, 92
Combo, 553
Comdex (event), 148
Comecon, 227
Comedy Central, 42
Come-stay-play model, 126
Comics, 4, 20, 25, 146, 207–208, 212, 215, 444
Command & Conquer, 551

Commandos series, 300
Commandos: Behind Enemy Lines (game), 123, 529
Commodore (company), 196, 221, 223, 428
Commodore (magazine), 566
Commodore 64 (computer), 29, 53, 108, 141, 161, 220, 221, 223, 225, 227, 228, 229, 233, 306, 386, 393, 453, 565–566, 584, 599
Commodore 64, Peek-Poke & Machine Code (book), 566
Commodore 128 (computer), 393, 453
Commodore AMIGA (computer), 14, 72, 161, 180, 229, 306, 386–388, 453, 455, 566–568, 584, 585, 599
Commodore PET (computer), 598–599
Commodore Plus4 (computer), 229
Commodore Power/Play (magazine), 604
Commodore VIC-20 (computer), 161, 386, 387, 565, 599
Commodore Világ (*Commodore World*) (magazine), 230
Communism, 146–147, 156, 197–200, 220, 227–228, 400, 410–411, 413, 439, 441, 614
Company of Heroes (game), 110
Compgamer Company, 545–546, 559, 561
Compuphiliacs (company), 568–569
Computer Adventure Studio, 405–406
Computer Entertainment Rating Organization (CERO), 336
Computer Entertainment Suppliers Association (CESA), 336
Computer Game Museum (Berlin), 205
Computer Gaming World (magazine), 604
Computer Input (magazine), 387
Computer Mania (magazine), 231
Computer Space (game), 593
Computerspielemuseum, 205
Comunidad INDIE, 36
CONACyT, 345
Conde Brothers, 37
Cóndor (game), 38
Condor Rush (game), 122
Coney Island (arcade chain), 347
Confrontation: Peace Enforcement, 442
CONI (Comitato Olimpico Nazionale Italiano), 312
Conquering Bytes (company), 75

consol.at (online magazine), 81
consol.MEDIA Publishing (company), 81
Console games, 17, 19–20, 29, 32, 36, 41, 47, 51–53, 78–79, 82, 121–122, 124, 127, 131–132, 137, 154–155n3, 207–211, 213–215, 251, 253, 346, 351–355, 384–385, 393–394, 434, 481, 495, 497, 505, 536, 545–547, 552, 556, 565, 584, 586
Consul 2717 (computer), 147
Contacto PSX (magazine), 345, 351
Continuum Entertainment, 89
Contra (game), 209, 471
Control Magazine, 360, 362–363, 365–366, 373
Conway, John, 580
Coop (store chain), 462
Coordinating Committee for Multilateral Export Controls (CoCom), 197–200
Copy protection, 131. *See also* Encryption; Piracy
Cordes, Agustin, 41
COREAR (Concurso de composicion de musica para videojuegos), 36
Core design, 586
Corneliussen, Hilde, 463
Corners, 442
Corous (company), 480
Corporation (1990 game), 585
Corre Chasqui Corre (*Run Chasqui Run*), 396
Counter-Strike (game), 83, 95, 123, 251–252, 460, 483, 508, 551, 619
Counter-Strike Online (game), 212
CoVboy (László Kiss), 230, 231
Cowboy Guns (game), 139
Cows vs. Aliens (game), 213
Crack Down (game), 210
Cradle to Cradle: Islands (game), 369
Crane, David, 595
CraneBalls (company), 155
CRASH, 230
Crash, of stock market in 1987, 388
Crash of 1977, 594, 597
Crash of 1983, 2, 332–333, 582, 594, 597–599

Crawford, Chris, 604
Crazy Cart, 32
Crazy Combi, 397
Crazy Panda (company), 440, 443
Crazyracing Kartrider, 509, 512, 559
CREA DIGITAL Grant, 140–141
Creative Connection (company), 143
Crime Cities, 416
Cristiano Ronaldo Freestyle (game), 433
Croft, Lara, 586
CrossFire, 510, 512
Cross Gate (game), 213
Crosstown trilogy, 455
Croucher, Mel, 583
Crowds Vote, 33
Crowe, Russell, 313
Crown IT (company), 33
Crowther, Will, 453
CryEngine 3, 558
Cryo Interactive Entertainment (company), 184
Crytek (company), 195, 201, 574
Császár, András, 221, 222
Cseri, István, 223
Csokonai, 228
Cthulhu Mythos, 457
CTO, 309
Cúchulainn, 299
Cuisset, Paul, 181, 182, 189
Culpa Innata (game), 572
Cultures Online (game), 202
Cupid Bistro (game), 212, 215–216
Cupid Bistro 2, 212
Curupira, 89
Customizable avatars, 561
Cut-scenes, 570
Cuy (Guinea Pig), 397
CVC Gameline, 597
CyberAgent, 335
Cybercafés. *See* Internet cafés
Cyber Empires (game), 109
CyberJuegos (company), 36, 40

CyberPlanet Interactive (PCL) Ltd. (Public), 556
Cyber-psychosis, 461
Cybersports and Online Gaming Association, 486
Cybertime (company), 75
Cybexlab (company), 153
Cyborg, The, 453
Cygert, Henryk, 406
CYPHA Entertainment, 25
Czech Republic, 8, 145–158, 227

Dabney, Ted, 593
Dacia (car), 227
Dactyl Nightmare, 600
Dactyl Nightmare 2: Race for the Eggs!, 600
DADIU, 463
Daewoo (company), 497, 498
Dafoe, Willem, 185
Daglish, Ben, 223
Daily Mirror, 221
Dakorah, Justin, 20
D'Alessandro, Jaime, 314
Damascus, 30
Damoria, 32
Danadata (company), 453
Dance Dance Revolution (DDR), 333, 546, 604
Dance Dance Revolution series, 601
Dancing Monster, 220
Danda gilli ("Poor man's cricket"), 236
Danish Game Awards, 461
Danmarks Radio, 461
Dante's Inferno (game), 310
Dapharen's Fear (game), 38
Dar Al-Fikr, 30, 32
Dardari Bros., 307
Dark Earth, 187
Darkness 2, The (game), 109
Darkness Within: In Pursuit of Loath Nolder (game), 572
Dark Orbit (game) 194
Dark Phantom, The: Dawn of the Darkness (game), 273, 275, 288
Dark Sector (game), 109

Dark Silence, 556
Darkwaves Games, 308
DASK (computer), 451
Data East (company), 595
Datel (company), 585
Datsun 280 Zzzap, 452
Dave & Busters (store chain), 349
David: In The Name of Crowds (game), 158n2
David Reid Electronics (company), 5, 384, 387
Davidson, Michael, 379, 384–385
Daylight Studios, 480
Daytona USA, 546
Dead Island (game), 201
Dead Island series, 416
Deadline Games (company), 455
Death Upon an Austrian Sonata, 484
de Blob (game), 59
Debuz (company), 545, 552, 561
Decadium Studios Game Developer, 90
Dedalord (company), 43
Dedektif Fırtına (*Detective Storm*), 568
Deep Silver (company), 201
Deep Silver Vienna (company), 74
Deer Hunter 2004, 90
Deer Hunter 2005, 90
Defender (game), 393, 594
Defenders (game), 426
Délirus, 91
Delphine Software (company), 182, 183
Demi-Gods and Semi-Devils Online (game), 215
Demographics, 320–321, 333–336
Demolition Operation (game), 282
Demon's Souls, 337
Demonware, 295, 296, 297, 298
Demoscenes, 2–3
DeNA, 320, 334–335
Dendej Sawanthat, 546
Dengen Chronicles, 311
Denmark, 11, 156
Denmark Girls (group), 460
Derby's Tycoon Online (game), 215

Der Clou! (game), 74
Der Industriegigant, 74
Derlien, Winrich, 196
Der Pferderennstall, 90
Der Standard (magazine), 82
DerStandard.at (online magazine), 82
Der verlassene Planet (*The Abandoned Planet*), 72–73
Desafino, 89
Desarrolladores Unidos de Videojuegos de América Latina (DUVAL), 43
Désilets, Patrice, 110
Detalion, 411, 417
Detetive Carioca, 91
Deuces Wild Casino Poker, 92
Deus Ex, 95
Deus Ex Machina, 583
Deutscher Computerspielpreis (German video games award), 204
Deutscher Kulturrat (German Culture Board), 204
de Vabres, Renaud Donnedieu, 190
Developers. *See* Game developers
Devil (game), 560
Devworks, 91
DF Interactive, 486
DGT—Deutsche Gamestage, 201
Dhruva Interactive (company), 242
Diablo (game), 311, 483, 506, 551
Diablo II (game), 213, 551
Diabolik: The Original Sin, 311
Diamond Dash (game), 201
Di Bello, Bonaventura, 307
DICE (Digital Illusions Creative Entertainment), 455, 456
Didaktik Gama (company), 147
Didaktik S (company), 147
Didi na Mina Encantada, 89
Die Pferdebande—Weiße Stute in Gefahr, 90
Die Ponyrancher, 90
Dier Pferdebande, 90
Die Völker (game), 74, 75
Digic Pictures 3D, 232
DigiPen Institute of Technology, 353

Digital Bros, 309
Digital content industry, 555
Digital Dreams (company), 566–567
Digital Extremes (company), 109
DigitalFun. *See* Viola, Fabio
Digital Games Research Association (DiGRA), xi, xiii
Digital Games Research Center (DIGAREC) of the University of Potsdam, 205
Digital Media Company, The (CMD), 47
Digital Reality, 224, 232
Digital Tales (company). *See* Bazzoni, Giovanni
DiGRA, 58, 170, 463
DiGRA Japan, xi, xiii
DikuMUD, 454–455, 581
Dinamic Multimedia (company), 527
Dinamic Software, 609
Dinç, Mevlüt, 566, 569–570
Dinç Interactive (later Sobee Studios), 570
Dingdong places, 251, 264n3
Dinky Kong, 387
Discovery/Hitek, 89
Discovery Software International, 453
Disney (company), 41, 43, 138, 222, 610
Disney Interactive, 310
Disney Online, 232
Distinctive Software (company), 108
Distribution, 4–5, 54, 76, 94, 127, 129, 147–151, 275, 279, 280, 286, 320–321, 336, 385, 388, 538, 569
Distribution anomalies, 385–387
DJ Hero 2, 296
DMA Design, 585. *See also* Rockstar North
DmC: Devil May Cry, 586
Docugame, 30
Dodoo, Albert, 20
Doghouse, The (arcade), 380
Do-it-yourself (DIY) tradition, 596
Domark, 221, 222, 586
Domo Studios, 126
Doña Gloria: The Game (game), 139
Donkey and Crow (game), 286

Donkeycat (company), 75
Donkey Kong (game), 53, 177, 209–211, 322, 471
Donkey Kong Country, 585
Donkey Kong Country Returns 3D, 312
Donohoe, Paschal, 294
Donsoft Entertainment, 88, 91
DontNod (company), 188
Doodle Jump, 604
DOOM (game), 19, 62, 148, 223, 460–461
Doom and Destiny, 308, 311
DOS, 109, 199–200, 276, 430, 551, 581, 599
Dot-com wave, 456
Double Negative Singapore, 480
Doupě (magazine), 149
Dovogame, 560
Downloadable content (DLC), 602
Dowrane Eftekhar (game), 282
Dracula (game), 385
Draft Marketin Esportivo, 89
Dragon, György, 226
Dragon 32 (computer), 582
Dragon Age (game), 109, 114
Dragon Ball (game), 210
Dragon Ball Z: Super Butōden, 549
Dragonfly (company), 511, 559
Dragon Nest (game), 212, 559
Dragon's Lair, 598
Drakar och Demoner, 459
Drankensang (game), 194
Dream-Build-Play competition, 615
Dreamcast. *See* SEGA Dreamcast
Dreamer: Musicstar Popstar, 91
Dreamfall: The Longest Journey, 455
Dream Gulong Online (game), 215
DreamHack (demo party), 460
Dreampainters (company), 308
Dream Traveler, 498
Driller, 583
Drive Me Bananas (game), 139
Dropfun (company), 284, 288

DrumMania, 546
DSI (company), 108
DSi (console), 15, 156
DSK Supinfocom (company), 244
Dual Blades (game), 570, 572
Dubai, 31–32
Dubinushka, 439
Duello Games (company), 574
Dun Darach, 299, 300
Dungeon and Fighter, 510
Dungeon Fighter Online, 504
Dungeon Master, 226
Dungeons & Dragons, 299, 300, 454, 458, 461
Dutch Game Awards, 360
Dutch Game Garden, 360, 371, 373
Dutch Games Association (DGA), 360, 362, 364, 370, 373
DVD players, 7, 52, 210, 264n5, 329, 331, 411, 542n4, 603
Dwango (company), 335
Dyack, Denis, 108
Dyer, Rick, 597–598
Dynacom (system), 51

EA Canada (company), 7, 108
Earl Grey and This Rupert Guy (game), 253
Earth 21xx series, 415
Earth Under Siege, 92
EA Sports, 559
Easter eggs, 595
E-books, 100
Ecco the Dolphin, 224–225
Ecole Nationale du Jeu et des Médias Interactifs Numériques (Enjmin), 191
Economic Development Board (EDB), 471, 472, 478
E-content publishing, 475, 484
EcoQuest: The Search for Cetus, 226
Eco Warriors, 311
Eden Online, 560
Edge (magazine), 149
Edia (game), 38
Edison, Thomas, 592

Edison parlors, ix, xii
Education, 17–18, 25, 28, 36, 37, 42–43, 46, 77, 81, 94–95, 99–100, 124, 141–142, 144, 152, 154, 157n19, 215, 345, 353, 442, 444, 481, 531, 587–588, 610, 612
Education and Culture Foundation of Minas Gerais (FUMEC), 98
Education of Neznaika, 440
EDU Games (system), 51
Edusoft (company), 38
Edutainment Croanak (game), 141
Efecto Studios, 137–138, 143
Egamesbox (game community), 260
Egenfeldt-Nielsen, Simon, 463
EGM en Español (magazine), 345, 352
Egypt, 19, 31, 33, 138, 594
Egypte (game), 184
Eidos Interactive, 110, 113–114, 123, 185, 435, 457, 529, 586
Eisler, Jan, 149
e-Learning Association of Thailand, 546
Electrical power, 3–4
Electric Dreams Software, 566
Electro Body, 410
Electromechanical games, 593, 439
Electronic Amusement Service Ltd., 380
Electronic Arts (EA) (company), 7, 41–42, 50, 71, 75, 108–110, 114, 151, 186, 196, 201, 202, 226, 299, 328, 330, 472, 474, 476, 480, 496, 502, 503, 538, 585
Electronic Entertainment Expo (E3), 125
Electronic Games (magazine), 604
Electronic Gaming Monthly (magazine), 345, 352
Electronic Plastic (book), 386
Electronic Sports League (ESL), 83
Electronics companies, 384–385
Electronic TV Game (Videoton), 220
Electro Sport, 598
Elektor (journal), 565
Elektronika (*Electronics*) (magazine), 149
Elite (game), 583
Ellersie Borough Council, 384
Elmadinah, 33

Elo Elektronik, 566
ELS Italia (e-sport association), 312
ELTE (Eötvös Lóránd University), 219
Eltern-LAN Workshops, 82
Eludamos (online magazine), 79–80
E-Ludo, 309, 313, 316
Em Busca dos Tesouros, 89
eMedia (periodical), 81
Emoak (company), 75
Emotion Engine, 331
Emperor (game), 560
Encryption, 131, 557. *See also* Piracy
Encyclopedia of Video Games, xv, 29
EndWar franchise, 187
England. *See* United Kingdom
Englen, 455
Enterprise (computer), 229
Entertainment Establishments Control Law (Japan), 335
Entertainment Software Ratings Board (ESRB), 9
Envisage Reality, 476
Eötvös Lóránd University (ELTE), 219
Epic Games (company), 109, 399, 410, 417
Epyx, 223, 584
Eranor (game), 354
Erdély, Dániel, 225, 233
Ere Informatique 180, 181
Ergen, Ahmet, 566
Eriksson, Kimmo and Viggo, 453
Erinia, 90
Escape from Alcatraz (game), 232
Esmann, Jan, 461
Esnaola Horacek, Graciela, 42
Espacio Byte, 46
Espaço Informática, 89, 90, 91
ESPN, 43
E-sport competition, 559
E-sports, 72, 79, 82, 460, 478, 508–509
E-sports (Italian), 312. *See also* NGI
eSports Association Austria, 82
Espris Pouya Nama (company), 272

ESRA, 280
Estelar Software, 89
E.T.: The Extraterrestrial, 10
E.T. Armies (game), 273, 275, 277
Eternal Darkness: Sanity's Requiem, 109
ETROM: The Astral Essence, 311
EUCROMA, 463
Eureka, 221
Europa Universalis, 456
Europe 2045 (game), 157
European Federation of Game Archives, Museums, and Preservation Projects (EFGAMP), 205
European Wars: Cossacks XVI–XVIII Centuries, 442
EVA. *See* Argentine Videogames Expo (EVA)
EVE Online, 457, 464
EverdreamSoft (company), 537–538
EverQuest, 455, 602–603
Evoga (company), 353, 354
Evoke (event), 205
Evolution (game), 108
Evolution Championship (tournament), 348
Evoluxion (company), 36, 40
Ewe tribe of Ghana, 20
Excalibur (magazine), 149
Excel Saga, 320
Exhibition of Argentinean Games (EXARGA), 36
Exidy, 596
Eyedentity Games, 212, 559
Eye of the Beholder, The, 226
EyeToy, 586

F1 2011, 586
F1 Racing Championship, 19
Fábián, István, 226
Fable, 586
Facebook, 20, 28, 33, 42, 47, 94, 139, 155, 201, 253, 397, 440, 561, 603
Fahrenheit (game), 185
Fairchild Channel F, 594
Fairchild Semiconductor, 594

Falafel Games, 29, 31
Falkland Islands, 39
Falling Fred Z (game), 43
Fallout 3, 10, 57, 63–64, 296, 300
Fallout Tactics: Brotherhood of Steel, 59
Famerama (game), 194
Famicom. *See* Nintendo Famicom
Famicom clones, 122
Fanafzar Sharif (company), 272, 287
Fandiño, Julio Enrique Aguilera, 141
Fantasy Earth Zero (game), 212, 213
Far Cry (game), 195, 201
Far Cry 2, 300
Fares al Ghad (company), 32
Fargas, Joaquin, 45
Farm 51, The, 419
Farmer, the Wolf, the Goat, the Cabbage, The, 220
Farmer Jane (game), 433
Farming Simulator series, 194
FarmVille, 28, 33, 94, 170, 253, 433, 602
FarmVille 2, 603
Fasoulas, Stavros, 162
Fatfoogoo (company), 75
Fayuca, 346. *See also* Piracy
Fazekas High School, 219
FBI, 177
FEC Company, 553
Federal Institute of Goiania (IF Goiano), 98
Federal Office for the Positive Assessment of Computer and Console Games (Bundesstelle für die Positivprädikatisierung von Computer- und Konsolenspielen (BuPP), 71, 78–79, 82
Federal Software (company), 129
Federal Technology University of Paraná (UTFPR), 98
Federal University of Bahia (UFBA), 98
Federal University of Mato Grosso do Sul (UFMS), 98
Federal University of Paraiba (UFPB), 98
Federal University of Piauí (UFPI), 98
Fehér, Gábor, 232
Feist (game), 538

Femina Bellica (group), 460
Fengyun (game), 130
Fénix, 396
Fenton, R. R. (Bill), 385–386
Fernandez, Agustin Perez, 43
Fernlund, Arne, 453
Ferranti (company), 579
Festival of Games, 360
Fídler, Miroslav, 153
FIFA 12: UEFA EURO 2012 (game), 41
FIFA 13, 311–312
FIFA 2012, 396
FIFA Online, 483, 503, 512
FIFA Online 2, 559
FIFA series, 20, 47, 53, 311, 352, 587
Fight Club (arcade gamers' organization), 209
Fighting Fantasy, 221
Figueiredo, António Dias, 428
Final Fantasy (game), 209–210
Final Fantasy II (game), 549
Final Fantasy VII (game), 312, 330–331, 352
Final Fantasy VIII (game), 210
Final Fantasy XII (game), 299
Final Fantasy Online (game), 212
Final Fantasy series, 214
Final Fight, 549
Fingle (game), 368
Finland, xi, xiii, 11, 156, 159–174, 213
Firebird Software, 566
Firedog Computer Entertainment, 212
Firedog Studio, 212, 215
Firefox (game), 385
Fireman, Fireman (game), 595
Firemint Studios, 61
First-person shooters (FPS), 551, 558
First Samurai (game), 569
Fist of Fu, 506, 507
Flair Software, 74
Flamin' Lab, 139–140
Flaregames, 201

Flick Fishing, 604
Flight Control, 61
Flip Flop, 442
Flipt (game), 20
Floppy disks, 599
Florido, Estanislao, 45
Flösser, Janos, 456
flOw, 604
Flower, 604
Flying Hamster (game), 188
Flying Heroes (game), 154
Flying Wild Hog (company), 419
FM4.at (online magazine), 82
FM Towns Marty (console), 599
Fnatic (group), 460
FNIV (Federazione Italiana Videogiocatori) (e-sport association), 312
Foerderverein fuer Jugend und Sozialarbeit e.V. (FJS), 203
Follis, Greg, 300
Football (handheld game), 595
Football Manager (game), 311, 570, 582
FootboCity (game), 574
Forbes (magazine), 125, 332
Force Unleashed II, The, 484
Fore One (company), 215
Forever Young Studio, 480
Forfás, 294
Forge-Reply (company), 311
Forrai, Gábor, 227
Fortugno, Nick, 50
Fotros (company), 272
Fountain (company), 5
Fountain Force 2 (console), 385
Fox Who Followed the Voice, The (game), 286
France, 24, 40, 144, 175–192, 213, 486
France Image Logiciel, 180
Francisco Matelli Matulovic, 91
Franco's dictatorship, 521
Franko: The Crazy Revenge, 408, 409
Franzani, Santiago Javier, 43

Frasca, Gonzalo, 98. *See also* Uruguay
Freddy Hardest in South Manhattan, 609
Frederiksen, Peter Ole, 453
Fred Skiing (game), 43
FreeStyle, 505, 512
FreeStyle 2, 505
FreeStyle Football, 505
Free-to-play model, 126, 132, 139, 195, 369
Freeverse Inc., 604
"French touch," 180–184, 188
Friedman, Thomas, 235
Friere, Paulo, 614
Frogger (game), 322, 347
Frogster Interactive Pictures, 201
Fruit Ninja (game), 61, 213
Fruzzle, 91
Fuentes, Enrique, 138
Fujitsu, 599
Fuka, František, 153
Fulco, Ivan, 314
Full-motion video (FMV), 599
Full Speed, 395–396
Funatics (company), 202
Funcom (company), 293, 455, 456
Funtree (company), 560
Fútbol Deluxe 96 (game), 37–38, 40
Futbolín, 522
Fute Bolon Line, 91
Futsim, 90
Future Cops (film), 208
Futurezone.at (online magazine), 82
Fuxoft (company), 153
Fuzzball, 455
FX Labs (company), 237

G4TV, 232
Gadnell, Stefan, 455
Gado (African artist), 18
Gaelco (company), 522, 525
Gaelic Games Football, 299

Galaxian, 471
Galaxy Edu Hitech, 560
Galaxy Game, 593
Galaxy Group Thailand Company, 546, 552, 559–560
Galaxy Patrol, 426
Galkin, Dmitri, 446
Galvez, Ernesto, 138, 142
Gamania (company), 212, 510
Gambys (game), 430
G/A/M/E (journal), 315
Game & Watch series, 17, 53, 211, 251, 322, 324–325, 536, 595
Game Atelier, The (company), 188
Game Boy. *See* Nintendo Game Boy
Game Boy Advance. *See* Nintendo Game Boy Advance
Game Boy Color. *See* Nintendo Game Boy Color
Gamebrasilis, 87, 101n2
Gamebryo, 558
G.A.M.E. Bundesverband der Computerspielindustrie (Association of Game Developers), 201, 203–204
Game City (event), 71, 78–80, 83
Game Cluster, 97
Game Con, 459, 560
GameCube. *See* Nintendo GameCube
Game design, 35, 40, 46, 81, 126, 145, 561, 556, 567, 572–573, 575
Game developers, 20, 57–58, 60–61, 121–123, 125, 129, 137, 142–143, 154, 498, 552, 558, 561
Game Developers' Association of Australia, 60
Game Developers Conference (GDC), 40, 138, 144
Game development, 20, 24–28, 37, 39–42, 47–48, 51, 54, 71–72, 74–75, 79–80, 83, 87–88, 92, 101, 130–132, 153–155, 271–277, 285–288, 377, 394–397, 469, 471, 488, 495, 498, 501, 505, 565, 567, 570, 574–575, 577, 579, 587, 622–623
Game.doc (film), 577
GameDuell (company), 202
Game engines, 5, 9, 20, 31, 116, 138–139, 141, 195, 225, 237, 272, 276, 278, 297, 331, 394, 411, 416–417, 433–434, 537, 539, 556, 557–558, 570, 572

Game Exchange Alliance, 480
Game Express (store chain), 350
Gamefest (event), 201
Gameforge (company), 194, 195, 201, 202
Game Gear (console). *See* SEGA Game Gear
Gamegenetics (company), 195
Game Genie, 585
Game Gestalt (company), 75
GAMEINDY (company), 545, 559, 561
Gamekings.tv, 360
Gamela (company), 350
GameLab (trade fair), 530
Game Lab Cologne (GLC), 205
Gameland (company), 443
Gameloft (company), 36, 40, 213
GameMaster (magazine), 352
Game of Life (1970 game), 580
Gameone (company), 215–216
Game on! El arte en juego (event), 51
Game-Orange (company), 125
Game Oven (company), 368
Game Page (television show), 155–156
Gameplay styles, 605
GamePro, 604
Gameprog.it, 310
Game Research for Training and Entertainment (GATE), 371
Gamerland (television show), 314
Gamer News, 561
Gamer.nl, 360
Gamers.at (online magazine), 81
GamersMotion, 81
Gamers' organization, 209
Games 4 Resilience Lab, 80
Games Academy, 205
Gamescollection.it, 314
Gamescom, 201
Games Convention, 201
Games for Health Europe, 360
Games Ireland Gathering (GIG), 295

Games Machine, The (magazine), 306, 317n6
Games per Computer (research group), 98
Gamesquare Interactive Co., Ltd. (Public), 556
GameStar (magazine), 149
Game Station (magazine), 250
Gamester (company), 40, 51
Gamester (website), 43
Games That Matter Productions (company), 74
Gameworks (store chain), 349
GameXP, 443
Gamexpo, 616
GamezArena (company), 71
Gamification, 46, 284, 288, 290, 308–309, 345, 361, 372–373
Gaming VS Reality series, 81
GamingXP (magazine), 82
Gamistry (company), 368
Gamos (company), 440, 442–443
Gang Xa, 553
Gap, The (magazine), 82
Garcia, Cesar, 348
Garcia, Juan, 38
García-Canclini, Néstor, 621
Gardens of Time (game), 41
Garena On-line Pte Ltd., 480
Gargoyle Games (company), 300
Garriott, Richard, 599
Garshasp (game), 273–275, 287
Gate2Play (company), 33
Gathering, The (demo party), 460
GCE/Milton Bradley Vectrex (console), 7, 161, 385, 536, 596, 602
GDC Europe, 201
Gears of War (game), 64, 399, 417
Gears of War: Judgment, 312
Geewa (company), 155
Gender. *See* Player demographics
Gender swapping, 581
Generala (game), 37
General Home Products (company), 593
General Instruments (company), 198, 580, 593

Génération 4 (magazine), 189
Generation X, 602
Genesis 3D engine, 31
Geometry Wars: Retro Evolved (game), 190
Georgia (country), 422
GEOS, 231
Geralt of Rivia (character), 418
Gerçeğin Ötesinde (Beyond the Truth), 568
German Business Simulations (genre), 72
German Computer Game Award, 204
German Democratic Republic (GDR), 229
German Parliament (Deutsches Parlament), 204
Germany, 41, 71–72, 74, 76–77, 81–82, 148, 177, 193–206, 227, 346, 594, 442, 500, 508
Gerry the Germ Goes Body Poppin', 566
Gesellschaft für Konsumforschung (GFK), 200
Gesellschaft für Medienwissenschaft (Society for Media Studies), 205
Gesellschaft für Sport und Technik, 198
GEVO Entertainment, 486
GfK-Rus, 446–447
GG ("Good Game" portal), 559
Ghajini (game), 237
Ghana, 20, 25, 28
GhettOut! (game), 158n3
Ghossoub, Vince, 29
Ghost Recon franchise, 187
GIANTS Engine, 537
GIANTS Software (company), 537
Gibbons, David, 457
Gielen, Jaro, 386
GIER, 452
Giften, 455
Gigaslave, 559
Giochipreziosi, 307
Gioventù Ribelle, 310
Gjedde, Albert, 461
GlobalFun, 40–41
Global Game Designers Guild (GGDG), 137, 140, 143–144
Global Game Jam, 25, 313, 371, 575, 611

Globalization, ix, 175, 184, 302n6, 444, 486, 503, 510, 513, 516, 553, 586, 614, 621, 623
Global marketplace, 9–10, 27–28, 33–36, 40–41, 54, 72, 83, 92, 94, 119, 123, 128, 377, 558, 570, 572, 586
Globant (company), 36, 41, 47
Globetrotter series, 455
Gloersen, Erik, 455
Gloof (company), 510
Gloom (game), 388
Glory Destiny Online, 560
GNI Soft, 559
Go (traditional game), 594
Go-Bang, 176
Godager, Gaute, 455, 457
God of War, 64. *See also* Violence
God of War: Ascension, 312
GodsWar, 545, 561
GOG.com, 417
Gohyman, Eduardo, 41
Goldensoft, 545–546, 559–560
Golden Triangle (group), 153
Golds of Virtual Boards, 91
Go-Maku, 176
Gomez, Jose Alberto, 616
Gonzalez, Martin, 43
Good Old Games, 417
Google, 143, 365
Google Play Store, 24
Gorbachev, Mikhail, 228, 439
Gorky series, 412
Gorminium, 453
Goth Con, 459
Gothic II, 73–74
Gothic series, 194
Gourier, Bruno, 180
Gouveia, Patrícia, 435
Grabbity (game), 138
Gradiente (company), 122
Graduation, 336
Graffiti (company). *See* Milestone

Graham Gooch Test Cricket, 582
Grand Chase, 506–507
Grandchase Chaos, 483
Grand Prix Circuit (game), 108
Grandslam, 580
Grandstand, 385, 387
Grand Theft Auto: San Andreas (game), 53, 586, 588
Grand Theft Auto: Vice City (game), 74, 619
Grand Theft Auto III (game), 61, 74, 586
Grand Theft Auto IV (game), 8, 110 294, 300, 469
Grand Theft Auto V (game), 11
Grand Theft Auto series, 110, 170, 586, 588, 605–606
Gran Turismo (game), 330
Gran Turismo 2 (game), 330
Graphical user interface (GUI), 599
Gravitation, 604
GRAVITY (company), 212, 495, 507, 551
Gray market, 250, 254–255, 261, 263, 346, 349, 351
Graziani, Alejandro, 41
Great Britain. *See* United Kingdom
Great Giana Sisters, The (game), 195
Great Legends: Vikings (game), 41
Great South Road, 386
Great Trader (game), 125
GREE, 320, 334–335
Greece, 594
Greek mythology, 20
Greenfield, Patricia Marks, 94, 99, 604
Greenland Studios, 90
Greentube (company), 71, 75
Gremlin (company), 184
Gremlin Interactive, 586
Grendel Games (company), 369
Greve Graphics, 453
Greytown, New Zealand, 381
Griffith, Mike, 297
Grog's Revenge (game), 108
Grønbech Søren "Sodan," 453
Groove the Worm (game), 141
Growing Games (company), 371

Grupo Alfas: Ambientes Lúdicos facilitadores de aprendizaje, 42
Grupo Avatar (Avatar Group), 395
Grupo Carso Telecom, 615
Grupo Forum, 429–430
Grupo Televisa, 352
Gryphon Rider (game), 369
G-Star Game Conference, 501
GT Interactive (company), 184
GT Racing Motor Academy (game), 213
Gualeni, Stefano, 309
Guardian, The (newspaper), 222
Guarini, Massimo, 305, 308, 309
Guarinoni, Cesar, 41
Guerilla Games (company), 364, 368, 370
Guild Wars, 509
Guild Wars 2, 311
Guimo, 89
Guip (game), 37
GuitarFreaks series, 601
Guitar Hero, 53
Guitar Idol, 91
Gumi Asia (company), 480
Gümüş, Ozan, 573
Gundam Extreme VS (game), 211
Gundam vs. Gundam Next (game), 209
Guru, 231
Gussy (game), 37
Gustafsson, Olof, 455
Gustavinho em O Enigma da Esfinge, 89
Gutiérrez, Andrés, 396

Habbo Hotel, 165
Hacettape University, 576
Hachette, 188
Hackerence (group), 460
Hack 'n' slash gameplay, 453–454, 458
Hades 2, 89
Haidarian, Alborz "HeMaN," 460
Haki, 20, 24
Haki 2, 20, 24

Halcyon system, 597–598, 601
Halfbrick Studios, 61, 213
Half-Life (game), 63, 551
Halo (game), 469
Halo series, 603
Hammer, Sebastian, 454
Hammerfist (game), 566
Hamza, Mohammed, 30
Hanafuda, 319, 322
Hanbitsoft (company), 498, 501
Hançer (The Dagger) (game), 566–567, 577
h.a.n.d. Global Singapore (company), 480
Handheld games, 17, 19–20, 151, 207, 209, 211, 213, 214, 384–386, 545, 556, 595, 602, 605
Hangame (company), 498, 501
Hanna-Barbera, 222
Hanuman: Boy Warrior (game), 238
Happy Birthday (game), 440, 442
Haptic interface, 397
Hård, Lars, 453
Hård, Niels, 453
Hardball! (game), 108
Hard Drivin', 600
Harry Potter and the Order of the Phoenix (film), 297
Harry Potter franchise, 95
Harvest, The (game), 20–21
Harvest Moon (game), 253
Hasbro (company), 601
Hasbro Interactive, 184
Hase und Wolf (game), 198
Hate the Sin, Love the Sinner (game), 273, 275, 285
Haunted House, 583
Havok (company), 296–297
Hawkins, Trip, 328, 599
Hawkquest, 388
Hawx franchise, 187
Head-to-Head game series, 595
HeartBit Interactive, 308, 311
Heartlight, 410
Heavy Rain (game), 185

Hein, Piet, 452
Hejlsberg, Anders, 463
Hellcopters (game), 353
Helle, Merete Pryds, 455
Hellwig, Johann C. L., 193
Herman, Leonard, 604
Heroes in Marshes: Tales of Upholding Justice (game), 123, 130
Heroes in Wulin (game), 123
Heroes of Gaia, 32
Hesperian Wars (game), 154
Heunis, Herman, 20
Hexa, 498
Hezbollah, 31
Hidden & Dangerous (game), 154
Hidden Treasure (game), 272, 282
Hideout (magazine), 149
HiE-D (Human Interface Electronic Device), 142
High School Musical (film), 355
High-score tables, 605
HipStreet, 559
Hirai, Kazuo, 332
Hirtenberg, 72
HistoryCity, 472
HitFox Game Ventures, 196
Hitman 2: Silent Assassin, 236
Hitman Absolution, 462
Hitman movie, 456
Hitman series, 456, 464
Hive Division, 313, 317n5
Hiyashi, Yosuke (*Ninja Gaiden*), 336
Hobbit, The (game), 58–59. *See also* Beam Software
Holden, Colin and Beverley, 383
Hollywood cinema, 6, 600
Holub, Radovan, 150
Holz, Frank, 553, 556
Homebrew, 387
Home PONG (console), 161, 178–179
Homeworld (1999), 110
Homeworld 2 (2003), 110
Hong Dong Hee, 498
Hong Kong, 207–218, 386, 485, 594
Hoplon Infotainment, 91
Horizon Riders (game), 40
Horror games. *See* Survival horror
Hot Circuits: A Video Arcade, 604
Hot Coffee mod, 586, 588
House 3, 443
Housemarque (company), 162
Houthoff Buruma—The Game (game), 369
Howard, Robert E., 458
Howard Dean for Iowa Game, The, 610
How the Urals Saved the Battle of Borodino, 442
HTC Middle East, 31
Hudson Soft (company), 209
Huevocartoon, 355
Huizinga, Johan, x, xii, 98, 620
Hülsbeck, Chris, 195
Humanisterna, 461
Humor, Comics and Games (research group), 98
Hunchback, 580
Hungarian Computer Technology Coordination Institute (SZKI), 233
Hungarian Museum of Science, Technology and Transport, 233
Hungarian Technological University, 229
Hungarian Trade Commission, 220
Hungarian University of Arts and Design, 233
Hungary, 219–234
Hunter on the Road, 440
Hutchison 3D Austria, 82
Hutterer, Gerd, 198
Hypaa, 560
Hypertext, 100
Hysteria Project (game), 188
Hyundai, 497, 498

IAGTG, 313
IBM PC games, 31, 33, 36, 41, 47, 52–53, 74, 121–123, 126, 127–130, 131, 138, 141, 146, 176, 197, 212, 252, 386, 429, 439, 451, 453, 568, 584–586, 598–599, 601

I Can Football (game), 570
IC Card Games, 334
Ice Age franchise, 109
Icon Games (company), 90–92
IDC (International Data Corporation), 119
IDDQ, 233
IDEA Games, 154
Időrégész (Time Archeologist), 226
Id Software, 203
idTech 5, 558
IEM Consulting Company, 556
IGDA, 25, 82, 143, 51, 542n8, 611
IGN (online magazine), 351
Ignis Games, 90
IJsfontein (company), 369
ilikescifi (company), 75
I.L. L'intrus (game), 179–180
Illusion Softworks, 154. *See also* 2K Czech
Il rosso e il nero—The Italian Civil War, 310
Imagic (company), 595
Imaginations FZ, LLC, 31
IMAX Siam Paragon Department Store, 559
Immersion Games (company), 19, 137–138, 143
Immersion Software and Graphics. *See* Immersion Games
Immersive Play Spaces Pte Ltd., 480
Immigrant population, 592
Immortals: The Heavenly Sage, 477
Impacto Alpha, 90
Imperium Galactica, 224
Imperium Galactica II: Alliances, 232
Imports, 6–7, 42, 96, 121–122, 132, 147, 149, 151, 177, 210, 214, 346, 350, 379–380, 394, 483, 505, 583–584, 609
Impossible Mission (game), 223, 584
Impossible Mission 2 (game), 223, 225
Inazuma Eleven series, 311
Incentive Software, 583
Incestibleach, 43
Incidente em Varginha, 89
Independent developers, 8–9, 30, 33, 36, 43, 45, 71, 81, 114–115, 137, 145, 154, 237, 239, 337, 359, 362, 364, 366, 368, 369, 475, 476, 479, 486, 532–533, 538, 595

India, 235–247
Indiagames (company), 237
Indie games. *See* Independent developers
Indigent—UFBA (research group), 98
Indogamers (game community), 260–261
Indonesia, 3, 10, 249–270, 485
Indonesian Warnet Association, 252
Indowebster (game community), 260
Industria VG (company), 43
Infernum Productions, 201
Infiniminer (game), 458
Infocom (company), 108
Infocomm Asia Holding (IAH), 477
Infocommunication Development Authority (IDA), 471, 478–479
Infogrames (company), 7, 59, 164, 179–181, 183–184, 237, 412, 414, 527, 586, 601
Informal Learning, 613, 618, 619, 621–622, 623. *See also* education
Infrastructures, 3–5, 18–19, 120, 144, 147–148, 596
In-game items, 553, 560
Initial D Arcade Stage 4 (game), 209
Initial D3, 547
Inka Madness, 396
Innogames, 195, 202
Insane Media, 92
Insausti, Eugenio, 41
Inside Games (magazine), 82
Institution of Italian Producers within Confindustria, 310. *See also* Carbone, Raoul
Intel (company), 197
Intel 286 processor, 394
Intenium (company), 202
Interactive comics, 316n1
Interactive Entertainment Association of Australia, 57
Interactive Fiction: The Computer Storygame "Adventure," 604
Interactive Games and Entertainment Association, 60
Interactive Games Association of Ireland (IGAI)/ Games Ireland, 294
Interama Games, 91

Interfaces, 30, 54, 83, 142, 433, 573, 600–601, 604
International Arcade Museum (IAM), 605
Internet cafés, 25–26, 100, 124–125, 129, 131–132, 211–212, 216, 241, 560–561, 613–620, 623
Interplay (company), 108
Intershop, 200
Interton (company), 196
Interton VC4000 (console), 197
Invaders from Space, 385
Invaders of the Lost Tomb, 223
Invaders of the Mummy's Tomb, 595
Invasão ET, 92
INVENTAD, 616
Inventarte, 397
Invention of video games, 592
INVEX (the International Fair of Information and Communication Technologies), 148
Invictus Games, 232
Invizimals (game), 531
IO Interactive, 456
iOS, 33, 40, 139, 238, 365, 368, 396–397, 441, 611
iPad, 139, 226, 433, 602–603, 611
Iparraguirre, Alejandro, 36
iPhone, 15, 139, 141, 157n17, 213, 226, 432, 440, 443, 537, 573
iPod, 602–603
IPS Computer Group, 413, 414–415
IQ 151 (computer), 146
Iran, 271–292
Iran Computer & Video Games Foundation, 272, 277, 279, 286, 287
Ireland, 293–304
Irem (company), 393
I, Robot (game), 600
Iron Byte, 609, 611
Iron Curtain, 219, 233n2
Ironhide Games, 611
Irrompibles (magazine), 54
iSlash (game), 574
Isometric graphics, 583
Israel, 594

İstanbul Efsaneleri: Lale Savaşçıları (*Legends of Istanbul: Tulip Warriors*), 567–568, 577
İstanbul Kıyamet Vakti (*IKV, Istanbul Apocalypse Time*), 570, 577
İstila (*Invasion*) (game), 573
Italian developers, 308–309. *See also* AESVI; AESVI4Developers
Italy, 41, 305–318, 508, 594
Itareritsukuseri, x–xi
ITE Media, 454
ITP, 313
IT Territory (company), 443
iWarrior (game), 20, 23
Iwata, Satoru (President), 326

Jack Orlando: A Cinematic Adventure, 415
Jacobo, Monica, 44
Jade Empire (game), 109
Jaguar Taller Digital, 143
Jakobsson, Mikael, 463
Janáček, Michal, 153
Japan, x–xi, xiii, 11–12, 29–30, 43, 47, 51, 119, 121, 126, 145–146, 177, 207–208, 210–211, 213–215, 259, 319–344, 348–349, 352, 380, 483, 485–486, 497, 500, 505, 506, 508, 510, 511, 536, 570, 594, 598–600, 602
Japan Social Game Association (JASGA), 335
Järvinen, Aki, 463
Java (island), 252
Jazzecity, 43
JCE (company), 501, 504
Jedi Alliance, 484
Jenin: Road of Heroes, 31
Jensen, Keld, 453
Jeopardy (game), 108
Jessen, Carsten, 461, 463
Jessy: Ein Zirkuspferd in Not, 90
Jetpack Joyride, 61
Jet Set Willy, 582, 583
Jianxia games, 123
Jianxiaqingyuan (game), 123, 126, 130
Jianxiaqingyuan 2 (game), 130

Jietou bawang (game), 208
Jinpan Electronic Corporation, 122
Jinyong Online (game), 212
Jogo do Banquinho do Raul Gil, 90
Johansson, Olle E., 453
John von Neumann Computer Society, 230
Jolt (company), 296–297
Jordan, 31–33
Jordanian Gaming Task Force, 33
Jørgensen, Kristine, 463
Jørgensen, Nina Klinker, 454
Journey through Europe, 440
JoWooD Entertainment AG (company), 71–75
Joy Box (company), 31
Joy Game (company), 574
Joymax (company), 501, 503
Joypad (magazine), 189
Joystick (magazine), 189
JTS Inc., 601
Juego de Escoba 15 (game), 37
Juego de Truco (game), 37
Jump Bug, 384
Just Dance (game), 311
Just Dance 4 (game), 311
Juul, Jesper, 98, 463
Jynx Playware, 90

K (magazine), 306
kabaddi, 236
Kabus 22 (*Nightmare 22*) (game), 570–572
Kalah, 442
Kalevala, 159
Kalisto Entertainment (company), 187
Kalypso (company), 308
Kanaga, David, 43
Kanazú S. A. S. (company), 141
Kane & Lynch series, 456
Kangas, Sonja, 160
Kanizsai, Zoltán, 223
Kankluay (movie), 552

Kankluay: The Last Battle, 552
Kantana Animation Company, 553
Kaplan, Larry, 595
Kaptajn Kaper i Kattegat, 453
Karma Revo, 559
Kasetsart University, 545
Kawaii characters, 333
KC series (home computer), 198
Kele Line, 453
Keloğlan (game), 566
Kemény, John, 219
Kenny, Enda, 294
Kenya, 18, 21, 25, 28
Kenyon, Les, 385–386
Képes, Gábor, 233
Kerbal Space Program, 355
Kerr, Aphra, 293, 295
Khan, Shah Rukh, 239
Khan Wars, 32
Kick It: Road to Brazil (game), 141
Kidd, Alex (mascot), 328
Kiesow, Ulrich, 194
Kijiji (game), 20
Kikuchi, Teruhide, 350, 357
Kılıçay, Özden, 566
Killer List of Videogames, 605
Killers (game), 37
Killerspiele, 203
Killzone series (game), 370
Kiloo Games, 459
Kim Bum Wong, 498
Kim Jeng Joo, 499
Kinect, 45, 75, 141–142, 188, 240, 309, 356, 475, 603
King Abdullah II, 33
King Arthur, 221
King Arthur: The Roleplaying Wargame, 224, 232
King Digital Entertainment plc. *See* Zacconi, Riccardo
Kingdom of Winds, 495
Kingdom Rush: Frontiers, 611
Kingdom Rush series, 611

King Naresuan Online (KNO), 553, 556
King Naresuan the Great, 553
King of Fighters, The (game), 208, 215, 348, 353, 546
King of Kings (game), 124–125, 551
Kingsoft Company, 122, 130, 196, 560
Kiss, Donát, 221, 222
Kitronix (company). *See* Kitset Electronics (company)
Kitset Electronics (company), 5, 378, 384
Klastrup, Lisbeth, 463
Klevjer, Rune, 463
Knight Foundation, 610
Knight Lore, 583
Knightrider (video parlor), 383
Knights of Glory, 29, 31
Knights of the Old Republic series, 109
Knutepunkt (Knudepunkt, Knutpunkt, Solmukohta), 460
Koch, Franz, 201
Koch Media, 74, 201, 310
KOEI (company), 214, 473, 483
Kojima, Hideo, 313
Kolář, Jarek, 153
Kompu Gacha (Complete Gacha), 334–35
Komunitas Gamers Indonesia (game community), 260
Konami (company), 209, 214, 393, 480, 601
Konix Multisystem, 584
Koramgame, 560
Korea, 125. *See also* North Korea *and* South Korea
Korea Creative Content Agency, 497
Korean Advanced Institute of Science and Technology (KAIST), 499
Korean Creative Content Agency (KOCCA), 501
Korean Game Development Institute (KGDI), 500–501
Korean Game Promotion Center, 499, 501
Korean IT Industry Promotion Agency, 500
Koridor (Corridor) (game), 573
Kosieradzki, Michał, 417
Koskimaa, Raine, 463
Koster, Raph, 50
Kotagames (online portal), 253
Kotak Game (game community), 260–261

Kotaku (online magazine), 351
Kotaku.com, 74
Kötü Kedi Şerafettin (Şerafettin the Bad Cat), 573
Kovács, Mihály, 219
Krajowa Agencja Wydawnicza, 404–405
Krajowe Wydawnictwo Czasopism, 404–405
Krautscape (game), 538
Krea Medie, 454
Kremlin, 439
Kriegsspiel (war game), 193
KRON Simulation Software 194
Krull, 10
Krygel, Roby, 40
Kuepper, Marc, 553, 556
Kuluya, 20, 24
Kulvakari, Seppo, 463
Kunana Island (game), 355
Kundratitz, Klemens, 201
Kung Fu Panda franchise, 109
Kuti, Fela, 18
Kuwait, 30
Kyd, Jesper, 456, 460

Laber, Nikki, 72
Laboratorium Komputerowe Avalon, 406, 407, 410–411, 413
Lagos, 25
L.A.I. Singapore Pte Ltd., 480
Lakshya Digital (company), 237, 243
Lamanna, Nicolás, 36, 41
La Mecca del Videogioco, 314
Land Games (magazine), 443
Lanetin Hikayesi (The Story of the Curse), 573
Lang, Fritz, 458
LAN game parties, 24, 460
Lange, Andreas 193–206
Langlois, Béatrice, 180
Langlois, Jean-Luc, 180
Lankhor (company), 180, 187
LAN networks, 148
LA Noire, 57, 60, 296

Lara Croft. *See* Croft, Lara
Lara Croft franchise, 95
L'Arche du Captain Blood (*Captain Blood's Ark*) (game), 181
Larger (company), 125
Large-scale integrated circuits (LSIs), 593
Larsen, Torben, 453
Larva Game Studio, 356
Laser (game), 425
Laserdisc games, 598, 601
Laserhawk, 388
Last Bullet, The (game), 282
Last Day on Earth (game), 356
Last Ninja, The, 223, 224, 583
Last Ninja 2: Back with a Vengeance, 566
Last of Us, The (game), 311, 312
László Kiss (CoVboy), 230, 231
LatDev Games & Tech (company), 36, 40
LatinGamers (website), 43
Law for Preventing Unjustifiable Extra or Unexpected Benefit and Misleading, 335
Leader Group (company), 307
League of Legends, 312
Lebanon, 11, 31, 33
Lee Hye Jing, 499
Lees, Anthony, 223
Left 4 Dead 2 (game), 64–65, 296
Legacy of Kain series, 109
Legacy of the Xan, 457
Legend (1991 game), 585
Legend: Legacy of the Dragons, 443
Legend of Mir, The (game) 503
Legend of Mir II, The (game), 125
Legend of Sword and Fairy, The (game), 122, 126
Legend of Zelda, The (game), 209
Legend of Zelda series, 211, 599
Legend of Zord, 31
Legends of Zork, 298
LEGO, 41
LEGO Group, The, 11–12
LEGO Indiana Jones games, 41
LEGO Pirates of the Caribbean, 586
LEGO Star Wars games, 41, 53
LEGO video games, 12, 16, 41, 53, 586
Leisure Suit Larry series, 154
Leisure time, 4
Leisure Vision (console), 384
Leiverouw, Leah A., 614
Lemmings, 585
Lengste Reisen. *See Longest Journey, The*
Les Lapins Crétins (*Raving Rabbids*) (game), 187
Les passagers du vent (*The Passengers of the Wind*) (game), 180
Leti Games, 20, 23–24, 28
Level (magazine), 149
Level 9 (company), 226
Level majstrov (*Level of Masters*) (television show), 155n3
Lewis, Jeremy, 298
Lex Venture, 91
Li, Arthur, 215
Liar: Legend of Sword, 498
Licensing, 322, 325, 328, 379, 385–386, 388
Lideroth, Jonas, 463
Liebe, Michael, 193–206
Life on the Screen (book), 95
Liga Game (game community), 260–261, 263
Ligachev, Yegor, 228
Lighthouse Interactive, 572
Light Shock Software, 307
Lila (Indian: "divine play"), 236
Liliegren, Fredrik, 455
Lima Game Jam, 395
Lima Sky, 604
Limbo, 458, 463, 464
Lime Odyssey, 560
Liminowicz, Mirosław, 406
Lineage (game), 212, 495, 499, 502–503, 506
Linietsky, Juan, 36, 38–39
Lionstork Studios, 486
Lisbon World Exposition, 429
Lischka, Konrad, 205
Lissy—Und Ihre Freund, 91
Little Big Adventure (game), 181
LittleBigPlanet, 586

Little Chicken Game Company (company), 369
Liverpool (store chain), 349
Livingstone, Ian, 221
Loaded (magazine), 54
LOCALGAMEHIST (e-mail discussion list), 1
Local Game Histories, 1
Localization, 31, 120, 123, 195, 208, 252, 295, 299, 302n6, 364, 380, 402, 416, 454, 474, 487, 499, 503, 510, 535–536, 574
Logical Design Works, 401
LOGO (computer language), 179
Lolita Syndrome, 322
Lone Wolf, 311
Longest Journey, The, 455, 456
Loop (festival), 142
Lord, Philip, 386
Lord of the Rings, The, 458
Lord of the Rings Online, The, 443
Loriciel, 180, 182
Lost Eden (game), 184
Lotfali Khan Zand (game), 272
Lotus Esprit Turbo Challenge, 585
Lovecraft, H. P., 458
Love Path (game), 273, 275
Løvind, Simon, 456
Low-Priced Legal Software Campaign, 130
LucasArts, 584, 585
Lucasfilm, 474, 610
Lucca Comics and Games, 314
Lucha libre, 348, 355
Lucha Libre: Heroes del Ring (game), 355
Lucha Libre AAA: Heroes del Ring (game), 138
Ludologica, 315. *See also* Bittanti, Matteo
Luma Arcade (company), 20–21, 26
Lumiere and Nycteris (game), 43
Lure of the Temptress, 584
Luz, Filipe Costa, 435

M3 Racer, 24
Macadam Bumper (game), 181
Mac App Store, 365

MacBook Pro, 15
M.A.C.H. 3, 598
Machinárium (game), 154, 157n18
Machinima, 45, 309
Machu Picchu, 397
Madagascar 3: Race across Europe (game), 369
MadByte (company), 574
Madden NFL (series), 587
Madfinger Games (company), 155
Madrid (game), 610
Madsen, Tom, 454
Maduro, Nicholás, 614
Maestro (game), 188
Mafia (game), 440
Mafia: The City of Lost Heaven (game), 154
Mafia Wars, 602
Mafia Wars 2, 603
Maghrabi, Joseph, 87
Magical Kingdom of Hugo (musical), 454
Magical Shopping Arcade Abenobashi, 320
Magic Chest, 440
Magic Dream, 440, 442
Magic Game, 440
Magma Studio Pte Ltd., 480, 486
Magnavox, 580, 593
Magnavox Odyssey, 178, 320, 322, 323, 346, 393, 593–594, 536
Magnavox Odyssey clone, 609
Magnetic Labyrinth, 442
Magnetic Scrolls (company), 226
Magnus og Myggen, 454
MagTrap (game), 139
MaguMagu, 510, 512
Mahidol University, 552
Mahjong, 37, 170
Mahjong Max, 92
Mail.ru Group (company), 440, 443
Majesco Entertainment, 434
Making Games Talents (company), 201
Makkoya, 499
Maktoob (company), 32

Malaysia, 213, 485
Maldark: Conqueror of All Worlds (game), 41
MALIYO Games, 20, 24
Malvinas 2032 (game), 38–40
Malzak (game), 378
MAME, 378
Mancala, 18
Mandorf, Ulf, 455
Manga, 319, 320, 321, 323
Mangatar, 311
Manhole, The, 599
Manhunt 2, 74
Manipulator, 456
Manoir de Mortevielle, Le (*Mortville Manor*) (game), 180
Manzur, Ariel, 36
MapleStory, 483, 495, 510, 512
Maquinitas, 347
Maragno, Luca, 314
Marble Blast Mobile, 20
Marc Ecko's Getting Up: Contents Under Pressure, 61
Maria Sharapova Tennis (game), 237
Marienbad (game), 175
Mario Bros. (game), 37, 208–209, 211, 322, 471, 499
Mario franchise, 51, 53, 137, 311, 595, 598
Mario Kart DS (game), 211
Mario Kart Wii (game), 210
MarketCity, 443
Markt & Technik (company), 72
Márquez, Pablo, 39
Martin-Barbero, Jesús, 614, 618, 620, 621–622
Marvel superheroes, 570
Más fútbol (game), 47
*M*A*S*H*, 10
Mass Effect (game), 14, 109, 114
Mass Effect 3 (game), 232
Massi, Nicolas, 39
Massively multiplayer online games (MMOGs), 20, 41, 75, 433, 441, 443, 476, 559–560, 581
Massively multiplayer online role-playing games (MMORPGs), 10–11, 32, 37, 41, 94, 124–126, 188, 257, 311, 444, 455, 457, 458, 481, 495, 503, 505–506, 509–512, 553, 557–561, 570, 602–603
Master Multimidia, 89
Master of Alchemy, 308
Master System. *See* SEGA Master System
Mastertroni (company), 221
Matchball Tennis, 90
MatchMove Games, 486
Mato, Fernando, 39
Matrix, The, 165
Mattel (company), 179, 307, 595, 597
Mattel Intellivision, 7, 161, 179, 306, 385, 536, 595, 597
Mattrick, Don, 108
Max Design (company), 71–74
Max Payne (game), 74, 165
Max Payne 2: The Fall of Max Payne (game), 73–74
Max Payne series, 110
Maysalward (company), 33
MBC (Middle Eastern TV network), 29
McCormick, Allen, 388
McDonald, Keza, 327
McGregor, Stephen, 380
Meantime (company), 91
Mediadesign Hochschule, 205
Media Development Authority (MDA), 471, 475–476, 478–479, 482–484, 486
Media Digital (company), 396
MediaXP (company), 81–82
Medienboard Berlin-Brandenburg, 205
Medrano, Antonio, 348
Mega Corp. (company), 595
Mega Drive. *See* Sega Mega Drive; Sega Genesis
Megarace (game), 184
MegaRamp (game), 433
Megatoon (company), 109
Mehrwertsteuer, 227
MelonDaiquiri (game development framework), 41
Memotest (game), 37
Menéndez, Paco, 524
Menghuanxiyou (game), 125

Mercenaries 2: World in Flames, 616
MercurySteam (company), 531
Meridian 59, 602–603
Merit, 560
Mérő, László, 225
Mertens, Matthias, 205
Metal Gear Solid: Philanthropy, 312
Metro: Last Light, 443
Metro 3D (company), 570
Metrogames (company), 36, 42
Metropolis Software (company), 411–412, 417
METU Game Technologies (GATE), 575
METUTECH Animation Technologies and Game Development Center (METUTECH-ATOM), 574–575
Méwilo (game), 182
Mexico, 3, 5, 8, 48–49, 100, 138, 144, 345–358, 594
Mexico City, 347, 348, 350
Mi'pu'mi Games (company), 75
Miąsik, Maciej, 410–411
Mica, Salvatore, 313
MICA (Mercado de Industrias Culturales Argentinas/Market of Argentine Cultural Industries), 42, 51
Michael Jackson: The Experience, 92
Michoacan, 347
Microcomputers, 386–387
Micro Forté (company), 59, 61
Microïds (company), 180
Micro Instrumentation and Telemetry Systems (company), 593
MicroProse, 225, 585
Micro Scooter Challenge, 89
Micros Genius (company), 122
Micro Sistemas (magazine), 87, 89
Microsoft, 32, 50, 52, 71, 75, 100, 151, 188, 189, 196, 197, 210, 214, 215, 255, 310, 326, 331–332, 352, 598–599, 603
Microsoft C#, 560
Microsoft Windows, 365
Microsoft Xbox. *See* Xbox (console)
Microsoft Xbox 360. *See* Xbox 360 (console)
Microsoft Xbox One. *See* Xbox One (console)

Microvision (console), 595, 602
Midas (game), 61
Midcoin (company), 306
Middle East Technical University (METU), 575–576
Midway Games (company), 7, 177, 471, 601
Midwinter, 585
Miechowski, Grzegorz, 411–412
Miffy's World (game), 434
Might 3D (company), 31
Mikoishi (company), 480
Mikrobáze (*Microbase*) (magazine), 149
MikroBitti (magazine), 161
Mikromat, 220
Mikrovilág (*Micro-World*) Christmas, 229
Milestone (company), 307, 310
Miller, Ed, 595
Millionaire (game), 123
Milton Bradley, 595
Minecraft, 458
Miners' Strike (1984–1985), 582
Mini #37, 20
Miniclip, 538
Minter, Jeff, 583
Mirage Software, 406–407, 413–414
Mirinae Software, 499
Mir-Mahna (game), 272, 273, 282
MirrorMoon, 308
Mirror's Edge, 457
Mirrorsoft, 220
Mis Ladrillos (game), 41
Missile Command (game), 293
Mission Bicentennial (game), 141
Mission Deep Sea (game), 573
MIT, 592
MIT-Gambit Game Lab, 488
Mithra's Planet (game), 272, 273, 275, 287
MIT Media Lab, 577
Mixed Reality Lab, 431
Mixi (company), 320, 334–35
Miyamoto, Shigeru, 187, 189–190, 322, 323, 326

MMOGs. *See* Massively multiplayer online games (MMOGs)
MMORPGs. *See* Massively multiplayer online role-playing games (MMORPGs)
MO5.com, 189
Mobage, 320, 335
Mobile games, 20, 24, 28, 33, 40–41, 54, 120–121, 126, 138–139, 141, 143, 145, 154–155, 207, 211, 213–215, 253, 359, 361, 377, 431, 476, 480–481, 497, 573–575
MobileXP (magazine), 82
Modems, 596, 599
MoiMir@Mail.ru (social network), 443
Mojang (company), 458
Møllehaug, Stein A. J., 460
Molleindustria. *See* Pedercini, Paolo
Molnár, József, 225
Molyneux, Peter, 189, 585
Momentum DTM (company), 572
Monde, Le (newspaper), 175
Money On-line, 475
Mônica no Castelo do Dragão, 89
Monkey Island series, 153
Monochroma (game), 574
Monopoly (game), 108
Monster & Me (game), 125
Monster Hunter Online (game), 213
Monster Madness: Battle for Suburbia (game), 138
Monsters Inc. franchise, 109
Monster World (game), 201
Monte Bello, 92
Moonga (game), 537
Moon Megatron (game), 426
Moon Patrol, 393
Moore, Alan, 457
Moorhuhn (game), 75, 202
Moraba, 24
Morais, Marco, 429
Moraldo, Horacio Hernán, 36, 38–39, 41
Moraldo Games (company), 36, 41
Moraldo Tech, 41
Morayra, Jian Carlo Joan, 396

Morgolock, Gulfas, 39
Mørkrid, Olav, 455
Morningstar, Chip, 237
Morocco, 24
Morpion (game), 175
Morskoi Boi (*Sea Battle*), 439, 441
Mortal Kombat (game), 53, 62, 601
Mortal Kombat series, 25
Mortensen, Torill Elvira, 463
Moshi Monsters, 586
Motion-Sports, 309
MotoGP series, 308
Motor de Juegos (company), 345
Mountain Sports, 75
Mount & Blade (game), 572, 574
Mount & Blade: Warband, 572–573
Mount & Blade: With Fire and Sword, 572
Mouse and the Snake, The (game), 286
Movimento contro la disinformazione sui videogiochi, 314
Mózgprocesor, 405–406
Mr. CEO (game), 20
Mr. Goodliving (company), 168
Mr. Gulp (game), 426
MS-DOS. *See* DOS
MSX (console), 30, 37, 565
MUD (*Multi-User Dungeon*) (game), 454, 580–581, 596, 599
MUDs, 123–125
Muhtar (game), 567
Müller, Mihály (Getto Kis), 231
Multicarts (console system), 122
Multiplayer.it, 314
Munch Time (game), 368
Mundo Gaturro (virtual world), 47
Mundo Nintendo (store chain), 350
MU Online, 503, 506, 511–512, 551
Murasaki Baby, 309
Murder in Tehran's Alleys 1933 (game), 273, 275, 288
Murray, Janet, 95, 98
Muzyka, Ray, 109
m.wire, 202

MXit, 20
My Brute (game), 188
My Fitness Coach Club (game), 75
Mylstar (company), 598
Myst, 599–600
Mystique (company), 595
mySugr GmbH (company), 75

Nabi Studios Pte Ltd., 480, 483–484, 486
Nader's Sword (game), 272, 273, 275, 282
N-Age Online, 551
Namco Bandai Games, 211, 251, 474
Namco, x, xiii, 37, 138, 177, 208, 210, 214, 293, 319, 321–22, 333, 393, 560
Namco Wonder Park, 214
Nam In-Hwan, 498
NA.P.S. Team (company), 307
Narco Police, 609
Naruto Shippuden Ultimate Ninja Storm 3 (D1 edition), 312, 475
Naska Digital (company), 143
NASSCOM (National Association of Software Services and Companies), 235, 241
National 863 Project, 125, 132
National Group DRIDKO, The, 47
National Semiconductor (company), 593
National University of Singapore, 472, 487–488
Nation Newspaper (Kenya), 20
NationStates 2 (game), 298
Naughty Sheep (game), 275
Naver (company), 498, 501
NBA All-Star Challenge, 549
NCSOFT (company), 212, 495, 499, 501–504, 509–510
NDi Media, 138
NDiTeravision, 138
NEC (company), 601
NEC PC-Engine/Turbografx-16, 121, 599
Necrovision, 63
Nehir, Hakan, 573
Nehru Place, New Delhi, 242

Neil, Ian, 455
Neitzel, Britta, 205
Nemen vs Llovaca (game), 37
Neocoregames, 232
Neo•Geo MVS, 348
Neoludica: Game Art Gallery, 313, 315. *See also* E-Ludo
NeoPle, 504
neo Software (company), 72, 74, 75
Neowiz (company), 501–502, 504, 509, 510
NES. *See* Nintendo Entertainment System (NES)
NESA-Pong (*PONG* clone), 349, 353
Net generation, 132–133
Netherlands, the, 3, 359–376
Netmarble (company), 501, 511
Nets (channel), 475
Neumann, John von, 219
Neverwinter Nights franchise, 109
Newcomer, 227
Newsgaming.com, 610
New Zealand Game Developer Association, 377
New Zealand, 3, 5, 7, 377–392, 597
New Zoo (company), 305, 308
NexGen Studio, 24, 476, 480, 486
Nexon Corporation (company), 212, 501, 504, 507, 509–510, 559
Next Level Conference, 205
Nexway (company), 474
Nezal Entertainment, 33
Ngā Taonga Sound & Vision, 389
NGD Studios (company), 36–37, 40–41
NGI, 312. *See also* E-sports
NHL series, 170
NHN (previously Hangame and Naver), 501–502, 504, 509–511, 560
NHN Japan, 335
NHN Singapore, 480
Nickelodeon, 138
Nicolas Eymerich: Inquisitor, 311
Nieto, Jairo, 142–143
Nievas, Naomi Marcela, 37

Nigeria, 20, 25, 28
Night Life, 322
Nightmare Creatures (game), 187
Night Trap, 62
Night Vigilante (game), 356
Nik (cartoonist), 47
Nikita (company), 440, 442–443
Nikita Online (company), 440, 442–443
Nim (traditional game), 159, 176, 451
Nimbi, 452
NIMROD (computer), 451, 579
Ninja Gaiden, 336
Ninja Gaiden 3: Razor's Edge, 66
Nintendo (company), 4, 6–7, 37, 40, 43, 51–53, 71, 75, 155n3, 161, 182, 188, 189, 196, 201, 208–211, 214, 229, 233, 251, 255, 310, 323–327, 348–353, 388, 394, 439–440, 484, 536, 585, 591, 594–595, 599, 601
Nintendo 3DS (console), 15, 53
Nintendo 64 (N64), 19, 52, 210, 326, 599, 601
Nintendo DS (NDS) (console), 40, 53, 92, 211, 312, 320, 325, 434
Nintendo DS Lite (console), 15
Nintendo DSi, 15, 156
Nintendo Entertainment System (NES), 6, 15, 17, 51, 121–122, 209–211, 216, 223, 322–323, 327–328, 349–351, 388, 394, 536, 547, 584, 598
Nintendo Famicom. *See* Nintendo Entertainment System (NES)
Nintendo Game Boy Advance, 38, 40, 53, 211, 570
Nintendo Game Boy Color, 53, 211
Nintendo Game Boy, 17, 53, 211, 214, 324–325, 584
Nintendo Game Gear, 53
Nintendo GameCube, 52, 210, 326, 331, 603
Nintendogs (game), 211, 320
Nintendo of America, 322
Nintendo Player (magazine), 189
Nintendo Super Scope, 547–548
Nintendo Wii U, 6, 326–327, 603
Nintendo Wii, 13, 40, 52, 92, 156, 210, 226, 311, 326, 332–333, 354–355, 556, 603

Nissan (company), 355
Nitro Chimp (game), 139
Nival Online (company), 443
Niv Studio (company), 272
Nixtron Interactive, 89
Nobbs, Andrew, 380
Nobunaga's Ambition, 484
No Cliché (company), 184–185, 187
Nohr, Rolf, 205
Nokia, 164–165
Nokia Indonesia (company), 253
No Limite (game), 89
Nomad Soul, The (game), 185
Noob (slang), 581
No One Lives Forever 2 (game), 236
Nordasia Partners Pte Ltd., 480
Nordic DiGRA, 463
Nordic Games GmbH, 75
Nordic Games Holding, 75
North Korea, 145
Norway, 485
Novaleaf (company), 553, 556, 559–560
Novarama (company), 531
Novint (company), 397
Novi Sad, 229
Novotrade (company), 220, 221, 222, 223, 225, 228
NoWhere Studios (company), 574
NRL Rugby League, 377
Ntreev Soft, 499
NTSC, 179
Nucleosys Digital Studio, 41
Nuku (game), 38
Nusantara Online (game), 10, 257–259
NUSSE (computer), 451
Nyboe, Katja, 454
Nygaard, Kristen, 462

Object Software Limited (company), 122
Obscure Pixels (website), 384
Occupations, 561

Ocean Software (company), 184, 221, 583, 585, 586
Octapolis (game), 162
Oddone, Andrés, 45
Oden (computer), 453
O Enigma dos Deuses, 89
Office for the Classification of Film and Literature, 57, 62–64
Official PlayStation Magazine (magazine), 149
OGame (game), 194, 202
Oi (company), 94
Older players, 561
Oldgamesitalia.it, 314. *See also* Retrogaming
Old Master Q, San-T (film), 208
O'Leary, Barry, 298
Oliveira, José, 425
Olivetti (company), 306
Ollmann, Klaus, 196
OLPC XO-1 (computer), 610
Olsen, Fred, 428
Olumhense, Benedict, 25
OMAC Industries, 298
One Laptop Per Child (project), 610
One Piece, 475
One Piece ARcarddass (game), 214
Oniria (company), 90–92
Oniric Games (company), 43
Online games, 24, 29, 31–32, 39, 47, 54, 82, 119–121, 123–126, 132–133, 151, 195, 201–202, 207, 212, 213–215, 250, 253, 256, 308, 359, 361, 368, 440–441, 443, 447–448, 469, 478, 480–481, 495–496, 507, 513, 545, 546, 551, 570
Open beta testing, 560
Opera Soft (company), 523
Operation Arabia, 33
Operation Flashpoint: Cold War Crisis (game), 154
Operation Flashpoint: Red River (game), 237
Orange Gum (company), 486
Ordenador, El (game), 37
Oric Atmos (computer), 181
Origin Systems (company), 584, 602

Orion's Belt (game), 433
Orkut, 42, 94
Ortoleva, Peppino, 315
Oscilloscope, 452
OSMOS (game), 190
Otaegui, Javier, 36, 38–40
Other Side Outscoring, 232
Otherworld, 484
Ourgame Online Game World (platform), 124
Ourworld (virtual world), 124
Outgun, 91
OutLand (game), 272
Outlive—A Era da Sobrevivência, 89
Outrun 2 (game), 547
Outrunner Special, 547
Outsourcing, 12, 41
Outspark (company), 507
Overlord I (game), 367
Overlord II (game), 367
Overplay (company), 91
Ovi Store, 24
Ovos (company), 75
Ovosonico. *See* Guarini, Massimo
OXM (*Official Xbox Magazine México*) (magazine), 345, 352
Özol, Özgür, 567
OZ Virtual (company), 458

Pachinko, 214, 319, 321, 329, 332
Pachinko Pictures, 61
Pacific Storm, 443
Pac-Man (game), x–xi, xiii, 12, 17, 37, 52–53, 208–209, 347, 393, 471, 595
Paddle Battle, 452
Painkiller, 399, 417
Pajatso, 159
Pakku-Man (*Pac-Man*), 322
PAL, 179
Palacios, Gabriel, 345
Pandemic Studios, 615
Pan European Game Information (PEGI), 9, 71, 152, 462

Panix, 378
Pannonia Film Studio, 222
Panoramical (game), 43
Pantoła, Roland, 406, 410–411
Paperama, 92
Paradigm (company), 184
Paradise Café (game), 427
Parasite Eve, 471
Paratext, 619
Para ve Güç (*Money and Power*) (game), 566
Parcheesi (traditional game), 236
Pariwana Studios, 396
Parkan, 440, 442
Parker Brothers (company), 595
Parra, Luis, 137, 139, 143
Participatory culture, 133
Party, The (demo party), 460
Parvaneh: Legacy of the Light's Guardians, 276, 278
Passage, 604
Pastagames (company), 188
Path of Go, The (game), 537
Pathway to Glory, 168
Pau Chu Kang, 473
Pawlo, Mikael, 462
Pazdan, Tomasz, 406, 411
Pazhitnoz, Alexey, 439
PBS, 138
PC Bangs, 4–5, 252, 497, 499, 505, 508
PC Fútbol, 528
PC Games (magazine), 443
PC games. *See* IBM PC games
PC Player (magazine), 189
PC XT (computer), 148
PC XT clone, 147
Peace of Cake? (game), 369
Peak Games (company), 574
Pebolim do Sao Paulo Futebol Clube, 91
Pedercini, Paolo, 309
Peggle HD, 299
Pelc, Janusz, 406, 410

Pelit, 161
Pendulo Studios (company), 528
Pentagon, 31
Pentavision, 559–560
People Can Fly, 399, 417
People's Net (website), 131
Peppiatt, John, 388
Pequeño, Matías, 41
Perceptum e Canal Kids, 89
Perceptum Informática, 89
Perestroika (game), 439
Perfect World, 443, 503
Perla de Oriente, 347
Perry, John, 387
Personae Studio, 480
Personal Gamer (e-sport association), 312
Perspective Software (company), 387
Perspex, 380
Persson, Johan, 456
Persson, Magnus "Notch," 458
Persuasive Games (company), 610
Persuasive Gaming in Context (PGiC), 372
Peru, 393–398
Perucchi, Christian F., 43
Petka and Vasily Ivanovich Save the Galaxy, 443
Pet Society, 483
PetsWar, 545, 561
Pettus, Sam, 327–328
Pfeiffer, Christian, 203
Phantom System (console system), 122
Pharaoh, 551
Phenomedia 201
Philippines, the, 485
Philips (company), 89, 599
Philips CD-i, 359–360
Philos Laboratories, 232
Phoenix (game), 393, 522, 546
Phoenix: The Fall and Rise of Videogames, 604
Picnic Hippo (company), 574
PillowFight, 91

Pilot Brothers: On the Trail of the Striped Elephant, 442
Pilot Brothers: The Case of the Serial Maniac Sumo, 442
Pinball, 159, 455
Ping-O-Tronic, 306
Pinho, Luís Paulo, 430
Pinto, Fernão Mendes, 429
Piracy, 7–8, 52–53, 120–123, 129–133, 147, 151, 209–210, 213–214, 242, 254–256, 312–313, 346, 348–349, 351, 440, 550, 557, 614, 619–20, 621
Piranha Bytes, 194
Piri the Explorer Ship (game), 569
Pis Yedili (game), 566
Pitfall! (game), 53
Pitts, Bill, 593
Pixar (company), 610
PIXEL (conference), 80
PixelHeroes: The First Fifty Years of Computer Games (book), 233
Pixeline, 454
Pixlers Entertainment (company), 71
Pix 'n Love Rush (game), 188
Pixofun (company), 573–574
PixOwl (company), 40
Plan Ceibel, 610
Planet Arkadia Pte Ltd., 481
Planeta Vermelho, 89
Planet Entertainment, 499
Planet Rakus, 20, 24
Plarium (company), 440, 443
Platogo (company), 75
Play, ix, 119
Play:Vienna (event), 82
PlayBrowser technology, 561
PlayCable service, 597
Playdead (company), 458
Playdom (company), 41
Player Barometer series, 170
Player behavior, 77, 545, 561–562
Player demographics, 35, 47, 57–58, 62, 150–152, 207–209, 212, 214, 244, 249, 445–447, 552

Player motivation, 445
Player One (magazine), 189
Playful Interactions (workshop), 577
PlaygroundSquad, 463
Play It Again project, 389
Playlogic (company), 364, 370
Playnus (company), 501
Playpark.com, 559
PlayStation (PS), 13, 17, 52–53, 161, 183, 189, 210, 214, 250, 312, 323, 329–330, 394, 441, 586, 601
PlayStation 2 (PS2), 52, 53, 156, 210, 212, 214, 331, 326, 351, 603
PlayStation 3 (PS3), 33, 52–53, 92, 109, 138, 156, 185, 210–211, 254, 311–312, 326, 603
PlayStation 4 (PS4), 6, 332, 603
PlayStation generation, 307
PlayStation Move, 44
PlayStation Network (PSN), 32, 332
PlayStation Portable (PSP), 40, 42, 53, 92, 156, 211, 312, 332, 441, 463, 557
PlayStation Portable Go (PSP Go), 15
PlayStation Portable Vita (PSP Vita), 309
PlayStation Vita (PS Vita), 53, 332
PlayStation X (PSX), 345, 351–352
Play Thai, 552, 553
Playware Studios, 481, 486
PM Studios, 311
PMD 85 (computer), 147
PoChickenPo, 91
Počítačové hry: Informace pro uživatele mikropočítačů (Computer Games: Information For Microcomputer Users) (magazine), 155
Pocket Gamer (magazine), 82
Pocket God, 604
Pokémon, 53, 311, 325–326
Pokémon Aka Midori (*Pocket Monsters: Red and Green*), 325
Poker Shark, 443
Poland, 122, 146, 227, 399–424, 428
Pole Position, 600
Poly (computer), 386

Poly Play (arcade machine), 198–199
PONG (game), ix, xii, 17, 37, 51, 53, 146, 176, 178, 220, 321, 347, 349, 425, 452, 471, 593
PONG clones, 51, 176, 349, 353, 384, 565
Pons, Ricardo, 44
Pool Live Tour (game), 155
PopCap Games (company), 295, 296, 299
Popcorn Adventure (game), 356
Popular Electronics (magazine), 1, 604
Popular Memory Archive, 389
Populous, 585
Populous II: Trials of the Olympian Gods, 580
Popwan (company), 123
Porko vs Dex (game), 38
Pornography, 462
Portidata (company), 430
Portnow, James, 50
Portugal, 7, 425–438
Portugal 1111 (game), 427, 430–431
Portuguese Society of Videogames Sciences, 435
Posada, Miguel, 138
Pou (game), 11, 33
Powerful Robot Games (company), 609–610
Power Glove, 547–548
Powerpuff Girls franchise, 41
PowerSlide, 59
Power Unlimited, 360
Poznański, Andrzej, 417
PP-01 (computer), 147
PP-06 (computer), 147
Prabhu, Shailesh, 238
Pratchett, Rhianna, 457
Pratchett, Terry, 457
Preloud (company), 90–91
Preservation, 2, 189, 205, 389, 604–605
"Press game," 547
Pressimage, 188
Press Start Studios, 139–140, 143
Prince of Persia (game), 187
Print Club (purikura), 329, 333
Prius Online, 510

Pro 3 Games (company), 75
"Pro" devices, 549
Prodigy (game), 566
Pro Evolution Soccer, 311
Proexport Colombia (company), 144
Programa Conectar Igualdad, 46
Programma 101 (computer), 101
Prograph (company), 307
Prohibition of video games, 616–617
Project Natal, 142
Promit Production, 553
PromptNow Ltd., 553, 556
Propaganda, 126, 145, 154
Proposed law n. 5093, 309
Protection of minors, 195, 202–203
Protégé Production, 476
Prototype (game), 110
Prototype 2 (game), 110
Proxy model, 126
Przedsiębiorstwo Zagraniczne "Karen," 401, 409–410
PS. *See* PlayStation
PS2. *See* PlayStation 2
PSP. See PlayStation Portable (PSP)
PSP Go. See PlayStation Portable Go (PSP Go)
PSP Vita. See PlayStation Portable Vita (PSP Vita)
PS Vita, 15
PSX. See PlayStation X (PSX)
Psygnosis, 430, 586
Pterodon Software, 153–154
Puiggrós, Federico Joselevich, 45
Pulga, La (game), 522
Pulqui II (game), 44
Puskás, Ferenc, 219
Pusu: Uyanış (*The Ambush: Awakening*), 570–571
Puszka Pandory, 403–404
Putzgrilla, 89
Pyro Studios (company), 529

QANTM/SAE Institute, 205
QB9 (company), 36, 42, 47
Qiandao Software (company), 122–123, 126, 130

QQ Audition, 506, 507
Quake (game), 19
Quaker Oats, 595
Qualcomm, 88
Qual é a Música, 89
Quantic Dream, 185
Quantum of Solace (film), 297
Quarter Horse (game), 598
Québec, 106, 109–113
Queen Sirikit, 553
Quevedo, Francisco de, 524
Quicksilva (company), 221, 223
Quirkat (company), 31, 33
Quo Vadis (conference), 201
Quraish (game), 10, 29–30, 32

R18+, 62, 64–66
R2Beat, 559
Rabbit Fable (game), 43
Racer, 20–21
Racine, Rémi, 109
Racing simulator, 455
Radarscope, 322
Radical Entertainment (company), 108, 110
Radical Studios, 346, 353, 354
Radikal Bikers (game), 526
Radio Shack, 599
Radio Shack TRS-80 (computer), 386, 581, 599
RADIOSOFT, 230
Radon Labs, 194
Ragnarok Online (game), 32, 212, 251–252, 259, 483, 495, 507, 551, 559, 561
Raiders of the Lost Ark, 10
Rainbow Arts, 195
Rainbow Media Pte Ltd., 481
Rainbow Six franchise, 187
Rait (company), 5
Rait Electronics, 378
Rake in Grass (company), 157n17
Rally-M, 439
Ramallo, Fernando, 43

Ramonovo kouzlo (*Ramon's Spell*) (game), 154
Ranj (company), 369
Rao, Rajesh, 236
Ra-One Genesis (game), 239
RapeLay, 336
Rare (company), 585
Rarebyte OG (company), 75
Ras Games (company), 287
Rasmussen, Laila Ingrid, 461
Raspberry Pi (computer), 588
Ratbag Games, 59
Ratbag Studios, 7
Ratings systems, 9, 11, 61, 63–65, 71, 76–78, 107, 152, 203–204, 279–281, 335–336, 462, 483, 500–501, 536, 543n15
Rátkai, István, 226
Ratloop Asia, 481, 486
Rats!, The (game), 583
Rau, Anja, 463
Raven Project, The (game), 184
Raven Software, 226
Raving Rabbids (game), 187
Raylight (company), 308
Rayman (game), 183, 185
Rayman Jungle Run (game), 188
Raynal, Frédérick, 181, 183–184, 189–190
Razer Pte Ltd., 481
RDI Halcyon. *See* Halcyon system
Reah: Zmierz się z nieznanym, 411
Reality (company), 486
Reality Engine, 138
Reality Pump (company), 195, 415, 419
Realm of Magic Online (game), 215
Real Racing, 61
Real U (company), 478
Recolector, El (game), 37
Recorcholis (store chain), 347, 348
Red Bull Crashed Ice Kinect (game), 356
Red Dead Redemption, 8, 606
Rede Multimídia (company), 90
RED Entertainment Distribution, 31
RedLynx (company), 168

Red Moon (game), 125
Redsteam Gameloft, 481
Reflex (magazine), 150, 156
Regatta, 442
Regnum (game), 37–38, 41
Regnum 2 (game), 37
Regnum Online (game), 37, 41
Regulations, 71, 76, 127–129, 132, 256–257, 383–384. *See also* Censorship; Ratings systems
Reis, Piri (Turkish admiral), 569
Relic Entertainment (company), 108, 110
REM das Computermagazin (radio show), 198
Remedy Entertainment (company), 165
Renren (company), 126
Rényi, Gábor, 220, 222
Reo-Tek (company), 573
Research Lab of Graphic Computing and Digital Games (research group), 98
Resident Evil series, 20
Retro Gamer, 225
Retrogaming, 314, 602
Reusswing, Jurgen, 553, 556
REV (game), 20
Révbíró, Tamás, 221, 228
Revoca-game (Recall-game), 395
Revolution Software, 584
Reynolds, Cher, 387
Reynolds, Hugh, 296
Rez (game), 190
RFT TV Special (console), 197
Rhaon (company), 212
Riachuelo Games, 92
Riccarton Borough Council, 384
Ridge Racer, 471
Ridiculous Fishing (game), 368
Riding Ground, 91
Riegler, Harald, 75
Riley, Hugh, 223, 233
Riot Games (company), 42, 50, 574
Risen franchise, 201
Rittirong Keawvichien, 545
Riven (game), 600
Roach n Roll (game), 139
Roadshow Entertainment (NZ), 388
Road to Jerusalem, 32
Road to the Final (game), 369
Robbo, 406, 407, 410
Robin Hood (game), 41
Robin of Sherwood (Robin Hood) (television show), 300
Rocketbirds: Hardboiled Chicken, 478
Rock Nano Private Ltd., 481
RockStar (game), 431
Rockstar Canada, 110
Rockstar Games, 7, 11, 71, 74, 110, 203
Rockstar London, 74
Rockstar Toronto (company), 110, 114
Rockstar Vancouver (company), 110
Rockstar Vienna, 71–73, 74–75
Rodriguez, Felipe, 139
Rodriguez, Gustavo "Gus," 350, 357
Rohan (game), 483
Rohrer, Jason, 604
Rojas, Eiver Arlex, 138–139
Rolocule (company), 239
Rolodex, 596
Romance of Three Kingdoms Online, 470
ROM Check Fail, 61
Ronimo Games (company), 367
Rosa, Jorge Martins, 435
Rosen, David, 214
Rosen Enterprises. *See* SEGA
Rötzer, Florian, 205
Roux, Anthony, 189
Rovio (company), 168, 213, 368
Royagaran Narmafzar (company), 272, 275, 287
RSK Entertainment (company), 273, 288
Rubik, Ernő, 219, 233
Rubik's Cube, 219, 222
Rubi-Ka, 457
Rubinstein, Marcelo, 38

Rud, Gabriel, 45
Rule of Rose (game), 314
Run Zombie Run (game), 356
RuneQuest, 459
RuneScape, 581, 586
Runes of Magic, 32
Running Fred (game), 43
Russia, 122, 439–450, 500
Russian Roulette (video game), 443
Rutz, Lajos, 230
Ruzsa, Ferenc, 225
Ryan, Marie-Laure, 463
Rylk, Tomáš, 153
Ryudragon, 90

S4 League, 559
Saarikoski, Petri, 160
Sabarasa Entertainment (company), 36, 39–40, 354
Saboteur, The, 300
Sacred franchise, 201
Saed, Mohamed, 31
SAE Institute, 345
Safari, 439
Saintfun, 560
Saint Seiya Senki (game), 211
Saints Row IV, 66
Sakhr (computer), 30
Sakura Wars (game), 123
Salameh, Paul, 33
Salazar, Oscar, 141
Sam & Max: Season One Wii, 74
Samart Corporation, 552
SAMMY (company), 214
Sam Power Footballer, 92
Samsung, 497, 498
Samurai Shodown, 455
Sanchez, Gonzalo "Phill," 345, 353, 355
Sandbox, The (game), 40
Sanders Associates, 592
Sandy & Júnior Ameaça Virtual, 90

SANK (computer), 452
Santa Ragione, 308
Santi Lothong, 546
Sanxion, 162
Sarafina! (film), 18
Saucedo, Jose M., 345, 351
Saudi Arabia, 33
Save the Turtles (game), 40
Saving Private Sheep (game), 188
Savitskaja, 444
Saygıner, Semih, 570
SBT Multimídia, 89
Scandinavia, 3, 451–468, 442
Scangames Norway, 455
Scarabaeus 223, 224
S.C.A.R.S. (game), 569
Sceptre of Goth, 596
Schaap, Frank, 463
Schäfer, Fritz, 196
Schein (company), 75
Schiller, Friedrich, 194
Schizm series, 411
Schmalz, James, 109
School shootings, 79, 83, 170
Schrödinger's Cat (book), 572
Schwarze Auge, Das (game), 194–195
Science of Cambridge, 581. *See also* Sinclair Research
Scooby-Doo franchise, 41, 109
Scoop Software, 90
Score (magazine), 149
Scotland Yard, 75
Scramble, 378, 385
Scratches (game), 41
Scratches: Director's Cut (game), 41
Screamer, 308
Screen Australia, 60
Screen Sign Arts, 380
Scrunff (game), 37
SD Gundam Online (game), 212, 215
SEA Gaming Pte Ltd., 481

Sea Rainbow Holding Corporation, 510
Sears, 580
SEBİT Education and Information Technologies, 569
SECAM, 179
Second Life, 478, 581
Second Self, The (book), 95
Secret Files franchise, 201
Secret World, The, 458, 464
SEGA (company), 17, 88, 122, 155n3, 177, 182, 184, 188, 196, 210, 214, 224, 229, 297, 348, 352, 387, 388, 389n2, 440, 473, 591, 596, 601
SEGA (SAMMY HOLDINGS), 327–29
SEGA CD, 52
Sega Computer (magazine), 387
SEGA console clones, 122
SEGA Dreamcast, 52, 88, 184–185, 210, 329, 331, 601–602
SEGA Enterprises, 560
SEGA Game Gear, 226
SEGA Genesis (Mega Drive), 17, 37, 51–53, 122, 210, 224–225, 328, 536, 584, 599
Segal, Marcus, 298
SEGA Master System, 51, 88, 328, 388, 584, 598
SEGA Mega Drive. *See* SEGA Genesis (Mega Drive)
SEGA Mega Jet, 327
SEGA Rosso (company), 209
SEGA SAMMY, 214
SEGA Saturn, 52, 210, 328–329
SEGA SC-3000, 387
SEGA SG-1000, 327
Sega Survivors (collective), 387
Seifert, Hannes, 72–73
Seifert, Michael, 454
Seiler, Gerhard, 75
Selener, Pablo, 38
SEMAPHORE (company), 33
Sember, Jeff, 108
Semih Saygıner ile Magic Bilardo (*Magic Billiards with Semih Saygıner*), 570
Senscape (company), 41
SEP, 345

September 12th: A Toy World, 610
Serious games, xi, 4, 24, 28, 71, 139, 359, 361, 363, 364, 368–369, 372, 610
Serrato, René, 139, 143
Service Games: The Rise and Fall of Sega (book), 327
Settlers, The (game), 194, 201
Settlers of Catan, The (board game), 194
Sexy Poker, 63
Shaban (game), 273, 275
Shadowlands, 457
Shadows on the Vatican, 310
Shaffer, Tim, 50
Shanda Interactive Entertainment Limited, 125, 501
Shang Tao, 25
Shape CD (company), 90
Sharara. *See* Nievas, Naomi Marcela
Sharkworld (game), 369
Shekaste Hasr (game), 282
Sherlock Holmes 2: Checkmate (game), 369
Shiny Entertainment, 184
Shockwave, 42
Shoot Bubble (game), 213
Show Biz Pizza Fiesta, 347, 353
Show Biz Pizza Place, 347
Show do Milhão, 89
Shutdown Law, 501
Sibly, Mark, 387–388
Sidam, 306
SID chip, 233
Sidhe Interactive, 377
Sierr, José "Pepe," 350
Sierra On-Line, 184, 221, 226, 584, 585
Signetics 2650A (computer), 196
Silent Hill: Homecoming, 64
Silents, The (group), 455
Silicon Graphics (company), 111
Silicon Knights (company), 108
Siliconworx, 567–568
Silkroad Online, 503
Silverdale Investments, 378

SimCity, 580
Sims, The, 311
Simulations and Games in Education (SIMGE), 576
Simulator Developer Co. (company), 272, 273, 287
Simulmondo, 307, 310, 315n1
Sinclair, Clive, 581
Sinclair (company), 149, 181, 196
Sinclair clones, 147, 609
Sinclair Research, 581, 584
Sinclair Z88, 386
Sinclair ZX Spectrum (computer), 30, 153, 181, 220, 221, 228, 229, 299–300, 386, 393, 425–428, 430, 453, 565, 581–582, 584
Sinclair ZX80, 581
Sinclair ZX81, 387, 393, 425, 565, 581
Sinc Studios, 20, 22
Sine Mora, 232
Sine Requie, 311
Singapore, 7, 9, 469–494, 500, 508, 511
Singstar Bollywood (game), 238
Siri, Santiago, 36–37, 39–40
Šisler, Vít, 152
Sittichai Theppaitoon (BankDebuz), 545, 561
Sizin Amstrad (magazine), 566
Skærmtrolden Hugo, 454
Skate (game), 114
SK Gaming, 460
Ski Challenge, 75
Skidmarks, 388
Skoar! (magazine), 237
Skoda (car), 227
Skool Daze, 582
Skripkin, Nikita, 439, 442
SK Telecom, 498
Sky Cat, 442
Skyrama, 75
Slang (company), 354, 355
Sloan, Patrick David, 380, 382
Slot machines, 439
Slugger, 510, 512

Smartphones, ix, xii, 20, 35, 40, 47, 151, 213, 263, 441, 448, 573, 579, 587
SMIL (computer), 452
Smilinguido em Desafio na Floresta, 90
Smith, Jonas Heide, 463
Smith, Matthew, 583
Smok Wawelski, 404
Snail Racers, 91
Snake, 169
Snake Battle, 442
SNES. *See* Super Nintendo Entertainment System (SNES)
Sniper, 439, 441
Sniper: Ghost Warrior series, 417
SNK (Shin Nihon Kikaku), 208, 348, 353, 549, 601
Sobee Studios, 570, 574
Sobor, 442
Social inclusion, 613, 614, 618
Social networks, 441, 443–444, 447
Social Quantum (company), 440, 443
Socialspiel Interactive Family Entertainment GmbH (company), 75
Softbank, 496
Softmax, 212, 499
Softstar Entertainment Inc., 122, 126
SoftView, 223, 225
Software Industry Promotion Agency (SIPA), 560
Software piracy, 401–402, 413–415
Soft-World International Corporation, 122, 130
Solaris 104, 89
Soldier One, 453
Solitaire, 170
SOMOS, 143
Son Işık Game Studios (company), 570
Song Jae-Kyung, 499
Songs of Impending Doom, 43
Sonic the Hedgehog (game), 210
Sonic the Hedgehog (mascot), 53, 323, 328, 330
Sonnori, 499
Sony, 7, 17, 32, 40, 50, 52, 151, 196, 201, 209, 211, 214, 310, 329–332, 351–352, 394, 440, 463, 586, 591, 599

Sony Computer Entertainment, 71, 75
Sony Computer Entertainment Europe, 586
Sony Computer Entertainment of America, 352
Sony PlayStation (PS). *See* PlayStation (PS)
Sony PlayStation Portable (PSP). *See* PlayStation Portable (PSP)
Sorvete Games e Nix & Huntta, 91
Sotnikov, E., 442
Soul (game), 440
Soulos, Cristian, 37
Sound Records, Video and Game Product Act, 501
Sourena Game Studio (company), 273
South Africa, 18–20, 24, 26, 28, 122
South Korea, 3, 119, 123, 126, 201, 211–212, 214, 252, 259, 485–486, 495–520, 574
Southlogic Studios, 89–90
South Ossetia, 442
Soviet Union, 145–146, 156, 228
Soyuzmultfilm, 222
Space Attack, 384
Space Harrier, 600
Space Invaders (game), 17, 37, 53, 208–209, 233, 306, 321, 322, 347, 471, 546, 594
Space Panic (game), 378
Space Pioneers (game), 202
Space Shooter, 90
SpaceTec, 378
Spacewar!, 592
Space Wars, 596
"Spacies," 378, 383
Spain, 40, 49, 123, 347, 521–534
Spanish folklore, 528
Special Force (2003 game), 31
Special Force (2004 game), 253, 504, 512, 559
Special Force 2: Tale of the Truthful Pledge, 31
Special Force series, 31
Special Studies Group/SSG, 59 (*see also Across the Dnepr: Second Edition*)
Spec Ops: The Line (game), 195, 201
Spectrum Holobyte, 233
Spectrum Világ (*Spectrum World*) (magazine), 230
Speed Freak (game), 596, 600
Speed Freaks (game) 293
Speed in the City (game), 272–273
Speed Punks (game), 293
Speed Up (game), 272
Sphere (game), 443
Spidron, 233
spielebox (institute), 82
Spielwerk GmBH (company), 75
Spil Games (company), 364, 368
Spilmuseet, 461
Spinvector (company), 308
Spirit of Khon, The (game), 10, 553–554, 560
Spirit Tales, 560
Splinter Cell franchise, 187, 313
Sponsorpay (company), 196
Spontaneo (game), 43
Sportronic (console), 384
Sprint 2, 452
Sproing (company), 71
Sproing Interactive Media GmbH (company), 75
Squad (company), 355
Square (company), 209, 210, 212
Square Enix (company), 42, 50, 113, 212, 214
Sripathum University, 545, 552, 560
Staengler, Ferenc, 226
Stærfeldt, Henrik, 454
Stanzani, Matteo, 309
StarCraft (game), 123, 250, 483, 508, 551
StarCraft II (game), 213, 483
StarCraft II: Wings of Liberty, 496
StarCraft League, 508
Stardoll, 32
Star Fox, 585
Stargate Online (game), 215–216
Starglider 2, 585
Star Trek, 458
Star Wars (game), 596
Star Wars: The Old Republic, 8–9

Star Wars Galaxies, 8
Statens Filmtilsyn, 461
State of Decay, 66
Steam, 41, 336, 365, 370
Stein, Robert, 220, 221, 233
Stern Pinball, 560
Steve Hyuga (game), 43
Stevenson, Bob, 223
Stickets, 61
Sticky Studios (company), 369
Stillalive Studios (company), 75
Stöckert, Gabor, 233
Stone Age (game), 125
Stone Throwers, 30
Stormfall: Age of War, 443
StormRegion, 232
Storyteller (game), 43–44
Strategy First (company), 572
Street Fighter (game), 208, 122, 215, 499
Street Fighter II (game), 53, 208, 210, 236, 546
Street Master, 499
Street Racer (1994 game), 569
Street Rod series, 410
Street Soccer Battles, 20
Streets of Rage, 17
Stroustrup, Bjarne, 463
Studio 21 (cinema chain), 251
Studio Evil, 308, 310
Studio Radiolaris (company), 75
S.T.U.N. Runner, 600
Sturmark, Christer, 461
Suárez, Gonzo, 529
Suarez, Jorge, 356
Suárez, Paco, 522
S.U.B., 226
Subelectro, 580
SUBOTRON Arcademy, 80, 82
Sudden Attack, 503, 511–512
Sudoku, 170
Sulake (company), 165

Sundquist, Göran, 452
Sunflowers Interactive Entertainment Software, 72, 194
Suntendy Interactive Multimedia Co., Ltd., 123
Suominen, Jaakko, 160
Superbike, 308
Super Boy, 499
Super Bubble Bubble, 499
Supercan, 453
SuperCan (game), 570
Super Cobra, 393
Super Falling Fred (game), 43
Super Famicom. *See* Super Nintendo Entertainment System (SNES)
Super FX chip, 585
Superman, 10
Super Mario, 53
Super Mario Bros. (game), 209, 322–323, 499
Super Mario Galaxy 2 (game), 210
Super Mario Land, 323
Super Mario series, 20
Super Mario World (game), 210, 323, 549
Super Mini Racing, 89
Super Nintendo Entertainment System (SNES), 17, 51–53, 121, 210, 250, 352, 536, 549, 584, 585
Super Programas em BASIC e Código Maquina (*Super Programs in BASIC and Machine Code*) (book), 426
Super Skidmarks, 388
Super Stardust HD, 164
Superstar Online, 483
Super Vôlei Brasil, 91
Super Vôlei Brasil 2, 92
Supreme Snowboarding, 164
Surreal games, 583
Survival horror games, 583
Survival Race, 34
Suter, Beat, 537, 542
Švelch, Jaroslav, 2, 152
SvilupParty, 310. *See also* Studio Evil; Ticonblu
Swalwell, Melanie, 2
Sweden, 40–41, 156

Swedish Video Game Awards, 461
Sweeney, David, 294
S.W.I.N.E., 232
Switzerland, 535–544, 594
Swordmen's Legend (game), 124
Sword of Sodan, 453
Sword of Sygos, 20
Sybo Games, 459
Syder Arcade, 308
Sydney Development Corporation, 108
SymMek (event), 460
Syndicat National du Jeu Video (SNJV), 187
Syria, 29–31
System, The (game), 311
System 3 Software (company), 223, 224, 566
System 80 (TRS-80 clone), 386
Szakács, Gábor (Sakman), 232
Székely, László, 230, 231
Szellőzőművek (Ventilation Works), 230
Szentesi, József, 223, 225
Szenttornyai, László, 225

T3 Entertainment, 212, 507, 559, 561
Tadeo in the Lost Inka Empire, 396
Taganrog, 444, 446
TAGS (The Awesome Game Studio) (company), 241
Tahadi (company), 32
Taikodom, 91
Tainá—Uma Aventura na Amazônia, 90
Taito, 37, 177, 208, 321–322, 378, 380, 382, 387, 393
Taiwan, 122, 211–213, 485, 506, 511
Tajemnica statuetki, 411–412
Tajemství oslího ostrova (*The Mystery of Donkey Island*), 153–154
Take a Champ, 24
Take-Two Interactive, 74, 110, 154, 201, 538
Takis Air Challenge (game), 355
TakTek (company), 33
Tales Runner (game), 212
TaleWorlds Entertainment, 572–574

Tamagotchi, 33
Tandy 1000 (computer), 599
Tandy Color Computer (computer), 599
Tank Destroyer, 442
Tankodrom, 439
Tank Racer (game), 41
Tapanimäki, Jukka, 162
Tariq's Treasure, 31
Tavella, Santiago, 45
Tawia, Eyram, 20, 25
Tax breaks for the video game industry, 60, 112–114, 177, 294, 309, 371, 379, 476, 587–588
Taylor, Chris, 50
TC 2048 (computer), 428
Teacher-in-a-Box (game), 369
Team Bondi (company), 57, 60. See also LA Noire
Team One (company), 346
Team Vienna (company), 75
Techland, 415–416
Techniat3D, 30–31
Technical University of Budapest, 219
Tecmo KOEI Games Singapore
Tecnopolis, 50
Tectoy, 87–89, 91–92
Teenage Mutant Ninja Turtles, 484
Teenagent, 412
Teenage Warriors, 397
Teixeira, Luís Filipe, 435
Tekes, 168
Tekken (game), 300, 471
Tekken 2 (game), 210
Tekken 6 (game), 508
Tekken series, 25
Tel, Jeroen, 233
Telefonica, 94
Telegraph, 592
Telemacht (company), 51
Telematch de Panoramic, 609
Telephones, 4–5, 596, 602
TeleSUR, 615

Televisa (company), 138
Television, 4, 6, 37, 43, 51, 53–54, 121, 126, 128, 131, 145, 155n3, 187–188, 215, 249, 297, 300, 314, 326, 433, 443, 454, 461, 472, 508, 559, 562, 571–573, 592, 600
Tembac, 44
Tempest, 596
Temple, Magnus, 233
Tencent (company), 507
Teravision, 616, 617
Teravision Games, 138
Terminator 3: Rise of the Machines (film), 232
Terramarque, 162
Terratools (company), 197, 200
Testa, Fernando, 39, 41
Test Drive (game), 108
Tetris (game), 37, 162, 219, 233, 324, 439
Tetris: From Russia with Love, 233
Teuber, Klaus, 194
Texas Instruments TI99/4a computer, 596, 599
Thai Game Software Industry Association, 545, 553, 555, 561
Thailand, 486, 506, 545–564, 507, 511
Thailand Planet, 560
Thai SMEs, 558
Thalamus Europe, 226
Tharaldsen, Jørgen, 293
Thatcher, Margaret, 582
The9 Limited (company), 125
Theme Park, 586
TheMobileGamer Pte Ltd., 481
Theocracy, 232
Think! Studio, 481
Thor (company), 123
Thorsen, Jens, 455
THQ (company), 7, 110, 188, 297
Three-dimensional graphics, 19, 32, 330, 433, 446, 512, 576, 583, 585, 596, 600
Three Kingdoms: Fate of the Dragon, The (game), 123
Three Melons (company), 11, 36, 41, 43
Tianlongbabu (game), 125, 483

Tiara Concerto (game), 212
Ticonblu (company), 308, 310, 311
TIC TOC (game), 43
Tie-in sponsors, 557
TIGA, 587
Tiger Mission, 453
Tihor, Miklós, 226
Tilt!, 314. *See also* Retrogaming
TILT (magazine), 188
TILT Club, 87
Time Crisis (game), 210, 471
Timeline Interactive (company), 19, 138
Time of War (game), 43
Times, The (newspaper), 222
Time Voyage (company), 486
Time-Warner Inc., 11
Timex Corporation, 7
Timex Portugal, 428–429
Timex Sinclair TS1000 (computer), 428
Timex Sinclair TS1500 (computer), 428
Time Zero (company), 443
Timezone (arcade chain), 251
TimSoft, 407
Tin, Harvey Kong, 386, 388
Tin Man Games (company), 61
Tir Na Nog (Arabesque) (game), 299–300
Titans (group), 460
Tito, Rui, 426
TK90X (computer), 609
T-Mek, 600
TN3STUDIO, 140–141
TNT (TV channel), 440, 443
Tobler, Andreas, 74
TOCA Race Driver 3 (game), 237
Todou (website), 131
Toei, 222
Tolkien, J. R. R., 153, 300
Tomb Raider, 123, 312, 330, 586, 600
Tomb Raider: Chronicles, 301
Tomb Raider series, 20, 30

Tomorrow, 230
Tomsk State University, 446
Tonka, 327
Topo-Tops (game), 356
Top Sector, 362, 371
TopWare Interactive, 412, 415
Toribash, 483
Torkamanchay (game), 282
Tørnquist, Ragnar, 457
Torre de Hanoi, 91
Torre Inteligente, 91
Tosca, Susana Pajares, 463
Total Overdose, 457
Toth, Lajos, 228, 229
Touch Dimension Interactive, 486
Tovar, Ivonne, 139, 143
Toxic Balls (game), 355
Toy Catz, 397
Toy Commander (game), 184, 187
Toy Racer (game), 187
Toys, 320, 325, 327, 328, 329
Toy Shop (game), 434
TQ Digital Entertainment, 125
Trabant, 227
Track Mania, 83
Traffic, 223
Train2Game, 588
Tramiel, Jack, 222
Tramis, Muriel, 181
Transmediale, 205
Transmedial franchises, 598
Trasante, Diego, 38
Traulian, 91
Traveller's Tales (company), 11
Travian (game), 194
Travian Games (company), 32, 194, 195, 202
T. R. C. (company), 153
TR-DOS, 581
Trecision, 307
Trenz, Manfred, 195
Tribal Wars (game), 202

Trimarchi, Simone "Akira," 312
Triumph Studios (company), 367
Trivial Pursuit Unhinged! (game), 108
Troff (film), 179
Trojan Chicken (company), 610
Tron (film), 179
Tronstad, Ragnhild, 463
Trophy Hunter, 90
TRS-80. *See* Radio Shack TRS-80
Trubshaw, Roy, 580
Truckers: Transport Company, 443
True Corporation, 559
True Digital Plus Co., Ltd, 559
Truth about the Ninth Company, The, 442
Tsang, Donald, 215–216
Tuck, Hugh, 593
TumGame (company), 545, 552, 561
Tunix (console), 384
Tunnels and Trolls, 300
Turath, 31
Turbine Entertainment Software, 602
Turbografx-16. *See* NEC PC-Engine/Turbografx-16
Turkey, 565–578
Turkle, Sherry, 95, 99
Turk Telecom, 574
Turrican (game), 195
Turtle, 228
Turtleship, 498
Tuzex (store chain), 145, 147
TV2 (television station), 452
TV NOVA, 155n3
TV Ping-Pong (game), 176
Twiddy, John, 223, 233
Twigger, 440, 442
Two Fish (company), 232
TXC Corporation (company), 122
Tyler Projects Pte Ltd., 481, 486
Typewriter, 596

Ubisoft, 7, 71–72, 75, 109–110, 112–114, 116, 138, 180, 183–188, 297, 364, 474, 485

Ubisoft Morocco, 19, 24
Ubisoft São Paulo, 92
Ubisoft Singapore, 481
Ubisoft Studios Milan, 308, 309
UCube, 556–557
U-GATE (the Utrecht Center for Game Research and Technology), 371
UK. *See* United Kingdom
Ulrich, Philippe, 189
Ultima Online, 124, 455, 581, 602–603
Ultima series, 584, 599
Ultima VII: The Black Gate, 39
Ultimate Play the Game, 583. *See also* Rare
Umut Tarlaları (Fields of Hope), 567–568
Uncharted, 311
Uncharted 2: Among Thieves, 296
Under Ash, 30
Under Ash II, 30
Undercover (game), 432
Undercover 2: Merc Wars (game), 432
UnderGarden, The (game), 108
Under Siege (Middle Eastern game), 30. *See also Under Ash II*
Under Siege (Portuguese game), 427, 434
Under Siege: Golden Edition, 30
Under Siege: Path to Freedom, 30
Under Siege: Remnant of Human, 30
Unearthed: The Trail of Ibn Battuta, 33
United Arab Emirates, 31, 32
UnitedGames (company), 364
United Kingdom, 3, 11, 17, 39, 149, 155n3, 177, 184, 386–388, 426, 500, 566, 569–570, 579–590, 594
United Nations, 30
United Socialist Party of Venezuela, 614
United States Marine Corps, 154
United States of America, ix, 3, 6, 11–12, 29, 36, 40–41, 49, 95, 100, 105–106, 108–110, 123, 138, 139–140, 145–146, 148, 175–177, 184, 208, 211, 213, 259, 346–350, 353, 348, 387, 428, 483, 485, 500, 506, 508, 536, 557, 569–570, 591–611
Unity (game engine), 139, 558
Universal, El (newspaper), 346, 348

Universidad Católica del Perú (PUCP), 395
Universidad Nacional Autónoma de México, 353
Universidad ORT Uruguay, 612
University of ELTE, 225
University of Louisiana, 546
University of Mainz, 204
University of Utah, 592
Universomo (company), 168
Universum (museum), 353
Unreal (game), 109
Unreal Engine 3, 558
Unreal series, 109
Urban Assault (game), 197
Űrhódító, 233
Uribe, Sebastian, 39
Uruguay, 609–612
Uruguayan National Videogame Creation Contest, 611
Urustar, 309
U.S. Gold, 583, 586
USK (Unterhaltungssoftware Selbstkontrolle/Entertainment Software Self-Rating), 71, 76–77, 203
Utopia (game), 298

Vaca Maia, 92
Vaca Vitória, 92
Vajna, Andrew G., 232
Valeur, Michael, 455, 456
Vallejo, Hermann, 139
Valve (company), 551
Valvular (ECC83) (game), 43
Vampire Season (game), 142
Vampiromania, 90
Van der Veer Games, 481
Vasco da Gama: A Grande Viagem (game), 430
VAT, 90
VBS1 (Virtual Battlespace Systems 1) (trainer), 154
VEB Microelektronic Mühlhausen, 198, 199
VEB Polytechnik, 199
Vector games, 596
Vectorbeam (company), 596
Vectrex. *See* GCE/Milton Bradley Vectrex (console)

Velazquez, Armando, 348
Velázquez, Diego, 524
Velvet Assassin, 63
Venezuela, 138, 613–628, 594
Venezuelan politics, 614–617
Venezuelan telecommunications industry, 615
Venezuelan video game industry, 616–617
Venice Biennale, 313
Versailles 1685: Complot à la Cour du Roi Soleil (game), 184
VG Map Gamejoint: Connecting Independent Developers, 43
VH1, 42
VHS, 393
VIC-20, 220
Victory Castle, 31. See also *Zoya: A Warrior from Palmyra*
Videkull, Kalle "Frostbeule" Mortlund, 460
Video Arcade Preservation Society (VAPS), 605
Video Entertainment System. See Fairchild Channel F
Video Game Classification in Australia, 61–66. See also Censorship; Office for the Classification of Film and Literature
Video Games Exhibition in Argentina (EVA-ADVA), 40
Video Games Indonesia (game community), 260–261
Video Games in the Model 1-on-1 International Symposium of Edutainment (conference), 42
Videogames Party (show), 312
Video Giocchi (magazine), 306
Videomaster Home T.V. Game, 580
Videopac (console), 179
Videospielkultur (Association for video game culture), 205
Videosport MK2, 580
Videotech, 378
Videoton Sportron 101, 220
Videotronics, 378
Video Vend Automatics, 378
Videoworld, 378
VidPlay (console), 384
Vienna Game AI Conference, 80
Vienna Game Conference—Future and Reality of Gaming (FROG), 79–80

Vienna Town Hall, 71
Viennot, Eric, 189
Vienom (company), 75
Vietcong (game), 154
Vietcong 2 (game), 154
VIGAMUS: The Videogame Museum of Rome, 313–314. See also Accordi Rickards, Marco; AIOMI
Viking-Con, 459
Vikings, The, 453
Villette, Antoine, 189
Vinacur, Nicolas, 38
Vinova Pte Ltd., 481
Viola, Fabio, 309, 316
Violence, 19, 62–64, 76–77, 121, 127, 335–336, 539–540
Virgin (company), 221
Virgin Games, 585
Virgin Games USA, 181
Virgin Interactive, 203
Virtua Fighter Cool Champ (game), 214
Virtua Fighter, 329, 330
Virtual Endosuite (game), 369
Virtual images, 561
Virtuality (company), 600
Virtual Russia, 440
Virus (1988 game). See Zarch
Visualization, Simulation and Digital Games (research group), 98
Vive Digital Plan, 144
Vivendi (company), 110, 184, 201
Vivo, Patricio Gonzalez, 45
Vkontakte (social network), 443
Vlambeer (company), 368
Vlček, Petr, 153
Vocanic (company), 481
Vochozka, Petr, 154
Vochozka Trading, 153
Vodafone, 94
von Reisswitz, Georg Leopold, 193
Vostu (company), 40, 42
Voyageurs du temps, Les (*The Time Travelers*) (game), 181–182

V-Play GmbH (company), 75
Vroom (game), 180 187
VUD, 203

Wadi Basheer, 31
Waei. *See* Wayi (company)
Waigua (program), 125
Walther, Bo Kampmann, 463
Wanmei (game), 125
Wanted! Monty Mole, 582
War73 (game), 30
Warbots, 91
WarCraft franchise, 95
Warcraft III, 460
Warcraft: Orcs & Humans, 551
Ward, Luca, 313
Warface, 443
WarFlow (WF), 560
War game (Kriegsspiel) (board game), 193
Warhammer 40,000: Dawn of War (game), 110
Warhammer: Mark of Chaos, 232
War Masters (game), 353
Wärn, Annika, 463
Warner Bros., 610
Warner Bros. Interactive Entertainment, 11
Warner Communications, 594
Warnets, 4, 252
War of Thrones, 443
Warrior, 596
Warriors, The (game), 110
Wartburg, 227
Watchmen (comics) 457
Watchmen (film), 297
Watchmen: The End Is Nigh, 457
Wavefront Technologies (company), 111
Wayi (company), 125
Way of the Exploding Fist, The, 69
WEA Musik, 196
Weapons of the Gods Online, 560
Web 3.0, 229

Web Gold System (WGS), 125
Webzen (company), 501, 503
Wedding Snake (game), 574
Weibo (company), 126
Wemade (company), 501, 503
Wencel, Lucjan, 401, 409–410
We Need a Hero (game), 275
Western Hills Residence Studio (company), 122–123, 126
West Germany, 177
Westra, Ans, 383
Wet (game), 109
Whale's Voyage, 74
WhiteAfrican blog, 20
Whitehead, Bob, 595
White Rabbit Interactive (company), 75
Wiedźmin series, 412, 416–417
Wiener, Norbert, 452
Wii. *See* Nintendo Wii
Wii Fit, 326
Wii Sports Resort (game), 210
Wii U. *See* Nintendo Wii U
Wild Races, 31
Wild Snake, 442
Wilhelmsson, Ulf, 463
Williams Electronics, 393, 594
Windows (operating system), 122, 226, 430, 560, 599, 603
Winguel, 91
Winner (game), 452
Winner Online Company Limited (WOCL), 559–560
Winning Eleven (Pro Evolution Soccer) series, 53
Wixel Studios, 33
Wizards (arcade), 380, 382
Wizards Productions, 33
Wizkid, 584
Wolfenstein 3D (game), 76, 148
Wonder Wizard, 593
Wooga (company), 201–202
WooWorld Pte. Ltd., 481
Wordfeud, 459
World Cyber Games (WCG), 50, 79, 83, 312, 395–396, 508

World of Legend, The (game), 125, 506
World of Temasek, 487
World of Warcraft (game), 121, 125, 212–213, 294, 311, 455, 462, 483, 508, 581, 603
World of Warcraft: Mists of Pandaria, 311
World Software, 408
World Trade Organization (WTO), 497, 513
World War II, 457
Worms (series), 584
Wozniak, Steve, 593
Wright, Will, 580
Wunderkind, 440, 442

Xaio Tiancai (company), 122
Xbox (console), 52, 74, 100, 210, 212, 250, 255, 331, 396, 603
Xbox 360 (console), 52–53, 92, 138, 156, 210, 215, 311, 312, 326, 332, 351–352, 560, 587, 603
Xbox Live, 32, 603
Xbox Live Arcade, 226 537, 560
Xbox LIVE Indie Game Channel, 475
Xbox One (console), 6, 603
Xendex (company), 71
Xenon 2: Megablast, 584
Xerox Star, 599
Xiaobawang (console system), 121–122, 129, 133n1
Xiaoping, Deng, 124
XIII Century: Rusich, 441
Xinxing (company), 122, 129
XL (mobile phone provider), 253
xLand, 410
X-Legend, 560
XMG Studio, 213
XNA technologies, 560
XO Laptop computer, 610–611
XOR (company), 143
X-rated games, 595
Xshort, 560
Xu, Jingchen, 208
Xuan-Yuan Sword series, 123, 126
Xuxa e os Duendes 2, 90

Yager Development, 195, 201
Yahoo!, 20, 32
Yahoo! Groups, 39
Yakuza, 321–322, 332
Yalazkan, Derya, 566
YAMI: Mechanical Invasion (game), 570–571
YDreams, 431
Yellow Monkey Studios (company), 239, 242
YetiZen Accelerator Program, 139–140, 144
Yip, Augustine, 109
Yo, Matías series, 37
Yoddha: The Warrior (game), 237
Yoga: The first 100% Experience Wii, 74
Yoğurt Technologies (company), 570
Yokoi, Gunpei, 322, 325
Young and Dangerous Online (game), 215
Younger gamers, 561
Youth culture, 382
Yudhoyono, Susilo Bambang, 257
Yugoslavia, 227

Z3 (computer), 193
Z-80 (computer), 37
Zaccaria, 306
Zacconi, Riccardo, 309
Zanussi, 306
Zaporozhye, 443
Zarch, 585
Zaxxon, 583
Zealot Digital Pte Ltd., 481
Zeebo Extreme Baja, 92
Zeebo Extreme Bóia Cross, 92
Zeebo Extreme Corrida Área, 92
Zeebo Extreme Jet Board, 92
Zeebo Extreme Rolimã, 92
Zeebo F. C. Foot Camp, 92
Zeebo Interactive Studios (ZIS), 88, 92, 243
ZekRealm Interactive, 481
Zemina, 499
Zentrum für Kunst und Medien (ZKM), 205

Zeschuk, Greg, 109
Zhangtu (game), 125
Zhiguli, 227
Zibumi (company), 574
Zingmobile Pte. Ltd. (company), 481
ZIO Studios (company), 143
Z-machine, 222
Zoetrope Interactive, 572
Zona Franca de Manaus, 100–101
Zorax, 89
Zork (game), 180
Zork: The Great Underground Empire—Part I (game), 108
Zotov, Stepan, 440
Zoya: A Warrior from Palmyra, 30
Zuccarino, Pablo (Paul), 37, 41
Zuse, Konrad, 193
Zwan, 309
ZX80. *See* Sinclair ZX80
ZX81. *See* Sinclair ZX81
ZX Magazín (magazine), 155
ZX Spectrum. *See* Sinclair ZX Spectrum
Zynga (company), 202, 296, 298, 433, 603
Zyron's Escape, 453
ZZAP! (Italian magazine), 306
Zzap!64 (magazine), 223, 230